STP 1156

Composite Materials: Fatigue and Fracture, Fourth Volume

Wayne W. Stinchcomb and Noel E. Ashbaugh, editors

ASTM Publication Code Number (PCN)
04-011560-33

ASTM
1916 Race Street
Philadelphia, PA 19103

ASTM Publication Code Number (PCN): 04-011560-33
ISBN: 0-8031-1498-2
ISSN: 1040-3086

Photocopy Rights

Peer Review Policy

Each paper published in this volume was evaluated by three peer reviewers. The authors addressed all of the reviewers' comments to the satisfaction of both the technical editor(s) and the ASTM Committee on Publications.

The quality of the papers in this publication reflects not only the obvious efforts of the authors and the technical editor(s), but also the work of these peer reviewers. The ASTM Committee on Publications acknowledges with appreciation their dedication and contribution to time and effort on behalf of ASTM.

Printed in Ann Arbor, MI
June 1993

Foreword

This publication, *Composite Materials: Fatigue and Fracture, Fourth Volume,* contains papers presented at the Fourth Symposium on Composite Materials: Fatigue and Fracture, which was held in Indianapolis, Indiana on 6–7 May 1991. The symposium was sponsored by ASTM Committees D-30 on High Modulus Fibers and Their Composites, E-9 on Fatigue, and E-24 on Fracture Testing. Wayne W. Stinchcomb, Virginia Polytechnic Institute and State University, and Noel Ashbaugh, University of Dayton, presided as symposium chairmen and were the editors of this publication.

Contents

Overview

Fatigue and fracture are topics of primary importance to the development of damage-tolerant composite materials, to the design of high-performance composite components and structures, and to the certification of composite products. The ASTM Symposium on Composite Materials: Fatigue and Fracture, held in Indianapolis, Indiana in May 1991, was the fourth in a series of ASTM symposia to address current issues related to fatigue and fracture of composites. This special technical publication (STP) contains peer-reviewed and approved papers presented at the Indianapolis symposium. The STP is one of more than 25 ASTM STPs, dating back to 1973, which either focus on or address the subjects of fatigue and fracture of composites and related issues.

The Fourth Symposium on Composite Materials: Fatigue and Fracture was sponsored by ASTM Committees D-30 on High Modulus Fibers and Their Composites, E-9 on Fatigue, and E-24 on Fracture Testing. Wayne Stinchcomb of Virginia Polytechnic Institute and State University and Noel Ashbaugh of the University of Dayton Research Institute served as chairmen of the symposium. Doug Ward of General Electric Aircraft Engines, Ted Nicholas of Wright-Patterson Air Force Base, Steven Lubowinski of BASF Structural Materials, James Whitney of the University of Dayton Research Institute, Charles Bakis of the Pennsylvania State University, and W. Steven Johnson of NASA Langley Research Center served as session chairmen. The authors of the STP papers present their experimental and analytical work on fatigue and fracture of composites conducted in the United States, Canada, Japan, the United Kingdom, and Austria.

Papers in the first section on strength and failure modes present results on the failure process in composites, including the initiation and growth of matrix cracks and delaminations and fractures. The papers describe failures of materials, components (tubes and notched laminates), and structural elements (stiffened panels). Results are presented for axial, transverse, and biaxial loading conditions.

The second section contains papers on the measurement, analysis, and modeling of damage, three elements essential to the development of relationships between the state of the material and its performance. The results presented in these papers reflect current progress in the understanding and modeling of damage and its consequences. The polymer matrix and glass-ceramic matrix composites discussed in this section are indicators of current interests in high-performance and high-temperature composites.

Intralaminar and interlaminar fracture, the subjects of the third section, continue to be topics of particular interest as new materials and material configurations are developed and evaluated for damage tolerance and fracture toughness. The papers in this section address Mode I and Mode II fracture, environmental effects and test methods. The importance of using proper experimental and analytical procedures is emphasized, and several practical recommendations are offered.

Micromechanics and interfaces add new dimensions to the topics of fatigue and fracture of composites and are the themes of the fourth section. These subjects are the focus of much of the current work on long-term behavior and provide much-needed information on the understanding and analysis of failure processes. Several micromechanics models are presented to represent the local behavior of matrices, fibers, and interfaces. Two test methods are also presented to measure and evaluate the quality and strength of interfaces.

Fatigue of polymer matrix composites, presented in the fifth section, continues to be a subject of primary interest and importance as higher performance requirements for composite materials and structures must be satisfied. Data are presented showing the effects of important factors such as laminate thickness, notches, interleaves, and stress ratios on the fatigue response of polymer matrix composites. The papers also discuss the development of damage, including matrix cracks and edge and local delaminations, and its influence on fatigue response.

The final section of the STP contains papers on the fatigue response of metal matrix and glass-ceramic matrix composites and an aramid fiber-reinforced epoxy/aluminum laminate (ARALL). The papers address the need for development and characterization of advanced, high-performance materials which can survive in aggressive and hostile environments. The papers in the sixth section present results of experimental work to evaluate the thermomechanical and tension-compression fatigue response of metal matrix composites, tension-tension fatigue damage in glass-ceramic matrix tubes, and high-cycle fatigue behavior of ARALL laminates.

The technical results contained in this STP provide many practical insights on new developments in areas of continuing interest, such as damage and failure analysis and test methods. They also provide new directions for emerging technology areas, such as micromechanics and interfacial analysis, which are applicable to fatigue and fracture of composites. The materials addressed are thermosetting and thermoplastic polymer matrix composites, metal matrix composites, and ceramic matrix composites, as well as a specialty laminate. Furthermore, the results were obtained for several configurations of composites ranging from laboratory specimens to subcomponents to structural elements.

One measure of the value of a technical publication is the applicability of new information contained therein to the solution of current problems. Another measure of value is the significance of new directions, methodologies, and innovations to help achieve further understanding and solutions. By both measures, this STP has value to all who are concerned with fatigue and fracture of composites.

The full spectrum of activities necessary to organize and hold a technical symposium and to publish an STP requires coordination and cooperation by a team of many people. To the ASTM staff, the sponsoring technical committees, the session chairman, the authors, the reviewers, and the clerical staffs of our organizations, we express our sincere appreciation for your professional efforts and jobs well done.

Wayne Stinchcomb
Virginia Polytechnic Institute and State University and the U.S. Air Force Academy; symposium chairman and editor

Noel Ashbaugh
University of Dayton, Research Institute, Dayton, OH; symposium chairman and editor

Strength and Failure Modes

Adam J. Sawicki,[1] Michael J. Graves,[2] and Paul A. Lagace[3]

Failure of Graphite/Epoxy Panels with Stiffening Strips

REFERENCE: Sawicki, A. J., Graves, M. J., and Lagace, P. A., "**Failure of Graphite/Epoxy Panels with Stiffening Strips**," *Composite Materials: Fatigue and Fracture, Fourth Volume, ASTM STP 1156*, W. W. Stinchcomb and N. E. Ashbaugh, Eds., American Society for Testing and Materials, Philadelphia, 1993, pp. 5–34

ABSTRACT: Nine graphite/epoxy panels were tested in uniaxial tension to examine the ability of stiffening strips to redirect propagating damage. The materials used were Hercules A370-5H/3501-6 prepreg fabric and AS4/3501-6 prepreg unidirectional tape. The layup of the unstiffened regions of the panels was four plies of fabric with the quasi-isotropic layup $[0_f/45_f]_S$. Three different four-ply unidirectional tape stiffener layups parallel to the applied load were tested. The specimen width was 203 mm with stiffener widths of either 48 or 64 mm. Slits of three lengths, 51, 71, and 102 mm (1.0 mm wide), were precut into the panels perpendicular to the direction of loading. The panels were tested under load control. Stress-strain and photoelastic data were taken. The finite element code ADINA was used to examine the two-dimensional response of the panels. The three slit lengths were modeled to gain an understanding of the local stress and strain response ahead of the slit in the stiffened region. Contour plots of maximum strain and the orientation of these maximum values with respect to the load direction near slits and stiffened regions were generated. These analytical models showed higher strains in the stiffened regions ahead of the slit compared to the strains in the unstiffened regions. The local orientation of the maximum tensile strains in the stiffened regions ahead of the slit remained perpendicular to the slit and parallel to the applied load. At failure, propagating damage progressed directly across the stiffener either perpendicular to the applied load or at 45° with respect to the applied load. It is concluded that the maximum tensile strain and its orientation ahead of the slit, as derived from the analysis, plays a key role in the prediction of subsequent damage propagation in these laminates. It is also concluded that the structural and material couplings, due to the unsymmetric layups in the stiffener regions, have little, if any, affect on the ability of the stiffener to redirect damage propagation, since in all cases damage progressed through the stiffeners.

KEY WORDS: graphite/epoxy fabric composites, stiffened panels, stiffening strips, stress concentrations, damage propagation, two-dimensional finite-element analysis

In recent years, the aerospace industry has seen an increase in the use of composite materials in both primary and secondary aircraft structures. Two characteristics which make these materials attractive compared to conventional metallic designs are: first, the composites are relatively low in density and second, they can be tailored through the stacking sequence to provide high strength and stiffness in directions of high loading. However, due to their anisotropic, laminated nature, composite materials have a more complicated response to loading and more complex failure modes than do isotropic materi-

---

[1]Aeronautics engineer, Boeing Helicopter, Philadelphia, PA.

[2]Boeing assistant professor of aeronautics and astronautics, Technology Laboratory for Advanced Composites, Massachusetts Institute of Technology, Cambridge, MA.

[3]Associate professor, Technology Laboratory for Advanced Composites, Massachusetts Institute of Technology, Cambridge, MA.

als. For example, whereas isotropic materials tend to fail primarily through cracking, composite failure is often a combination of matrix cracking, fiber breakage, and interply delamination. Consequently, one of the challenges facing designers who wish to utilize composites in primary structures is damage propagation and how the material and structure interact to attenuate the potentially destructive effects.

An important problem in aircraft structures is damage propagation initiating from stress concentrations such as those associated with doors, attachment holes, and windows. A great deal of wing and fuselage design has concentrated on preventing catastrophic failure due to rapid propagation of damage. In a typical metallic aircraft fuselage, catastrophic failure is prevented using frame reinforcements bolted to the skin. These reinforcements provide additional bending stiffness to the fuselage and act as crack arrestors. Such structures are known as being damage tolerant.

Damage tolerance of composite materials and the issues facing designers can be found in several references [1–3]. From these references it can be concluded that three methods of damage-tolerant design are considered: structures can be redundant, they can be made of materials having high fracture toughness, and they can be augmented with structural reinforcements. Reinforcements can either decrease the force driving damage propagation or increase the local fracture resistance. In this investigation, the ability of stiffening strips to influence the path of propagating damage in composites was examined. These stiffening strips were used to influence the local strength and stiffness and thereby influence the force driving damage propagation. The effects of stiffening strip layup and placement relative to slit tip upon damage propagation in composite panels were examined. In particular, the membrane and bending states of stress and strain near a slit tip and how they influence the direction of damage propagation were investigated.

The influence of stiffening and/or softening strips on the failure of composite panels has been addressed by several investigators [4–6]. These studies were primarily concerned with predicting panel failure loads and, as such, did not address directly the question of damage propagation.

Bhatia and Verette [4] tested [0/ ± 45/90]$_s$ graphite/epoxy panels under uniaxial tension. They replaced two sections of the 0 and 90° plies with ± 45° ply strips. Slits and holes were machined into the panels. They found that the direction of damage propagation changed when the initial flaw size was above a particular threshold. They also reported that the damage propagated directly toward the arrestment strip and then aligned itself with the strip material axes prior to arrestment. They concluded that linear elastic fracture mechanics (LEFM) provided a reliable tool for predicting initial fracture strength. They did not provide any detailed description of the strain state ahead of the damage nor any influence this had on subsequent damage propagation paths.

Sendeckyj [5] examined boron/epoxy and graphite/epoxy panels similar to those of Bhatia and Verette. LEFM was again used to predict panel failure loads for various initial flaw types. Sendeckyj found that delaminations produced during damage arrest can affect the ultimate strength of the arrestment panel, and that buffer strips with low in-plane stiffness and high fracture toughness were optimal.

Porter and Pierre [6] considered the effects of stiffening ratio, skin laminate layup, tear strap attachment, and strap spacing on the fracture response of graphite/epoxy skin panels. They concluded that a combination of finite element stress analysis and a point stress fracture theory provided accurate predictions of panel failure. They did not address directly the question of the influence of the straps on the path of damage propagation.

The above-mentioned investigations on reinforced composite panels under tensile loading concluded that initial flaw size and in-plane stiffness and fracture resistance contribute significantly to damage arrest and ultimate strength. The present investigation focused on the influence of in-plane stiffness as well as local bending stiffness on the damage propagation characteristics of panels with stiffening strips. In particular, these flat panel tests were compared to equivalent layups in pressurized cylinder tests in which propagating damage was shown to be redirected by the stiffening strips [7]. For all tests reported herein the failure stress versus flaw size was found to be in good agreement with the Mar-Lin correlation [8–11].

Experimental Procedure

Nomenclature

The panels manufactured in this investigation were composed of fabric plies and tape stiffeners. The 0° direction of the fabric refers to the warp direction, while the 90° direction refers to the fill direction. Plies cut at an angle of 45° to the fill direction are referred to as 45° plies. The subscript "f" refers to fabric plies. The layup of the unstiffened regions of the panels was four plies of fabric with the layup $[0_f/45_f]_S$. The stiffened regions had the following layups: Type A, $[0_4/(0_f/45_f)_S]_T$; Type B, $[0_2/0_f/45_f]_S$; and Type C, $[0_f/0/45_f/0]_S$.

Materials

Two graphite/epoxy material systems were used. The material used for fabricating the unstiffened sections of the panels was Hercules Corporation A370-5H/3501-6 graphite/epoxy fabric, a five-harness satin weave (4.3 yarns/mm) impregnated with the 3501-6 resin system. The material was furnished in 1245-mm-wide continuous rolls of preimpregnated fabric or "prepreg." The material used for fabricating the stiffeners was Hercules Corporation AS4/3501-6 graphite/epoxy tape.

Specimen Manufacture

Nine panel specimens, 711 mm long and 203 mm wide, were fabricated (see Table 1). Three panels were manufactured for each of the layup Types A, B, and C. Stiffener widths were either 48 or 64 mm with slit lengths of 51, 71, or 102 mm.

The Type A specimens were fabricated with the stiffeners running the length of the specimen. The stiffener length was shorter in the Types B and C specimens to provide a more level bonding surface across the width of the specimen for fiberglass loading tabs. However, terminating the stiffeners near the edge of the loading tab was found to produce a stress concentration that could lead to premature failure. To prevent failure near the

TABLE 1—*Stiffener width and slit length in manufactured panel specimens.*

Panel Number	Stiffener Width, mm	Slit Length, 2a, mm
A1	48	51
A2	48	102
A3	64	71
B1	48	51
B2	48	102
B3	64	71
C1	48	51
C2	48	102
C3	64	71

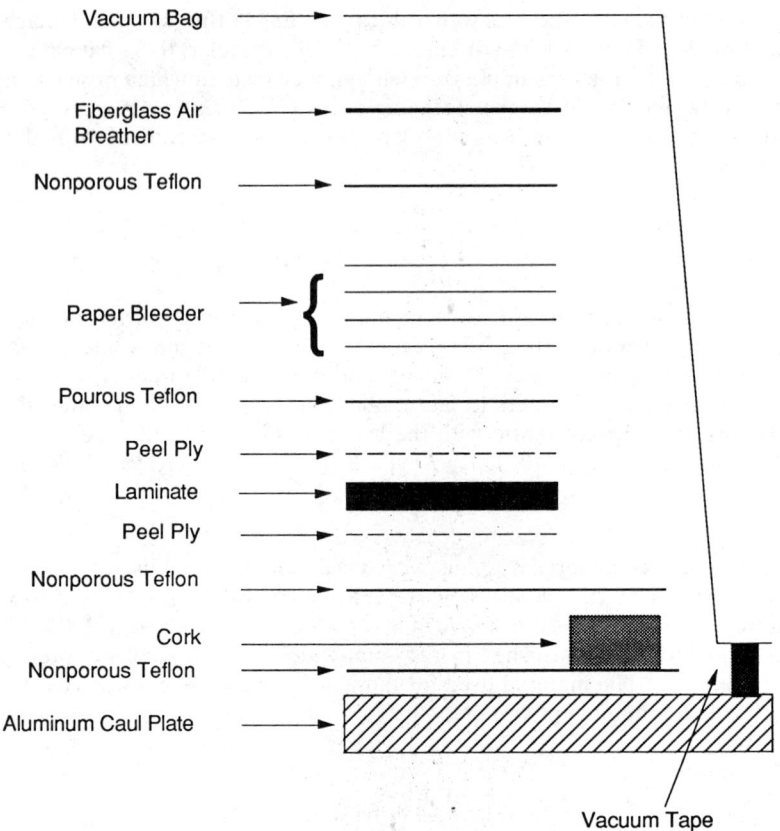

FIG. 1—*Graphite/epoxy cure setup.*

tabs, the area was reinforced with four plies of wet layup plain-weave Boatex 7781 fiberglass with a Volan finish on each side of the panel prior to testing. Specimens B2, B3, C1, C2, and C3 had fiberglass reinforcements applied, whereas Specimens A1, A2, A3, and B1 did not.

The panels were cured using the standard 3501-6 resin cure cycle. The curing assembly is illustrated in Fig. 1. All panels were fabricated using a flat aluminum caul plate. (See Fig. 2 for an illustration of the resulting structural asymmetry of the cured panels, which contributes to the local bending near the stiffeners and consequently must be included when discussing damage propagation.)

The slits were cut by hand using a jeweler's saw (0.2-mm nominal thickness) and a metal rule as a guide. The end of each slit was sharped with a jeweler's saw ground to a point. All slits were cut perpendicular to the loading direction and had a nominal width of 1.0 mm.

To provide uniform load introduction into the panels, precured fiberglass/epoxy loading tabs were bonded to the ends. The tabs used were 3M Type 1002 with a layup of 15 plies of alternating 0 and 90° plies and a thickness of 3.2 mm. American Cyanamid FM-123-2 film adhesive was used to bond the loading tabs to the laminates.

Once the fiberglass tabs had been bonded to the panels, holes were drilled in the tabs and laminates such that the panels could be mounted in specimen-loading jigs. The dimensions of the jig plates, the placement of the holes, and the order of nut torquing are shown

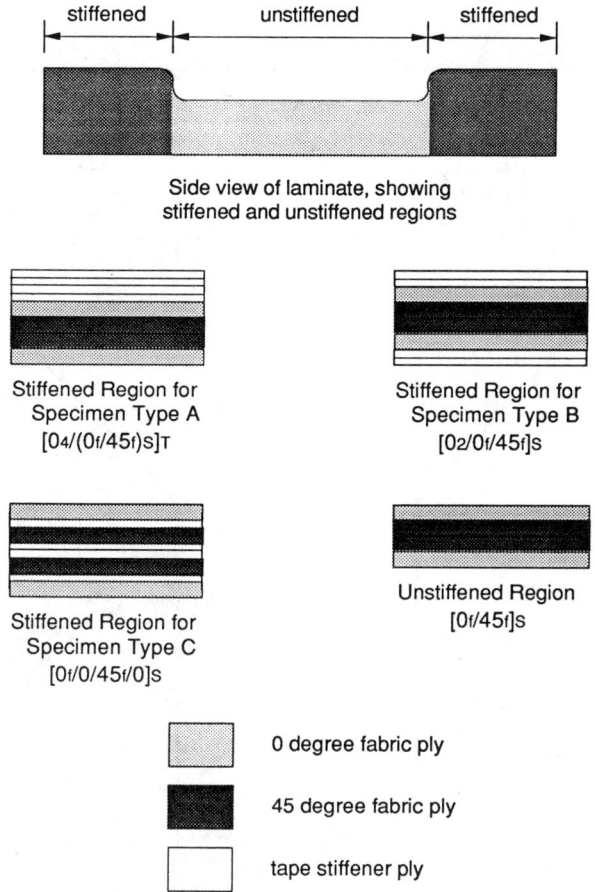

stiffened   unstiffened   stiffened

Side view of laminate, showing
stiffened and unstiffened regions

Stiffened Region for
Specimen Type A
[0₄/(0f/45f)s]ᴛ

Stiffened Region for
Specimen Type B
[0₂/0f/45f]s

Stiffened Region for
Specimen Type C
[0f/0/45f/0]s

Unstiffened Region
[0f/45f]s

0 degree fabric ply

45 degree fabric ply

tape stiffener ply

FIG. 2—*Edge-on view of panel layups.*

in Fig. 3. All nuts in each jig were first tightened by hand, then torqued to 160.9 N/m using a torque wrench. The final torque of 187.8 N/m was applied using a compressed air wrench. The dimensions of the manufactured panels, including the fiberglass reinforcements, are shown in Fig. 4. A side view of a panel secured in a jig is shown in Fig. 5.

Specimen Instrumentation

Micro Measurements EA-06-125AD-120 strain gages were used to measure the far field strains in the panel. The gage locations are shown in Fig. 6. All panels had a longitudinal and transverse gage on each side, with the exception of Panel A1, which also had longitudinal gages located in the stiffened section on each side of the panel to verify that the far-field longitudinal strain remained constant across the width of the panel.

Photoelastic stress tests were also performed on each panel to examine the full-field laminate strains near a slit. Photoelastic sheets of varying thickness were cut to size (254 mm long by 203 mm wide) and preslitted using a drill and coping saw (see Table 2). The sheets were bonded to the flat side of the panels using Measurements Group Inc.— Photoelastic Division photoelastic epoxy. PC-1 resin and PCH-1 hardener were mixed in a

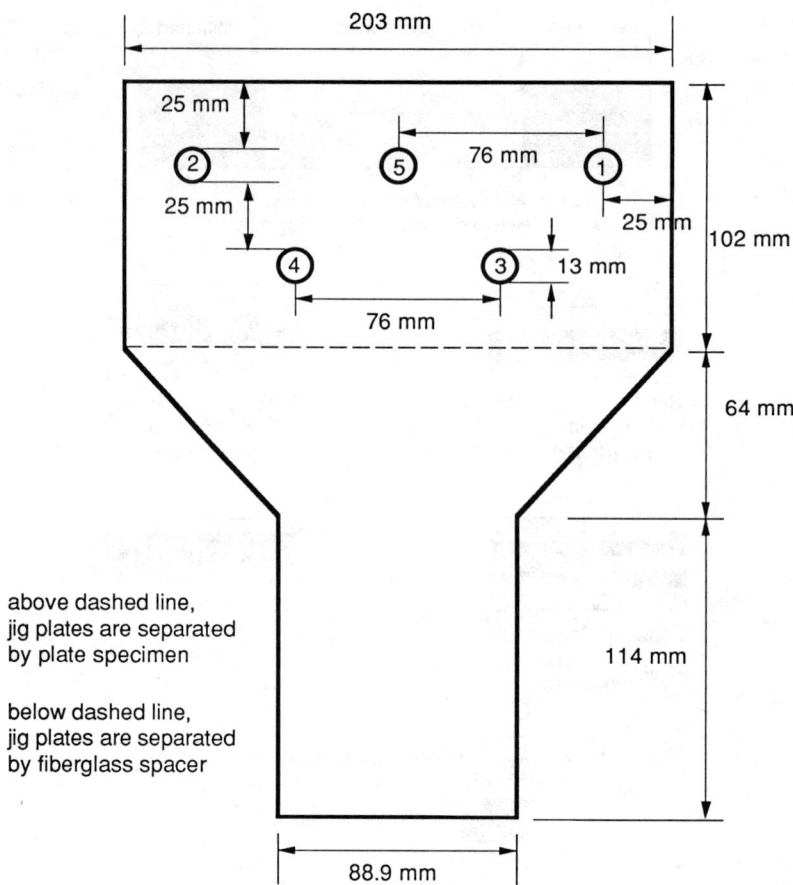

FIG. 3—*Jig plate dimensions and bolt-tightening sequence.*

ratio of 10:1. A piece of TFE-fluorocarbon was used to align the sheet and laminate slits during bonding. The elastic modulus of the photoelastic coating after cure was approximately 3 GPa, about 5% of the modulus of the graphite/epoxy laminates. The location of the photoelastic sheet on the laminates is shown in Fig. 6.

Specimen Test Procedure

Panel tension tests were performed in an MTS 810 universal test machine with hydraulic grips. Strain gage data were taken using Vishay 2120 strain gage amplifiers. The data were recorded using a Digital Equipment Corporation PDP-11/34 data acquisition system. The tests were conducted under load control with a ramp rate of 200 s to apply 220 000 N, yielding a loading rate of 1100 N/s. Strain data was recorded every 0.5 s. The loads were halted every 4400 N from 0 to 66 000 N, and then every 2200 N from 66 000 N until failure. At each halt, the specimen was photographed under load using a Polariscope Model 031 manufactured by Measurements Group, Inc. The resulting photographs of loaded panels showed colored bands which corresponded to the difference in principal strains within the panel.

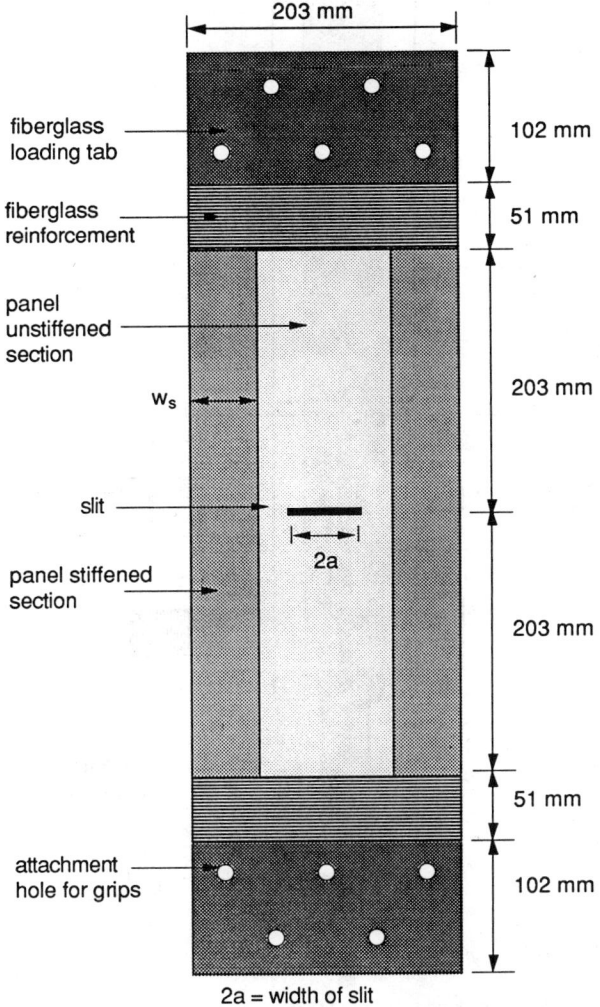

2a = width of slit
w$_s$ = width of stiffeners after milling

FIG. 4—*Manufactured specimen dimensions.*

The relationship between photoelastic fringe color and strain intensity is given by

$$\bar{\epsilon}_1 - \bar{\epsilon}_2 = fN_f \tag{1}$$

where

$\bar{\epsilon}_1, \bar{\epsilon}_2$ = principal strains,
N_f = the fringe order number, and
f = the fringe order-microstrain relationship.

The fringe order-microstrain relationships for the various photoelastic sheets used are

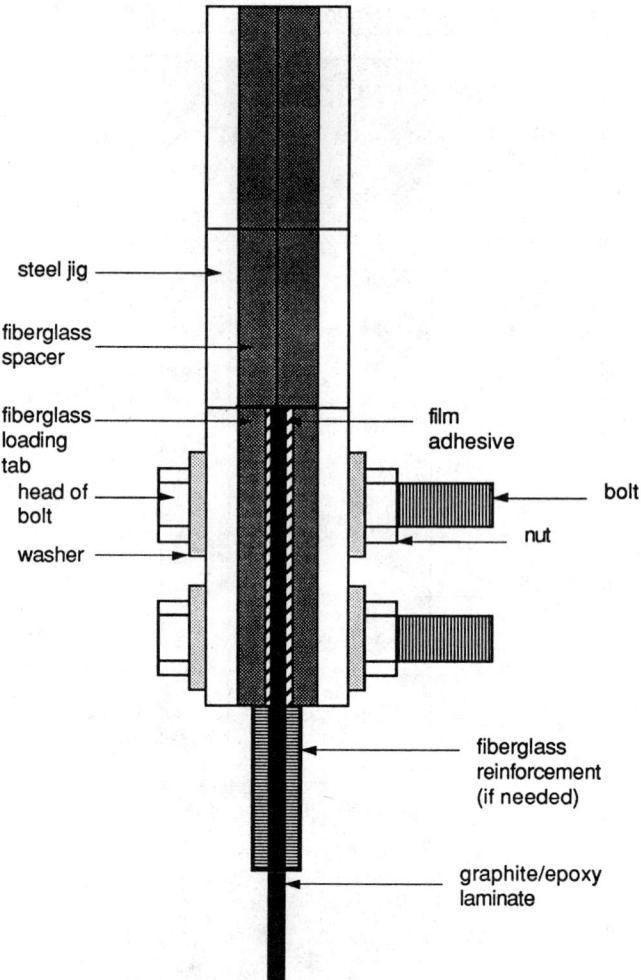

steel jig

fiberglass
spacer

fiberglass
loading
tab

film
adhesive

head of
bolt

bolt

washer

nut

fiberglass
reinforcement
(if needed)

graphite/epoxy
laminate

FIG. 5—*Side view of specimen secured in jig.*

given in Table 2, while the color-fringe order correspondence is given in Table 3. During loading, delamination of the photoelastic coating from the panel near the slit tip could be heard and can be identified in the resulting photographs. No attempt was made to repair this damage.

Theory and Analysis

Mechanical Properties of Materials

The elastic constants (Young's moduli, shear moduli, and Poisson's ratios) for individual fabric and tape plies are given in Tables 4 through 6. In this nomenclature, the "1" and "L" subscripts refer to the longitudinal (load) direction and the "2" and "T" subscripts refer to the transverse direction in the panels.

The elastic constants for the unstiffened and stiffened laminates were determined using

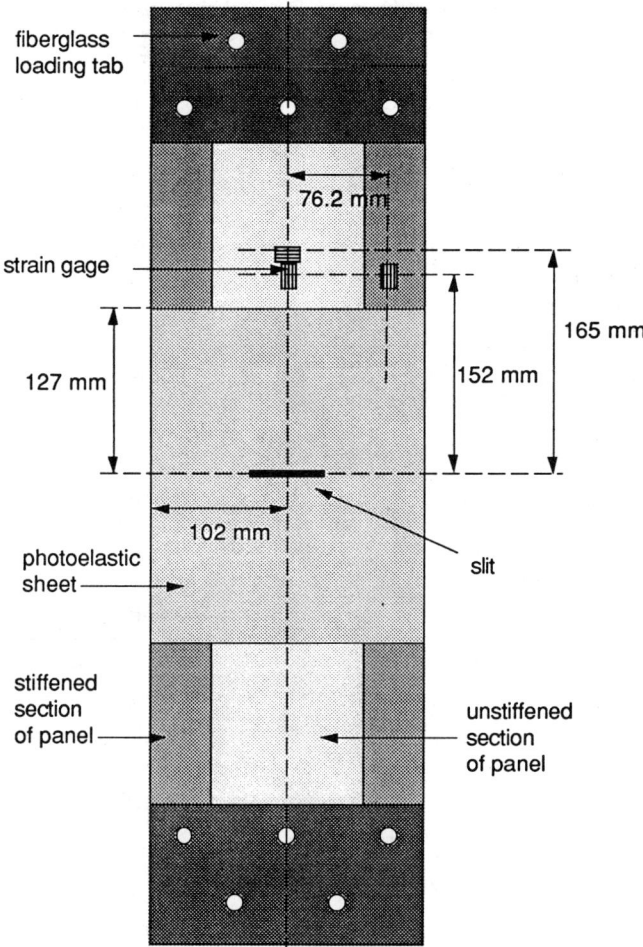

FIG. 6—*Location of strain gages and photoelastic sheet.*

TABLE 2—*Properties of photoelastic sheets.*

Panel Number	Thickness of Sheet, mm	f, microstrain/fringe
A1	1.016	1890
A2	1.016	1890
A3	1.016	1890
B1	1.016	1890
B2	0.965	1990
B3	0.965	1990
C1	0.965	1990
C2	1.016	1890
C3	0.508	3790

TABLE 3—*Isochromatic fringe characteristics.*

Color	Fringe Order Number, N
Black	0.00
Gray	0.28
White	0.45
Pale yellow	0.60
Orange	0.80
Dull red	0.90
Purple (tint of passage)	1.00
Deep blue	1.08
Blue-green	1.22
Green-yellow	1.39
Orange	1.63
Rose red	1.82
Purple (tint of passage)	2.00
Green	2.35
Green-yellow	2.50
Red	2.65
Red-green transition	3.00
Green	3.10
Pink	3.65
Pink-green transition	4.00
Green	4.15

the laminated plate theory. The stress-strain relations for the case of plane stress can be given in both stiffness and compliance form:

Stiffness form:

$$
\begin{bmatrix} \sigma_1 \\ \sigma_2 \\ \sigma_6 \end{bmatrix} = \begin{bmatrix} Q_{11} & Q_{12} & Q_{16} \\ Q_{12} & Q_{22} & Q_{26} \\ Q_{16} & Q_{26} & Q_{66} \end{bmatrix} \begin{bmatrix} \epsilon_1 \\ \epsilon_2 \\ \epsilon_6 \end{bmatrix}
\tag{2}
$$

Compliance form:

$$
\begin{bmatrix} \epsilon_1 \\ \epsilon_2 \\ \epsilon_6 \end{bmatrix} = \begin{bmatrix} S_{11} & S_{12} & S_{16} \\ S_{12} & S_{22} & S_{26} \\ S_{16} & S_{26} & S_{66} \end{bmatrix} \begin{bmatrix} \sigma_1 \\ \sigma_2 \\ \sigma_6 \end{bmatrix}
\tag{3}
$$

where

$$\sigma_1, \sigma_2, \sigma_6 = \text{in-plane stress,}$$
$$\epsilon_1, \epsilon_2, \epsilon_6 = \text{in-plane strains,}$$
$$Q_{11}, Q_{22}, \text{etc.} = \text{in-plane stiffness, and}$$
$$S_{11}, S_{22}, \text{etc.} = \text{in-plane compliance.}$$

The in-plane laminate stiffness and compliance properties are given in Table 6. All of the stiffened layups had the same laminate in-plane stiffness and compliance properties.

The calculated bending stiffness was determined using the laminated plate theory (see Tables 7 through 9). The bending stiffness is referenced to the geometric midplane for both

TABLE 4—*In-plane ply material properties used in the analysis.*

	AS4/3501-6	A370-5H/3501-6
E_L	142.00 GPa	72.50 GPa
E_T	9.81 GPa	72.50 GPa
G_{LT}	6.00 GPa	4.43 GPa
v_{LT}	0.30	0.059
t_{PLY}	0.134 mm	0.35 mm

TABLE 5—*Unstiffened region laminate in-plane material properties.*

Stiffness, GPa		Compliance, GPa^{-1}	
Q_{11}	57.9	S_{11}	0.0194
Q_{22}	57.9	S_{22}	0.0194
Q_{66}	19.3	S_{66}	0.0517
Q_{12}	19.2	S_{12}	−0.0064
Q_{16}	0.0	S_{16}	0.0000
Q_{26}	0.0	S_{26}	0.0000

TABLE 6—*Stiffened region laminate in-plane material properties.*

Stiffness, GPa		Compliance, GPa^{-1}	
Q_{11}	81.4	S_{11}	0.0131
Q_{22}	44.6	S_{22}	0.0238
Q_{66}	15.6	S_{66}	0.0639
Q_{12}	14.7	S_{12}	−0.0043
Q_{16}	0.0	S_{16}	0.0000
Q_{26}	0.0	S_{26}	0.0000

TABLE 7—*Unstiffened region laminate bending stiffness.*

Stiffness, GPa · mm^3	Actual, $[0_f/45_f]_S$	Based on Smeared Properties
D_{11}	15.8	13.2
D_{22}	15.8	13.2
D_{66}	1.9	4.4
D_{12}	1.8	4.4
D_{16}	0.0	0.0
D_{26}	0.0	0.0

the stiffened and unstiffened laminates and, as such, does not account for the structural asymmetry illustrated in Fig. 2. For comparison, the bending-stretching coupling terms for laminate Type A are given in Table 9, and bending stiffness based on orthotropic or smeared laminate properties is included in Table 8.

TABLE 8—*Stiffened region laminate bending stiffness.*

Stiffness, GPa · mm³	Type A, $[0_4/(0_f/45_f)_S]_T$	Type B, $[0_2/0_f/45_f]_S$	Type C, $[0_f/0/45_f/0]_S$	Based on Smeared Properties
D_{11}	62.8	69.5	47.6	48.5
D_{22}	56.0	19.5	36.6	26.6
D_{66}	15.4	4.1	5.0	9.3
D_{12}	15.2	2.9	4.7	8.8
D_{16}	0.0	0.0	0.0	0.0
D_{26}	0.0	0.0	0.0	0.0

TABLE 9—*Bending-stretching coupling stiffness.*

Stiffness, GPa · mm²	Type A, $[0_4/(0_f/45_f)_S]_T$
B_{11}	-36.2
B_{22}	-55.3
B_{66}	-18.1
B_{12}	-18.4
B_{16}	0.0
B_{26}	0.0

Far-Field Load-Stress Relations

The in-plane stiffness and compliance given in Tables 5 and 6 were used to estimate the stress ratio between the unstiffened and stiffened regions of the panel. Assuming the far-field longitudinal strain was the same in the stiffened and unstiffened regions, the ratio of longitudinal stresses was given as

$$\frac{\sigma_{stiff}}{\sigma_{unstiff}} = \frac{(S_{11})_{unstiff}}{(S_{11})_{stiff}} \tag{4}$$

The loads applied over the stiffened and unstiffened sections were given by

$$P_{unstiff} = (w_{panel} - 2w_{stiff})(t_{unstiff})\sigma_{unstiff} \tag{5}$$

$$P_{stiff} = (w_{stiff})(t_{stiff})\left(\frac{\sigma_{stiff}}{\sigma_{unstiff}}\right)\sigma_{unstiff} \tag{6}$$

where

w_{panel} = the width of the panel,
w_{stiff} = the width of the stiffeners,
$t_{unstiff}$ = the thickness of the unstiffened region, and
t_{stiff} = the thickness of the stiffened regions.

The total load applied to the panel was

$$P_{total} = P_{unstiff} + 2P_{stiff} \tag{7}$$

Therefore, the relationship between longitudinal stress and applied load was

$$\frac{\sigma_{\text{unstiff}}}{P_{\text{total}}} = \left[(w_{\text{panel}} - 2w_{\text{stiff}})(t_{\text{unstiff}}) + 2(w_{\text{stiff}})(t_{\text{stiff}})\left(\frac{(S_{11})_{\text{unstiff}}}{(S_{11})_{\text{stiff}}}\right) \right]^{-1} \tag{8}$$

For panels with 48-mm-wide stiffeners, $\sigma_{\text{unstiff}}/P_{\text{total}} = 2352.74$ m^{-2}, whereas for panels with 64-mm-wide stiffeners, $\sigma_{\text{unstiff}}/P_{\text{total}} = 2118.72$ m^{-2}. Therefore, all experimental load data was multiplied by these factors to produce stress-strain plots for strains recorded in the unstiffened regions.

Finite Element Analysis

The panels were modeled in two dimensions using ADINA [12], a displacement-based finite element code which was run on a DEC MicroVAX. The analysis was performed to determine the effects of slit size and stiffener width upon the stress and strain fields in the

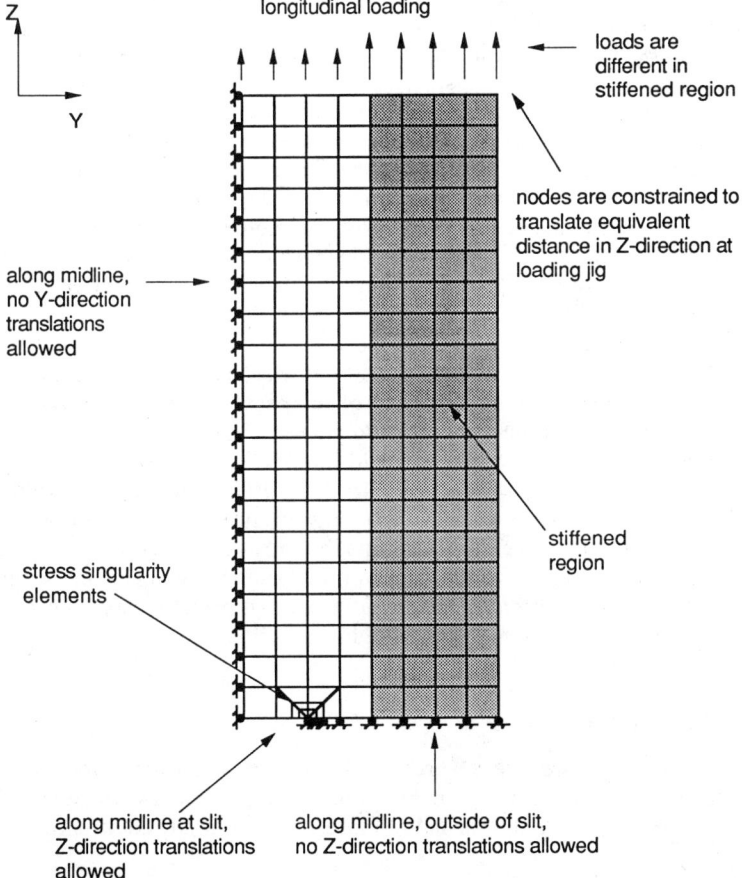

FIG. 7—*ADINA finite element model of panel specimens (quarter model): boundary conditions and applied loads.*

panels. The analysis was static and did not include any dynamic effects associated with damage propagation.

The elements used in the panel analysis were two-dimensional eight-node isoparametric plane stress elements. Each element node was allowed to translate in the Y direction (the transverse direction) and the Z direction (the longitudinal direction) as defined in the ADINA input specifications. Three-by-three Gaussian integration was used. A typical quarter model of the panel is shown in Fig. 7.

The typical square elements used were 12.5 mm on a side. Near the slit tip, special triangular elements were used to model the assumed singular stress distribution. Such elements had a side collapsed to one node. The side nodes were placed one quarter of the side length away from that node to obtain a singular stress distribution near the slit tip, as recommended by Bathe [13].

The finite element method was used to examine the load response of the three types of panels. The slit length and stiffener width of each panel analyzed were: Case 1—slit length = 50 mm, stiffener width = 50 mm; Case 2—slit length = 100 mm, stiffener width = 50 mm; Case 3—slit length = 75 mm, stiffener width = 62.5 mm. The purpose of analyzing the three cases was to examine the influence of the stiffened region upon the local stress and strain distributions around the slit and upon the global load response of the panel. Case 1 examined a panel with stiffeners a distance away from slit tips. Case 2 examined a panel with stiffeners close to the slit tips. Case 3 examined the effect of changing stiffener width upon stiffeners close to slit tips.

Boundary conditions were then applied to the elements. As shown in Fig. 7, all of the nodes along one side of the model were constrained to move in the Z-direction only to model the symmetry of the specimen across the Z axis. Along an adjacent side, all nodes except those alongside the slit were constrained to move in the Y-direction only to model the symmetry of the specimen across the Y axis. The nodes along the sides of the slit were allowed to move in both the Y and Z directions. Along the loaded side of the panel, all nodes were constrained to move an equivalent distance in the Z direction to simulate the constraint of the specimen jig.

Along the loaded side, in-plane stresses were applied to the mesh. The loadings were different for the unstiffened and stiffened regions. It was assumed that the far-field strain in the Z-direction was constant along the width of the panel. Therefore, to model the stress distribution in the panels, the far-field longitudinal stress, $\sigma_{unstiff}$, was set to a value of 0.1 N/m² for the unstiffened section. The far-field longitudinal stress in the stiffened region was then found using Eq 4. This σ_{stiff} of 0.148 09 N/m² was then applied to the stiffened section of the panel. The strains obtained from the analysis were transformed into principal strains to determine the magnitudes of maximum tensile strain and the orientation of maximum tensile strain relative to the direction of loading. The principal strain data was also used to determine the strain intensity (difference of principal strains) so that a comparison could be made with the photoelastic data.

Failure Stress Predictions

Predictions of panel failure stress were based upon the Mar-Lin failure correlation. The Mar-Lin correlation uses a fracture toughness parameter coupled with an assumed power law relationship to determine the far-field failure stress of a flawed structure. The correlation has the form

$$\sigma_f = H_c(2a)^{-m} \tag{9}$$

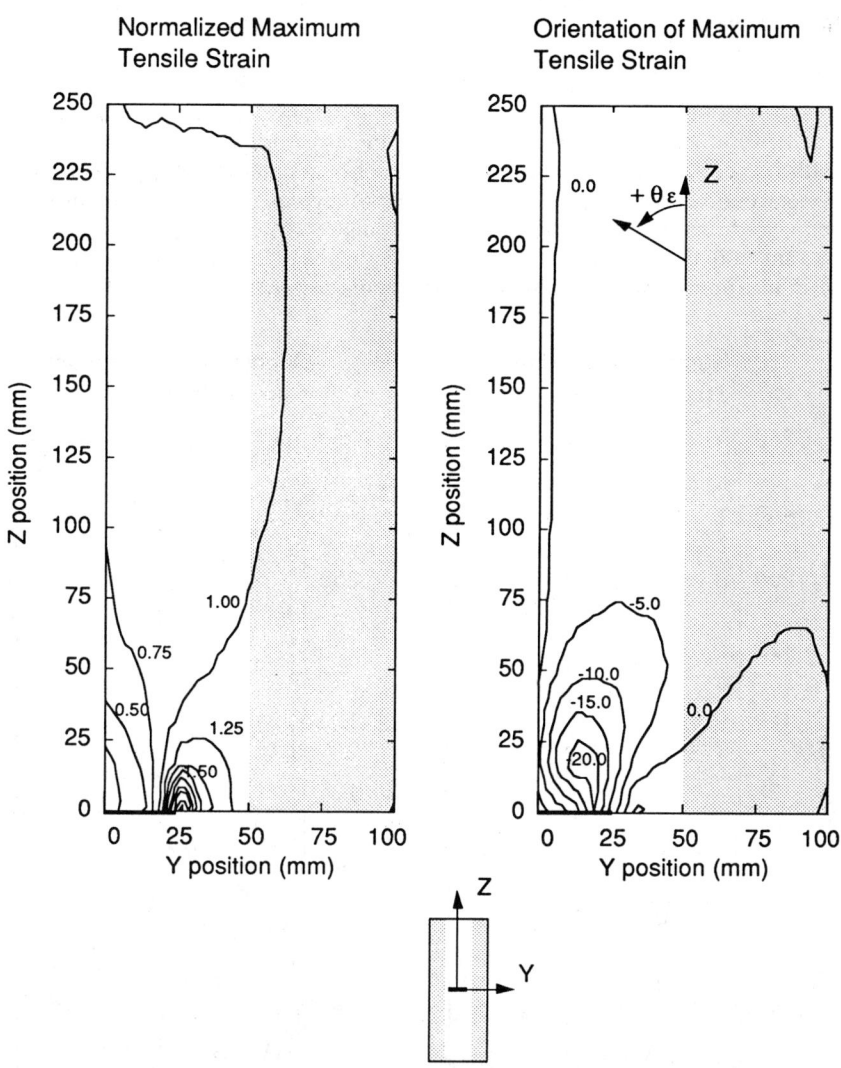

FIG. 8—*Normalized maximum tensile strain and orientation of maximum tensile strain for panel Model 1 with slit length = 50 mm and stiffener width = 50 mm.*

where

σ_f = the far-field fracture stress,
H_c = a composite fracture parameter,
m = the critical singularity responsible for fracture, and
$2a$ = the notch size perpendicular to the maximum tensile stress.

For the A370-5H/3501-6 graphite/epoxy fabric, the value of H_c is 738, and m is theoretically taken to be 0.28. The Mar-Lin correlations do not take into account the finite width of the panel.

Analytical Results

Maximum Tensile Strains

Figures 8 through 10 show maximum tensile strains normalized to the far-field unstiffened region maximum tensile strain and angles of orientation for maximum tensile strain with respect to the Z-direction for the three panel models. All three cases showed strain concentrations near the slit tip and a general increase in maximum tensile strain in the clockwise direction about the slit tip. The angles of orientation between maximum tensile strain and the Z-direction were near zero ahead of the slit tip. There was no increase in maximum tensile strain across the boundary between the far-field unstiffened

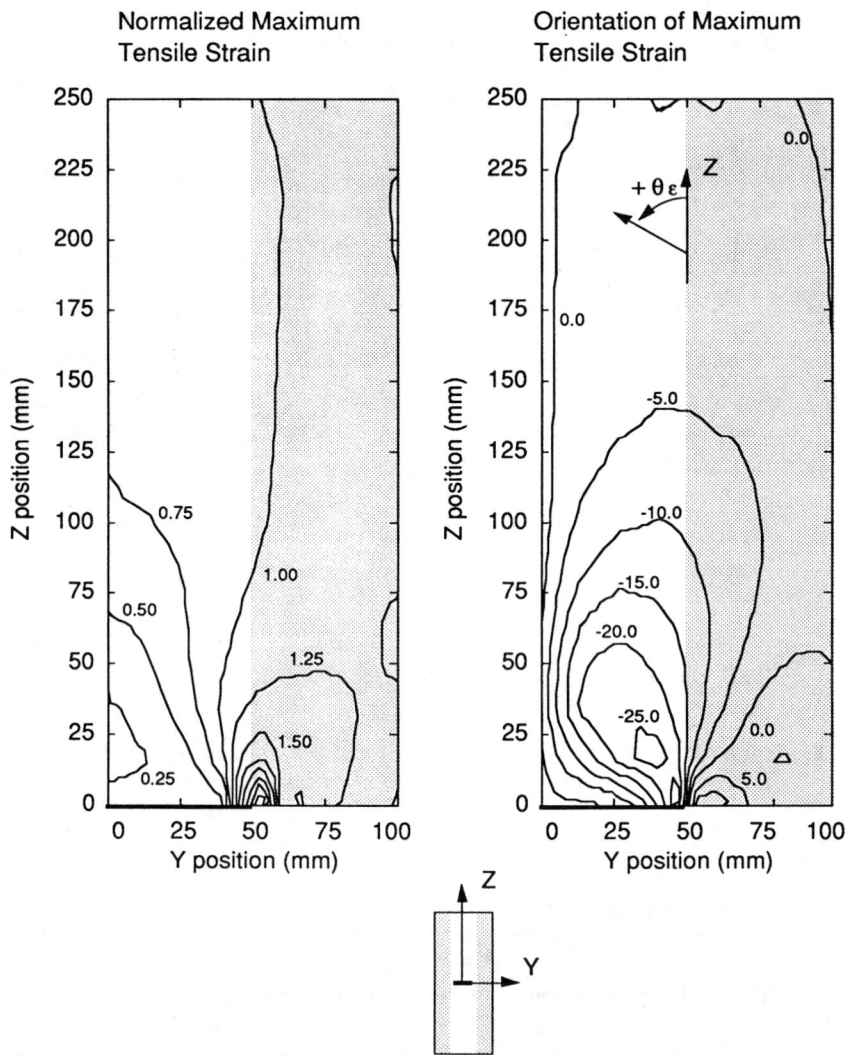

FIG. 9—*Normalized maximum tensile strain and orientation of maximum tensile strain for panel Model 2 with slit length = 100 mm and stiffener width = 50 mm.*

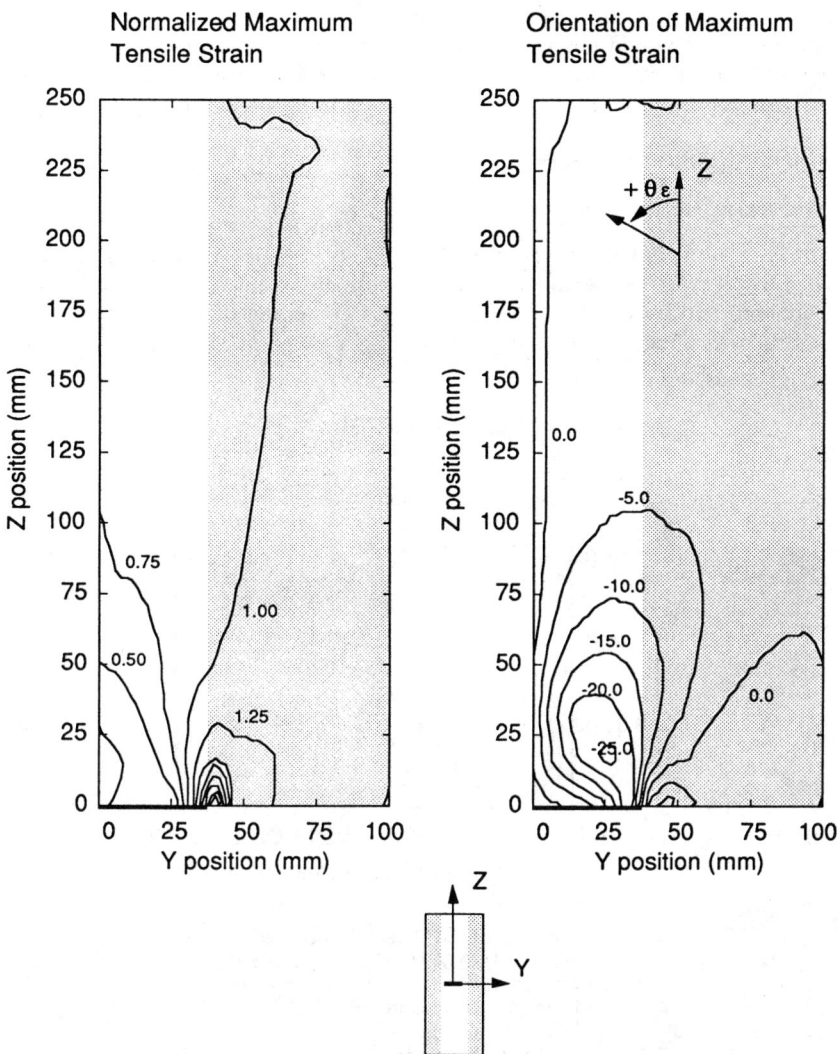

FIG. 10—*Normalized maximum tensile strain and orientation of maximum tensile strain for panel Model 3 with slit length = 75 mm and stiffener width = 62.5 mm.*

and stiffened regions. This verified the constraint that the far-field longitudinal strain in the stiffened region must be the same as the strain in the unstiffened region.

The major difference in the results of the different models was the magnitude of strains in the stiffened region near the slit tip. In Fig. 8, where the slit did not extend to the stiffened region, the intense strain contours near the slit tip did not extend into the stiffened region, and the strains did not intensify in the stiffened region. In Fig. 9, where the slit did extend to the stiffened region, the intense strain contours surrounding the slit tip extended into the stiffened region, and strains intensified in the stiffened region.

In all cases, it was evident that the region ahead of the slit tip showed strain magnitudes greater than those in the far-field unstiffened region. The angles of orientation for maximum tensile strain remained close to zero in the region ahead of the slit tip. Therefore, propagating damage would enter regions of high tensile strain in the direction of loading, and since the angles of orientation for maximum tensile strain remained near zero, the most likely direction of damage propagation would be through the stiffened region.

Experimental-Analytical Comparison

Far-Field Stress-Strain Behavior

The stress-strain behavior of typical A, B, and C-type panel specimens is shown in Figs. 11 through 13. The stress-strain behavior on both the front (the side with photoelastic coating applied) and back sides of the panels was compared to predicted behavior.

The stress-strain behavior of the panel specimens was linear to failure. The Type A

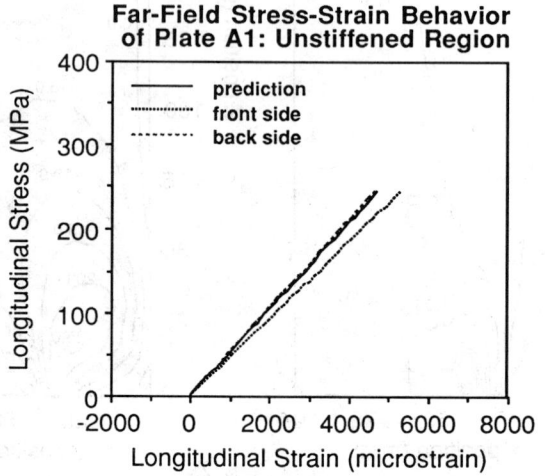

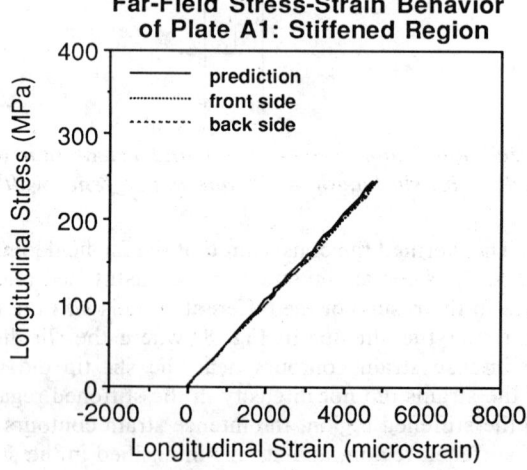

FIG. 11—*Comparison of experimental and analytical far-field stress-longitudinal strain relationships for unstiffened and stiffened regions of panel Type A.*

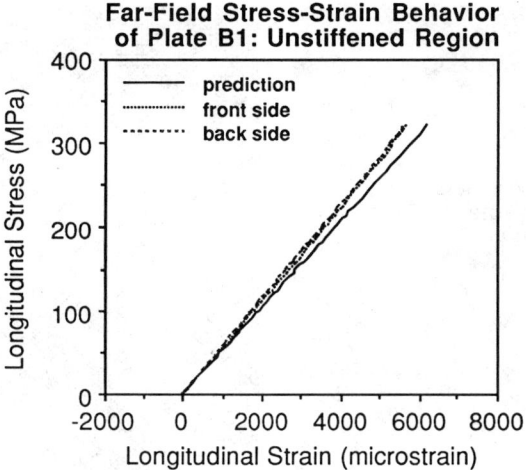

FIG. 12—*Comparison of experimental and analytical far-field stress-longitudinal strain relationships for unstiffened region of panel Type B.*

specimens exhibited greater differences between front and back surface stress-strain behavior than did the Types B and C specimens; this was expected of the Type A laminates, which exhibited bending-stretching coupling.

Photoelastic Data

Typical photoelastic data and computer-enhanced images for the panel specimens were compared to finite element predictions of strain intensity (difference of principal strains) in Figs. 14 through 25. The finite element models were used to predict the strain intensity in the upper right quadrant of the photoelastic data. The digitized photographs, computer-

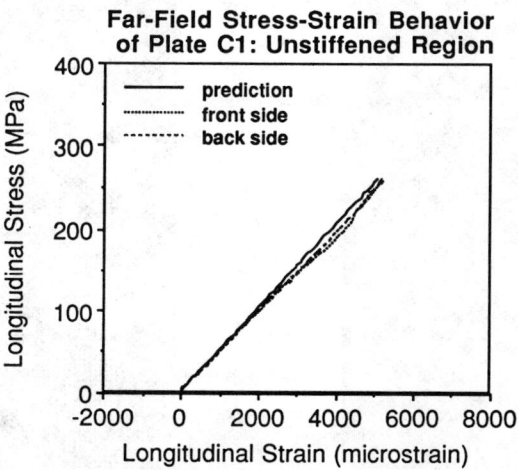

FIG. 13—*Comparison of experimental and analytical far-field stress-longitudinal strain relationships for unstiffened region of panel Type C.*

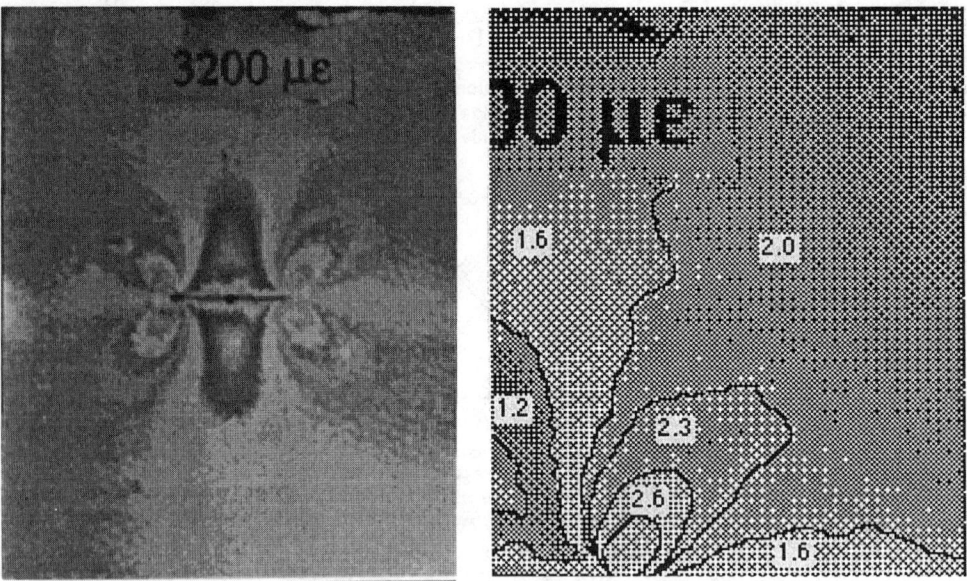

FIG. 14—*Digitized photograph and computer-enhanced image of photoelastic test result for Panel A1 at far-field longitudinal strain of 3000 microstrain.*

enhanced images, and contour plots show the fringe patterns associated with the strain intensity at a far-field longitudinal strain of 3000 microstrain. The numeric magnitudes refer to the fringe order numbers of the photoelastic fringes. The fringe order numbers of computer image fringes for panel specimens B2, B3, C1, and C3 were modified to facilitate

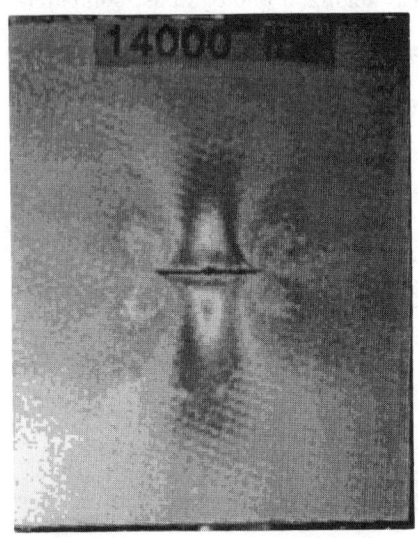
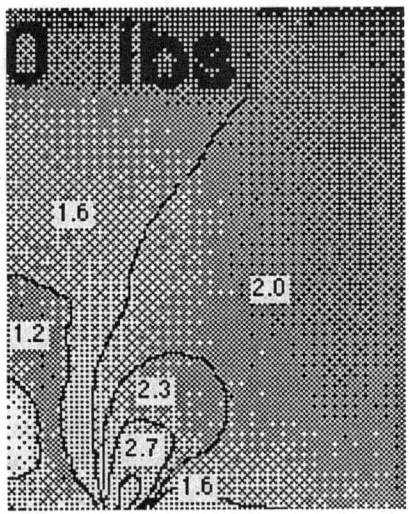

FIG. 15—*Digitized photograph and computer-enhanced image of photoelastic test result for Panel B1 at far-field longitudinal strain of 3000 microstrain.*

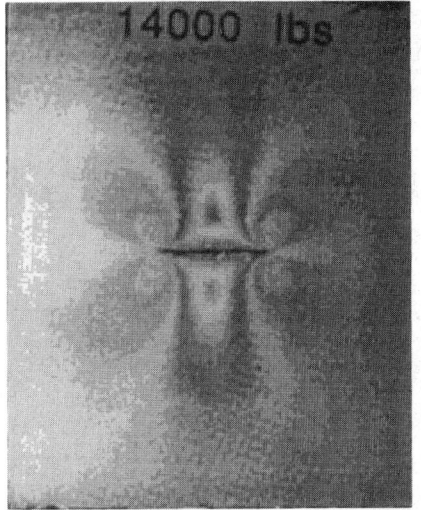 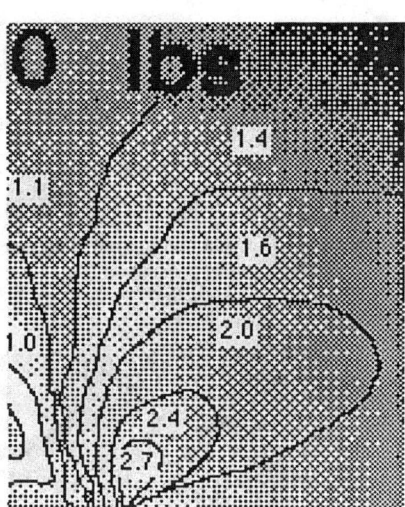

FIG. 16—*Digitized photograph and computer-enhanced image of photoelastic test result for Panel C1 at far-field longitudinal strain of 3000 microstrain.*

comparison to the other panels (this was necessary because of differences in photoelastic sheet thickness).

All panels exhibited regions of high strain intensity near the slit tips. The strain intensity decayed to the far-field value away from the slit. Regions of low strain intensity were present above and below the slit away from the tips.

Panels A2 and A3 had additional regions of low strain intensity near the slit tips not

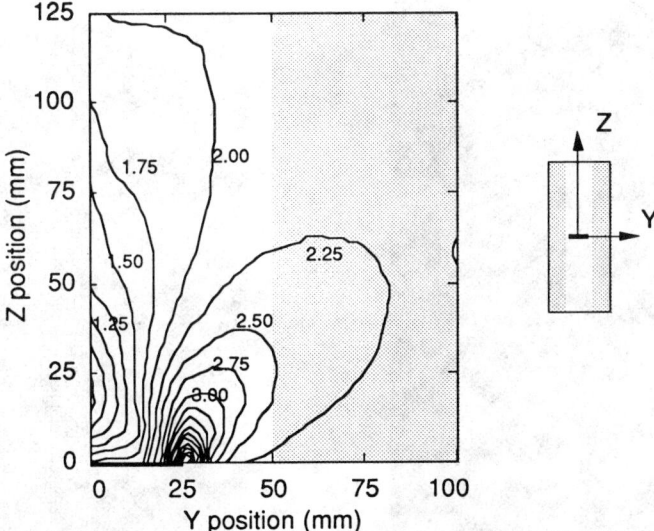

FIG. 17—*Analytical prediction of photoelastic patterns for Panels A1, B1, and C1 at far-field longitudinal strain of 3000 microstrain.*

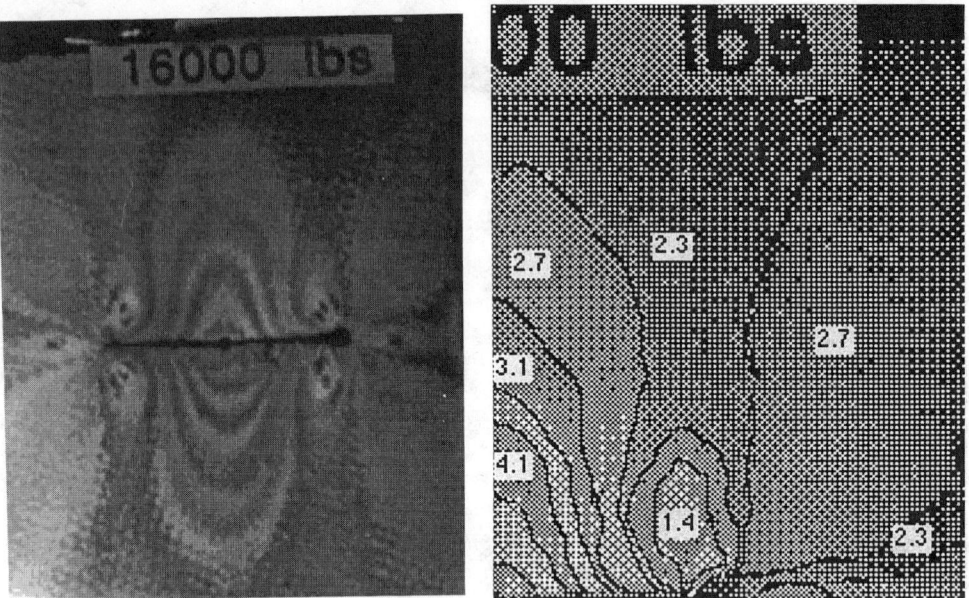

FIG. 18—*Digitized photograph and computer-enhanced image of photoelastic test result for Panel A2 at far-field longitudinal strain of 3000 microstrain.*

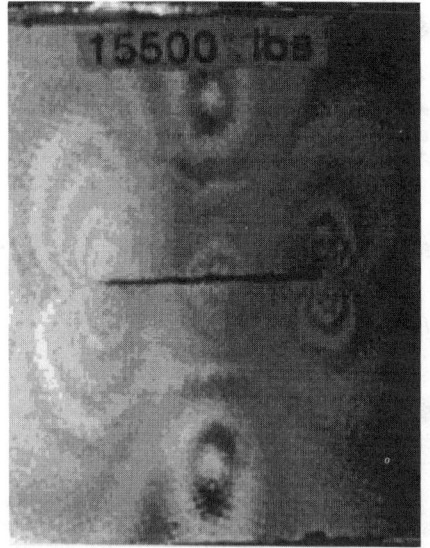 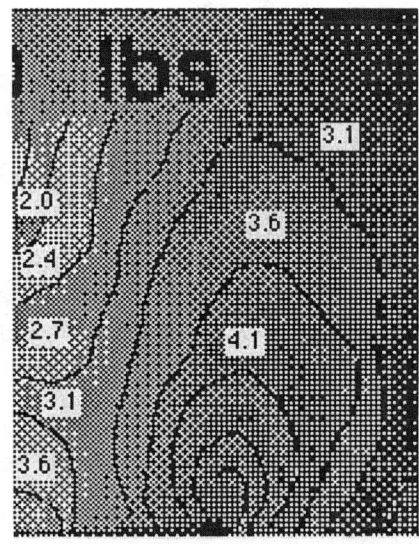

FIG. 19—*Digitized photograph and computer-enhanced image of photoelastic test result for Panel B2 at far-field longitudinal strain of 3000 microstrain.*

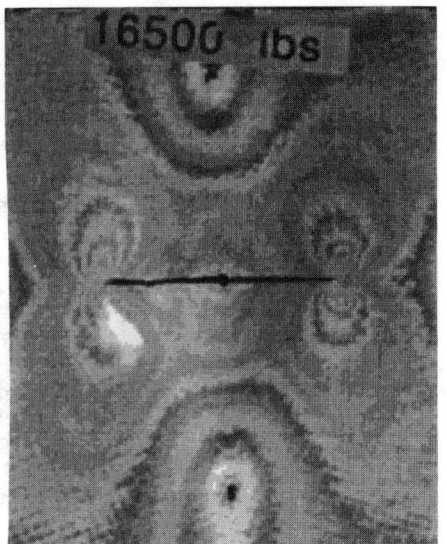 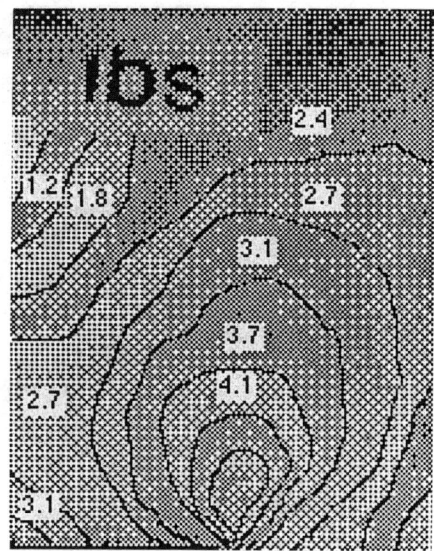

FIG. 20—*Digitized photograph and computer-enhanced image of photoelastic test result for Panel C2 at far-field longitudinal strain of 3000 microstrain.*

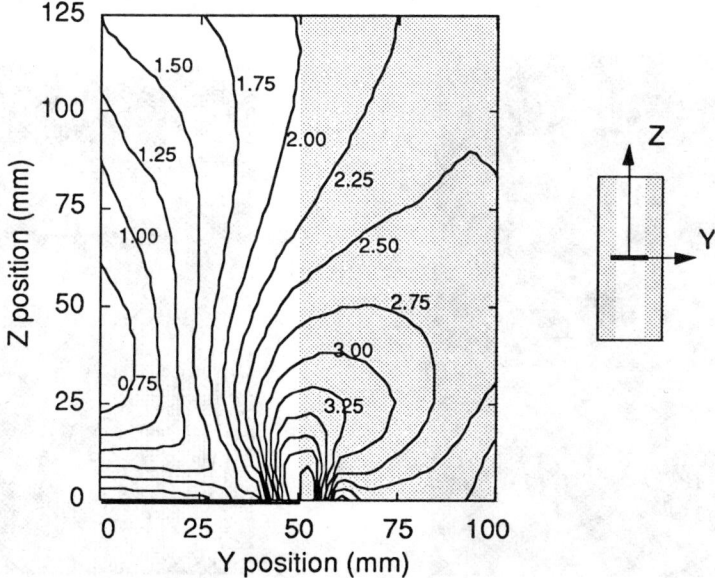

FIG. 21—*Analytical prediction of photoelastic patterns for Panels A2, B2, and C2 at far-field longitudinal strain of 3000 microstrain.*

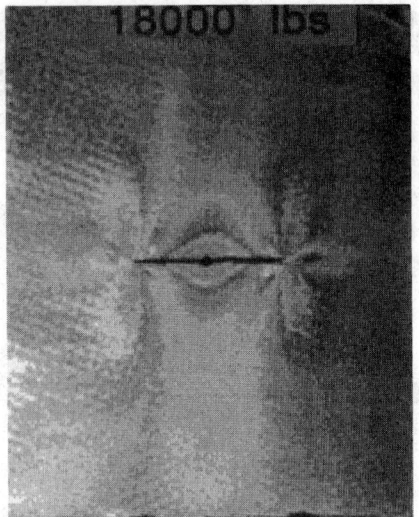
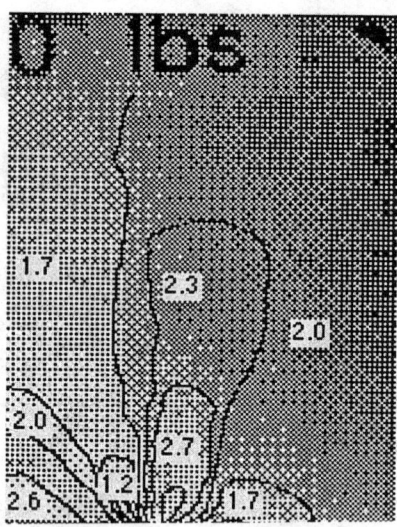

FIG. 22—*Digitized photograph and computer-enhanced image of photoelastic test result for Panel A3 at far-field longitudinal strain of 3000 microstrain.*

present in the other panels. These regions developed in the Type A panels (the panels with bending-stretching coupling) when the slits extended to the stiffened regions. The Types B and C panels with large slits exhibited similar bending-influenced photoelastic effects above and below the slits. In addition to their different photoelastic patterns, these panels exhibited large bending displacements, such that the slit appeared to bluge out-of-plane.

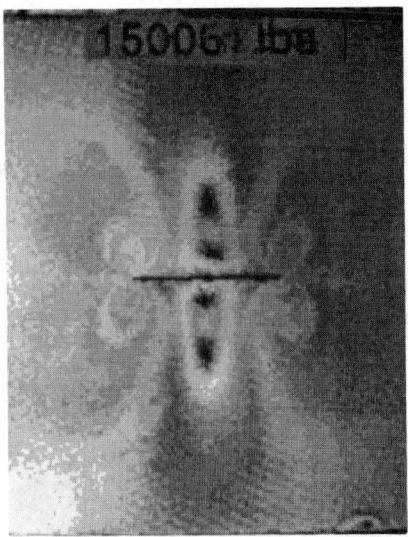
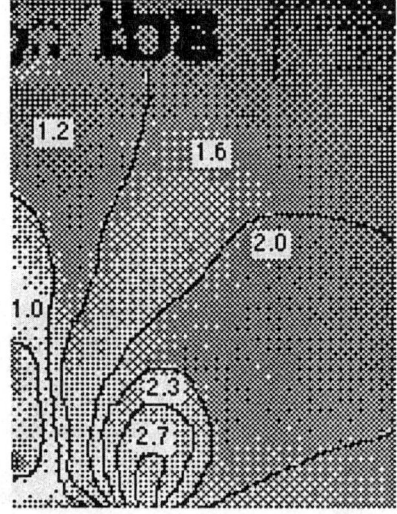

FIG. 23—*Digitized photograph and computer-enhanced image of photoelastic test result for Panel B3 at far-field longitudinal strain of 3000 microstrain.*

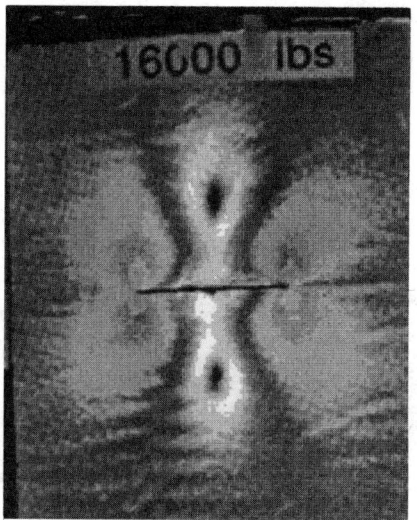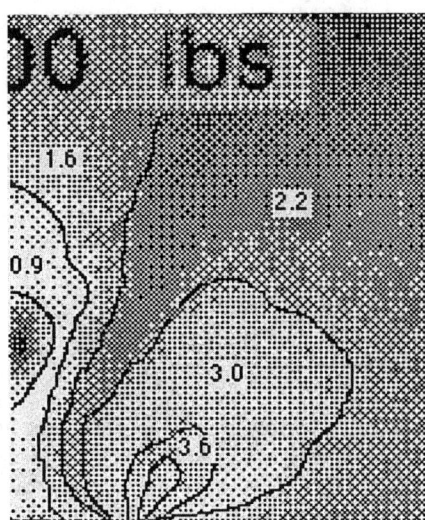

FIG. 24—*Digitized photograph and computer-enhanced image of photoelastic test result for Panel C3 at far-field longitudinal strain of 3000 microstrain.*

Failure Stresses

The experimental failure loads and the corresponding unstiffened region far-field longitudinal failure stresses for the panel specimens are given in Table 10.

Failure stresses of the panel specimens are plotted versus slit length in Fig. 26. The experimental data are compared to predictions based upon the Mar-Lin failure correlation.

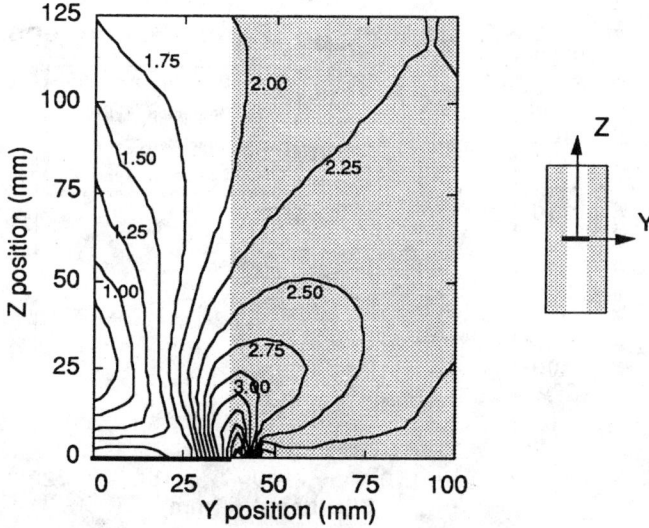

FIG. 25—*Analytical prediction of photoelastic patterns for Panels A3, B3, and C3 at far-field longitudinal strain of 3000 microstrain.*

TABLE 10—*Failure loads and unstiffened region failure stresses for panel specimens.*

Panel Designation	Slit Length, 2a, mm	Failure Load, kN	Failure Stress, MPa
A_1	51	108.14	254.42
A_2	102	103.02	242.38
A_3	71	156.20	330.94
B_1	51	118.82	279.55
B_2	102	99.68	234.52
B_3	71	129.05	273.42
C_1	51	113.92	268.02
C_2	102	76.32	179.56
C_3	71	113.48	240.43

The failure stresses of the panel specimens decreased exponentially with increasing slit size, as was predicted. The failure stresses of the panels were generally higher than predicted, but still within experimental scatter. Specimen A3, which was expected to have bending-stretching coupling in the stiffening strip, was the notable exception to the predicted response.

Failure Modes

Photographs of the failed panels are shown in Figs. 27 through 29. Damage propagated from the slit tips in all panels except Panel B1. Panel B1 failed at the loading tab due to local stress concentrations produced through termination of stiffener plies.

The Type A specimens exhibited damage propagation both along the 45° plies and perpendicular to loading. Their stiffener plies showed a tendency to disbond from the panel at failure. The Types B and C specimens all exhibited damage propagation perpen-

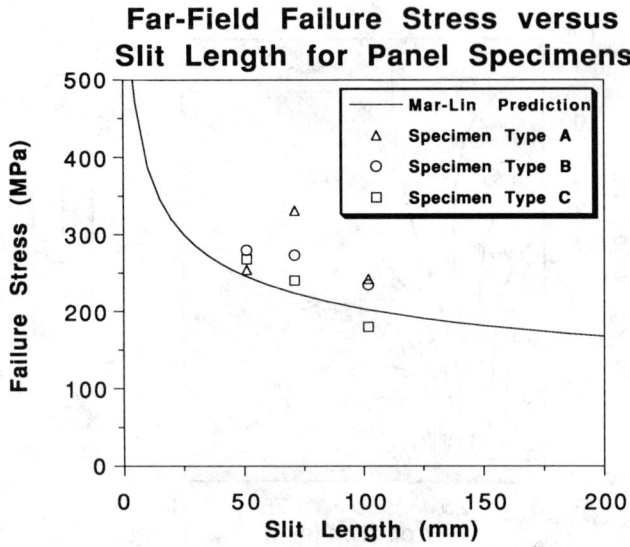

FIG. 26—*Mar-Lin predictions and experimental failure stresses versus slit size for panel specimens.*

FIG. 27—*Post-failure photographs of Panels A1 (slit length 51 mm), A2 (slit length 102 mm), and A3 (slit length 71 mm), from top to bottom.*

dicular to loading. These differences may be attributed to the bending-stretching coupling inherent in the nonsymmetric A laminate which exacerbated the failure in the relatively weak epoxy layer, separating the unstiffened panel and the stiffening strips.

None of the specimens exhibited damage redirection or arrest upon reaching the stiffened regions. This result was consistent with the finite element plate models, which indicated that the maximum tensile strains in the stiffening strips were aligned with the loading direction.

Summary and Conclusions

Predictions of far-field longitudinal failure stress agreed well with established failure correlations for notched composites [8–11]. What makes these results interesting, however, is the implication that the presence of the stiffeners does not affect the failure initiation behavior—it is the stiffener influence on the orientation of the principal strain field in the region of the flaw that affects the damage propagation path.

The two-dimensional plane stress finite element models were consistent with the photoelastic behavior of panels that did not exhibit significant bending response to loading (the Types B and C panels with short slits). Bending effects were not accounted for in the finite element models and, therefore, the models did not accurately predict the

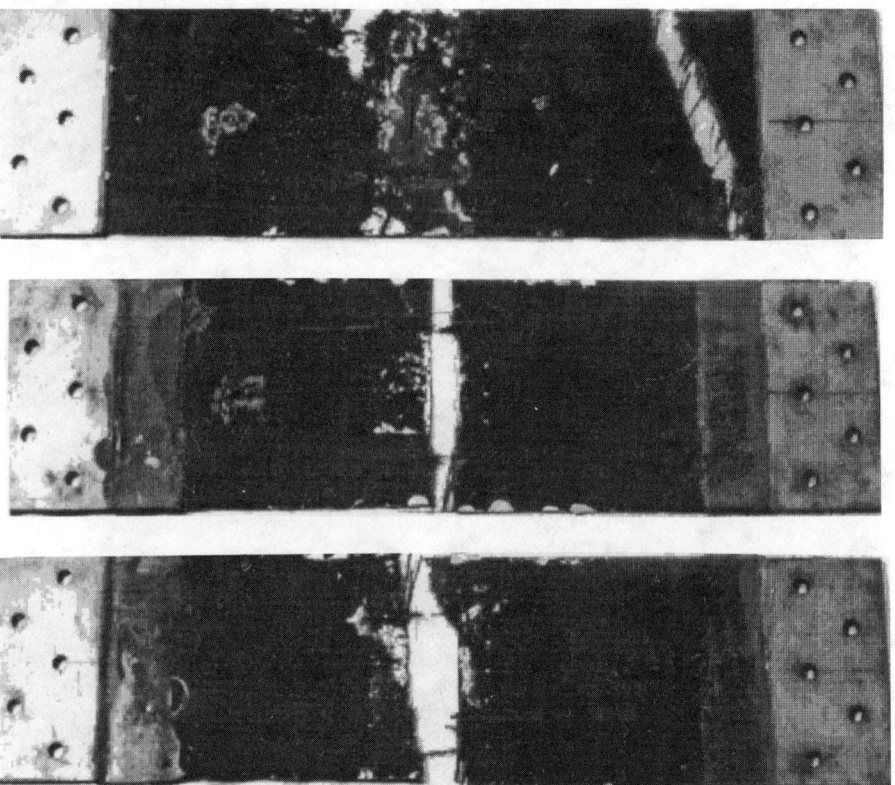

FIG. 28—*Post-failure photographs of Panels B1 (slit length 51 mm), B2 (slit length 102 mm), and B3 (slit length 71 mm), from top to bottom.*

photoelastic behavior of the panels that exhibited significant bending-stretching coupling. The models, however, did predict the overall magnitudes of strain intensity of the photoelastic images.

The Type A specimens exhibited damage propagation both along the 45° plies and perpendicular to loading. Their stiffening plies showed a tendency to disbond from the panel at failure. All of the Types B and C specimens exhibited damage propagation perpendicular to loading. These differences may be attributed to the local bending-stretching coupling inherent in the nonsymmetric A laminate layup as evidenced by the photoelastic data and the relatively weak epoxy layer separating the unstiffened panel and the stiffening strips. As can be seen in Fig. 10, the magnitude of the maximum tensile strain increases by a factor of at least two directly in front of the slit in the stiffener region, while the orientation of the strain remains perpendicular to the slit. It is hypothesized that damage propagates perpendicular to a principal strain direction, so that, in the case of these stiffened panels, damage propagated in a direction parallel to the slit.

Data and analysis from identical layups used in 304-mm-diameter pressurized cylinder tests [7] show that both the magnitudes and orientation of these strains ahead of the slit and into the stiffened region can dramatically affect the damage propagation path. In the case of the cylinders, the stiffening strips redirected the damage propagation whereas, in the flat panels, the damage continued through the stiffened region. It is concluded that the

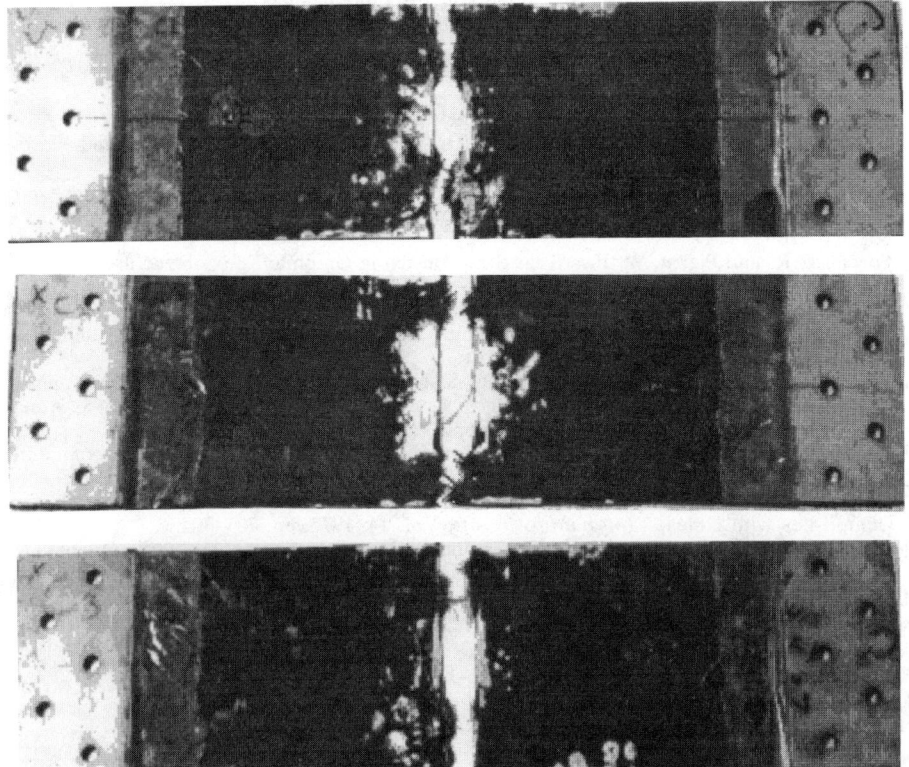

FIG. 29—*Post-failure photographs of Panels C1 (slit length 51 mm), C2 (slit length 102 mm), and C3 (slit length 71 mm), from top to bottom.*

maximum tensile strain and its orientation ahead of the slit plays a key role in the prediction of subsequent damage propagation in these laminates. It is also concluded that the structural and material couplings due to the unsymmetric layups in the stiffener regions of these flat panels have little, if any, effect on the ability of the stiffener to redirect damage propagation since, in all cases, damage progressed through the stiffeners. This is in contrast to the inherent structural coupling in shell structures which, for even symmetrical layups, have been shown to produce dramatic damage path redirection due to the stiffening strips.

Acknowledgments

This work was performed in the Technology Laboratory for Advanced Composites (TELAC) of the Department of Aeronautics and Astronautics at the Massachusetts Institute of Technology. This work was sponsored by NASA Research Grant NAG-1-991.

References

[1] Horton, R. E., Whitehead, R. S., et al., "Damage Tolerance of Composites—Vols. I. II, and III," AFWAL-TR-87-3030, July 1988.

[2] "Aircraft Structural Integrity Program Airplane Requirements," MIL-STD-1530A (11) USAF, 11 December 1975, superseded MIL-STD-1530 USAF, 11 September 1972.

[3] Kanninen, M., Mills, E., Hahn, G., et al., "A Study of Ship Hull Crack Arrester Systems," Technical Report SSC-265, Naval Systems Command, Department of the Navy, Washington, DC, 1976.

[4] Bhatia, N. M. and Verette, R. M., "Crack Arrest of Laminated Composites," *Fracture Mechanics of Composites, ASTM STP 593*, American Society for Testing and Materials, Philadelphia, 1976, pp. 200–214.

[5] Sendeckyj, G. P., "Concepts for Crack Arrestment in Composites," *Fracture Mechanics of Composites, ASTM STP 593*, American Society for Testing and Materials, Philadelphia, 1976, pp. 215–226.

[6] Porter, T. R. and Pierre, W. F., "Tear Strap Design in Graphite/Epoxy Structure," NASA TM 84116, Vol. II, January 1981, pp. 265–280.

[7] Sawicki, A. J., "The Failure of Integrally Stiffened Composite Plates and Cylinders," master's thesis, TELAC Report 90-17, Department of Aeronautics and Astronautics, Massachusetts Institute of Technology, September 1990.

[8] Lagace, P. A., "Notch Sensitivity and Stacking Sequence of Laminated Composites," *Composite Materials: Testing and Design (Seventh Conference), ASTM STP 893*, American Society for Testing and Materials, Philadelphia, 1986, pp. 161–176.

[9] Mar, J. W. and Lin, K. Y., "Fracture of Boron/Aluminum Composites with Discontinuities," *Journal of Composite Materials*, Vol. 11, 1977, pp. 405–421.

[10] Mar, J. W. and Lin, K. Y., "Fracture Mechanics Correlation for Tensile Failure of Filamentary Composites with Holes," *Journal of Aircraft*, Vol. 14, 1977, pp. 703–704.

[11] Fenner, D. N., "Stress Singularities in Composite Materials with an Arbitrarily Oriented Crack Meeting an Interface," *International Journal of Fracture*, Vol. 12, 1975, pp. 705–721.

[12] "Automatic Dynamic Incremental Nonlinear Analysis-Input: User's Manual," Report ARD 87-4, ADINA R. and D., Inc., 1987.

[13] Bathe, K. J., *Finite Element Procedures in Engineering Analysis*, Prentice-Hall, Inc., Englewood Cliffs, NJ, 1982.

G. D. Chu[1] *and C. T. Sun*[1]

Failure Initiation and Ultimate Strength of Composite Laminates Containing a Center Hole

REFERENCE: Chu, G. D. and Sun, C. T., **"Failure Initiation and Ultimate Strength of Composite Laminates Containing a Center Hole,"** *Composite Materials: Fatigue and Fracture, Fourth Volume, ASTM STP 1156,* W. W. Stinchcomb and N. E. Ashbaugh, Eds., American Society for Testing and Materials, Philadelphia, 1993, pp. 35–54

ABSTRACT: A failure analysis method that includes the free edge effect was proposed for failure prediction in composite laminates containing a center hole. The location of onset of failure in the hole was determined by comparing the strengths of off-axis coupon specimens with the hoop stress around the hole. The subsequent failure progression was performed by the ply-by-ply classical laminate failure analysis. Experiments on $[0/90/\pm 45]_S$ and $[0/90]_S$ laminates and their off-axis specimens were performed to verify the present method.

KEY WORDS: failure, composite, laminate, hole, free edge, hoop stress

The strength of laminates containing cutouts has been a subject of immense interest as evidenced by the long list of publications. Various methods have been proposed for strength prediction of these laminates. A good review of papers on this subject was given by Awerbuch and Madhukar [1].

Generally speaking, the existing strength prediction methods can be categorized into three groups. The first method is called the stress fracture model, which was initially proposed by Whitney and Nuismer [2,3]. In this model, a critical length from the cutout must be chosen so that the point stress or the average stress over the critical length is equal to the ultimate stress of the unnotched laminate. Tan [4,5] extended this concept for more general situations. The main drawback in this method is that the critical length is basically a curve-fitting parameter which is dependent on the stacking sequence, the geometry of the cutout, and the loading condition [6].

The second method can be called the equivalent crack method since a crack is used to represent the initial flaw [7] or to describe the damage progression in the notched laminate [8]. As the form of damage in laminates is quite different from that of a crack, this representation with a crack is questionable. At most, the fracture mechanics parameters such as stress intensity factor and strain energy release rate are used as curve-fitting parameters.

The third method is basically a ply-by-ply failure progressive analysis using lamination theory [9,10] to predict the failure initiation and ultimate strength of notched laminates. Certain failure criteria were utilized to determine failure and failure modes of each lamina, and some rules were employed to estimate the degradation of the mechanical properties in the damaged region. Because of the limitation of the 2-D nature in lamination theory, the

[1]Graduate student and professor, respectively, School of Aeronautics and Astronautics, Purdue University, West Lafayette, IN 47907.

effect of interlaminar stresses on the failure of laminates was not included in the failure analysis.

Since free edge stresses are present near the cutout, failure often emanates from the boundary of the cutout. This phenomenon was observed by Shalev and Reifsnider [11] as well as by O'Brien and Raju [12], who studied laminates containing a circular hole. O'Brien and Raju [12] analyzed interlaminar stresses around the hole and calculated the mixed mode strain energy release rate for an assumed crack in every interface. Delamination was assumed to occur when the Mode I strain release rate reached a critical value. They found that the delamination location is not at the location of maximum hoop stress.

In this study, a new method is proposed for predicting the strength of laminates containing a circular hole. Initial failure location around the hole is determined based on the strengths of off-axis specimens cut from the laminate. The subsequent damage progression is analyzed using the classical ply-by-ply failure procedure in addition to the laminate strength at the edge of the hole. This approach is different from that of Chang [10]. The analytical predictions are compared with experimental results for $[0/90/\pm45]_s$ and $[0/90]_s$ laminates.

Specimen and Test Procedure

Specimens

Two types of laminates were tested in this study. The first type was $[0/90/\pm45]_s$ quasi-isotropic laminate, and the second was $[0/90]_s$ cross-ply laminate. Off-axis specimens were cut from 30 by 30-cm laminate panels.

Both unnotched and notched specimens were tested for strength. All the unnotched specimens were 25.4 mm wide and 300 mm long. End tabs 3.8 cm in length were used.

Two different specimen sizes corresponding to two hole sizes, 6.35 and 12.7 mm in diameter, were used for the $[0/90/\pm45]_s$ laminate. The specimen containing a 6.35 mm hole was 25.4 mm wide. The one containing a 12.7-mm hole was 50.8 mm wide. Only one hole size (6.35 mm) was considered in the $[0/90]_s$ laminate.

During the tension test, some of the off-axis (7.5 and 15°) $[0/90/\pm45]_s$ specimens failed along the straight edges instead of the hole edge, indicating that the straight edge of the coupon specimen could be weaker than the hole edge. To ensure that failure would occur at the hole, shallow notches were created along the straight edges using a diamond wheel. These edge notches increased the edge strength, as found by Sun and Chu [13].

Tension Test

All specimens in this study were mechanically loaded at a stroke rate of 0.0254 cm/min on an MTS machine. Hydraulic grips were used to clamp the specimens. All tests were performed at room temperature.

For unnotched specimens, four replicas were tested for strength. For specimens containing a center hole, five replicas were used.

Damage Inspection

Attention was focused on damage initiation and progression in notched specimens. In general, the specimen was loaded to a number of load levels before final failure. At each load level, the specimen was removed from the loading machine, and a radiograph was taken by using techniques of diiodobutane (DIB)-enhanced X-ray examination. These radiographs were used to determine the damage initiation location and to observe the subsequent damage propagation.

Five notched specimens for each loading angle (0, 7.5, and 15°) were tested in tension. The first specimen was loaded monotonically to final failure to obtain ultimate strength. For the $[0/90/\pm 45]_S$ laminate, the second specimen was loaded at small load steps starting at 20% of ultimate strength with 5% increments until reaching 40% of ultimate strength. Initial failure occurred below the 40% load level. The subsequent load levels were 60, 80, 90, and 95%, which would reveal the damage propagation characteristics. With the damage initiation load level obtained from the second specimen, load levels for the remaining three specimens were chosen to be 35, 40, 60, 80, 90, and 95%.

The notched $[0/90]_S$ specimens were used mainly to determine failure initiation location and damage progression behavior. The load steps were chosen as 60, 80, 90, and 95% of the ultimate strength of the notched specimen.

It should be noted that it was difficult to define accurately the location of damage initiation based on the radiographs because of the small size of the hole and the finite size of the damage. Thus, the range in angular position along the hole boundary was indicated.

Failure Initiation

Experimental results indicate clearly that failure in the notched laminate specimens initiated from the boundary of the hole. Figures 1 and 2 show the radiographs of the specimens loaded to 90% of the ultimate strength. Common to all specimens are the matrix cracks present in the surface plies. For the 0° specimen, there are four surface cracks running in the 0° direction. These cracks are symmetrically situated with respect to the center of the hole. For the off-axis specimens, only a pair of surface matrix cracks are present. Again, these two cracks are symmetrically located with respect to the center of the hole. The failure initiation location was estimated based on the radiographs at 40% ultimate strength for the $[0/90/\pm 45]_S$ laminate and at 60% ultimate strength for the $[0/90]_S$ laminate. The failure initiation locations in the first quadrant for all specimens are listed in Tables 1 and 2. It is evident that failure does not initiate at the maximum hoop stress location.

It is then obvious that free edge stresses at the hole boundary must have played an important role in failure initiation. In fact, many researchers [14–16] have found that many coupon specimens of composite laminates fail as a result of interlaminar stresses at the free edge. To predict free edge failure, the 3-D stress field near the free edge must be obtained first. In addition, failure criteria based on these free edge stresses must be available. Such a procedure would be extremely expensive computationally. Moreover, said analysis may not necessarily be accurate.

An alternative to performing the 3-D failure analysis is to resort to a series of off-axis coupon specimen tests. Near the hole boundary, the laminate is under a state of "unidirectional" hoop stress. The strength of the laminate at a point along the hole may be estimated by testing the uniaxial strength of a coupon specimen with the same stacking sequence at the location of interest. This would allow the use of 2-D analysis in which the laminate is regarded as a 2-D plane stress homogeneous orthotropic solid.

Unnotched Laminate Strength

The laminate strength along the circular free edge boundary of the hole was estimated by measuring the strengths of unnotched off-axis laminate coupon specimens. The off-axis angles considered were 0, 7.5, 15, and 22.5° for the $[0/90/\pm 45]_s$ laminate and 0, 7.5, 15,

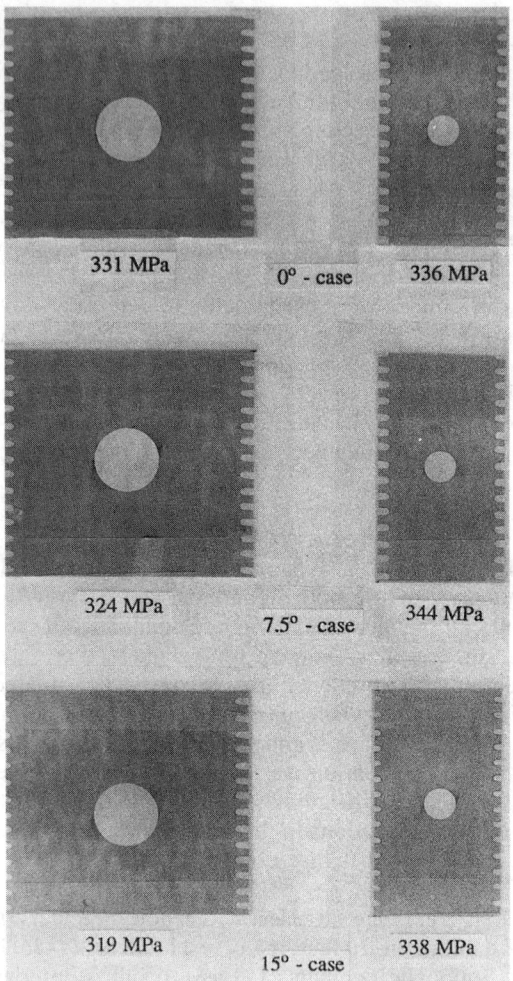

FIG. 1—*Radiographs of damage around the hole for [0/90/±45]$_s$ laminate with different loading angles at 90% failure strength.*

22.5, 30, 37.5, and 45° for the [0/90]$_s$ laminate. The average tensile strength based on four specimens and the standard deviation are listed in Table 3. The experimental results indicate that the strengths of both laminates depend highly on the loading angle ψ (see Fig. 3).

Strictly speaking, to cover the entire boundary of the hole, the off-axis angle of the unnotched specimen must cover the range from 0 to 90°. However, the failure initiation locations observed experimentally were all located within the range ψ = 0 to 15°. To facilitate the analysis, the experimental results were extended as a periodic function of ψ, (Fig. 4). Thus, for the [0/90]$_s$ laminate, the period is 90°, and for the [0/90/±45]$_s$ laminate, the period is 45°. This periodic behavior was observed by Sun and Zhou [14] to be more or less valid for the [0/90/±45]$_s$ laminate.

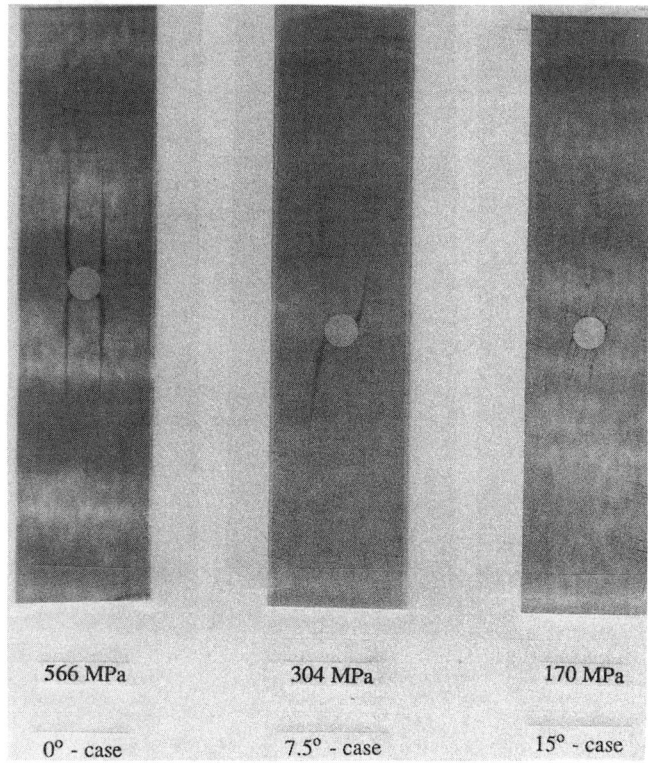

566 MPa   304 MPa   170 MPa

0° - case   7.5° - case   15° - case

FIG. 2—*Radiographs of damage around the hole for [0/90]ₛ laminate with different loading angles at 90% failure load.*

TABLE 1—*Locations of failure initiation in the first quadrant of the [0/90/±45]ₛ laminate containing a center hole under on- and off-axis loadings.*

	Location of Failure Initiation, θ					
	0°		7.5°		15°	
Loading Angle Specimen	Experimental	Predicted	Experimental	Predicted	Experimental	Predicted
50.8 mm width	12°		8°		4°	
	|	10°	|	5°	|	2.5°
12.7-mm hole	9°		3°		0°	
25.4 mm width	11°		8°		3°	
	|	10°	|	5°	|	2.5°
6.35-mm hole	9°		3°		0°	

TABLE 2—*Locations of failure initiation in the first quadrant of the [0/90]ₛ laminate containing a center hole under on- and off-axis loadings.*

	Location of Failure Initiation, θ	
Off-axis Angle	Experimental	Predicted
0°	16°–13°	15°
7.5°	12°–8°	10°
15°	7°–3°	6°

Prediction of Failure Initiation

Consider a notched laminate subjected to uniform stress σ_N^∞ as shown in Fig. 5. At a point θ on the hole boundary, the loading angle ψ is given by (see Fig. 5)

$$\psi = \theta + \phi \tag{1}$$

It is assumed that failure would occur if

$$\sigma_{ult}(\psi) = \sigma_{\theta\theta} \tag{2}$$

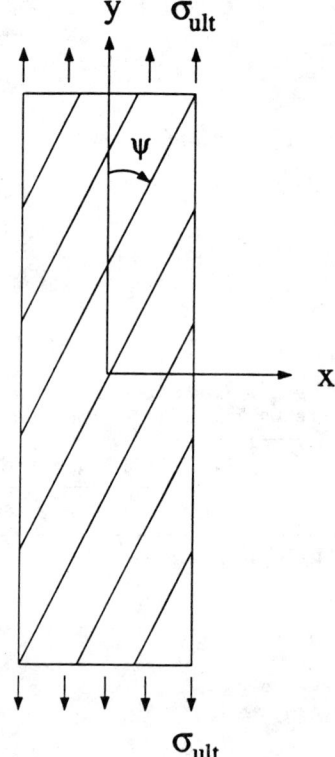

FIG. 3—*Off-axis specimen under uniaxial loading.*

TABLE 3—*Strengths of unnotched off-axis laminate specimens.*

	Failure Strength, MPa			
	$[0/90/\pm45]_s$		$[0/90]_s$	
Off-axis Angle	Average	Deviation	Average	Deviation
0°	647	11.3	915	15.0
7.5°	428	12.4	450	22.0
15°	394	10.3	270	12.0
22.5°	378	13.1	206	7.0
30°	...	...	166	5.0
37.5°	...	...	150	5.0
45°	...	...	137	4.0

where $\sigma_{\theta\theta}$ is the hoop stress at that point. Note that the actual state of stress around the hole is 3-D in nature. However, in this study, the in-plane failure stress (σ_{ult}) was measured from coupon specimens that included the effect of the 3-D free edge stresses. Thus, the 2-D hoop stress was used in Eq 2 for failure predictions. To determine the failure initiation location, we consider

$$\frac{\sigma_{ult}/\sigma_0}{\sigma_{\theta\theta}/\sigma_N^\infty} = k(\theta, \phi) \tag{3}$$

and find the θ that produces the minimum value of k for a given ϕ. From Eqs 2 and 3, it is evident that the load at which failure initiates is given by

$$\sigma_{Ni}^\infty = \sigma_0 k_{min} \tag{4}$$

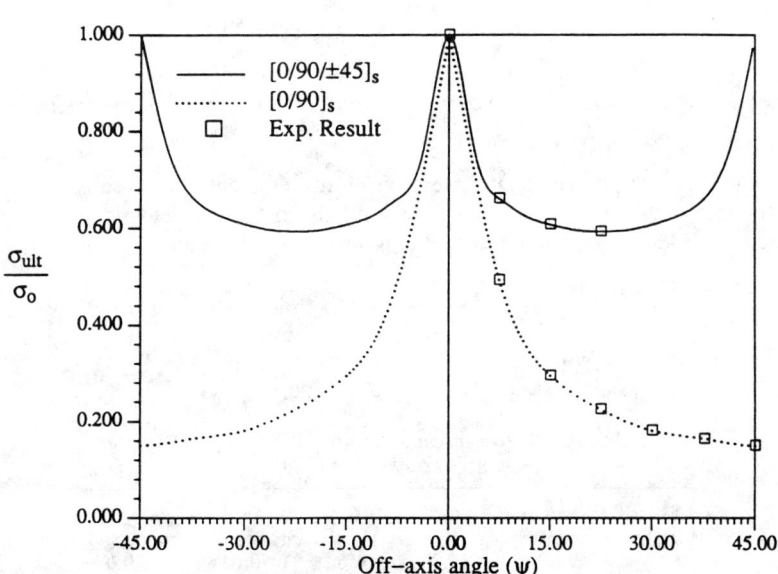

FIG. 4—*Strength ratio σ_{ult}/σ_0 of unnotched $[0/90/\pm45]_s$ and $[0/90]_s$ laminates subjected to off-axis loadings (σ_0 is the strength of 0° specimen).*

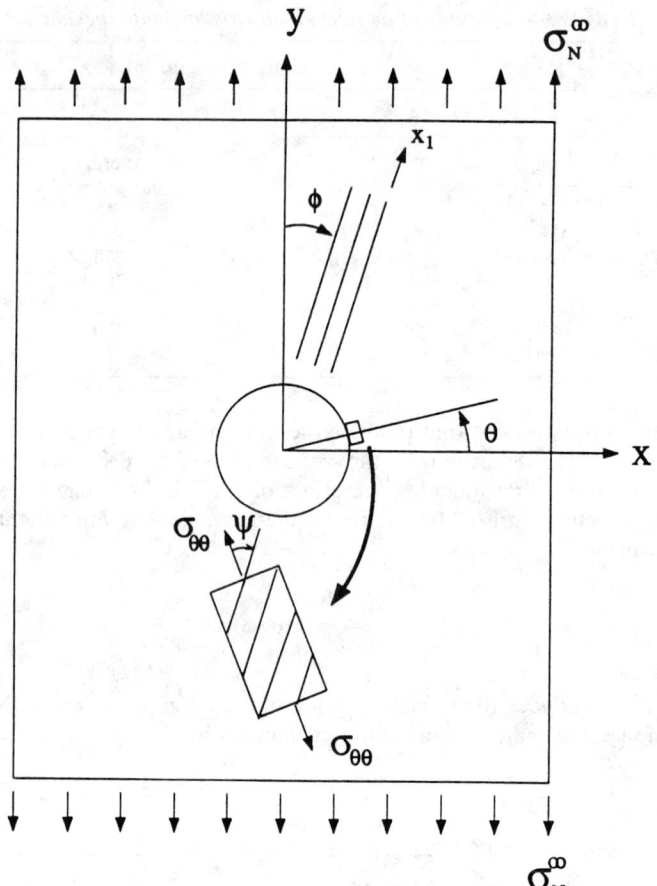

FIG. 5—*Coordinates for composite laminate containing a center hole.*

A four-noded isoparametric 2-D plane stress finite element was used to perform stress analysis to obtain the hoop stress, $\sigma_{\theta\theta}$, around the hole. The elastic properties of the unidirectional AS4/3501-6 graphite/epoxy composite are given as

$$E_1 = 139\,\text{GPa}, \quad E_2 = 9.86\,\text{GPa}, \quad G_{12} = 6.21\,\text{GPa}, \quad \nu_{12} = 0.30 \tag{5}$$

The ply thickness is 0.127 mm.

TABLE 4—*Effective modulus of $[0 + \psi/90 + \psi]_s$ laminates (units expressed in GPa).*

ψ	E_x	E_y	G_{xy}	ν_{xy}	$\eta_{xy,x}$
0°	74.8	74.8	6.62	0.04	0.0
7.5°	64.8	64.8	7.00	0.17	−0.108
15°	47.5	47.5	8.32	0.40	−0.222

Effective elastic properties were calculated using classical laminate theory. The [0/90/±45]$_s$ laminate can be considered as an isotropic plate with the effective moduli given as

$$E = 55.0 \text{ GPa}, \quad G = 21.3 \text{ GPa}, \quad \nu = 0.30 \qquad (6)$$

The off-axis specimens cut from the [0/90]$_s$ laminate can be regarded as [ψ/90 + ψ]$_s$ laminates with respect to the x-y coordinate system. The effective moduli for $\psi = 0, 7.5,$ and 15° are listed in Table 4.

Figures 6 and 7 show the strength distribution function $k(\theta, \phi)$ for the [0/90/±45]$_s$ and [0/90]$_s$ laminates, respectively. For the 0° specimen for both laminates, it is possible to have two failure initiation locations in the $-90° \leq \theta \leq 90°$ range. For other off-axis specimens, a single location is found in the same range.

The predicted initial failure locations and the experimental results are compared in Tables 1 and 2. The predictions are quite accurate.

Table 5 lists the experimental and predicted failure initiation loads for the off-axis specimens of the [0/90/±45]$_s$ laminate. For the [0/90]$_s$ laminate, only the predicted values are presented because the first load increment in the test was set at 60% of the ultimate strength, which was higher than the failure initiation load.

From Table 5, it is seen that the predicted failure initiation loads agree with the experimental data fairly well. The predictions are slightly lower. This may be due to the fact that the load step used in the experiment was not small enough to catch the precise moment failure occurred.

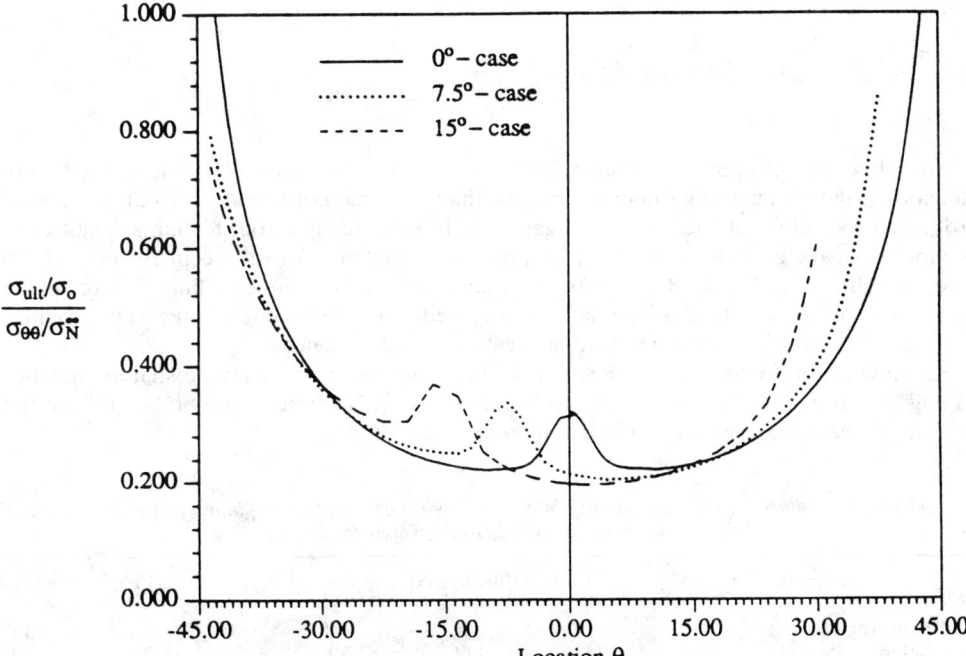

FIG. 6—*Strength distribution* k(θ) *around the hole (6.35 mm diameter) in the [0/90/±45]$_s$ laminate for different loading angles.*

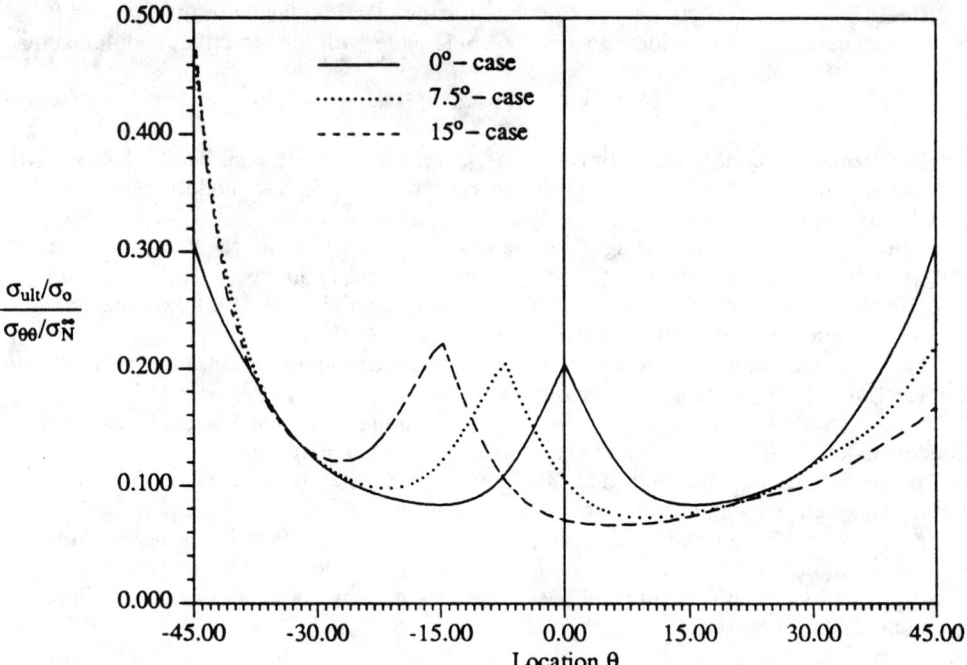

FIG. 7—*Strength distribution* k(θ) *around the hole (6.35 mm diameter) in the [0/90]ₛ laminate for different loading angles.*

Damage Progression and Final Failure

Experimental Result

In a laminate coupon specimen with a center hole, there are two competing failure locations; the straight edge may be weaker than the hole edge, and final failure may be triggered by failure at the straight edges [14]. Indeed, results from testing straight edge coupon specimens of the [0/90/±45]ₛ laminate indicate that off-axis specimens failed from the straight edge. Figure 8 presents the radiographs taken close to final failure of the specimen. It is easy to see that failure occurred along the straight edge. The fracture surface of the broken specimen also suggested the same scenario.

In view of the foregoing, all [0/90/±45]ₛ specimens were cut to create shallow notches along the straight edges. According to Sun and Chu [12], these edge notches reduce the effect of free edge stresses and hence increase the strength.

TABLE 5—*Failure initiation loads* σ_{Ni}^{∞} *(MPa) of [0/90/±45]ₛ and [0/90]ₛ laminates containing a center hole under on- and off-axis loadings.*

Laminate	[0/90/±45]ₛ			[0/90]ₛ		
Off-axis Angle	0°	7.5°	15°	0°	7.5°	15°
Predicted, MPa	143	139	130	76	66	60
Experimental, MPa Avg	165	149	140	...	...	...
Experimental, MPa deviation	8.1	5.1	4.7	...	...	...

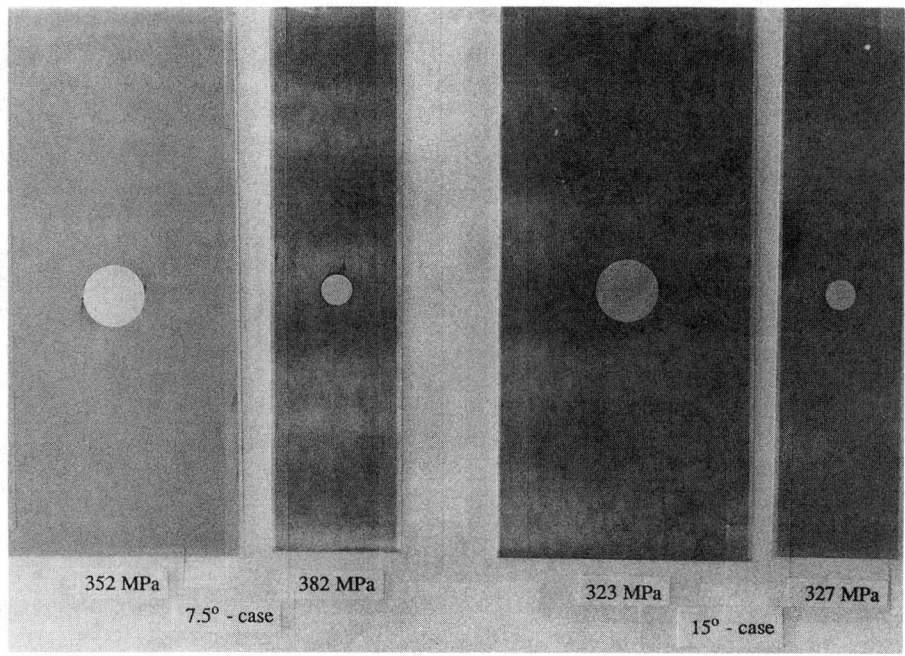

352 MPa   382 MPa   323 MPa   327 MPa

7.5° - case   15° - case

FIG. 8—*Radiographs of edge failure in 7.5° and 15° specimens of notched [0/90/±45]$_s$ laminates with straight edges.*

Table 6 lists the ultimate stresses of [0/90/±45]$_s$ and [0/90]$_s$ specimens containing a center hole. One group of the [0/90/±45]$_s$ specimens had straight edges and the other had both edges notched. Comparing the strengths of both groups, we note that the strengths of the off-axis specimens with notched edges are higher. From the broken specimen, it was also confirmed that failure of the specimen with notched edges emanated from the hole.

It is interesting to note that all the [0/90/±45]$_s$ off-axis specimens have a similar strength although their corresponding unnotched strengths (see Table 3) are significantly different.

The damage progression in the off-axis [0/90]$_s$ specimens is shown in Fig. 9. In the 0° specimen, four matrix cracks in the surface laminas propagated in a stable manner in the 0° direction. The final fracture failure, however, occurred near the hole.

TABLE 6—*Failure stresses (MPa) of [0/90/±45]$_s$ laminate containing a center hole under on- and off-axis loadings.*

		[0/90/±45]$_s$						[0/90]$_s$		
		Edge Unnotched			Edge Notched			Edge Unnotched		
Specimen										
Off-axis Angle		0°	7.5°	15°	0°	7.5°	15°	0°	7.5°	15°
50.8 mm width	Average	373	353	337	376	376	373	⋯	⋯	⋯
12.7-mm hole	Deviation	12.1	13.2	11.5	19.7	5.6	6.1	⋯	⋯	⋯
25.4 mm width	Average	375	344	331	373	384	379	614	314	204
6.35-mm hole	Deviation	17.4	10.2	9.0	7.9	7.7	5.7	10.2	14.1	2.1

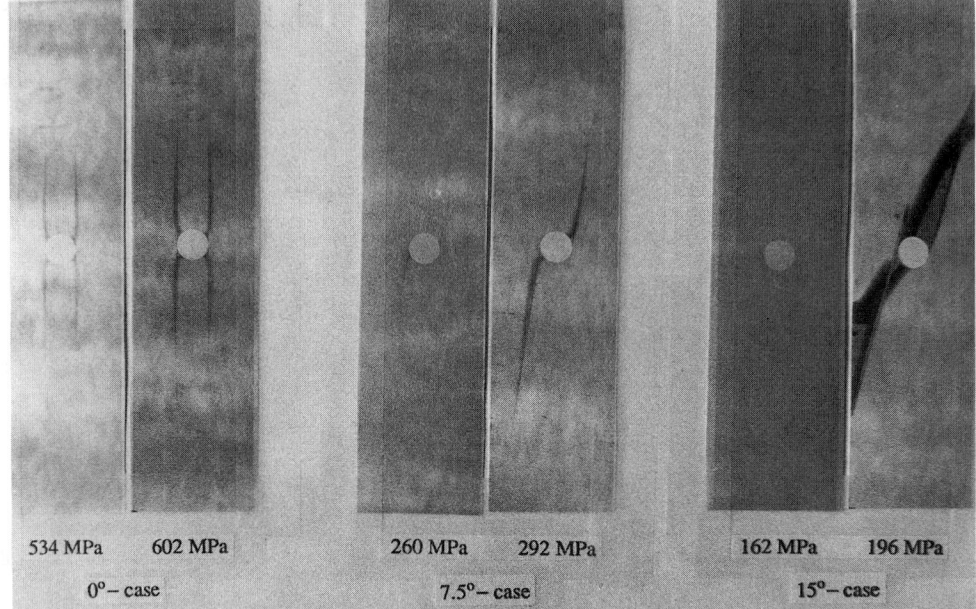

534 MPa   602 MPa   260 MPa   292 MPa   162 MPa   196 MPa

0° – case   7.5° – case   15° – case

FIG. 9—*Radiographs showing damage in [0/90]ₛ off-axis specimens at 80 and 95% ultimate strength.*

For the 7.5 and 15° specimens, two surface matrix cracks appear on each side of the specimen (Fig. 9). The cracks propagated along the fiber direction of the surface layer. The final fracture surface seemed to coincide with these surface cracks.

In the 15° specimen, the surface cracks appeared before the load reached 60% ultimate strength. The cracks did not grow significantly as the load increased. Rapid crack growth occurred only at the loads close to the final failure load.

Figures 10 through 12 show the damage before final failure and the fracture surface in the [0/90/±45]ₛ specimens. Surface cracks were evident. However, they did not propagate as far as those in the [0/90]ₛ laminate. The fracture surface was very messy. Nevertheless, the final failure zone seems to have been confined in a horizontal band enclosing the hole.

Numerical Simulation

As observed from the experiment, the damage progression in notched laminates is very complicated. The failure modes are 3-D in nature. To render the analysis manageable, the following procedures were taken:

1. 2-D plane stress finite elements formulated based on classical lamination theory were used for stress analysis.
2. At the failure initiation load, the finite element at the failure initiation location was assumed to have failed. This was accomplished by setting all ply stiffnesses to zero.
3. After allowing initial failure, surface matrix cracking in the top and bottom layers was assumed to exist in the element adjacent to the initially failed element along the fiber direction. Within this element, the top and bottom layers were assumed to have delaminated.

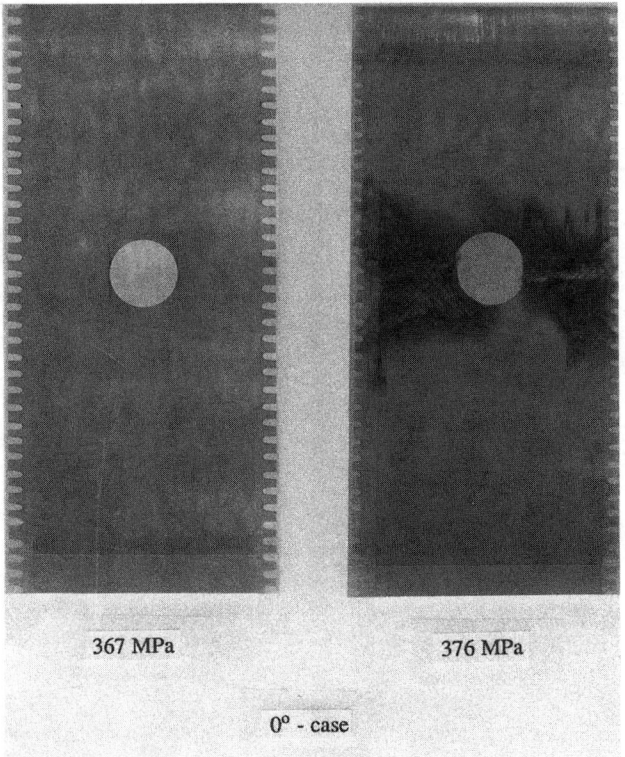

FIG. 10—*Radiographs showing damage in [0/90/±45]ₛ 0° specimens at 95% ultimate strength and final failure.*

4. The subsequent failure analysis consisted of two parts. The first part was for the elements at the hole edge. For these elements, failure was assumed to occur if the hoop stress ($\sigma_{\theta\theta}$) reached the ultimate strength of the corresponding unnotched coupon specimen. Thus, the edge effect on strength was included in this part. Note that, in this analysis, the material properties of failed elements were modified by setting the top and bottom ply stiffnesses to zero and setting E_2 and $G_{12} = 0$ for the remaining plies. This was based on the experimental observation. The second part of the analysis was for the elements covering the interior portion of the laminate. In this part, the edge effect for the element is not important, and failure was assumed to be dominated by in-plane stresses. Thus, the classical ply-by-ply stiffness discounting procedure using Hill-Tsai criterion and classical laminate theory was adopted.

In using the Hill-Tsai criterion for failure analysis of the interior portion of the laminate, the modes of failure were determined by comparing the ratios σ_{11}/X, σ_{22}/Y, and σ_{12}/S (X, Y, and S are the longitudinal, transverse, and in-plane shear strengths of the unidirectional composite, respectively). The maximum ratio indicates whether failure is due to fiber breakage or to matrix cracking. For fiber breakage, we set $E_1 = 0$; for transverse matrix failure, $E_2 = 0$; and for in-plane shear failure, $E_2 = G_{12} = 0$.

The proposed failure analysis procedure requires many iterations in order to perform ply

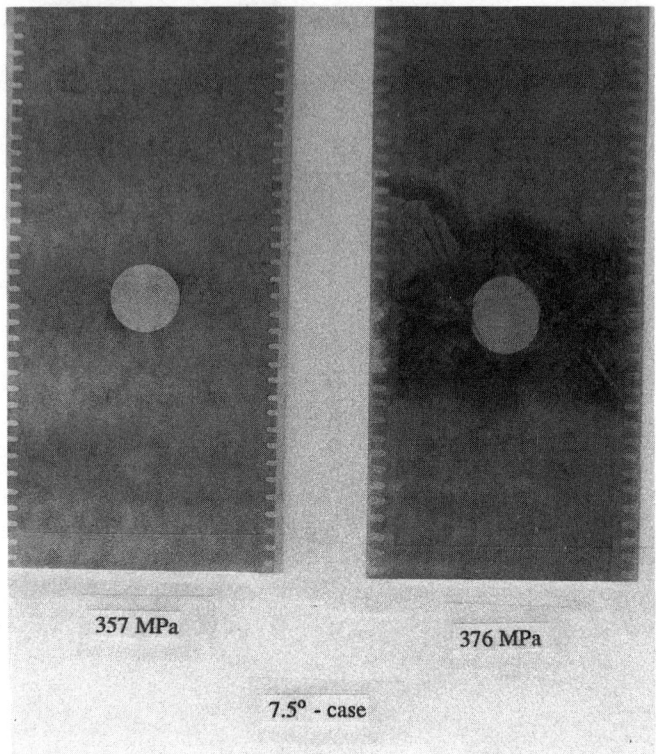

357 MPa

376 MPa

7.5° - case

FIG. 11—*Radiographs showing damage in [0/90/±45]ₛ 7.5° off-axis specimens at 95%*
ultimate strength and final failure.

stiffness reduction during damage progression. To facilitate these calculations, several
material damage codes corresponding to different combinations of ply failure modes were
established. They are listed in Table 7.

Type 8 (for [0/90/±45]ₛ) and Type 4 (for [0/90]ₛ) were used to formulate the element at
the failure initiation site.

Type 5 for [0/90]ₛ laminate and Type 9 for the [0/90/±45]ₛ laminate were used to model
the surface crack. This material type incapacitates the surface ply after matrix cracking
based on the experimental observation that the surface ply would separate from the rest of
the laminate in the vicinity of the surface crack. The surface crack was made to advance in
the surface layer fiber direction if that element degenerated into material Type 5 (for [0/
90]ₛ) and Type 9 (for [0/90/±45]ₛ).

In the numerical analysis, the elastic constants given by Eq 5 and the following ply
strength properties obtained by Sun and Zhou [14] were used.

$$X = 2.14 \text{ GPa}, \quad Y = 56.5 \text{ MPa}, \quad S = 110.3 \text{ MPa} \qquad (7)$$

Finite element analyses were performed for both [0/90/±45]ₛ and [0/90]ₛ laminates containing a 6.35-mm hole. The off-axis angles 0, 7.5, and 15° were considered. Figure 13
shows the simulated failure progression in the 7.5° off-axis [0/90/±45]ₛ specimen. The
shaded area indicates the surface crack location. The dark area indicates the region where

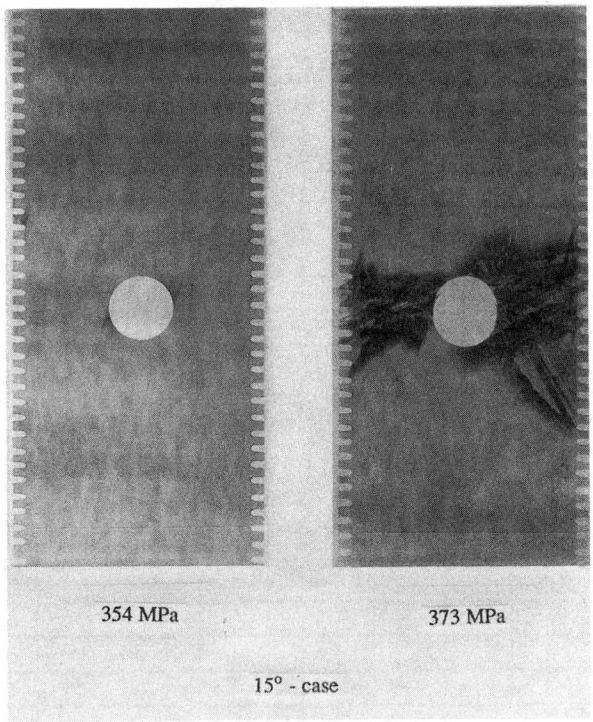

FIG. 12—*Radiographs showing damage in [0/90/±45]ₛ 15° off-axis specimens at 95% ultimate strength and final failure.*

the elements have degenerated into material Type 8, which could not take additional loads. Such a failure path seems to agree with the experimental observation.

Figure 14 presents the finite element result for the 7.5° off-axis [0/90]ₛ specimen. The predicted surface crack is seen to have extended substantially away from the hole. The failure path is indicated by the elements that degenerated into material Type 4. Again, the prediction seems to agree with the experimental result.

TABLE 7—*Material properties in the finite element analysis for the plies of [0/90/±45]ₛ and [0/90]ₛ laminates.*

Laminate Material	[0/90/±45]ₛ			[0/90]ₛ		
	$E_1 = 0$	$E_2 = 0$	$G_{12} = 0$	$E_1 = 0$	$E_2 = 0$	$G_{12} = 0$
1	⋯	⋯	⋯	⋯	⋯	⋯
2	⋯	90°	90°	⋯	90°	90°
3	⋯	90°, 45°	90°	⋯	0°, 90°	0°, 90°
4	⋯	90°, −45°	90°	0°	0°, 90°	0°, 90°
5	⋯	90°, ±45°	90°, ±45°	0°	0°	0°
6	⋯	45° or −45°	⋯	⋯	⋯	⋯
7	⋯	0°, 90°, 45°	0°, 90°, 45°	⋯	⋯	⋯
8	0°	0°, 90°, ±45°	0°, 90°, ±45°	⋯	⋯	⋯
9	0°	0°	0°	⋯	⋯	⋯

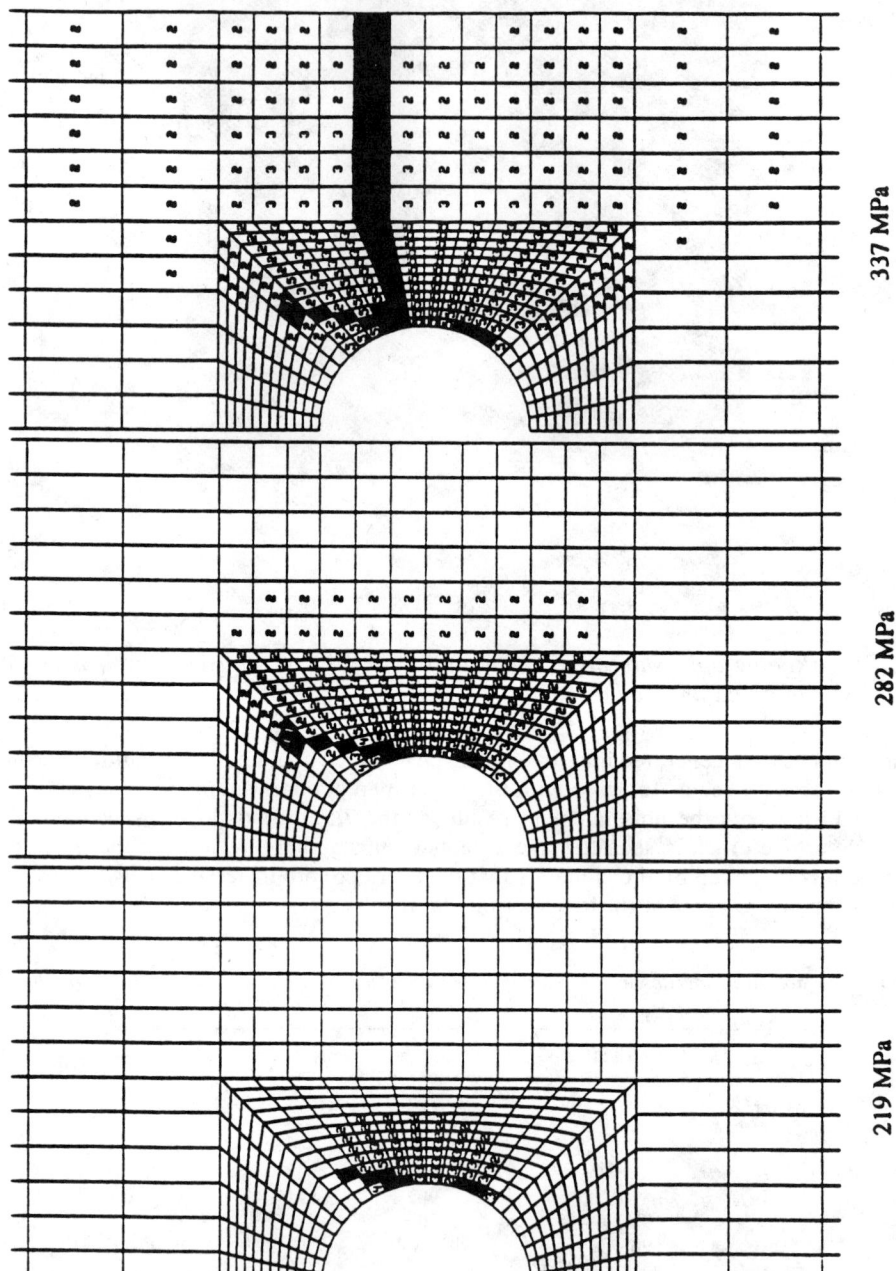

219 MPa   **282 MPa**   **337 MPa**

FIG. 13—*Predicted damage in [0/90/ ±45]$_s$ 7.5° off-axis specimens at 60, 80, and 95% ultimate strength.*

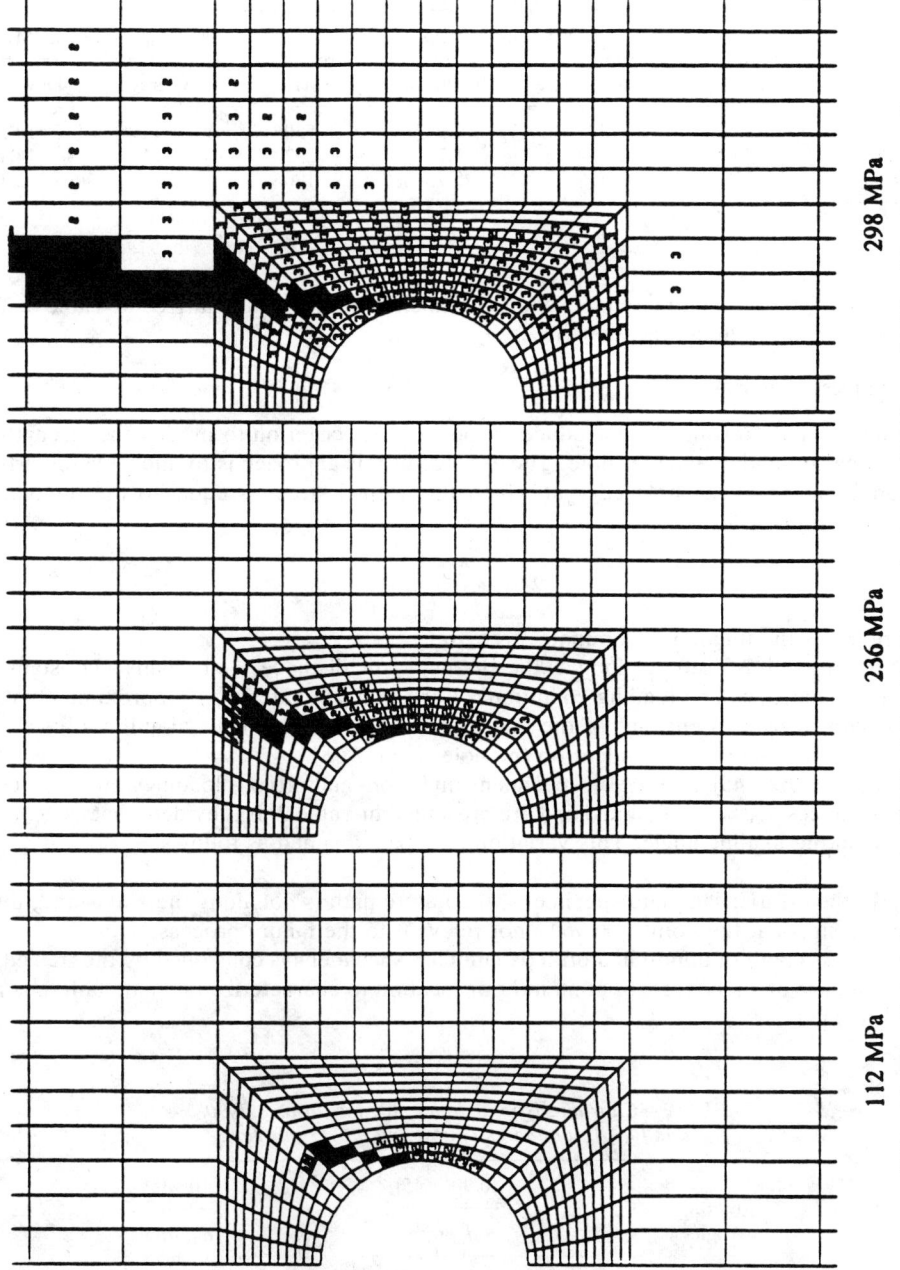

112 MPa   236 MPa   298 MPa

FIG. 14—Predicted damage in $[0/90]_s$ 7.5° off-axis specimens at 40, 80, and 95% ultimate strength.

TABLE 8—*Ultimate stresses of [0/90/ ±45]ₛ and [0/90]ₛ laminates containing a 6.35-mm hole under on- and off-axis loadings.*

| Off-Axis Angle | [0/90/ ±45]ₛ | | | [0/90]ₛ | | |
| | Predicted, MPa | Experimental, MPa | | Predicted, MPa | Experimental, MPa | |
		Average	Deviation		Average	Deviation
0°	337	373	7.9	550	614	10.2
7.5°	334	384	7.7	292	314	14.1
15°	329	379	5.7	190	204	2.1

The predicted ultimate strengths and experimental data are summarized in Table 8. The worst error in prediction is about 13%.

Point Stress Criterion

Whitney and Nuismer [2] introduced a point stress criterion to predict the strength of laminates containing a center hole. The essence of this approach is to find a characteristic distance, d_0, from the hole edge at which the normal stress is equal to the unnotched laminate strength, i.e.,

$$\sigma_{yy|y=0, x=d_0} = \sigma_{\text{ult}} \tag{8}$$

where σ_{ult} is the unnotched laminate strength.

This approach is useful as long as the characteristic distance, d_0, remains a constant for different hole sizes and loading conditions. For some laminates, d_0 is approximately constant with respect to changes of the hole size [2,3]. However, several authors have also found that d_0 could be dependent on the hole size [17,18].

For the [0/90/ ±45]ₛ and [0/90]ₛ specimens under on- and off-axis loadings, the characteristic distances, d_0, were obtained and are presented in Table 9. It is evident that d_0 depends highly on the loading angle. This variation may be explained as follows:

1. In the off-axis laminate specimens, the failure path is not along the *x*-axis and, thus, the stress at the point $x = d_0$ is not relevant to the failure process.
2. The strength of unnotched off-axis laminate specimens is controlled by the free edge. The use of such strength at an interior location to characterize laminate failure is not justified.

TABLE 9—*Characteristic distance, d_0, for off-axis [0/90/ ±45]ₛ and [0/90]ₛ laminates containing a 6.35 mm hole.*

| Loading Angle | [0/90/ ±45]ₛ | [0/90]ₛ |
	d_0, mm	d_0, mm
0°	1.12	1.96
7.5°	3.56	1.81
15°	4.38	2.56

Conclusions

In this study, a procedure for determining failure initiation in laminates containing a center hole was presented. In addition, a finite element simulation for the subsequent failure progression was performed. Based on the comparison between the analytical and experimental results, the following conclusions were obtained:

1. The free edge strength along the hole in a composite laminate can be estimated from the strengths of unnotched off-axis specimens cut from the laminate in various directions. Failure occurs at a point on the hole boundary when the hoop stress is equal to the corresponding free edge strength.
2. The classical ply-by-ply stiffness discounting procedure using the Hill-Tsai criterion seems to be adequate for describing the damage propagation in regions away from the hole edge. This procedure is capable of capturing the main characteristics of damage propagation and estimating the final failure load of the notched laminate.
3. The characteristic distance d_0 in Whitney and Nuismer's point stress criterion seems to depend highly on the loading direction and cannot be regarded as a material constant.

The effect of hole size is an important issue to be addressed. To carry out this study, the size effect of the coupon specimen used in measuring the free edge–controlled laminate strength must be established first. Results of this aspect of research will be reported in a future paper.

Acknowledgment

This work was supported by ONR Grant No. N00014-90-J-1666 with Purdue University. Y. Rajapakse was the technical monitor.

References

[1] Awerbuch, J. and Madhukar, M. S., "Notched Strength of Composite Laminates: Predictions and Experiments—A Review," *Journal of Reinforced Plastics and Composites,* Vol. 4, January 1985, pp. 3–159.
[2] Whitney, J. M. and Nuismer, R. J., "Stress Fracture Criterion for Laminated Composites Containing Stress Concentrations," *Journal of Composite Materials,* Vol. 8, July 1974, pp. 253–265.
[3] Nuismer, R. J. and Whitney, J. M., "Uniaxial Failure of Composite Laminates Containing Stress Concentrations," *Fracture Mechanics of Composites, ASTM STP 593,* American Society for Testing and Materials, Philadelphia, 1975, pp. 117–142.
[4] Tan, S. C., "Effective Stress Fracture Models for Unnotched and Notched Multidirectional Laminates," *Journal of Composite Materials,* Vol. 22, 1988, pp. 322–340.
[5] Tan, S. C., "Laminated Composites Containing an Elliptical Opening. I. Approximate Stress Analyses and Fracture Models," *Journal of Composite Materials,* Vol. 21, 1987, pp. 925–948.
[6] El-Zein, M. S. and Reifsnider, K. L., "The Strength Prediction of Composite Laminates Containing a Circular Hole," *Technology and Research,* JCTRER, Vol. 12, No. 1, Spring 1990, pp. 24–30.
[7] Waddoups, M. E., Eisenmann, J. R., and Kaminski, B. E., "Macroscopic Fracture Mechanics of Advanced Composite Materials," *Journal of Composite Materials,* Vol. 5, 1971, pp. 446–454.
[8] Backlund, J. and Aronsson, C. G., "Tensile Fracture of Laminates with Holes," *Journal of Composite Materials,* Vol. 20, July 1986, pp. 259–307.
[9] Sandhu, R. S., Gallo, R. L., and Sendeckyj, G. P., "Initiation and Accumulation of Damage in Composite Laminates," *Composite Materials: Testing and Design (6th Conference), ASTM STP 787,* I. M. Daniel, Ed., American Society for Testing and Materials, Philadelphia, 1982, pp. 163–182.

[*10*] Chang, F. K. and Chang, K. Y., "A Progressive Damage Model for Laminated Composites Containing Stress Concentrations," *Journal of Composite Materials,* Vol. 21, 1987, pp. 834–855.

[*11*] Shalev, D. and Reifsnider, K. L., "Study of the Onset of Delamination at Holes in Composite Laminates," *Journal of Composite Materials,* Vol. 24, 1990, pp. 42–71.

[*12*] O'Brien, T. K. and Raju, I. S., "Strain-Energy-Release Rate Analysis of Delamination Around an Opening Hole in Composite Laminates," *25th AIAA Structures, Structural Dynamics and Materials Conference,* Palm Springs, CA, May 1984, pp. 526–536.

[*13*] Sun, C. T. and Chu, G. D., "Reducing Free Edge Effect on Laminate Strength by Edge Modification," *Journal of Composite Materials,* Vol. 25, February 1991, pp. 142–167.

[*14*] Sun, C. T. and Zhou, S. G., "Failure of Quasi-Isotropic Composite Laminates with Free Edges under Off-axis Loading," *Journal of Reinforced Plastics and Composites,* Vol. 7, November 1988, pp. 515–557.

[*15*] Reifsnider, K. L., Henneke, E. G., and Stinchcomb, W. W., "Delamination in Quasi-Isotropic Graphite Epoxy Laminates," *Composite Materials: Testing and Design (4th Conference), ASTM STP 617,* American Society for Testing and Materials, Philadelphia, 1977, pp. 93–105.

[*16*] Pipes, R. B. and Daniel, I. M., "Moire Analysis of Interlaminar Shear Edge Effect in Laminated Composites," *Journal of Composite Materials,* Vol. 5, 1971, pp. 255–259.

[*17*] Karlak, R. F., "Hole Effects in a Related Series of Symmetrical Laminates," *Proceedings of Failure Modes in Composites, IV,* The Metallurgical Society of AIME, Chicago, IL, 1977, pp. 105–117.

[*18*] Pipes, R. B., Wetherhold, R. C., and Gillespie, J. W., Jr., "Notched Strength of Composite Materials," *Journal of Composite Materials,* Vol. 12, 1979, pp. 148–160.

Paul A. Lagace,[1] Narendra V. Bhat,[1] and Ahmet Gundogdu[1]

Response of Notched Graphite/Epoxy and Graphite/PEEK Systems

REFERENCE: Lagace, P. A., Bhat, N. V., and Gundogdu, A., "**Response of Notched Graphite/Epoxy and Graphite/PEEK Systems,**" *Composite Materials: Fatigue and Fracture, Fourth Volume, ASTM STP 1156*, W. W. Stinchcomb and N. E. Ashbaugh, Eds., American Society for Testing and Materials, Philadelphia, 1993, pp. 55–71.

ABSTRACT: A test program was conducted to compare the tensile behavior of laminates made with graphite fibers and a low strain-to-failure matrix system (epoxy) versus a high strain-to-failure matrix system (PEEK), in the presence of notches. Graphite/epoxy (AS4/3501-6) and graphite/PEEK (AS4/APC-2) laminates were utilized in three different layups: $[(\pm45)_2/0/90]_S$, $[45/0_2/-45/0_2/90]_S$, and $[\pm45/0]_S$. These laminates represent "soft," "stiff," and "medium" configurations, respectively, in regard to their longitudinal stiffness as manifested by the percentage of 0° plies in each. Two different hole configurations, open and filled, were utilized with three different hole diameters: 6.35, 9.53, and 12.7 mm. All specimens were tested to failure in tension with X-ray photographs taken at various loading intervals with the aid of a dye penetrant. Semi-empirical relations were successfully used to correlate the data for both material systems. The graphite/epoxy system showed earlier and more extensive matrix damage than the graphite/PEEK system. In the case of the "soft" $[(\pm45)_2/0/90]_S$ laminate, this led to a lower failure stress for the graphite/epoxy laminate as compared to the graphite/PEEK laminate; whereas, this led to a higher failure stress for the "stiff" $[45/0_2/-45/0_2/90]_S$ graphite/epoxy laminate. This is related to the number of major load-carrying plies (0°) in the laminate and the load redistribution caused by the damage. These results indicate that the in-plane "toughness" of a composite laminate is a structural property and cannot be directly correlated to the "toughness" or strain to failure of the matrix system but must account for the damage progression which occurs in the specific laminate configuration.

KEY WORDS: graphite/epoxy, graphite/thermoplastic, fracture strength, notch sensitivity, damage progression

High strain-to-failure matrix systems have been developed in an attempt to increase design allowables, as the composite systems made with these matrices exhibit increased delamination damage resistance and tolerance characteristics, especially to impact events [1]. However, these material systems may exhibit damage modes which are different than those of conventional "brittle" (i.e., low strain-to-failure) epoxy systems. For example, there is evidence that damage modes which occur in epoxy systems can increase their resistance to damage, such as notches, via stress-relieving damage mechanisms [2]. Work on high strain-to-failure systems shows that damage can play a similar role [3].

It is therefore necessary that the damage modes of composite materials made with these high strain-to-failure matrix systems be characterized and understood as to their role in the failure of the composite material. The overall behavior of such composite materials can then be understood. Currently, the higher strain-to-failure systems, materials such as

[1]Technology Laboratory for Advanced Composites, Dept. of Aeronautics and Astronautics, Massachusetts Institute of Technology, Cambridge, MA 02139.

graphite/PEEK, are not well understood either in empirical terms or in terms of the damage which occurs before, and results in, final in-plane failure. Much of the work on these materials has concentrated on their through-thickness properties and resistance to delamination [4–6]. However, the in-plane damage mechanisms control failure in many applications. An understanding of these items is essential before these new material systems can be confidently applied in primary load-carrying structures.

A considerable amount of information exists in the literature pertaining to the response of composite laminates in the presence of notches [7]. This contains literature pertaining essentially to "brittle" systems, while there is comparatively little literature available on high strain-to-failure systems, such as the AS4/PEEK system, which is the subject of the current work. This particular material has been documented with respect to its resistance to delamination and impact [1,6,8,9], but little data are available in terms of its tensile response in the presence of stress concentrators like holes [3,9,10].

The objectives of the current work are to document the damage mechanisms which occur in a high strain-to-failure matrix graphite system, to determine how the progressive damage contributes to final failure in the presence of notches, and to compare this to the behavior of a typical epoxy system. A general understanding of the response in these cases will help establish a better understanding of the response of the high strain-to-failure matrix systems to general in-plane loading and the progressive damage which can result.

Approach

The AS4/3501-6 graphite/epoxy system and the AS4/APC-2 graphite/PEEK (poly-etheretherketone) system were utilized. The former represents a typical "first generation" epoxy matrix system, while the latter represents the class of "tougher" (i.e., higher strain-to-failure) matrix systems. The same fiber type was utilized so that the basic unidirectional stiffness properties, shown in Table 1, were virtually the same for the two systems. Failure strains for the unidirectional ply of the two systems are also shown, indicating a factor of two increase in the transverse tensile strain to failure for the graphite/PEEK system over the graphite/epoxy system.

Three different laminates were used: $[(\pm45)_2/0/90]_S$, $[45/0_2/-45/0_2/90]_S$, and $[\pm45/0]_S$. The first represents a "soft" laminate configuration due to the fact that up to 75% of the laminate is made of 45 and 90° plies. It also represents some typical aircraft skin configurations. The second laminate is representative of the "stiff" laminate family, due to its containing more than 50% of its plies in the 0° orientation. The third configuration was chosen due to the abundance of data for this laminate with respect to brittle matrix systems [11]. This is an intermediate configuration with respect to stiffness, with 33% of the fibers in the 0° direction.

Two different hole configurations (open and filled) were utilized, with three different

TABLE 1—*Nominal material properties for the two systems.*

Property	AS4/3501-6	AS4/APC-2
E_L	138.0 GPa	134.0 GPa
E_T	9.4 GPa	8.9 GPa
G_{LT}	6.0 GPa	5.1 GPa
ν_{LT}	0.28	0.28
t_{ply}	0.134 mm	0.125 mm
ε_L^{ult}	15 000 μstrain	14 500 μstrain
ε_T^{ult}	5000 μstrain	10 000 μstrain

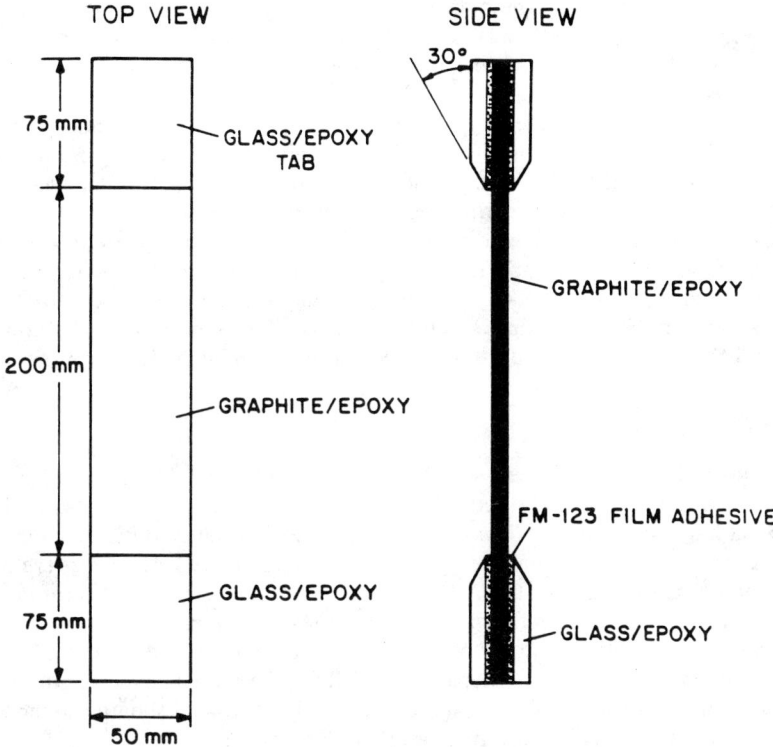

FIG. 1—*Tensile coupon geometry.*

hole diameters: 6.35, 9.53, and 12.7 mm. The open hole allows data to be taken of a basic configuration, while the filled configuration models the constraint in the case of a hole with an inclusion (such as a bolt). Three specimens, of the geometry shown in Fig. 1, were made and tested for each configuration. This represents a total of 54 specimens of each material as shown in Table 2.

In the case of specimens tested with filled holes, a hardened steel bolt machined to the nominal diameter of the hole was used as filler material. In most cases, the actual dimensions of the hole did not conform to the exact hole dimensions. To ensure a tight fit of the

TABLE 2—*Test matrix.*

	Laminate					
	$[(\pm 45)_2/0/90]_S$		$[45/0_2/-45/0_2/90]_S$		$[\pm 45/0]_S$	
Hole Diameter, mm	Open[a]	Filled	Open	Filled	Open	Filled
6.35	3[b]	3	3	3	3	3
9.53	3	3	3	3	3	3
12.7	3	3	3	3	3	3

[a]Hole condition.
[b]Number of specimens of each material tested.

bolt, it was necessary to wrap the bolt in aluminum foil before inserting the bolt into the hole. The aluminum foil used was 0.03-mm-thick standard foil, which was wrapped smoothly around the bolt. The number of wraps was varied by trial and error until a tight fit of the bolt was achieved.

The first specimen in each category was strain gaged around the hole and in the far-field region before testing, as shown in Fig. 2. This gage, with an element size of 6.4 by 3.2 mm, was placed as close to the edge of the hole as possible. The onset of damage at the hole edge was indicated by the nonlinearities and discontinuities in the strain data recorded by these gages, as was seen in the load versus strain plots. This information was utilized in the tests on the remaining two specimens to set points during the test where loading was stopped, returned to zero, and X-ray photographs taken, utilizing dye penetrant, to determine the damage state. The testing was then continued until the next load increment was reached, and the previous procedure was repeated up to the final failure of the specimen.

Experimental Techniques

The composite laminates were layed up and processed as 350 by 305-mm plates. In the case of the graphite/epoxy, the standard manufacturer's cure cycle of a 1-h flow stage at 116°C and a 2-h set stage at 177°C was utilized. This was conducted in an autoclave under vacuum with a 0.59-MPa external pressure. The graphite/epoxy laminates were postcured in an oven at 177°C for 8 h. For the AS4/APC-2, a soldering iron was utilized in the layup process in order to "tack" together neighboring plies. These plates were consolidated in an autoclave at a plate temperature of 390°C under a vacuum and at 0.34 MPa external pressure. A series of temperature steps were used for the autoclave setting to insure that the caul plate and the thermoplastic laminate exceeded the consolidation temperature of 375°C. The plates were cooled at a rate of about 40°C/min.

All coupons were cut to their final width of 50 mm using a water-cooled diamond wheel. All specimens showed a thickness within 10% of the nominal per ply thickness of 0.134

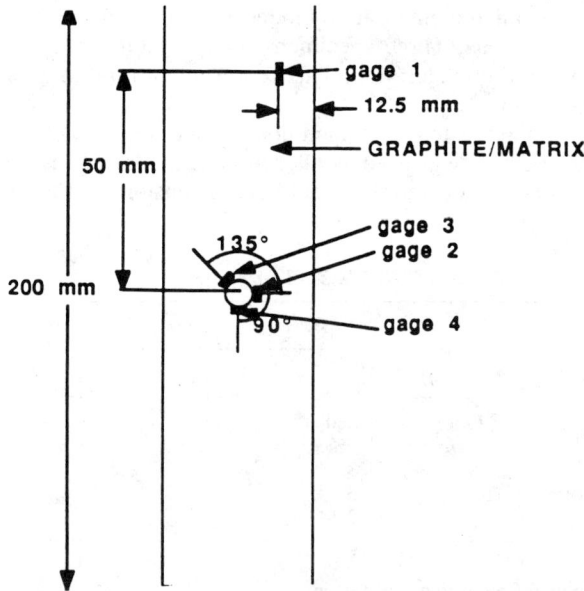

FIG. 2—*Locations of strain gages around hole.*

mm for the graphite/epoxy and 0.125 mm for the graphite/PEEK. Holes were drilled in the specimens using a two-step procedure. A slightly undersized diamond grit-coated drill was first used to create a hole. A reamer of the exact hole size and with a finer diamond grit was utilized to give the hole edge a smooth surface finish. Glass/epoxy tabs with a 0/90 configuration were bonded to the specimens in a secondary operation. This yielded the final specimen as shown in Fig. 1. Strain gages were placed on one specimen of each configuration in the locations previously shown in Fig. 2.

All specimens were tested in a Material Test System MTS 810 testing machine equipped with hydraulic grips. A stroke rate of 1.1 mm/min was utilized. This yields a strain rate of approximately 5000 microstrain/min in the test section. Load, strain, and stroke data were recorded automatically through a data acquisition system.

The first specimen of each configuration, equipped with strain gages, was tested monotonically to failure. The load-strain plots obtained from the strain gages in the vicinity of the hole were examined to determine the nonlinearities and discontinuities, which indicate the possible occurrence of damage. This information was used to determine specific loads at which the subsequent tests were interrupted to nondestructively examine the specimen for damage. At these specified loads, the specimen was removed from the testing machine and all the edges swabbed using a cotton swab dipped in dye penetrant. The dye penetrant used was 1,4-diodobutane (DiB). Care was taken to clean up all the excess dye penetrant on the surface of the specimen. An X-ray photograph was taken and the specimen replaced in the testing machine and loaded to the next specified interval. This process was repeated up to failure. Although it was not proven, it is unlikely that this test interruption and restart had any significant effect on damage progression and final failure.

Notched Strength Results

Two different techniques were utilized to correlate the notched strength data. The first is the correlation by Mar and Lin [12,13] as modified by Lagace [11]. This correlation is of the form

$$\sigma_f = H_c(2r)^{-m} \tag{1}$$

where σ_f is fracture stress, $2r$ is notch length, H_c is the "composite fracture parameter," and m is defined as the value of the stress singularity at the tip of a discontinuity lying at a fiber/matrix interface. This problem was solved by Fenner [14], and the value is dependent upon the Poisson's ratio and shear modulus of the fiber and matrix. The calculated value of m for both systems used in this work is 0.28. This value is the same for both systems due to the fact that the elastic properties on which this value is dependent are not significantly different for the two matrix systems. Thus, the two systems should vary only in their value of H_c. The experimental fracture stresses are correlated with this equation by calculating the average value of H_c for each laminate for each notch size and condition. The value of the composite fracture parameter is then averaged over all the notch sizes for that particular laminate and hole condition.

The second technique is the point stress correlation proposed by Whitney and Nuismer [15,16]. They proposed that failure would occur when the stress at some characteristic distance, d_0, ahead of a notch reached the unnotched failure stress, σ_0. The operative equation is

$$\frac{\sigma_f}{\sigma_0} = 2/\{2 + \xi^2 + 3\xi^4 - (K_\infty^T - 3)(5\xi^6 - 7\xi^8)\} \tag{2}$$

where

$$\xi = r/(r + d_0) \tag{3}$$

and K_∞^T is the stress concentration at the hole edge. This has been modified [11] to provide a two-parameter fit by replacing the unnotched failure stress by a "notched fracture stress parameter," σ_p. An average value for this notched fracture stress parameter can then be found for each laminate along with the characteristic dimension.

The results of these correlative techniques are shown for the three laminates in Tables 3

TABLE 3—*Reduced fracture stress data summary for $[(\pm 45)_2/0/90]_S$ laminate.*

	Open Holes		Filled Holes	
Value	AS4/3501-6	AS4/APC-2	AS4/3501-6	AS4/APC-2
$H_c{}^a$	511 (1.7%)^b	595 (1.3%)	527 (1.7%)	599 (4.2%)
σ_p	603 MPa	631 MPa	605 MPa	486 MPa
d_0	1.41 mm	2.17 mm	1.60 mm	4.29 mm
R^c	0.84	0.96	0.98	0.99

^{a}Units are MPa(mm)$^{0.28}$.
^{b}Values in parentheses are coefficients of variation.
^{c}Correlation coefficient of the Whitney-Nuismer curve fit.

TABLE 4—*Reduced fracture stress data summary for $[45/0_2/-45/0_2/90]_S$ laminate.*

	Open Holes		Filled Holes	
Value	AS4/3501-6	AS4/APC-2	AS4/3501-6	AS4/APC-2
$H_c{}^a$	1238 (2.5%)^b	1148 (2.9%)	1218 (7.8%)	1155 (4.2%)
σ_p	1837 MPa	2110 MPa	1557 MPa	1327 MPa
d_0	0.80 mm	0.35 mm	1.25 mm	1.70 mm
R^c	0.80	0.42	0.87	0.82

^{a}Units are MPa(mm)$^{0.28}$.
^{b}Values in parentheses are coefficients of variation.
^{c}Correlation coefficient of the Whitney-Nuismer curve fit.

TABLE 5—*Reduced fracture stress data summary for $[\pm 45/0]_S$ laminate.*

	Open Holes		Filled Holes	
Value	AS4/3501-6	AS4/APC-2	AS4/3501-6	AS4/APC-2
$H_c{}^a$	796 (5.8%)^b	823 (6.0%)	830 (5.8%)	775 (2.5%)
σ_p	804 MPa	1024 MPa	849 MPa	928 MPa
d_0	2.44 mm	1.20 mm	2.44 mm	1.41 mm
R^c	0.80	0.72	0.94	0.95

^{a}Units are MPa(mm)$^{0.28}$.
^{b}Values in parentheses are coefficients of variation.
^{c}Correlation coefficient of the Whitney-Nuismer curve fit.

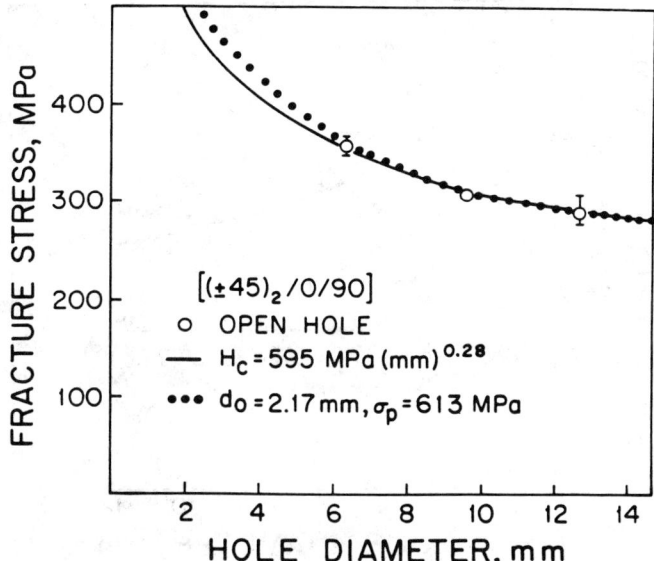

FIG. 3—*Notched strength correlations and data for graphite/PEEK [(±45)₂/0/90]ₛ laminate with open holes.*

through 5. A sample plot of these correlations and the data is shown in Fig. 3 for the case of the graphite/PEEK [(±45)₂/0/90]ₛ laminate with open holes. It can clearly be seen that both correlative techniques provide equally good fits to the data. However, the correlative data in Tables 3 through 5 show that the value of the characteristic dimension in the point stress criterion is very sensitive to the data and thus varies widely across the various configurations. In the Mar-Lin correlation, the calculated value of m is 0.28 for both cases. Thus, the value of H_c provides a better comparative factor. Therefore, this is used for further comparison.

All the fracture stress data are presented in Figs. 4 through 6, along with the respective Mar-Lin correlations, for the [(±45)₂/0/90]ₛ, [45/0₂/−45/0₂/90]ₛ, and [±45/0]ₛ laminates, respectively. In these figures, the value of H_c averaged for the open and filled hole configurations is utilized.

Several observations can be made concerning the correlations and the fracture data. One, the Mar-Lin correlation works very well in all cases. The sensitivity of the two material systems to the presence of notches are similar, varying not in the shape of the curve (determined by the parameter m) but in the magnitude (determined by the parameter H_c). Two, this fact also holds true for all the laminate configurations as the correlation works as well for all three laminate cases. Three, the hole condition has little effect on the final fracture stress as the value of H_c is virtually the same for the two cases of open and filled hole for each material and laminate configuration. Four, there is a difference between the values of the composite fracture parameter between the graphite/epoxy and graphite/PEEK laminates. However, the difference is dependent upon laminate configuration.

Further observation shows that the difference in H_c between the two material systems correlates with the percentage of 0° plies in the laminate. The stiffest laminate, [45/0₂/−45/0₂/90]ₛ, with 50% of the plies in the 0° direction, shows the AS4/APC-2 to have an 8% lower H_c than the AS4/3501-6. The next stiffest laminate, [±45/0]ₛ, with 33% of the plies in the 0° direction, has virtually no difference between the two material systems with

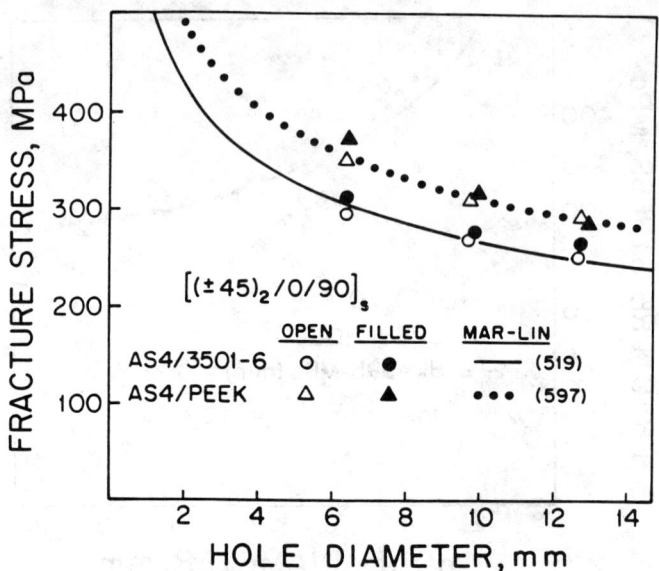

FIG. 4—*Fracture stress data and Mar-Lin correlations for [(±45)₂/0/90]ₛ laminates.*

the data showing a 2% larger value of H_c for the graphite/epoxy system. The "softest" laminate, $[(\pm 45)_2/0/90]_s$, with only 17% of the plies in the 0° direction shows a 13% larger value of H_c for the graphite/thermoplastic system over the graphite/epoxy system. Whether or not this observation can be linked to physical phenomena is discussed in the following section on damage progression.

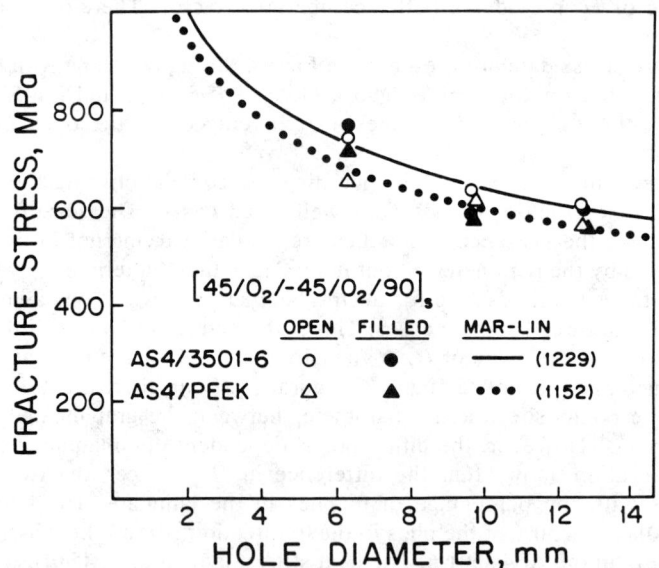

FIG. 5—*Fracture stress data and Mar-Lin correlations for [45/0₂/−45/0₂/90]ₛ laminates.*

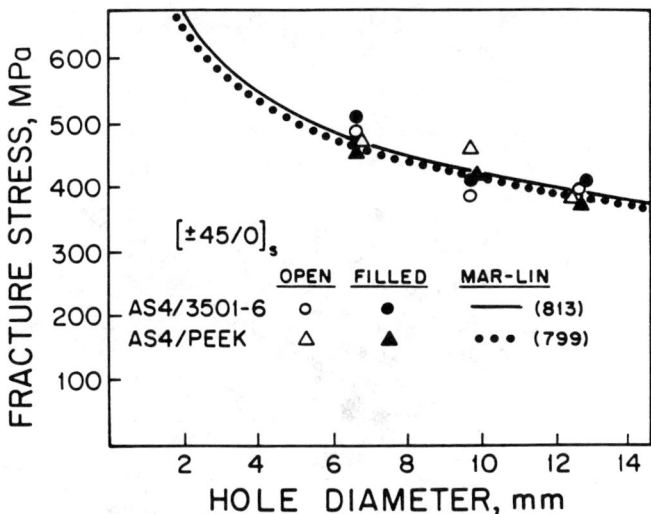

FIG. 6—*Fracture stress data and Mar-Lin correlations for [±45/0]ₛ laminates.*

Damage Progression

The descriptions of the observed progression of damage are made for each laminate configuration.

[(±45)₂/0/90]ₛ Laminate

The graphite/epoxy laminate with open holes shows cracking in the 90° plies at about 80% of the ultimate load as can be seen in Fig. 7. Some splits also appear in the 0° plies, and delamination occurs at the interface of the splits in the 0 and 90° plies. Very near failure, 45° splits occur, and these are evident in the final failure shown in Fig. 7. The same type of damage, somewhat reduced in magnitude, occurs in the graphite/epoxy laminates with filled holes. The final failure, shown in Fig. 8, is virtually identical to the case of the laminates with open holes.

In the graphite/PEEK laminates of this configuration, cracking also begins in the 90° plies, but at higher percentages of the ultimate loads (above 90%) than in the graphite/epoxy case. The magnitude of this cracking is somewhat reduced, as can be seen in Fig. 9, and no 0° ply splitting and subsequent interface delamination is observed. The final failures, shown in Figs. 9 and 10, are very similar to the graphite/epoxy case, with, once again, the behavior being virtually identical for both the cases of filled and open holes.

[45/0₂/−45/0₂/90]ₛ Laminate

Splitting in the 0° plies is the dominant damage mechanisms in these graphite/epoxy laminates in both the cases of open and filled holes, as shown in Figs. 11 and 12. This splitting develops at approximately 80% of the ultimate load and grows in the direction of the loading. Delamination develops at the +45/0 or −45/0 interface and bridges the two splits. This splitting and delamination continues to grow as the load is increased and some cracking occurs in the 90° plies. The splitting and delamination mechanism is evident in the final failure.

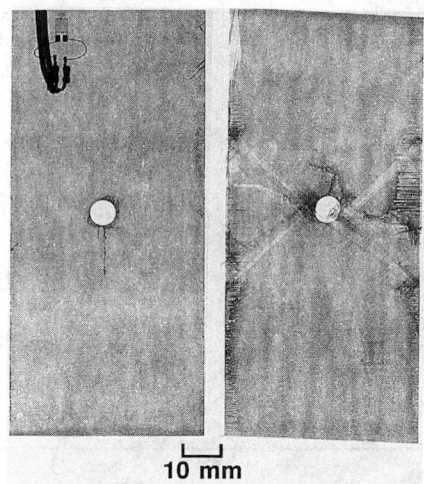

FIG. 7—*X-ray photographs of damage sequence in [(±45)$_2$/0/90]$_s$ graphite/epoxy specimen with 6.35-mm-diameter open hole* (left) *at 80% of ultimate load and* (right) *after failure.*

This splitting in the 0° plies does occur in the graphite/PEEK laminates, but it occurs at a much higher load (generally above 90% of the ultimate load) and to a lesser extent. This can be seen for the open hole case in Fig. 13 and for the filled hole case in Fig. 14. There is little growth of these splits in the 0° plies, and no delamination develops to bridge the splits at the +45/0 or −45/0 interface(s). Some cracking does occur in the 90° plies. The final failures of the graphite/PEEk specimens are similar to the graphite/epoxy specimens except that the two predominant 0° splits are not evident and less delamination is involved.

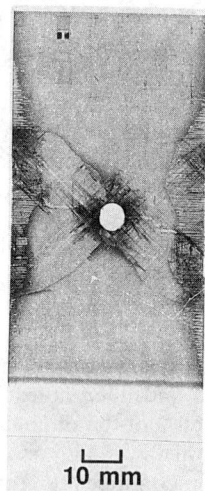

FIG. 8—*X-ray photograph of damage after failure in [(±45)$_2$/0/90]$_s$ graphite/epoxy specimen with 6.35-mm-diameter filled hole.*

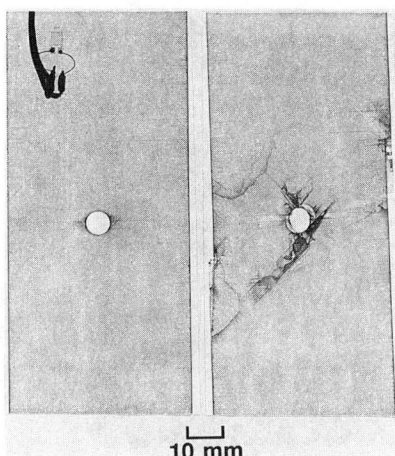

FIG. 9—*X-ray photograph of damage sequence in [(±45)₂/0/90]ₛ graphite/PEEK speci-men with 6.35-mm-diameter open hole (left) at 98% of ultimate load and (right) after failure.*

[±45/0]ₛ Laminate

Very little damage is observed prior to failure for both material systems and for both hole conditions for the [±45/0]ₛ laminates. Some splits do develop in the +45° and −45° plies, as shown in Fig. 15, in the graphite/epoxy laminates, but these show little growth and are very small in size compared to the damage observed for the cases of the previous two laminates. Final failure for all the configurations of the [±45/0]ₛ laminate, as seen in Figs. 15 through 18, show delamination at the 45° ply interface and 0° ply fiber failure along +45° and/or −45° lines.

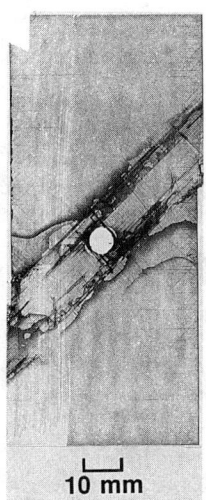

FIG. 10—*X-ray photograph of damage after failure in [(±45)₂/0/90]ₛ graphite/PEEK specimen with 6.35-mm-diameter filled hole.*

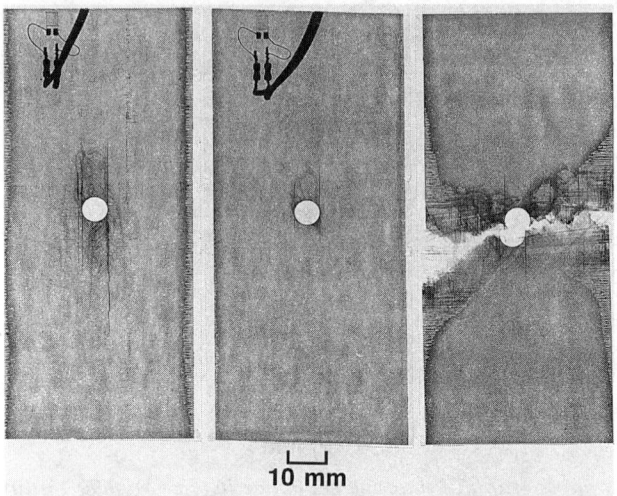

FIG. 11—*X-ray photograph of damage sequence in [45/0₂/−45/0₂/90]ₛ graphite/epoxy specimen with 6.35-mm-diameter open hole* (left) *at 95% of ultimate load,* (center) *at 81% of ultimate load, and* (right) *after failure.*

Discussion

The notched strength fracture data are well correlated for both material systems, independent of hole condition, using a technique developed for graphite/epoxy systems. However, the relative strength of the two material systems is dependent upon the laminate configuration. It was earlier observed that the difference in relative strength correlated with the percentage of 0° plies in the particular laminate. The 0° plies are the major load-carrying plies, and their failure represents the ultimate failure of these laminates. It is

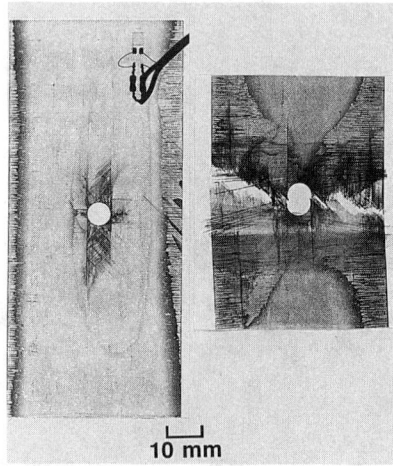

FIG. 12—*X-ray photograph of damage sequence in [45/0₂/−45/0₂/90]ₛ graphite/epoxy specimen with 6.35-mm-diameter filled hole* (left) *at 90% of ultimate load and* (right) *after failure.*

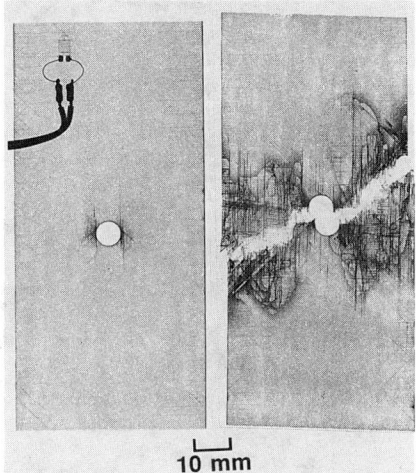

FIG. 13—*X-ray photograph of damage sequence in [45/0₂/ −45/0₂/90]ₛ graphite/PEEK specimen with 6.35-mm-diameter open hole* (left) *at 98% of ultimate load and* (right) *after failure.*

therefore not surprising that the notched strength is linked to the percentage of 0° plies in the laminate. In this particular case, the stiffest laminate shows a greater notched strength for the graphite/epoxy system, while the laminate with the smallest percentage of 0° plies shows a greater notched strength for the graphite/PEEK system. This can be explained given the observations on damage progression.

   In the case of the laminate with the smallest percentage of 0° plies, the $[(\pm 45)_2/0/90]_S$

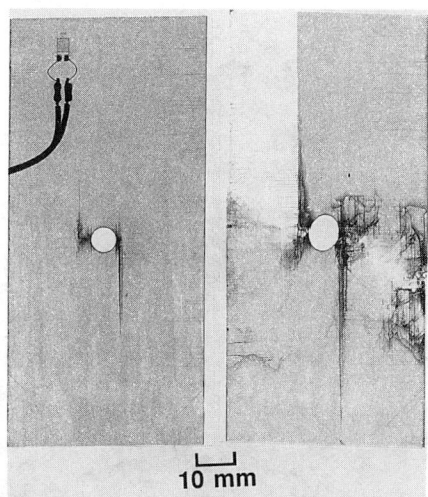

FIG. 14—*X-ray photograph of damage sequence in [45/0₂/ −45/0₂/90]ₛ graphite/PEEK specimen with 6.35-mm-diameter filled hole* (left) *at 81% of ultimate load and* (right) *after failure.*

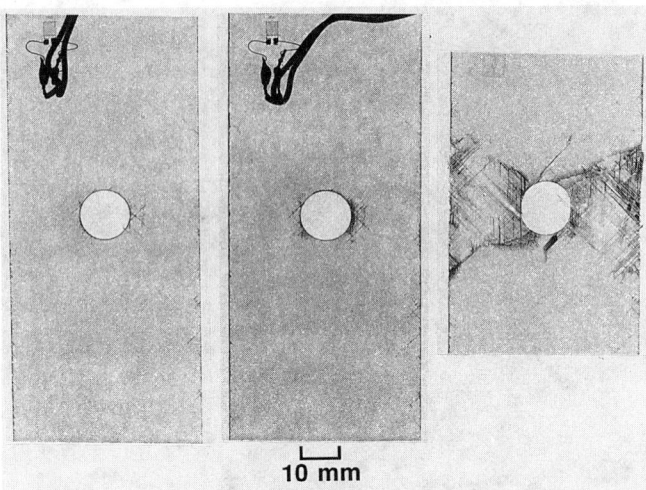

FIG. 15—*X-ray photograph of damage sequence in [±45/0]ₛ graphite/epoxy specimen with 12.7-mm-diameter open hole* (left) *at 78% of ultimate load,* (center) *at 93% of ultimate load, and* (right) *after failure.*

laminate, the main damage observed prior to final failure is cracking in the 90° plies. This occurs earlier, and to a greater extent, in the graphite/epoxy laminate than in the graphite/PEEK laminate. Since the latter is a higher strain-to-failure system and 90° ply cracking is matrix controlled, it is expected that 90° plies would crack at a lower load in the graphite/epoxy laminates. This 90° ply cracking leads to 0° ply splitting and delamination between the 0° and 90° plies, which contributes to the final failure process. Subsequent 0° ply splitting and delamination does not occur in the graphite/PEEK case. The graphite/epoxy

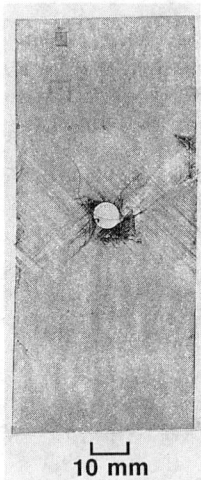

FIG. 16—*X-ray photograph of damage after failure in [±45/0]ₛ graphite/epoxy specimen with 6.35-mm-diameter filled hole.*

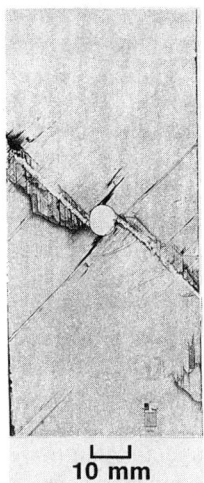

FIG. 17—*X-ray photograph of damage after failure in [±45/0]ₛ graphite/PEEK specimen with 6.35-mm-diameter open hole.*

[(± 45)₂/0/90]ₛ laminate thus fails at a lower load than the graphite/PEEK configuration. The final failure mode also shows more delamination for the graphite/epoxy case, which underscores the importance of the 90° cracking and subsequent delamination in the failure process. This scenario implies that the cracking in the 90° plies and subsequent splitting and delamination cause a higher stress in the 0° plies. It has previously been reported that such ply "uncoupling" does cause a reduction in load-carrying capability [17]. In this case, the 0° plies, and thus the laminate, fail at a lower load. The damage which occurs in this graphite/epoxy laminate is thus detrimental to ultimate strength for this configuration.

At the other end of the scale is the [45/0₂/ − 45/0₂/90]ₛ laminate with the highest percentage of 0° plies, 50%. Once again, the first damage occurs in the form of matrix damage.

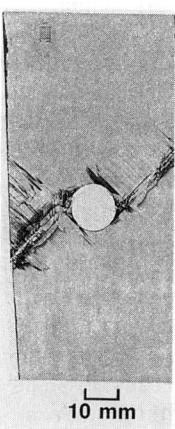

FIG. 18—*X-ray photograph of damage after failure in [±45/0]ₛ graphite/PEEK specimen with 12.7-mm-diameter filled hole.*

This occurs earliest in the graphite/epoxy case. However, the damage now first occurs in the 0° plies in the form of splitting. This propagates in the graphite/epoxy case, and delamination forms between the two splits at the neighboring ply interface. Very little of this occurs in the graphite/PEEK $[45/0_2/-45/0_2/90]_s$ laminate.

A similar damage has been observed in previous work [2,17]. It has been noted that the splitting and subsequent delamination is able to cause a stress redistribution around the hole such that the 0° plies do not see the full stress concentration at the notch. Therefore, the laminate is able to carry increased load due to the reduction in the stress in the fibers adjacent to the notch. In the graphite/PEEK case, the damage does not fully develop, thus the stress concentration is not reduced and failure occurs at a lower load. It would therefore appear that the damage which occurs in the graphite/epoxy laminate is beneficial in terms of ultimate load-carrying capability in this case.

The third laminate, $[\pm45/0]_s$, represents an intermediate case where virtually no damage occurs prior to failure. In this case, the graphite/epoxy and graphite/PEEK configurations fail at virtually the same stress. Since final failure occurs when the major load-carrying plies, the 0° plies, fail, it would be expected that there would be no difference between the two material systems if no damage occurs to cause stress redistribution since the critical path, the fibers, is unchanged between the two cases.

The phenomenon of a damage zone at notches in composite laminates, and subsequent redistribution of local stress, has been previously noted [3,17]. The current results illustrate that the existence of damage may be detrimental or beneficial depending upon the laminate, the loading situation, and the subsequent progression of damage. The introduction of higher strain-to-failure matrix systems may therefore degrade in-plane composite performance, particularly where high loads must be carried and larger percentages of 0° plies are likely to be employed.

Summary

The current work shows that the notched strength of composite systems with high strain-to-failure (a.k.a. "tough") matrix systems can be correlated using the same semi-empirical techniques utilized for the typical epoxy matrix systems. However, the improved delamination resistance of these systems does not necessarily translate into improved in-plane performance. The matrix damage which can occur in the vicinity of a hole can be beneficial by acting as a stress-relieving mechanism on the major load-carrying-plies or can be detrimental by causing increased stress concentration on these plies. A key parameter is the percentage of major load-carrying plies, in this case 0° plies, in the laminate and the subsequent type of damage which develops. In this case, splitting of the 0° plies and subsequent delamination occurred in the laminate with the largest percentage of 0° plies. This damage acted as a stress-relieving mechanism. In contrast, the laminate with the smallest percentage of 0° plies was first damaged by cracks in the 90° plies. This caused an increase in stress concentration in the 0° plies and an earlier failure.

The matrix damage always occurred at lower loads in the lower strain-to-failure system. In some cases this led to a lower failure load, in other cases to a higher failure load, as compared to the laminates made with the higher strain-to-failure system. This clearly demonstrates that in-plane "toughness" is not a matrix property nor a composite property, but is configuration dependent. Thus, in-plane "toughness" is a structural property, and a higher "toughness" matrix material does not guarantee a higher "toughness" performance and may even yield a lower "toughness" performance.

These results indicate that, in order to characterize the response of composite laminates to general loading and their damage tolerance capabilities, research should be concentrated on damage mechanisms and subsequent failure rather than on the characterization of bulk properties. Furthermore, the characterization of matrix properties will not necessarily allow the representation of overall composite properties. Damage models need to be developed, based on physical reality, to incorporate the matrix and fiber properties in determining the overall composite laminate response.

References

[1] Leach, D. C., Curtis, D. C., and Tamblin, D. R., "Delamination Behavior of Carbon Fiber/ Poly(etheretherketone) (PEEK) Composites," *Toughened Composites, ASTM STP 937*, American Society for Testing and Materials, Philadelphia, 1987, pp. 358–380.

[2] Lagace, P. A. and Nolet, S. C., "The Effect of Ply Thickness on Longitudinal Splitting and Delamination in Graphite/Epoxy under Compressive Cyclic Load," *Composite Materials: Fatigue and Fracture, ASTM STP 907*, American Society for Testing and Materials, Philadelphia, 1986, pp. 335–360.

[3] Simonds, R. A. and Stinchcomb, W. W., "Response of Notched AS4/PEEK Laminates to Tension/Compression Loading," *Advances in Thermoplastic Matrix Composite Materials, ASTM STP 1044*, American Society for Testing and Materials, Philadelphia, 1989, pp. 133–145.

[4] Hinkley, J. A., Johnston, N. J., and O'Brien, T. K., "Interlaminar Fracture Toughness of Thermoplastic Composites," *Advances in Thermoplastic Matrix Composite Materials, ASTM STP 1044*, American Society for Testing and Materials, Philadelphia, 1989, pp. 251–263.

[5] Hibbs, M. F., Tse, M. K., and Bradley, W. L., "Interlaminar Fracture Toughness and Real-Time Fracture Mechanism of Some Toughened Graphite/Epoxy Composites," *Toughened Composites, ASTM STP 937*, American Society for Testing and Materials, Philadelphia, 1987, pp. 115–130.

[6] O'Brien, T. K., "Fatigue Delamination Behavior of PEEK Thermoplastic Composite Laminates," *Proceedings*, American Society for Composites First Technical Conference, Dayton, OH, October 1986, Technomic, Lancaster, PA, pp. 404–420.

[7] Awerbach, J. and Madhukar, M. S., "Notched Strength of Composite Laminates: Predictions and Experiments—A Review," *Journal of Reinforced Plastics and Composites*, Vol. 4, January 1985, pp. 3–159.

[8] Carlile, D. R., Leach, D. C., Moore, D. R., and Zahlan, N., "Mechanical Properties of the Carbon Fiber/PEEK Composite APC-2/AS-4 for Structural Applications," *Advances in Thermoplastic Matrix Composite Materials, ASTM STP 1044*, American Society for Testing and Materials, Philadelphia, 1989, pp. 199–212.

[9] Carlile, D. R. and Leach, D. C., "Damage and Notch Sensitivity of Graphite/PEEK Composite," *Proceedings*, 15th SAMPE National Technical Conference, Cincinnati, OH, October 1983, SAMPE, Corina, CA.

[10] Tan, S. C., "Tensile and Compressive Notched Strength of PEEK Matrix Composite Laminates," *Journal of Reinforced Plastics and Composites*, Vol. 6, July 1987, pp. 253–266.

[11] Lagace, P. A., "Notch Sensitivity and Stacking Sequence of Laminated Composites," *Composite Materials: Testing and Design (Seventh Conference), ASTM STP 893*, American Society for Testing and Materials, 1986, pp. 161–176.

[12] Mar, J. W. and Lin, K. Y., "Fracture of Boron/Aluminum Composites with Discontinuities," *Journal of Composite Materials*, Vol. 11, 1977, pp. 405–421.

[13] Mar, J. W. and Lin, K. Y., "Fracture Mechanics Correlation for Tensile Failure of Filamentary Composites with Holes," *Journal of Aircraft*, Vol. 14, 1977, pp. 703–704.

[14] Fenner, D. N., "Stress Singularities in Composite Materials with an Arbitrarily Oriented Crack Meeting an Interface," *International Journal of Fracture*, Vol. 12, 1975, pp. 705–721.

[15] Nuismer, R. J. and Whitney, J. M., "Uniaxial Failure of Composite Laminates Containing Stress Concentrations," *Fracture Mechanics of Composites, ASTM STP 593*, American Society for Testing and Materials, Philadelphia, 1975, pp. 117–142.

[16] Whitney, J. M. and Nuismer, R. J., "Stress Fracture Criterion for Laminated Composites Containing Stress Concentrations," *Journal of Composite Materials*, Vol. 8, 1974, pp. 253–265.

[17] Harris, C. E. and Morris, D. H., "A Fractographic Investigation of the Influence of Stacking Sequence on the Strength of Notched Laminated Composites," *Fractography of Modern Engineering Materials: Composites and Metals, ASTM STP 948*, American Society for Testing and Materials, Philadelphia, 1987, pp. 154–173.

Deirdre A. Hirschfeld[1] and Carl T. Herakovich[2]

Failure Analysis of Notched Unidirectional Graphite/Epoxy Tubes Under Combined Loading

REFERENCE: Hirschfeld, D. A. and Herakovich, C. T., **"Failure Analysis of Notched Unidirectional Graphite/Epoxy Tubes Under Combined Loading,"** *Composite Materials: Fatigue and Fracture, Fourth Volume, ASTM STP 1156,* W. W. Stinchcomb and N. E. Ashbaugh, Eds, American Society for Testing and Materials, Philadelphia, 1993, pp. 72–85.

ABSTRACT: Crack growth from notched unidirectional graphite/epoxy tubes subjected to axial, torsional, and combined loading conditions was studied experimentally and theoretically. Tubes with fiber orientations of 2.5, 15, 45, and 87.5° were notched with slots 0.5 cm long, parallel to the tube axis.

The stress state at the notch of the thin tubes was determined using an elasticity analysis of an infinite plate with an elliptical hole. The normal stress ratio theory was used to predict the far-field stresses and the direction of crack growth at crack initiation. The slots in the tubes were modeled as an ellipse having the same notch tip radius as the slot, an ellipse having the same length and width, or as a circular hole depending on the loading regime. The normal stress ratio theory correctly predicted the direction of crack growth parallel to the fibers, but did not provide good correlation with the experimental far-field stress for crack initiation in all cases. It is believed that this lack of correlation for the critical stresses is due to the different notch geometries in the theoretical model and actual tubes.

The fracture surfaces of the failed tubes were examined using a scanning electron microscope. It was found that failure occurred primarily with the cracks growing from the notch either fully within the matrix or along the fiber/matrix interface depending upon the far-field loading.

KEY WORDS: notched tubes, graphite/epoxy, combined loading, torsion, failure analysis, composites, tension, compression, crack growth, fracture

Nomenclature

a Semi-axis of elliptical hole along the x-axis
b Semi-axis of elliptical hole along the y-axis
a_{ij} Anisotropic elastic compliance coefficients
s_1, s_2 Nonconjugate roots of the characteristic equation
Z_1, Z_2 Complex variables
Φ_0, Ψ_0 Complex potentials
$\sigma_x, \sigma_y, T_{xy}$ Stress components
$\sigma_x^\infty, \sigma_y^\infty, T_{xy}^\infty$ Far-field stress components
ϕ Angle of radial plane around the crack tip or of position along elliptical boundary

[1]Assistant professor, Center for Advanced Ceramic Materials, Virginia Polytechnic Institute and State University, 301 Holden Hall, Blacksburg, VA 24061-0256.
[2]Henry L. Kinnier, Professor of Civil Engineering and Director of Applied Mechanics, Thorton Hall, University of Virginia, Charlottesville, VA 22903-2442.

$R(\phi)$ Normal stress ratio
$\sigma_{\phi\phi}$ Normal stress acting on plant ϕ
$T_{\phi\phi}$ Tensile strength normal to plane ϕ
β Angle between plane ϕ and fiber orientation
X_t Unidirectional composite tensile strength parallel to the fibers
Y_t Unidirectional composite tensile strength
r_s Distance from crack tip
y Fiber orientation
Ω Radial direction bounded by tangent to the ellipse at each position, ϕ

The widespread use of composite materials, particularly as critical structural members, has made it necessary to fully understand the failure modes of these heterogeneous materials. Once failure has been fully characterized, mechanics-based theories can be developed to reliably predict failure under arbitrary loading, thus avoiding overly conservative safety factors.

The first step in achieving this goal is to examine the fundamental failure problem of crack propagation in a unidirectional lamina. This is especially important because fibrous graphite/epoxy composites have proven to be sensitive to defects that occur during both manufacture and service. Characterization of the crack growth from these flaws in laminae is necessary for the understanding and subsequent modeling and prediction of laminate failure modes.

Various theories have been evaluated on the basis of their ability to predict crack growth and failure loads [1]. Of these, the best theories are based upon available analytical solutions that determine the stress distribution in the vicinity of idealized cracks in homogeneous, anisotropic materials for idealized loading and geometric configurations. In particular, the normal stress ratio (NSR) theory [1–3] has proven to be one of the more successful theories in its ability to predict crack growth and failure loads in notched unidirectional composites and, hence, was chosen for use in this study.

Previously, the NSR theory has been used to study crack growth in unidirectional laminae for tensile loading conditions [1,4] and for shearing conditions using off-axis tensile coupons and Iosipescu tests [5,6]. An in-depth study of the influence of flaw shapes and far-field loading on crack growth from notches is given in Ref 7. In this study, the NSR theory is further evaluated by monitoring crack propagation in unidirectional tubular specimens subjected to various loading conditions. The use of tubular specimens avoids the effects of a finite width specimen and allows various combinations of axial and torsional loading.

Analytical Approach

To provide a location for crack initiation, notches were machined into the tubes. The notches were neither ellipses nor cracks, but rather slots having straight sides and semicircular ends as shown in Fig. 1a.

Three approaches have been used to approximate the notch geometry. One method is to model the notch as a sharp crack. This method has been shown to yield partial agreement with experimental results for the critical far-field stresses [2]. However, this approach is limited in three ways. First, it can only be used for comparison with experiments where the actual notch is truly sharp. This can be accomplished only in unidirectional composites with the notch aligned parallel to the fibers, and even then with difficulty. Another limitation of the sharp crack solution is that the stresses are infinite at the crack tip, which

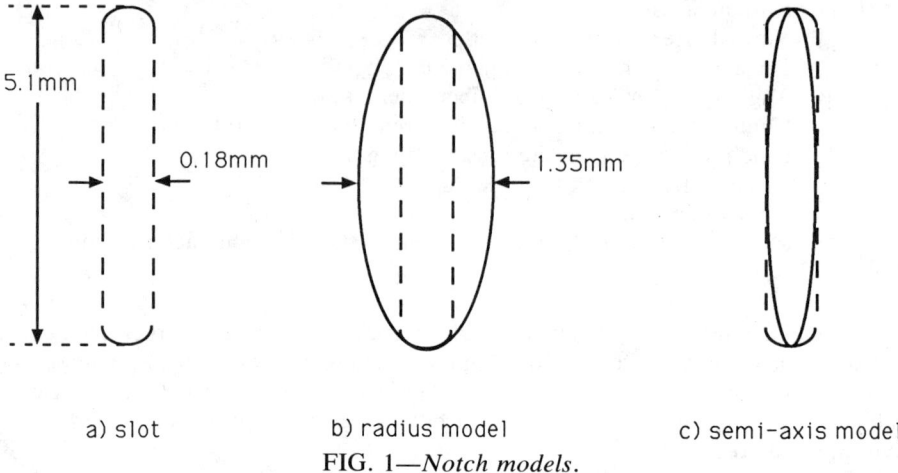

a) slot   b) radius model   c) semi-axis model

FIG. 1—*Notch models.*

necessitates the selection of a critical distance from the crack tip for failure analysis. The third difficulty is that the solution is applicable only to loading configurations where the crack does not close. In this study, the notches are aligned parallel to the tube axis, which results in crack closure for some loading configurations. Also, the notch is not aligned parallel to the fiber direction. Accordingly, the sharp crack model is not appropriate for this study.

The second approach is to model the slot as an ellipse. When modeling a slot as an ellipse, there are two extremes that envelope the slot geometry. First, the ellipse can be of the same length and have the same minimum radius of the notch tip. By matching the radius and length, the width of the ellipse is greater than that of the notch. This model (Fig. 1*b*) will be referred to as the radius model in this study. The other extreme is to match the semi-axes of the ellipse with the length and width of the slot. This approximation (Fig. 1*c*) is called the semi-axis model. For the notches used in the experimental study, the notch models have a/b ratios of 0.035 for the semi-axis model and 0.265 for the radius model. The third approach considered in this study models the notch as a circular hole ($a/b = 1.0$) since the ends of the notch are semicircular. This model is referred to as the circle model.

Since there is no tractable analytical technique for evaluating the stresses in a notched anisotropic cylinder, a notched infinite plate theory approach is used. Such an approach is valid for the thin tubes used in this study. Lekhnitskii [8] and Savin [9] developed a solution for a traction-free elliptical hole (semi-axis ratio of a/b) at the center of an infinite homogeneous anisotropic plate (Fig. 2) utilizing a complex potential formulation. Sih, Paris, and Irwin [10] use a similar approach to derive expressions for the near-crack-tip stresses for cracks.

The stress components near the hole are:

$$\sigma_x = \sigma_x^\infty + 2Re\{s_1^2\Phi_0'(Z_1) + s_2^2\Psi_0'(Z_2)\}$$
$$\sigma_y = \sigma_y^\infty + 2Re\{\Phi_0'(Z_1) + \Psi_0'(Z_2)\}$$
$$T_{xy} = T_{xy}^\infty - 2Re\{s_1\Phi_0'(Z_1) + s_2\Psi_0'(Z_2)\}$$

(1)

where the primes denote the derivative of the complex potentials, Φ_0 and Ψ_0, with respect

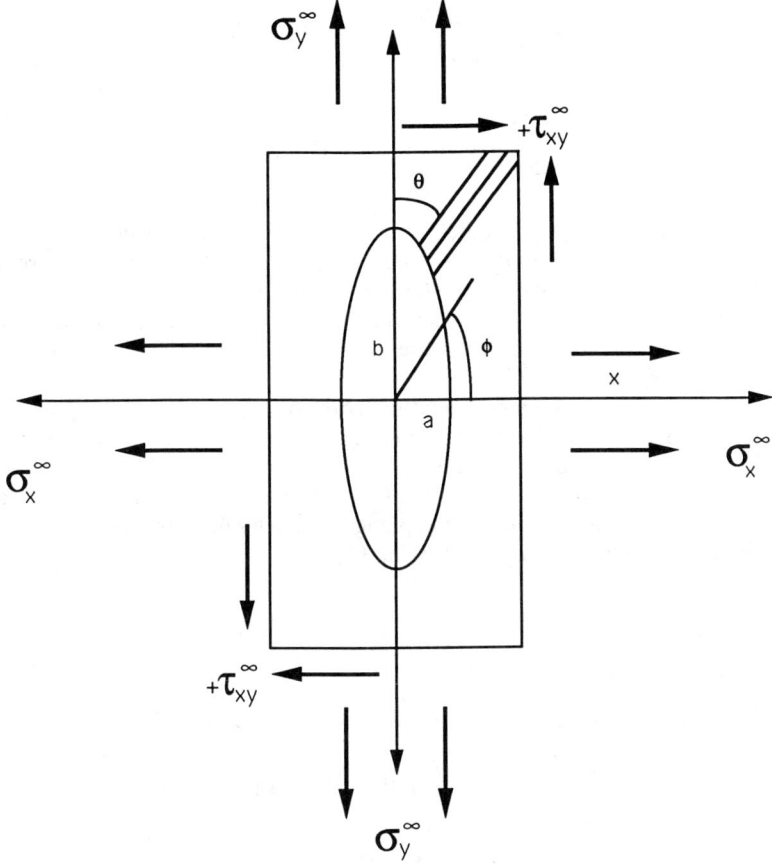

FIG. 2—*Elliptical notch in unidirectional lamina under biaxial loading.*

to the complex arguments Z_1 and Z_2, the far-field applied stresses are identified by the superscript infinity, and s_1 and s_2 are the nonconjugate roots of the characteristic equation

$$a_{11}s^4 + 2a_{16}s^3 + (2a_{12} + a_{66})s^2 - 2a_{26}s + a_{22} = 0 \qquad (2)$$

which depends on the anisotropic elastic compliance coefficients, a_{ij}. The complex variables, Z_1 and Z_2, are defined in Cartesian coordinates as

$$Z_1 = x + s_1y \quad \text{and} \quad Z_2 = x + s_2y \qquad (3)$$

The derivatives of the complex functions, Φ_0 and Ψ_0, are

$$\Phi_0'(Z_1) = A_1 \left(1 - \sqrt{\frac{Z_1^2}{Z_1^2 - (a^2 + s_1^2b^2)}} \right) \qquad (4)$$

$$\Psi_0'(Z_2) = A_2 \left(1 - \sqrt{\frac{Z_2^2}{Z_2^2 - (a^2 + s_2^2b^2)}} \right)$$

where

$$A_1 = \frac{-i(a - is_1b)}{2(s_1 - s_2)(a^2 + s_1^2b^2)} [\sigma_x^\infty b + \sigma_y^\infty ias_2 + T_{xy}^\infty(bs_2 + ia)]$$

$$A_2 = \frac{i(a - is_2b)}{2(s_1 - s_2)(a^2 + s_2^2b^2)} [\sigma_x^\infty b + \sigma_y^\infty ias_1 + T_{xy}^\infty(bs_1 + ia)]$$
(5)

This analytical solution for an infinite plate with an elliptical hole yields real stresses on the edge of the notch. If the a/b ratio equals 1, then the notch is a circular hole. Furthermore, it has been shown when the a/b ratio is 0.005 or less for a notch aligned parallel to the loading axis, the stresses approach those calculated by the crack-like notch model [7]. It follows that the stresses in the vicinity of the elliptical hole are significantly affected by the a/b ratio.

The NSR theory is a phenomenological criterion originally developed by Buczek and Herakovich [2] to predict the direction of crack growth in composites. The theory assumes that cracks grow due to local crack-tip tensile normal stress only. The direction of crack extension corresponds to the maximum value of the ratio, $R(\phi)$, which is defined as the ratio of the normal stress acting on a radial plane defined by the angle around the crack tip, ϕ, to the tensile strength normal to that plane ($T_{\phi\phi}$)

$$R(\phi) = \frac{\sigma_{\phi\phi}}{T_{\phi\phi}}$$
(6)

The tensile strength normal to the ϕ plane is assumed to have the form

$$T_{\phi\phi} = X_T \sin^2\beta + Y_T \cos^2\beta$$
(7)

where

β = the angle between the radial plane, ϕ, and the fiber direction, θ,
X_T = unidirectional composite tensile strength parallel to the fibers, and
Y_T = unidirectional composite transverse tensile strength.

The geometry of β, $T_{\phi\phi}$, and $\sigma_{\phi\phi}$ is shown in Fig. 3.

The NSR theory has been applied by Beuth and Herakovich [3] to predict the critical applied far-field stresses for crack growth. The critical stresses are defined as the smallest far-field stresses causing $R(\phi) = 1.0$ at some point near an existing flaw with r_s being the distance from the crack tip. It is obvious that the NSR theory is a variation of the maximum stress theory for isotropic materials [11] modified to account for anisotropic strength. In particular, $T_{\phi\phi}$ provides the directional strength dependence expected for anisotropic materials.

In this study, the normal stress ratio is evaluated along the elliptical boundary where the stresses are not singular as in the sharp crack model. Figure 3b illustrates the procedure used to apply the normal stress ratio to an elliptical notch as developed by Gurdal and Herakovich [7]. The NSR is evaluated over a range of radial directions at each point along the elliptical boundary. The radial direction, Ω, is varied over 180° bounded by the tangent to the ellipse at each position defined by the angle ϕ. The predicted crack extension direction Ω_c occurs in the direction of maximum normal stress ratio at a point on the

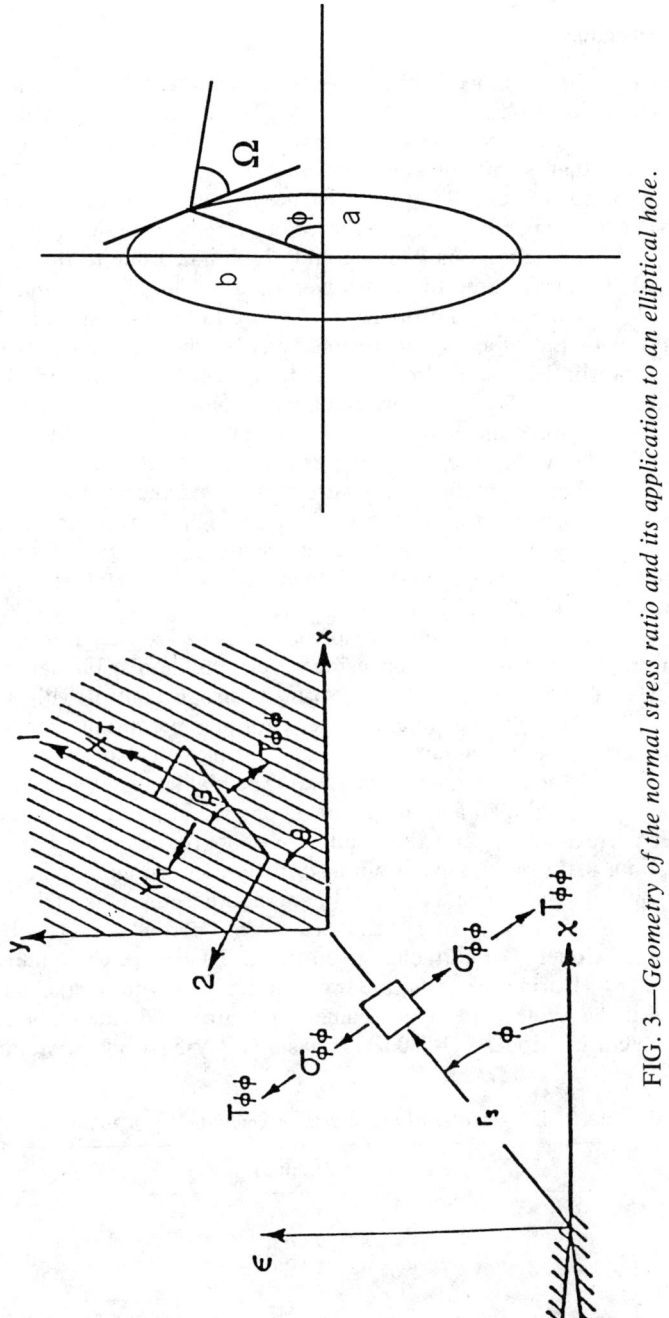

FIG. 3—*Geometry of the normal stress ratio and its application to an elliptical hole.*

ellipse (ϕ_c). Thus both the site of origin of crack extension and the direction of crack growth are obtained.

Experimental Procedure

Filament-wound tubes, nominally 5.0 cm outside diameter (OD), 0.2 cm thick, and 25.4 cm long, consisting of 16 plies of AS4/3501-6 graphite/epoxy, were manufactured by Hercules, Inc., Magna, UT. Special handling was required to ensure that the tubes were truly unidirectional, that is, the tubes were wound in one direction only. Tubes with fiber orientations of 2.5, 15, 45, and 87.5°, as measured from the axial direction, were fabricated. The 2.5 and 87.5° specimens were studied because they were the closest approximation to 0 and 90° fiber orientations which could be obtained due to the limitations of the fabrication capability at the time the tubes were made. The fiber volume fraction of the composite was 74%. The material properties [12] used in the analysis are listed in Table 1. Notches aligned with the tube axis were machined in the tubes at NASA-Langley Research Center (Branson Ultrasonic Impact Grinder Model UAM-20) using steel shim stock as a cutting tool. The corners of the shim stock were rounded in an attempt to avoid sharp corners. The notches were measured after machining and were found to have nominal dimensions of 0.018 cm wide by 0.5 cm long. It is noted that the tubes were ultrasonically C-scanned before and after notching to ensure that no significant damage was present in the tubes. The tubes were carefully prepared for gripping [12] by drying and cleaning, and then the tube surfaces were abraded and cleaned again prior to gluing to the plugs with a high-strength adhesive. A fixture was used to maintain alignment of the tube and plug axes. The load train is illustrated in Fig. 4.

The tubes were tested in tension, compression, negative and positive torsion, and combined compression/negative torsion using a servo-hydraulic biaxial testing machine (Instron Model 1350). The axial tests were run in stroke control with a strain rate of approximately 0.1%/min while maintaining zero torque, permitting free rotation of the tube. The torsion tests were conducted in rotary control with a shear strain rate of 0.1 %/min with free axial displacement (zero axial load). The biaxial tests were conducted in load control with both the axial load and torque being applied at a rate that corresponded to 7 MPa/min. (This corresponds to an axial strain rate of approximately 0.5%/min and a shear strain rate of about 0.1%/min.) Two strain gage rosettes were mounted on the outside of the tubes: one on the side opposite the notch, and another approximately 2 cm below the notch. Strains, loads, and axial/rotary feedback were recorded for all the tests (Solatron Orion Data Logger Model 3530F attached to an IBM AT Personal Computer) and analyzed using MATPAC [13]. During the tests, the formation and growth of cracks from the notch were monitored using a telescope, video camera, monitor, and video tape recorder (Sony Videocassette Recorder Model VO-5800H, Panasonic WV-5470 Video Monitor, Panasonic

TABLE 1—*Material properties of graphite/epoxy tubes.*

AS4/3501-6 Graphite/Epoxy
E_1 = 136.3 GPa
E_2 = 9.9 GPa
G_{12} = 6.5 GPa
v_{12} = 0.276
S = 82.8 MPa
X_T = 1.8 GPa
Y_T = 49.7 MPa

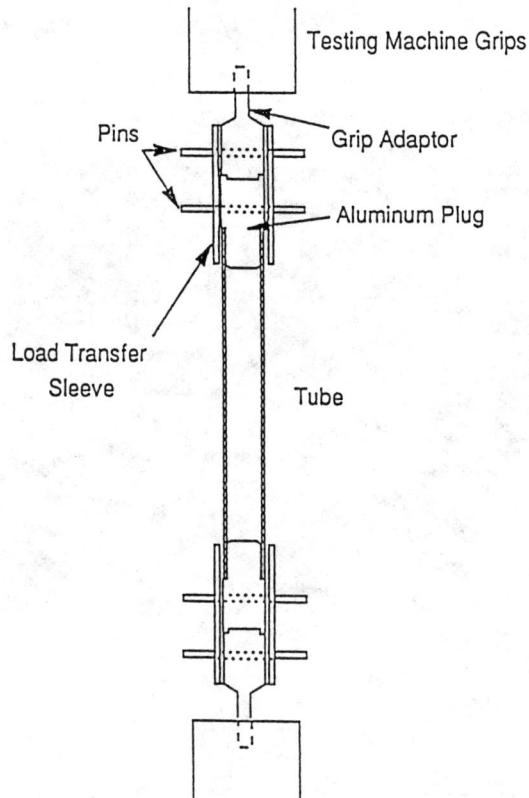

FIG. 4—*Complete tube assembly and load train.*

TV camera Model WV1800, and Questar QM1 telescope). Also, a microphone (Realistic Stereo Electret Microphone Model 33-1065) was positioned near the specimen to record any sound emitted during the test. After each tube failed, care was taken in removing the specimen from the testing machine to preserve the fractured surface for analysis using a scanning electron microscope (Phillips Model SEM-550).

Results and Discussion

Crack Initiation

Most tubes failed catastrophically with cracks initiating at the tangent point where the straight side of the slot meets the semicircular ends independent of loading. The normal stress ratio theory did correctly predict the side of the notch where cracks initiate; however, the exact location of crack initiation was not accurate because the exact geometry of the slot was not modeled. The cracks formed symmetrically on opposite sides of the notch and grew parallel to the fibers as shown in Fig. 5.

The NSR theory correctly predicted crack growth parallel to the fibers with angles less than 2% different from the fiber orientation angle for all loading cases except one (Table 2). The exception was the 87.5% tube loaded in compression. In this case, the predicted direction of crack growth using the radius model was 6° off the fiber orientation. The

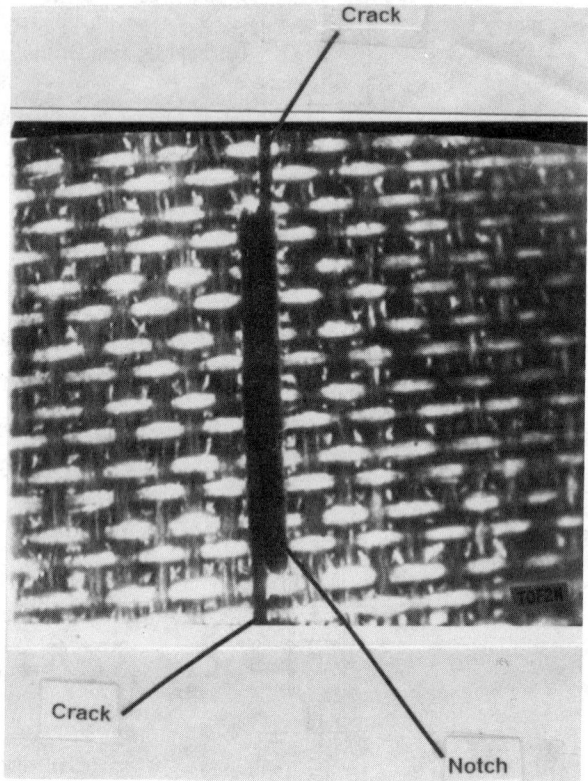

FIG. 5—*Failure of a 2.5° tube in negative torsion.*

predictions using the circle and semi-axis models were 9 and 33° from the fiber orientation, respectively. This may be due to the very small values of $\sigma_{\phi\phi}$ at the tip of the notch for the 87.5° tube in compression.

Table 3 summarizes the predicted and experimental critical far-field stresses for all specimens that exhibited crack growth from the notch. It should be noted that 2.5° tubes did not fail when loaded axially (either in tension or compression), even though the predicted critical failure stresses were less than the machine capacity. The predicted critical stress for tensile loading using the radius model for the 15° tubes corresponded with the experimental value. For the 45° tubes, the experimental failure stress was between the predicted stresses from the circle and radius models. The 87.5° tube tested in tension failed near the grips instead of at the notch.

Positive torsion tests (where the tubes were torqued in the positive fiber orientation direction as shown in Fig. 2) were conducted for all four fiber orientations; however, the 45 and 15° tube tests were unsuccessful as crack growth tests. The 45° tubes did not fail when loaded to the testing machine capacity, and two 15° tubes failed approximately 1.0 cm away from the notch. No flaws were evident on either the fracture surfaces or the ultrasonic C-scans of the 15° tubes to account for this behavior. The 2.5° tube failed catastrophically at a stress much larger than predicted. The 87.5° tube exhibited slow crack growth, with the crack initiating at 40.7 MPa, which agrees well with the circle model prediction. The predicted critical far-field stresses for the 45° tube were less than the testing machine capacity for the semi-axis and radius models; however, the circle model

TABLE 2—*Experimental and predicted crack growth directions.*

Loading	Fiber Orientation, θ	Tube	Experimental Crack Growth Direction	Predicted Circle Model, Ω_c	Semi-Axis Model, Ω_c	Radius Model, Ω_c
Tension	15°	15C1B	15°	17°	16°	17°
		15A2	15°			
Tension	45°	45F3	225°	44°	44°	44°
		45D3	45°			
Compression	45°	45E3	45°	47°	48°	47°
Compression	87.5°	90L1A	87.5°	79°	55°	274°
Positive torsion	2.5°	0F	2.5°	1°	1°	1°
Positive torsion	87.5°	90J	267.5°	269°	269°	269°
Negative torsion	2.5°	0F2N	2.5°	4°	4°	4°
Negative torsion	45°	45B2N	45°	45°	45°	45°
		45F1N	45°			
		45E2N	45°			
Negative torsion	87.5°	90KN	267.5°	87°	87°	87°
Biaxial[a]	2.5°	OF1N	2.5°	3°	4°	4°
Biaxial[a]	15°	15B2	15°	16°	16°	16°
Biaxial[a]	45°	45F2	45°	45°	45°	45°
Biaxial[a]	87.5°	90K1	87.5°	85°	86°	86°

[a]Biaxial compression/negative torsion, $\sigma_y = \tau_{xy}$.

TABLE 3—*Experimental and predicted critical stresses.*

Loading	Fiber Orientation, θ	Average Experimental Failure Stress, MPa	Predicted Critical Stress Circle Model, MPa	Semi-Axis Model, MPa	Radius Model, MPa
Tension	15° (i)[a]	182.2	95.2 (−48%)[b]	292.6 (+60%)[b]	196.0 (+7%)[b]
Tension	45° (i)	45.9	32.9 (−28%)	76.6 (+67%)	58.1 (+27%)
Compression	45° (m)[c]	−211.8	−138.7 (+35%)	−165.6 (+22%)	−153.2 (+28%)
Compression	87.5° (i)	−164.9	−431.9 (−162%)	−46.2 (+72%)	−412.6 (−150%)
Positive torsion	2.5° (m)	55.9	33.6 (−40%)	10.8 (−81%)	34.6 (−38%)
Positive torsion	87.5° (m)	40.7[d]	33.6 (−17%)	3.2 (−92%)	17.0 (−58%)
Negative torsion	2.5° (m)	−60.0	−29.7 (+50%)	−8.3 (+87%)	−28.6 (+52%)
Negative torsion	45° (m)	−18.3	−18.4 (−1%)	−2.2 (+88%)	−11.7 (+36%)
Negative torsion	87.5° (i)	−30.4[d]	−29.7 (+2%)	−2.8 (+91%)	−15.4 (+49%)
Biaxial[e]	2.5° (i)	−60.0	−30.4 (+49%)	−8.3 (+87%)	−27.5 (+54%)
Biaxial[e]	15° (i)	−26.2	−27.0 (−3%)	−4.2 (+84%)	−19.4 (+26%)
Biaxial[e]	45° (i, m)	−22.8	−26.6 (−17%)	−2.2 (+90%)	−12.6 (+45%)
Biaxial[e]	87.5° (i, m)	−96.6	−60.1 (+38%)	−3.2 (+97%)	−20.6 (+79%)

[a]"i" indicates interface as dominant failure mode.
[b]Values in parentheses indicate percent difference of predicted stress with respect to experimental stress.
[c]"m" indicates matrix as dominant failure mode.
[d]Far-field stress at crack initiation prior to slow crack growth.
[e]Biaxial compression/negative torsion, $\sigma_y = \tau_{xy}$.

predicted a critical stress of over 138 MPa, which is more than the testing machine capacity of 104 MPa.

Tubes with fiber orientations of 2.5, 45, and 87.5° were loaded in negative torsion. Again, the 87.5° tube failed by slow crack growth, with the crack initiating at 30.4 MPa. The 45° tubes failed in a unique manner when subjected to negative torsional loading: the fibers did not break, but were pulled out of the tubes bridging the gap emanating from the notch. For the 2.5° tube, the stresses predicted using the circle and radius models were about half the experimental value. For the 45 and 87.5° tubes, the experimental stresses agreed well with the NSR theory circle model predictions.

Notched tubes for the four fiber orientations were tested biaxially in combined compression-negative torsion with load ratios such that σ_y^∞ is equal to T_{xy}^∞. All tubes failed with cracks growing parallel to the fibers in an unstable manner. Both the 15 and 45° tubes exhibited fibers bridging the crack before final failure. The predicted critical stresses using the circle model agreed very well with the experimental failure stresses for the 15 and 45° tubes loaded biaxially. However, the circle model critical stress prediction for the 2.5° tube was approximately half that of the experimental value, and the prediction for the 87.5° tube was only 62% of the experimental failure stress.

For compression loading, the predicted critical stresses could not be correlated with a particular model. The 45° experimental failure stresses were larger than those predicted, and the 87.5° results were considerably different from all three model predictions.

In general, the correlation between predicted and experimental far-field stresses at crack initiation varied with the circle model, having good agreement in five of the twelve cases. These cases of good agreement between the stress predictions and experimental results occurred when there was torsional loading of 45 and 87.5° tubes, and biaxial loading of 45 and 15° tubes. Several factors may affect the correlation. Primarily, the discontinuity in the notch where the straight sides intersect the tangent of the semicircular ends is a stress concentration point where the cracks initiate. None of the models used accounts for this exact geometry. Also, micromechanical effects were not considered. Another factor is that linear elastic material response was assumed in the analysis.

Fractography

Fractographs of the notched specimens were taken from a representative area within 0.1 cm of the notch. The fracture surfaces of all specimens were examined; however, only representative fractographs are presented.

The fracture of notched unidirectional 15 and 45° tubes in tension and a 87.5° tube in compression is characterized primarily by fiber-matrix interfacial failure, while the 45° tube in compression is characterized by matrix failure. Figure 6 shows fractographs of 45° tubes failed in tension and compression. When fiber-matrix interfacial failure is the dominant failure mode for axial loading, the angle of the hackles in the matrix are almost perpendicular to the fibers. This is indicative of a stress state other than pure shear between adjacent fibers and may be due to a large tensile force acting between fibers near the notch. In the compression case, the matrix appears crushed between the fibers with few or no hackles present.

The tubes loaded in torsion exhibited matrix failure, except for the 87.5° tube in negative torsion, which was dominated by fiber-matrix interfacial failure. In this case, the hackles associated with matrix failure tended to be at 45° with respect to the fibers, which is indicative of a pure matrix shear failure mode [14–16].

In the case of biaxial loading, failure generally occurred along the interface with the matrix crushed between fibers (2.5 and 15° tubes) or a combination of interfacial and

FIG. 6—*Fractographs of 45° tubes failed in tension and compression.*

matrix failure (45 and 87.5° tubes) as shown in Fig. 7. Even though crack growth was parallel to the fibers, a few fibers were broken by shearing and/or crushing near the notch in the 87.5° tube. The angle of the hackles varied due to the complex state of stress as failure progressed.

Conclusions

Initiation of crack growth from slot-like notches in thin, unidirectional graphite/epoxy tubes subjected to axial, torsional, and combined load states has been studied experimentally and theoretically. It is concluded that the critical far-field stresses for crack initiation are dominated by local notch geometry effects. Cracks tended to initiate at local discon-

FIG. 7—*Fractographs of 2.5 and 45° tubes failed in compression/negative torsion.*

tinuities in the slot geometry. The normal stress ratio theory generally provided good correlation with experimental result for direction of crack growth. The predictions for critical stresses were less satisfactory. Of three notch geometry models considered, modeling the notch as a circle provided the best correlation with the experiment. The circle model predictions were excellent in five out of twelve cases. Unfortunately, the correlation was poor in the remaining cases.

Analysis of the fracture surfaces near the notch revealed that failure occurred either along the fiber/matrix interface or primarily within the matrix, depending upon the far-field loading.

References

[1] Beuth, J. L. Jr. and Herakovich, C. T., "On Fracture of Fibrous Composites," *Composites '86: Recent Advances in Japan and the United States,* S. Cadet, S. Umekawa, and A. Kobayashi,

Eds., *Proceedings*, Japan-U.S. Conference on Composite Materials III, Tokyo, 1986, Publisher: Japan Society for Composite Materials, Tokyo, pp. 267–277.

[2] Buczek, M. B. and Herakovich, C. T., "A Normal Stress Criterion for Crack Extension Direction in Orthotopic Composite Materials," *Journal of Composite Materials*, Vol. 19, 1985, pp. 544–578.

[3] Beuth, J. L. Jr. and Herakovich, C. T., "Analysis of Crack Extension in Anisitropic Materials Based on Local Normal Stress," *Theoretical and Applied Fracture Mechanics*, Vol. 11, 1989, pp. 27–46.

[4] Beuth, J. L. Jr., Gregory, M. A., and Herakovich, C. T., "Crack Growth in Unidirectional Graphite-Epoxy Under Biaxial Loading," *Experimental Mechanics*, Vol. 24, No. 3, September 1986, pp. 245–253.

[5] Gregory, M. A., Beuth, J. L. Jr., Barbe, A., and Herakovich, C. T., "Application of the Normal Stress Ratio Theory for Predicting the Direction of Crack Growth in Unidirectional Composites," *Fracture of Fibrous Composites*, AMD Vol. G00294, C. T. Herakovich, Ed., ASME, New York, 1985, pp. 33–42.

[6] Herakovich, C. T., Gregory, M. A., and Beuth, J. L. Jr., "Crack Growth Direction in Unidirectional Off-Axis Graphite-Epoxy," *Mechanical Characterization of Load Bearing Fibre Composite Laminates*, Cardon and Verchery, Eds., Elsevier Applied Science Publishers, New York, 1985, pp. 97–114.

[7] Gurdal, Z. and Herakovich, C. T., "Effect of Initial Flaw Shape on Crack Extension in Orthotopic Composite Materials," *Theoretical and Applied Fracture Mechanics*, Vol. 8, 1987, pp. 59–75.

[8] Lekhnitskii, S. G., *Theory of Elasticity of an Anisotropic Elastic Body*, Mir Publishers, Moscow, 1981.

[9] Savin, G. N., *Stress Concentration Around Holes*, Pergammon Press, New York, 1961.

[10] Sih, G. C., Paris, P. C., and Irwin, G. R., "On Cracks in Rectilinerarly Anisotropic Bodies," *International Journal of Fracture Mechanics*, Vol. 1, No. 3, 1965, pp. 189–203.

[11] Erodgan, F. and Sih, G. C., "On the Crack Extension in Plates under Plane Loading and Transverse Shear," *ASME Journal of Basic Engineering*, Vol. 85, 1965, pp. 519–527.

[12] Hirschfeld, D. A., "Failure Analysis of Notched Graphite-Epoxy Tubes," Ph.D. dissertation, Virginia Polytechnic Institute and State University, Blacksburg, VA, 1990.

[13] Hidde, J., "Materials Testing Package," Composite Mechanics Group, Department of Engineering Science and Mechanics, Virginia Polytechnic Institute and State University, Blacksburg, VA, 1987.

[14] Purslow, D., "Some Fundamental Aspects of Composites Fractography," *Composites*, Vol. 12, No. 4, October 1981, pp. 241–247.

[15] Purslow, D., "Matrix Fractography of Fibre-reinforced Epoxy Composites," *Composites*, Vol. 17, No. 4, October 1986, pp. 289–303.

[16] Donaldson, S. L., *Fracture Toughness Testing of Graphite/Epoxy and Graphite/PEEK Composites*," *Composites*, Vol. 16, No. 4, April 1985, pp. 103–112.

Sheng Liu,[1] Zafer Kutlu,[1] and Fu-Kuo Chang[1]

Matrix Cracking-Induced Delamination Propagation in Graphite/Epoxy Laminated Composites Due to a Transverse Concentrated Load

REFERENCE: Liu, S., Kutlu, Z., and Chang, F.-K., "**Matrix Cracking-Induced Delamination Propagation in Graphite/Epoxy Laminated Composites Due to a Transverse Concentrated Load,**"*Composite Materials: Fatigue and Fracture, Fourth Volume, ASTM STP 1156*, W. W. Stinchcomb and N. E. Ashbaugh, Eds., American Society for Testing and Materials, Philadelphia, 1993, pp. 86–101.

ABSTRACT: An investigation was performed to study the delamination propagation induced by matrix cracking in graphite/epoxy laminated composite beams subjected to transverse concentrated loading. The major focus of the study was to fundamentally understand the interaction between matrix cracking and delamination propagation in laminated composites resulting from transverse loadings. A nonlinear finite element analysis combined with a failure criterion and fracture mechanics was developed for modeling the damage initiation and propagation in composite beams subjected to a concentrated line load. A contact analysis based on the Lagrange multiplier technique was also adopted for modeling the contact-slip condition of delamination and matrix crack surfaces during loading. Results of the calculations based on the analysis were compared favorably with the existing data and the data generated during the investigation.

KEY WORDS: delamination, matrix cracking, stable and unstable crack propagation, laminated composites, transverse loading

It is well known that laminated composites are vulnerable to transversely concentrated loading such as low-velocity impact, which can cause significant internal damage such as matrix cracks and delaminations. These matrix cracks and delaminations apparently accompanied each other experimentally in impact tests, strongly indicating the existence of a strong interaction between matrix cracking and delamination during impact. It was also found experimentally that similar results could be produced in low-velocity impact tests by quasi-static loading [1–7].

Recent studies of two-dimensional impact tests by Choi and Chang [8–11] showed that there were two types of matrix cracks which could initiate interface delaminations: shear cracks embedded inside the material and bending cracks which appeared in the outermost plies of the laminate. Previous impact studies by Choi et al. [12] and others [7,13–14] also found that the delamination always propagated more extensively in the fiber direction than in the transverse direction of the fibers of the bottom layer next to the delaminated interface. Figure 1 shows the measured delamination lengths of specimens subjected to two-dimensional line-loading impact [8–11]. For [0/90/0] composites, once damage oc-

[1]Graduate student, post-doctoral student, and associate professor, respectively, Dept. of Aeronautics and Astronautics, Stanford University, Stanford, CA 94305-4035.

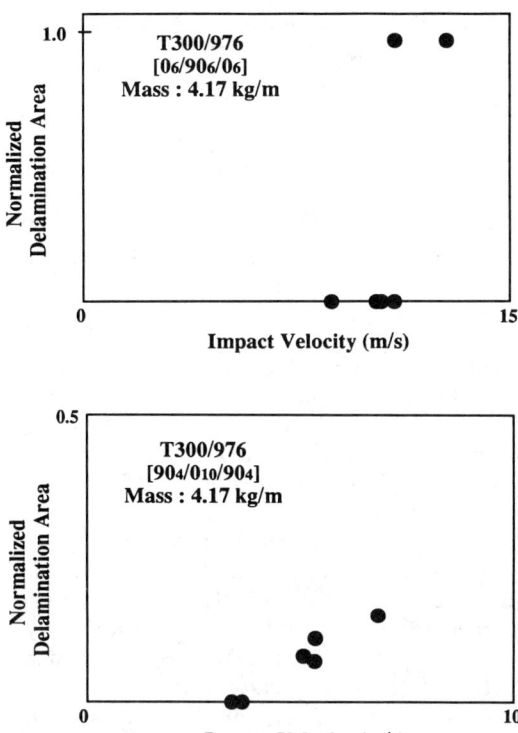

FIG. 1—*Measured delamination area of laminated composite plates resulting from two-dimensional line loading impact at different impact velocities.* Top: *the measured delamination area within [0₆/90₆/0₆] composites.* Bottom: *the measured delamination area within [90₄/0₁₀/90₄] composites.*

curred the delamination propagated immediately to the boundary, independent of the impact velocity. However, for [90/0/90] composites, the extent of the delamination was proportional to the impact velocity after damage had initiated.

Apparently, the interaction of matrix cracking and delamination plays an important role in damage initiation and development in laminated composites subjected to transverse loading. It is believed that in order to fundamentally understand the impact damage in laminated composites, the knowledge of such interaction must be fully developed. Therefore, the objective of this investigation was to study the damage mechanism and mechanics of laminated composites subjected to transverse concentrated loading. To simplify the problem, only laminated composite beams subjected to transverse concentrated line loading were considered, and the load was applied quasi-statically.

Experiments

T300/976 graphite/epoxy prepregs were selected for fabricating the specimens. Composite panels were manufactured according to a standard cure cycle and sliced into specimens. The dimensions and ply orientations of the specimens are listed in Table 1. Each specimen was X-rayed before the test to examine any internal damage resulting from curing and cutting. No apparent damage was found in any of the specimens.

TABLE 1—*Ply orientations and geometry of test specimens*

Ply Orientation, Flat Panels	Span Length (L), in.[a]	Width (W), in.	Thickness (H), in.	Number of Specimens
$[0_6/90_3]_s$	1.5	0.955	0.098	4
$[0_4/90_4]_s$	1.5	1.030	0.0896	4
$[0_4/90_3/0_2]_s$	1.5	0.920	0.098	5
$[0_3/90_2/0_3/90]_s$	1.5	0.930	0.098	5

[a] 1 in. = 0.0254 m.

Figure 2 shows the schematic of the testing fixture for the specimens. During the test, a line-nosed cylindrical load head was applied downward at the center line of the specimen. An MTS machine was utilized for all the tests. The rate of the load-head increment was 0.002 in./s (5.08×10^5 m/s). A strain gage mounted at the center of the back surface of each specimen was used to record the strain history as a function of the applied load. The results of the tests are summarized in the following.

The test results of $[0_6/90_3]_s$ and $[0_4/90_4]_s$ specimens recorded from a strain gage are shown by open circles in Fig. 3. As can be seen from the figure, the relationships between the applied load and the measured strains for both ply orientations were quite linear until the maximum load was reached, after which the load dropped sharply. Below the maximum load, no visual damage was observed during the test. However, significant damage, including multiple matrix cracks and delaminations, was found right after the load drop, indicating that the initiation of the damage and its growth occurred simultaneously and that the growth of the delaminations was quite unstable. Apparently, once the damage was initiated, it immediately propagated extensively inside the material and resulted in catastrophic failure.

A photograph of a close side view of a damaged specimen near the central loading region is presented in Fig. 4. It seems that a pair of matrix cracks located in the 90° ply group near the loading area branched into interface delaminations. The cracks were inclined in an angle from the loading direction. It was also noticed that the same damage pattern as shown in Fig. 4 appeared also in the similar specimens tested by impact loading [8–11].

The effect of the stacking sequence on the transverse strength was also investigated. Figure 5 presents the final failure load of the tested specimens as a function of ply orientation. The thicknesses of the laminates were kept constant. Solid symbols with the

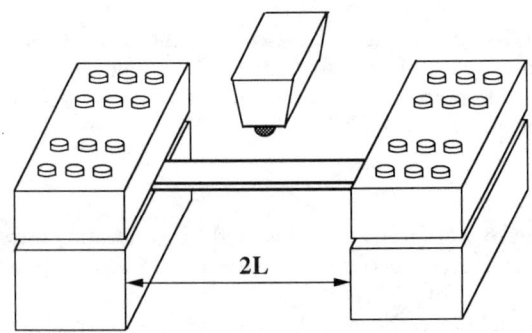

FIG. 2—*The schematic of the testing fixture used.*

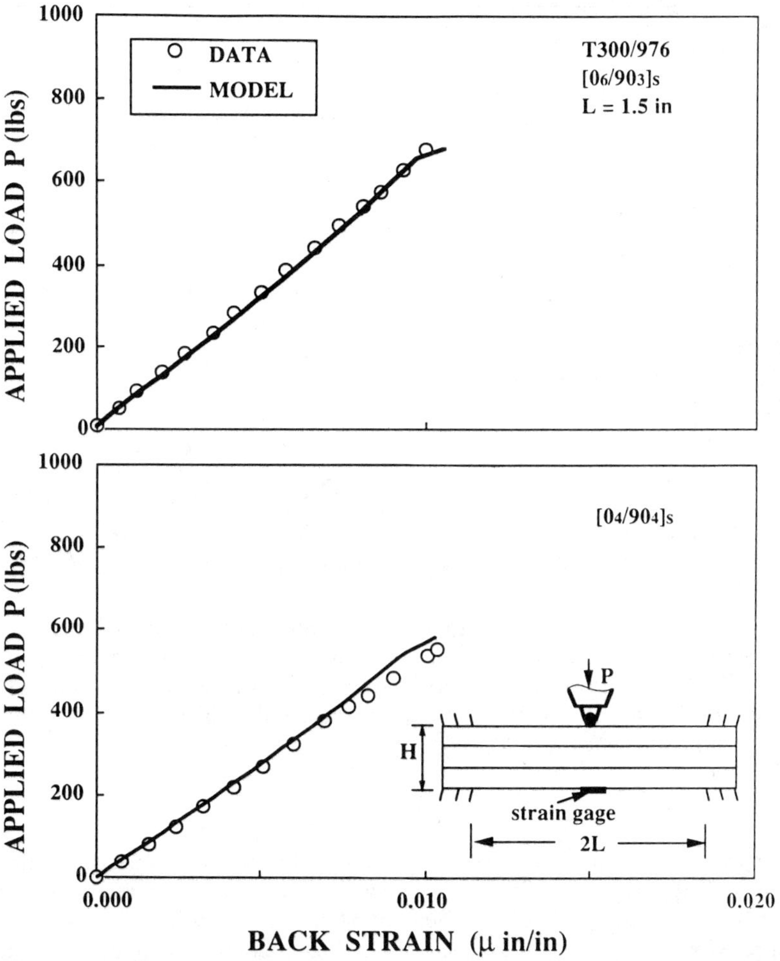

FIG. 3—*The responses of [0$_6$/90$_3$]$_s$ and [0$_4$/90$_4$]$_s$ composites between the applied load and the strain measured at the center of the back surface of the laminates. Comparisons between the experimental data and the model predictions (1 in. = 0.0254 m, 1 lbf = 4.45 N).*

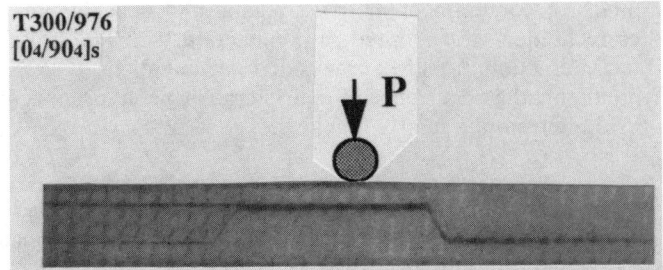

FIG. 4—*A photograph of a close side-view near the loading area of a damaged [0$_4$/90$_4$]$_s$ composite resulting from a transversely concentrated line load.*

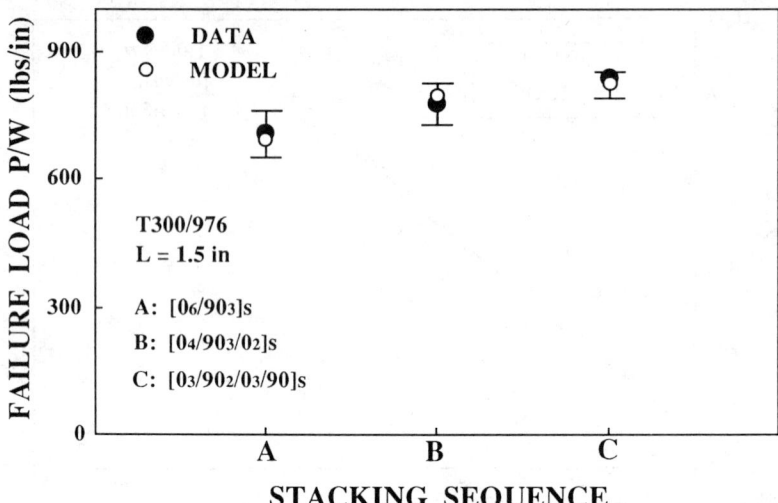

FIG. 5—*Effect of stacking sequence on the final failure load of cross-ply composites subjected to a transversely concentrated line load (1 in. = 0.0254 m, 1 lbf = 4.45 N).*

error bars denote the test data. Apparently, the transverse strength of the material increased as the plies with the same fiber orientation dispersed more uniformly across the thickness. A similar trend was also observed for the material subjected to impact loading. The damage patterns for all the ply orientations were quite similar and resembled that of the $[0_6/90_3]_s$ composites shown in Fig. 4.

Analysis

In this investigation, an analytical model was proposed for analyzing delamination growth induced by matrix cracking in a laminated composite beam resulting from transverse concentrated loading. The development of the model was based on the analysis developed previously by Kutlu and Chang [16–18] for modeling delamination propagation in laminated composites. The major modification of the model was the consideration of matrix cracking and its interaction with delamination. Basically, the model consists of three portions: a stress analysis, a contact analysis, and a failure analysis. The stress analysis was developed for calculating the stresses and the deformations of the composite laminate containing matrix cracks and delaminations. The contact analysis was developed to deal with the interface condition of matrix cracking and delamination growth. The Lagrange multiplier technique was adopted in conjunction with the stress analysis for modeling the surface interaction of matrix cracks and delaminations. The failure analysis was proposed for predicting the occurrence of matrix cracking and for modeling delamination propagation resulting from the matrix cracking.

Stress Analysis

It was assumed that the width of the laminate was considerably larger than its thickness, hence the problem was analyzed two dimensionally. A two-dimensional generalized plane strain condition in the x_1-x_3 plane was adopted. Therefore, as a first approximation the displacement field of the laminate can be assumed in the following form [16–18]

$$u_1 = u_1(x_1, x_3)$$

$$u_2 = \alpha \times x_2 \tag{1}$$

$$u_3 = u_3(x_1, x_3)$$

where α is a constant. Consequently, the strain E_{22} will be constant, but may not be zero. However, the free edge effect on the response of the plate was neglected.

Accordingly, the total potential energy of the laminate can be expressed in the initial configuration as [18,19]

$$\Pi = \sum_{m=1}^{N} \int_{^0\Omega^m} W^m(E_{ij})\ ^0d\Omega - \int_{^0S} {^0\bar{T}_i} \cdot u_i\ ^0dS \tag{2}$$

where

N = the total number of plies in the laminate,
$^0\Omega^m$ = the cross-sectional area of the mth layer of the laminate,
$\bar{T}$ = the surface tractions,
^0S = the boundary where the surface tractions $\bar{T}^i$ are applied,
W = the strain energy function, and
E_{ij} = the components of Green strain based on the finite deformation theory [19].

The strain energy function, W, for the mth layer can be expressed as

$$\frac{\partial W^m(E_{ij})}{\partial E_{ij}} = S_{ij} = C_{ijkl}^m E_{kl} \tag{3}$$

where S_{ij} = the components of the second Piola-Kirchhoff stress [19], and C_{ijkl}^m = the moduli of the mth layer.

Contact Analysis

The contact analysis proposed in this investigation was a direct extension of the contact analysis previously developed by Kutlu and Chang [16–18] for modeling the interface condition of laminated composites containing multiple delaminations. Due to the presence of matrix cracks and delaminations, the condition of the surfaces of both the matrix cracks and the delaminations had to be specified in the analysis. A local coordinate system (s,r) was introduced along a surface of each cracked or delaminated interface as shown in Fig. 6. The directions of s and r are tangent and normal to the surface, respectively. Thus, the other surface of the cracked interface at time t (deformed configuration) can be described with respect to the prescribed surface in terms of the local coordinates as

$$^tr = {^tf}(^ts) \tag{4}$$

As the laminate deforms, the local coordinates change with respect to the global coordinate system. Therefore, Eq 4 can be rewritten in terms of the global displacements as

$$^tr = {^tF}(u_i) \tag{5}$$

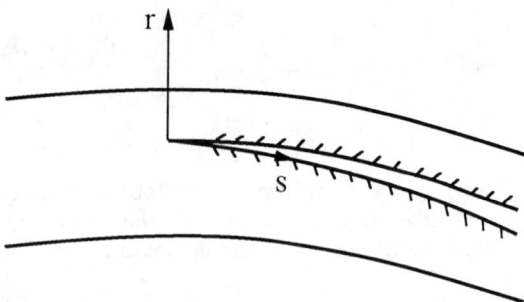

FIG. 6—*A local coordinate system along the surfaces of a matrix crack or delamination used in the analysis.*

In order to prevent the interply and intraply crack surfaces from overlapping, a constraint condition has to be satisfied in the deformed configuration along the interface. The surfaces of each cracked interface should satisfy the following condition

$$'F(u_i) \geq 0 \qquad (6)$$

If the upper and lower surfaces of the cracked interface are in contact, the above condition reduces to

$$'F(u_i) = 0 \qquad (7)$$

and for the open region of the cracked interface, the normal surface traction, $'T_r$, on the cracked interface must be equal to zero, i.e.,

$$'T_r = 0 \qquad (8)$$

Accordingly, by combining Eqs 6 and 8, the cracked interface condition can be expressed as

$$'F(u_i) \cdot 'T_r = 0 \qquad (9)$$

Note that, based on the proposed interface condition, the two surfaces of each cracked interface can slide against each other, but interface friction was not considered in the analysis.

By imposing the interface condition into the total potential energy using the Lagrange multiplier technique, the total potential energy given in Eq 2 can be modified as

$$\Pi(u_i,\lambda) = \sum_{m=1}^{N} \int {}^0\Omega^m \, W^m(E_{ij}) \, {}^0d\Omega - \int {}^0S \, {}^0\bar{T}_i \cdot u_i \, {}^0dS + \int {}^0S_c \, {}^0\lambda F(u_i) \, {}^0dS \qquad (10)$$

where $^0\lambda$ is the Lagrange multiplier, and ^0S_c is the crack interface.

In order to solve Eq 10, a nonlinear finite element method was developed. The total potential energy including the crack interface condition was expressed in terms of nodal displacements and nodal Lagrange multipliers, and was then minimized to obtain the equations of equilibrium.

Failure Analysis

Matrix cracking and delamination growth of laminated composite beams subjected to transverse concentrated loading were the major concern of this study. Fiber breakage was not considered in this study. Failure criteria were proposed for predicting the occurrence of the initial failure, and fracture mechanics based on linear elastic fracture mechanics were developed for modeling crack propagation once the damage was initiated.

Prediction of Initial Failure

In order to predict initial damage, failure criteria previously developed by Chang et al. [11] were adopted. For matrix cracking, the matrix cracking criterion can be described as

$$\left(\frac{\sigma_{yz}}{S_i}\right)^2 + \left(\frac{\sigma_{yy}}{Y_t}\right)^2 = e_M^2 \tag{11}$$

where the subscript y is the local coordinate of the 90° ply group normal to the fiber direction, the subscript z is the out-of-plane direction, S_i is the in-situ interlaminar shear strength within the laminate under consideration, Y_t is the in-situ ply transverse tensile strength [20], and σ_{yz} and σ_{yy} are the interlaminar shear stress and normal stress, respectively. The occurrence of matrix cracking is predicted when the value of e_M is equal to or greater than unity.

The delamination failure criterion previously developed by Chang et al. [21] can be expressed as

$$\left(\frac{\sigma_{zz}}{Y_t}\right)^2 + \left(\frac{\sigma_{yz}^2 + \sigma_{xz}^2}{S_i^2}\right) = e_d^2 \tag{12}$$

where σ_{zz}, σ_{yz}, and σ_{xz} are the normal stress and interlaminar stresses, respectively. The occurrence of delamination at an interface is predicted when the value of e_d is equal to or greater than unity.

Whenever the combined state of stresses satisfies either one of the criteria, initial failure is predicted. The corresponding failure criterion indicates the initial mode of failure. Once the initial failure is predicted, fracture mechanics is then introduced to simulate the growth of the local damage as the applied load continues to increase.

Modeling of Crack Propagation

In order to simulate crack propagation, a small initial crack length was introduced immediately after the occurrence of the initial failure, depending upon the type of the failure mode. For matrix cracking, a vertical crack was immediately generated in the failed ply at the location where the matrix cracking was predicted. The size of the crack was assumed to be equal to the thickness of the cracked ply group, fully extended to the interfaces of the neighboring plies, which have a different ply orientation. However, if delamination was predicted as the initial failure mode, an initial delamination length about one to two plies thick, a commonly suggested initial flaw size would then be introduced along the delaminated interface. The stresses and deformations of the laminate would then be recalculated at the same load, and fracture mechanics analysis was then applied to determine the growth of the initial failure and to model the damage propagation.

The mixed mode fracture criterion was adopted to predict the initiation of a crack propagation and can be expressed as follows [16]

$$\left(\frac{G_I}{G_{IC}}\right)^\alpha + \left(\frac{G_{II}}{G_{IIC}}\right)^\beta = e_D \qquad (13)$$

where G_{IC} and G_{IIC} are the critical strain energy release rates corresponding to Modes I and II fractures, respectively. It was assumed that G_{IC} and G_{IIC} did not change with delamination length. $\alpha = 1$ and $\beta = 1$ was selected for this study because it was found previously [16,18] to provide the best fit to the experiments.

The onset of crack growth is predicted when the value of e_D is greater than or equal to unity ($e_D \geq 1$). The strain energy release rates, G_I and G_{II}, for Mode I and Mode II fractures, respectively, can be expressed based on the linear elastic fracture mechanics as

$$G_I = \lim_{\Delta a \to 0} \left\{ \frac{1}{2\Delta a} \int_0^{\Delta a} [u_r^+(a + \Delta a) - u_r^-(a + \Delta a)]\sigma_{rr}\, ds \right\} \qquad (14)$$

and

$$G_{II} = \lim_{\Delta a \to 0} \left\{ \frac{1}{2\Delta a} \int_0^{\Delta a} [u_s^+(a + \Delta a) - u_s^-(a + \Delta a)]\sigma_{rs}\, ds \right\} \qquad (15)$$

where Δa is the crack extension, σ_{rr} and σ_{rs} are the stresses at the crack tip associated with a crack length of a, and u_r^+, u_s^+, and u_r^-, u_s^- are the displacements of the upper and lower surfaces of the crack associated with a crack length of $a + \Delta a$, respectively.

It is well documented in the literature [23,24] that there exists an oscillatory characteristic of the stress singularity near the crack tip at the interface of dissimilar materials. Mathematically, as Δa approaches zero, the definition of strain energy release rates is not well defined. However, the region of stress oscillation is believed to be very small [25]. Hence, it has been argued by Wang and Crossman [26] that, in practice, Δa should be chosen large enough so as to be consistent as the effective ply continuum assumption for inhomogeneous media such as fiber-reinforced composites. For graphite epoxy, for instance, Δa should be in the order of several fiber diameters [27]. For any value smaller than that, the values of G_I and G_{II} may have to be redefined for the inhomogeneous media. Sun and Jih [28] also suggested that a finite length of Δa should be used always when evaluating the G_I and G_{II} values at the interface for fiber-reinforced composites. Furthermore, Zafer and Chang [17,18] have recently demonstrated that the values of G_I and G_{II} calculated from a finite length of Δa correlated very well with their experiments.

Discussion

Numerical simulations of the mechanical response of the specimens are also presented in Fig. 3 by the solid lines. The material properties used in the calculations are listed in Table 2. Clearly, the predictions agreed with the test data very well from initial loading to final failure. The predicted final failure load also coincided well with the test results. Matrix cracking due to interlaminar shear stress and inplane transverse tensile stress in the 90° ply group located near the loading area was predicted to be the initial failure mode. Immediately after the occurrence of matrix cracking, the model predicted a significant load

TABLE 2—*Material properties for T300/976 graphite/epoxy prepreg tape.*

T300/976 Graphite/Epoxy

Ply longitudinal modulus	E_x =	20 200	ksi[a]
Ply transverse modulus	E_y =	1 410	ksi
Out-of-plane modulus	E_z =	1 410	ksi
In-plane shear modulus	G_{xy} =	810	ksi
Out-of-plane shear modulus	G_{xz} =	810	ksi
Out-of-plane shear modulus	G_{yz} =	546	ksi
Poisson's ratio	ν_{xy} =	0.29	
Poisson's ratio	ν_{xz} =	0.29	
Poisson's ratio	ν_{yz} =	0.40	
Ply longitudinal tensile strength	X_t =	167	ksi
Ply longitudinal compressive strength	X_c =	163	ksi
Ply transverse tensile strength	Y_t =	5.5	ksi
Ply transverse compressive strength	Y_c =	25.1	ksi
Ply shear strength	S =	14.5	ksi
Ply longitudinal thermal expansion coefficient	α_x =	0.3	μin./in. °F[b]
Ply transverse thermal expansion coefficient	α_y =	16	μin./in. °F
Critical strain energy release rate for Mode I	G_{IC} =	0.50	lbf/in.
Critical strain energy release rate for Mode II	G_{IIC} =	1.80	lbf/in.

[a]1 ksi = 6.89 MPa.
[b]100 °F = 37.78 °C.

drop due to an extensive delamination propagation which occurred at the interface between the bottom 0° and the central 90° ply groups. Accordingly, the load at which matrix cracking occurred corresponded to the maximum load.

Numerical simulations of the deformed configurations of the $[0_4/90_4]_s$ composite as a function of the applied load are shown in sequence in Fig. 7. Only one half the beam was simulated because of its symmetric condition. A matrix crack located a distance away from the center of beam was predicted first in 90° plies near the loading area. The matrix crack immediately initiated delaminations from its crack tips along the upper and lower interfaces of the cracked 90° ply group. The delamination along the lower interface grew away from the loading area toward the boundary. The growth of the lower interface delamination was very unstable, hence, it propagated instantly to the boundary. The simulations agreed well with the observed damage pattern in $[0_4/90_4]_s$ composites from the experiments shown in Fig. 4.

Figure 8 shows the predicted mechanical response of a $[0_4/90_4]_s$ composite as a function of the load head displacement during loading. The upper figure presents the relationship between the applied load and the displacement, and the lower figure shows the predicted delamination length along the lower interface as a function of the displacement. Clearly, the applied load dropped significantly when the displacement reached a critical value at which matrix cracking occurred. Apparently, the growth of the delamination induced by the internal shear matrix crack is very unstable and catastrophic.

Due to unstable delamination growth, the value of e_D in the delamination growth criterion, $(G_I/G_{IC}) + (G_{II}/G_{IIC})$, was always greater than unity. Both G_I and G_{II} contributed considerably to the growth of the delamination, indicating a mixed mode crack growth. However, as the delamination propagated a distance away from the matrix crack during growth, the ratio of G_{II}/G_{IIC} was found to be increasing and becoming a dominant factor for the unstable delamination propagation. Apparently, both Mode I and Mode II fracture

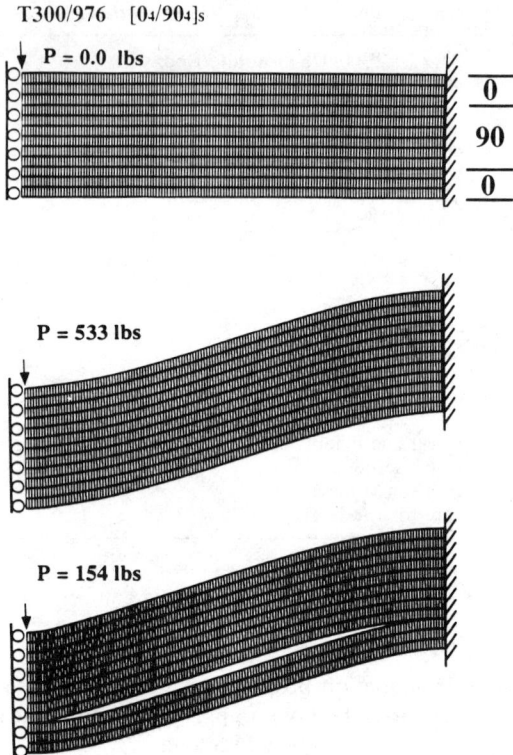

T300/976   [0₄/90₄]ₛ

FIG. 7—*Numerical simulations of deformed configurations of a [0₄/90₄]ₛ composite as a function of the applied concentrated load (1 lbf = 4.45 N).*

energy release rates, especially the latter one, are important for governing the delamination initiation and growth. The predicted final failure loads of the test specimens with three different ply orientations given in Fig. 5 are also presented in the same figure by open symbols. Again, the predictions agreed with the test data very well.

However, a completely different response can be produced from the model by simply placing 90° plies at the outermost surfaces. For instance, a [90₈/0₄]ₛ composite beam was considered in the analysis. Numerical simulations of the deformed configurations of the specimens corresponding to various loading stages are presented in Fig. 9. The model predicted that a surface matrix crack located at the bottom 90° ply group directly beneath the applied load initiated the damage, corresponding to the initial failure mode. The matrix cracking was due to an excessive transverse tensile stress normal to the fiber direction in the 90° plies as a result of bending. Such a bending crack could produce an interface delamination between the bottom 90° and the central 0° ply groups.

Figure 10 shows the predicted delamination length of a [90₈/0₄]ₛ composite as a function of the load-head displacement. Unlike the shear matrix crack, the delamination induced by the bending crack grew very stably and gradually with the increase of the applied load. No catastrophic failure was predicted for this type of ply orientation. The calculated strain energy release rates as a function of the load-head displacement are presented in Fig. 11. Apparently, Mode I fracture dominates the growth of the delamination induced by the

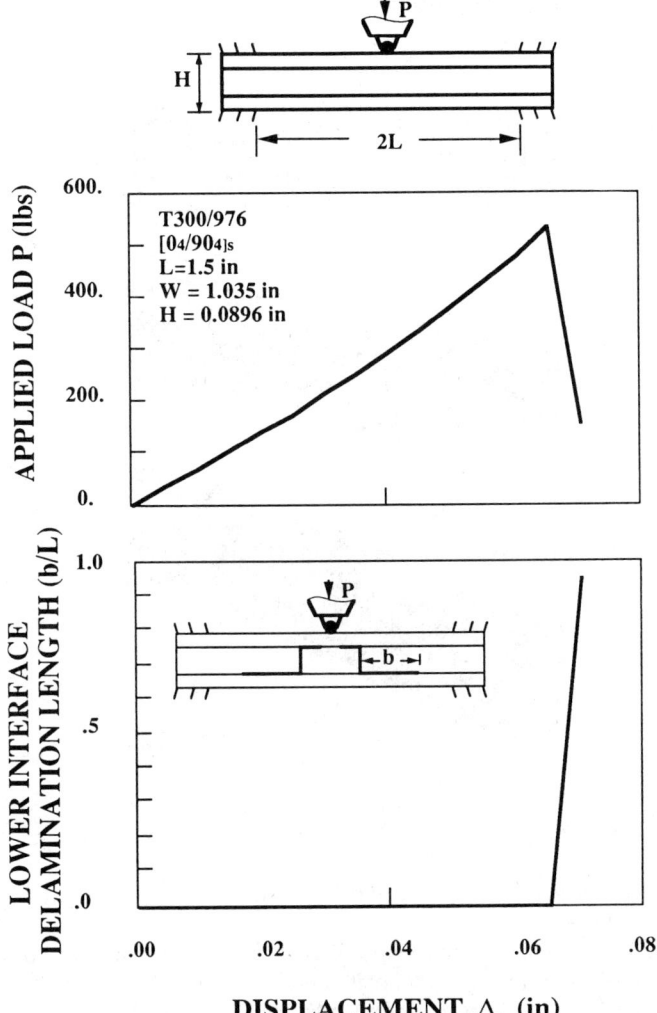

FIG. 8—*The predicted responses of a [0₄/90₄]ₛ composite as a function of the central deflection at the location where the load was applied. Top: the calculated applied load as a function of the displacement. Bottom: the calculated delamination length located at the lower 90/0 interface as a function of the displacement. (1 in. = 0.0254 m, 1 lbf = 4.45 N).*

bending crack. Similar results were also obtained by Sun et al. [*15*] on the AS4/3501 composite. Numerical calculations were also performed to compare with the test data generated by Sun et al. [*15*]. Excellent agreements were also obtained between the model predictions and Sun's test data. The results of the comparison are given in Ref *22*.

Accordingly, based on the study it is believed that both the embedded and surface matrix cracks correspond to the initial damage which can produce interface delaminations in composites subjected to transverse concentrated loading. However, the growth of the delamination induced by the embedded cracks apparently is very unstable and cata-

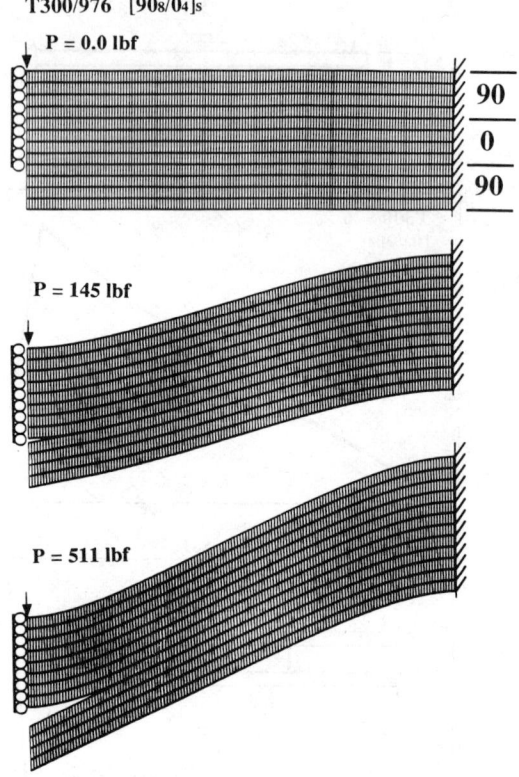

FIG. 9—*Numerical simulations of deformed configurations of a [90₈/0₄]ₛ composite as a function of the applied concentrated load (1 lbf = 4.45 N).*

strophic, while the surface crack-induced delamination grows progressively and stably and proportionally to the applied load.

Based on the above comparisons, we see that the model can adequately simulate the transverse response of laminated composite beams subjected to quasi-static concentrated loading. Based on the calculations, Mode I fracture dominates the onset and the growth of the interface delamination induced by the bending crack. However, both Mode I and Mode II fracture toughnesses are important, especially Mode II fracture, for the initiation and the growth of the delamination induced by the shear matrix crack.

Conclusions

An experimental and analytical investigation was performed to study the damage mechanism and mechanics of laminated composite beams subjected to transverse concentrated loading. An analytical model was developed for simulating the mechanical response of the composite beams subjected to a concentrated line load, from initial loading to final failure. Based on the study, the following remarks can be made for laminated composites with graphite fibers and a brittle matrix system due to transversely concentrated loading:

1. Matrix cracking is the initial failure mode.
2. Matrix cracking can result in interface delaminations.

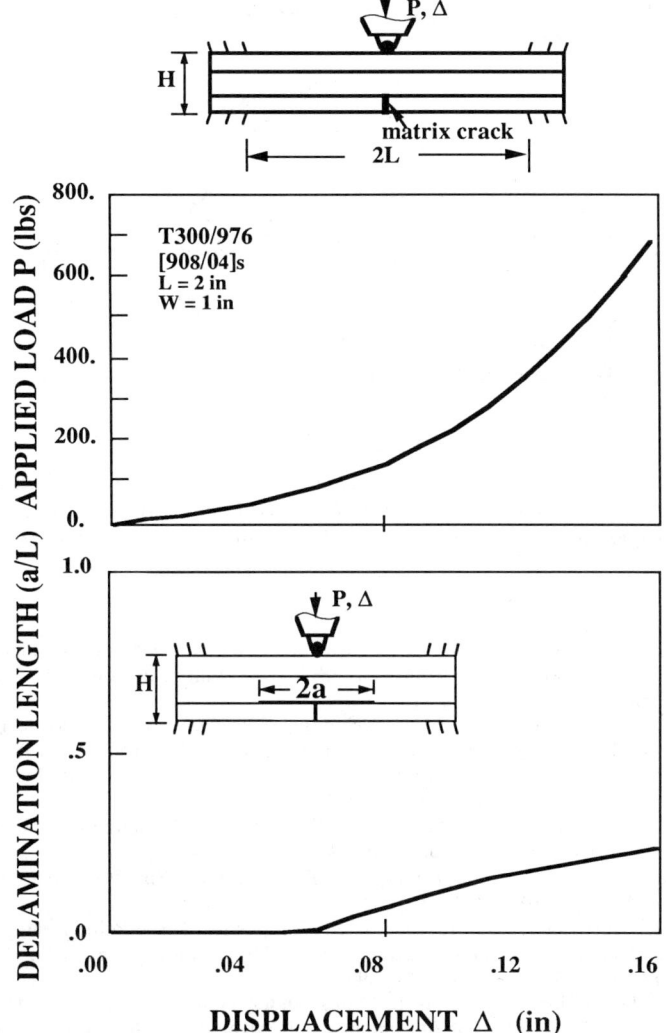

FIG 10—*The predicted delamination length at the lower 0/90 interface of a [90₈/0₄]ₛ composite as a function of the central deflection measured at the location where the load was applied (1 in. = 0.0254 m, 1 lbf = 4.45 N).*

3. Delamination growth induced by a matrix bending crack is stable and progressive.
4. Delamination growth induced by a matrix shear crack is unstable and catastrophic.
5. Ply orientation significantly affects the failure mode and the extent of the damage.

Acknowledgments

The support of the Air Force Office of Scientific Research on this project is gratefully acknowledged. Lt. Col. George A. Haritos was the contract monitor.

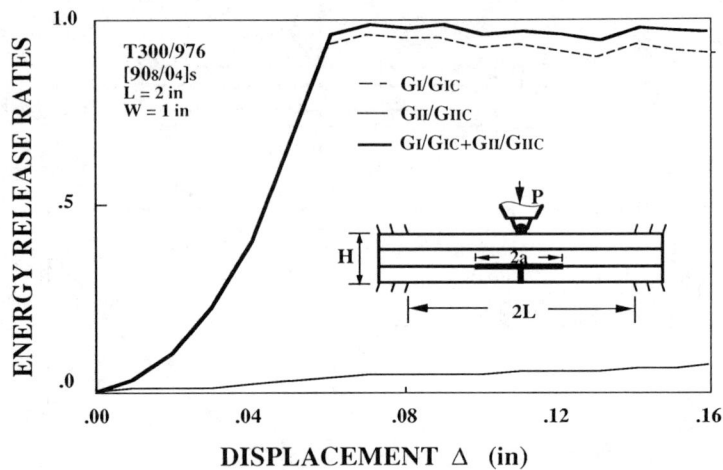

FIG. 11—*The predicted strain energy release rates of a matrix crack-induced delamination in a [90₈/0₄]ₛ composite as a function of the central deflection measured at the location where the transversely concentrated line load was applied (1 in. = 0.0254 m).*

References

[1] Bostaph, G. M. and Elber, W., "Static Indentation on Composite Plates for Impact Susceptibility Evaluation," *Proceedings,* U. S. Army Symposium on Solid Mechanics, Cape Cod, MA, September 1982.

[2] Elber, W., "Failure Mechanics in Low-Velocity Impacts on Thin Composite Plates," NASA technical paper TP 2152, NASA-Langley Research Center, Hampton, VA, 1983.

[3] Sjöblom, P. O. and Hartness, J. T., "On Low-Velocity Impact of Composite Materials," *Journal of Composite Materials,* Vol. 22, January 1988, pp. 30–52.

[4] Joshi, S. P. and Sun, C. T., "Impact-Induced Fracture in a Laminated Composite," *Journal of Composite Materials,* Vol. 19, 1986, pp. 51–66.

[5] Poe, C. C. Jr., "Simulated Impact Damage in a Thick Graphite/Epoxy Laminate Using Spherical Indenters," NASA TM-100539, 1988.

[6] Starbuck, M., "Damage States in Laminated Composites Three-Point Bend Specimens—An Experimental/Analytical Correlation Study," Ph.D. dissertation, Virginia Polytechnic Institute and State University, Blacksburg, VA, 1990.

[7] Finn, S., "Delaminations in Composite Plates Under Transverse Static or Impact Loading," Ph.D. dissertation, Department of Aeronautics and Astronautics, Stanford University, Stanford, CA, 1991.

[8] Chang, F. K., Choi, H. Y., and Downs, R. J., "A New Approach Toward Understanding Damage Mechanisms and Mechanics of Laminated Composites Due to Low-Velocity Impact, Part I—Experiments," *Journal of Composite Materials,* Vol. 25, August 1991, pp. 992–1011.

[9] Chang, F. K., Choi, H. Y., and Wu, H. Y. T., "A New Approach Toward Understanding Damage Mechanisms and Mechanics of Laminated Composites Due to Low-Velocity Impact, Part II—Analysis," *Journal of Composite Materials,* Vol. 25, August 1991, pp. 1011–1038.

[10] Choi, H. Y., Wang, H. S., and Chang, F. H., "Effect of Laminate Configuration and Impactor's Mass on the Initial Impact Damage of Composite Plates Due to Line-Loading Impact," *Journal of Composite Materials,* Vol. 26, No. 6, 1992, pp. 804–827.

[11] Chang, F. K., Choi, H. Y., and Jeng, S. T., "Characterization of Impact Damage in Laminated Composites," *Proceedings,* 34th International SAMPE Symposium and Exhibition, Reno, NV, May 1989.

[12] Choi, H. Y., "Damage in Graphite/Epoxy Laminated Composites Due to Low-Velocity Impact," Ph.D. dissertation, Department of Mechanical Engineering, Stanford University, Stanford, CA, 1990.

[13] Wu, H. T. and Springer, G. S., "Impact Induced Stresses, Strains, and Delaminations in Composites Plates," *Journal of Composite Materials,* Vol. 22, 1988, pp. 533–560.

[14] Liu, D., "Impact Induced Delamination—A View of Bending Stiffness Mismatching," *Journal Of Composite Materials,* Vol. 22, 1988, pp. 674–692.

[15] Sun, C. T. and Manoharan, M. G., "Growth of Delamination Cracks Due to Bending in a [$90_s/0_s/90_s$] Laminate," *Composite Science and Technology,* Vol. 34, 1989, pp. 365–377.

[16] Kutlu, Z. and Chang, F. K., "Modeling Compression Failure of Laminated Composites Containing Multiple Through-the-Width Delaminations," *Journal of Composite Materials,* Vol. 26, 1992, pp. 350–387.

[17] Chang, F. K. and Kutlu, A., "Collapse Analysis of Composite Panels Containing Multiple Delaminations," *Proceedings,* AIAA/ASME/ASCE/AHS 30th Structures, Structural Dynamics and Materials (SMD) Conference, Mobile, AL, April 1989.

[18] Kutlu, Z., "Compression Failure of Laminated Composite Panels Containing Multiple Delaminations," Ph.D. dissertation, Department of Mechanical Engineering, Stanford University, Stanford, CA, 1991.

[19] Bathe, K. J., *Finite Element Procedures in Engineering Analysis,* Prentice-Hall, Inc., Englewood Cliffs, NJ, 1982.

[20] Chang, F. K. and Lessard, L. B., "Damage Tolerance of Laminated Composites Containing an Open Hole and Subjected to Compressive Loading: Part I—Analysis," *Journal of Composite Materials,* Vol. 25, No. 1, January 1991, pp. 2–43.

[21] Chang, F. K. and Springer, G. S., "The Strengths of Fiber Reinforced Composite Bends," *Journal of Composite Materials,* Vol. 20, No. 1, January 1986, pp. 30–45.

[22] Liu, S., Kutlu, Z., and Chang, F. K., "Delamination Propagation Induced by Matrix Cracking in Laminated Composites Subjected to Transverse Loading," AFOSR Technical Report of 1991 for Contract No. 89-0554.

[23] Edogan, F., "Stress Distribution in a Nonhomogeneous Elastic Plane with Crack," *Journal of Applied Mechanics,* Vol. 30, 1963, pp. 232–236.

[24] Cominou, M. and Chang, F. K., "The Interface Crack in a Shear and Bending Field," *Advances in Aerospace Structures and Materials, AD-01,* S. S. Wang and W. J. Renton, Eds., 1981, pp. 287–291.

[25] Anderson, G. P., Devries, K. L., and Williams, M. L., "Finite Element in Adhesion Analysis," *International Journal of Fracture,* Vol. 9, 1973, pp. 421–435.

[26] Wang, A. S. D. and Crossman, F. W., "Initiation and Growth of Transverse Cracks and Edge Delamination in Composites, Part I—An Energy Method," *Journal of Composite Materials,* Vol. 14, 1980, pp. 71–87.

[27] Kybicki, E. F. and Pagano, N. J., "A Study of the Influence of Micro-Structure on the Modified Effective Modulus Approach for Composite Laminates," *Proceedings,* First International Conference on Composite Materials, AIMMPE, 1976.

[28] Sun, C. T. and Jih, C. J., "On the Strain Energy Release Rates for Interfacial Cracks in Bi-material Media," *Engineering Fracture Mechanics,* Vol. 28, 1987, pp. 13–27.

Damage: Measurement, Analysis, and Modeling

Roderick H. Martin[1] and Wade C. Jackson[2]

Damage Prediction in Cross-Plied Curved Composite Laminates

REFERENCE: Martin, R. H. and Jackson, W. C., **"Damage Prediction in Cross-Plied Curved Composite Laminates,"** *Composite Materials: Fatigue and Fracture, Fourth Volume, ASTM STP 1156,* W. W. Stinchcomb and N. E. Ashbaugh, Eds., American Society for Testing and Materials, Philadelphia, 1993, pp. 105–126.

ABSTRACT: This paper details the analytical and experimental work required to predict delamination onset and growth in a curved cross-plied composite laminate subjected to static and fatigue loads. The composite used was AS4/3501-6, graphite/epoxy. Analytically, a closed-form stress analysis and 2-D and 3-D finite element analyses were conducted to determine the stress distribution in an undamaged curved laminate. The finite element analysis was also used to determine values of strain energy release rate at a delamination emanating from a matrix crack in a 90° ply. Experimentally, transverse tensile strength and fatigue life were determined from flat 90° coupons. The interlaminar tensile strength and fatigue life were determined from unidirectional curved laminates. Also, Mode I fatigue and fracture toughness data were determined from double cantilever beam specimens. Cross-plied curved laminates were tested statically and in fatigue to give a comparison to the analytical predictions. A comparison of the fracture mechanics life prediction technique and the strength-based prediction technique is given. Generally, both prediction techniques gave good comparisons with the experimental results.

KEY WORDS: composite material, delamination, curved laminate, strain energy release rate, fracture toughness, matrix crack

Nomenclature

a_0 — Initial delamination length
a — Delamination length
A,B,C,D — Constants in curve fit expression
E_{11} — Elastic modulus in the fiber direction
E_{22} — Elastic modulus transverse to fiber direction
E_{33} — Elastic modulus through thickness
f — Frequency
G — Total strain energy release rate
G_c — Critical value of strain energy release rate
G_{Ic} — Mode I static interlaminar fracture toughness
G_{Imax} — Maximum Mode I cyclic strain energy release rate
L — Moment arm length
m — Constant in DCB compliance expression
n — Exponent in DCB compliance expression
P — Load

[1]Analytical Services and Materials, Inc., MS 188E, NASA Langley Research Center, Hampton, Virginia, 23665-5225.
[2]U.S. Army Aerostructures Directorate, MS 188E, NASA Langley Research Center, Hampton, Virginia, 23665-5225.

P_{max}	Maximum cyclic load
P_{min}	Minimum cyclic load
R	Radius or load ratio
R_{inner}	Inner radius
t	Thickness
w	Width
z	Width-wise coordinate
α	Delamination angle counter clockwise
β	Delamination angle clockwise
δ_{max}	Maximum cyclic displacement
δ_{min}	Minimum cyclic displacement
Θ	Angle around curved portion of laminate
σ_r	Radial stress
σ_θ	Tangential stress
σ_{2c}	Transverse static strength
σ_{2max}	Transverse cyclic stress
σ_{3c}	Interlaminar static strength
σ_{3max}	Interlaminar cyclic stress
$\tau_{r\theta}$	Interlaminar shear stress
ν_{12}	Poisson's ratio

Many laminated composite structures, such as an angle bracket, a co-cured web, or a frame have a loaded curved portion [1] (Fig. 1). The final failure in such structures may be a complex progression of ply cracking, delamination, and fiber failure. Delamination may initiate from radial stresses caused by the bending. Also, any tangential stresses present may cause matrix cracks to develop in an off-axis ply. If a crack occurs, singular interlaminar stresses will be created where the matrix crack meets the adjacent plies. These

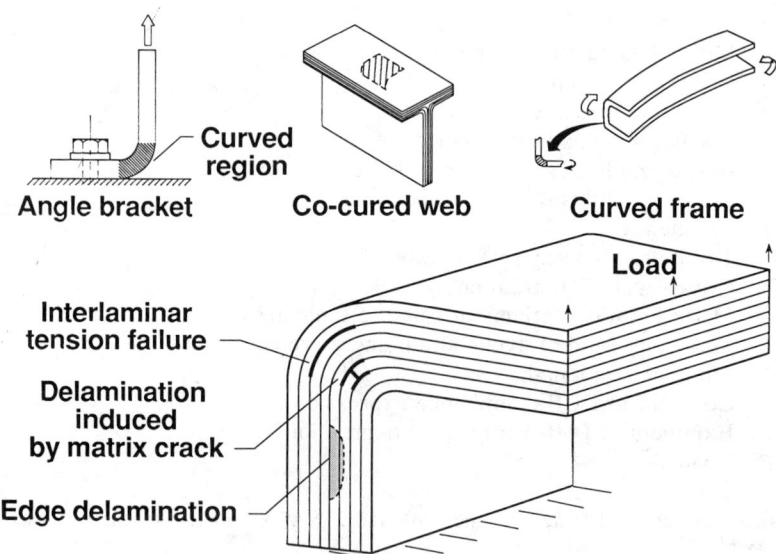

FIG. 1—*Structural configurations and damage modes for curved laminates.*

interlaminar stresses may create local delaminations, shown schematically in Fig. 1. The difference in material properties between adjacent plies of different orientation may also cause mathematically singular stresses at the free edges which may initiate edge delaminations. These edge delaminations may interact with the previously mentioned failure modes and complicate delamination onset predictions using classical strength based failure criteria. Therefore, the need to understand the stress distribution and damage mechanisms in a curved laminate is important in aiding the structural designer to predict the strength of such a component.

To extend strength prediction techniques to account for singular stresses from matrix cracks, delaminations, material defects, edges, or any other discontinuity, a fracture mechanics prediction capability is necessary. Interlaminar fracture mechanics based failure criteria offer a technique to predict the onset and growth of delamination in a component with a singular stress source [2–8]. For an interlaminar fracture mechanics based failure criterion, a value of strain energy release rate, G, must be determined at every potential delamination source. This value of G may be termed the critical value, G_c, and will cause delamination onset and growth when it equals the interlaminar fracture toughness of the composite. The techniques to determine the G_c may depend on the source of the potential delamination.

Figure 2 gives some examples from the literature of how G_c has been determined for different structures. For edge delamination in a flat multidirectional laminate, G increased from zero to a plateau within a few ply thicknesses from the edge (Fig. 2a) [2]. The value of G at the plateau was considered as G_c and used to predict delamination onset at the edge. In a tapered laminate, the value of G at the peak was considered to be G_c and was used to predict delamination onset (Fig. 2b) [3]. For delamination growth from a matrix crack in a curved laminate [6], the point of inflection in the G versus delamination length curve was postulated to give the critical value of G required to predict static delamination onset (Fig. 2c). The effect of the free edge on the growth of local delaminations initiating from a matrix crack in a flat laminate subjected to tensile loads was investigated in Ref 7.

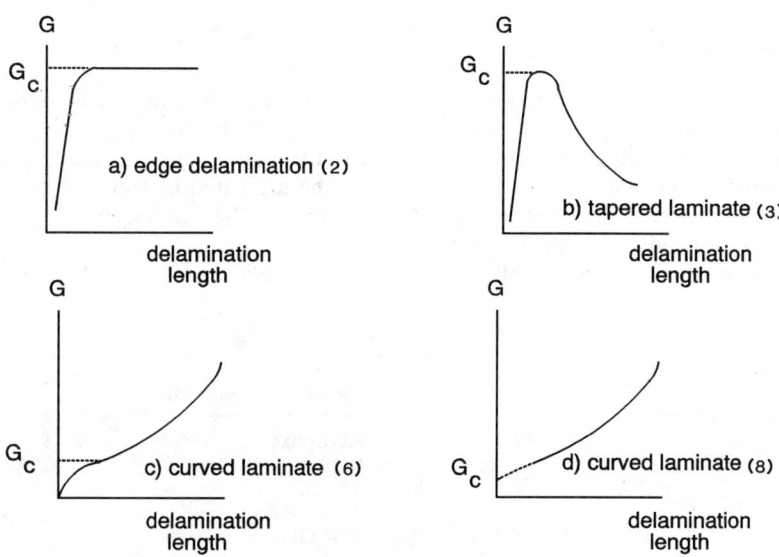

FIG. 2—*Shapes of G curves for different structures.*

For a straight delamination front perpendicular to the edge, the value of G increased to a plateau value as the delamination grew from the matrix crack. This plateau was assumed to give the critical value of G similar to the edge delamination case (Fig. 2a). A curved laminate was analyzed with a delamination emanating from a matrix crack in Ref 8. The G versus delamination length curve was extrapolated to zero delamination length to determine a critical value of G (Fig. 2d).

The purpose of this paper is to predict damage in a curved laminate subjected to static and fatigue loads using a strength and a fracture mechanics based failure criterion. The stress distribution in an undamaged cross-plied curved laminate was determined using a closed form solution and 2-D and 3-D finite element analyses (FEA). The G distribution for a delamination initiating from a matrix crack in a set of 90° plies was determined from the 2-D and 3-D finite element analyses. Static and fatigue tests were conducted using flat 90° laminates to determine the transverse strength, σ_{2c}, and the fatigue life. Unidirectional curved laminates were used to determine interlaminar tensile strength, σ_{3c}, and the fatigue life. The Mode I interlaminar fatigue and fracture toughness of the composite material were determined using a unidirectional double cantilever beam (DCB) specimen. Attempts to predict interlaminar tension delamination of the cross-plied curved laminate were made by comparing the maximum radial stress to the interlaminar tensile strength and life data for the composite. Attempts to predict matrix cracking were made by comparing the maximum radial and tangential stress in a 90° ply to the transverse and interlaminar strength and life data. Finally, attempts to predict delamination onset from a matrix crack were made by comparing the appropriate values of G to the fatigue and fracture toughness data. Static and fatigue tests were also conducted on a cross-plied curved laminate to determine the damage modes and static and fatigue strength to compare with the predictions.

In addition to the curved laminate being used as a structural component, it has been considered as a possible specimen to determine σ_{3c} for composite materials [9]. Therefore, it is a secondary objective of this paper to determine whether or not the multidirectional curved laminate is suitable for σ_{3c} determination considering the additional damage modes that may be present and the difficulty in determining stresses at singularities.

Experimental Procedure and Results

This section describes the materials, the experimental procedure, and experimental results. All specimens were fabricated from Hercules AS4/3501-6 graphite/epoxy. The specimens were cured in an autoclave according to the manufacturer's recommendations. Material elastic moduli were obtained from Ref 4 and are listed in Table 1.

The ply thicknesses were nominally 0.125 mm thick. The volume fractions for each specimen type are given in their corresponding subsections below. Prior to testing, all specimens were dried using the following cycle: 1 h at 95°C, 1 h at 110°C, 16 h at 125°C,

TABLE 1—*Material properties used in the analyses.*

AS4/3501-6 Graphite/Epoxy
$E_{11} = 140$ GPa
$E_{22}, E_{33} = 11.0$ GPa
$G_{12}, G_{13}, G_{23} = 5.84$ GPa
$\nu_{12}, \nu_{13}, \nu_{23} = 0.3$

and 1 h at 150°C. Following the drying cycle, the specimens were stored in a desiccator cabinet until tested.

Transverse Tension Strength Tests

Flat 90°, unidirectional, 24-ply specimens were fabricated for the transverse tension tests. The specimen dimensions were 150 by 33 mm with the fibers oriented in the 33-mm width direction. The specimens had a fiber volume fraction of 65.0%. The specimens were subjected to static tests and to load-controlled fatigue tests at an R-ratio (P_{min}/P_{max}) of $R = 0.1$ and at a frequency of 5 Hz. The tests were conducted until failure occurred. If the specimen had not failed after at least 1×10^6 cycles, the test was stopped. In most of these tests no loading tabs were used, and most specimens failed at or near the grips. In a few specimens glass epoxy loading tabs were used. However, the failures were still at or near the grips, so the use of tabs was discontinued. The static failure load or the number of cycles to failure was recorded, and the results are shown in Fig. 3. The straight lines were drawn by eye as approximate fits of the lower and upper bound curves between 10^2 and 10^6 cycles. The use of these fits will be described later. The arrows on the fatigue data points in Fig. 3 and subsequent figures indicate that the specimen did not fail.

Interlaminar Tension Strength Tests

Unidirectional, 0° curved laminate specimens were fabricated for interlaminar tensile strength determination. The specimens had 24 plies and their configuration and dimensions are given in Fig. 4. The prepreg was laid over a solid aluminum block, and the panels were cured in an autoclave. The specimens had an average volume fraction in the curved region of 54.7%. The static strengths of these specimens were taken from Ref 4. Fatigue tests were conducted using the test fixture shown in Fig. 5. The fatigue tests were conducted under load control at an R-ratio of 0.1 and a frequency of 5 Hz. The number of cycles to the onset of delamination was noted. The onset of delamination also corresponded to a rapid loss of bending stiffness of the curved portion, causing the specimen to open. When the specimen opened, the machine hydraulics and the cyclic counter switched off. From Ref 4 the maximum normalized radial stress was determined to be ($\sigma_r w/P$) = 4.37 mm^{-1},

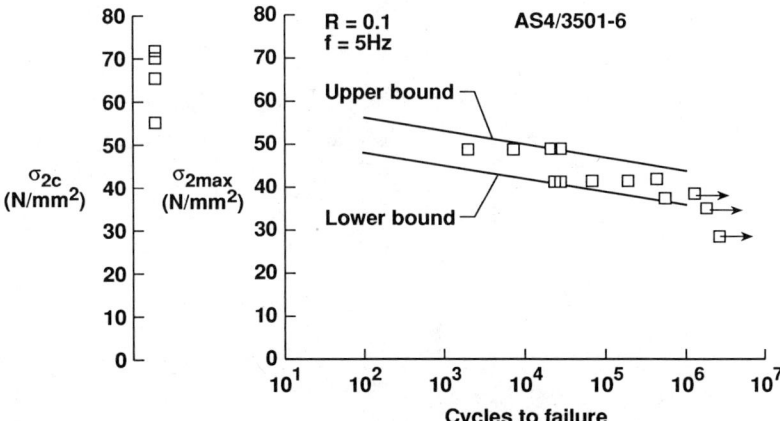

FIG. 3—*Transverse tension static and fatigue strength of AS4/3501-6 from 90° flat laminates.*

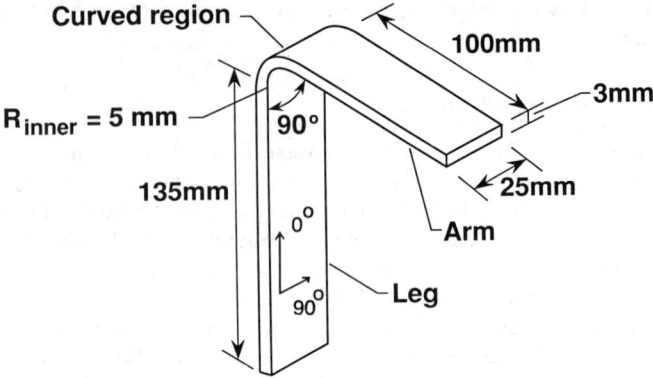

FIG. 4—*Unidirectional and cross-plied curved laminate configuration.*

occurring near the midthickness of the curved region. The tangential and in-plane shear stresses were negligible at this location. This value of radial stress was used with the applied loads to determine σ_{3max} (or σ_{3c}) using Eq 1.

$$\sigma_{3max} = \left(\frac{\sigma_r w}{P}\right)\left(\frac{P_{max}}{w}\right) = 4.37\left(\frac{P_{max}}{w}\right) \tag{1}$$

The fatigue results from this work and the static results from Ref *4* are shown in Fig. 6.

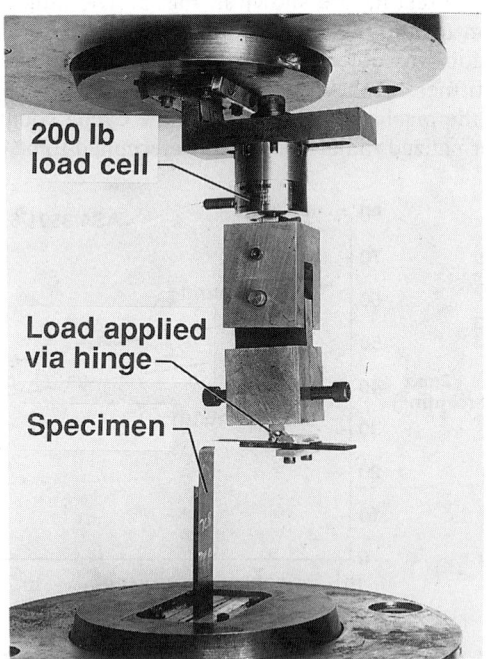

FIG. 5—*Curved laminate test fixture.*

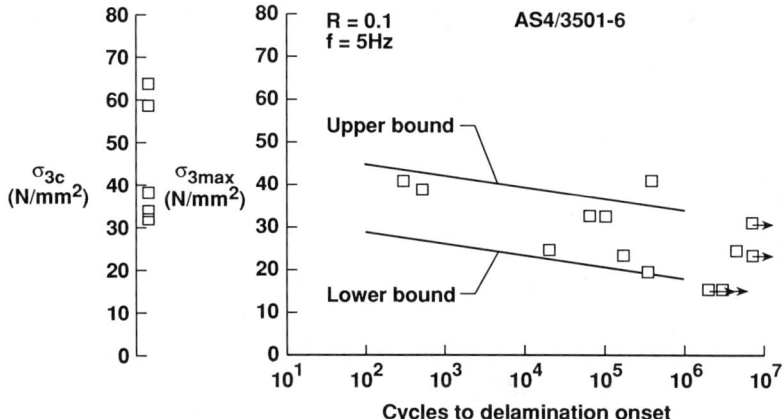

FIG. 6—*Interlaminar tension static and fatigue strength of AS4/3501-6 from unidirectional curved laminates. Static results from Ref 4.*

The straight lines were drawn by eye as an approximate fit to the lower and upper bound curves between 10^2 and 10^6 cycles. The use of these fits will be described later.

Double Cantilever Beam Tests

Double cantilever beam (DCB) specimens were fabricated to determine the Mode I fatigue and static fracture toughness of the composite. The specimens were 24-ply, 0° unidirectional laminates. A 13-μm nonadhesive Kapton film was placed at the midplane at one end prior to curing to simulate a delamination. The specimen dimensions were 100 by 25 mm. The specimens had a volume fraction of 55.9%. Piano hinges were bonded to the surface to allow the load to be transferred to the beam. The static tests were conducted under displacement control at a loading rate of 0.5 mm/min. Fracture toughness, G_{Ic}, was determined for a delamination initiating from the end of the thin insert [10]. The fatigue tests were conducted under displacement control at an R-ratio ($\delta_{min}/\delta_{max}$) of $R = 0.1$, at a frequency of 5 Hz at various maximum cyclic displacements [10,11]. The number of cycles to delamination onset was determined by monitoring the maximum cyclic load, P_{max}. If a 1% decrease in P_{max} was observed, then the delamination was assumed to have grown. Delamination onset was also monitored by visual inspection of the tip of the insert. The fracture toughness and the maximum cyclic strain energy release rate, G_{Imax}, versus the number of cycles to delamination onset are given in Fig. 7, where G_{Imax} was calculated from

$$G_{Imax} = \frac{nP_{max}\delta_{max}}{2wa_0} \qquad (2)$$

where n is the exponent in the static compliance expression $\delta/P = ma^n$ (an average value of $n = 2.58$ from three specimens was used in Eq 2), and a_0 is the initial delamination length. The visual and 1% load drop methods of determining the number of cycles to delamination onset gave inconsistent results. The visual detection method gave lives greater, less than, and similar to the 1% load drop method. In some specimens a 1% load drop was not detected after visual delamination growth had been observed. Hence, the visual data were used in the predictions presented later.

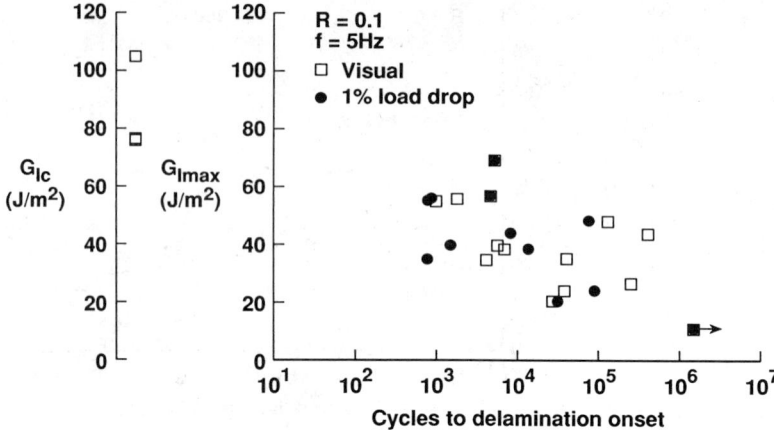

FIG. 7—*Interlaminar static and fatigue fracture toughness of AS4/3501-6 using the DCB specimen.*

Cross-Plied Curved Laminate Tests

Static and fatigue tests were conducted on cross-plied curved laminates. The cross-plied specimens were laid up to the dimensions given in Fig. 4 using the same procedure described for the unidirectional curved laminates above. The layup used was $[0_4/90_3/0_5]_s$ and will be referred to as Layup A. This particular layup was chosen because it was anticipated that a matrix crack would occur in the tension-loaded 90° plies and that delamination would grow from the matrix crack. The layup was not intended to represent a viable layup for a structural component. The specimens had a volume fraction of 59.5% in the curved region. During the static and fatigue tests the first audible sign of damage corresponded to complete loss of bending stiffness of the curved portion of the specimen.

The results of the static and fatigue tests on Layup A are shown in Fig. 8. The fatigue results are plotted as the maximum applied load per unit width versus number of cycles to onset of damage. The static results of Ref 6 are also shown in Fig. 8. The layup used in that

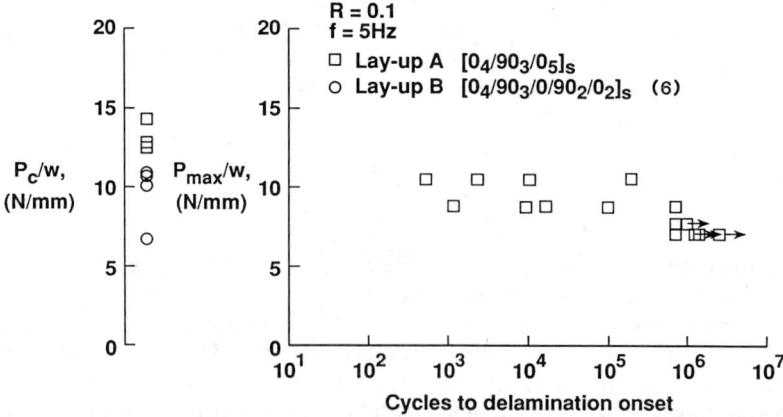

FIG. 8—*Static and fatigue strength of cross-plied curved laminates.*

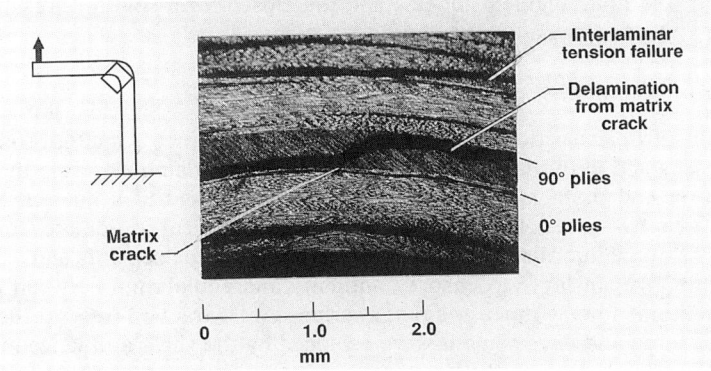

FIG. 9—*Damage in curved laminate, Layup A.*

work was $[0_4/90_3/0/90_2/0_2]_s$ and is referred to as Layup B. The failure paths for Layups A and B are shown in Figs. 9 and 10, respectively. In Fig. 9, an oblique matrix crack can be seen in the tension loaded $[90]_3$ plies with delaminations emanating from this crack. Other delaminations consistent with an interlaminar tension failure can be seen in the 0° plies. In Fig. 10, a straight matrix crack can be seen in the tension-loaded $[90]_3$ plies and also in the $[90]_2$ plies above. This second matrix crack, which was not analyzed in Ref 6, was another reason Layup A was chosen for this work, i.e., to eliminate the complications of two matrix cracks in the analysis.

Analysis

This section details the analysis conducted on the cross-plied curved laminates to determine the stress distribution with no damage present and the strain energy release rate distribution with matrix ply cracking and delamination present. The stress analysis consists of a closed-form multilayer theory. These results are compared with a 2-D and a 3-D finite element analysis. The finite element analysis was also used to determine the G

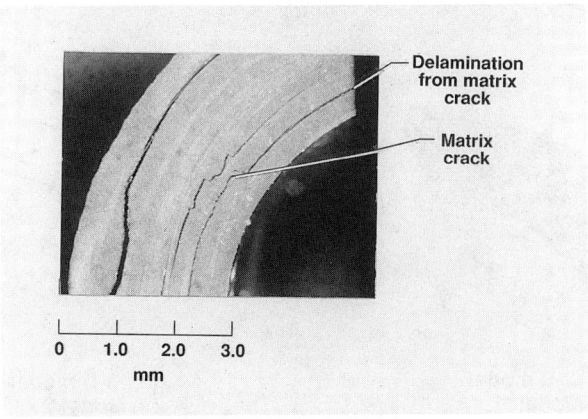

FIG. 10—*Damage in curved laminate, Layup B.*

distribution. In the finite element models and the closed-form solution, residual thermal and moisture stresses were not considered.

2-D Closed-Form Solution

Classical anisotropic elasticity theory was used to construct a multilayered theory to calculate stresses and deformation fields around the curved laminate using the method in Ref *12*. Both an end moment and an end force were applied to a quarter section of a circular beam. An Airy stress function was written in cylindrical coordinates for each layer in the beam. At every interface between layers, both the displacements and stresses were matched to the adjacent layer to ensure continuity and equilibrium between layers. The boundary conditions for the inner and outer surfaces were traction free. At the end of the beam, the end force and end moment were balanced by the shear and tangential stresses, respectively. The total stress in the beam is a summation of the stresses caused by the end force and the moment. Once the constants in the Airy stress function were known, the stresses were determined at any location (R,Θ) in the beam using the expressions in Ref *12*.

Finite Element Analysis

The finite element package MSC/NASTRAN [*13*] was used for the analysis. A 2-D finite element model of the complete specimen was created and the mesh of the curved portion is shown in Fig. 11*a*. This model was used to determine the stress distribution around the radius and through the thickness for comparison with the closed form solution and to determine the global variation of G with delamination length. The model used four-noded isoparametric elements and was a plane strain analysis. Around the curved region there was one element per ply thickness and one element per one degree sweep. Larger elements were used in the legs and the arms. The final model had approximately 8100 degrees of freedom. A matrix crack was modeled through the group of [90]$_3$ plies nearest to the smallest radius ($R = 5$ mm) at an angle of 45° around the radius. This crack location was consistent with the crack observed experimentally in Figs. 9 and 10. Delaminations were

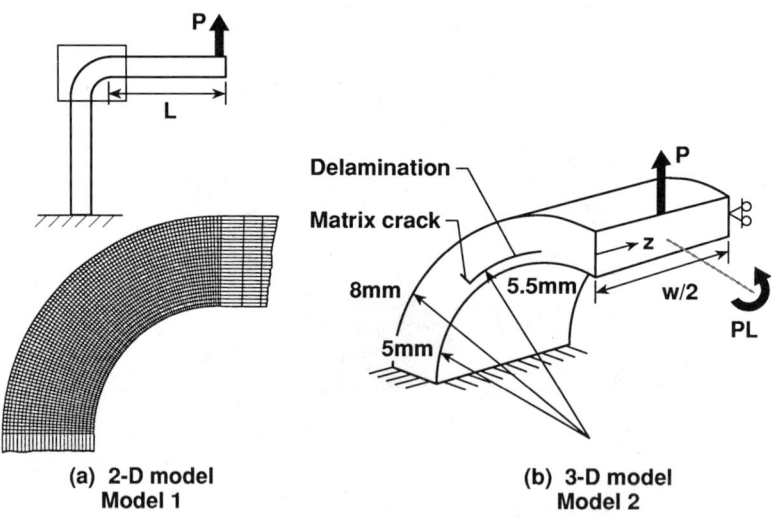

(a) 2-D model
Model 1

(b) 3-D model
Model 2

FIG. 11—*Models used in finite element analyses.*

modeled emanating in each of the four directions from the matrix crack in one direction at a time (Fig. 11b). The notation for the paths of these cracks is given in (Fig. 12). To simulate the matrix crack and delamination in the model, a free surface was included by the use of coincident nodes. The coincident nodes were restrained together using multi-point constraints (MPCs). By releasing the appropriate MPCs in different analysis cases, several delamination lengths could be modeled using one finite element mesh. Strain energy release rate was determined using the virtual crack closure method [14].

A full 3-D model of only the curved region was also created (Fig. 11b). The model was rigidly supported at one end. Only half the width, w, was modeled because of symmetry. The total width was 25 mm. A matrix crack was modeled in the same location as the 2-D model. To reduce the number of degrees of freedom, only the delamination growing clockwise from the bottom of the matrix crack was modeled, that is, along Path b1 (Fig. 12).

The 2-D model showed that a delamination was most likely to form along this path. Near the matrix crack and the free edge, the elements were 1/16 of a ply thickness square by one ply thickness wide (z-direction), where one ply thickness equals 0.125 mm. This refinement was used to capture any variations in the value of G in the close vicinity of the matrix crack. The model had approximately 30 000 degrees of freedom and used six- and eight-noded, solid, modified isoparametric elements. This model was also used to determine the effects of the free edges on the stress distribution assuming an undamaged laminate.

Analytical Results

Stress Analysis

Figure 13 shows the variation of the radial and shear stress and Fig. 14 the variation of the tangential stress with Θ in Layup A. The plots are in the first tension-loaded 90° ply as indicated by the arrow on the figures. The stresses were determined from the 2-D model and the closed-form solution, and Θ is defined in Fig. 12. In Fig. 13 the two solutions agree reasonably well around much of the curved portion and diverge towards Θ = 0° and Θ = 90° because of the differences in boundary conditions. In Fig. 14 the agreement is poor but the trend of the variation with theta is similar. Between 25° < Θ < 75° the stresses vary by only a small amount, and failure might initiate anywhere in this region.

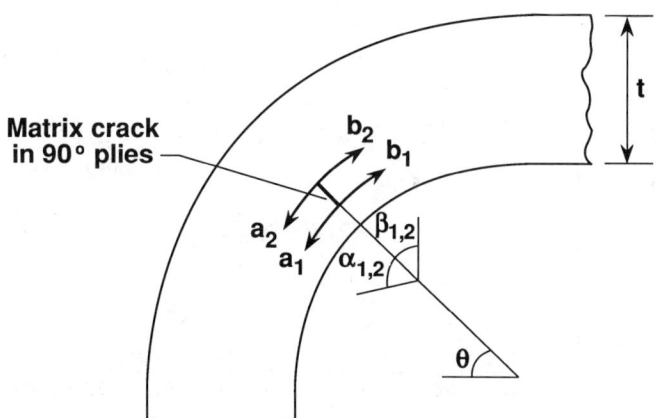

FIG. 12—*Notation for delamination growth from matrix crack.*

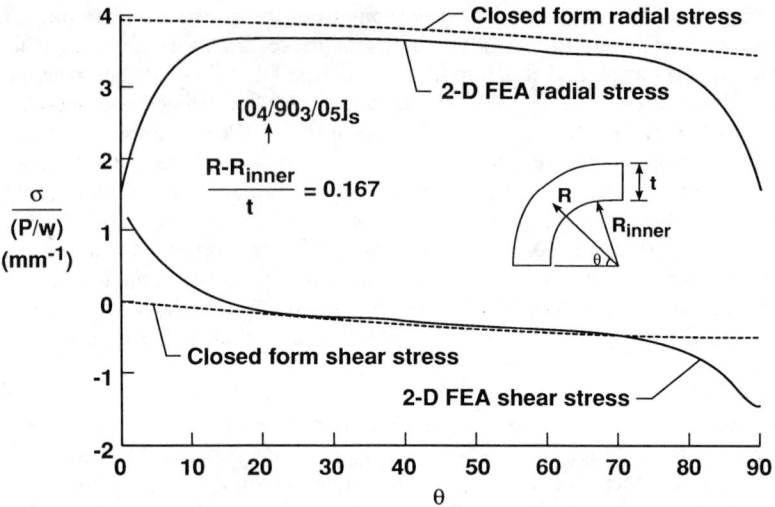

FIG. 13—*Radial and shear stress variation with theta.*

Figure 15 shows the radial and shear stress components plotted across the width along the first 0/90 interface determined using the 3-D model. Also, plotted for comparison, are the results from the closed-form solution. Figure 15 shows that the radial stress is largest at the free edge, indicative of the singular nature of the interlaminar radial stress at the free edge in a 0/90 interface. No attempt was made to further refine the finite element mesh at the edges to show more clearly the singular stresses because of the large number of degrees of freedom of the current model. The closed-form solution compared well to the FEA solution away from the free edge. The shear stress FEA results in Fig. 15 did not agree well with the closed-form solution. However, the difference is small compared to the magnitude of the radial stress.

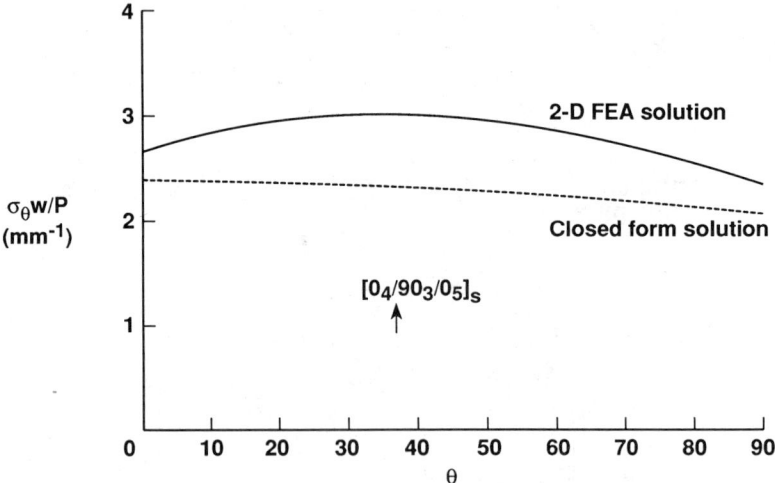

FIG. 14—*Tangential stress variation with theta.*

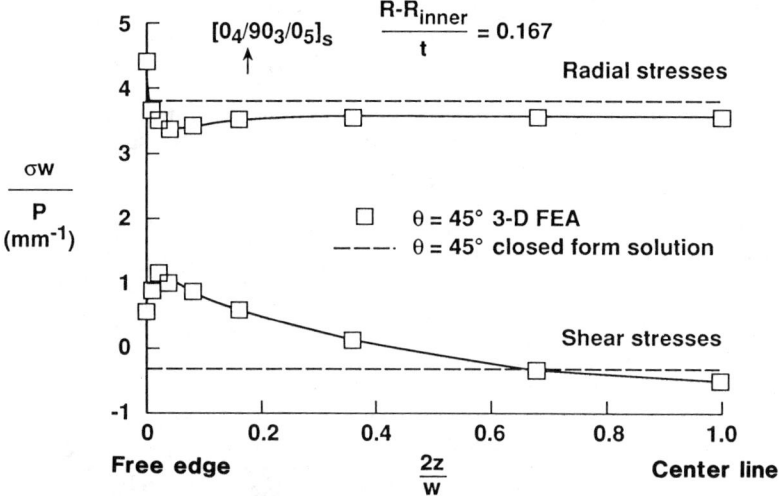

FIG. 15—*Radial and shear stress variation across width.*

Figures 16, 17, and 18 show the variation of the radial, tangential, and shear stresses, respectively, through the thickness in Layup A as determined from the 2-D model, the 3-D model, and the closed-form solution and are plotted at $\Theta = 45°$. The results from the 3-D model are plotted at $2z/w = 0.0$ (the free edge) and $2z/w = 1.0$ (the center line). In Fig. 16 the 2-D FEA results and the closed form solution agree reasonably well for the radial stress. The radial stress reaches a maximum towards the center of the laminate in the 0° plies. In the 3-D FEA at the center line, the results compare well with the closed-form solution. However, at the free edge the two solutions vary, demonstrating the singular

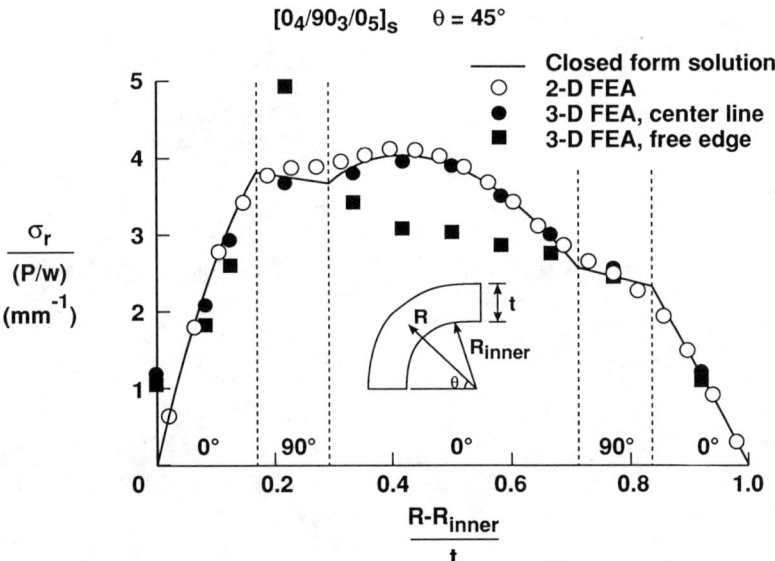

FIG. 16—*Radial stress through thickness.*

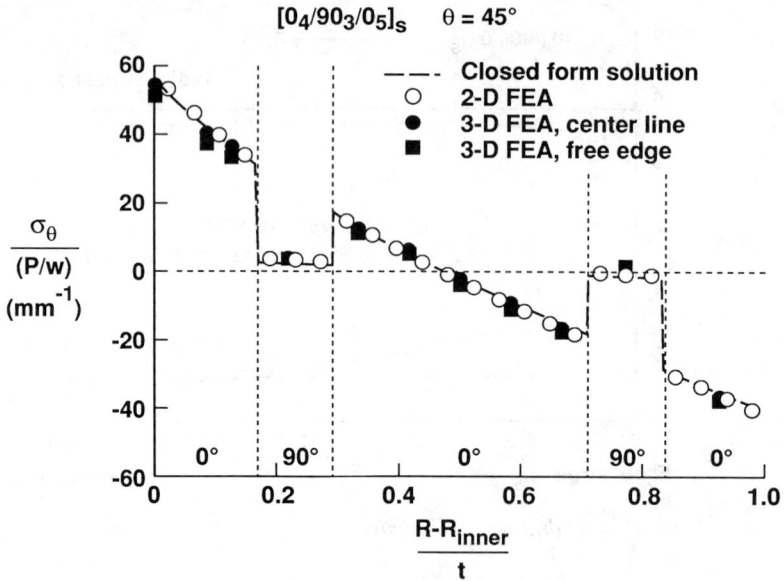

FIG. 17—*Tangential stress through thickness.*

nature of the stresses at the free edge. In Fig. 17 the tangential stress reaches a maximum at the inside edge, $(R - R_{inner}/t = 0.0)$ where there is a 0° ply. For the 90° plies, the tangential stress has a maximum tension value in the 90° ply closest to the inner radius, $(R - R_{inner}/t) = 0.167$. The tangential stress at this location may contribute to matrix cracking in these plies. The results for all three solutions agree reasonably well. In Fig. 18, the shear

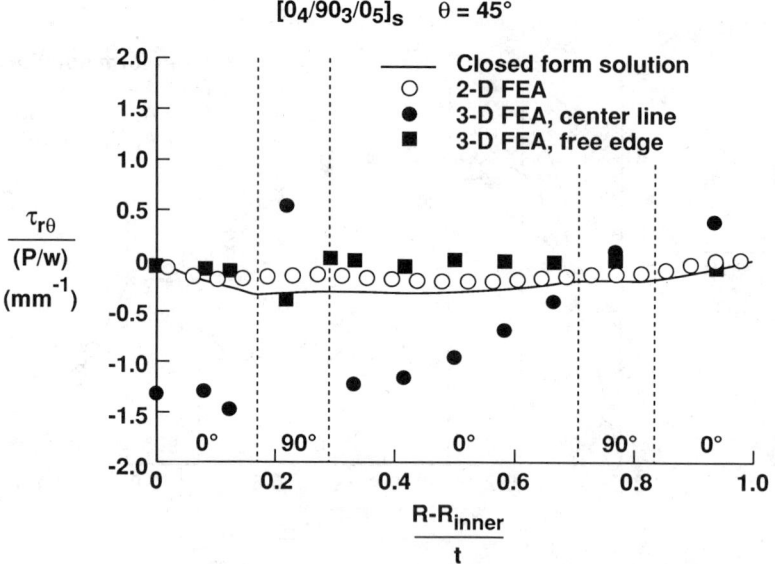

FIG. 18—*Shear stress through thickness.*

stress results from the closed-form solution and the 2-D FEA solution agree well, and the shear stress is small. From the 3-D solution in the center, the results are inconsistent with the 2-D solutions. The cause for this is not known presently, but may be caused by the large aspect ratio of the elements in the center width of the model.

Strain Energy Release Rate Analysis

In Ref 6, the variation in G with delamination length for delamination growth from a matrix crack along each of the four paths shown in Fig. 12 was determined for Layup B. The value of G along Path b1, G_{b1}, was usually higher than the other three values (G_{a1}, G_{a2}, and G_{b2}) (Fig. 19). Therefore, it was assumed delamination would initiate along Path b1. In Ref 4 a technique for predicting the delamination growth path for a delamination with two fronts was given. The technique consisted of comparing the values of G at each front and growing the delamination at the location of the highest G. Following a similar method the next highest value to G_{b1} was G_{a2}. By assuming growth along Path b1 and comparing values of G_{a2} to G_{b1}, it was observed that G_{b1} was always larger than G_{a2} around most of the curved portion. Therefore, delamination should grow along Path b1 followed by delamination along Path a2, as observed in Fig. 10. Figure 20 shows the variation of G_{b1} with β for Layup B. Because G_{b1} is continually increasing, delamination growth should be unstable once initiated, as observed experimentally. Also, shown in Fig. 20 is the Mode I component of strain energy release rate, G_1. Because of the oscillatory nature of the stress fields between plies of different orientation, the Mode I value may not converge [15]. However, with the element size used, the delamination remained largely Mode I around much of the curved portion due to the high radial stresses present. The above analyses were not repeated for Layup A. A similar result is expected since the matrix crack and delamination paths in Figs. 9 and 10 are similar.

Figure 21 shows the distribution of G_{b1} across the width at several delamination lengths (angle β) from the matrix crack for Layup A using the 3-D model. Across 80% of the width, G was constant and reached a maximum at the free edge. The results are very similar for Layup B in Ref 6. Figure 21 indicates that the delamination front may not remain straight and perpendicular to the edge as it grows. Delamination front curvature

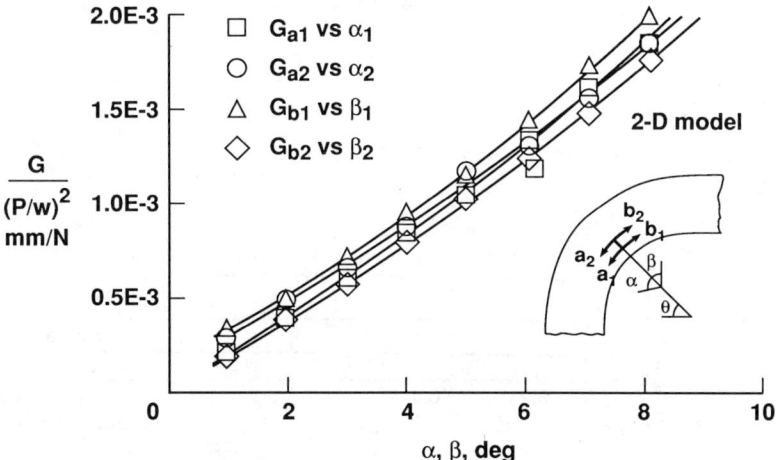

FIG. 19—*G variation with delamination growth along four paths for Layup B* [6].

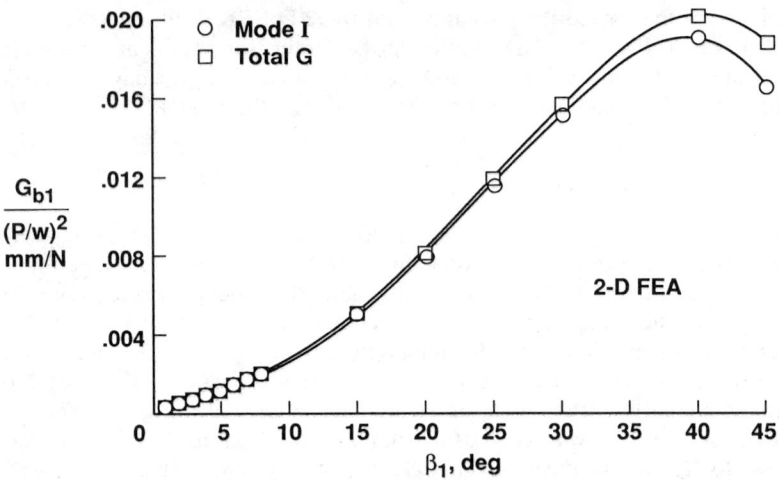

FIG. 20—G *variation along Path b1 for Layup B* [6].

was not considered in the analysis. Figure 22 shows the variation of G with β using the 3-D model. As delamination grows, the rate of change of G initially decreases and then begins to increase with a point of inflection in between. Close to the matrix crack, G is very small. Also shown in Fig. 22 are the results using the 2-D model (solid circular symbols). As expected, the values of G from the 3-D model in the interior are similar to those obtained from the 2-D model. A polynomial curve fit expression to the G distribution on the free edge and the center line is shown in Fig. 23. The first part of this curve, where $dG/d\beta$ is decreasing, may be analogous to the distributions for edge delamination [2], Fig. 2a, or for delamination from dropped plies [3], Fig. 2b, where the singular stress source, the edge and the dropped ply, respectively, dominates the G distribution. However, for the curved

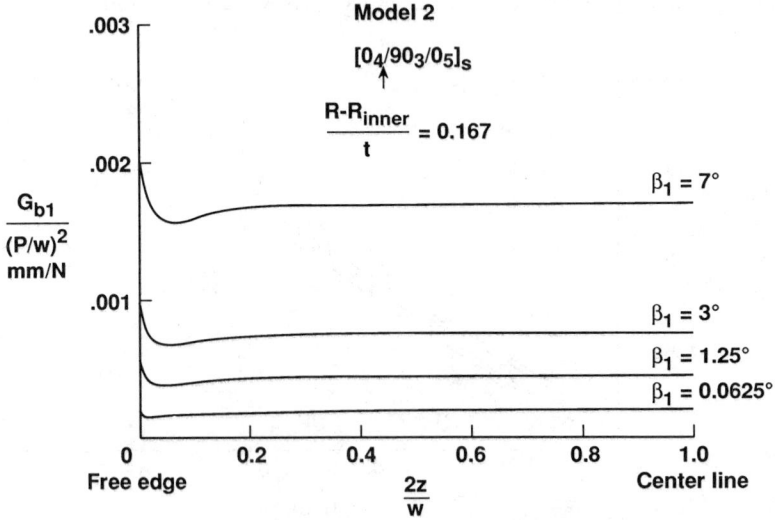

FIG. 21—G *variation across width at various delamination lengths.*

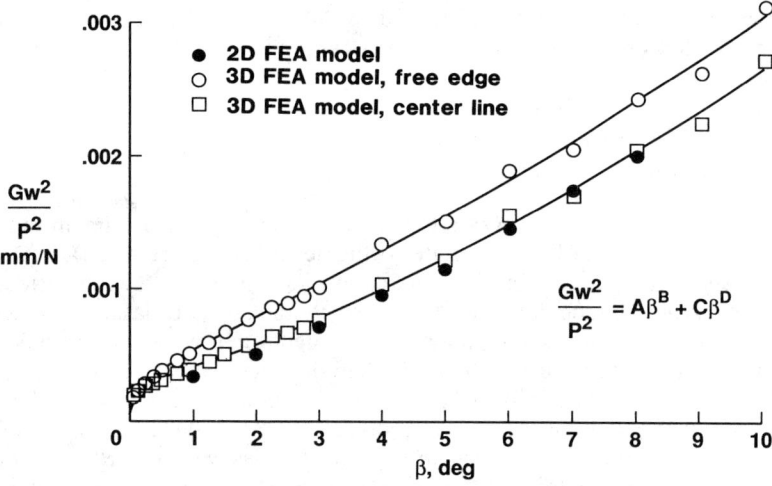

FIG. 22—*G variation with delamination growth close to matrix crack.*

laminate, there is a high radial force which also contributes to G and causes $dG/d\beta$ to increase with no peak or plateau, as observed in Refs 2 and 3. It was postulated in Ref 6 that the point of inflection may be used to determine a critical value of G to predict delamination growth much as the plateau or peak value in Refs 2 and 3. At the free edge in Layup A, the point of inflection ($d^2G/d\beta^2 = 0$) was determined to be at $\beta = 2.13°$ (1.6 ply thicknesses from the matrix crack), yielding a critical value of $(Gw^2/P^2) = 8.08$ E-4 mm/N. At the center of the width the point of inflection was determined to be at $\beta = 1.08°$ (0.83 ply thicknesses from the matrix crack), yielding a critical value of $(Gw^2/P^2) = 4.17$ E-4

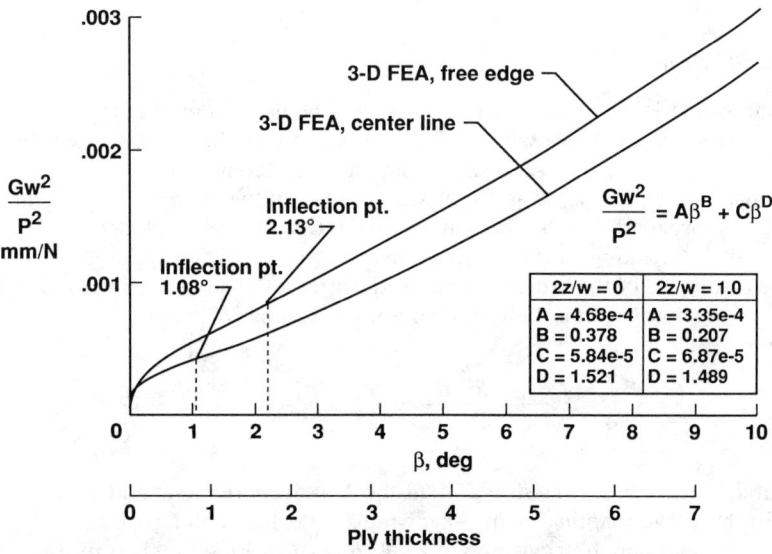

FIG. 23—*Curve fit of G variation with delamination growth.*

mm/N. These values of G may be compared to the static and fatigue toughness of the composite to predict delamination onset for the curved laminate.

Damage Onset Predictions

Delamination from Radial Stresses

In Refs *16* and *17* delamination was predicted by comparing the maximum radial stress to the transverse strength of the composite. If the delamination occurs in a 0/90 interface and the analysis used to determine the maximum radial stress ignores the effects of the free edge, this prediction technique may be incorrect, and any correlation between experimental failure loads and predictions may be coincidental. However, if the delamination occurs between plies of the same orientation, then this technique is valid. Figure 9 shows an interlaminar tension failure in the 0° plies as well as a matrix crack with delaminations emanating from it. It is not known which occurred first. From the closed-form solution in Fig. 16, the maximum normalized radial stress in the 0° plies was $(\sigma_r \, w/P) = 4.04 \text{ mm}^{-1}$ at $(R - R_{inner}/t) \approx 0.42$. At this location the tangential and shear stresses were negligible. By comparing this value of radial stress to the interlaminar tensile strength results shown in Fig. 6, a prediction for Layup A can be made using Eq 3

$$\frac{P_{max}}{w} = \frac{\sigma_{3max}}{\left(\dfrac{\sigma_r w}{P}\right)} = \frac{\sigma_{3max}}{4.04} \tag{3}$$

and is shown in Fig. 24. Individual experimental data points from Fig. 6 have been used for the predictions. The predictions are close to the experimental failures and are generally conservative. From Fig. 10 no interlaminar tension delaminations occurred alone in Layup B, and the criterion in Eq 3 is not relevant.

Transverse Ply Cracking Prediction

A strength-based failure criterion could possibly be used to predict the onset of matrix cracking in the 90° plies. However, the 3-D stress analysis showed the presence of stress singularities at the free edge. Also, the strength of the composite in the 90° flat tension specimens may not be the same as the in-situ strength of the 90° plies in the laminate [*18*]. Therefore, any predictions must be purely qualitative and are included in this paper to show that matrix cracking in the 90° plies may occur. In the 90° plies there is a biaxial stress state composed of the radial and tangential stresses. The Tsai-Hill criterion has been derived for predicting first ply failure in composite lamina [*19*] as

$$\frac{\sigma_{11}^2 - \sigma_{11}\sigma_{22}}{X^2} + \frac{\sigma_{22}^2}{Y^2} + \frac{\tau_{12}^2}{S^2} = 1 \tag{4}$$

where X and Y are the tensile strengths in the 1 and 2 directions and S is the 12 shear strength. In the curved laminates the shear ratio is negligible compared to the radial and tangential ratios and may be neglected in Eq 4. Rewriting Eq 4 in terms of the normalized radial and tangential stresses and omitting the shear stress terms, a predicted load per unit

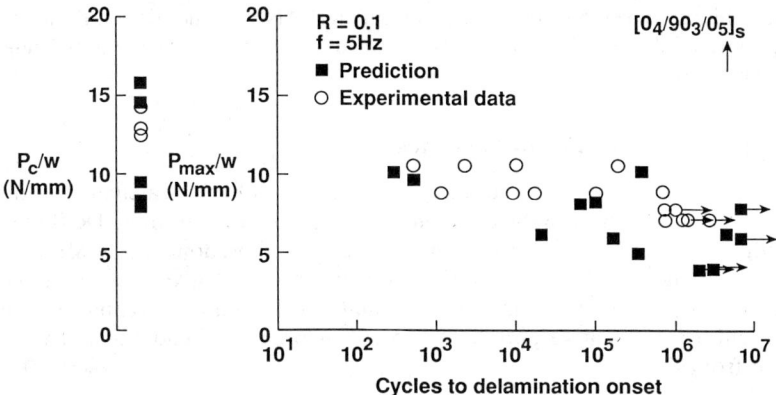

FIG. 24—*Prediction of interlaminar tension failure between 0° plies.*

width may be determined as

$$\left(\frac{w}{P_{max}}\right)^2 = \frac{\left(\dfrac{\sigma_r w}{P}\right)^2 - \left(\dfrac{\sigma_r w}{P}\right)\left(\dfrac{\sigma_\theta w}{P}\right)}{(\sigma_{3max})^2} + \frac{\left(\dfrac{\sigma_\theta w}{P}\right)^2}{(\sigma_{2max})^2} \tag{5}$$

From the closed form solutions in Figs. 16 and 17, the maximum radial and tangential stresses in the 90° plies were $(\sigma_r\, w/P) = 3.91$ mm^{-1} and $(\sigma_\theta\, w/P) = 2.35$ mm^{-1}, respectively. Because individual data points may not be used in Eq 5 for σ_{2max} and σ_{3max}, straight-line lower and upper bound curves were fit to the transverse tension and interlaminar tension fatigue data in Figs. 3 and 6, respectively. These curves were drawn for demonstration purposes only and are not meant to characterize the material property. Using these straight line fits in Eq 5, lower and upper bound predictions can be made and are shown in Fig. 25. The prediction curves indicate that matrix cracking may occur prior to final failure of the curved laminate. These prediction curves do not take into account the singular nature of the stresses at the free edge nor the residual thermal stresses, both of

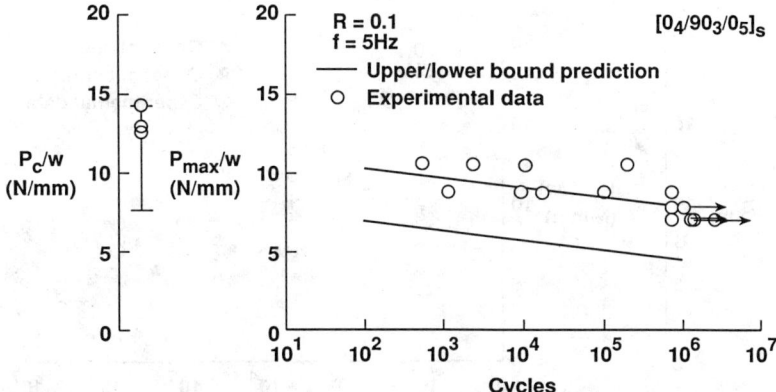

FIG. 25—*Prediction of transverse matrix cracking using the Tsai-Hill criterion.*

which will effectively reduce the fatigue life further. Also, as mentioned in the experimental work, some of the 90° flat specimens failed in the grips and hence may underestimate the transverse strength of the composite.

Delamination Onset from the Matrix Crack

If a matrix crack is assumed to already exist, the values of G calculated at the point of inflection may be compared to the delamination data obtained from the DCB in Fig. 7. The G distribution in Fig. 20 showed that delamination was predominantly Mode I. If the G distribution had significant Modes II and III components, a mixed-mode failure criterion would be necessary. However, assuming a total G criterion and comparing it to Mode I data will yield a conservative approach [3,5,10]. A predicted load per unit width may be determined from Eq 6

$$\frac{P_{max}}{w} = \sqrt{\frac{G_{Imax}}{\left(\frac{Gw^2}{P^2}\right)}} \tag{6}$$

The results of this prediction are shown in Fig. 26 using the values of normalized G at the free edge and in the center. The predictions at the free edge are conservative, and the predictions at the center are close to the experimental data. This prediction does not include the number of cycles to form the matrix crack and hence will generally be a conservative prediction. In the analysis the matrix crack was assumed to be straight; it is not clear how an angled crack would affect the G distribution or the Mode I/Mode II mix of G. However, if a Mode II component was present, it would be anticipated that the predicted number of cycles to delamination onset would be higher because of the higher fatigue and fracture toughness in Mode II [10]. It is probable that the matrix crack occurred before the delamination, although it was not detected prior to final failure in the experimental work.

Discussion

If the stresses that cause delamination are not singular, then damage onset or delamination may be predicted using a strength criterion. This was demonstrated for interlaminar

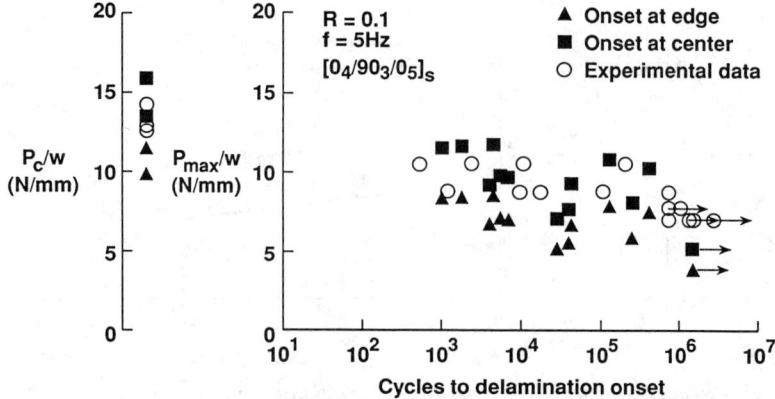

FIG. 26—Prediction of delamination onset using fracture mechanics criterion.

tension delaminations initiating in the 0° plies. However, if the stresses are singular, such as at a free edge or a discontinuity, a fracture mechanics prediction technique represents a means for predicting damage onset and growth. This was demonstrated for delaminations initiating from a matrix crack. Because closed-form and 2-D FEA solutions do not account for free edge singularities, strength predictions may be inaccurate and, more importantly, potentially unconservative using these analyses techniques. Therefore, it is important to determine where damage initiates before selecting a particular criterion to predict it. However, 2-D analyses are useful for initial design because damage initiation depends on layup. For example, matrix cracks can be avoided if a layup is chosen so that the tangential stresses in the 90° plies are not highly tensile. Also, free edge delamination may be minimized if a layup is chosen so that the radial stresses are low at 0/90 interfaces (or other perpendicular interfaces, e.g. $+45/-45$).

Because of matrix cracking and interface affects at free edges, multidirectional laminates should not be used to determine interlaminar tensile strength, σ_{3c}, unless it is certain that interlaminar tension failure occurs within a group of plies of the same orientation. The simplest means to ensure this is in a unidirectional curved beam [4]. In Ref 4, a 2-D closed-form analysis of a unidirectional curved laminate determined that the stresses were purely radial at the failure location in the laminate.

Summary

This paper details the analytical and experimental work required to predict delamination onset in a curved composite laminate subjected to static and fatigue loads. The composite used was AS4/3501-6, graphite/epoxy. Analytically, a closed-form stress analysis and a 2-D and a 3-D finite element analyses were conducted to determine the stress distribution in an undamaged curved laminate. The finite element analyses were also used to determine values of strain energy release rate at a delamination emanating from a matrix crack in a 90° ply. Experimentally, transverse tensile strength and fatigue life were determined from flat 90° coupons. The interlaminar tensile strength and fatigue life were determined from unidirectional curved laminates. Also, Mode I fatigue and fracture toughness data were determined from double cantilever beam specimens. The analysis and the strength and toughness data were used to predict the static and fatigue strength of cross-plied curved laminates. The prediction for interlaminar tension delamination in the 0° plies was in reasonable agreement with the experimental results for the curved laminate. The interlaminar fracture mechanics approach compared the critical value of strain energy release rate at the free edge, and in the center, to the fatigue and fracture toughness of the composite. The predictions at the free edge were conservative, and the predictions at the center were in agreement with the experimental data. This prediction does not include the number of cycles to form the matrix crack and hence will generally be a conservative prediction.

References

[1] Kedward, K. T., Wilson, R. S., and McLean, S. K., "Flexure of Simply Curved Composite Shapes," *Composites*, Vol. 20, No. 6, November 1989, pp. 527–536.
[2] O'Brien, T. K., "Mixed-Mode Strain-Energy-Release Rate Effects on Edge Delamination of Composites," *Effects of Defects in Composite Materials, ASTM STP 836*, D. J. Wilkins, Ed., American Society for Testing and Materials, Philadelphia, 1984, pp. 125–142.
[3] Murri, G. B., Salpekar, S. A., and O'Brien, T. K., "Fatigue Delamination Onset Prediction in Tapered Composite Laminates," *Composite Materials: Fatigue and Fracture (Third Volume), ASTM STP 1110*, T. K. O'Brien, Ed., American Society for Testing and Materials, Philadelphia, 1991, pp. 312–339.
[4] Martin, R. H., "Delamination Failure in a Unidirectional Curved Composite Laminate," *Com-

posite Materials: Testing and Design, Tenth Volume, ASTM STP 1120, G. C. Grimes, Ed., American Society for Testing and Materials, Philadelphia, 1992.

[5] O'Brien, T. K., "Local Delamination in Laminates with Angle Ply Matrix Cracks: Part II— Delamination Analysis and Characterization," this publication. Also published as NASA TM 104076, 1991.

[6] Martin, R. H., "Analysis of Delamination Onset and Growth in Curved Laminates," Mechanics Computing in 1990's and Beyond, H. Adeli and R. L. Sierakowski, Eds., American Society for Civil Engineers, New York, 1991, pp. 922–927.

[7] Salpekar, S. A. and O'Brien, T. K., "Combined Effect of Matrix Cracking and Stress-Free Edge on Delamination," Composite Materials: Fatigue and Fracture, Third Volume, ASTM STP 1110, T. K. O'Brien, Ed., American Society for Testing and Materials, Philadelphia, 1991, pp. 287–311.

[8] Sun, C. T. and Kelly, S. R., "Failure in Composite Angle Structures—Part II: Onset of Delamination," Journal of Reinforced Plastics and Composites, Vol. 7, May 1988, pp. 233–244.

[9] Paul, P. C., Saff, C. R., Sanger, K. B., Mahler, M. A., Kan, H. P., and Kautz, E. F., "Analysis and Test Techniques for Composite Structures Subjected to Out of Plane Loads," Composite Materials: Testing and Design, Tenth Volume, ASTM STP 1120, G. C. Grimes, Ed., American Society for Testing and Materials, Philadelphia, 1992.

[10] Murri, G. B. and Martin, R. H., "Effect of Initial Delamination on Mode I and Mode II Interlaminar Fracture Toughness and Fatigue Fracture Threshold," this publication.

[11] Martin, R. H. and O'Brien, T. K., "Characterizing Mode I Fatigue Delamination of Composite Materials," Proceedings, American Society for Composites, Fourth Technical Conference, Blacksburg, Virginia, October 1989, Technomic Publishing Co., Lancaster, PA, pp. 257–266.

[12] Ko, W. L. and Jackson, R. H., "Multilayer Theory for Delamination Analysis of a Composite Curved Bar Subjected to End Forces and End Moments," Composite Structures 5: Proceedings of the 5th International Conference, Paisley, Scotland, 24–26 July, 1989, pp. 173–198.

[13] NASTRAN Users Manual, Version 66B, MacNeal-Schwendler Corporation, Los Angeles, November 1989.

[14] Shivakumar, K. N., Tan, P. W., and Newman, J. C. Jr., "A Virtual Crack-Closure Technique for Calculating Stress Intensity Factors for Cracked Three Dimensional Bodies," International Journal of Fracture, Vol. 36, 1988, pp. R43–R50.

[15] Sun, C. T. and Jih, C. J., "On the Strain Energy Release Rates for Interfacial Cracks in Bi-Material Media," Engineering Fracture Mechanics, Vol. 28, 1987, pp. 13–27.

[16] Chang, F. K. and Springer, G. S., "The Strengths of Fiber Reinforced Composite Bends," Journal of Composite Materials, Vol. 20, January 1986, pp. 30–45.

[17] Sun, C. T. and Kelly, S. R., "Failure in Composite Angle Structures Part I: Initial Failure," Journal of Reinforced Plastics and Composites, Vol. 7, May 1988, pp. 220–232.

[18] Hart-Smith, L. J., "Some Observations About Test Specimens and Structural Analysis for Fibrous Composites," Composite Materials: Testing and Design (Ninth Volume), ASTM STP 1059, S. P. Garbo, Ed., American Society for Testing and Materials, Philadelphia, 1990, pp. 86–120.

[19] Tsai, S. W., "Strength Theories of Filamentary Structures" Fundamental Aspects of Fiber Reinforced Plastic Composites, R. T. Schwartz and H. S. Schwartz, Eds., Wiley Interscience, New York, 1968, pp. 3–11.

Manon Bolduc[1] and Clermont Roy[2]

Evaluation of Impact Damage in Composite Materials Using Acoustic Emission

REFERENCE: Bolduc, M. and Roy, C., "**Evaluation of Impact Damage in Composite Materials Using Acoustic Emission,**" *Composite Materials: Fatigue and Fracture, Fourth Volume, ASTM STP 1156,* W. W. Stinchcomb and N. E. Ashbaugh, Eds., American Society for Testing and Materials, Philadelphia, 1993, pp. 127–138.

ABSTRACT: The Defence Research Establishment, Valcatier (DREV) has been studying the application of composite materials for use in a lightweight anti-tank recoilless gun. Since thermoset resins are very sensitive to impact, it is important to determine the severity of damage using a nondestructive technique in order to establish the serviceability of the tube. Acoustic emission (AE) was identified as a very promising method to measure the level of damage accumulation due to low-velocity impact, and a study was undertaken. The experimental program consisted of impacting 28 of 36 tubes with energies varying from 3 to 18 J. These tubes, fabricated by a wet winding process using carbon fiber/epoxy resin, were 25.4 cm long, and had a 9-cm internal diameter and a 3.6-mm wall thickness. The winding angle was ±71°. The AE was measured while pressurizing these tubes to rupture. Results indicate the potential of acoustic emission as a means of estimating the serviceability of a composite tube.

KEY WORDS: acoustic emission, filament wound tube, carbon fiber, epoxy resin, low-velocity impact

During the last ten years, the Defence Research Establishment (DREV) has been studying the application of composite materials for a generic anti-tank recoilless gun. To achieve the desired weight reduction, a graphite-epoxy composite material was used to manufacture the gun tube and venturi. Although the ballistic viability of such a design was well proven during tests and evaluations, one major concern that remains is the sensitivity of the thermoset resin (epoxy) to the low-level damage accumulation due to a mixture of normal loading and rough handling in the field. Such damage is generally barely visible, and it is therefore important to develop an adequate NDT inspection technique to permit testing as close as possible to an operational environment. Acoustic emission was identified as a very promising method to measure the damage level accumulation due to low-velocity impact, and a study was undertaken.

The experimental program was carried out using a wet wound tube of nine layers, 25.4 cm long and with a 9-cm internal diameter and a 3.6-mm wall thichness. The winding angles for these tubes were ±71°, the same as for the light recoilless gun. Thirty-six graphite epoxy tubes were manufactured using Grafil Courtaulds E-XAS 12K graphite fibers in conjunction with D.E.R. 383 epoxy resin.

[1]Defense scientist, Defence Research Establishment Valcartier, 2459 Pie XI Blvd. North, Courcelette, Quebec, Canada G0A 1R0.
[2]Professor, Applied Sciences faculty, Mechanical Engineering Dept., Sherbrooke University, Sherbrooke, Quebec J1K 2R1.

Twenty-eight of these tubes were impacted at energy levels varying from 3 to 20 J, using a simple drop tower designed at DREV. Following these impacts, the tubes were pressurized to rupture, and acoustic events were recorded during pressurization. A plastic liner was placed inside all these tubes to avoid weeping. Similar recordings were also carried out on nonimpacted tubes for comparison purposes.

Experimentation

Fabrication Process

The sample's design criteria were based on the high pressure held by the weapon, which could reach 60 MPa in service, and on the dimensions of the muzzle, which is the most susceptible to be damaged by impact.

The winding process is well adapted to the composite tube fabrication, allowing the high fibers content needed to support the high pressure. The tubes were manufactured using the Defence Research Establishment, Valcartier (DREV) wet winding machine. They were wound as twelve 1-m-long sections that were subsequently cut in three sections to obtain 36 samples of 25.4-cm-long specimens with a 9-cm internal diameter with a wall thickness of 3.6 cm. The tubes were fabricated with Grafil Courtaulds E-XAS 12K graphite fiber added of D.E.R. 383 epoxy resin. They were wound at an angle of $\pm 71°$ and cured in an oven under atmospheric pressure for 2 h at 90°C followed by 4 h at 150°C.

Experimental Program

The experimental program was established to evaluate the ability of the acoustic emission technique to detect damages in composite tubes and indicate their severity.

The experiments were done in two parts so that we could establish the impact range and become familiar with the acoustic emission technique in the first part and then complete the experimental program with more accuracy. Therefore, the samples were split in two groups of 10 and 24 besides the two tubes used to verify the setup. From the first ten, seven were impacted at an energy level varying from 3 to 18 J, and the last three were not damaged. They were all pressurized up to the failure point to record acoustic emission data. A plastic liner was added inside the tubes to avoid weeping. Because the results indicated that the impact range was high enough to generate damages, but not so high as to cause a complete penetration through the tubes, the remaining 24 tubes were tested with the same approach. Finally, out of the 36 samples, 28 were damaged at seven different energy levels varying from 3 to 18 J, six were not impacted, and two were used to verify the setup.

Impact Tests

The setup used to impact the tubes is shown in Fig. 1. The drop tower designed at DREV for this purpose consisted of a translucent acrylic tube for guiding the impactor to the specimen and an instrumented impactor to determine the level of energy initiating damage in the composite material. The impactor was a Dynatup 15.58-KN instrumented striking tup connected to an ETI-630 data acquisition system. The striking tup had a half sphere head of 1.27 cm in diameter that concentrated the energy at one point. The impactor was dropped from a fixed height of 1.4 m by which the variation of energy was achieved by increasing the mass with some lead discs. The tube was fixed on a V-shaped base with two metal belts.

The velocity of the impactor was measured just prior to impact by a photoelectric cell

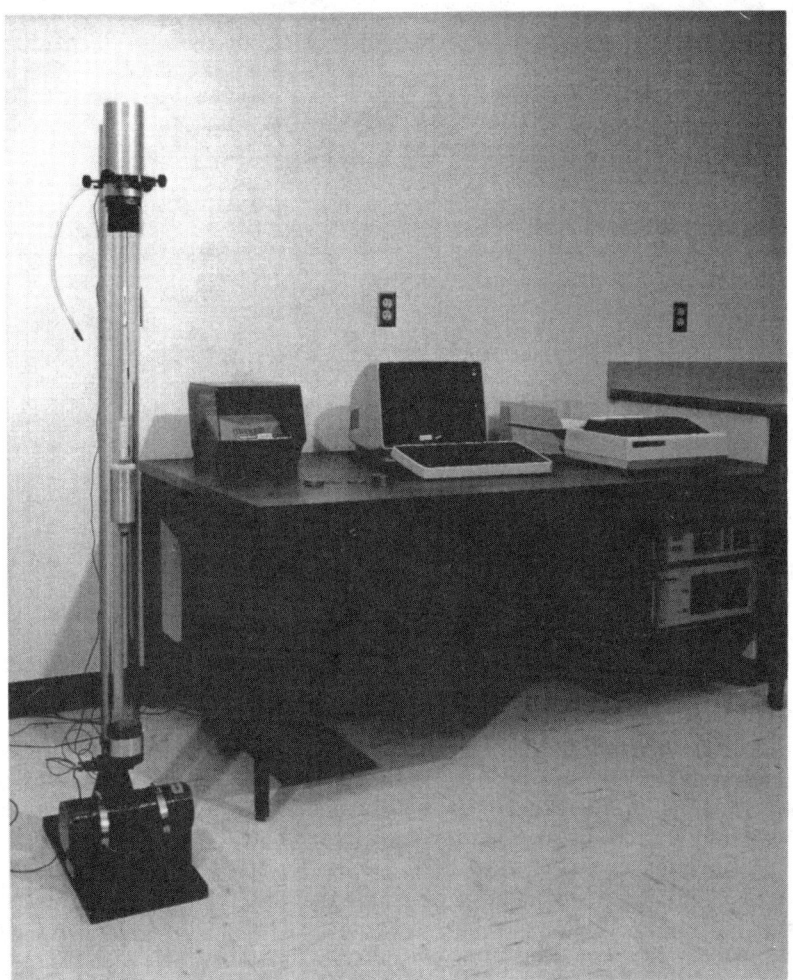

FIG. 1—*Impact tubes setup.*

activated by a flag fixed on the impactor. This, together with the mass of the impactor, gave the energy of the impactor. The absorbed energy was calculated by the data acquisition system from the impactor strain gage records.

Figure 2 shows the correlation between the energy absorbed by the specimen and the impact energy for some samples. The line is a quadratic fit. Although the object of the study was not to analyze the damage effects in tubes, we were concerned by the fact that if the line were continued, the energy absorbed would reach 100% between 18 and 20 J. Some researchers [1] noticed a mode damage modification from intralaminar to interlaminar at this turning point for a 16-layer composite tube. We intend to examine thoroughly this field in future studies.

Prior to and after impact, all tubes were examined by a C-scan technique at the National Aeronautical Establishment (NAE) in Ottawa. This indicated some manufacturing flaws and damages in the tubes due to the impact. It is pointed out that acoustic emission is an in-situ monitoring technique while the C-scan is quasi-static.

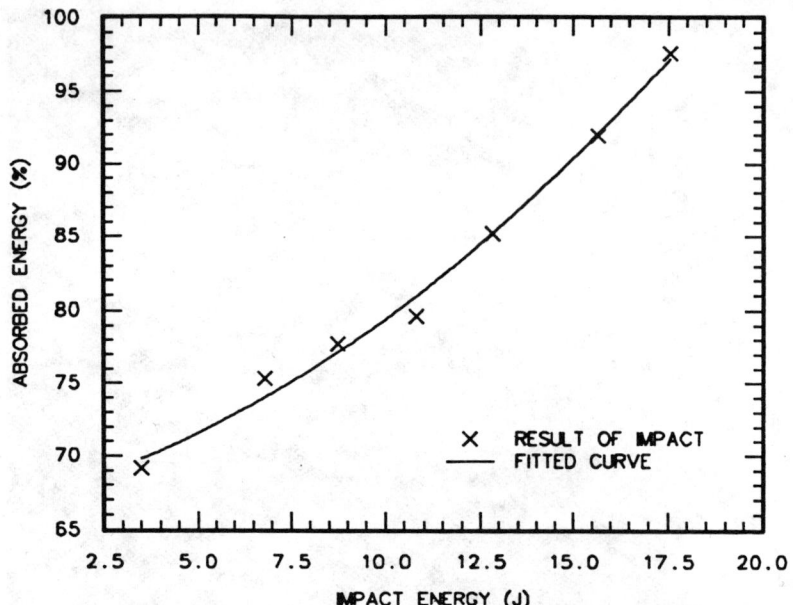

FIG. 2—*Curve of absorbed energy versus impact energy.*

Acoustic Emission

"Acoustic Emission is the generation, propagation and detection of transient stress waves in materials as they undergo deformation on fracture" [2]. "In composites, acoustic emission is generated by cracking of the matrix, debonding of the matrix from the fibres, laminate separation, fibre pullout and breakage of the fibres" [3].

Figure 3 shows a schematic view of the acoustic emission principle. Under stress condition, an acoustic emission source is created that generates mechanical waves in the structure. These waves are detected by sensitive piezoelectric transducers attached to the surface by which they are transformed in electric signals called acoustic events. The

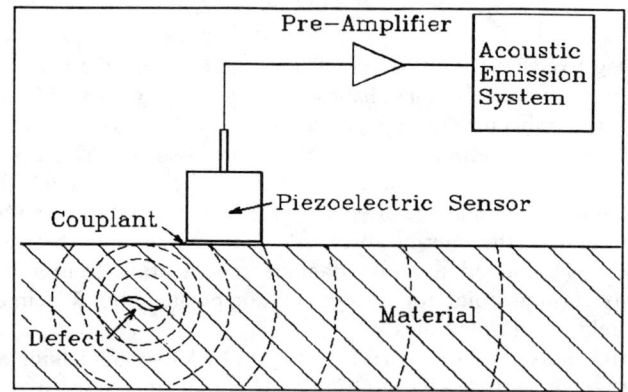

FIG. 3—*Schematic view of the acoustic emission principle.*

signals are then filtered, amplified, and analyzed to extract their parameters and recorded for analysis by the acoustic emission system. A couplant is used to obtain a good contact, thus ensuring a better elastic wave transmission.

To be recorded, an acoustic signal must be above a threshold that is set to eliminate noise. Parameters extracted from the signal are showed in Fig. 4. They are the number of counts in an event, maximum amplitude of an event, rise time to the peak amplitude from the start of the event, event duration, time at the beginning of the event, etc. All of these parameters are recorded in real time. An integrated software package is used for post analysis of all the data.

The setup used for the experimental program is shown in Fig. 5. The high pressure pump outlet (1) was connected to one end of the protection chamber (2) where the tube (3) was fixed. The pressure was measured by a transducer (5) at the top of the chamber. The pressurizing medium was a mixture of cutting fluid and water. A piezoelectric sensor (4) with a resonant frequency of 175 kHz was fixed on the composite tube. Signals received through the sensor were conditioned and amplified by 40 dB through the preamplifier (6). The signals were amplified again and processed using an AET 5500 acoustic emission analyzing and recording system (7). This equipment consists of a signal processing unit connected to a terminal (8) for information display.

The procedures followed in the experimental program were based on the ASTM Practice for Acoustic Emission Examination of Reinforced Thermosetting Resin Pipe (RTRP) (E 1118–86) and on the guide presented by D. J. McNally [3]. Even if the pressure limit set by ASTM E 1118 was exceeded (maximum permitted was 35 MPa, maximum reached was 63 MPa), the general procedure was followed. Thus, the pressurization in cycle with loading, load-hold periods, unloading was respected. Some preliminary tests had been done before the beginning of the experimental program to eliminate all background noises, so the unload period before each experiment was shorter than the one suggested in norm ASTM E 1118. The loading sequence used is shown in Fig. 6.

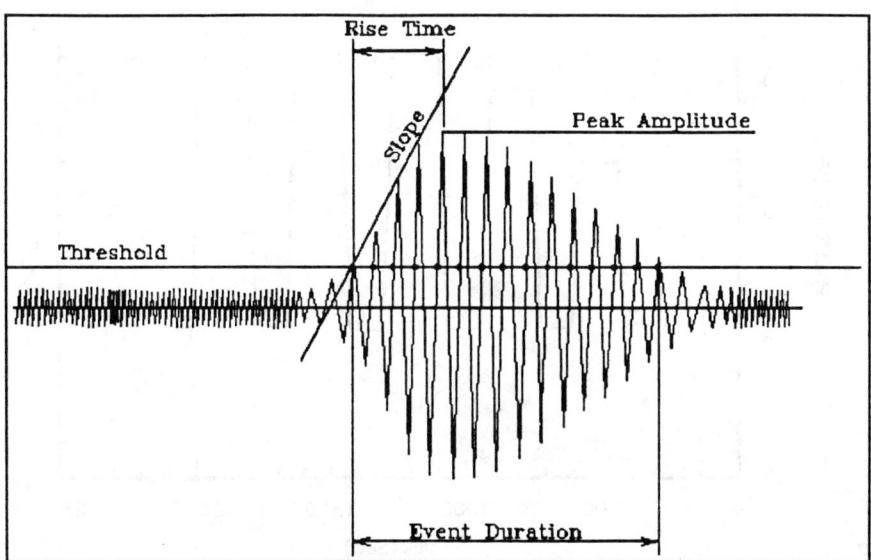

FIG. 4—*Parameters extracted from acoustic emission signal.*

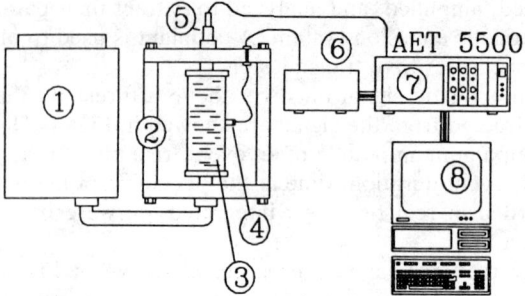

1 hydraulic pump
2 safety pressure chamber
3 composite tube
4 acoustic emission sensor
5 pressure transducer
6 pre-amplifier
7 acoustic emission system
8 terminal / display

FIG. 5—*Acoustic emission setup.*

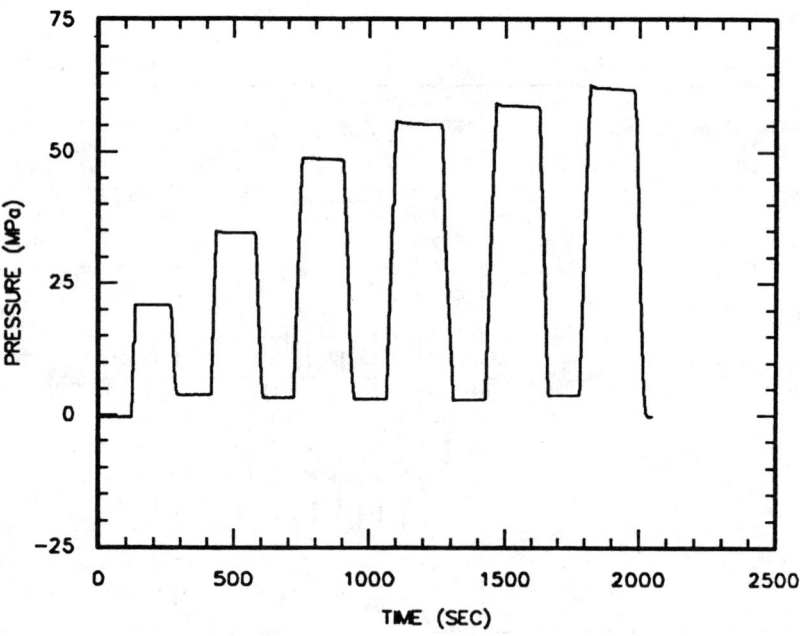

FIG. 6—*Loading sequence.*

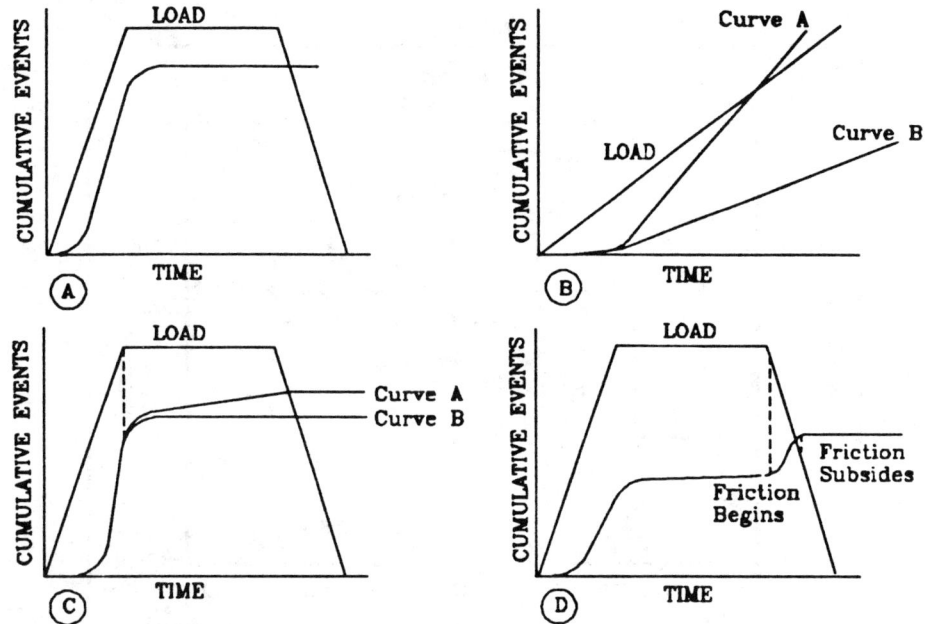

FIG. 7—*Cumulative events versus time with the load history superimposed on the plot.*

Results

Of all the data available from acoustic emission, only a few are of practical interest. D. J. McNally [3] has presented a good interpretation of some. One of the most useful forms of the data has been the plot of cumulative events versus time with the load history superimposed on the plot (see Fig. 7A). We can follow the acoustic emission activity response during pressurization, load-hold, and depressurization for each cycle. In examining the plot of cumulative events versus time during the rising load, one should note the slope of the curve (Fig. 7B). A steeper curve (as in Curve A) indicates more acoustic emission activity, and this means a more rapid damage progression. The same can be said for the load-hold period (Fig. 7C). If there is no acoustic emission activity during this period, then the tube is of good quality. Another section of interest is where the drop in event rate appears after the load is held constant (Fig. 7C, the knee following the dot line). This gives a useful indicator of material response to step load. For example, Curve A indicates a less desirable material property than Curve B. During the depressurization, the sudden burst of acoustic emission indicates the point at which two adjacent delaminated surfaces came in contact with each other, causing frictional noise (Fig. 7D).

The plots of cumulative events versus time with the load history superimposed were traced from data of all experiments. As an example, the curve of cumulative events versus time for Tube 11B is shown in Fig. 8A. Between the first and the last loading step, we can observe an increase in slope between each step during the load-hold period. During the depressurization, we see the sudden burst of acoustic emission for the last three loading steps, indicating delamination. If we examine the plots for Tube 10B, we have no sign of delamination during the depressurization (Fig. 8B). If we compare the curves of total events for those two tubes (Fig. 9), we observe the large difference in magnitude of total

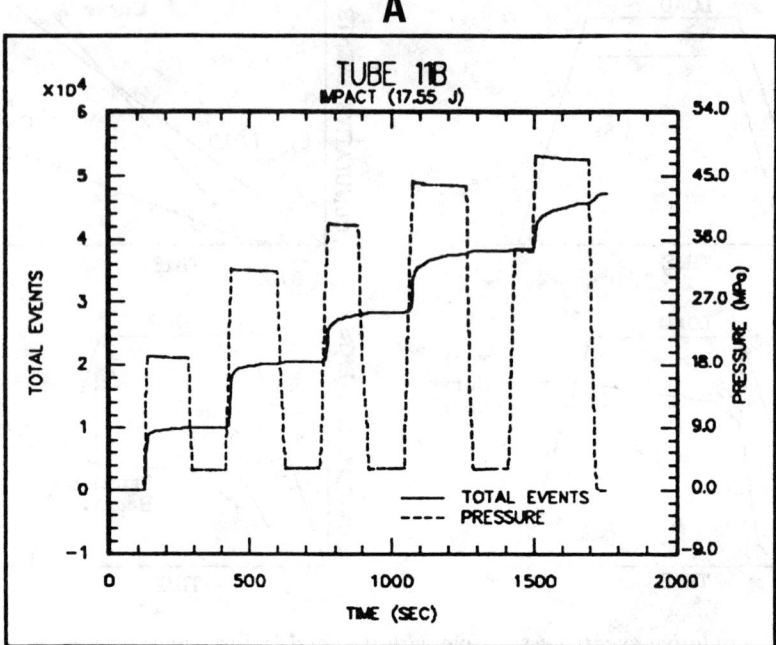

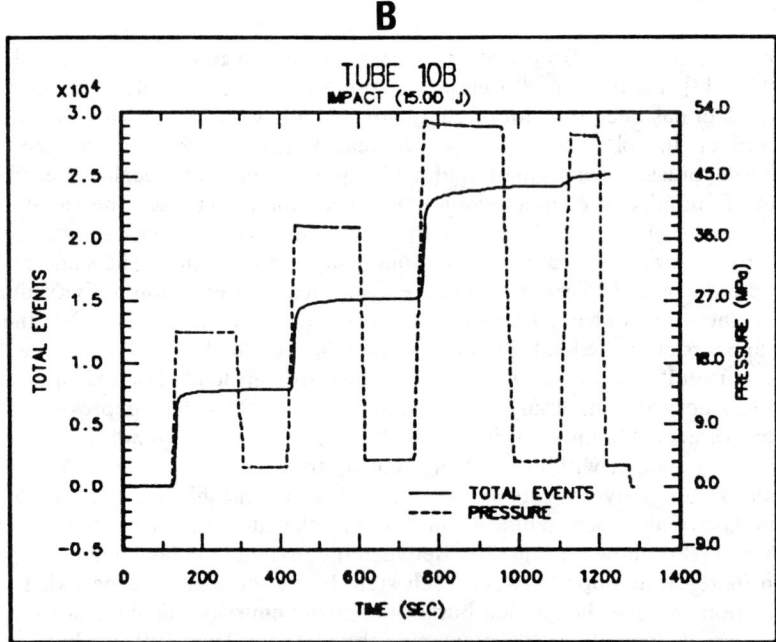

FIG. 8—*Cumulative events versus time for Tubes 11B and 10B.*

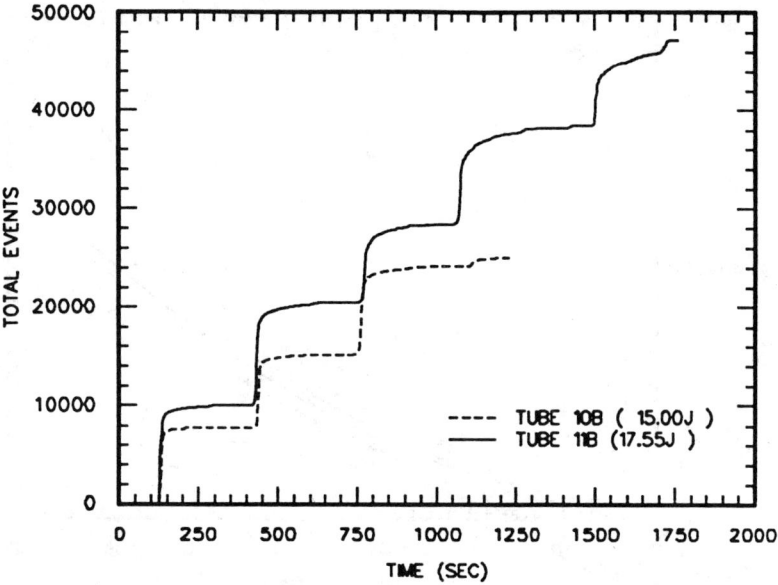

FIG. 9—*Comparative curves of total events versus time for tubes 11B and 10B.*

events. Also the slope of the curve for Tube 10B during the constant load is flatter. Thus, we can say that Tube 11B is more damaged than Tube 10B.

Another interesting curve is the total events versus pressure for tubes that had experienced different impact energies. A quadratic approximation was done to the data to

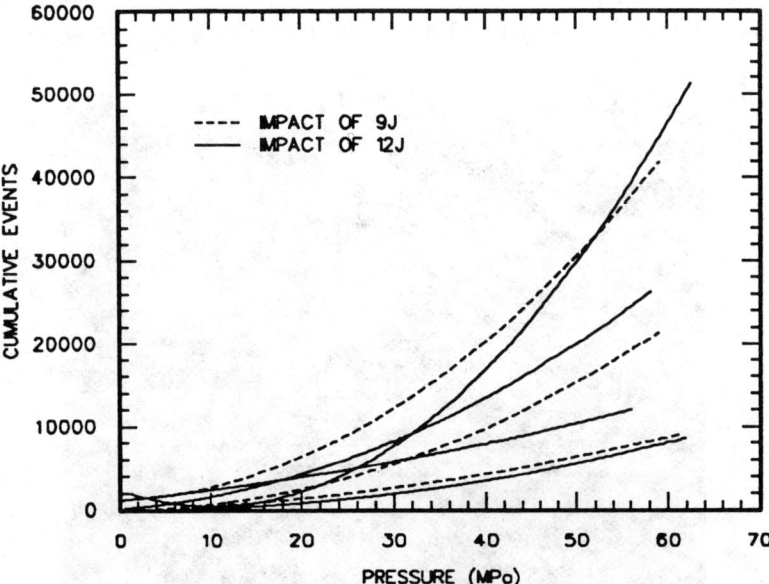

FIG. 10—*Total events versus pressure for tubes impacted at 9 and 12 J.*

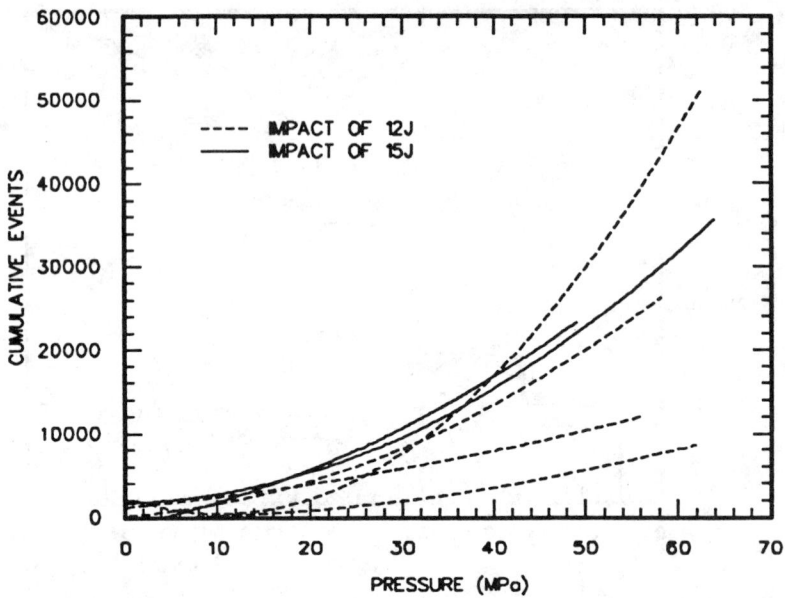

FIG. 11—*Total events versus pressure for tubes impacted at 12 and 15 J.*

facilitate the comparison. Curves for each energy level were compared with curves from each of the other energy levels on a one-to-one basis: 3 versus 6 J, 3 versus 9 J . . . 3 versus 18 J, 6 versus 9 J . . . 6 versus 18 J and so on. It should be noted that the quoted energy levels represent nominal values. For example, 3-J curves were in fact scattered from 2.47 to 3.50 J. From 3 to 12 J, no clear distinction appeared between the different energy levels. Figure 10 presents a typical example of the situation: the cumulative events curves versus pressure for 9 and 12 J energy levels are scattered and cross each other

FIG. 12—*Photomicrograph of Sample 7B (magnification ×25).*

within the same boundaries. However, past the 15 J level, as evident in Fig. 11, which presents the curves for the 12 and 15 J levels, it seems that a threshold was crossed and, in most cases, the curves for impact of 15 J and over can be differentiated from those for lower energy levels. The above can be explained by two rationales: either the difference in damages from 3 to 12 J can be considered minor for this type of solicitation, which shows that there is no fiber breakage, or defects from the fabrication process affected more the strength of the product than the impact itself. The first explanation is supported by the observation that the burst pressure didn't seem to be affected by either these or other defects produced by impacts up to 12 J.

To further verify these possibilities, a few samples were cut and observed on a microscope. As shown in Fig. 12, we can notice a point where two separate plies come into contact with one another. Research done by M. A. Hamstad [4] confirmed that this defect may affect the strength. Furthermore, curves of total events versus pressure for tubes that have experienced no impact are scattered, too (Fig. 13). This confirms that some sample had minor defects coming from the fabrication process.

We exclude lines with highest acoustic activities because these samples are most likely to contain defects other than those produced by the impact. Figure 14 presents some representative curves and shows that the acoustic activities increase with impact energy. Following these experiments and the results of their analyses, it is believed that if we do a pressure test on a similar tube that has experienced an unknown level of damage, we will be able to estimate an equivalent level of energy and determine if the tube is still serviceable or not.

Conclusion

The purpose of this project is to verify the acoustic emission as a nondestructive evaluation technique to detect and qualify damages in composite tubes. The experimental

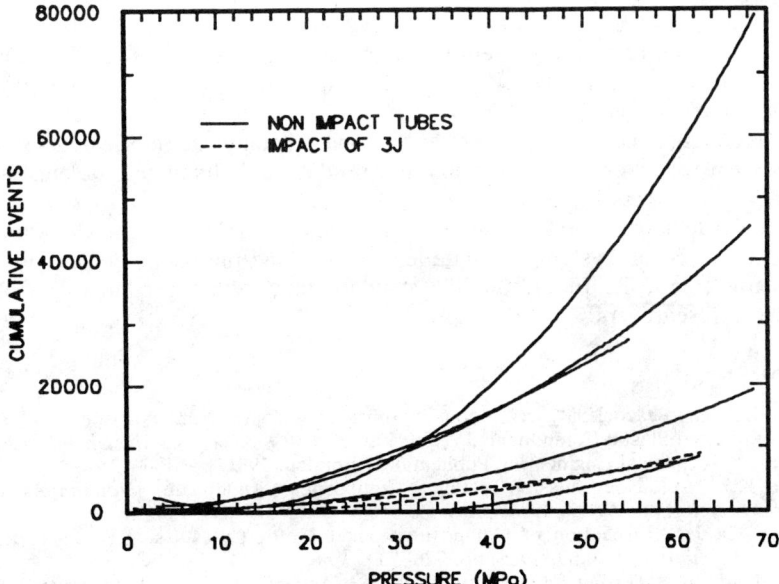

FIG. 13—*Total events versus pressure for nondamaged tubes with 3-J impact tubes.*

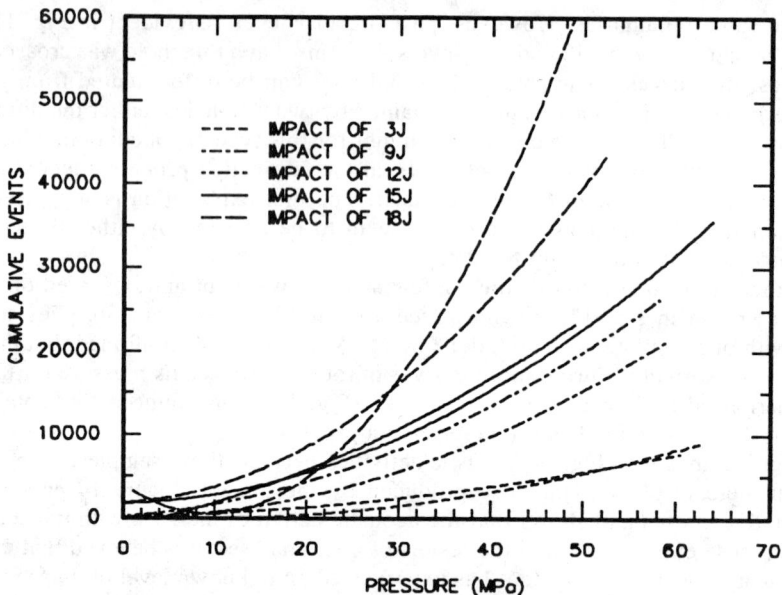

FIG. 14—*Best plots of all samples.*

program showed that interesting results were obtained from curves of cumulative events versus time with the load history superimposed on the plot and from total events versus pressure. The scattering of some of the results point out a problem that could be related to the fabrication process. The best plots indicate the potential of acoustic emission as a mean to estimate the serviceability of a composite tube.

The following conclusions can be reached:

1. Impacts under 12 J do not seem to affect seriously the tubes burst pressure, but a transition level exists between 12 and 15 J. Also, over 15 J, the strength of the tubes is clearly influenced.
2. It is recommended that tests should be done on composite fabricated with prepreg fibers that would ensure a more uniform product and eliminate problems due to the fabrication process.
3. A standard should be established that takes into consideration the calibration of the acoustic emission system, the data acquisition, and the results validation for high-pressure composite tubes. Such a standard may help the comparison of results between researchers.

References

[1] Davies, J. R. and Peacock, M. J., " Acoustic Emission: a Complementary Inspection Methode for Fibre-Reinforced Plastic Components," *GRP Vessels and Pipework for the Chemical Processing Industries*, Mechanical Engineering Publications, London, 1983, pp. 106–128.
[2] Fowler, T. J., "Acoustic Emission Testing of Composites—An Update," *National Association of Corrosion Engineers*, Vol. 7, 1986, pp. 7–20.
[3] McNally, D. J., "Inspection of Composite Rocket Motor Cases Using Acoustic Emission," *Materials Evaluation*, No. 43, 1985, pp. 728–739.
[4] Hamstad, M. A., "Testing Fiber Composites with Acoustic Emission Monitoring," *Journal of Acoustic Emission*, Vol. 1, No. 3, 1982, pp. 151–164.

Lynn Boniface,[1] Stephen L. Ogin,[2] and Paul A. Smith[2]

The Change in Thermal Expansion Coefficient as a Damage Parameter During Thermal Cycling of Crossply Laminates

REFERENCE: Boniface, L., Ogin, S. L., and Smith, P. A., "**The Change in Thermal Expansion Coefficient as a Damage Parameter During Thermal Cycling of Crossply Laminates,**" *Composite Materials: Fatigue and Fracture, Fourth Volume, ASTM STP 1156,* W. W. Stinchcomb and N. E. Ashbaugh, Eds., American Society for Testing and Materials, Philadelphia, 1993, pp. 139–160.

ABSTRACT: An experimental and theoretical study of the effect of 90° ply cracking on the thermal expansion coefficients of crossply laminates has been carried out. It has been found experimentally that reductions in the coefficient of thermal expansion of up to 50% are caused by 90° ply cracks induced mechanically, although considerable care is needed in the experimentation. This behavior was modeled using a simple shear-lag analysis, and the resulting analytical expressions are compared with other approaches available in the literature. The growth of matrix cracks in a model GFRP system under severe thermal cycling (77 to 373 K) is investigated. The changes in expansion coefficient are affected by the growth of 0° ply cracks in addition to the 90° ply cracks. The crack growth rate/cyclic strain energy release rate range data are compared with those reported previously for mechanical fatigue cycling of similar material. The two data sets are consistent if plotted in terms of a fracture mechanics parameter which aims to account for the temperature dependence of material properties.

KEY WORDS: transverse ply crack, thermal expansion coefficient, shear-lag model, strain energy release rate, thermal cycling, crossply laminate, GFRP, CFRP, crack growth rate, Paris relation

The effect of matrix cracks on the properties of composite laminates, particularly of the crossply type, has been researched extensively over the years. It is well known that laminate elastic properties, such as longitudinal modulus, Poisson's ratio, and shear modulus are reduced by the presence of 90° cracks. The extent of the reduction depends on the material type and the volume fraction of 90° plies [1–7].

More recently, there has been work concerning the effect of cracking on the longitudinal and transverse thermal expansion coefficients of crossply laminates [8–12]. Much of this work has been analytical: Bowles [8], Herakovich and Hyer [9], and Lim and Hong [11] studied the problem using finite element analysis; Lim and Hong also employed shear-lag analysis; Hashin [10] and McCartney [12] used more rigorous mechanics methods. These studies all suggest that 90° ply cracks produce very large percentage changes in the thermal expansion coefficients in crossply laminates, and, in the limiting case, could lead to the sign of the expansion coefficient changing in CFRP systems. It may be attractive to make

[1]Research fellow, Department of Materials Science and Engineering, University of Surrey, Guildford, Surrey GU2 5XH, UK.
[2]Lecturer, Department of Materials Science and Engineering, University of Surrey, Guildford, Surrey GU2 5XH, UK.

use of this to develop a technique for monitoring crack development in crossply laminates in certain situations, e.g., under thermal cycling conditions. However, experimental data to test the models are limited, although Bowles [8] presents some results for CFRP laminates which agree well with his theoretical model.

The aim of the present work is to generate data for the degradation of the thermal expansion coefficient with transverse crack density in GFRP and CFRP crossply laminates and to compare the data with theoretical predictions. The simple theoretical model used here is based on a shear-lag analysis but, before applying it to the data, we compare its predictions with those of other models. The experimental data for the thermal expansion coefficient as a function of crack density is generated under mechanical loading and, for GFRP, thermal cycling conditions. During the latter experiments it was also possible to generate crack growth rate data (as a function of applied strain energy release rate range) which can be compared with data reported previously for mechanical fatigue cycling.

Analysis

The geometry of a $(0/90)_s$ laminate containing a regular array of transverse ply cracks spaced $2s$ apart is shown in Fig. 1. The longitudinal plies each have thickness b, and the transverse ply has a total thickness of $2d$. The origin of the coordinate system used in the simple shear-lag stress analysis is at the center of the transverse ply, midway between two cracks. To find expressions for the thermal expansion coefficients of the cracked laminate, we need to consider the stress-strain relations in each ply. For purely thermal loading (i.e., no applied mechanical stress), a one-dimensional shear-lag analysis (as used in Refs 4,7,13) gives the longitudinal stress in the 0° ply as a function of crack spacing, σ_1:

$$\sigma_1 = -\sigma^T\left(1 - \frac{\cosh \lambda y}{\cosh \lambda s}\right) \tag{1}$$

In this equation $-\sigma^T$ is the thermal stress in the longitudinal ply at a temperature T when the laminate is uncracked. This can be calculated using the laminated plate theory, LPT [14]. The sign convention used is such that T is less than the laminate cure temperature, and hence σ_1 is negative (i.e., compressive). The quantity λ is given by

$$\lambda^2 = \alpha G_{23}(b + d)E_0/d^2bE_1E_2 \tag{2}$$

where E_1 and E_2 are the moduli of a lamina parallel and perpendicular to the fibers, respectively; G_{23} is the out-of-plane shear modulus of a lamina; and E_0 is the rule-of-mixtures modulus of the $(0/90)_s$ laminate [$E_0 = (bE_1 + dE_2)/(b + d)$]. The value of α depends on the assumptions made in the shear lag analysis and generally is between 1 and 3 [7].

The longitudinal stress in the transverse ply is found from equilibrium and is simply $-\sigma_1(b/d)$. If we write the transverse stress in the 0° ply as σ_2, then the transverse stress in the 90° ply is $-\sigma_2(b/d)$, and the stress/strain relations for the plies can now be written. For the 0° ply

$$\epsilon_L(y) = \frac{\sigma_1}{E_1} - \frac{\nu_{21}\sigma_2}{E_2} + \alpha_1\Delta T \tag{3}$$

$$\epsilon_T(y) = \frac{\sigma_2}{E_2} - \frac{\nu_{12}\sigma_1}{E_1} + \alpha_2\Delta T \tag{4}$$

and for the 90° ply

$$\epsilon_L(y) = -\frac{\sigma_1(b/d)}{E_2} + \frac{\nu_{12}\sigma_2(b/d)}{E_1} + \alpha_2\Delta T \tag{5}$$

$$\epsilon_T(y) = -\frac{\sigma_2(b/d)}{E_1} + \frac{\nu_{21}\sigma_1(b/d)}{E_2} + \alpha_1\Delta T \tag{6}$$

In Eqs 2 to 6 $\epsilon_L(y)$ and $\epsilon_T(y)$ are the longitudinal and transverse strains in the laminae; α_1 and α_2 are the coefficients of thermal expansion of a lamina (parallel and perpendicular to

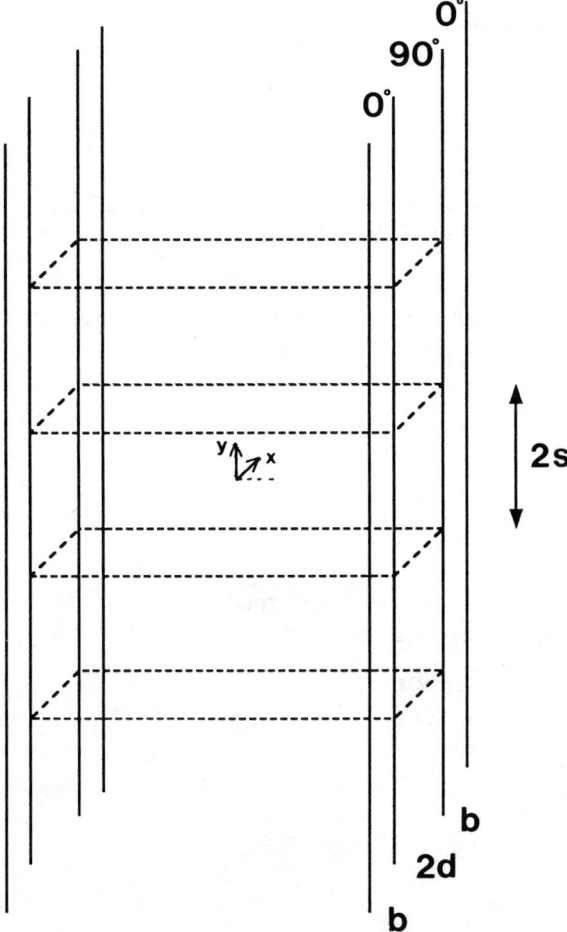

FIG. 1—*Geometry of (0/90)$_s$ laminate containing an array of transverse ply cracks, spaced 2s apart.*

the fiber direction, respectively); ν_{12} and ν_{21} are the major and minor Poisson's ratios of a lamina; and ΔT is the temperature change to which the laminate has been subjected. By equating Eqs 4 and 6, an expression for σ_2 can be obtained

$$\sigma_2 = -\frac{(b + d)\nu_{12}E_2}{(dE_1 + bE_2)}\sigma^T\left(1 - \frac{\cosh \lambda y}{\cosh \lambda s}\right) + \frac{dE_1E_2(\alpha_1 - \alpha_2)\Delta T}{(dE_1 + bE_2)} \qquad (7)$$

Substituting Eq 7 into Eq 3 gives an expression for $\epsilon_L(y)$

$$\epsilon_L(y) = -\frac{\sigma^T}{E_1}\left(1 - \frac{\nu_{12}\nu_{21}E_1(b + d)}{bE_2 + dE_1}\right)\left(1 - \frac{\cosh \lambda y}{\cosh \lambda s}\right) + \left[\alpha_1 + \frac{d\nu_{21}E_1(\alpha_2 - \alpha_1)}{bE_2 + dE_1}\right]\Delta T \qquad (8)$$

The mean longitudinal strain in the laminate is given by

$$\epsilon_m = \frac{1}{s}\int_0^s \epsilon_L(y)\, dy \qquad (9)$$

Substituting from Eq 8 and integrating gives

$$\epsilon_m = -\frac{\sigma^T}{E_1}\left(1 - \frac{\nu_{12}\nu_{21}E_1(b + d)}{bE_2 + dE_1}\right)\left(1 - \frac{\tanh \lambda s}{\lambda s}\right) + \left[\alpha_1 + \frac{d\nu_{21}E_1(\alpha_2 - \alpha_1)}{bE_2 + dE_1}\right]\Delta T \qquad (10)$$

But, by definition

$$\epsilon_m = \alpha_L(s) \cdot \Delta T \qquad (11)$$

where α_L is the longitudinal coefficient of thermal expansion of the cracked laminate. So combining Eqs 10 and 11, we have

$$\alpha_L(s) = -\frac{(\sigma^T/\Delta T)}{E_1}\left(1 - \frac{\nu_{12}\nu_{21}E_1(b + d)}{bE_2 + dE_1}\right)\left(1 - \frac{\tanh \lambda s}{\lambda s}\right)$$

$$+ \left[\alpha_1 + \frac{d\nu_{21}E_1(\alpha_2 - \alpha_1)}{bE_2 + dE_1}\right] \qquad (12)$$

Equation 10 can be rewritten in a more readily usable form by noting that as $2s \to \infty$, $\alpha_L(s)$ must tend to the appropriate uncracked value, α_L. Using this condition we obtain

$$\alpha_L(s)/\alpha_L = \left\{ 1 - \frac{\left[\alpha_1 + \dfrac{dv_{21}E_1(\alpha_2 - \alpha_1)}{bE_2 + dE_1} \right]}{\alpha_L} \right\} \left(1 - \frac{\tanh \lambda s}{\lambda s} \right)$$

$$+ \left[\alpha_1 + \frac{dv_{21}E_1(\alpha_2 - \alpha_1)}{bE_2 + dE_1} \right] \bigg/ \alpha_L \tag{13}$$

By substituting Eq 7 into Eq 4 and integrating to obtain an expression for the mean transverse strain, a similar expression to Eq 13 can be found for the normalized transverse coefficient of thermal expansion of the crossply laminate as a function of crack spacing, $\alpha_T(s)$

$$\alpha_T(s)/\alpha_T = \left\{ 1 - \frac{\left[\alpha_2 - \dfrac{dE_1(\alpha_2 - \alpha_1)}{bE_2 + dE_1} \right]}{\alpha_T} \right\} \left(1 - \frac{\tanh \lambda s}{\lambda s} \right)$$

$$+ \left[\alpha_2 - \frac{dE_1(\alpha_2 - \alpha_1)}{bE_2 + dE_1} \right] \bigg/ \alpha_T \tag{14}$$

Comparison with Other Models

Before presenting any experimental data, it is useful to compare the predictions of Eqs 13 and 14 with those of some of the other models in the literature. Table 1 shows the basic lamina properties used by Hashin [10], McCartney [12], and Bowles [8] in their modeling of GFRP and CFRP (0/90)$_s$ laminates. Figure 2 shows the prediction of McCartney for the normalized longitudinal expansion coefficient as a function of transverse ply crack density along with the prediction of Eq 13 for various values of the shear-lag constant, α. Very reasonable agreement between the predictions is obtained when a value of 1 is used for the shear-lag constant, α. Figure 3 shows a similar plot, this time for the transverse coefficient of thermal expansion; again agreement is best (this time with Eq 14) for an α value of 1. Figure 4 shows predictions for the longitudinal expansion coefficient of a (0/90)$_s$ GFRP laminate (which has very different properties to its CFRP counterpart) as a function

TABLE 1—*Lamina properties for various glass/epoxy and carbon/epoxy systems used for modeling expansion coefficient changes in crossply laminates.*

	Material		
	CFRP, T300/5208 [8]	GFRP [10]	CFRP [12]
E_1, GPa	132	41.7	208.3
E_2, GPa	10.8	13	6.5
v_{12}	0.24	0.3	0.255
v_{23}	0.49	0.42	0.413
G_{12}, GPa	5.7	3.4	1.65
G_{23}, GPa	3.4	4.58	2.30
α_1, K^{-1}	-0.11×10^{-6}	6.72×10^{-6}	-0.3×10^{-6}
α_2, K^{-1}	27.2×10^{-6}	29.32×10^{-6}	28.1×10^{-6}

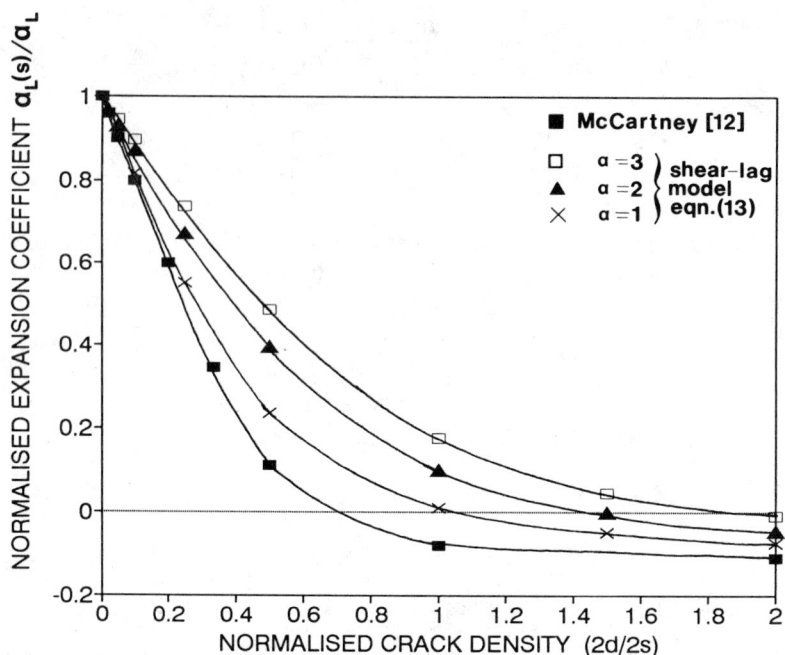

FIG. 2—*Longitudinal thermal expansion coefficient as a function of normalized transverse ply crack density for a (0/90)$_s$ CFRP laminate. Comparison between the prediction of McCartney [12] and shear-lag model (Eq 13).*

of transverse ply crack density, this time from Hashin [10]. Agreement in this case is also satisfactory with a value for α of 1.

It is interesting also to compare the shear-lag model predictions with those of finite element analysis. This comparison is seen in Fig. 5 for crossply CFRP laminates of three different transverse ply thicknesses from the T300/5208 system, with the finite element results taken from Bowles [8]. For these results agreement is excellent with an α value of 3 (which results from assuming a parabolic profile of the longitudinal displacement across the thickness of the transverse ply). This is not unexpected, since a shear-lag analysis with a value for α of 3 has been found to give similar predictions for longitudinal stiffness reduction of a crossply laminate as a function of crack density to the predictions of a finite element analysis [15]. However, it does suggest that the models of Hashin [10] and McCartney [12] predict a slightly larger effect of transverse cracks on the longitudinal expansion coefficient than finite element analysis.

Experimental

Materials and Damage Monitoring Techniques

Two crossply laminates were investigated, a transparent glass/epoxy system (processed as described in an earlier work [13]) and carbon fiber/PMR15 laminates (fabricated at Rolls-Royce, Derby). The basic material properties and laminate geometries are summarized in Table 2. Coupons, 220 by 20 mm wide, were cut from the laminates using a water-

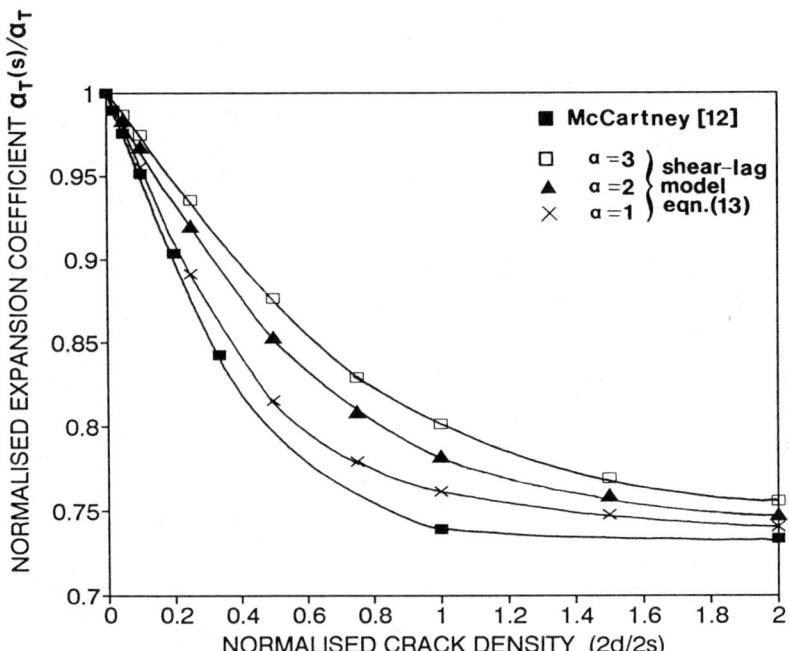

FIG. 3—*Transverse thermal expansion coefficient as a function of normalized transverse ply crack density for a $(0/90)_s$ CFRP laminate. Comparison between the prediction of McCartney [12] and shear-lag model (Eq 14).*

lubricated diamond saw. Aluminum alloy tabs 50 mm long and 1.5 mm thick were bonded to the ends of the coupons, and 10-mm-long strain gages oriented in the longitudinal and transverse laminate directions were bonded at the center of the coupon. In the transparent GFRP system, transverse ply cracks could be seen and counted directly. For the CFRP system, two methods were used: both edges of each coupon were polished carefully prior to testing so that cracks could be counted using an optical microscope; for some specimens the crack count determined in this way was verified using dye-penetrant enhanced X-radiography.

Mechanical Testing

All mechanical tests were carried out using an Instron 1196 testing machine, and the output from the strain gages was taken with the load signal to a chart recorder. Specimens were loaded to progressively higher strains to produce higher crack densities. Between loadings, the longitudinal thermal expansion coefficient (CTE) was measured as described in the next section.

Measurement of Thermal Expansion Coefficient

The basis of the experimental technique used for the measurement of the CTE has been reported by other workers [17]. The strain in the composite was monitored as a function of temperature, and temperature-induced effects on the gage response were compensated by

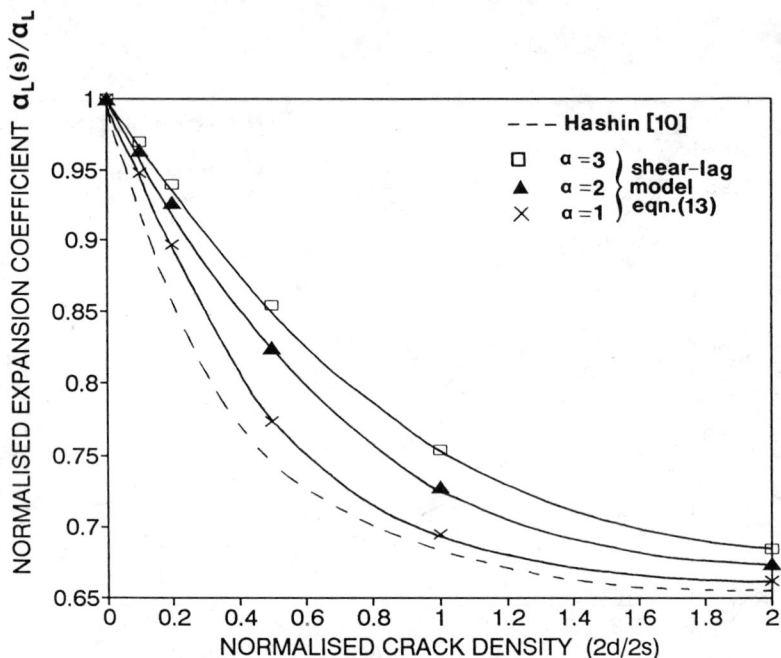

FIG. 4—*Longitudinal thermal expansion coefficient as a function of normalized transverse ply crack density for a (0/90)ₛ CFRP laminate. Comparison between the prediction of Hashin [10] and shear-lag model (Eq 13).*

using an identical gage bonded to a reference sample [*18*]. In order to try to eliminate any effect of water uptake on the CTE measurements, all specimens were dried for approximately 30 h at 50°C in a vacuum oven (and stored in a desiccator between tests) before the initial measurement of the CTE in the uncracked laminate. The CTE was measured over a temperature range of either 20 to 100°C (for the CFRP) or 20 to 70°C (for the GFRP) using a thermal cycling cabinet (thermal shock test chamber model Checkmate 320/T/S from Lenton Thermal Design Ltd.) with a Eurotherm type 818P controller/programmer. The procedure used is outlined briefly below.

The coupon was placed in the thermal chamber, and the strain gages (TML PL-10 type) on the composite coupon and the quartz reference sample were attached in separate three-wire quarter-bridge circuits. The temperature was then increased from ambient to 100°C in steps of 10°C at a heating rate of 5°C/min, holding at temperature for 10 min before taking the strain reading. The output from each gage was measured to give the composite strain, ϵ_S, and the quartz reference sample strain, ϵ_R. These are related to the composite CTE, α_S; the reference sample CTE, α_R; and the temperature difference, ΔT by the expression

$$\alpha_S - \alpha_R = (\epsilon_S - \epsilon_R)\Delta T \qquad (15)$$

Hence the CTE for the composite can be found from the sum of the slope of a plot of (ϵ_S − ϵ_R) against temperature and the CTE of the reference sample. For each sample the measurements were repeated at least three times (at each crack density) to obtain an average of the strain-temperature slope.

The values obtained with the technique were consistent with measurements made by

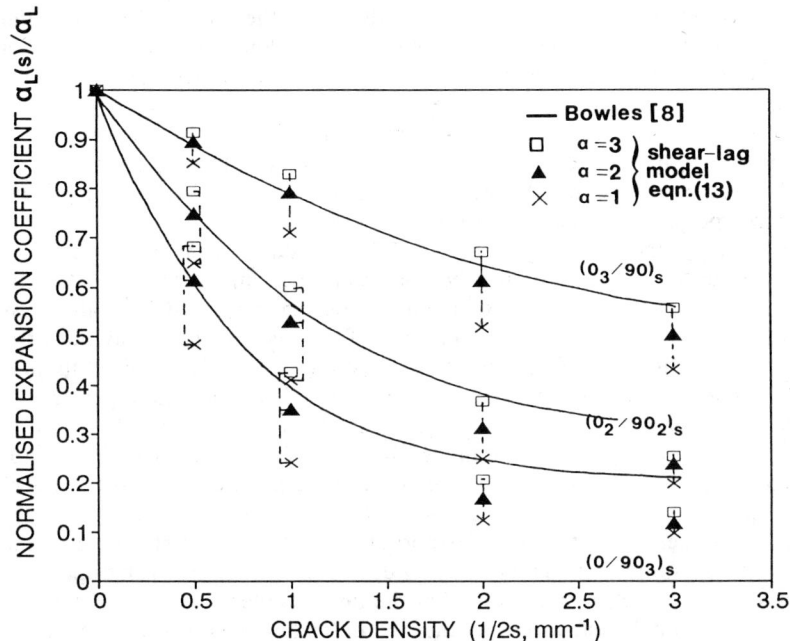

FIG. 5—*Longitudinal thermal expansion coefficient as a function of normalized transverse ply crack density for $(0_3/90)_s$, $(0_2/90_2)_s$, and $(0/90_3)_s$ CFRP laminates. Comparison between the finite element based predictions of Bowles [8] and shear-lag model (Eq 13).*

Mulheron [*19*] on unidirectional GFRP using thermo-mechanical analysis. His values for α_1 and α_2 were $9 \times 10^{-6} \text{ K}^{-1}$ and $33 \times 10^{-6} \text{ K}^{-1}$ at 20°C compared with values in the present study of $8.7 \times 10^{-6} \text{ K}^{-1}$ and $34.3 \times 10^{-6} \text{ K}^{-1}$.

Thermal Cycling

For a preliminary study of thermal cycling in the GFRP system, two simple cycles were chosen. The first cycle was between temperature limits of liquid nitrogen (77 K) and room temperature (293 K). This entailed immersing the specimen in liquid nitrogen for 30 s and then holding for 1 min at room temperature. A total of 250 cycles of this type were applied to the coupon, and then the same coupon was subjected to a second, more severe, cycle between temperature limits of 77 and 373 K. The upper limit was achieved by placing the specimen in an air-oven for 1 min. The extent of cracking was recorded at regular intervals by taking photographs of the (transparent) coupon, and the CTE was measured at transverse ply crack densities comparable to those introduced by mechanical loading.

Results and Discussion

Changes in Thermal Expansion Coefficient for Laminates with Cracks Introduced Mechanically

The mean values of the CTE for the uncracked crossply laminates are shown in Table 2. Using LPT in conjunction with the unidirectional CTE data gives predicted CTE values for the crossply laminates in both the longitudinal and transverse directions of 15.4×10^{-6}

K^{-1} for the GFRP system and 1.8×10^{-6} K^{-1} for the carbon/PMR15 system. This represents satisfactory agreement with the mean values obtained experimentally, although there is some variation in measurements between specimens, especially for carbon/PMR. Possible reasons for this are discussed below.

Figure 6 shows typical strain difference/temperature data (Eq 15) for nominally identical runs on uncracked and cracked GFRP and CFRP systems. The effect of cracking on the CTE can be seen by noting that the cracked laminates (Figs. 6b and 6d) show lower slopes than the corresponding uncracked systems (Figs. 6a and 6c).

Figures 7 and 8 show the CTEs inferred from data such as Fig. 6 plotted against transverse ply crack density, D (number of cracks per millimetre), for the GFRP and CFRP systems, respectively. The predictions of the shear-lag model are also shown. The experimental scatter could be due to a number of factors. The gage may not be perfectly aligned with regard to the principal material directions, and this may lead to an overestimate of the expansion coefficient compared to the nominal situation of perfect alignment. The crack density values used are average values for the entire length of the coupon— measured from photographs by counting the number of cracks intersecting a grid of lines drawn parallel to the length of the specimen—and the local crack density under a strain gage may differ from this, especially at low crack densities.

Overall, the agreement between the experimental data and the shear-lag predictions is satisfactory. Interestingly, the data are better described with higher values of the shear-lag parameter α (i.e., $\alpha = 2,3$), suggesting that the finite element model of Bowles [8] would have described the data well, whereas other theoretical models [10,12] might have overestimated the CTE reduction.

Transverse Ply Crack Growth under Thermal Cycling

The initial thermal cycle range of 77 to 293 K was not severe enough to lead to fully developed matrix cracks (i.e., cracks spanning the width and thickness of the coupon) on

TABLE 2—*Properties of glass/epoxy and carbon/PMR laminates.*

	Material	
	Glass/Epoxy	Carbon/PMR15
Unidirectional lamina properties:[a]		
E_1, GPa	43	119
E_2, GPa	13	8.7
ν_{12}	0.3	0.3
G_{12}, GPa	4	4
G_{23}, GPa	4.64[b]	3.11[b]
α_1, K^{-1}	8.7×10^{-6}	-0.53×10^{-6}
α_2, K^{-1}	34.3×10^{-6}	26.4×10^{-6}
0/90/0 Laminate geometry and properties:		
b, mm	0.3	0.125
d, mm	0.3	0.125
E_0, GPa	28.0	69.1
α_L, K^{-1}	14.4×10^{-6}	1.7×10^{-6}
α_T, K^{-1}	14.6×10^{-6}	2.1×10^{-6}

[a]Unidirectional carbon/PMR15 data taken from Ref 16, GFRP data measured in the present study.
[b]Determined assuming transverse isotropy, $G_{23} = E_2/2(1 + \nu_{23})$, with ν_{23} taken as 0.4.

the first cycle. However, incipient transverse ply cracks and longitudinal ply splits are visible near the edges of the coupon after about 100 cycles (Fig. 9a). These propagate with further cycling, and additional longitudinal ply splits are observed away from the free edges (Figs. 9b and 9c). These observations are consistent with those reported by other workers testing continuous fiber and woven CFRP laminates [20,21].

For the same coupon under the more severe thermal cycle (77 to 373 K), Fig. 10, further cracks initiated and existing transverse and longitudinal cracks propagated across and along the coupon, respectively, more rapidly than in the previous thermal cycle.

From photographs of the cracked laminate at a number of cyclic intervals (Figs. 9 and 10), it is possible to construct plots of transverse ply crack length as a function of number of thermal cycles for the two different cycle types investigated. Typical plots are shown in Fig. 11 for some of the cracks indicated in Figs. 9 and 10. To a good approximation the graphs are linear, suggesting that the fracture mechanics parameter governing crack growth rate is independent of the length of the transverse ply crack; such behavior has been discussed in detail previously for mechanical fatigue cycling [13,22,23]. It was also found, in accordance with mechanical fatigue, that the crack growth rates are affected by the presence of neighboring cracks, which effectively shield a growing crack from some of the applied stress. This is shown by the crack length versus number of cycles data for Crack c (Fig. 11b) where the crack growth rate (i.e., the slope) is reduced after 100 cycles when Crack c starts to interact with an internally initiated crack (Fig. 10). This stress shielding can be modeled using shear-lag analysis and thus incorporated into fracture mechanics-based analyses leading to expressions for the strain energy release rate associated with a growing transverse ply crack [22–25].

Fracture Mechanics Analysis of Transverse Ply Crack Growth

For a model array of transverse ply cracks where a third crack grows between two existing cracks (Fig. 12), the expression for the average strain energy release rate associated with the growth of the third crack under combined mechanical and thermal loading is given by [23]

$$G = \left(1 + \frac{b}{d}\right)\left(\sigma\frac{E_2}{E_0} + \sigma^T\right)^2\left(\frac{d}{b}\right)\frac{E_0}{E_1 E_2}\frac{1}{\lambda}\left[\tanh\left(\frac{\lambda s_{AC}}{2}\right)\right.$$

$$\left. + \tanh\left(\frac{\lambda s_{BC}}{2}\right) - \tanh\left(\frac{\lambda s_{AB}}{2}\right)\right] \quad (16)$$

(It should be noted that the approaches in Refs 24 and 25 are very similar to that in Ref 23, which leads to Eq 16; they differ only significantly in the form of stress analysis used to model the shear-lag effect.)

If transverse ply crack growth under thermal cycling can be described by fracture mechanics in a similar way to mechanical fatigue cycling [23], we might expect that the crack growth rate (da/dN) would be related to the applied strain energy release rate range via a Paris-type relation

$$da/dN = A(\Delta G)^m \quad (17)$$

where A and m are material constants. The left-hand side of Eq 17 is simple to determine from the slope of plots of crack length against number of cycles such as those shown in Fig. 11. The right-hand side could be calculated from Eq 16, noting that the applied

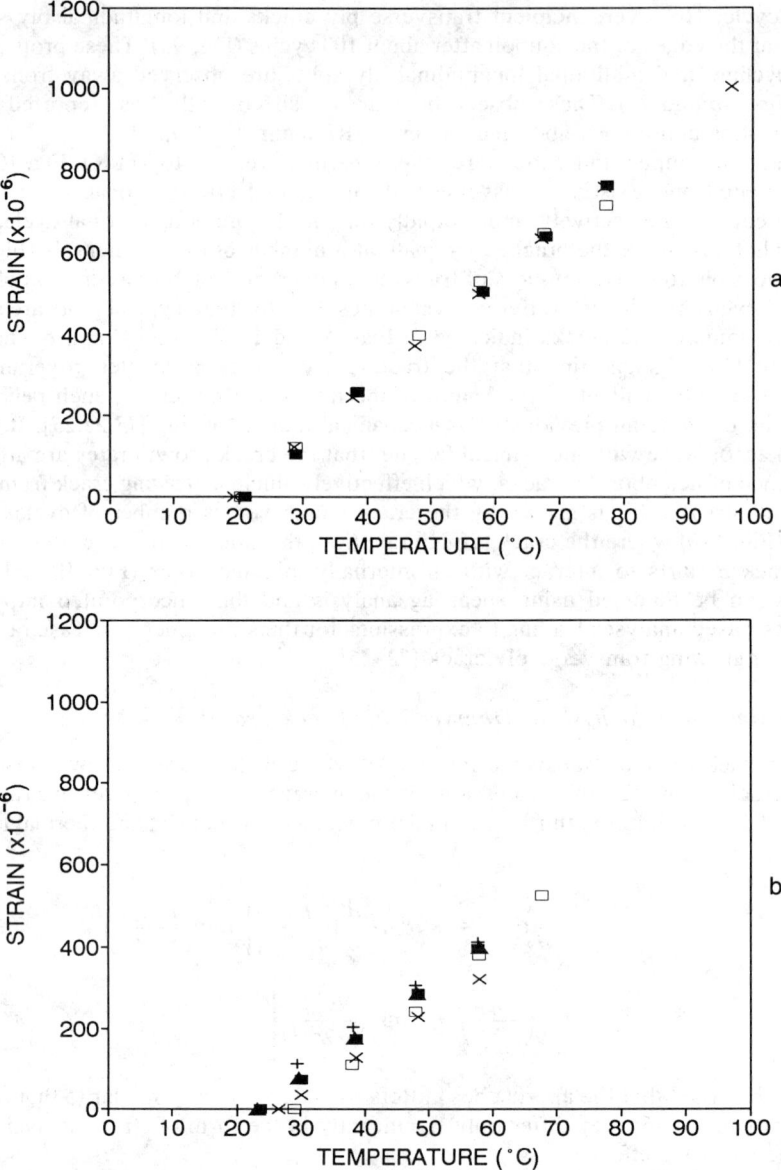

FIG. 6—*Typical plots of strain against temperature enabling CTE of crossply laminates to be measured: (a) GFRP, uncracked; (b) GFRP, D = 0.78 mm^{-1}; (c) CFRP, uncracked; and (d) CFRP, D = 2.08 mm^{-1}.*

mechanical stress, σ, is zero. However, such an approach is complicated by the change in the elastic material properties of the system (most notably E_2, G_{23}, and α_2) and the crack growth resistance of the transverse ply (G_c) with temperature during the cycle.

In an attempt to develop a general way of allowing for the variation in material properties during a cycle (and hence being able to rationalize thermal cycling and mechanical

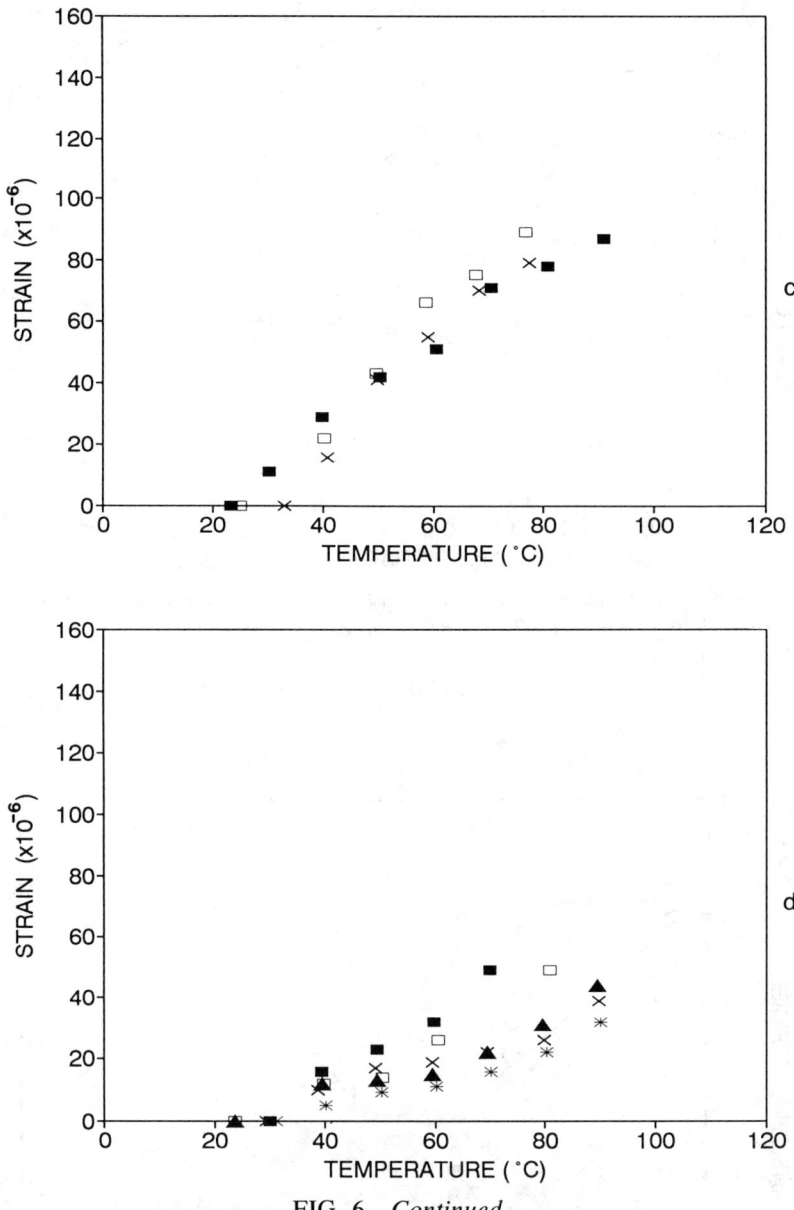

FIG. 6—*Continued*.

fatigue data), the following procedure is proposed. In metals, crack growth data for different materials can be superposed by normalizing the applied stress intensity range by the Young's modulus, i.e., the Paris relation is modified to the form [26]

$$da/dN = B(\Delta K/E)^n \tag{18}$$

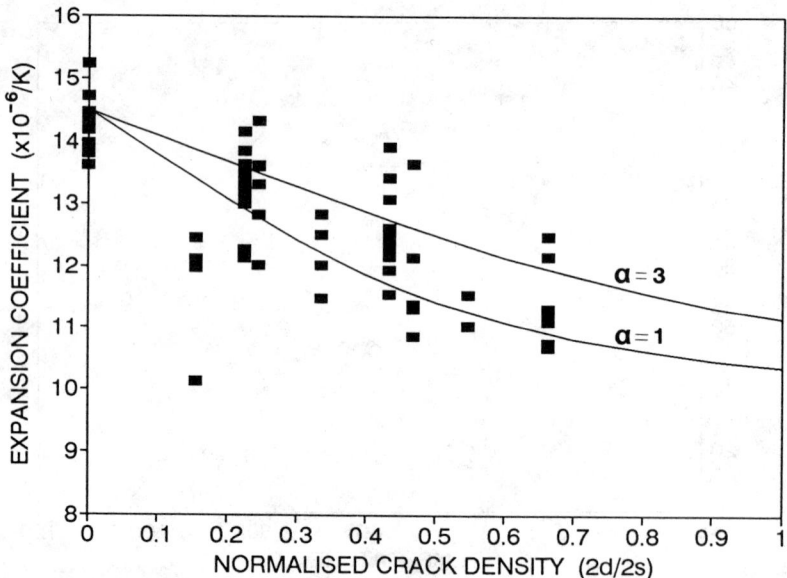

FIG. 7—*Longitudinal thermal expansion coefficient as a function of normalized transverse ply crack density for a (0/90)ₛ GFRP laminate. Comparison between experimental data and shear-lag model (Eq 13).*

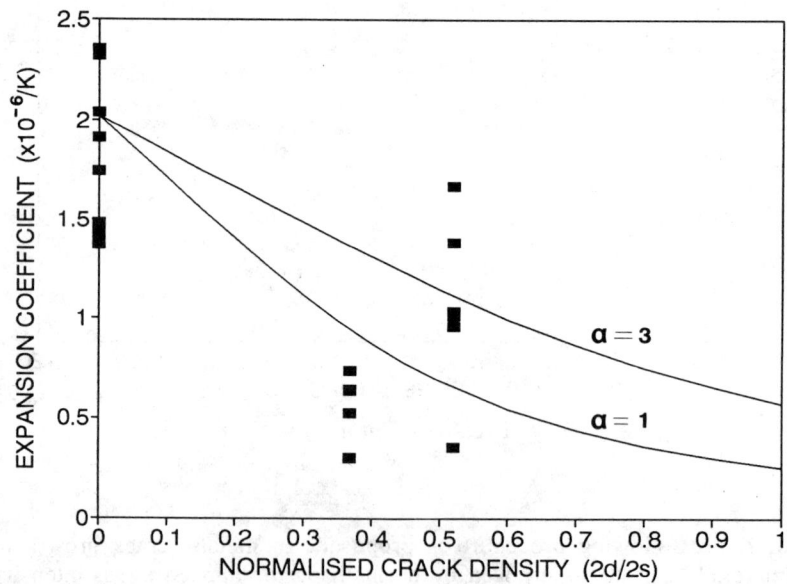

FIG. 8—*Longitudinal thermal expansion coefficient as a function of normalized transverse ply crack density for a (0/90)ₛ CFRP laminate. Comparison between experimental data and shear-lag model (Eq 13).*

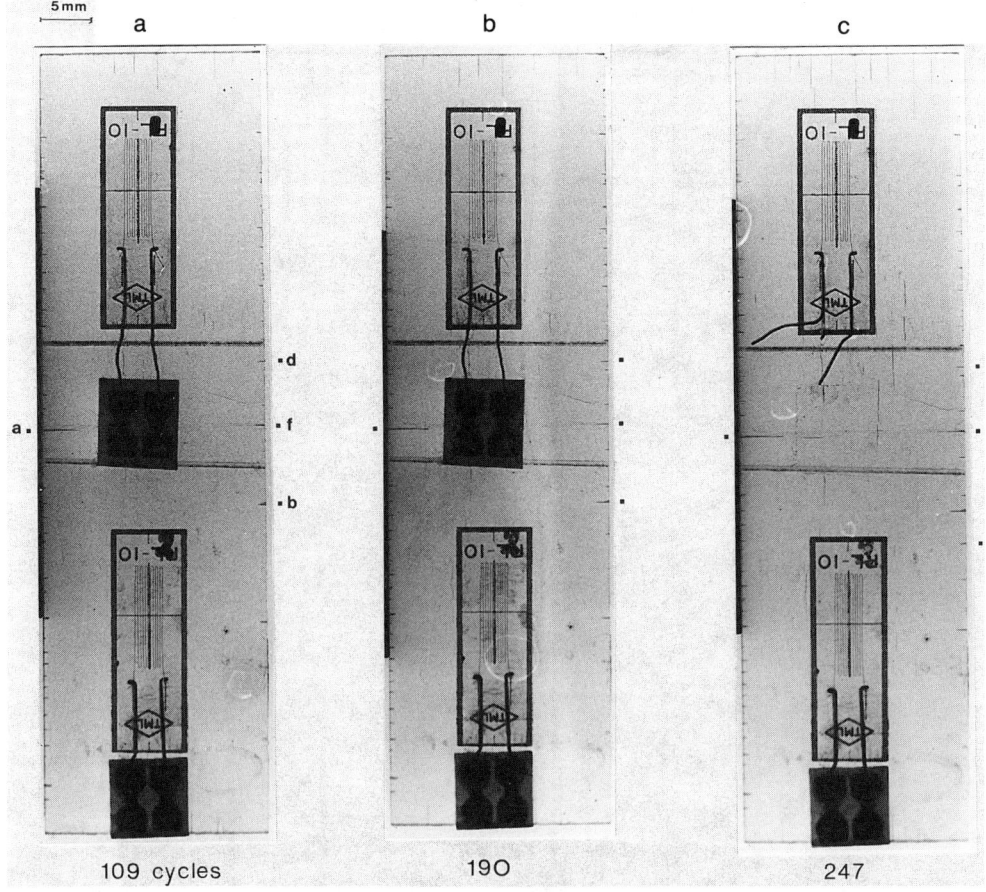

FIG. 9—*Transverse and longitudinal ply matrix cracking in crossply GFRP laminate subjected to 77 to 293 K thermal cycle. Photographs show cracking after* (a) *109,* (b) *190, and* (c) *247 cycles.*

Rewriting Eq 18 in terms of strain energy release rate, G, and incorporating the effect of transverse modulus variation in the thermal cycle gives

$$da/dN = C(\Delta\{[G/E_2]^{1/2}\})^p \qquad (19)$$

Equation 19 can be applied to a thermal cycle by calculating the quantity $[G/E_2]^{1/2}$ at the upper and lower temperatures and subtracting (note that E_2 will be constant during a mechanical fatigue cycle).

For propagating cracks during the thermal cycles of the present study we calculate $\Delta\{[G/E_2]^{1/2}\}$ by making some simplifying assumptions. First, it is assumed that cracks interact only with the crack closest to them. Hence, for a crack at distance s_{AC} from another crack, $\tanh(\lambda s_{BC}/2) = \tanh(\lambda s_{AB}/2) = 1$ and Eq 16 for G becomes

$$G = \left(1 + \frac{b}{d}\right)\left(\sigma\frac{E_2}{E_0} + \sigma^T\right)^2\left(\frac{d}{b}\right)\frac{E_0}{E_1E_2}\frac{1}{\lambda}\left[\tanh\left(\frac{\lambda s_{AC}}{2}\right)\right] \qquad (20)$$

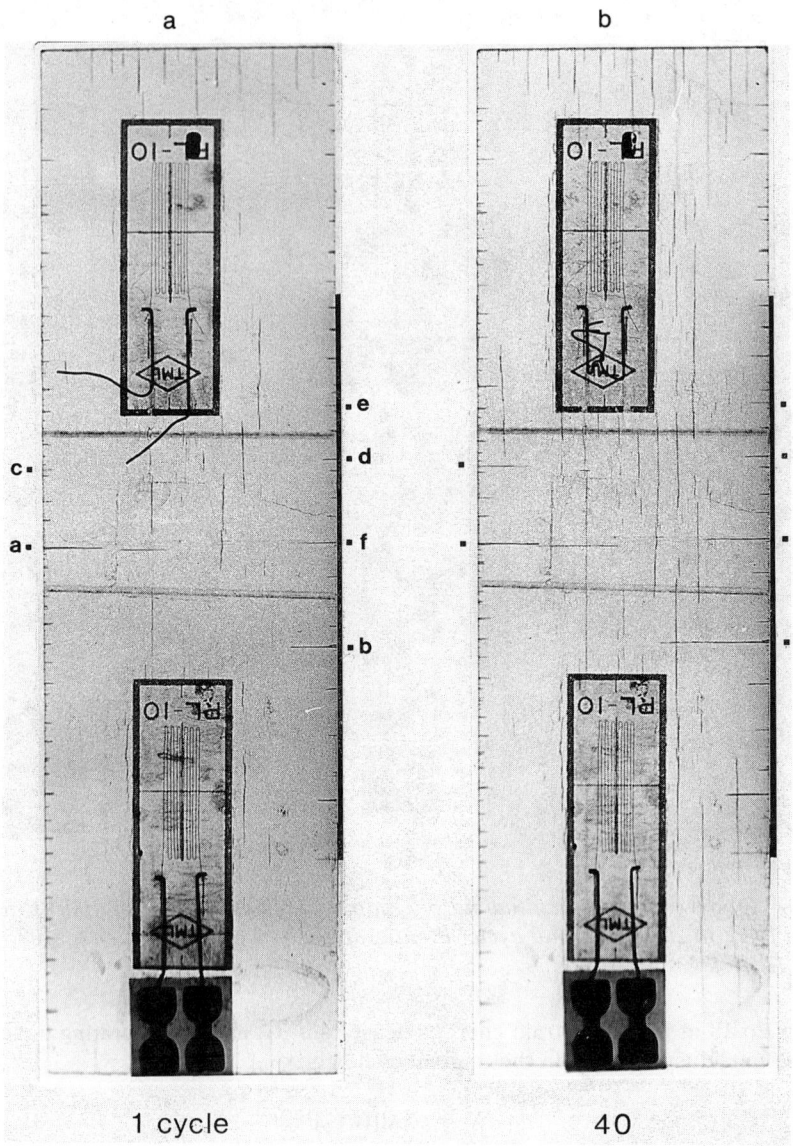

FIG. 10—*Transverse and longitudinal ply matrix cracking in crossply GFRP laminate subjected to 77 to 373 K thermal cycle. Photographs show cracking after (a) 1, (b) 40, (c) 100, and (d) 150 cycles.*

To calculate the thermal stress, room temperature properties are used in conjunction with LPT. In calculating G/E_2 we estimate values of E_2 at the extreme temperatures of the thermal cycles from results obtained by Hill [27] in the temperature range from 23 to 90°C, which we extrapolate linearly to the temperatures of interest (this gives $E_{2(77 \text{ K})} = 16.4$ GPa, $E_{2(293 \text{ K})} = 13.1$ GPa, $E_{2(373 \text{ K})} = 12.0$ GPa).

From the two thermal cycles the values of $\Delta\{[G/E_2]^{1/2}\}$ for about 50 propagating cracks

c   d

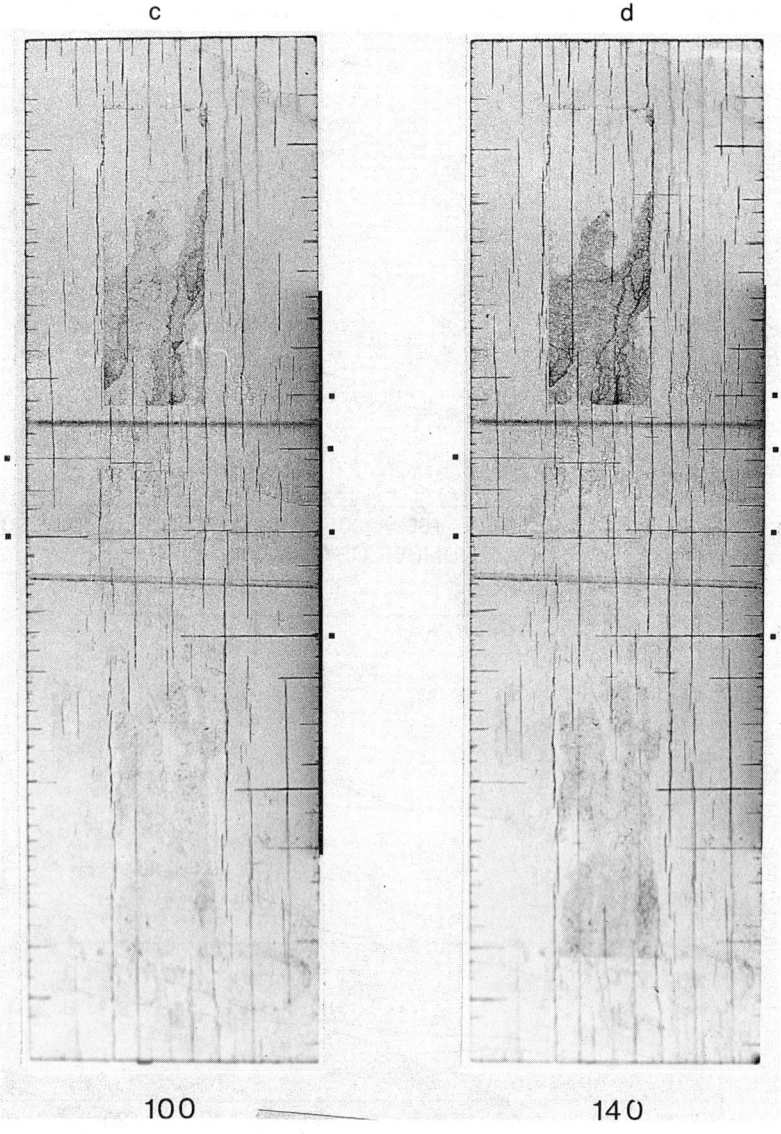

100 ———————— 140

FIG. 10—*Continued.*

were calculated (three cracks in the 77 to 293 K cycle and the remainder in the 77 to 373 K cycle). It was found that the parameter $\Delta\{[G/E_2]^{1/2}\}$ is insensitive to the value of s_{AC} for most of the cracks studied because tanh $(\lambda s_{AC}/2)$ in Eq 20 is generally close to unity.

Figure 13 is a log-log plot of crack growth rate against the strain energy release rate range parameter under mechanical fatigue cycling for a $(0/90_5)_s$ (taken from Ref *23*, modified to include the effect of room temperature thermal stresses) and the thermal cycling data for the $(0/90)_s$ laminate of the present study. There is reasonable consistency between the mechanical cycling and thermal cycling data. The data for the less severe thermal cycle follow the trends of the mechanical data very closely, while the data for the more severe

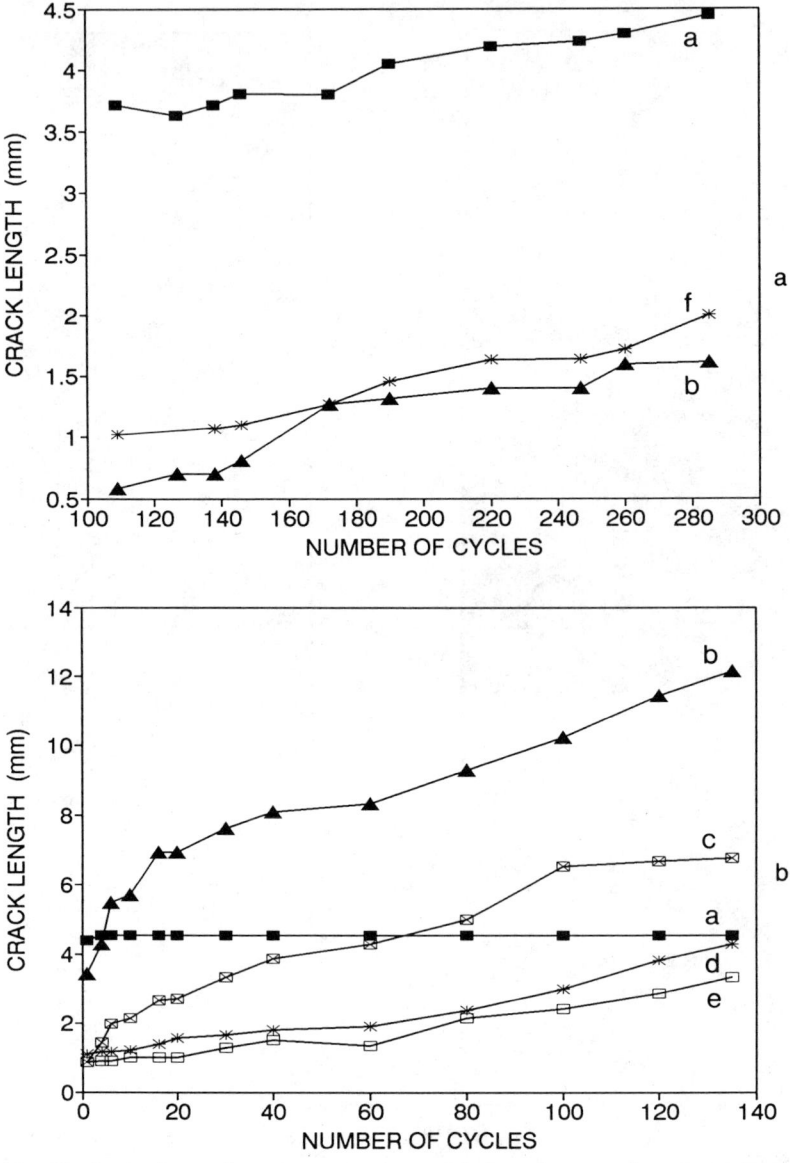

FIG. 11—*Typical plots of transverse ply crack length as a function of number of thermal cycles for (0/90)$_s$ GFRP laminate for cracks* a *to* f *labeled in Figs. 9 and 10:* (a) *77 to 293 K thermal cycle, and* (b) *77 to 373 K thermal cycle.*

thermal cycle show a range of crack growth rates comparable with that shown by the mechanical cycling data at a slightly lower value of $\Delta\{[G/E_2]^{1/2}\}$. The data from the few cracks in the more severe thermal cycle, which were calculated to have significant interaction with neighboring cracks, are also reasonably consistent. Although the agreement between the sets of mechanical and thermal data is good, it is important to consider some of the limitations of the comparison, and this is done below.

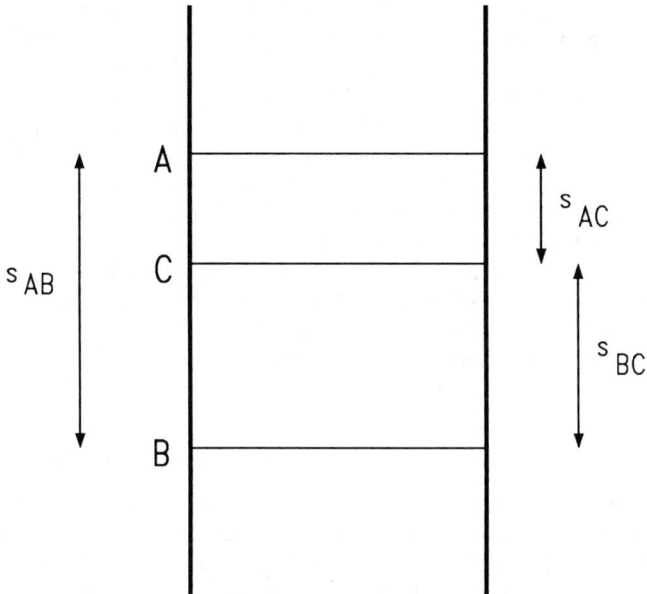

FIG. 12—*Front view of a crossply laminate showing schematically the growth of a model array of transverse ply cracks where Crack C grows between Cracks A and B.*

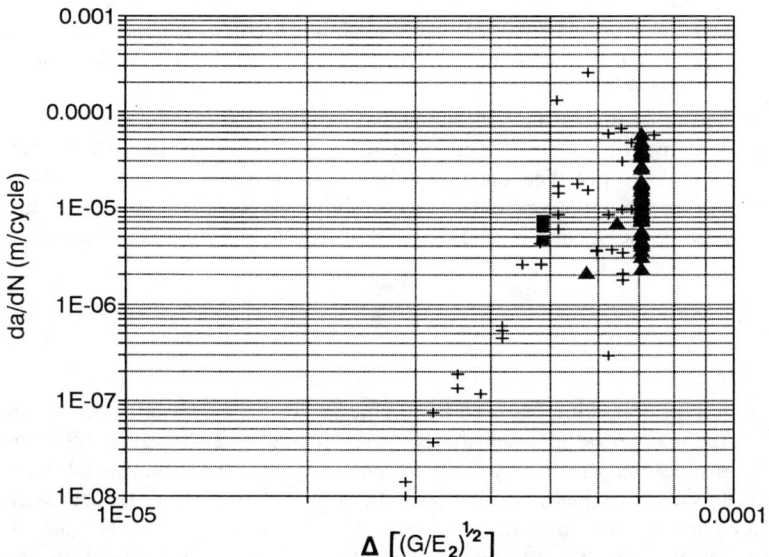

FIG. 13—*Log-log plot of transverse ply crack growth rate against fracture mechanics parameter* $\Delta[(a/E_2)^{1/2}]$. *Data shown are for mechanical fatigue of a* $(0/90_5)_s$ *GFRP laminate (from Ref 23) and the thermal cycling data of the present study for the* $(0/90)_s$ *GFRP laminate:* $+$ = *mechanical data,* ■ = 77 *to* 293 *K thermal cycle,* ▲ = 77 *to* 373 *K thermal cycle.*

The calculation of $\Delta\{[G/E_2]^{1/2}\}$ for the thermal cycle is limited because of the lack of temperature-dependent property data over the complete range of 77 to 373 K. Another consideration is that the modulus normalization of the fracture mechanics parameter, $\Delta K/E$, or in this case $\Delta\{[G/E_2]^{1/2}\}$, is only strictly applicable over the linear portion of a Paris plot; as G_c is approached, the crack growth behavior is difficult to predict. It may also be important that, although the 90° ply material is nominally the same for both the mechanical and thermal cycling data, the mechanical data were obtained using a laminate with a much thicker transverse ply (3.2 mm compared to 0.6 mm). It is possible that for the same value of $\Delta\{[G/E_2]^{1/2}\}$, a transverse crack will grow more rapidly in a laminate with a thick transverse ply, where there is less constraint, than in a thinner transverse ply. In other words, the relative contributions of the various crack opening modes to the total strain energy release rate may be different in laminates with very different transverse ply thicknesses, and this in turn may affect the growth rate.

Longitudinal Ply Split Growth Under Thermal Cycling and Its Effect on CTE

The growth rate of the 0° ply splits which initiated under thermal cycling was generally higher than that of the 90° ply cracks. This is because the constraint exerted by the 90° plies on the 0° plies is different from that exerted by the 0° plies on the 90° plies. Hence the expression used for the strain energy release rate calculation for 90° ply crack growth is not directly applicable to the 0° ply split growth. Only the crack growth rate data for the 90° ply cracks are plotted in Fig. 13.

Finally, we looked at the residual CTE of the cracked GFRP laminate after thermal cycling. This cannot be plotted meaningfully in Fig. 7 because of the concurrent growth of 0° ply splits along with the transverse ply cracks. The effect of the 0° splits is to reduce the CTE by a greater amount than 90° ply cracking alone [10]. This means that the value of the reduced longitudinal CTE does not characterize the cracking damage uniquely since different combinations of 0° ply splitting and 90° ply cracking could give the same reduced CTE. A method of modeling the simultaneous 0° and 90° ply cracking, suggested in Ref 8, is to use finite element analysis of a cracked laminate to deduce effective degraded elastic constants for a cracked ply. These can then be used in a laminated plate theory calculation for the response of a laminate cracked in more than one ply. Such an approach could also be attempted using the shear-lag analysis to deduce the effective constants for a cracked ply.

Conclusions

1. Transverse ply cracking leads to large changes in the CTEs of crossply laminates.

2. This behavior can be modeled fairly simply using an analysis based on a shear-lag model. The predictions are in reasonable agreement with other analytical approaches in the literature as well as the experimental data of the present study for transverse cracks introduced by mechanical loading.

3. Under thermal cycling, the occurrence of 0° ply splits in conjunction with 90° ply cracking also affects the CTE and means that cracking damage cannot be characterized simply from a single measurement of the reduced CTE.

4. The use of a modified strain energy release rate parameter is proposed to enable transverse ply crack growth rates as a result of thermal cycling to be compared with data obtained under mechanical fatigue cycling.

Acknowledgments

The authors thank Rolls-Royce plc for financial support during the course of this work and would like to thank S. Schofield and D. Lind of Rolls-Royce, Derby for helpful discussions. It is also a pleasure to acknowledge our colleague, R. Whattingham, for technical assistance.

References

[1] Highsmith, A. L. and Reifsnider, K. L., "Stiffness Reduction Mechanisms in Composites," *Damage in Composite Materials, ASTM STP 775,* K. Reifsnider, Ed., American Society for Testing and Materials, Philadelphia, 1982, pp. 103–117.

[2] Dvorak, G., Laws, N., and Hejazi, M., "Analysis of Progressive Matrix Cracking in Composite Laminates: I. Thermoelastic Properties of a Ply with Cracks," *Journal of Composite Materials,* Vol. 19, 1985, pp. 216–234.

[3] Talreja, R., "Transverse Cracking and Stiffness Reduction in Composite Laminates," *Journal of Composite Materials,* Vol. 19, 1985, pp. 355–375.

[4] Ogin, S. L., Smith, P. A., and Beaumont, P. W. R., "Matrix Cracking and Stiffness Reduction During the Fatigue of a (0/90)$_s$ GFRP Laminate, *Composite Science and Technology,* Vol. 22, 1985, pp. 23–31.

[5] Allen, D. H., Harris, C. E., and Groves, S. E., "A Thermomechanical Constitutive Theory for Elastic Composites with Distributed Damage—Part II: Application to Matrix Cracking in Laminated Composites," *International Journal of Solid Structures,* Vol. 23, 1987, pp. 1319–1338.

[6] Nuismer, R. J. and Tan, S. C., "Constitutive Relations of a Cracked Lamina," *Journal of Composite Materials,* Vol. 22, 1988, pp. 306–321.

[7] Smith, P. A. and Wood, J. R., "Poisson's Ratio as a Damage Parameter in the Static Tensile Loading of Simple Crossply Laminates," *Composite Science and Technology,* Vol. 38, 1990, pp. 85–93.

[8] Bowles, D. E., "Effect of Microcracks on the Thermal Expansion of Composite Laminates," *Journal of Composite Materials,* Vol. 17, 1984, pp. 173–187.

[9] Herakovich, C. T. and Hyer, M. W., "Damage-induced Property Changes in Composites Subjected to Cyclic Thermal Loading," *Engineering Fracture Mechanics,* Vol. 25, 1986, pp. 779–791.

[10] Hashin, Z., "Thermal Expansion Coefficients of Cracked Laminates," *Composite Science and Technology,* Vol. 31, 1988, pp. 247–260.

[11] Lim, S. G. and Hong, C. S., "Effect of Transverse Cracks on the Thermomechanical Properties of Crossply Laminated Composites," *Composite Science and Technology,* Vol. 34, 1989, pp. 145–162.

[12] McCartney, L. N., "Theories of Stress Transfer in a Crossply Laminate Containing a Parallel Array of Transverse Cracks," NPL Report DMA(A)189, National Physical Laboratory, Teddington, England, March 1990.

[13] Boniface, L. and Ogin, S. L., "Application of the Paris Equation to the Fatigue Growth of Transverse Ply Cracks," *Journal of Composite Materials,* Vol. 23, 1989, pp. 735–754.

[14] Jones, R. M., *Mechanics of Composite Materials,* Scripta-McGraw-Hill, Washington, DC, 1975.

[15] Caslini, M., Zanotti, C., and O'Brien, T. K., "Study of Matrix Cracking and Delamination in Glass/Epoxy Laminates," *Journal of Composite Technology Research,* Vol. 9, 1987, pp. 121–130.

[16] Elmes, D. A. and Gilbert, D. G., "Thermal Stress Analysis in PMR-15 Carbon Fibre Laminates," *Extending the Limits,* Proceedings of the PRI 3rd International Conference on Fibre Reinforced Composites, Liverpool, Plastics & Rubber Institute, London, 1988, pp. 28/1–28/12.

[17] Corvi, A. and Reale, S., "Engineering Evaluation of the Thermal Expansion Characteristics of Fibrous Composite Laminates," *Extending the Limits,* Proceedings of the PRI 3rd International Conference on Fibre Reinforced Composites, Liverpool, Plastics & Rubber Institute, London, 1988, pp. 33/1–33/9.

[18] "Measurement of Thermal Expansion Coefficient Using Strain Gages," Note TN-513, Measurement Group Inc., Technical, Raleigh, NC, 1986.

[19] Mulheron, M. J., "A Study of Thermal Strains in Glass Fibre Reinforced Plastics," Ph.D. thesis, University of Surrey, 1984.

[20] Adams, D. S., Bowles, D. E., and Herakovich, C. T., "Thermally Induced Transverse Cracking in Graphite-Epoxy Crossply Laminates," *Journal of Reinforced Plastics and Composites,* Vol. 5, 1986, pp. 152–169.

[21] Owens, G. A. and Schofield, S. E., "Thermal Cycling and Mechanical Property Assessment of Carbon Fiber Fabric Reinforced PMR-15 Polyimide Laminates," *Composite Science and Technology,* Vol. 33, 1988, pp. 177–190.

[22] Boniface, L., Ogin, S. L., and Smith, P. A., "Fracture Mechanics Approaches to Transverse Ply Cracking in Composite Laminates," *Composite Materials: Fatigue and Fracture, ASTM STP 1110,* T. Kevin O'Brien, Ed., American Society for Testing and Materials, Philadelphia, 1990, pp. 9–29.

[23] Boniface, L., Ogin, S. L., and Smith, P. A., "Strain Energy Release Rates and the Fatigue Growth of Matrix Cracks in Model Arrays in Composite Laminates," *Proceedings of the Royal Society of London A,* Vol. 432, 1991, pp. 427–444.

[24] Laws, N. and Dvorak, G. J., "Progressive Transverse Cracking in Composite Laminates," *Journal of Composite Materials,* Vol. 22, 1988, pp. 900–916.

[25] Nairn, J. A., "The Strain Energy Release Rate of Composite Microcracking: A Variational Approach," *Journal of Composite Materials,* Vol. 23, 1989, pp. 1106–1129.

[26] Hertzberg, R. W., *Deformation and Fracture Mechanics of Engineering Materials,* John Wiley and Sons, New York, 1976.

[27] Hill, M. C., University of Surrey, private communication, 1991.

Bradley A. Lerch[1] and James F. Saltsman[1]

Tensile Deformation of SiC/Ti-15-3 Laminates

REFERENCE: Lerch, B. A. and Saltsman, J. F., "Tensile Deformation of SiC/Ti-15-3 Laminates," *Composite Materials: Fatigue and Fracture, Fourth Volume, ASTM STP 1156,* W. W. Stinchcomb and N. E. Ashbaugh, Eds., American Society for Testing and Materials, Philadelphia, 1993, pp. 161–175.

ABSTRACT: The damage mechanisms of a laminated, continuous SiC fiber-reinforced Ti-15-3 composite were investigated. Specimens consisting of unidirectional as well as cross-ply laminates were pulled in tension to failure at room temperature and 427°C and subsequently examined metallographically. Selected specimens were interrupted at various strain increments and examined to document the development of damage. When possible, a micromechanical stress analysis was performed to aid in the explanation of the observed damage. The analyses provided average constituent microstresses and laminate stresses and strains. The damage states were found to be highly influenced by the fiber architecture.

KEY WORDS: SiC/Ti-15-3, composite, damage modes, tensile properties

Metal matrix composites (MMC) are currently being developed for high-temperature applications in engines and the structures of hypersonic flight vehicles. Both applications require a stiff, lightweight material capable of carrying significant thermal and mechanical loads. Conventional monolithic materials are currently operating at or near their limits and do not offer the potential for meeting the demands of the next generation of engines and flight vehicles.

A candidate material now under active consideration by many investigators is SiC fiber-reinforced Ti-15V-3Cr-3Sn-3Al (Ti-15-3) composite. This composite material offers the potential for meeting the above requirements but presents a host of potential problems to be addressed. The SiC fiber and the Ti matrix have significantly different mechanical and thermal properties, and it is necessary to understand how these differences affect the behavior of the composite.

The purpose of this study is to investigate the damage mechanisms produced by tensile loads at room temperature (RT) and 427°C (800°F). The damage is determined by metallographic methods, and a micromechanical stress analysis program is used to further understand the observed damage. A more detailed discussion of the damage mechanisms is presented in a companion NASA publication [1].

Experimental Procedure

Material

The composite was fabricated from alternating layers of Ti-15V-3Cr-3Sn-3Al (Ti-15-3) foils and continuous SiC (SCS-6) fibers. The various fiber layups were consolidated at

---

[1]Research engineers, NASA Lewis Research Center, Mail Stop 49–7, 21000 Brookpark Rd., Cleveland, OH 44135.

elevated temperatures, yielding eight-ply material which contained a nominal fiber volume fraction of 34%. Detailed descriptions of the microstructure have been given previously [2]. These composite plates were manufactured in 1987 and contained a Ti-wire weave which was used to keep the SiC fibers parallel during consolidation. Specimens were cut from one of four plates, depending on the desired orientation. Dogbone-shaped specimens having a 15.2-cm length and 1.3-cm width and a straight gage section (3.8 by 0.8 cm) were used for most tests.[2] The orientations were: $[0°]_8$, $[90°]_8$, $[90°/0°]_{2s}$, $[0°/90°]_{2s}$, $[\pm 30°]_{2s}$, $[\pm 45°]_{2s}$, and $[\pm 60°]_{2s}$.

Tension Testing

The majority of the specimens were heat-treated for 24 h at 700°C in a vacuum prior to testing. The purpose of this heat treatment was to partially stabilize the alpha-beta Ti structure [3]. The remainder of the specimens were tested in the as-received condition. No differences were observed in the mechanical properties of the specimens from the two heat treatments.

Tension tests were conducted with a servohydraulic test machine. Strain was measured with a 1.3-cm gage length, clip-on extensometer. Alumina probes were used to contact the specimen and permit strain measurement at elevated temperatures. To determine Poisson's ratio, 0°/90° strain gages were mounted on a few $[0°]_8$, $[90°]_8$, and $[90°/0°]_{2s}$ specimens. Tension specimens were loaded in strain control at a strain rate of 10^{-4}/s unless otherwise indicated. Stress-strain data were recorded digitally as well as on analog recorders. Specimens which were tested at 427°C were heated with an induction coil.

Metallography

Damage modes for both failed and interrupted tests were examined with the use of optical metallographic techniques which had been developed previously [2] for this material. A sufficient number of untested sections were similarly polished to ensure that the different types of damage were not a result of metallographic preparation.

Tension Test Results

Room Temperature Tests

Typical stress-strain curves for room-temperature (RT) tests are presented in Fig. 1 for all the fiber orientations tested as well as for the unreinforced matrix. As expected, the $[0°]_8$ orientation is the strongest, followed by, in order of decreasing strength, the $[90°/0°]_{2s}$ and $[0°/90°]_{2s}$, $[\pm 30°]_{2s}$, the matrix, $[\pm 45°]_{2s}$, $[90°]_8$, and $[\pm 60°]_{2s}$ orientations. Note that the $[90°/0°]_{2s}$ and $[0°/90°]_{2s}$ specimens have comparable mechanical properties. Also, the $[\pm 60°]_{2s}$ and $[90°]_8$ specimens had approximately the same strength. In general, the degree of nonlinear stress-strain behavior increased as the strength decreased. The average tensile properties for each orientation are given in Table 1.

Excluding the $[0°]_8$ orientation, all layups contained a second linear portion in the stress-strain curves, and an example of this is shown for a $[0°/90°]_{2s}$ specimen in Fig. 2. The slope of this second linear portion is indicated in Table 1 as E_s. The E_s value is always significantly less than the initial modulus, E_i, and this is due to debonding of the fiber/matrix interface [1,4].

---

[2] Six other specimen types were used in this study. The mechanical properties were not dependent on the specimen geometry. Information on the dimensions of these specimens can be found in Ref *1*.

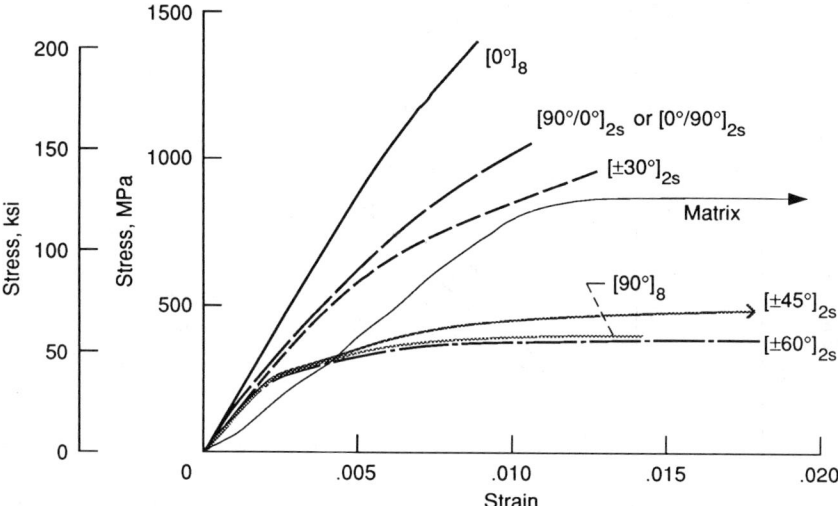

FIG. 1—*Room temperature tensile curves for various laminate orientations. Note that the [±45°]₂ₛ specimen continues to strains larger than 0.020.*

Failure strain (defined as strain at complete separation) is smallest for the [0°]₈ orientation and equal to 0.90%. Failure strains for the other orientations are nominally the same (at least for the few specimens tested) and range from approximately 1 to 2%. Due to the large specimen-to-specimen variation in failure strain for all orientations (except [0°]₈), this property is not given in Table 1.

Tests at 427°C

The [0°/90°]₂ₛ, [90°/0°]₂ₛ, and [± 30°]₂ₛ orientations were tested only at room temperature (RT). The remaining orientations were tested at RT and 427°C. Test results show [1] that there are no significant temperature-induced variations in mechanical properties at RT and 427°C for the [0°]₈ and [90°]₈ orientations. Temperature-induced variations were observed for the [± 45°]₂ₛ and [± 60°]₂ₛ orientations. Typical stress-strain behavior for [0°]₈ and [90°]₈ orientations tested at RT and 427°C are shown in Fig. 3, and the curves lie within the

TABLE 1—*Average room temperature properties.*

Laminate Code	Ultimate Tensile Strength (UTS), MPa	Initial Modulus of Elasticity, E_i, GPa	Secondary Modulus of Elasticity, E_s, GPa	Poisson's Ratio, ν_{12}
Fiber	. . .	395	. . .	. . .
[0°]₈	1390	180	NA	0.28
[90°/0°]₂ₛ, [0°/90°]₂ₛ	1020	145	115	0.21
[± 30°]₂ₛ	995	150	85	. . .
Matrix	870	90	NA	. . .
[± 45°]₂ₛ	530	115	35	. . .
[90°]₈	420	120	30	0.17
[± 60°]₂ₛ	390	115	25	. . .

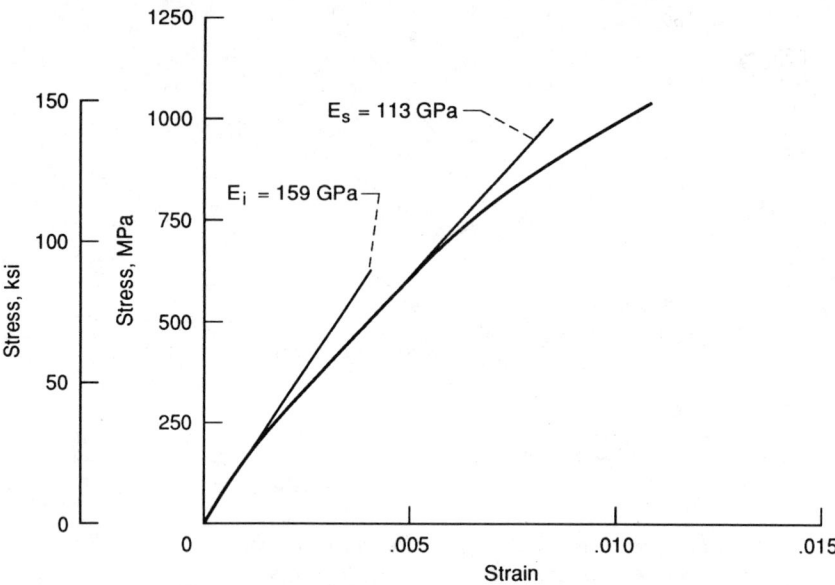

FIG. 2—*Tensile curve for [0°/90°]₂ₛ specimen. Note the bilinear response at lower strains indicated by tangent lines.*

observed scatter of specimen-to-specimen variation. In Fig. 3, there is a large variation in failure strain between RT and 427°C for the [90]₈ orientation. However, this orientation typically exhibits a large scatter in failure strains, and, since only one duplicate test was run per temperature, it is not known if test temperature influences the failure strain. Additional tests are necessary to determine this. The difference in the tensile behavior between RT and 427°C for the [±45°]₂ₛ orientation is shown in Fig. 4. Specimens are less stiff and weaker at 427°C than the ones tested at RT. Tangent moduli at selected stress levels, corresponding to E_I and E_S, for the [±45°]₂ₛ orientation are also shown in Fig. 4. The properties of the [±60°]₂ₛ orientation show a similar temperature dependence [1].

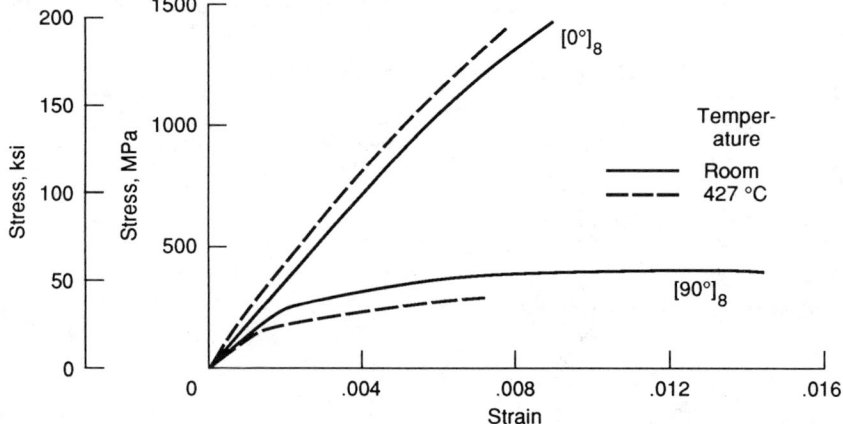

FIG. 3—*Tensile curves for [0°]₈ and [90°]₈. Strain rate is 10^{-4} s^{-1}.*

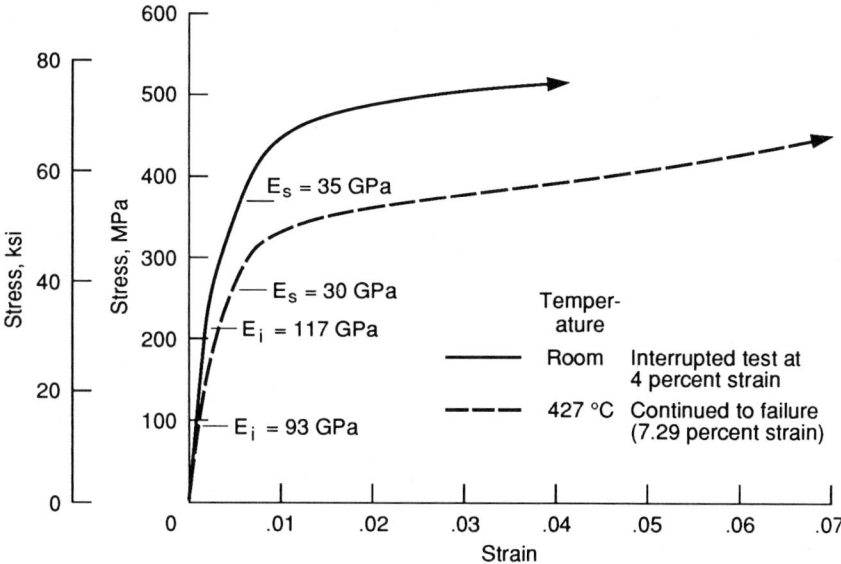

FIG. 4—*Tensile curves for [±45°]₂ₛ. Strain rate is* 10^{-4} s^{-1}.

Specimens having $[0°]_8$, $[90°]_8$, and $[\pm 45°]_{2s}$ orientations were tested at 427°C at different strain rates. The $[0°]_8$ specimens were pulled at 10^{-3}, 10^{-4}, and 10^{-5} s^{-1}, and the stress-strain curves for these tests are shown in Fig. 5. No consistent strain rate effects were observed. Likewise, the $[90°]_8$ orientation showed no strain rate effects for the order-of-magnitude change in strain rate examined (Fig. 6)[3]. The stress-strain curves for the $[\pm 45°]_{2s}$ specimens showed similar tensile properties for both strain rates used (Fig. 7). However, the waviness of the stress-strain curve for the specimen tested at the slower rate indicates that some rate-dependent behavior may be present. This waviness is believed to be actual material behavior (as will be shown later) and not some test artifact.

Discussion of Damage

Damage development and the cause of nonlinearities in the stress-strain curve was investigated by using data from specimens tested to failure as well as from specimens which were interrupted before failure. The specimens were examined metallographically for cracking, debonding, and evidence of matrix flow. Three orientations are discussed here. The remaining orientations are discussed in Ref *1*.

To better understand the relationship between the stress-strain curves and the observed damage mechanisms, the composite stress-strain behavior was modeled by using the laminate program AGLPLY [*5*]. This code is based on the fiber vanishing-diameter model and incorporates the elastic-plastic response of the matrix. The fiber was assumed to be elastic. Laminate behavior as well as average constituent microstresses can be described. The input values used for the constituents along with an estimate of residual stresses induced during consolidation are given in the Appendix of Ref *1*.

---

[3]The jog in the curve for the specimen tested at a strain rate of 10^{-5} s^{-1} was a result of material cracking. This occurred on some specimens and was not dependent on fiber layup or test conditions.

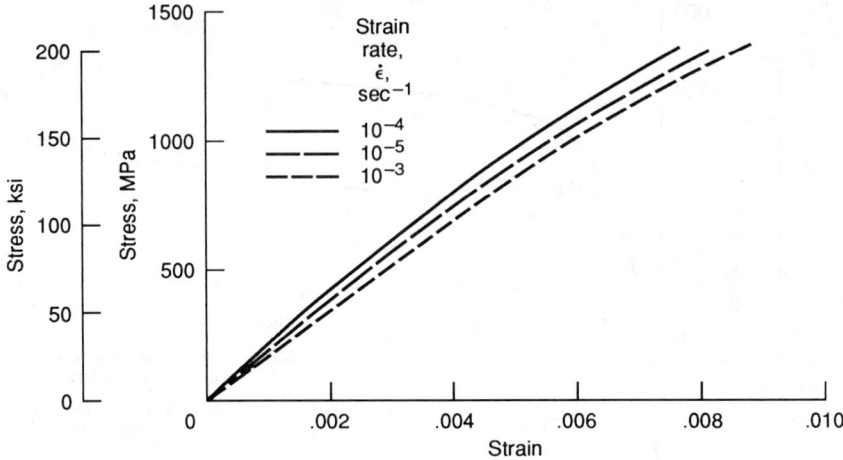

FIG. 5—*Tensile curves for [0°]₈ specimens tested at 427°C.*

[0°]₈ Laminate

The predicted longitudinal stress-strain curve is given for the [0°]₈ layup in Fig. 8 along with the experimental results from two specimens. Also included in this figure is the stress-transverse-strain response. Figure 8 indicates good agreement between the predicted and experimental curves, especially for the longitudinal strain. The AGLPLY calculations are linear to a laminate stress of ≈900 MPa. The matrix stress at this point is ≈710 MPa and corresponds to the proportional limit of the matrix [1]. Thus, the deviation from linearity at a laminate stress of ≈900 MPa is predicted on the basis of the yielding of the matrix. The stress-transverse-strain also indicates this point of yielding.

In order to investigate the cause of the nonlinearities, two interrupted tension tests were performed at RT. One specimen was interrupted at a longitudinal strain of 0.5%, which represents the first signs of nonlinearity (Point A in Fig. 8), and no damage was metallographically observed in the composite. The second specimen was strained just short of

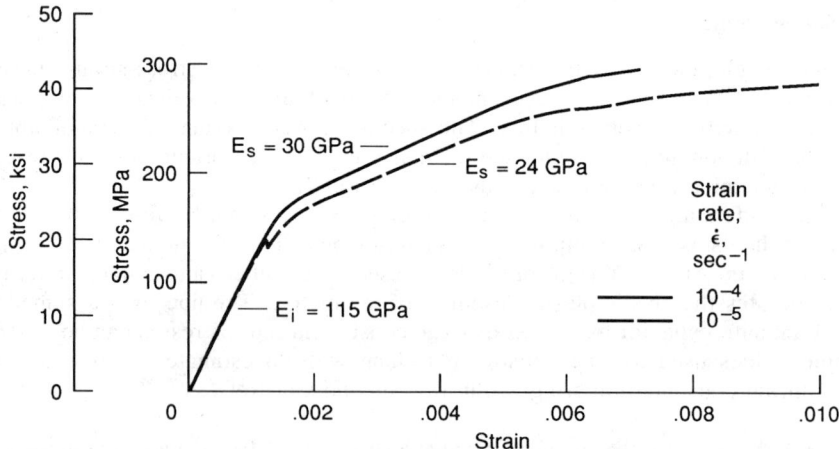

FIG. 6—*Tensile curves for [90°]₈ specimens tested at 427°C.*

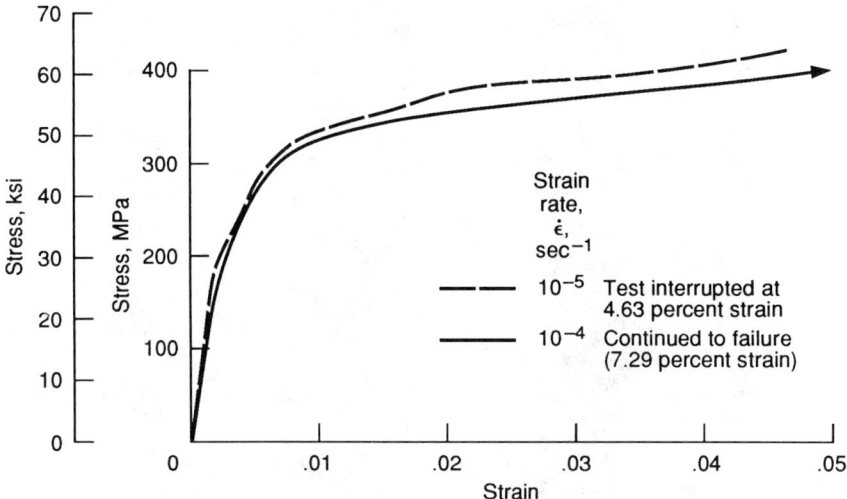

FIG. 7—*Tensile curves for [±45°]₂ₛ specimens tested at 427°C.*

failure (Point B in Fig. 8) to a longitudinal strain of 0.85%. This strain is 94% of the average failure strain of 0.90% for the [0°]₈ orientation. No matrix or fiber cracks were observed, nor was there any evidence of delamination. However, some debonding was evident (Fig. 9). No metallographic evidence of slip bands indicating matrix plasticity could be found in either specimen, although AGLPLY predicted yielding of the matrix at these stresses.

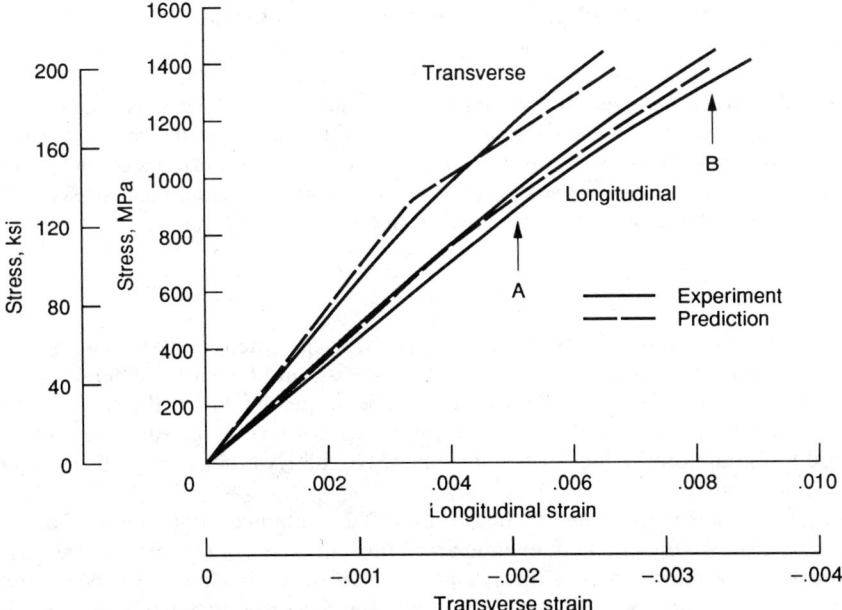

FIG. 8—*Room temperature tensile curves for [0°]₈ specimens. Data from two specimens are plotted for longitudinal strain.*

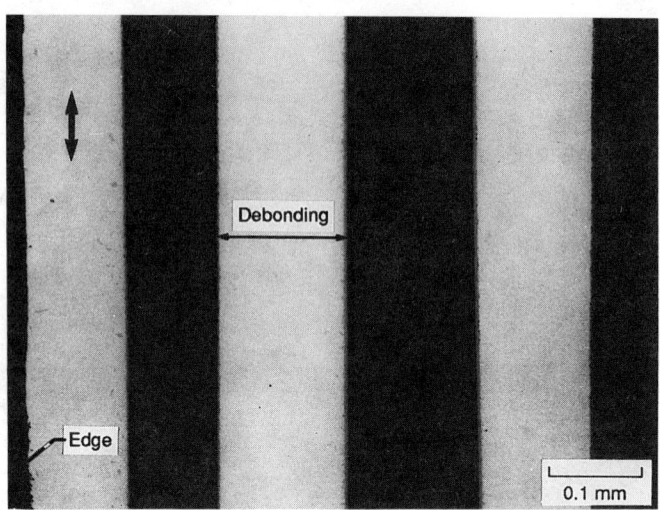

FIG. 9—*Face section of [0°]₈ specimen tested to a strain of 0.85% at room temperature. Double-pointed arrow indicates loading direction.*

Chemical etchants failed to reveal slip bands. In addition, since the matrix has very limited strain-hardening capabilities, microhardness indentations in the in-situ matrix were also ineffective in determining whether or not plasticity occurred and subsequently caused the nonlinearity. It is not known if the debonding observed in Fig. 9 is sufficient to cause such nonlinearities. The only conclusive statement which can be made about the nonlinearities in the [0°]₈ specimens is that it is *not* caused by fiber or matrix cracks.

The [0°]₈ specimens show little evidence of damage up to nearly the failure point. The only observable change due to straining is some debonding. A quantitative measure of debonding is impossible (regardless of the fiber layup) because it is observed as a line defect, and there is a resolution problem when the interfacial gap is small. This also inhibits defining when (i.e., at what strain) debonding starts. Furthermore, the interface does not debond completely, but only separates along certain portions of the fiber length and circumference, which depends on the local bond strengths and local stresses. Nevertheless, large interfacial gaps can be identified in both transverse and longitudinal sections.

[±45°]₂ₛ Laminate

Due to the high strains achieved in this layup, these specimens contained unique damage modes. The RT specimen was interrupted at 4% strain. Extensive fiber cracking and debonding were observed. The fiber cracks were perpendicular to the fiber axis. Some matrix cracking was observed in through-thickness sections (Fig. 10), and these cracks appeared to initiate on the face of the specimen. No matrix cracks were observed beyond the first fiber row.

Specimens which were pulled to failure at 427°C contained large amounts of damage. Fibers were extensively cracked, even in areas far removed from the fracture surface (Fig. 11). The fiber cracks were not always perpendicular to the fiber axis but were bent, approaching a direction parallel to the fiber axis. Large gaps associated with debonding, as well as extensive matrix cracking, were also observed (Fig. 11). The matrix in several debonded areas has locally necked down and has resulted, in some cases, in a reduction in

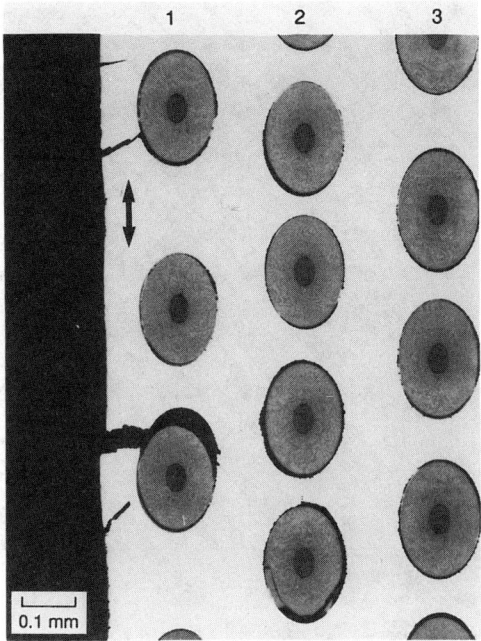

FIG. 10—*Through-the-thickness section of [±45°]$_{2s}$ specimen. Room temperature test interrupted after 4% strain. Double-pointed arrow indicates loading direction.*

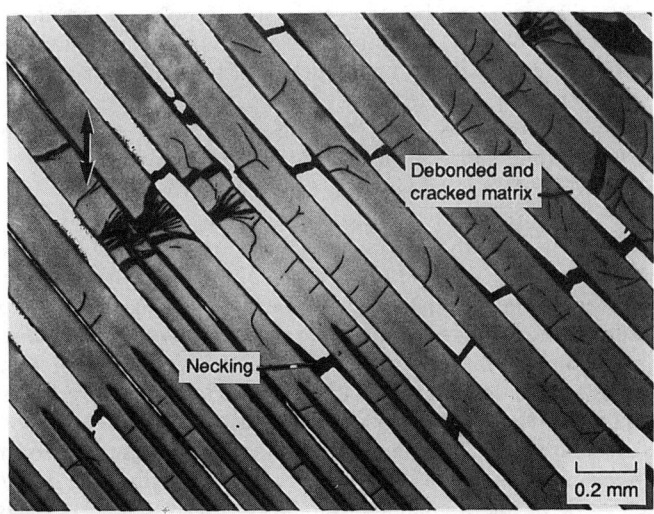

FIG. 11—*Face section of [±45°]$_{2s}$ specimen pulled to failure at 427°C. Double-pointed arrow indicates loading direction.*

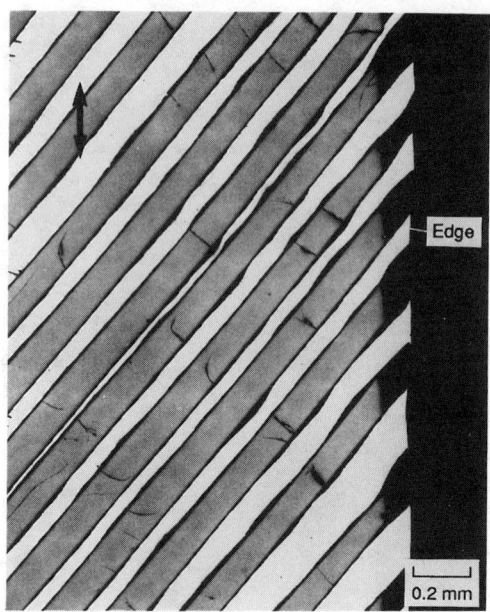

FIG. 12—*Face section of [±45°]$_{2s}$ specimen tested to failure at 427°C. Double-pointed arrow indicates loading direction.*

area of 43%. Figure 12 shows the edge of the specimen at which the matrix has been stretched past the plane of the fiber ends by approximately one fiber diameter. This figure also shows a waviness in both the fibers and the matrix slivers between fibers. This waviness is due to the large amounts of damage.

Testing at a slower strain rate of 10^{-5} s^{-1} at 427°C resulted in observable slip bands in the matrix (Fig. 13). Slip could not be resolved in any of the other tested specimens. These slip bands are believed to be due to dynamic precipitation of alpha-titanium caused by the slower straining rates, thus causing the waviness in the stress-strain curve observed in Fig. 7.

The [±45°]$_{2s}$ layup has a unique stress-strain response in that the laminate failure strains are significantly larger than those of any of the other layups ($\approx$7% compared to $\approx$1%). Prediction of the RT [±45°]$_{2s}$ stress-strain curve is shown in Fig. 14. There is good agreement between the predicted and experimental laminate stress-strain curve in the linear elastic region. However, there is no agreement for the rest of the stress-strain curve. The modeling of the [±45°]$_{2s}$ (and all other cross-ply laminates) was based upon experimental data from tension tests on the [90°]$_8$ orientation. For the [90°]$_8$ orientation, the stress at which debonding occurred could be experimentally verified and was equal to 275 MPa. To account for the debonding analytically, the transverse fiber modulus was reduced at the applied stress where debonding occurred as recommended by Johnson [6].

After debonding occurred, the secondary modulus, E_s, was determined to be 29 GPa from a linear regression analysis of the experimental stress-strain data for the [90°]$_8$ layup. The transverse fiber modulus needed to give a secondary modulus of 29 GPa was found to be 13 GPa, which is 3% of the original fiber modulus. The value of 13 GPa is substituted in AGLPLY for the original transverse modulus in each layup when debonding occurs.

The predicted results for the [45°]$_{2s}$ layup using this debonding criterion did not repro-

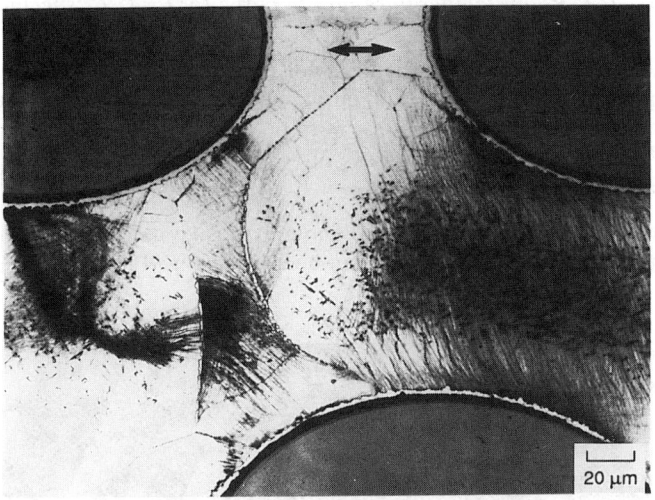

FIG. 13—*Sections of [±45°]₂ₛ specimen showing slip bands: test temperature, 427°C; strain rate, 10⁻⁵ s⁻¹; etched. Double-pointed arrows indicate loading direction.*

duce the experimental data. Both the secondary modulus and the applied stress, at which debonding occurred, were too high. Examination of the calculated fiber-matrix interfacial stresses at the point of actual debonding for a laminate stress of 250 MPa indicates that the radial stresses, σ_{33}, and the fiber-matrix interface shear stresses, σ_{13}, are 240 and 125 MPa, respectively, Since σ_{13} is a large percentage of σ_{33}, it was assumed that this shear stress would also play a role in the debonding process since the shear stress could also crack the interface. Therefore, an arbitrary effective stress criterion for debonding was used in which $\sqrt{\sigma_{33}^2 + \sigma_{13}^2} = 275$ MPa.

Attainment of this criterion for the [±45°]₂ₛ specimen occurred at a predicted applied laminate stress of 250 MPa, which agrees with the actual break in the stress-strain curve

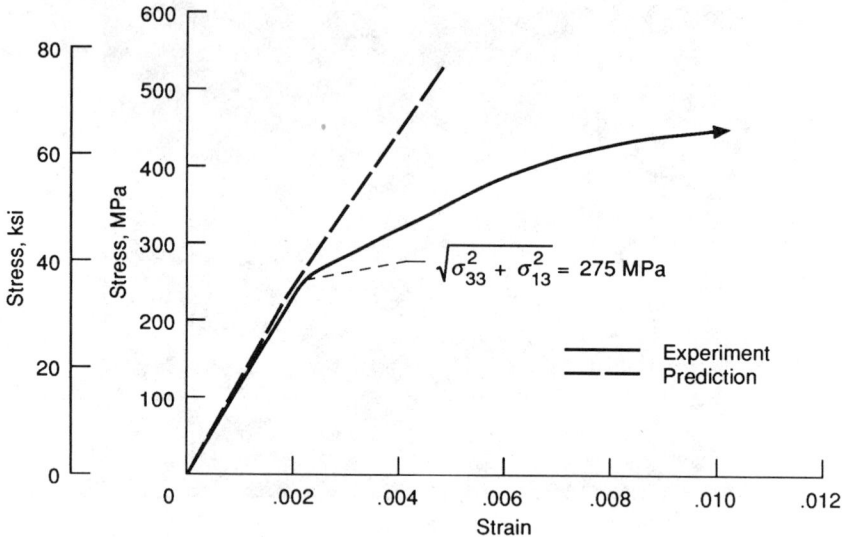

FIG. 14—*Room temperature tensile curve for [±45°]₂ₛ specimen.*

(Fig. 14). However, use of this criterion did not improve AGLPLY's predictive capabilities for the remainder of the stress-strain curve which contains both the secondary modulus and the nonlinear behavior.

[±30°]₂ₛ *Laminate*

Specimens which were tested to failure show extensive damage near the fracture surface. The fracture profile in the first ply can be seen in Fig. 15. The fibers, as well as the

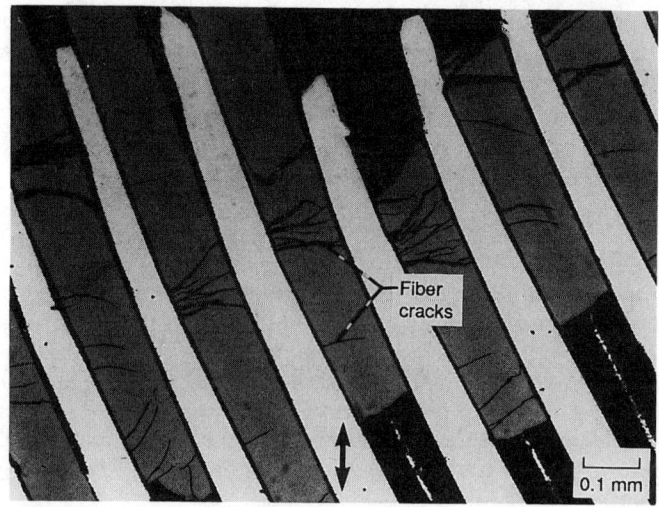

FIG. 15—*Face section of [±30°]₂ₛ specimen tested at room temperature. Fiber rotation is accommodated by fiber cracking. Double-pointed arrow indicates loading direction.*

matrix slivers between the fibers, have rotated as if to align themselves with the load axis. Extensive fiber cracking is observed near the fracture surface, with less occurring away from it. Fiber cracks are seen initiating on one side of the fiber. Some of these cracks do not propagate completely across the fiber diameter. Debonding was also observed along most fibers, whereas matrix cracks were rarely observed.

To investigate the development of damage, one specimen was strained to 0.8%, well into the nonlinear range (see Point A in Fig. 16). Examination of this specimen showed considerable debonding, as indicated in Fig. 17. In some areas of debonding, the matrix slivers between fibers had begun to neck. Some matrix cracks were observed, but these were limited to the outer matrix sheet. There was also one isolated area in which cracks (a total of five cracks) were found in four fibers. This occurred on the edge of the specimen and in the second ply. No other fiber cracks were observed.

The behavior of the $[\pm 30°]_{2s}$ orientation cannot be predicted by using AGLPLY, as seen in Fig. 16. Although the linear elastic portion of the stress-strain curve can be accurately predicted, the rest of the curve cannot. The first point of debonding can be adequately modeled by using the arbitrary stress criterion for debonding, as discussed previously for the $[\pm 45°]_{2s}$ laminate. However, when the reduced transverse fiber modulus is substituted, the predicted secondary modulus, E_s is much higher than the experimental value and in fact shows little deviation from the linear elastic line. Additionally, the primary break from linearity in Fig. 16 occurs at about 660 MPa and was due to matrix yielding. The cause of the nonlinearity could be a result of local plasticity as a consequence of debonding and as observed by the necking of the matrix (Fig. 17). Global plasticity, surface cracks, or a combination of these mechanisms could also contribute to the nonlinearity, none of which can be accounted for in AGLPLY. It is unlikely that fiber cracking contributed to the nonlinearity, since few fiber cracks were observed at strains <1%.

The findings indicate that the damage occurring in this composite system is complex and highly dependent upon the fiber layup. Methods to handle matrix plasticity, constituent cracking, and fiber/matrix debonding must be included in any model which will be used to accurately predict the tensile stress-strain behavior of this system.

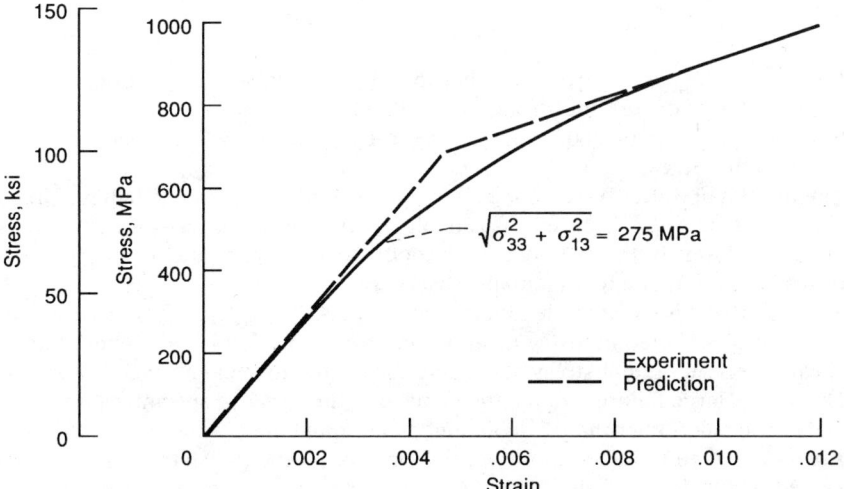

FIG. 16—*Room temperature tensile curve for [±30°]₂ₛ specimen.*

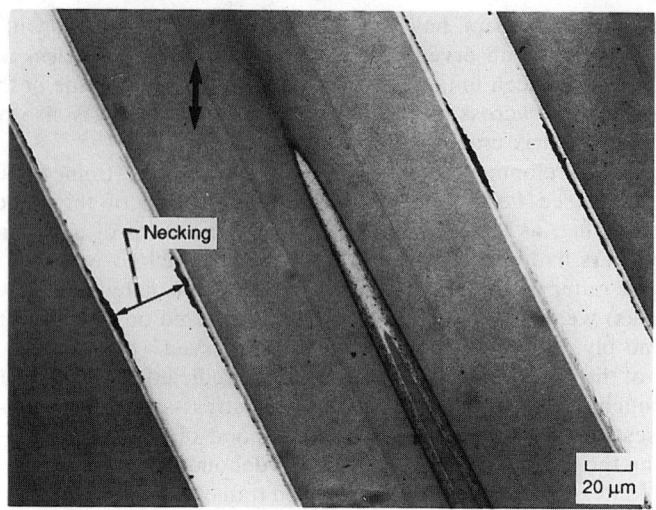

FIG. 17—*Face section of* [±30°]₂ₛ *specimen tested at room temperature to a strain of 0.8%. Double-pointed arrow indicates loading direction.*

Summary

Laminates of SiC/Ti-15-3 composite were pulled in tension at RT and 427°C (800°F). Laminate orientations tested were [0°]₈, [90°]₈, [0°/90°]₂ₛ, [90°/0°]₂ₛ, [±30°]₂ₛ, [±45°]₂ₛ, and [±60°]₂ₛ. After testing, detailed metallography was performed to identify the damage modes. Tests which were interrupted before failure allowed damage development to be followed. A laminate analysis was used to determine the constituent microstresses during straining. These analyses, coupled with the metallographic observations, allowed a description of the tensile behavior of the various laminates. A summary of the results is highlighted below.

1. RT stress-strain curves revealed that the [0°]₈ orientation is the strongest, followed by, in order of decreasing strength, the [0°/90°]₂ₛ and [90°/0°]₂ₛ, [±30°]₂ₛ, the matrix, [±45°]₂ₛ, [90°]₈, and [±60°]₂ₛ. In general, the degree of nonlinear behavior increased as strength decreased.
2. The [0°]₈ layup showed only some fiber/matrix debonding prior to failure. No constituent cracking was observed. The nonlinearity in the stress-strain curve was predicted analytically through matrix plasticity. However, no metallographic evidence of plastic flow was experimentally observed.
3. The point of fiber/matrix debonding in the [±45°]₂ₛ and [±30°]₂ₛ orientations was accurately predicted by using an arbitrary stress criterion in which both the average shear stress and radial stress at the interface acted to separate the fiber and matrix.
4. Due to the large failure strains, the damage in the [±45°]₂ₛ specimens was extensive and consisted of fiber/matrix debonding, matrix and fiber cracking, necking of matrix slivers between the fibers, and possible matrix plasticity. Matrix plasticity was observed in the form of slip bands when tested at a slower strain rate (10^{-5} s^{-1}) at 427°C.

Acknowledgments

The authors would like to thank R. Shinn for his technical help with the strain gage experiments and R. Corner for performing the fractography and assisting with the tensile experiments. A special thanks goes to T. Leonhardt, M. Miller, and W. McCort for their diligence in preparing the seemingly endless number of metallographic samples. The authors would also like to thank Dr. W. S. Johnson of NASA Langley for supplying AGLPLY and for his valued technical advice regarding its use.

References

[1] Lerch, B. A. and Saltsman, J. F., "Tensile Deformation Damage in SiC Reinforced Ti-15-3," NASA TM-103620, 1991.

[2] Lerch, B. A., Hull, D. R., and Leonhardt, T. A., "Microstructure of a SiC/Ti-15-3 Composite," *Composites,* Vol. 21, No. 3, May 1990, pp. 216–224.

[3] Lerch, B. A., Gabb, T. P., and MacKay, R. A., "Heat Treatment Study of the SiC/Ti-15-3 Composite System," NASA TP-2970, 1990.

[4] Johnson, W. S., Lubowinski, S. J., and Highsmith, A. L., "Mechanical Characterization of Unnotched SCS/Ti-15-3 Metal Matrix Composites at Room Temperature," *Thermal and Mechanical Behavior of Metal Matrix and Ceramic Matrix Composites, ASTM STP 1080,* J. M. Kennedy, H. Moeller, and W. S. Johnson, Eds., American Society for Testing and Materials, Philadelphia, 1990.

[5] Bahei-El-Din, Y. A. and Dvorak, G. J., "Plasticity Analysis of Laminated Composite Plates," *Journal of Applied Mechanics,* Vol. 49, No. 4, December 1982, pp. 740–746.

[6] Johnson, W. S., private communication, NASA Langley Research Center.

Chian-Fong Yen[1] and Kent W. Buesking[2]

Material Modeling for Unidirectional Glass and Glass-Ceramic Matrix Composites with Progressive Matrix Damage

REFERENCE: Yen, C.-F. and Buesking, K. W., "**Material Modeling for Unidirectional Glass and Glass-Ceramic Matrix Composites with Progressive Matrix Damage,**" *Composite Materials: Fatigue and Fracture, Fourth Volume, ASTM STP 1156,* W. W. Stinchcomb and N. E. Ashbaugh, Eds., American Society for Testing and Materials, Philadelphia, 1993, pp. 176–194.

ABSTRACT: Material models have been developed for characterizing the nonlinear stress-strain response of unidirectional glass and glass-ceramic matrix composites associated with progressive transverse matrix cracks perpendicular to fibers and ultimate composite failure. The models utilize a combined micromechanical and statistical approach. A simple micromechanical model has been used to evaluate the fiber and matrix stresses in the presence of transverse matrix cracks, fiber-matrix debonding, and residual stresses. Utilizing the chain-of-bundles approach, statistical models for matrix failure and fiber failure have been introduced to relate the density of transverse matrix cracks to the applied composite stress and to compute the final composite strength, respectively. Procedures to determine the required statistical and mechanical material parameters from experimental data have been established. The material models that are developed are applied to a glass matrix composite (C/borosilicate) and a glass-ceramic matrix composite (SiC/LAS).

KEY WORDS: glass matrix composite, ceramic matrix composite, micromechanics, statistical failure model, progressive matrix damage

A variety of continuous fiber-reinforced glass and glass-ceramic matrix composites (GCMCs) have been successfully developed [*1–4*]. In the GCMCs, brittle matrices are usually reinforced with high-strength carbon or SiC fibers to enhance the composite strength. When a unidirectional composite is subjected to an axial tensile load, the low-strain capability of the brittle matrix usually results in matrix cracking initiated at a low stress level. The matrix cracking occurs as microcracks that form perpendicular to the fibers throughout the specimen's cross section. The unbroken fibers, with or without interface debonding, will bridge the cracks and take up the extra stress released by the matrix. The microcracks occur at multiple matrix sites and are somewhat equally spaced. Eventually at one site the microcrack widens to form a primary crack which exhibits considerable fiber pullout before ultimate fiber fracture [*5*].

In a GCMC, several different "strength" levels may be of interest to the designer for various structural applications. These include the onset of local damage, the proportional limit, and the ultimate strength. While the stress level at which overall failure occurs will define an ultimate load, the levels associated with the progression of matrix cracks and interface debonding may be of interest as this may cause the overall tensile modulus to become reduced below required stiffness levels. Furthermore, recent work by Kim and

[1]Materials Science Corp., 500 Office Center, Suite 250, Fort Washington, PA 19034.
[2]Currently at MSNW, Inc., P.O. Box 715, Blue Bell, PA 19422.

Pagano [6] has shown that matrix cracking begins at a stress level much lower than the proportional limit. Although the apparent linear response between the onset of matrix cracking and the proportional limit may be attributed to the very low crack density, the evolution of the matrix damage in this region may play an important role on the residual strength for GCMCs used in thermostructural applications. This is because the surrounding atmosphere penetrating through the microcracks may cause fiber degradation under high temperatures and significantly reduce the strength of the composites [7].

Attention has been focused on modeling the matrix cracking of GCMCs [8–12]. In these references, the predicted first matrix crack stress level may be recognized as a lower bound steady-state value; however, no consideration is given to the effects of the distribution and progression of the remaining matrix cracks on the composite response. Since glass matrices are known to have pronounced statistical distributions of flaws or imperfections, loading beyond the microcracking threshold causes the matrix crack density to increase. On the other hand, a statistical chain-of-bundles failure model which considers the cumulative damage of composites has been proposed by Rosen [13]. The use of this model for materials having dispersed regions of fiber failure was treated in Refs 13 through 17 for epoxy matrix composites. In the recent work of Sutcu [18] and Curtin [19], models are developed to predict the ultimate tensile strength of ceramic matrix composite materials in terms of statistical fiber strength. These models consider the effect of multiple matrix cracking on the composite ultimate strength. However, the influence of the progression of matrix damage on the overall stress-strain response has not been reported. Probabilistic models have also been developed to characterize the crack growth in the matrix, e.g., Ref 20. These models emphasize transverse cracking in mainly 90° layers of epoxy composites.

In the work presented in this paper, material models have been developed for the stress-strain response of unidirectional GCMCs associated with: (1) onset, progression, and saturation of transverse matrix cracks perpendicular to fibers; (2) ultimate composite failure due to fiber failure; and (3) statistical variations in matrix and fiber strengths. To accomplish this we have utilized a combined micromechanical and statistical approach.

A micromechanical model given by Aveston and Kelly (A&K) [9] has been used to evaluate the fiber, matrix, and interface shear stresses with the presence of transverse matrix cracks and fiber-matrix debonding. Statistical models utilizing the approach provided by Rosen [13] and Zweben [14] have been generalized to account for the statistical strength distributions of fibers and matrix. The material models which are developed have been applied to a glass matrix composite (C/borosilicate) and a glass-ceramic matrix composite (SiC/LAS) with satisfactory results.

This paper starts with a brief description of the A&K model with the consideration of residual stresses. The statistical models for the progressive matrix failure and the final fiber failure are then formulated by utilizing the chain-of-bundles model. Finally, the models are used to characterize the stress-strain response of C/borosilicate and SiC/LAS.

Micromechanical Model

Consider a unidirectional composite being stressed in the fiber direction. The axial Young's modulus of the composite, E_c, may be computed from the axial moduli of the fiber and matrix, E_f and E_m, respectively, by the rule of mixture relation

$$E_c = V_f E_f + V_m E_m \tag{1}$$

where V_f and V_m are the volume fractions of the fiber and matrix, respectively. The matrix stress, σ_m, and fiber stress, σ_f, are then computed from the composite stress, σ_c

$$\sigma_m = \frac{E_m}{E_c} \sigma_c + \sigma_{mo} \tag{2a}$$

and

$$\sigma_f = \frac{E_f}{E_c} \sigma_c - \frac{V_m}{V_f} \sigma_{mo} \tag{2b}$$

where σ_{mo} is the residual stress in the matrix.

If the matrix fails at a stress, σ_{mu}, the first matrix crack in the composite is expected to be formed at the composite stress σ_c', from Eq 2a

$$\sigma_c' = \frac{E_c}{E_m} (\sigma_{mu} - \sigma_{mo}) \tag{3}$$

Corresponding to σ_c', the fiber stress, σ_f', can be computed from Eq 2

$$\sigma_f' = \frac{E_f}{E_m} \sigma_{mu} - \frac{E_c}{E_m V_f} \sigma_{mo} \tag{4}$$

At the matrix crack, the stress carried by the matrix, σ_{mu}, will be supported by the fibers that bridge the crack. The extra stress carried by the fibers is then σ_{mu}/V_f, which is transferred by interfacial shear back into the matrix over a distance x on each side of the crack. The interface shear stress at the crack can attain a high value. However, the interface bonding may remain intact until the debonding occurs at the interface when the shear value exceeds the interface strength.

The interface bonding-debonding behavior following the matrix crack is considered in the model of A&K. On the debonded interface, they assume that the fibers may be drawn through the matrix or that the matrix may slide over the fibers under a constant interface shear traction, τ', which may be much smaller than the interface shear strength, τ_u. Based on the A&K model [9], with the presence of a matrix crack and a debonded fiber length, l', from the crack surface, the stress distributions of fiber, σ_f, matrix σ_m, and interface shear, τ, expressed in terms of the distance from the crack along the fiber direction x are listed in Table 1. Note that, as an approximation, the matrix stress, σ_m, is assumed to be uniform in the transverse plane. It is seen from Table 1 that the assumption of a constant sliding shear, τ', results in the fiber and matrix stresses varying linearly in the debonded region, $0 \leq x \leq l'$, while exponential distributions are for σ_f, σ_m, and τ in the bonded region, $x' = x - l' \geq 0$.

If the matrix has a well-defined, single-value breaking stress, σ_{mu}, the additional stress carried by the fibers at the crack plane ($x = 0$) is

$$\Delta\sigma_0 = \frac{\sigma_c}{V_f} - \sigma_f' \tag{5}$$

where σ_c is the applied stress and σ_f' is the undisturbed fiber stress at the first matrix crack (Eq 4). Reduced by the interface shear traction $2\tau'l'/r_f$, the additional stress carried by the fibers at the debonded length, l', from the crack ($x' = 0$, Table 1) is

$$\Delta\sigma_0' = \Delta\sigma_0 - \frac{2l'\tau'}{r_f} \tag{6}$$

where r_f is the fiber radius. Since the interface shear stress is limited by the shear strength, τ_u, we have, from $\tau(x')$ in Table 1, $\tau_{max} = \tau(x' = 0) = \tau_u$. This leads to the relation

$$\Delta\sigma_0' = \frac{2\tau_u}{\sqrt{\phi}r_f} \tag{7}$$

From Eq 6, utilizing Eqs 4, 5, and 7, the debonded length can be computed from Eq 8.

$$l' = \frac{(\Delta\sigma_0 - \Delta\sigma_0')r_f}{2\tau'} \tag{8}$$

It can be seen from Eqs 7 and 8 that if debonding does not occur, then $l' = 0$ and $\Delta\sigma_0 = \Delta\sigma_0' < 2\tau_u/\sqrt{\phi}r_f$. The threshold stress, σ_c'', for the onset of debonding can be computed from Eqs 4, 5, and 7

$$\sigma_c'' = \frac{V_f E_f \sigma_{mu} - E_c \sigma_{mo}}{E_m} + \frac{2V_f\tau_u}{\sqrt{\phi}r_f} \tag{9}$$

From Eqs 3 and 9, the bonding between fibers and matrix will remain intact at first matrix crack if $\sigma_c'' > \sigma_c'$. That is, $2\tau_u/\sqrt{\phi}r_f \geq V_m\sigma_{mu}/V_f$.

By increasing the load beyond the level of initial matrix crack, the matrix will crack again in the bonded region. The distance, l'', at which $\sigma_m(x' = l'') = \sigma_{mu}$ can be obtained from $\sigma_m(x')$ in Table 1 is

$$l'' = -\frac{1}{\sqrt{\phi}} \ln\left[\frac{\Delta\sigma_0 - \dfrac{V_m}{V_f}\sigma_{mu}}{\Delta\sigma_0'}\right] \tag{10}$$

TABLE 1—Stress distribution of fiber, matrix, and interface shear in a unidirectional composite with the presence of a matrix crack and a debonded fiber length (l') from the crack plane [9].[a]

Stress Component	Debonded Region, $0 \leq x \leq l'$	Bonded Region, $x' = x - l' \geq 0$
Fiber, σ_f	$\sigma_f' + \Delta\sigma_0 - \dfrac{2\tau'}{V_f}x$	$\sigma_f' + \Delta\sigma_0' \exp(-\sqrt{\phi}x')$
Matrix, σ_m	$\dfrac{2\tau'}{r_f}\dfrac{V_f}{V_m}x$	$\dfrac{2\tau'V_f l'}{r_f V_m} + \dfrac{V_f}{V_m}\Delta\sigma_0'[1 - \exp(-\sqrt{\phi}x')]$
Interface shear, τ	τ'	$\dfrac{r_f}{2}\Delta\sigma_0'\sqrt{\phi}\exp(-\sqrt{\phi}x')$

[a]Where
$$\phi^{1/2} = \left(\frac{2G_m E_c}{E_f E_m V_m}\right)^{1/2}\frac{1}{r_f[\ln(R/r_f)]^{1/2}}, \quad R = r_f[\pi/(2\sqrt{3}V_f)]^{1/2}, \text{ from Ref 9,}$$
$\Delta\sigma_0 = \sigma_c/V_f - \sigma_f'$, and
$\Delta\sigma_0' = \Delta\sigma_0 - 2\tau'l'/r_f$.

The minimum crack spacing, l, at a stress level, σ_c, is then equal to $l' + l''$. The maximum crack spacing is $2l$ [9].

The average composite strain in a partially bonded composite with constant crack spacing can be determined from the fiber strain averaged over a representative region between two adjacent cracks. Let the crack spacing be δ, where $l \leq \delta \leq 2l$, and the average strain can be computed from the fiber stress, σ_f, given in Table 1. That is

$$\epsilon_c = \epsilon_1 + \epsilon_2 + \epsilon_2 \tag{11}$$

$$\epsilon_1 = \frac{\sigma'_c}{E_c}$$

$$\epsilon_2 = \frac{(\Delta\sigma_0 + \Delta\sigma'_0)l'}{\delta E_f}$$

and

$$\epsilon_3 = \frac{2}{\delta E_f} \int_0^\zeta (\sigma_f - \sigma'_f)\, dx = \frac{2\Delta\sigma'_0(1 - e^{-\sqrt{\phi}\zeta})}{\delta\sqrt{\phi}E_f}$$

where the bonded length is $2\zeta = \delta - 2l'$. Note that ϵ_1 is the strain for the undisturbed state, and ϵ_2 and ϵ_3 are, respectively, the additional strains in the debonded and bonded regions due to the matrix cracks with crack spacing, δ. In the case of no interface debonding, we have $l' = \epsilon_2 = 0$ and $\zeta = \delta/2$.

Increasing the applied load beyond the formation of the first matrix crack results in increasing the debonding length and decreasing the crack spacing. The matrix cracking is then expected to reach a saturation state when the matrix is completely debonded from the fibers. In loading the composite beyond the matrix crack saturation state, the fibers will be stretched further and will slip through the blocks of cracked matrix, which can take no further share of load. Consequently, the axial modulus of the cracked composite will be $E_f V_f$.

The minimum crack spacing, l_s, at the crack saturation can be derived from the expressions for l' and l''. By solving $\Delta\sigma_0$ from $d(l' + l'')/d(\Delta\sigma_0) = 0$ and substituting this $\Delta\sigma_0$ back into Eqs 8 and 10, we have

$$l_s = \frac{1}{\sqrt{\phi}}\left(1 + \ln\frac{\tau_u}{\tau'} - \frac{\tau_u}{\tau'}\right) + \frac{V_m}{V_f}\frac{\sigma_{mu}V_f}{2\tau'} \tag{12}$$

This establishes l_s as a function of τ_u/τ', σ_{mu}/τ', and $\sqrt{\phi}$, which is a parameter including the constituent moduli. Considering the special case of $\tau_u = \tau'$, we have $l_s = l_0 = V_m\sigma_{mu}V_f/2V_f\tau'$. This is the minimum crack spacing for an initially unbonded composite given in Ref 8. Since it is physically impossible to have an interface shear strength smaller than the sliding shear τ', we then conclude that $l_s \leq l_0$.

The influence of residual stress on the composite stress-strain response can be readily seen from Eq 3. A tensile residual stress in the matrix reduces the applied stress required for matrix cracking, while a compressive matrix residual stress delays the onset of matrix cracking. The residual stress also has significant influence on the unloading behavior. It can be readily shown that unloading from the stress-strain curve assuming zero interface shear traction, i.e., $\tau' = 0$, results in: (1) a positive residual strain of $\sigma_{mo}V_m/E_f V_f$ if the matrix is under a residual tension; and (2) the closure of matrix cracks at a tensile composite stress of $\sigma_{mo}E_c/E_m$ if the matrix is under a residual compression, [21].

The A&K model provides a simple approach for modeling brittle matrix composites by taking into account the interface bonding-debonding behavior. This model is well suited for the characterization of the glass and glass-ceramic matrix composites.

Statistical Model for Matrix Failure

Utilizing the chain-of-bundles approach, the model used to determine matrix failure in a GCMC subjected to a tensile load parallel to the fiber axis is given in Fig. 1. Instead of fiber bundle layers, the composite is considered to consist of a series of identical matrix layers. The axial dimension of a layer will be referred to as the "ineffective length," δ_m. Each layer is composed of representative matrix elements which are chosen based on the hexagonal arrangement of fibers as shown in Fig. 1b. Therefore, a composite of length L consisting of N fibers will have $M = L/\delta_m$ layers in the composite and N elements in each layer, for a total of MN matrix elements.

In this model, the matrix is treated as having a statistical distribution of flaws that can result in breaks of individual matrix elements at various stress levels. Because of the scattered flaws, the matrix elements break randomly throughout the body. However, due to the pronounced brittleness of the glass matrix, the failure of a matrix element in a layer usually results in crack propagation across the layer (see Fig. 1). For brittle matrices with statistical strength distributions, transverse matrix cracks can be initiated at low stress levels while the fiber-matrix interface may remain intact. Without interface debonding, fiber stress concentrations can occur where the transverse matrix crack tips intersect the fibers. Here, we consider that the fibers are strong enough to survive the stress concentration until the interface debonding occurs at higher stress levels.

Consider the strength of a representative volume element of matrix material ΔV_{mo} being characterized by a Weibull cumulative distribution function $F(\sigma_m)$ as

$$F(\sigma_m) = 1 - \exp\left[-\left(\frac{\sigma_m}{\alpha_m}\right)^{\beta_m} \right] \tag{13}$$

where α_m and β_m are the Weibull parameters for the matrix. When the most critical flaw in a layer starts to propagate under a certain load, we consider that the crack becomes unstable and can penetrate through the layer. The strength of a layer is thus determined by the strength of the weakest representative volume element (RVE). Consequently, the probability cumulative distribution function for the strength of a layer, $G(\sigma_m)$, with N elements in the layer, has the form

$$G(\sigma_m) = 1 - [1 - F(\sigma_m)]^N = 1 - \exp\left[-N\left(\frac{\sigma_m}{\alpha_m}\right)^{\beta_m} \right] \tag{14}$$

In a composite consisting of M layers, the number of layers, p, expected to fail under the matrix stress, σ_{mp}, based on the noninteractive assumption, e.g., Ref 14, can be estimated as

$$p = MG(\sigma_{mp}) = M\left\{ 1 - \exp\left[-N\left(\frac{\sigma_{mp}}{\alpha_m}\right)^{\beta_m} \right] \right\} \tag{15}$$

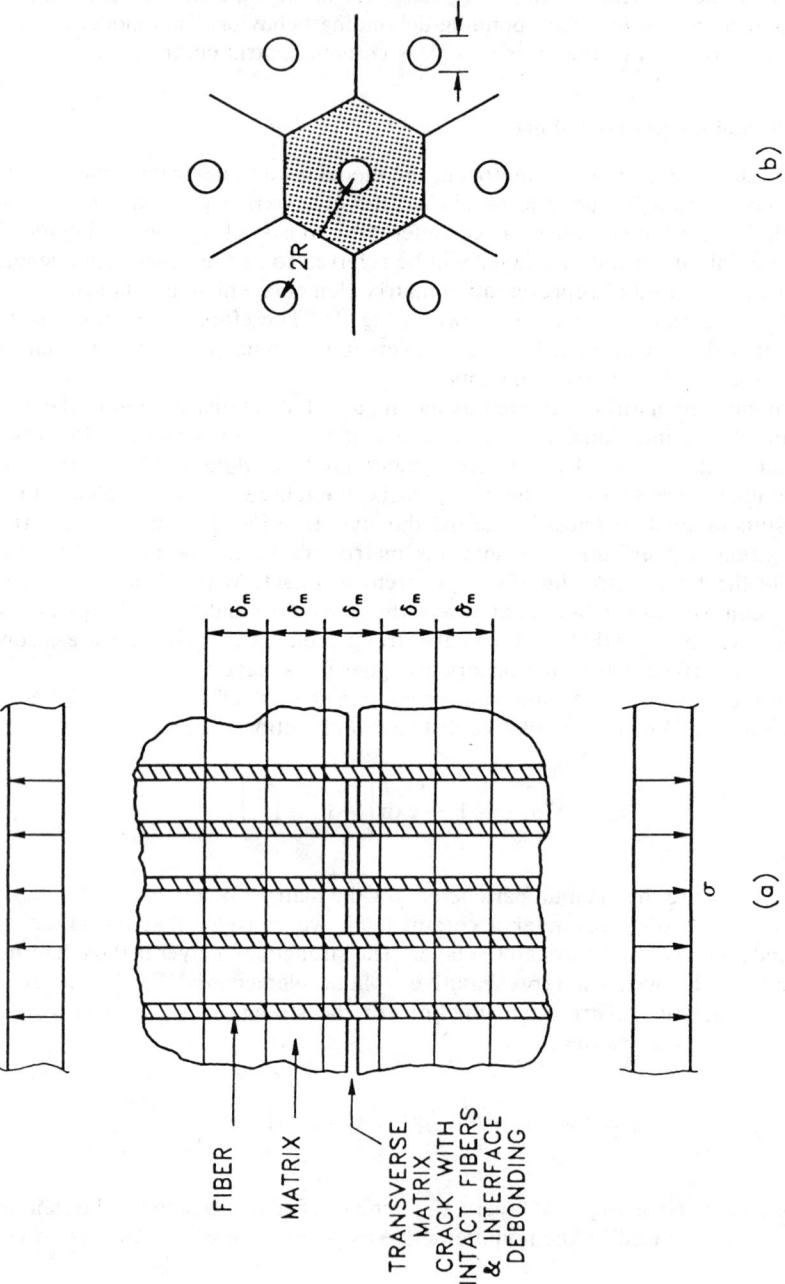

FIBER

MATRIX

TRANSVERSE
MATRIX
CRACK WITH
INTACT FIBERS
& INTERFACE
DEBONDING

(a)

(b)

FIG. 1—*Statistical model for matrix failure.*

where $1 \le p \le M$. By computing the composite stress, σ_{cp}, from the matrix stress, σ_{mp}, using Eq 3, Eq 15 provides the relationship between the matrix crack density and the applied stress.

Note that the Weibull parameters for matrix strength must be obtained before the statistical model can be used to characterize the composite stress-strain response. The constants α_m and β_m can be evaluated by using experimental strength versus length data, as suggested in Ref 13 for fibers. However, it is expected to be extremely difficult to duplicate the in-situ matrix behavior by conducting experiments using monolithic specimens. Alternately, the matrix Weibull parameters can be obtained from the crack density versus applied stress relation measured from tension tests of the composite (e.g., Ref 6). The layer length, δ_m, is then the average crack spacing in the gage length at the saturation state of matrix cracking. It is noted from Ref 22 that the theoretical value for the mean crack spacing is $1.34 l_s$ where l_s is the minimum crack spacing evaluated at the crack saturation.

The statistical model can now be used to simulate the stress-strain response of a glass matrix composite with progressive matrix cracks. A composite of gage length L is separated into M layers with N elements in each layer. The matrix stress at the onset of the first matrix crack, σ_{mi}, is obtained from

$$MG(\sigma_{mi}) = 1 \tag{16}$$

At a loading level at which the crack number is p, $1 \le p \le M$, the matrix breaking stress, σ_{mp}, can be computed from Eq 15. The corresponding values for composite stress, σ_{cp}, the threshold debonding stress, σ_{cp}'', and the debonding length, l_p', can be computed from Eqs 3, 9, and 8, respectively, with σ_{mu} replaced by σ_{mp}. If $\sigma_{cp} > \sigma_{cp}''$, debonding will occur, and the composite strain can be approximately computed as

$$\epsilon_{cp} = \epsilon_1 + \frac{p}{M}(\epsilon_2 + \epsilon_3) \tag{17}$$

where ϵ_1, ϵ_2, and ϵ_3 are given in Eq 11 with σ_f', $\Delta\sigma_0$, and $\Delta\sigma_0'$ computed from Eqs 4, 5, and 6, respectively, using the updated values of σ_{mp} and σ_{cp}. If $\sigma_{cp} \le \sigma_{cp}''$, the fiber/matrix bond remains intact. Then, we have $\epsilon_2 = 0$ and $l_p' = 0$.

Note that in Eq 17, the additional strain in the bonded region, ϵ_3, is computed only within a cracked layer, i.e., $\delta = \delta_m$ and $\zeta = \frac{1}{2}\delta_m - l_p'$ in Eq 11. This is because for most of the GCMCs considered here, the stress variation outside a cracked layer is negligible due to the exponentially decreasing fiber stress in the bonded region shown in Table 1. As an example, a C/borosilicate of 40% fiber volume fraction provides $r_f\sqrt{\phi} = 1.4$. The stress change in a fiber then drops below 1% of $\Delta\sigma_0'$ at a distance $x' = 5r_f$. Considering uniform stresses in the unbroken layers leads to the use of the noninteractive assumption to obtain the crack number in Eq 15. Furthermore, the interface shear strength, τ_u, is likely to have its own statistical distribution. However, for simplicity, τ_u is assumed to be constant in this paper. The matrix cracks are saturated at

$$MG(\sigma_{ms}) = qM \tag{18}$$

where q should approach 1; however, $q = 0.99$ has been used here. Beyond the crack saturation stress, the matrix carries no extra load and, thus, the composite elastic modulus is determined from the fiber elastic modulus as $E_{cs} = E_f V_f$.

Statistical Model for Fiber Failure

The model is formulated by assuming that the fiber failure occurs after the matrix cracks have been fully saturated. This assumption is justified for the glass matrix composites because the brittle matrices are always reinforced with strong fibers. Furthermore, the matrix cracks are usually deflected by the matrix/fiber interface because of the relatively weak interfacial bond strength in these composites. This prevents the matrix cracks from penetrating into the fibers and allows the strength of the fiber to be fully utilized.

We again utilize the chain-of-bundles approach. A composite is considered to consist of a series of identical fiber links (Fig. 2). The axial dimension of a link is characterized by the ineffective length, δ_f. As shown in Fig. 2, a layer of the link may consist of several layers of cracked matrix for which the mean crack spacing is given by the matrix failure model at the crack saturation state. By noting that the fiber-matrix interface is assumed fully debonded at crack saturation, the stress distribution along an unbroken fiber can be found using the A&K micromechanical model (see Fig. 2). Along a broken fiber, the fiber has zero stress at the break. If we utilize the same constant interface shear as assumed for the matrix failure, the fiber stress then increases linearly from zero at the fiber break to the average load level carried by an unbroken fiber at the layer boundary. The layer thickness, δ_f, can be obtained by balancing the fiber surface shear force of $2\pi r_f(\delta_f/2)\tau_i$ with the

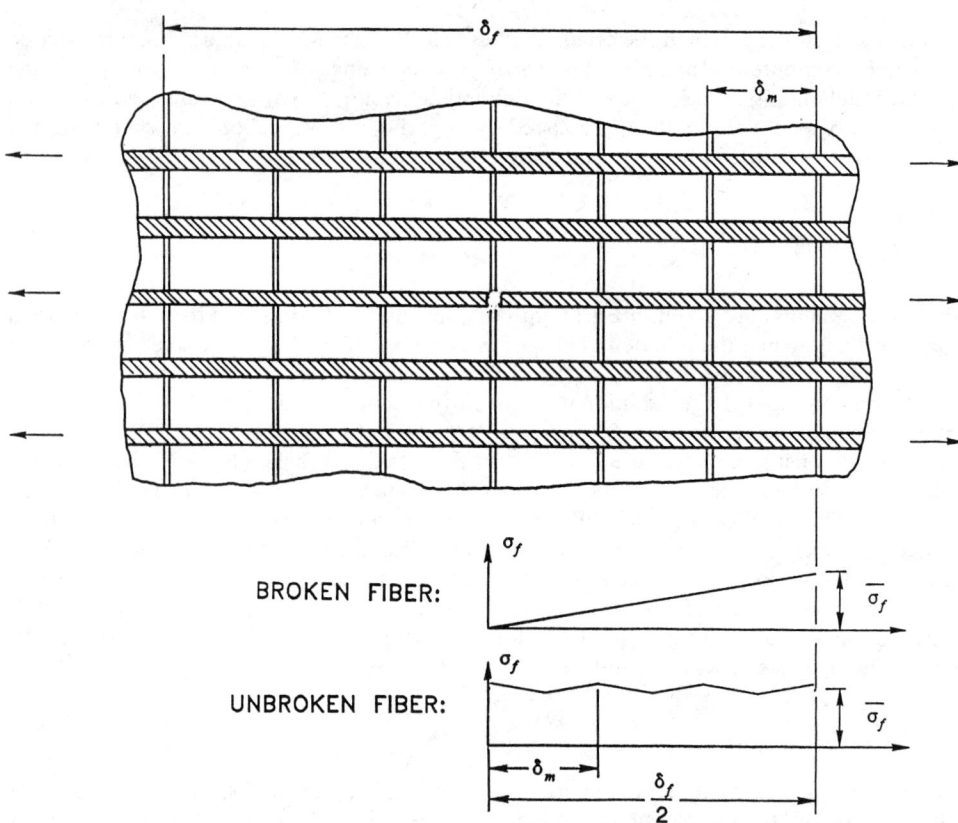

FIG. 2—*Statistical model for fiber failure.*

average fiber load of $\pi r_f^2 \overline{\sigma}_f$, where $\overline{\sigma}_f$ is the average fiber stress (see Fig. 2). That is,

$$\delta_f = \frac{r_f \overline{\sigma}_f}{\tau_i} \tag{19}$$

We assume that the strength of fibers of length L is characterized by a Weibull distribution

$$F(\sigma_f) = 1 - \exp\left[-\left(\frac{\sigma_f}{\alpha_f}\right)^{\beta_f}\right] \tag{20}$$

where α_f and β_f are the Weibull parameters for the fiber. The statistical distribution of strength for a fiber link, σ_f, can then be scaled as

$$F_{\delta_f}(\sigma_f) = 1 - \exp\left[-\frac{\delta_f}{L}\left(\frac{\sigma_f}{\alpha_f}\right)^{\beta_f}\right] \tag{21}$$

The strength distribution for a fiber link can be further generalized by taking into account the periodic distribution of fiber stress as given in Ref *21*. This, however, would lead to a complicated formulation that requires a numerical procedure to obtain the composite strength.

A simple estimate of the composite strength can be obtained by first replacing the periodic fiber stress profile by its average stress. We then consider that the composite load is evenly shared by the surviving fibers within a link of the fiber bundle. By neglecting the load carried by the broken fibers in the link, the load-carrying capacity of the composite is related to the average fiber stress, $\overline{\sigma}_f$, by

$$\frac{\sigma_c}{V_f} = \overline{\sigma}_f[1 - F_{\delta_f}(\overline{\sigma}_f)] + \frac{\tau'\delta_m}{2r_f} \tag{22}$$

where the second term represents the portion of stress carried by the cracked matrix blocks. The current probability of fiber failure, $F_{\delta_f}(\sigma_f)$, is given by Eq 21, and the layer length of a bundle link, δ_f, is given by Eq 19.

Maximizing Eq 22 with respect to $\overline{\sigma}_f$ yields the composite strength, σ_{cu}, as

$$\frac{\sigma_{cu}}{V_f} = \alpha_f\left[\frac{\tau'L}{r_f\alpha_f(\beta_f + 1)e}\right]^{\frac{1}{\beta_f + 1}} + \frac{\tau'\delta_m}{2r_f} \tag{23}$$

where e is the natural logarithm base. The corresponding maximum average fiber stress, $\overline{\sigma}_{fm}$, is

$$\overline{\sigma}_{fm} = \alpha_f\left[\frac{\tau'L}{r_f\sigma_f(\beta_f + 1)}\right]^{\frac{1}{\beta_f + 1}} \tag{24}$$

Equation 23 can now be used to compute the composite strength based on the fiber failure by knowing the Weibull parameters for fibers α_f and β_f, the value of interface shear traction, τ', the mean fiber radius, r_f, and the matrix crack spacing, δ_m, which is determined from the statistical model for matrix failure. Due to the neglect of the load carried by broken fibers, the result computed from Eq 23 is a lower bound.

For comparison, we also include the simple formulation provided by Curtin [19]. Rather than use the chain-of-bundles approach, Curtin [19] employs a Weibull expression for the fraction of fiber breaks occurring in a critical gage length, δ_c, to derive the composite strength in the form

$$\frac{\sigma_{cu}}{V_f} = \sigma_c \left(\frac{\xi}{\beta_f + 2}\right)^{\frac{1}{\beta_f + 1}} \frac{\beta_f + 1}{\beta_f + 2} + \frac{\tau'\delta_m}{2r_f}$$

and   (25)

$$\sigma_c = \left(\frac{\alpha_f^{\beta_f}\tau'L}{r_f}\right)^{\frac{1}{\beta_f + 1}}$$

where $\xi = 2$ if the load carried by the broken fiber is considered, and $\xi = 1$ by neglecting the load-carrying capacity of the broken fibers. Equation 25 yields an upper bound result for $\xi = 2$ and a lower bound result for $\xi = 1$. In Eq 25 we also include the portion of load carried by the cracked matrix which has been excluded in Ref 19. Comparisons with other statistical models are given in Ref 19.

The simplified result of Eq 25 provides an accurate approximation for a more rigorous but complicated approach proposed by Sutcu [18]. Numerical results indicate that the lower bound result of Eq 23 is bounded by the results of Eq 25 with $\xi = 1$ and 2 and is closer to that of $\xi = 1$. Both the lower bound of Eq 23 and the upper bound of Eq 25 ($\xi = 2$) will be applied to predict the strengths of GCMCs in the next section.

Applications of the Model

Considered here for correlation study are the unidirectional composites of borosilicate matrix reinforced with Hercules high-modulus carbon fibers (C) and lithium aluminosilicate (LAS) reinforced with Nicalon[3] (SiC) fibers. The C/borosilicate has a tensile residual stress in the matrix, while the SiC/LAS has a compressive matrix residual stress. Stress-strain curves of tension tests for the C/borosilicate of various fiber volume fractions are reported in Refs 6 and 23. Experimental results of SiC/LAS are given in Refs 5, 6, 24, and 25.

Applying the matrix and fiber failure models requires material parameters including Weibull parameters for matrix (α_m, β_m) and fiber (α_f, β_f), matrix elastic modulus (E_m) and Poissons' ratio (ν_m), fiber modulus (E_f), matrix residual stress (σ_{mo}), interface shear traction (τ'), and interface shear strength (τ_u). Among them, the constituent elastic properties E_f, E_m, and ν_m are commonly known. However, only a limited amount of mechanical measurements on GCMCs exist in the literature that provide the required in-situ fiber strength data and interface shear traction. There is no published data to our knowledge on the in-situ matrix strength distribution and the interface shear strength. Alternatively, the matrix Weibull parameters (α_m, β_m) and the interface shear strength (τ_u) were estimated from the experimental stress-strain curves utilizing the procedures described in the following paragraphs.

The residual stress, σ_{mo}, is computed by performing a thermoelastic micromechanical analysis [21] on the composite with a given range of stress-free temperature. Note that the cooling process analysis assumes a perfectly bonded fiber-matrix interface. The stresses at

---

[3]Nicalon is a trademark of Nippon Carbon Co. Ltd., Japan.

the onset of initial matrix cracking, σ_{ci}, and at the matrix crack saturation σ_{cs} are identified from the experimental stress-strain curve. The corresponding matrix stresses, σ_{mi} and σ_{ms}, are respectively computed from σ_{ci} and σ_{cs} using Eq 2 and the residual stress, σ_{mo}. From Eqs 14, 16, and 18, the Weibull parameters (α_m, β_m) can be determined. Finally, the proper value for τ_u that provides the best fit to the experimental stress-strain curve is chosen. Note that the total number of representative layers (M) in a gage length (L) is determined by the layer thickness δ_m, i.e., $M = L/\delta_m$. δ_m is chosen as the average crack spacing at the matrix crack saturation state (i.e., $\delta_m = 1.34\,l_s$, where l_s is computed from Eq 12 with σ_{mu} replaced by σ_{ms}). Furthermore, it will become clear later that the proportional limit of a stress-strain curve is significantly influenced by τ_u. Therefore, it is instructive that τ_u can be chosen based on the experimental proportional limit. The values for r' and τ_u can be further verified by the measured crack spacing.

C/Borosilicate

The experimental stress-strain results for $V_f = 42\%$ given in Ref 6 are shown in Fig. 3. The results for $V_f = 40\%$ and 55% from Ref 23 are plotted in Fig. 4. Typical stress-strain response can be observed from these figures. An initial linear response is followed by a nonlinear stage due to the matrix crack initiation, progression, and saturation. Once the matrix microcracking is completed, linear behavior is again resumed until the composite failure. Note that, since the specimen was unloaded before final failure, the maximum applied stress shown in Fig. 3 does not represent the composite ultimate strength.

The material properties and the test parameters are summarized in Table 2. We chose the parameters α_m, β_m, τ_u, and τ' that could provide the best fit of the theoretical curve to the experimental data in Fig. 3. The parameters were then used to predict the stress-strain curves given in Fig. 4. Over the temperature range from 20 to 600°C, the coefficient of

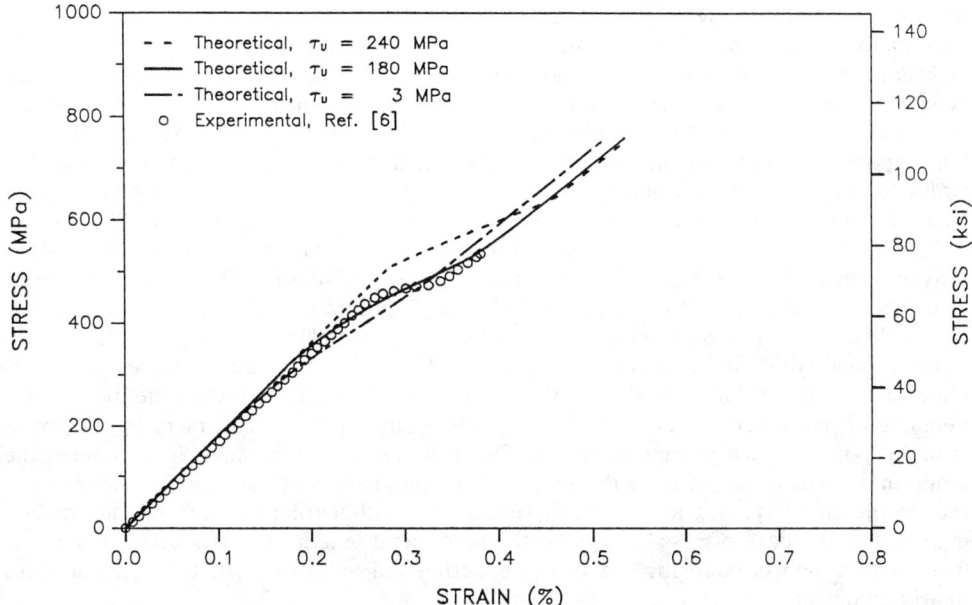

FIG. 3—*Axial stress-strain curves for unidirectional HMU/borosilicate with* $V_f = 42\%$, $E_f = 350$ *MPa,* $E_m = 63$ *MPa,* $\alpha_m = 1041$ *MPa,* $\beta_m = 8.0$, *and* $\tau' = 3$ *MPa.*

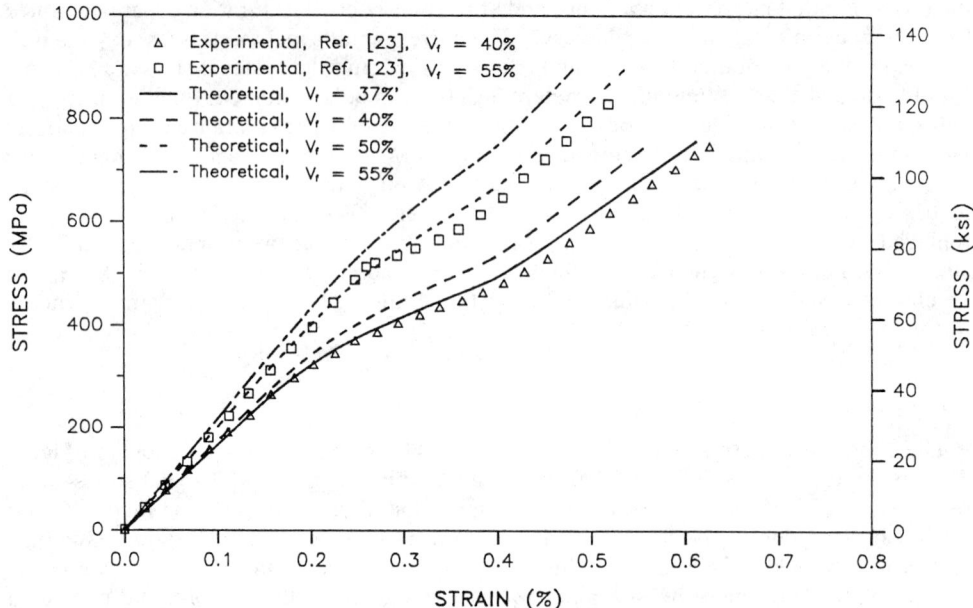

FIG. 4—*Axial stress-strain curves for unidirectional HMU/borosilicate with various fiber volume fractions, E_f = 350 MPa, E_m = 63 MPa, α_m = 1041 MPa, β_m = 8.0, τ_u = 180 MPa, and τ' = 3 MPa.*

thermal expansion (CTE) of the borosilicate matrix is approximately $3.2 \times 10^{-6}/°C$ [26], while the CTE of the high-modulus fiber is commonly known to depend on the temperature. The temperature-dependent CTEs of the fiber listed in Table 2 were obtained from the data for similar PAN fiber given in Ref 27. The computed matrix residual stress (σ_{mo}) is 77 MPa. The initial matrix microcracking occurs at 123 MPa, which was experimentally determined in Ref 6 from the first noticeable acoustic emission. Note that this microcracking onset stress (σ_{ci}) of 123 MPa is much lower than the 350 MPa proportional limit reported in Ref 6 (see also Fig. 3). Matrix cracking was considered to be saturated at stress (σ_{cs}) of 580 MPa. Based on the values of σ_{mo}, σ_{ci}, and σ_{cs}, the matrix stresses corresponding to the onset of cracking (σ_{mi}) and crack saturation (σ_{ms}) are 119 and 287 MPa, respectively. Due to the lack of experimental data, we assumed τ' = 3 MPa. This is compatible with the values measured for other GCMCs [25,28]. Finally, the interface shear strength (τ_u) of 180 MPa provided the best fit to the experimental data shown in Fig. 3. The Weibull parameters were computed as α_m = 1041 MPa and β_m = 8.

Theoretical stress-strain curves for τ_u = 3, 180, and 240 MPa are compared in Fig. 3. Note that the curve for $\tau_u = \tau'$ = 3 MPa replicates the results based on the model with unbonded fibers given by Aveston, Cooper, and Kelly [8]. The effect of τ_u on the stress-strain response is clearly demonstrated in Fig. 3. It is seen that increases in τ_u increase the strain increment resulting from the matrix cracking. This is because higher values of τ_u reduce the crack spacing and, in turn, reduce the load-bearing capacity of the cracked matrix. For the case of τ_u = 180 MPa, the theoretical length of matrix crack spacing is 0.140 mm, which is compatible with the experimental values reported for similar glass matrix composites [6,29].

Comparing of the theoretical curve to the experimental results in Fig. 3 also shows that τ_u affects the proportional limit significantly. Based on the current theory, numerical

TABLE 2—*Material properties and test parameters for C/borosilicate and SiC/LAS.*

	C/Borosilicate		SiC/LAS	
Parameters	Value	Reference	Value	Reference
Fiber modulus, E_f (GPa)	350	[4]	190 200	[24]
Matrix modulus, E_m (GPa)	63		86	
Poissons' ratio	0.22	[26]	0.17	[24]
Fiber radius, r_f (μm)	4	[6]	8	[24]
Matrix Weibull parameters,				
α_m (MPa)	1041	...[a]	1734	...[a]
β_m	8		4.3	
Interface shear traction, τ'				
(MPa)	3	...[a]	2, 3	...[a]
Interface shear strength, τ_u				
(MPa)	180	...[a]	172–193	...[a]
Stress-free temperature				
range, ΔT (°C)	− 600	[6]	− 600	...[a]
Fiber thermal expansion				
coefficient, 10^{-6}/°C	− 1.15 at 100°C			
	− 0.61 at 300°C			
	− 0.07 at 600°C	[27]	3.9	[6]
	0.23 at 800°C			
Matrix thermal expansion				
coefficient, 10^{-6}/°C	3.2	[26]	1.5	[6]
Cross-sectional area of				
composite, A_{sp} (mm × mm)	3 × 7.6	[6]	3 × 7.6	[6]
	2 × 8	[23]	2 × 8	[24]
Gage length, L (mm)	25	[23]	25	[24]

[a]Estimated value.

results indicate that before the debonding occurred the additional strain introduced by the bonded matrix cracks was negligible compared to the total composite strain at the same stress. However, when loading beyond the interface bonding strength, the stress released by debonding introduces a significant amount of additional strain that results in the nonlinear stress-strain response. Consequently, the stress at the debonding initiation can be recognized as the proportional limit. In the case of $\tau_u = 180$ MPa, the debonding starts at 330 MPa, where the crack density is about 18% of that at the final saturation state, while the additional strain from the bonded matrix cracking is only 0.3% of the total strain.

The parameters list in Table 2 were used to compute the theoretical curves shown in Fig. 4 to simulate the experimental results of $V_f = 40$ and 55%. In each case, the Weibull parameter, α_m, was rescaled to the new RVE of the matrix, which is determined by the fiber volume fraction and the layer length. Also included in Fig. 4 are the theoretical curves for $V_f = 37$ and 50%. Those fiber volume fractions were obtained by the best fit to the initial moduli of the experimental data for $V_f = 40$ and 55%, respectively. It is seen that the experimental results of $V_f = 40$ and 55% are well characterized by the theoretical curves of $V_f = 37$ and 50%, respectively. The discrepancies in fiber volume fractions are less than 10% and are possibly the result of fabrication variations of the specimens.

Nicalon/LAS

The second material considered here is the unidirectional SiC/LAS. The experimental stress-strain data for a LAS-III composite of 39% fiber volume fraction obtained from Ref 6 are shown in Fig. 5. Note that the specimen was unloaded at 400 MPa and then reloaded to failure. The data for a LAS-II composite of 46% fiber volume fraction reported in Ref *24* are shown in Fig. 6. While both LAS-II and LAS-III matrix materials consisted predominantly of standard LAS, they were tailored to be more compatible with the SiC-type fibers [*4*].

The material properties and the test parameters are summarized in Table 2. The parameters α_m, β_m, τ_u, and τ' together with the range of stress-free temperatures were estimated to provide the best fit to the experimental data in Fig. 5 and the measured lengths of crack spacing given in Refs *5* and *25*. The parameters were then used to predict the stress-strain curves given in Fig. 6.

To estimate the matrix residual stress (σ_{mo}), we first considered a stress-free temperature range of $-1000°C$ as cited in Ref *6*. A matrix residual stress of -147 MPa was computed by performing the thermoelastic analysis for the thermal mismatch between the matrix and fibers. Then, based on the measured stress for matrix crack initiation ($\sigma_{ci} = 200$ MPa) reported in Ref *6* and using Eq 2, the computed -147 MPa residual stress leads to the unlikely result that matrix cracking initiates at a compressive matrix stress ($\sigma_{mi} = -132$ MPa). Apparently, the use of a $-1000°C$ stress-free temperature overestimates the compressive residual stress and, in turn, results in the unreasonable compressive stress for matrix crack initiation. Due to the uncertainty in the stress-free temperature, the simulation is proceeded by recognizing the matrix residual stress as the maximum compressive initial stress in the matrix that could provide the best fit to the experimental stress-strain data. Following the aforementioned procedure and using $\sigma_{ci} = 200$ MPa and $\sigma_{cs} = 434$ MPa, we obtained the material parameters: $\sigma_{mi} = 44$ MPa, $\sigma_{ms} = 212$ MPa, $\sigma_{mo} = -90$ MPa, $\alpha_m = 1734$ MPa, $\beta_m = 4.3$, $\tau' = 2$ MPa, and $\tau_u = 172$ MPa. Note that $\tau' = 2$ MPa was chosen to provide an average crack spacing of 0.4 mm which was measured in

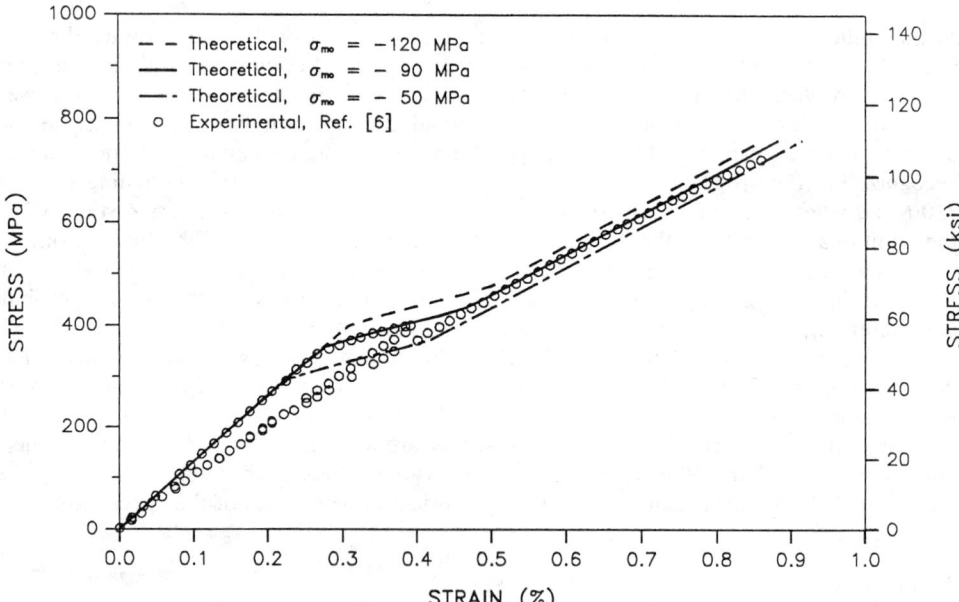

FIG. 5—*Axial stress-strain curves for unidirectional Nicalon/LAS-III with $V_f = 39\%$, $E_f = 200$ MPa, $E_m = 86$ MPa, $\alpha_m = 1734$ MPa, $\beta_m = 4.3$, $\tau_u = 172$ MPa, and $\tau' = 2$ MPa.*

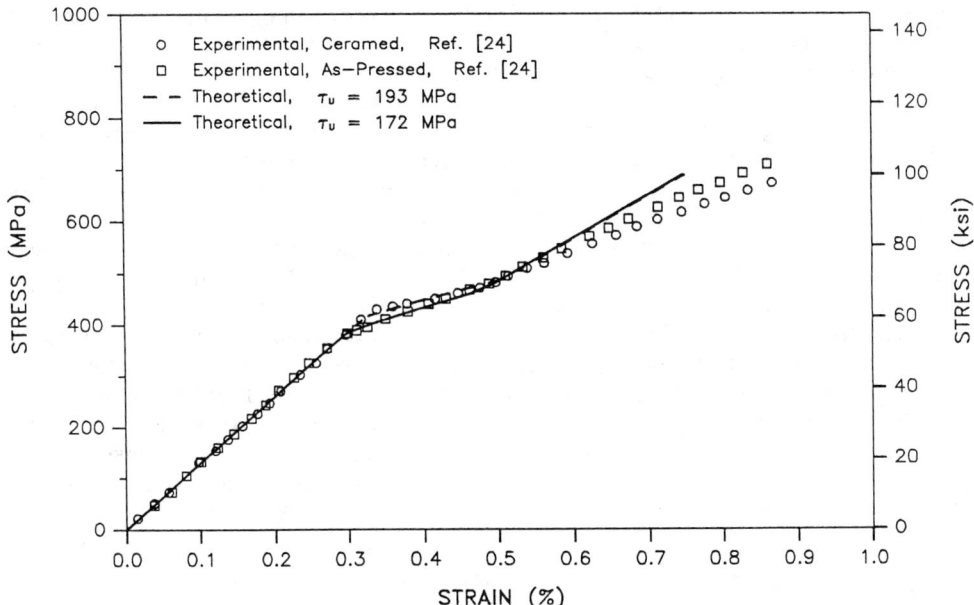

FIG. 6—*Axial stress-strain curves for unidirectional Nicalon/LAS-II with* V_f = 44%, E_f = 193 MPa, E_m = 86 MPa, α_m = 1734 MPa, β_m = 4.3, *and* τ' = 3 MPa.

Ref *25* for a SiC/LAS-III with V_f = 45%. The resulting theoretical stress-strain curve is plotted in Fig. 5.

Based on the thermoelastic analysis, -90 MPa residual stress corresponds to a range of stress-free temperature of $-600°C$ for the SiC/LAS-III with V_f = 39%. To demonstrate the effect of the residual stress on the stress-strain response, the curves for σ_{mo} = -120 and -50 MPa are included in Fig. 5. The increase in compressive matrix residual stress delays the onset of matrix crack and debonding and thus increases the proportional limit, Eq 9. The residual stress also affects the unloading stress-strain behavior for a cracked specimen. A compressive residual stress results in the closure of matrix cracks at a tensile composite stress. Based on the simple formulation of Ref *21* as discussed in the section of micromechanical model, the closure of matrix cracks occurs at 59 MPa, which is slightly higher than the value observed from Fig. 5 where the unloading stress-strain response rejoins the initial elastic loading path.

The material parameters estimated for the SiC/LAS-III were then used to correlate the data of SiC/LAS-II with some variations in the interface properties. First, τ' = 3 MPa with τ_u = 172 MPa were determined from the measured 0.3-mm average crack spacing and 300-MPa proportional limit reported in Ref *5* for a SiC/LAS-II with V_f = 40%. The estimated value τ' = 3 MPa is compatible with the measured τ' = 2 MPa for the SiC/LAS-II by the indentation method [28]. Utilizing τ' = 3 MPa and τ_u = 172 MPa, the experimental stress-strain data of the as-pressed specimen can be well correlated by the theory (Fig. 6). The creamed SiC/LAS-II, however, shows higher proportional limit, and it requires τ_u = 193 MPa (τ' = 3 MPa) to best match the experimental data.

Based on the above correlation studies for the SiC/LAS composites, the results indicate that the interface properties are influenced, to some degree, by the fabrication process and the slight variation in the constituents of the LAS matrices. Further research efforts are necessary to quantify these effects. It can also be observed from Fig. 6 that the theoretical curves can predict the nonlinear stress-strain data resulting from the matrix failure. However, the measured data are seen to deviate from the theoretical results after the saturation of matrix cracking. This is likely to be the result of the load-carrying capacity of fibers gradually degrading by the progression of fiber breakage. The ultimate composite failure can be estimated by the proposed statistical model of fiber failure.

In Ref 24, Nicalon fibers extracted from as-fabricated and creamed LAS-II matrix composites were carefully tested to determine the Weibull parameters (α_f, β_f) for the fiber strengths. The post-processed fiber strength data are shown in Table 3. The system parameters are listed in Table 2 with $\tau' = 3$ MPa and $\tau_u = 172$ MPa, which were obtained by the aforementioned correlation study of the matrix failure for the SiC/LAS-II system. The theoretical predictions based on the proposed lower bound model were computed from Eq 23, while the upper bound results of Curtin [19] were given by Eq 25 with $\xi = 2$. The theoretical results are compared with the experimental data in Table 3. It is seen that, while the prediction of Eq 25 provides good upper bounds to the experimental data, the prediction of Eq 23 provides satisfactory lower bound results. The only inconsistency is for Sample 2369(c), for which the lower bound result is slightly higher than the largest experimental value.

Concluding Remarks

The models developed in this paper have been used to characterize the stress-strain response of unidirectional GCMCs with satisfactory results. Utilizing the chain-of-bundles approach, statistical models for matrix failure and fiber failure have been introduced to relate the density of transverse matrix cracks to the applied composite stress and to compute the final composite strength, respectively. Based on the proposed models, the onset of matrix cracking can occur at a stress much lower than the proportional limit at which the fiber-matrix interface debonding is initiated. The apparent linear stress-strain response between the onset of matrix cracking and proportional limit is because the strain increment released by the bonded matrix cracks is negligible compared with the total strain. Loaded beyond the proportional limit, the increase in crack density and debonding length results in nonlinear stress-strain response until the saturation of matrix cracking. At this point the fiber-matrix interface is completely debonded. The stress-strain curve may resume its linear response until the ultimate fiber failure. The average matrix crack spacing

TABLE 3—*Tensile strength for SiC/LAS-II.*

| Sample | V_f, % | Fiber Weibull Parameters[a] | | Experimental[a] | Ultimate Strength, MPa | |
		a_m	b_m		Theoretical[c] Lower Bound	Theoretical[c] Upper Bound
2369 (p)[b]	46	1740	3.8	717–795	683	771
2369 (c)[d]	46	1740	2.7	641–681	690	804
2376 (p)[b]	44	1615	3.9	653–683	615	693
2376 (c)[d]	44	1632	3.1	664–686	626	719

[a]Experimental data from [24].
[b]As-pressed.
[c]$\tau' = 3$ MPa.
[d]Creamed.

after crack saturation is a function of the constituent elastic moduli, fiber volume fraction, matrix strength at crack saturation, interface shear strength, and shear traction. The final composite failure depends on the matrix crack spacing, interface shear traction, and Weibull parameters of the fibers. Procedures have been established to determine some in-situ material parameters including the matrix Weibull parameters, interface shear strength, and interface shear traction.

The models can also be applied to other continuous fiber-reinforced ceramic matrix composites (CMCs) such as the SiC/RBSN reported in Ref 30. The simple micromechanical model of A&K appears to be well suited for the GCMCs considered in this work. However, for a wide variety of CMCs, the interfacial sliding and debonding behaviors may be very different. In this case, the proposed statistical models can be readily incorporated with other micromechanical models, or even finite element methods, to properly characterize the interface behavior with the presence of transverse matrix cracks. In addition, the models can be further generalized to account for the interaction between the progressive matrix failure and fiber failure.

Through the current material modeling, we are able to greatly enhance the analytical understanding of the effect of micromechanical parameters on the failure characteristics of the GCMCs. This will enable material developers and structural designers to effectively control, improve, and implement the glass and glass-ceramic composites.

Acknowledgment

This work was supported by the Office of Naval Research (ONR) under Contract No. N00014-89-C-0211.

References

[1] Sambell, R. A., Bowen, D., and Phillips, D. C., "Carbon Fiber Composites With Ceramic and Glass Matrices, Part 2, Continuous Fibers," *Journal of Material Science,* Vol. 7, 1972, pp. 676–681.

[2] Prewo, K. M., Bacon, J. F., and Dicuss, D. L., "Graphite Fiber-Reinforced Glass Matrix Composites," *SAMPE Quarterly,* Vol. 10, No. 4, 1979, p. 42.

[3] Prewo, K. M. and Brennan, J. J., "High Strength Silicon Carbide Fiber-Reinforced Glass Matrix Composites," *Journal of Material Science,* Vol. 15, 1980, pp. 463–468.

[4] Prewo, K. M., Brennan, J. J., and Layden, G. K., "Fiber-Reinforced Glasses and Glass-Ceramics for High Performance Applications," *American Ceramic Society Bulletin,* Vol. 65, No. 2, February 1986, pp. 305–322.

[5] Mah, T., Mendiratta, M. G., Katz, A. P., Ruh, R., and Maziyasni, K. S., "Room Temperature Mechanical Behavior of Fiber-Reinforced Ceramic-Matrix Composites," *Journal of the American Ceramic Society,* Vol. 68, 1985, pp. C-27 to C-30.

[6] Kim, R. Y. and Pagano, N. J., "Crack Initiation in Unidirectional Brittle-Matrix Composite," *Journal of the American Ceramic Society,* Vol. 74, 1991, pp. 1082–1090.

[7] Prewo, K. M., "Silicon Carbide Fiber Reinforced Glass-Ceramic Composite Tensile Behavior at Elevated Temperature," submitted to the *Journal of Material Science.*

[8] Aveston, J., Cooper, G. A., and Kelly, A., "The Properties of Fiber Composites," Paper 1, *Conference Proceedings,* National Physical Laboratory, IPC Science and Technology Press Ltd., Surrey, England, 1971, pp. 15–26.

[9] Avenston, J. and Kelly, A., "Theory of Multiple Fracture of Fibrous Composites," *Journal of Material Science,* Vol. 8, 1973, pp. 352–362.

[10] Budiansky, B., Hutchinson, J. W., and Evans, A. C., "Matrix Fracture in Fiber-Reinforced Ceramics," *Journal of Mechanics and Physics of Solids,* Vol. 34, No. 2, 1986, pp. 167–189.

[11] Marshall, D. B., Cox, B. N., and Evans, A. G., "The Mechanics of Matrix Cracking in Brittle Matrix Fiber Composites," *Acta Metallurgica,* Vol. 33, No. 11, 1985, pp. 2013–2021.

[12] McCartney, L. N., *Proceedings of the Royal Society,* A409, 1987, pp. 329–350.

[13] Rosen, B. W., "Tensile Failure of Fibrous Composites," *AIAA Journal,* Vol. 2, No. 11, November 1964, pp. 1982–1991.

[*14*] Zweben, C., "Tensile Failure Analysis of Fibrous Composites," *AIAA Journal,* Vol. 6, No. 12, December 1968, pp. 2325–2331.

[*15*] Zweben, C. and Rosen, B. W., "A Statistical Theory of Material Strength With Application to Composite Materials," *Journal of Mechanics and Physics of Solids,* Vol. 18, 1970, pp. 180–206.

[*16*] Batdorf, S. B., "A New Micromechanical Theory for Composite Failure," *Proceedings,* 1980 JANNAF Rocket Nozzle Technology Subcommittee Meeting, Monterey, CA, October 1980, Chemical Propulsion Information Agency, Laurel, MD, pp. 227–238.

[*17*] Harlow, D. G. and Phoenix, S. L., "The Chain of Bundles Probability Model for the Strength of Fibrous Materials I: Analysis and Conjectures," *Journal of Composite Materials,* Vol. 12, April 1978, pp. 195–214.

[*18*] Sutcu, M., "Weibull Statistics Applied to Fiber Failure in Ceramic Composites and Work of Fracture," *Acta Metallurgica,* Vol. 37, No. 2, 1989, pp. 651–661.

[*19*] Curtin, W. A., "Theory of Mechanical Properties of Ceramic-Matrix Composites," *Journal of the American Ceramic Society,* Vol. 74, 1991, pp. 2837–2845.

[*20*] Wang, A. S. D., Chou, P. C., and Lei, S. C., "A Stochastic Model for the Growth of Matrix in Composite Laminates," *Journal of Composite Materials,* Vol. 18, 1986, pp. 239–254.

[*21*] Volk, D., Yen, C. F., and Buesking, K., "Structural Development of Fiber-Reinforced Glass Matrix Composites," MSC TFR 2112/8601, Contract No. N00014-89-C-0211, Office of Naval Research, Arlington, VA, June 1990.

[*22*] Kimber, A. C. and Keer, J. G., "On The Theoretical Average Crack Spacing in Brittle Matrix Composites Containing Continuous Aligned Fibers," *Journal of Materials Science Letters,* Vol. 1, 1982, pp. 353–354.

[*23*] Prewo, K. M. and Nardone, V. C., "Carbon Fiber-Reinforced Glass Matrix Composites for Space Based Application," UTRC Report R86-917161-1, ONR Contract N0014-85-C-0332, Office of Naval Research, Arlington, VA, September 1986.

[*24*] Prewo, K. M., "Tension and Flexural Strength of Silicon Carbide Fiber-Reinforced Glass Ceramics," *Journal of Material Science,* Vol. 21, 1986, pp. 3590–3600.

[*25*] Cao, H. C., Sbaizero, O., Ruhle, M., Evans, A. G., Marshall, D. B., and Brennan, J. J., "Effect of Interfaces on the Properties of Fiber-Reinforced Ceramics," *Journal of the American Ceramic Society,* Vol. 73, 1990, pp. 1691–1699.

[*26*] Prewo, K. M., "A Compliant, High Failure Strain, Fiber-Reinforced Glass-Matrix Composite," *Journal of Material Science,* Vol. 17, 1982, pp. 3549–3563.

[*27*] Yasuda, E., Tanabe, Y., Machino, H., and Takaku, A., "Thermal Expansion Behavior of Various Types of Carbon Fibers up to 1000°C," *Proceedings,* 18th Biennial Conference on Carbon, Worcester, MA, July 1987, The American Center Society, PA, pp. 30–31.

[*28*] Marshall, D. B., "An Indentation Method for Measuring Matrix-Fiber Frictional Stresses in Ceramic Composites," *Journal of the American Ceramic Society,* Vol. 67, 1984, pp. C259–C260.

[*29*] Phillips, D. C., Sambell, R. A., and Bowen, D. H., "The Mechanical Properties of Carbon Fiber-Reinforced Pyrex Glass," *Journal of Material Science,* Vol. 7, 1972, pp. 1454–1464.

[*30*] Bhatt, R. T. and Phillips, R. E., "Laminate Behavior for SiC Fiber-Reinforced Reaction-Bonded Silicon Nitride Matrix Composites," NASA TM-101350, 1988.

*Sailendra N. Chatterjee,[1] Edward C. J. Wung,[2] and
Chian F. Yen[2]*

Modeling Ply Crack Growth in Laminates Under Combined Stress States

REFERENCE: Chatterjee, S. N., Wung, E. C. J., and Yen, C. F., "**Modeling Ply Crack Growth in Laminates Under Combined Stress States,**" *Composite Materials: Fatigue and Fracture, Fourth Volume, ASTM STP 1156,* W. W. Stinchcomb and N. E. Ashbaugh, Eds., American Society for Testing and Materials, Philadelphia, 1993, pp. 195–217.

ABSTRACT: First ply failure in brittle matrix composites usually indicates generation of transverse cracks in a ply. In a laminate, these cracks cannot cause failure since they are constrained by fibers in adjoining laminae. Many investigators are studying stiffness loss due to such cracks under transverse tensile stresses. In this work, use is first made of a phenomenological approach that models the response of homogeneous materials that have a large number of possible microcrack planes with different orientations in space. Densities of cracks of similar orientation can be treated as internal variables in the strain (or complementary) energy, which is also a function of the average strains (or stresses) over a representative volume element. The energy formulation is employed to model the response of a lamina under combined stresses containing cracks parallel to fibers. Estimates of the change in complementary energy due to the dilute concentration of cracks in a ply as obtained from simple solutions and results reported in literature are utilized to calculate the increase in compliance. Crack initiation and growth are predicted by the use of criteria which are similar to those used for homogeneous media. Results are compared with test data for $(\pm \theta)_{ns}$ layups for small and moderate crack densities. A relationship is established between the phenomenological and fracture mechanics-based approaches for the particular case of cross ply layups with large crack densities. Results are compared with test data. Advantages and limitations of both approaches are discussed.

KEY WORDS: plycracks, microcracks, phenomenological models, internal variables, strain energy, complementary energy, increased compliance, energy release rate, fracture mechanics approach

The objective of this study was to investigate how a lamina in a laminate responds to increasing load under combined stress states. First ply failure in organic matrix composites often indicates generation of transverse damage in a ply in the form of microcracks parallel to the fibers. Due to the existence of fibers of other orientations in the adjacent plies, such damage cannot propagate catastrophically, and a kind of in-situ pseudo-plastic response of the damaged plies usually occurs as the load is increased. For these reasons, commonly utilized ply-by-ply failure analysis leads to prediction of final failure loads which may differ significantly from test data.

In general, ply or microcracks are generated due to the combined action of extensional as well as shear stresses. Also, results of tests on $(\pm \theta)$ laminates [1] show that loading to a fraction of ultimate load and unloading may leave some permanent strain. It is quite possible that some elastoplastic behavior occurs in addition to microcracking. However, it is likely that microcrack surfaces generated by the combined action of transverse normal

[1]Staff scientist and [2]research engineer, Materials Sciences Corp., 930 Harvest Dr., Blue Bell, PA 19422.

and shear stresses are not smooth, and thus they may provide resistance to the reverse process (sliding back freely) during unloading. It should also be noted that some residual stresses due to processing may be present at the lamina level, which can explain the residual strain after unloading. Effects of such stresses are not studied here, but may be included in the analysis models without any difficulty.

In what follows we will attempt to model the growth of damage and the nonlinear response during monotonic loading without any effects of plasticity. Even with these assumptions, the mechanical response is difficult to quantify based on tests alone, since tests on unidirectional material often cause premature failure due to catastrophic propagation of cracks parallel to the fibers. In the next section we review and modify some phenomenological approaches [2–4] for modeling microcrack growth in continuous media. Next we make use of energy estimates (based on mechanistic approaches) to choose some of the constants required for modeling stiffness loss due to ply cracks in dilute concentration. These models, along with the test data for one laminate, are then utilized to predict the response of different $(\pm\theta)_{ns}$ layups.

Recently, several authors have attempted to quantify stiffness loss in $(0_m/90_n)_s$ and other layups of graphite/epoxy and glass/epoxy composites [4–13]. In general all of these works (except Ref 4 which employs curve-fitting techniques) employ approximate solutions for evaluating stiffness loss due to ply cracks. Some of these authors [7,9,10,12] have also made use of an energy release rate approach for predicting the crack density for a given load. As crack density increases, the reduction in strain energy is assumed to be used in the creation of new crack surfaces. In a later section we discuss a modification of the phenomenological model to consider large crack densities and model crack growth, which is conceptually similar to the approaches discussed above.

Phenomenological Constitutive Models

Constitutive laws for modeling microcracking in a continuum are discussed in Refs 2 and 3. Such models have been found to be very successful in representing the behavior of

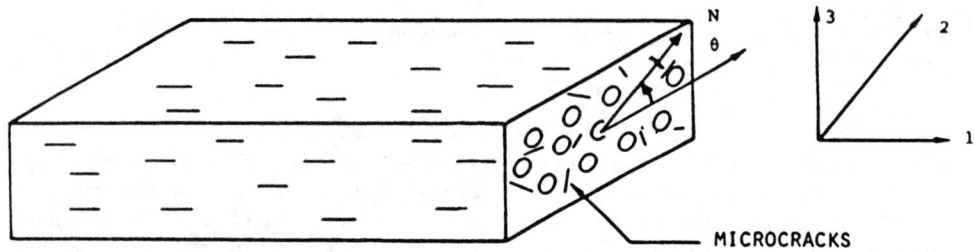

ALL CRACK PLANES PARALLEL TO X_1 AXIS

ORIENTATION = θ $0 < \theta < 2\pi$

$N = [0, \cos\theta, \sin\theta]$ – CRACK ORIENTATION VECTOR

$\omega(N)$ – SOME VOLUME AVERAGE OF AREA OF CRACKS WITH NORMAL N

FIG. 1—*Microcracks parallel to fibers in a lamina.*

brittle materials and make use of a phenomenological approach and internal variables. Considering possible microcracking parallel to fibers only (Fig. 1), one can adopt the formulation given in Ref 2 in the form given below where stress, strain, energy, and crack densities as well as the mechanical response denote the corresponding averages over a representative volume element (RVE) containing several microcracks.

$$\underline{\sigma} \cdot \partial\underline{\epsilon} - \rho\partial\psi \geq 0; \text{ Clausius-Duhem inequality}$$

$$\underline{\sigma} = \underline{\sigma}[\underline{\epsilon},(\omega^{\alpha}, \underline{N}^{\alpha}), \alpha = 1, 2 \ldots k]$$

ψ = Helmholtz free energy density

$$= \psi[\underline{\epsilon},(\omega^{\alpha}, \underline{N}^{\alpha})]$$

ρ = mass density

$$\sigma_i = \rho\frac{\partial\psi}{\partial\epsilon_i}; \sum_{\alpha=1}^{k} R^{\alpha}\partial\omega^{\alpha} \geq 0 \tag{1}$$

R^{α} = conjugate thermodynamic force, an energy release rate

$$= -\rho\frac{\partial\psi}{\partial\omega^{\alpha}}$$

k = number of crack plane orientations

ω^{α} = nondimensionalized density of microcracks with orientation $\underline{N}^{\alpha}$ (direction cosine vector)

The strain energy ($U = \rho\psi$) (Fig. 2) is expressed as the strain energy in the undamaged material, which is a quadratic function of the strain $\underline{\epsilon}$ less the decrease in strain energy due to cracks, which is a sum of the products of the crack densities ω^{α} ($\alpha = 1, 2, \ldots k$) and a function involving squares of the extensional and two shear strains on the crack planes. The function is written with three (or two in the case of isotropic materials) material constants C_i' (sometimes more constants are employed, as in Ref 4). Note that the decrease in strain energy is equal to $\Sigma R^{\alpha} \omega^{\alpha}$, and, therefore, R^{α} is a measure of the energy release rate due to an increase in crack density $\partial\omega^{\alpha}$, which occurs due to the increasing number or sizes of the set of cracks α. A power law type of relationship may be chosen to relate damage increment ω^{α} to R^{α}, which is assumed to be independent of other damage planes (i.e., dilute concentration of cracks and no interaction). This relation and the constants C_i' are chosen to fit experimental data.

The formulation in terms of strain may not be adequate to model the response of plies damaged by microcracks, since transverse tensile stress in absence of such tensile strain can cause initiation and growth of ply cracks. Further, if elastoplastic behavior is to be incorporated (by superposing the two phenomena as outlined in Ref 3), it is advantageous to have a formulation in terms of stress employing a complementary energy-type formulation (Fig. 2) as outlined below.

$$\partial W_c - \underline{\epsilon} \cdot \partial \underline{\sigma} \geq 0$$

$$\underline{\epsilon} = \underline{\epsilon}(\underline{\sigma}, (\omega^\alpha, \underline{N}^\alpha), \alpha = 1, 2 \ldots k)$$

$$W_c = W_c(\underline{\sigma}, (\omega^\alpha, \underline{N}^\alpha)) \tag{2a}$$

$$\epsilon_i = \frac{\partial W_c}{\partial \sigma_i}; \quad \sum_{\alpha=1}^{k} R^\alpha \partial \omega^\alpha \geq 0$$

$$R^\alpha = \frac{\partial W_c}{\partial \omega^\alpha}$$

where k = number of crack planes. Note that W_c is the complementary strain energy density expressed in terms of stresses $W_c + \rho\psi = \underline{\sigma} \cdot \underline{\epsilon}$. All other quantities are defined in Eq 1. For small crack densities one may assume that

$$W_c = \frac{1}{2}\left[S_{ij}^0 \sigma_i \sigma_j + \sum_{\alpha=1}^{k} \omega^\alpha (C_1 h_\alpha \sigma_\alpha^2 + C_2 \tau_{t\alpha}^2 + C_3 \tau_{a\alpha}^2) \right] \tag{2b}$$

where

$$
\begin{aligned}
\sigma_i (i = 1 \ldots 6) &= \text{stresses in layer coordinates in contracted notation,} \\
S_{ij}^0 &= \text{initial elastic compliances before damage initiates,} \\
\sigma_\alpha, \tau_{t\alpha}, \tau_{a\alpha} &= \text{extensional, transverse, and axial shear stresses on crack plane } \alpha; \\
&\quad \text{and}
\end{aligned}
$$

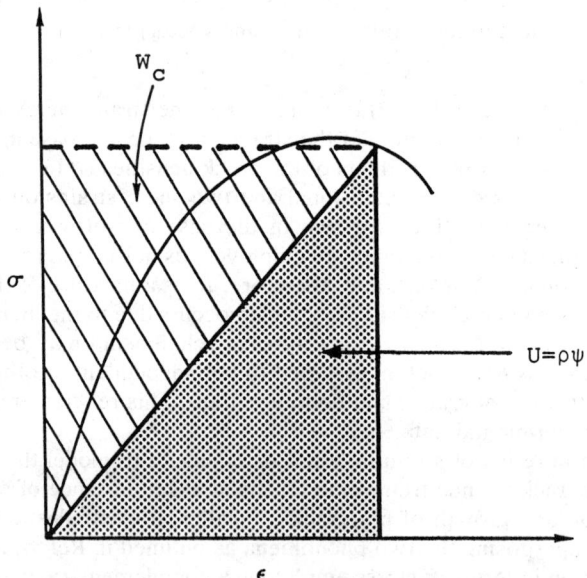

FIG. 2—*Stored strain and complementary energies in damaged state.*

C_i = constants which may be chosen from energy considerations discussed later.

Note that R^α, the measure of the energy release rate (discussed in the context of Eq 1), is given by half of the term within parentheses in Eq 2b and $R^\alpha \omega^\alpha$ is the increase in complementary energy density due to the set of cracks α.

Since the crack faces are expected to transfer compressive normal stresses, one has to choose

$$h_\alpha = 0 \qquad \sigma_\alpha \leq 0$$
$$= 1 \qquad \sigma_\alpha > 0 \qquad\qquad (2c)$$

When the crack sizes are small compared to the ply thickness and the crack densities are low, one can choose the values of C_1, C_2, and C_3 from the solution of a single crack (Fig. 3) in an infinite transversely isotropic medium as a first approximation, as discussed in the next subsection. Using the values in Eqs 2a,b and the damage growth model presented later, we performed calculations to determine stiffness loss and crack densities with several values of $k \geq 3$ (sets of crack planes) in $(\pm \theta)_{ns}$ graphite/epoxy laminates subjected to increasing tensile load in the 0° direction. The results indicate that the densities of cracks in all planes are small except for the density for the set of cracks which are perpendicular to the lamination planes. For simplicity we will therefore choose $k = 2$ for this study to consider two sets of crack planes, one of which is described above. The other set describes the crack planes parallel to lamination planes. In the initial stages, stresses on these planes are zero and no cracks will appear in this set. Therefore, $\omega^2 = 0$, and henceforth we will denote ω^1 by ω. Tests show that cracks in this set may start as microcracks but grow very rapidly in the x_3 direction to the full thickness, $2a$, of the ply (or

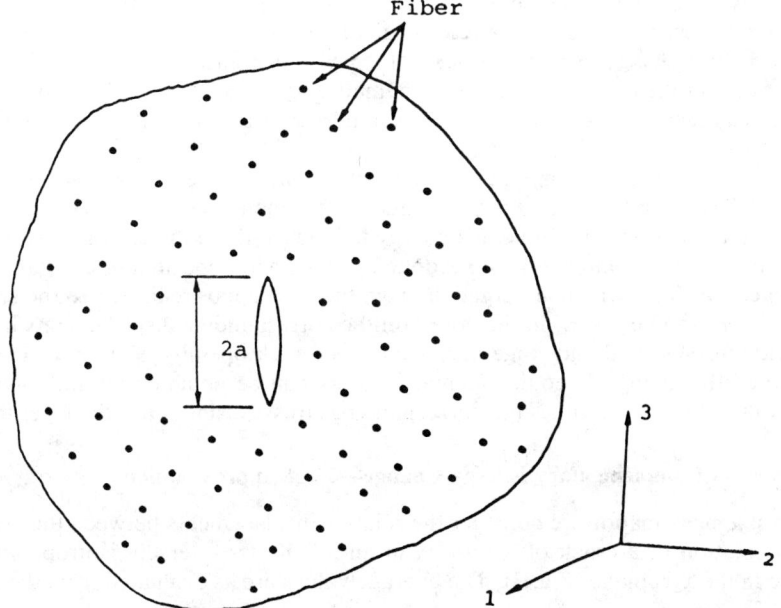

FIG. 3—*An infinite fibrous medium containing one crack.*

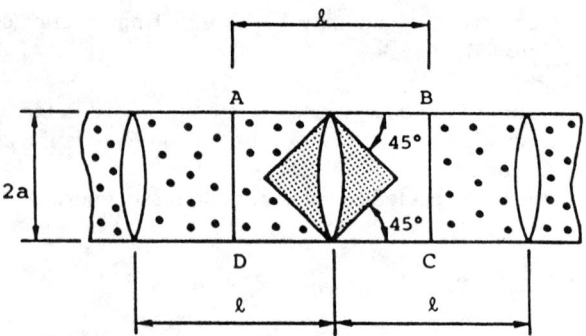

FIG. 4—*A fiber lamina with transverse cracks.*

layer as shown in Fig. 4) and become quite long in the x_1 (or fiber) direction, extending through the entire width of the test coupon (they are called ply cracks). The crack density increases as an increasing number of cracks appear. Following Refs 2, if we define the crack density, ω, to be the ratio of the assumed loss of load-bearing area (denoted by the shaded area in Fig. 4) per crack to the original area then

$$\omega = (\sqrt{2a})^2/2al = a/l \tag{3}$$

where $2a$ = the layer thickness, and l = the average crack spacing.

It may be noted that the loss in load-bearing area may be chosen in other ways if desired (a circle, an ellipse with a fixed ratio of the dimensions of the major axis to the minor one, or even a rectangle), yielding a higher value of crack density, i.e., $q\,a/l$ with $q > 1$. The expression for complementary energy in terms of ω given by Eq 3 will, however, remain the same if C_i in Eq 2b are reduced by the same factor. Therefore, without any loss in generality, we will choose the constants, C_i, from energy considerations described in the next paragraph based on the measure of ω given by Eq 3. We note, however, that ω (as defined) does not change in a continuous fashion as an increasing number of cracks appear (the process described above), and the mathematical description in Eq 2a involving derivatives of ω is an idealization, but it gives us a simple approach to model the phenomenon of ply crack growth.

Now we note that the change in complementary energy due to a single ply crack in a layer located inside a laminate which is sufficiently wide (as compared to laminate thickness) can be calculated from the crack energy in a quasi–three-dimensional problem. This energy is evaluated by finding the work done by equal and opposite but constant tractions on the crack surfaces (which are equal in magnitude but opposite in sign to the stresses $\sigma_2 = \sigma_{22}$, $\sigma_6 = \tau_{23}$, and $\sigma_3 = \tau_{12}$ in the layer) on the corresponding displacements in absence of far field stresses. If the average displacements (crack opening, sliding, or tearing) are known, then the change in complementary energy can be obtained by multiplying these average values by $\frac{1}{2} \times 2a$ times the corresponding stresses, since $2a$ is the layer thickness.

An Estimate of Complementary Energy Change—First Approximation

As a first approximation we consider the relative displacements between the crack faces from the solution of a crack of length $2a$ in an infinite transversely isotropic composite (isotropic in the x_2x_3 plane, Fig. 3). The average values are as evaluated from the results in Ref *14*.

For opening, $\overline{U}_2^* \approx \pi a h \sigma_2 / E_T^0$

For sliding, $\overline{U}_3^* \approx \pi a \sigma_6 / E_T^0$ \hfill (4)

For tearing, $\overline{U}_1^* \approx \pi a \sigma_3 / 2 G_A^0$

where h is defined in Eq 2c. Approximate equality signs are used since these results are valid when the fiber direction modulus (x_1) is very high and the problem is quasi–three-dimensional. The equality sign holds for the plane stress condition. Note that E_T^0 is the transverse Young's modulus and G_A^0 is the axial shear modulus of the composite in the undamaged state. The following expression is obtained for the crack energy (per unit width)

$$W^{cr} = \frac{\pi a^2 \sigma_2^2}{E_T^0} + \frac{\pi a^2 \sigma_6^2}{E_T^0} + \frac{\pi a^2 \sigma_3^2}{2 G_A^0} \tag{5}$$

We note that the derivative of each of the terms on the right side of Eq 5 with respect to Eq 2a yields the energy release rate at the two tips of the crack for each of the three modes, i.e.,

$$G_I = \pi \sigma_2^2 a / E_T^0, \; G_{II} = \pi \sigma_6^2 a / E_T^0, \; G_{III} = \pi \sigma_3^2 a / 2 G_A^0 \tag{6}$$

Thus the in-plane stresses $\sigma_2 = \sigma_{22}$ and $\sigma_3 = \tau_{12}$ produce Mode I (opening) and Mode III (tearing) type of deformations near the crack.

On the other hand, consider a fiber lamina with parallel slit cracks (transverse cracks) as shown in Fig. 4. The complementary energy in the representative volume element (the region ABCD with unit length in x_1 direction) is assumed to be (from Eq 2b with $k = 2$, $\omega^2 = 0$, $\omega^1 = \omega$)

$$V \cdot W_c = V \left[\frac{1}{2} S_{ij}^0 \sigma_i \sigma_j + \frac{1}{2} \omega (C_1 h \sigma_2^2 + C_2 \sigma_6^2 + C_3 \sigma_3^2) \right] \tag{7}$$

where $V = 2al$ and the change in complementary energy density due to the cracks is given by the second term within the brackets. Since $\omega = a/l$, the total change in the complementary obtained from Eq 7 will be identical to that from Eq 5 provided

$$C_1 = \frac{\pi}{E_T^0}, \; C_2 = \frac{\pi}{E_T^0}, \; C_3 = \frac{\pi}{2 G_A^0} \tag{8}$$

The stress-strain relations of the damaged solid can be obtained from Eqs 2 or 7 as

$$\epsilon_i = \frac{\partial W_c}{\partial \sigma_i} = (S_{ij}^0 + S_{ij}^*) \sigma_j \tag{9}$$

where the only three nonvanishing components of S_{ij}^* are

$$S_{22}^* = \omega C_1, \; S_{33}^* = \omega C_3, \; S_{66}^* = \omega C_2 \tag{10}$$

Thus, when cracks are introduced in dilute concentration, the coefficients S_{22}^0, S_{33}^0, and S_{66}^0 change to the following (these values are the effective compliances of a cracked layer).

$$S_{22} = \frac{1}{E_T} = \frac{1}{E_T^0} (1 + \omega C_1 E_T^0)$$

$$S_{33} = \frac{1}{G_A} = \frac{1}{G_A^0} (1 + \omega C_3 G_A^0) \qquad (11)$$

$$S_{66} = \frac{1}{G_T} = \frac{1}{G_T^0} (1 + \omega C_2 G_T^0)$$

while other coefficients remain unchanged, i.e., they remain equal to those of the un-cracked fibrous composite. Equation (11) implies that compliance changes are related linearly to $\omega = a/l$ when ω is small.

Similar changes were reported in Ref 5 where the problem of an infinite fibrous medium with aligned slit cracks parallel to the fibers was solved by the self-consistent method. Figure 5 shows changes in the three coefficients S_{22}, S_{33}, and S_{66} of a T300 graphite/epoxy composite calculated from Eq 11 with C_i from Eq 8 for given values of the crack density ω. Results reported in Ref 5 are also plotted on the same figure for comparison. Note that the crack density, β, defined in Ref 5 is equal to 2ω where ω is the crack density defined in Eq 3. The comparison in Fig. 5 shows that Eq 11, which is strictly valid for $\omega \rightarrow 0$, yields S_{33}

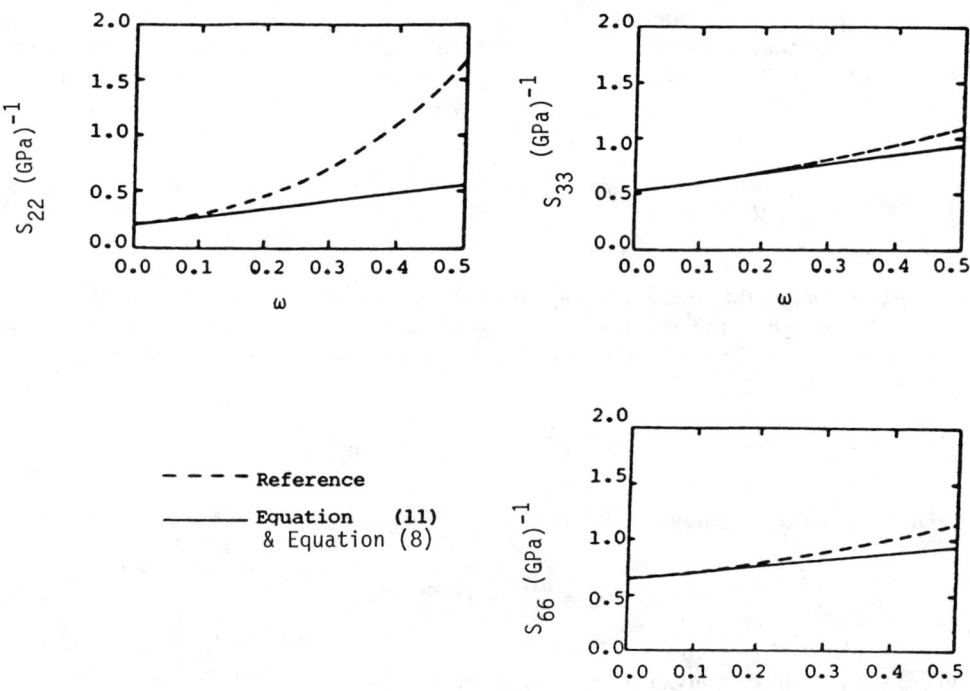

FIG. 5—*Compliance changes in T300 graphite/epoxy system caused by cracks of density ω, fiber volume = 20%.*

and S_{66}, which are close to those from the self-consistent method even when $\omega = 0.5$. The agreement in the case of S_{22} is good for $\omega \leq 0.2$, beyond which the self-consistent method shows that S_{22} increases much more rapidly with ω. It may be noted that the self-consistent method does not consider correctly the interaction between the cracks for the Mode I case for larger values of ω (on the order of 0.5). The rate of increase in S_{22} (or the rate of stiffness reduction) is expected to reduce (as will be discussed in a later section; also see Ref 15) as ω approaches 0.5, in contrast to what is predicted by the self-consistent method. The purpose of the comparison given here is only to show that the compliance changes almost linearly with ω for dilute concentrations (at least up to $\omega \approx 0.2$).

Better Estimates of Complementary Energy Change

To obtain correct estimates of C_i in the expression for the increase in complementary energy (Eq 7) for dilute concentration of ply cracks, it is necessary to determine the stress state in the whole laminate containing a transverse crack in a layer (Fig. 6). In this section we give some simple results which will be useful for studying the responses of $(\pm\theta)_{ns}$ or cross ply laminates under tension. We note that the necessary results can be obtained by solving the quasi–three-dimensional problem with known prescribed tractions on the crack surfaces as discussed in the previous section. If we consider the lamina to be located inside a $(\pm45)_{ns}$ system, then each of the constants C_1, C_2, and C_3 can be determined by solving a separate problem for each case. The constant C_2 is not required for the case under consideration because there are no transverse shear stresses at the location of the crack in the uncracked laminate. First we will obtain an estimate of C_3 associated with in-plane shear stress (tearing mode) in a cracked layer. The case of normal stress is discussed later.

For the case of $(\pm45)_{ns}$ laminate, the problem is that of a Mode III crack (see the last of Eqs 4 or 6) in a 0/90 system referred to in the principal material coordinates of the cracked layer with a crack located in one of the 90° layers. The crack faces are subjected to tractions equal in magnitude to the in-plane shear stress in the layer. If the axial and transverse shear moduli of the layers (G_A and G_T) are not much different, the energy release rate at the tips of a crack of length $2a'$ will be the same as that in an isotropic elastic strip of finite width. For a symmetrically located crack (with shear traction $= \sigma_3$) in

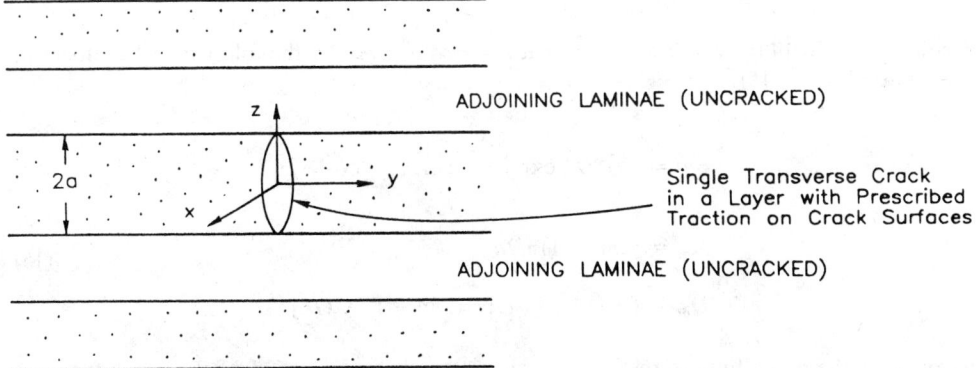

FIG. 6—*Problem to be solved to obtain better estimates of energy change.*

a strip of width $2b$, one has

$$G_{III} = \frac{\pi\sigma_3^2 a'}{2G} \left[(2b/\pi a') \tan(\pi a'/2b)\right] \tag{12}$$

where G is the shear modulus. When $a' \ll b$ and $G = G_A^0$, one obtains the result given by last of Eq 6. When $a' = a = b/2$, the numerical factor in the last of Eq 6 changes from $\pi/2$ to 2.0. The complementary energy change due to the cracks of length $2a$ ($\approx$ layer thickness) is given by

$$\Delta W_c = 2 \int_0^a G_{III}(a')\, da' \tag{13}$$

which indicates that the constant C_3 will be a little higher than $\pi/2G_A^0$ but less than $2/G_A^0$. If the total thickness of all of the outer layers on one side is greater than half of that of the cracked layer (Fig. 6), then $\pi/2G_A^0$ is a good estimate of C_3. The same estimate holds for a ply crack in an outer layer with half the thickness of the inner layer considered above.

Next we consider the plane problem when normal stress is prescribed on the crack surfaces. If the layers just next to the transversely isotropic (in the yz plane, Fig. 6) cracked 90° layer are very stiff (the Young's modulus in y direction of the adjacent 0° layers $= E_A \gg E_T$), a simple solution can be obtained by neglecting the transverse deformations (in the z direction). We will now consider the 90° strip with $v = 0$ along its edges (Fig. 6), since by assumption strains ϵ_{yy} in the adjacent 0° layer are negligible and satisfy the equilibrium equation in the y direction (called the shear analogy model because of its similarity with antiplane problems). The governing equations for plane stress are ($\sigma_x = 0$, note that the results will be practically identical if one assumes the case of plane strain $\epsilon_x = 0$ as discussed after Eq 4)

$$\frac{\partial \sigma_{yy}}{\partial y} + \frac{\partial \tau_{yz}}{\partial z} = 0$$

$$\sigma_{yy} = E_T \frac{\partial v}{\partial y} \tag{14}$$

$$\tau_{yz} = G_T \frac{\partial v}{\partial z}$$

A solution of Eq 14 for $y > 0$ with $v = 0$ at $z = \pm a$ is given by the following (the subscript T is omitted in Eq 15)

$$v = \sum_{m=1}^{\infty} D_m \exp\left(-\sqrt{G/E}\,\lambda_m y\right) \cos \lambda_m z$$

$$\lambda_m = (2m - 1)\pi/2a \tag{15}$$

$$D_m = (-1)^{m-1} 8\sigma_2 a / [\pi^2 (2m - 1)^2 \sqrt{GE}]$$

where σ_2 is the magnitude of the compressive normal stress applied on the crack faces. A similar solution holds for $y < 0$ and, evaluating the crack energy, one obtains

$$W^{cr} = 1.084\sigma_2^2 a^2/\sqrt{E_T G_T} \tag{16}$$

For $G_T = E_T/2 (1 + u_T)$ and $u_T \approx 0.3$, one finds that

$$C_1 \approx 1.74/E_T \tag{17}$$

Exact numerical solution for the plane strain ($\epsilon_x = 0$) problem of an isotropic strip (note that the 90° layer is isotropic in the yz plane) clamped at the edges [16] yields the following estimate for C_1

$$C_1 \approx 1.98/E_T \tag{18}$$

which is little higher than that in Eq 17. The numerical factor is expected to be higher if the flexibility of the layers adjacent to the cracked layer is considered in the analysis (discussed in a later section). However, we note that for most of the ($\pm \theta$) systems considered in the next section, a slight difference in the value of C_1 does not affect W_c significantly since in-plane shear stresses (tearing mode) dominate this quantity.

From the results given above, it appears that, for (± 45)$_{ns}$ systems, values of C_1 and C_3 from Eq 18 and the last of Eq 8, respectively, should yield good estimates of the change in W_c. It is noted that the change in complementary energy estimated this way yields the total change in the system, not necessarily in the cracked layer alone as postulated by Eq 7, especially for in-plane shear stresses. Also note that for other ($\pm \theta$)$_{ns}$ layups, the values of C_1 and C_3 may differ from those estimated for (± 45)$_{ns}$ systems. Further, coupling terms involving product of transverse normal stress and in-plane shear stress should also be included in the expression of Eqs 2 and 7 for W_c because of the loss in symmetry (or antisymmetry) conditions about the crack plane. These differences or modifications of the expression for W_c may be determined by performing stress analysis of some representative layups. Such calculations are, however, beyond the scope of the present work. It appears, though, that if θ is close to 45° (i.e., $30° \leq \theta \leq 60°$), the estimates obtained here should give reasonably accurate results.

Damage Function

Damage initiation and progression are characterized by the energy release rate as quantified by R (in Eq 2a) per unit volume of the representative volume element. Taking the derivative of W_c as given by Eq 7 (or Eq 2b) for the only active set of crack planes for ply cracks under consideration) with respect to ω, one obtains

$$R = \frac{\partial W_c}{\partial \omega} = \frac{1}{2} (C_1 h \sigma_2^2 + C_2 \sigma_6^2 + C_3 \sigma_3^2) \tag{19}$$

for dilute concentration of cracks, where the stresses are the ply stresses in each layer susceptible to cracking and the value of ω may vary from layer to layer. Following the formal approach used in damage mechanics [2], we assume that damage initiates when the energy release rate, R, reaches a critical value, R_0. For modeling the phenomenon of stable growth of crack density, ω (increasing number of ply cracks) for $R > R_0$, we assume a power law-type relation between R and ω (similar to that in Ref 2) in terms of the damage function $g(R, \omega)$

$$g = (R/R_0)^p - 1 - \omega/\Delta = 0 \tag{20}$$

where $\Delta = \pi/k$, k being the number of crack planes considered. For this study $k = 2$ and $\Delta = \pi/2$. The factor Δ is introduced to consider more crack planes if needed. Additional damage is possible only if $g = 0$ and $(dg/dR)dR > 0$.

The two material parameters, R_0 and p, in Eq 20 can be determined from test data for a selected layup of a given material system. To outline the procedure for evaluation of R_0 and p and of the nonlinear stress-strain response due to the increasing number of ply cracks, we consider results for a graphite/epoxy AS1/3501 system reported in Ref 1.

Correlations with Data Reported in Ref 1

The properties of the uncracked unidirectional composite are given in Table 1. The effective properties of a cracked lamina can be computed from Eq 11 with the values of C_i estimated in the previous subsection. The classical lamination theory can then be used to compute the reduction of the axial stiffness, E_0, of a $(\pm\theta)_s$ laminate as a function of the crack density ω (assumed to be the same in all the layers). Detailed results can be found in Ref 17. It may be noted that similar estimates of stiffness loss are reported in Ref 11 based on a different approach.

The evaluation of R_0 and p is made as follows: the elastic limit $\sigma_0^{(A)}$ (far field stress in $0°$ direction) and another stress point, $\sigma_0^{(B)}$, located in the nonlinear range are identified from the stress-strain data of a $(\pm\theta)_s$ tension test (Fig. 7). The crack density, $\omega^{(A)}$, associated with Point A is zero because the damage process has not yet started at this point. The crack density, $\omega^{(B)}$, at Point B (Fig. 7) which is needed to reduce the laminate stiffness from $E_0^{(A)}$ to $E_0^{(B)}$ is found from the stiffness loss versus ω data computed using the lamination theory (discussed above). One can also express the conjugate thermodynamic force (the same in each layer) using the lamination theory in terms of a function, $r(\omega)$, for a given layup (for details see Ref 17)

$$R = r(\omega)(\sigma_O^2) \tag{21}$$

For Points A and B we therefore have

$$R_0 = r(0)(\sigma_0^{(A)})^2 \text{ and } R^{(B)} = r(\omega^{(B)})(\sigma_0^{(B)})^2 \tag{22}$$

Substitution of the values of R_0 and $R^{(B)}$ from Eq 22 and $\omega^{(B)}$ in Eq 20 yields the following relation for evaluation p

$$p = \ln (1 + \omega^{(B)}/\Delta)/\ln \left(\frac{R^{(B)}}{R_0} \right) \tag{23}$$

where $\Delta = \pi/2$.

TABLE 1—*Material properties of uncracked unidirectional composite.*

AS1/3501 Graphite/Epoxy
$E_A^0 = 130 \text{ GPa} = 18.85 \text{ Msi}$
$E_T^0 = 10.5 \text{ GPA} = 1.52 \text{ Msi}$
$G_A^0 = 6.06 \text{ GPa} = 0.87 \text{ Msi}$
$G_T^0 = 4.06 \text{ GPa} = 0.58 \text{ Msi}$
$\nu_A^0 = 0.35$
$\nu_T^0 = 0.31$

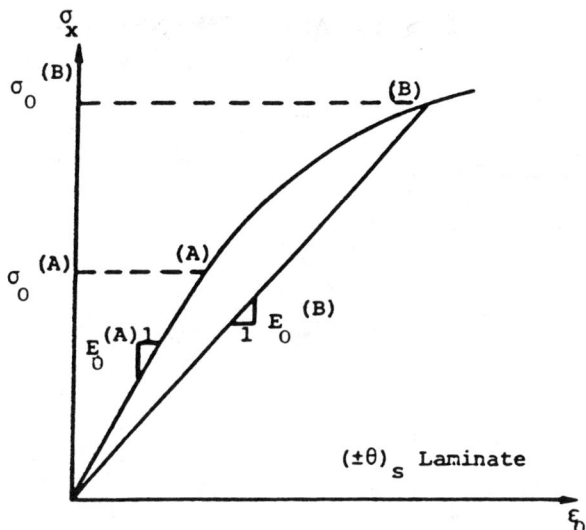

FIG. 7—*A schematic drawing of a typical stress-strain response of a* $(\pm\theta)_s$ *laminate in a tension test for the determination of* R_0 *and* p.

The above procedure was used to analyze the stress-strain data of an ASI/3501 $(\pm30)_s$ laminate reported in Ref *1*, and the following values were found

$$R_0 = 149 \times 10^3 \text{ Pa} = 21.6 \text{ psi and } p = 0.1152 \qquad (24)$$

Note that the presence of significant nonlinearity and nearly identical behavior of the two outer (θ) layers and the $(-\theta)_2$ layer make the use of data from such layups a very attractive method for determination of R_0 and p.

The damage analysis methodology (Eqs 19 and 20) and the procedure for computation of effective compliances (Eq 11) have been incorporated in a computer code developed earlier for nonlinear laminate analysis [*18*] which employs an iterative scheme to evaluate laminate response for increasing loads. To verify the proposed data interpretation scheme and the validity of the model, the damage parameters R_0 and p (Eq 24) obtained from the $(\pm30)_s$ tension test data (as discussed above) were used in this code to generate analytic predictions of the stress-strain curves for various $(\pm\theta)_s$ laminates with $\theta = 30, 35, 40, 45,$ and 50° in tension tests. Results for $\theta = 30°$ and 50° are compared in Fig. 8 with test data from Ref *1*. Other results can be found in Ref *17*. Reasonable agreement is observed in all cases. Maximum values of ω were of the order of 0.3. Since the same values of R_0 and p obtained from the $(\pm30)_s$ data are used for all other layups, this agreement shows that the model is simple and useful for dilute concentration of ply cracks.

Correlation with Data from Ref *17*

Results of tests on $(\pm15)_{ns}$, $(\pm30)_{ns}$, $(\pm45)_{ns}$, $(\pm60)_{ns}$, and $(\pm75)_{ns}$ (with $n = 1, 2,$ and 4) systems are reported in Ref *17*. We first determined R_0 and p for this AS4/3501-6 system for the $(\pm30)_s$ layup. The values were obtained by using material properties listed in Table

STRESS STRAIN RESPONSE

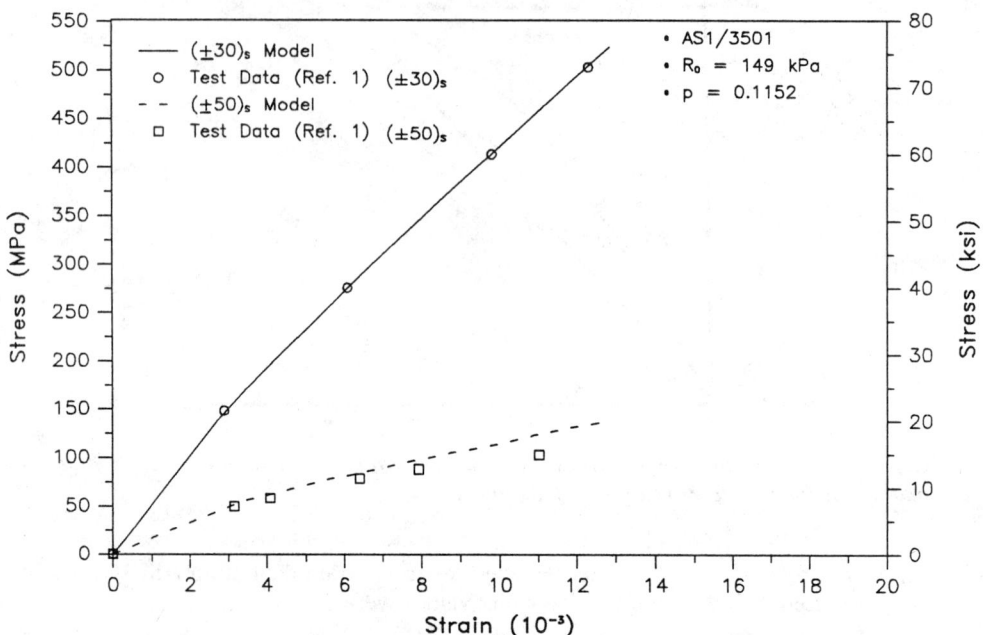

FIG. 8—*Stress-strain responses of AS1/3501 (±30)$_s$ and (±50)$_s$ laminates.*

1 and the procedure outlined in the previous section. We obtain

$$R_0 = 170 \times 10^3 \text{ Pa} = 24.7 \text{ psi}$$

$$p = 0.1073 \tag{25}$$

These values are slightly different from those of the AS1/3501-6 laminate (Eq 24). The (±60)$_{ns}$ and (±70)$_{ns}$ systems did not show much nonlinearity, and most of the specimens exhibited failure due to fractures in the surface plies parallel to fibers [17] and delaminations before the generation of many ply cracks. The (±15)$_{ns}$ systems showed an increase in stiffness (as in Ref 1) often noticed in such layups (and in unidirectional materials) due to a different mechanism. This stiffening can be attributed to an increase in fiber modulus at higher strain levels due to reorientation of graphite crystals [19]. The stiffening is, however, negligible in ($\pm\theta$)$_{ns}$ layups for $\theta \geq 30°$, and detailed correlations for such laminates reported in Ref 17 show good agreement with the model predictions. Some representative results are shown in Figs. 9 and 10.

For $n > 1$ in ($\pm\theta$)$_{ns}$ layups, we assumed R_0 to be twice the value of what is given by Eq 25 for the layers which are not at the center or on the surface. Since the thicknesses for these layers are half of those at the center, this result follows from fracture mechanics concepts discussed in the next section. Note that the layers at the center are two-ply thick. Crack spacing, l, in these layers can be found if desired as $l = a/\omega$, a being the ply thickness. The same result is valid for outer layers. For other layers in ($\pm\theta$)$_{ns}$ layups, the crack spacing will be given by $a/2\omega$. We have performed some correlations of crack

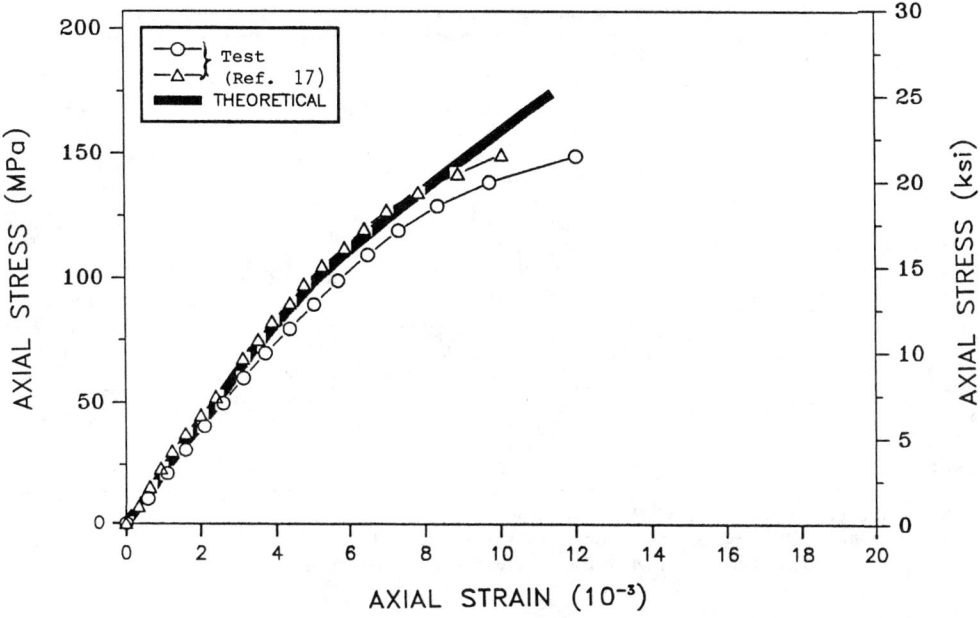

FIG. 9—*Axial stress versus axial strain for* [±45]$_{2s}$.

spacing reported in literature [4] for cross ply graphite epoxy laminates and the results are in reasonable agreement with those reported.

Off-axis tests reported in Ref 17 do not show much nonlinearity, and crack growths in these specimens appear to be quite fast. Results from these tests, as well as those from (±60)$_{ns}$ and (±75)$_{ns}$ systems can be utilized to check the usefulness of R_0 as the damage

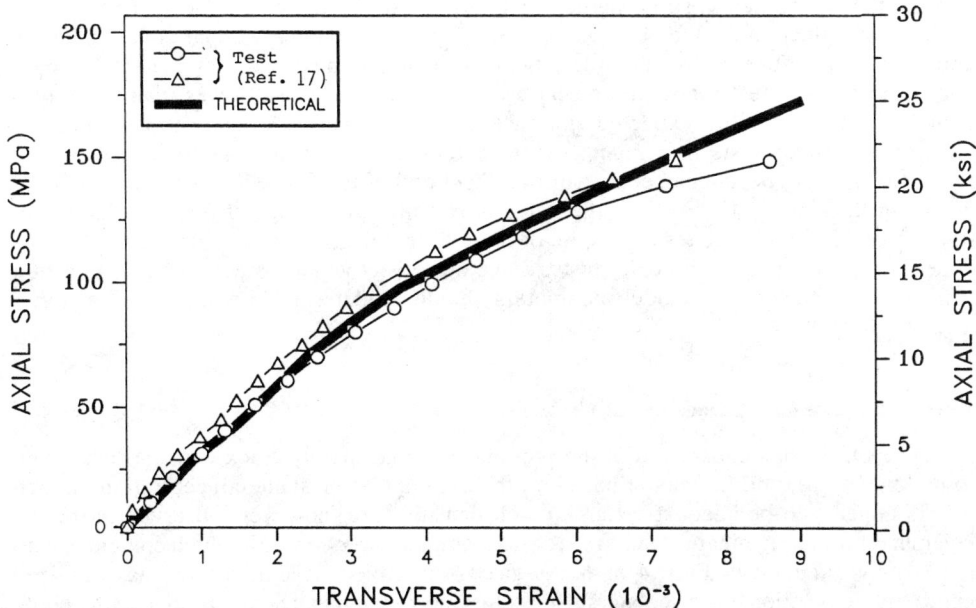

FIG. 10—*Axial stress versus transverse strain for* [±45]$_{2s}$.

TABLE 2—*Damage initiation and failure stresses in* $(\pm 60)_{ns}$, $(\pm 75)_{ns}$, *and off-axis tests.*[a]

| | | Damage Initiation Stress, MPa (ksi) | | Failure Stress, MPa (ksi) |
Tests	Layup	Calculated	Experimental (average)	Experimental (average)
$(\pm \theta)_{ns}$ [b]	$(\pm 60)_s$	50 (7.2)	55 (8.0)	70 (10.2)
$(\pm \theta)_{ns}$	$(\pm 60)_{2s}$	50 (7.2)	55 (8.0)	85 (12.3)
$(\pm \theta)_{ns}$	$(\pm 75)_s$	43 (6.3)	41 (6.0)	50 (7.3)
$(\pm \theta)_{ns}$	$(\pm 75)_{2s}$	43 (6.3)	41 (6.0)	49 (7.1)
Off-axis[c]	$(10)_8$	194 (28.2)	143 (20.8)	351 (50.9)
Off-axis	$(10)_{12}$	194 (28.2)	145 (21.0)	303 (43.9)
Off-axis	$(15)_8$	130 (18.9)	142 (20.6)	281 (40.8)
Off-axis	$(15)_{12}$	130 (18.9)	97 (14.12)	232 (33.6)
Off-axis	$(20)_8$	99 (14.3)	97 (14.0)	228 (33.0)
Off-axis	$(20)_{12}$	99 (14.3)	63 (9.1)	142 (20.6)
Off-axis	$(30)_8$	68 (9.8)	48 (6.9)	132 (19.2)

[a]Critical energy release rate = $R_0 \approx 24.7$ psi.
[b]$C_1 = 1.98/E_T$, $C_3 = \pi/2G_A$, based on constrained ply cracks.
[c]$C_1 = C_3 = \pi/E_T$, based on through transverse cracks in specimens.

initiation parameter. Some results are given in Table 2. It may be noted that R_0 in $(\pm 60)_{ns}$ and $(\pm 75)_{ns}$ systems are computed based on the same constrained crack growth concept discussed earlier. Thus the same values, C_1 and C_3, are utilized.

On the other hand, in off-axis tests, through-transverse cracks are likely to appear first. These cracks then increase in number as well as in size and later join together to cause failure. For this reason, C_1 and C_3 quantify the Mode I and Mode II effects and the appropriate values are $C_1 = C_3 = \pi/E_T^0$ (see the first two lines of Eq 8). It may be noted that the damage initiation stress agrees well with the observed results for the $(\pm 60)_{ns}$ and $(\pm 75)_s$ tests, since the value of R_0 was determined for initiation of similar constrained ply cracks in the $(\pm 30)_s$ system. However, the damage initiation stresses obtained experimentally from off-axis tests show more scatter. These values are also lower than those predicted. These results are expected since existing initial flaws are likely to influence the results for off-axis tests to a much greater extent. Also, damage initiation stresses are lower for thicker specimens because of the likelihood of the existence of more flaws (and larger) in thicker sheets. Further, the value of R_0 may itself be lower for the initiation of through cracks in off-axis tests. In spite of these differences, the results from off-axis tests appear to be consistent with $(\pm \theta)$ tests. Since our objective is to model the behavior of a lamina in a laminate, the model parameters should be determined from $(\pm \theta)_{ns}$ or other laminate tests.

Effects of Large Crack Densities in Cross Ply Laminates

For $(\pm \theta)_{ns}$ layups considered in the previous section, the ply crack densities are found to be low (≤ 0.3), and the model based on the assumption of dilute concentration of such cracks appears to be adequate. Large crack densities are, however, observed in the 90° layers in cross ply laminates loaded in tension in the 0° direction [13], which coincides with the x_2 or y direction in Figs. 4 or 6. As discussed earlier, several authors have utilized approximate solutions to calculate stiffness loss in such laminates due to increasing crack densities [5–13]. In this section we first postulate the forms of change in strain and complementary (stress) energy densities in a $(0_m/90_n)_s$ layup when the crack density defined by Eq 3 is large. We consider equally spaced cracks in the 90° layer ($2n$ plies thick) with crack spacing equal to l. Noting that the complementary energy in the representative volume $2al$ of the 90° layer containing one crack (Fig. 4) is expressed by Eq 7 for $\omega \rightarrow 0$, one can postulate the following expression for large ω

$$V \cdot W_c = V \left[\frac{1}{2} S_{ij}^0 \sigma_i \sigma_j + \frac{1}{2} \omega C_1 F(\omega) \sigma_2^2 \right]$$

$$V = 2al$$

$$F(\omega) = 1, \omega \rightarrow 0$$

(26)

The stresses are average, and they will be quite different from that in the uncracked state because of the large crack density. Note that the stress σ_2 is tensile and hence $h = 1$ as per Eq 2c. Also, the average stresses, σ_6 and σ_3, used in Eq 7 are zero in the problem under consideration. The second term within the bracket quantifies the effect of energy change in the whole laminate (not in the 90° layer alone) due to the cracks as discussed previously, and $F(\omega)$ is as yet an unknown function quantifying crack interaction effects. The corresponding expression for strain energy is

$$V \cdot U = V \left[\frac{1}{2} C_{ij}^0 \epsilon_i \epsilon_j - \frac{1}{2} \omega C_1 f(\omega)(\sigma_2^0)^2 \right]$$

(27a-d)

$$f(\omega) = 1, \omega \rightarrow 0$$

$$C_{ij}^0 = \text{initial stiffness}$$

$$\sigma_2^0 = \text{stress in the 90° layer in the uncracked laminate} = \delta E_T^0 \epsilon_2,$$

$\epsilon_2 =$ the laminate strain in the x_2 direction, $E_T^0 =$ the initial transverse modulus, and $\delta =$ a factor close to unity. Note that $\delta = 1$ if the Poisson effect is neglected. The second term in the bracket quantifies the crack energy (change in strain energy density due to the crack) in a representative volume element of the laminate with 0° and 90° layers with length l (Fig. 4). For wide laminates in the x_1 direction, this energy can be estimated (see discussions preceding Eq 4 for the case of dilute concentration) by finding the work done by constant pressure, σ_2^0, prescribed on the crack surfaces on the crack opening displacement in the quasi–three-dimensional problem. The quantity, $aC_1 f(\omega)\sigma_2^0$, is the average crack opening displacement ($2a =$ thickness of 90° layers) when the ends $y = \pm l/2$ of the RVE (of the laminate) are not allowed to displace in the x_2 direction but are free of shear stresses. Therefore, $C_1 f(\omega)$ can be obtained by solving the boundary value problem of the RVE described above. A similar procedure was suggested for evaluating C_1 for dilute concentration of cracks, $f(\omega) = 1$, and, in such a case, Eq 27a is of the form suggested in Ref 2. In the following we give the values of C_1 and approximate expressions for $f(\omega)$ obtained from: (1) the approximate shear analogy model presented earlier (but with $U_y = 0$ at the ends $y = \pm l/2$), which is valid for infinitely stiff 0° layers, and (2) finite element solutions for a $(0/90_3)_s$ glass/epoxy material. The transversely isotropic layer properties used are

$$E_A = 41.7 \text{ GPA}, E_T = 13 \text{ GPa, and } \nu_A = \nu_T = 0.3 \tag{29}$$

The shear analogy

$$C_1 = \frac{1.74}{E_T^0}; f(\omega) \approx \tanh\left(\frac{\beta}{\omega}\right);$$

$$\beta = \pi/4\sqrt{2(1 + \nu_T)} = 0.487 \tag{30}$$

The finite element solutions

$$C_1 = \frac{2.8}{E_T^0}; f(\omega) \approx \tanh\left(\frac{\beta}{\omega}\right); \beta = 0.32 \tag{31}$$

It may be noted that the value of C_1 given above is identical to that obtained by integrating the energy release rates calculated from Isida's solution [20] for a cracked strip (six 90° plies) bonded to two stiffeners (0° plies) on its edges. The value of β in Eq 31 was evaluated to fit the finite element results for $f(\omega)$ given in Table 3 for values of ω up to 2.0 (crack spacing up to one fourth the 90° layer thickness). The concept of ω as a measure of loss in a load-bearing area (see discussion before Eq 3) is meaningless for large ω, but we still use the mathematical definition (Eq 3).

It may be noted that for comparatively large ω, $f(\omega) \approx \tanh(\beta/\omega) \approx \beta/\omega$, and if the Poisson effect is neglected, the effective stiffness loss (reduction of the original transverse modulus E_T^0) as obtained from Eqs 27a and 27d is $\beta C_1 E_T^0$. Use of the results (Eq 31) shows that 10% of the original stiffness will remain for large ω (on the order of 2.0). The value obtained by using Eq 30 is 15%. It appears, therefore, that for a different problem, if C_1 is known (or estimated for a dilute concentration case), β, and hence $f(\omega)$ can be evaluated from

$$\beta C_1 E_T^0 \approx 0.9 \tag{32}$$

It may be noted that for $\omega \approx 0.5$, Eqs 31 and 27 imply that about 20% of the original stiffness is retained (effectively) by the 90° layers.

Direct evaluation of $F(\omega)$ in the expression for complementary energy (Eq 26) is more complicated. However, $F(\omega)$ can be obtained by the iterative method if $f(\omega)$ is known. If

TABLE 3—Values for f(ω) from finite element solution.

	Finite Element Results			
ω	≤0.1	0.3	0.5	2.0
$f(\omega)$	1.0	0.768	0.564	0.165

the Poisson effect is neglected, then Eqs 27a-d yield

$$\sigma_2 = E_T^0 \epsilon_2 (1 - e(\omega))$$

$$E_T^0 \epsilon_2 \approx \sigma_2 [1 + e(\omega)/(1 - e(\omega))] \qquad (33a\text{-}c)$$

where

$$e(\omega) = \omega C_1 f(\omega) E_T^0.$$

To make the strain derivable from Eq 26 identical to that in Eq 33b one must choose

$$F(\omega) = f(\omega)/[1 - e(\omega)] \qquad (34)$$

which, with Eq 26, implies that the effective compliance of the 90° layers for $\omega = 0.5$ will be about five times that of the original compliance $(1/E_T^0)$. We have calculated the stiffness loss in the $(0/90_3)_s$ glass/epoxy laminate using both strain and complementary energy formulations and the values of C_1 and $f(\omega)$ from Eq 30 as well as Eq 31. All the results are close to one another except, in the initial stages, where Eq 30 yields less stiffness loss in the beginning since C_1 is less than that in Eq 31. For strain energy formulation, the strain energy density in the 0° layer is expressible in terms of original stiffnesses, and finding the total strain energy density, U_L, in the laminate using Eq 27 one obtains

$$U_L = \frac{1}{2} E_0^* \epsilon_2^2 = \frac{1}{2} E_0 \epsilon_2^2 - V_{90} \left[\frac{1}{2} \omega C_1 f(\omega) \right] (\delta E_T \epsilon_2)^2 \qquad (35)$$

where

V_{90} = the thickness fraction of the 90° plies,
E_0 = the original effective modulus of the laminate,
E_0^* = the reduced modulus due to cracks, and
δ = 0.994.

The results of normalized stiffness, E_0^*/E_0, are plotted in Fig. 11 along with the analytical lower bound results from Ref 9 and test data reported in Ref 13. The present calculations yield slightly higher values (lower stiffness loss) from the lower bound estimates for high values of ω (0.2 to 0.5), which is expected. However, they are also higher than the test data when ω is on the order of 0.5. The reason for this difference is not clear, but may possibly be attributed to growth of some delaminations between 0° and 90° plies from the ply cracks for higher crack densities.

The strain energy release rate quantified in terms of conjugate thermodynamic force, R, can be evaluated from either the complementary energy or the strain energy formulation (see Eqs 1 and 2). The results are given as follows

$$R = \frac{\partial W_c}{\partial \omega} = \frac{1}{2} C_1 \sigma_2^2 [F(\omega) + \omega F'(\omega)] \qquad (36a)$$

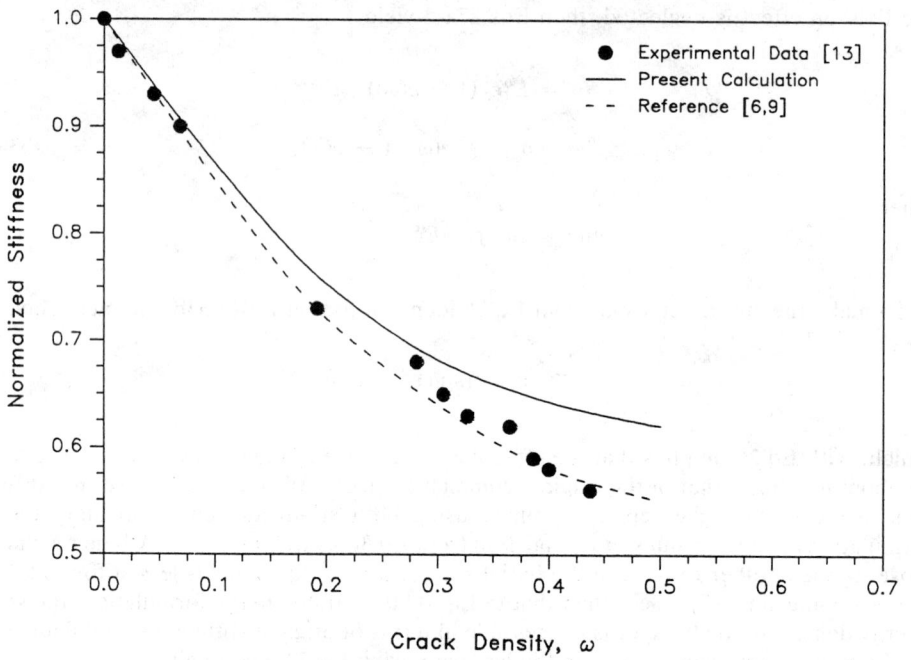

FIG. 11—*Stiffness reduction in (0/90₃)ₛ glass/epoxy.*

and

$$R = -\frac{\partial U}{\partial \omega} = \frac{1}{2} C_1(\sigma_2^0)^2[f(\omega) + \omega f'(\omega)] \tag{36b}$$

If the first crack appears at $\sigma_2 = \sigma_2^0 = \sigma_2^*$ then

$$R_0 = \frac{1}{2} C_1 \sigma_2^{*2} \tag{37}$$

For $\sigma_2^0 > \sigma_2^*$ using Eq 36b, ω can be found from the following relation (the calculation using 36a is more complicated)

$$f(\omega) + \omega f'(\omega) = \frac{R_0}{R_{app}} \tag{38}$$

where

$$R_{app} = \frac{1}{2} C_1(\sigma_2^0)^2 \tag{39}$$

σ_2^0 is given by Eq 27d for the known laminate strain, ϵ_2. Using Eq 31 for $f(\omega)$ reduces Eq 38 to

$$\tanh \beta/\omega - \beta/\omega \ \text{sech}^2 \ \beta/\omega = \frac{R_0}{R_{app}} \tag{40}$$

which we can use as the damage (crack) growth law instead of the power law-type relation (Eq 20).

Finally, we predict the growth of damage, ω, using Eq 40 and correlate it with measured crack densities ($1/l = \omega/a$, $a = 0.024$ in. $= 0.609$ mm) with an assumed value of $R_0 \approx 345 \times 10^3$ Pa $= 50$ psi. The results are plotted in Fig. 12 along with the predictions in Ref 9, which incorporates the effect of a residual stress of 1.97 ksi (13.6 MPa) in the 90° layers. Correlations of both predictions with test data [13] appear to be reasonable. We also note that R_0 may be related to the strain energy release rate, G_I^c, for transverse cracking by the following equation, which establishes a relationship between the phenomenological formulation and fracture mechanics-based approaches [9,10,12]

$$R_0 = G_I^c/a \tag{41}$$

where $2a =$ the thickness of the cracked plies, which in the present problem yields

$$G_I^c \approx 1.2 \ \text{in. lb/in.}^2 \approx 210 \ \text{J/m}^2 \tag{42}$$

This value is lower than that assumed in Ref 9 (330 J/m²) but comparable to that used in Ref 12 (200 J/m²). If the effects of residual stresses are considered, this value will be higher.

In this connection we note that the value of R_0 obtained in the previous section yields

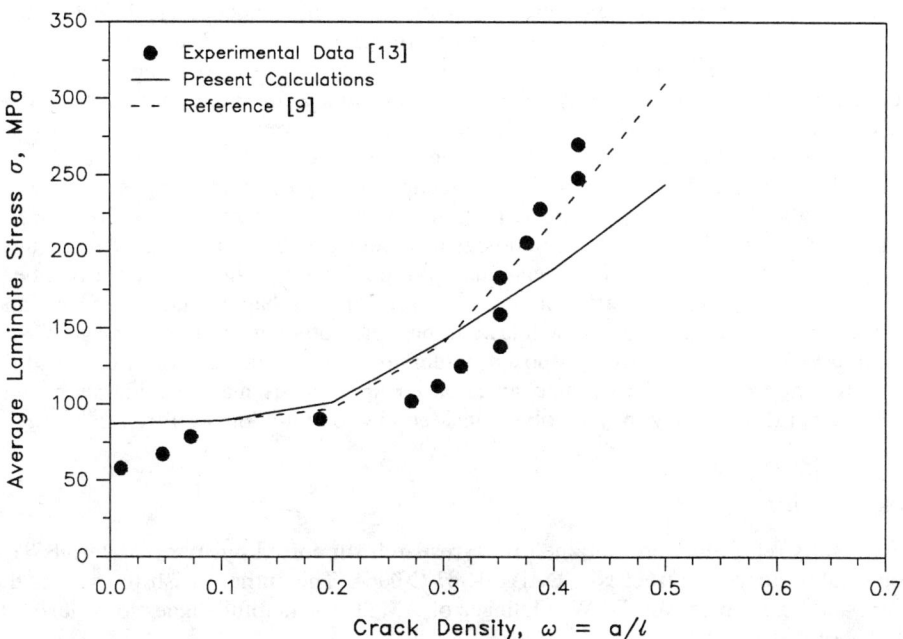

FIG. 12—*Crack density versus applied stress $(0/90_3)_s$ glass/epoxy.*

the following estimate, G^c, in AS4/3501-6 composites, since $R_0 \approx 170 \times 10^3$ Pa = 24.7 psi and $a = 0.127$ mm = 0.005 in.

$$G^c \approx 0.13 \text{ lb/in.} \approx 22 \text{ J/m}^2 \tag{43}$$

This value is very low as compared to the values of G_I^c used in Refs 10 and 12 for studying crossply graphite/epoxy systems with consideration to residual stresses. If we consider the existence of a residual transverse tensile stress of about 5.5 ksi in each of the layers of a (± 45)$_s$ system, then the estimate of G^c will increase to about 0.29 in. lb/in.^{2} (50 J/m^2), which still appears to be lower than the G_I^c assumed by other authors. It is possible that under mixed mode (Modes I and III) conditions (with Mode III dominating) the critical energy release rate is lower than that for pure Mode I. Alternatively, there exists a possibility that some other alternative mechanism (like plasticity or edge delaminations) causes the onset of nonlinearity. Edge delamination mechanism is, however, unlikely at such load levels and results from the (± 45) tests [*17*] with specimens of different widths show the onset of nonlinearity at similar stress levels. These points should be addressed in future studies with careful examinations for determining damage onset and growth.

Conclusions

We have shown that phenomenological approaches for dilute concentration of ply cracks ($\omega < 0.3$) provides a simple but useful tool for modeling the nonlinear response of angle ply graphite/epoxy laminates. The phenomenological formulation has been extended to consider the effects of cracks in nondilute concentration for the case of cross ply layups under tension. It has also been shown that a relationship exists between the critical conjugate thermodynamic force for crack initiation in the phenomenological model and the critical strain energy release rate used in fracture mechanics-based approaches. Further work is, however, needed for modeling such damage growth using both approaches since differences between critical energy release rates are found for single mode (Mode I) and mixed mode (Modes I and III) cases. The results for large crack densities appear to show that these models may not be accurate when transverse crack densities become high, possibly because delaminations begin to appear. Also, growth of transverse cracks of other orientations (in addition to those perpendicular to planes of lamination) have been observed in laminates. The phenomenological model can be easily extended to model growth of such cracks as well as delamination-type flaws by choosing more number of crack planes ($k > 2$). In fact, such models were developed to handle many damage planes. However, appropriate damage growth laws in presence of such complex damage patterns in composite laminates must be proposed, understood, and validated for effective utilization of this approach. Fracture mechanics-based approaches may provide such understanding even though they may involve considerable computational efforts.

Acknowledgments

The results reported were obtained from research supported by Army Materials Technology Laboratory, Contract No. DAAL04-87-C-0064. The authors wish to express their sincere appreciation to Mr. D. W. Oplinger of AMTL for helpful suggestions during the progress of this work.

References

[1] Lagace, P. A., "Nonlinear Stress-Strain Behavior of Graphite/Epoxy Laminates," *Proceedings, Part 1,* AIAA/ASME/ASCE/AHS 25th Structures, Structural Dynamics and Materials Conference, AIAA, New York, May 1984, p. 63.

[2] Ilankamban, R. and Krajcinovic, D., "A Constitutive Theory for Progressively Deteriorating Brittle Solids," *International Journal of Solids and Structures,* Vol. 23, No. 11, 1987, p. 1521. See also Krajcinovic, D., "Continuous Damage Mechanics Revisited: Basic Concepts and Definitions," *ASME Journal of Applied Mechanics,* Vol. 52, No. 4, 1985, p. 829.

[3] Chaboche, J. L., "Continuum Damage Mechanics: Part I and Part II," *ASME Journal of Applied Mechanics,* Vol. 55, No. 1, March 1988, pp. 59 and 65.

[4] Talreja, R., "Transverse Cracking and Stiffness Reduction in Composite Laminates," *Journal of Composite Materials,* Vol. 19, No. 4, July 1985, p. 355.

[5] Dvorak, G. J., Laws, N., and Hejazi, M., "Analysis of Progressive Matrix Cracking in Composite Laminates. I. Thermoelastic Properties of a Ply with Cracks," *Journal of Composite Materials,* Vol. 19, May 1985, p. 216.

[6] Hashin, Z., "Analysis of Cracked Laminates: A Variational Approach," *Mechanics of Materials,* Vol. 4, 1985, p. 121.

[7] Laws, N. and Dvorak, G. J., "Progressive Transverse Cracking in Composite Laminates," *Journal of Composite Materials,* Vol. 22, 1988, p. 900.

[8] Tan, S. C. and Nuismer, R. J., "A Theory of Matrix Cracking in Composite Laminates," *Journal of Composite Materials,* Vol. 23, 1989, p. 1009.

[9] Nairn, J. A., "The Strain Energy Release Rate of Composite Microcracking: A Variational Approach," *Journal of Composite Materials,* Vol. 23, 1989, p. 1106.

[10] Lim, S. G. and Hong, C. S., "Prediction of Transverse Cracking and Stiffness Reduction in Cross-Ply Laminated Composites," *Journal of Composite Materials,* Vol. 23, 1989, p. 695.

[11] Lee, J., Allen, D. H., and Harris, C. E., "Internal State Variable Approach for Predicting Stiffness Reductions in Fibrous Laminated Composites With Matrix Cracks," *Journal of Composite Materials,* Vol. 23, 1989, p. 1273. See also Allen, D. H., Groves, S. E., and Harris, C. E., "A Thermomechanical Constitutive Theory for Elastic Composites with Distributed Damage, Part II: Application to Matrix Cracking in Laminated Composites," *International Journal of Solids and Mechanics,* Vol. 23, 1987, p. 1379.

[12] Tan, S. C. and Nuismer, R. J., "A Theory of Progressive Matrix Cracking in Composite Laminates," *Journal of Composite Materials,* Vol. 23, 1989, p. 1029.

[13] Highsmith, A. L. and Reifsnider, K. L., "Stiffness Reduction Mechanisms in Composite Laminates," *Damage in Composite Materials, ASTM STP 775,* American Society for Testing and Materials, Philadelphia, 1982, p. 103.

[14] Sih, G. C. and Liebowitz, H., "Mathematical Theories of Brittle Fracture," *Fracture An Advanced Treatise, Vol. II,* H. Liebowitz, Ed., Academic Press, Inc., New York, 1968.

[15] Delameter, W. R., Herrman, G., and Barnett, D. M., "Weakening of an Elastic Solid by a Rectangular Array of Cracks," *Journal of Applied Mechanics,* Vol. 42, 1975, p. 74.

[16] Peterson, D. F., Prasad, S. N., and Chatterjee, S. N., "Complex Eigen Function Solution of Periodic Cracks in an Elastic Strip," *Journal of Applied Mechanics,* Vol. 40, 1973, p. 1126.

[17] Chatterjee, S. N., Wung, E. C. J., Ramnath, V., Yen, C. F., Adams, D. F., and Kessler, J., "Composite Specimen Design Analysis—Parts I and II," Army Materials Technology Laboratory, MTL TR 91-5, January 1991.

[18] Chatterjee, S. N. and Wung, E. C. J., "Software Development for Matrix Composite Material Evaluation," MSC TFR 1816/8304, Final Report, Contract No. N00164-87-C-0233 for Naval Weapons Support Center, January 1988.

[19] Beetz, C. P. Jr. and Budd, G. W., "Strain Modulation Measurements of Stiffening Effects in Carbon Fibers," *Review of Scientific Instruments,* Vol. 54, 1983, p. 1222.

[20] Isida, M., in *Mechanics of Fracture, Vol. 1,* G. C. Sih, Ed., Noordhoff, Groningen, Netherlands, 1973.

Xianqiang Lu[1] and Dahsin Liu[1]

Assessment of Interlayer Shear Slip Theory for Delamination Modeling

REFERENCE: Lu, X. and Liu, D., "**Assessment of Interlayer Shear Slip Theory for Delamination Modeling,**" *Composite Materials: Fatigue and Fracture, Fourth Volume, ASTM STP 1156,* W. W. Stinchcomb and N. E. Ashbaugh, Eds., American Society for Testing and Materials, Philadelphia, 1993, pp. 218–235.

ABSTRACT: This paper studied the effects of delamination on composite structure behavior. Composite beams made of glass/epoxy and pure glass were examined. Experimental results were compared with analytical solutions which were based on a previously developed theory, namely the interlayer shear slip theory (ISST). The ISST accounted for both interlaminar shear stress continuity and displacement discontinuity conditions on the delamination interface. A finite element program was developed for delamination simulation. It was found from the studies that central deflections and fundamental vibration frequencies were not significantly affected by the changes of delamination length and location except when delamination extended to beam ends. In addition, the changes of deflection and vibration were not sensitive to the delamination position in the beam thickness direction except at midplane. Since both beam ends and midplane had maximum transverse shear stresses, it is concluded from these results that the modeling of delamination should consider the transverse shear effect.

KEY WORDS: composite beam, central delamination, end delamination, interlaminar shear stress, interlayer shear slip, central deflection, and vibration frequency

Fiber-reinforced polymer-matrix composite laminates have some advantage to high performance structures which require materials with high stiffness and strength but low density. However, in contrast to their in-plane properties, the composite laminates are very vulnerable to transverse loading. As a consequence, delamination can easily occur on the composite interface. Since delamination can greatly reduce the integrity of composite structures, it has become an important issue in the composite laminate analysis.

In studying the behaviors of composite laminates with delamination, the classical laminate theory (CLT) [1–3] and the high-order shear deformation theory (HSDT) [4] have been widely used. Barbero and Reddy have also presented the generalized laminated plate theory (GLPT) for similar study [5]. Since these three theories do not consider the interlaminar stress continuity conditions on the composite interface, they have deficiencies in examining delamination. Some of them even experience difficulty [6] if the delamination is not located on the midplane of a composite laminate.

In addition to the laminate theories, the elasticity approach has also been used to study composite laminates with delamination. However, it is not a practical technique because analytical solutions only exist for a very few special cases [7,8]. Moreover, some conventional finite element methods have been employed for delamination analysis. It has been concluded that they require many elements to obtain accurate results [9,10]. Hence, it is

[1]Department of Material Sciences and Mechanics, Michigan State University, East Lansing, MI 48824.

the purpose of this study to extend a multiple layer theory developed by the authors [11], namely the interlayer shear slip theory (ISST), for delamination simulation. This quasi-two-dimensional technique can give accurate results for both deflections and stresses.

In this study, ISST is converted into a finite element program for both static and vibration analysis. The effects of delamination size, position, and location on central deflection and natural vibration frequency are studied. Experimental investigations are also performed. The comparison between the experimental results and the analytical solutions are used to assess the feasibility of using ISST for delamination modeling.

Fundamentals

As mentioned above, the analytical study is based on a previously developed interlayer shear slip theory [11]. For completeness, a brief discussion is summarized in the following section. Since finite element formulation for delamination modeling is the main subject of this study, it is presented in a later section.

Interlayer Shear Slip Theory

A composite beam composed of n layers, as shown in Fig. 1, is considered. A Cartesian coordinate system is chosen such that the midplane of the beam occupies a domain, Ω, in

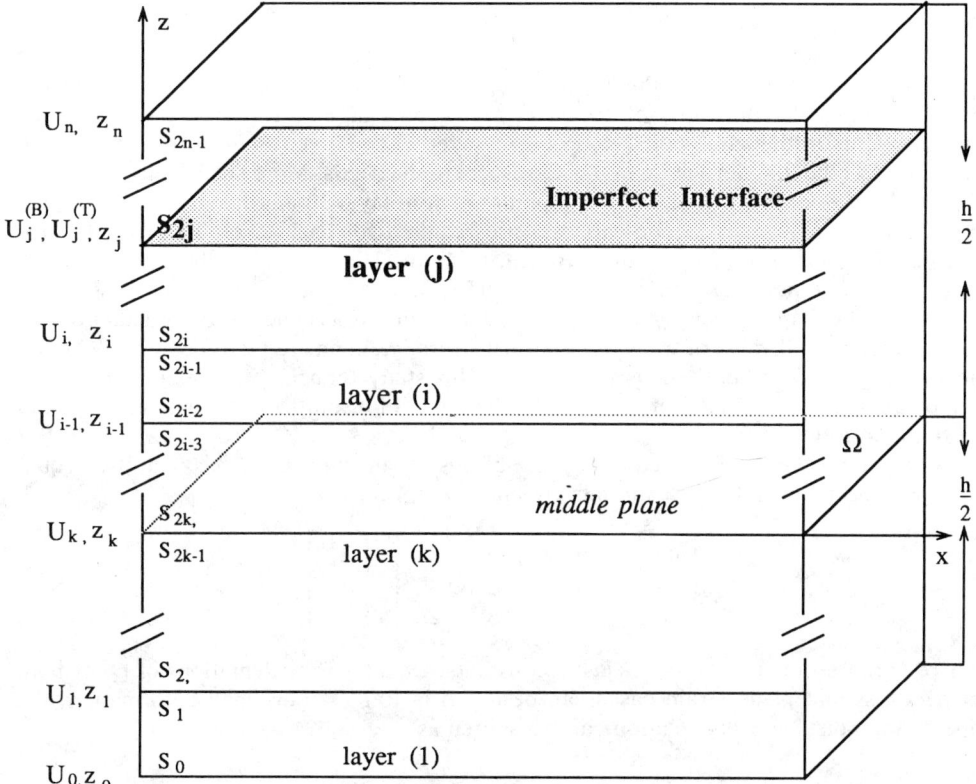

FIG. 1—*Nodal variables and coordinate system of an n-layer laminate.*

the x-y plane, and the z-axis is normal to this plane. The displacement field at a generic point (x,z) in the beam is assumed to be as follows

$$u(x, z, t) = \sum_{j=0}^{n} \{U_{i-1}^{(T)}(x, t)\phi_1^{(i)}(z) + U_i^{(B)}(x, t)\phi_2^{(i)}(z) + S_{2i-2}(x, t)\phi_3^{(i)}(z) + S_{2i-1}(x, t)\phi_4^{(i)}(z)\}$$

(1)

$$w(x, z, t) = w(x, t)$$

where t is the variable of time and $\phi_j^{(i)}$ are Hermite cubic shape functions which can be expressed as

$$\phi_1^{(i)} = 1 - 3[(z - z_{i-1})/h_i]^2 + 2[(z - z_{i-1})/h_i]^3$$

$$\phi_2^{(i)} = 3[(z - z_{i-1})/h_i]^2 - 2[(z - z_{i-1})/h_i]^3$$

$$\phi_3^{(i)} = (z - z_{i-1})[1 - (z - z_{i-1})/h_i]^2$$

(2)

$$\phi_4^{(i)} = (z - z_{i-1})^2([(z - z_{i-1})/h_1 - 1]/h_i)$$

if $z_{i-1} \leq z \leq z_i$, otherwise

$$\phi_1^{(i)} = \phi_2^{(i)} = \phi_3^{(i)} = \phi_4^{(i)} = 0$$

The superscript (i) denotes the layer number, i.e., the ith layer of the composite beam, and h_i denotes the thickness of the layer. As shown in Fig. 1, $U_i^{(B)}$ and $U_i^{(T)}$ are the displacements at the same point (x, z_i) but in layer (i) and layer $(i + 1)$, respectively. In addition, S is the first derivative of u with respect to the z-axis. More specifically, S_{2i-1} and S_{2i} represent nodal values of $\partial u/\partial z$ at a point (x, z_i) in layer (i) and layer $(i + 1)$, respectively. As a consequence, four variables are required for each layer, i.e., two U's for translation and two S's for rotation. Hence, using the Hermite cubic shape functions as given in Eq 2 to assemble the composite layers is justified [12].

In Eq 1, the displacement w in the thickness direction is assumed to be constant due to the relatively small value of transverse normal stress, σ_z, compared to other stress components [13,14]. Accordingly, σ_z is neglected in this study though it is questionable under some circumstance [15]. A study regarding the effect of σ_z on the composite analysis can be found in Ref 16.

In this study, fiber-reinforced composite beams are of interest. The constitutive equations for special orthotropic materials are employed, i.e.,

$$\begin{bmatrix} \sigma_x \\ \sigma_z \\ \tau_{xz} \end{bmatrix}^{(i)} = \begin{bmatrix} Q_{11} & Q_{13} & 0 \\ Q_{13} & Q_{33} & 0 \\ 0 & 0 & Q_{55} \end{bmatrix}^{(i)} \begin{bmatrix} \epsilon_x \\ \epsilon_z \\ \gamma_{xz} \end{bmatrix}^{(i)}$$

(3)

where Q represents functions of elastic constants. Details of the definitions of Q for both plane stress and plane strain cases can be found in Ref 14. For small deformation, the linear strain-displacement relation can be written as

$$\epsilon_x = \frac{\partial u}{\partial x}, \ \epsilon_z = \frac{\partial w}{\partial z}, \ \gamma_{xz} = \frac{\partial w}{\partial x} + \frac{\partial u}{\partial z}$$

(4)

In addition, both the displacements and interlaminar shear stress continuity requirements should be enforced on the composite interfaces, i.e.,

$$\Delta U_i = U_i^{(T)} - U_i^{(B)},$$
$$i = 1, 2, \ldots, n - 1 \tag{5}$$
$$\lim_{z \to z_i} \tau_{xz}^{(i+1)} = \lim_{z \to z_i} \tau_{xz}^{(i)}$$

where ΔU_i is the interlayer shear slip on the ith interface.

Perfect Interface—If the shear slip vanishes, $\Delta U_i = 0$, the interface between layer (i) and layer ($i + 1$) is called a perfect interface, and the bonding on the interface is called rigid bonding.

For perfect interface, the displacements on the composite interfaces draw the following conclusion

$$U_i^{(T)} = U_i^{(B)} = U_i \tag{6}$$

By substituting Eq 1 and Eq 4 into Eq 3, the stresses can be expressed in terms of displacements. With the use of continuity conditions of interlaminar stress in Eq 5, S_{2i-1} can be verified to be a function of S_{2i} and w

$$S_{2i-1} = \frac{Q_{55}^{(i+1)}}{Q_{55}^{(i)}} S_{2i} + \left[\frac{Q_{55}^{(i+1)}}{Q_{55}^{(i)}} - 1 \right] \frac{\partial w}{\partial x} \quad i = 1, 2, \ldots, n - 1 \tag{7}$$

Assume the shear tractions vanish on both top and bottom surfaces, i.e.,

$$\tau_{xz}\big|_{z = \pm(h/2)} = 0 \tag{8}$$

In the same fashion used in obtaining Eq 7, two more variables can be reduced by using Eq 8

$$S_0 = S_{2n-1} = -\frac{\partial w}{\partial x} \tag{9}$$

Therefore, it is concluded from Eqs 6, 7, and 9 that only two variables, U_i and S_{2i}, are required to express each nodal point. Consequently, the total number of independent variables is reduced to $2n + 1$. The reduced variables are assigned new notations and shown in Fig. 2. The displacement field can then be written as follows

$$u(x, z, t) = \sum_{j=0}^{n} U_j \Phi^j + \sum_{j=1}^{n-1} \overline{S}_j \Psi^j + \left[\sum_{j=1}^{n-1} \left(\frac{Q_{55}^{(i+1)}}{Q_{55}^{(i)}} - 1 \right) \Theta_j^{(1)} - \phi_3^{(1)} - \phi_4^{(n)} \right] \frac{\partial w}{\partial x}$$

$$w(x, z, t) = w(x, t) \tag{10}$$

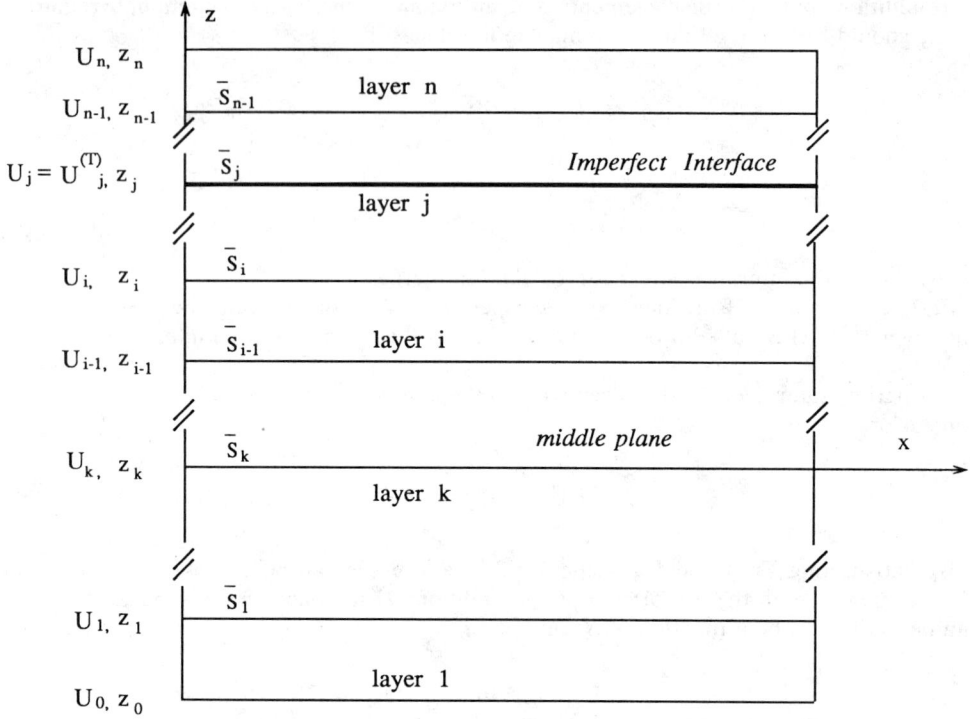

FIG. 2—*Reduced variables and coordinate system of an* n-*layer laminate.*

The shape functions, Φ, Ψ, and Θ, are

$$\Phi^j = \begin{cases} \phi_1^{(i)} & j = i - 1 \\ \phi_2^{(i)} & j = i \end{cases} \quad \text{layer } (i)$$

$$\Phi^j = 0 \qquad\qquad\qquad \text{others}$$

$$\Psi^j = \begin{cases} \phi_3^{(i)} & j = i - 1 \\ \dfrac{Q_{55}^{(i+1)}}{Q_{55}^{(i)}} \phi_4^{(i)} & j = i \end{cases} \quad \text{layer } (i) \qquad (11)$$

$$\Psi^j = 0 \qquad\qquad\qquad \text{others}$$

$$\Theta_j^{(1)} = \begin{cases} \phi_4^{(i)} & j = i & \text{layer } (i) \\ 0 & & \text{others} \end{cases}$$

Delaminated Interface—Assume the composite beam of interest has n_1 ($n_1 \leq n - 1$) delaminated interfaces which are located between layer (j_k) and layer ($j_k + 1$) as shown in Fig. 1. A set, Π, which contains all delaminated interfaces j_k, $k = 1, 2, \ldots, n_1$, is defined. In order to model the delaminated interface, a linear slip law presented by Newmark et al. [17] is employed, i.e.,

$$\Delta U_{j_k} = \mu_{j_k} \tau_{xz}^{(j_k)}; \ \mu_{j_k}(x) \geq 0. \quad \text{where} \quad J_k \in \Pi \qquad (12)$$

The coefficient μ is used to represent the interfacial bonding quality. For rigid bonding, μ is zero. However, μ goes to $+\infty$ if there is no bonding on the interface. In other words, if μ becomes infinite, the interface is delaminated. Since the slip ΔU should be a finite value, τ_{xz} must vanish on the delaminated interfaces. The values of μ between 0 and $+\infty$ correspond to nonrigid interfaces, by which is meant that the composite interface has the ability to transfer shear stress up to some extent [11].

Substituting Eq 12 into Eq 5, the displacements above and below the imperfect interface will have the following relation

$$U_{jk}^{(B)} = U_{jk} - \mu_{jk}\tau_{xz}^{(jk)}, \quad \text{where} \quad U_{jk} = U_{jk}^{(T)} \quad \text{and} \quad J_k \in \Pi \tag{13}$$

Therefore, it still requires two variables, U_i and S_{2i}, for each node. And the total number of independent variables remains $2n + 1$. The displacement field for a composite beam which includes n_1 delaminated bonded interfaces then becomes

$$u(x, z, t) = \sum_{j=0}^{n} U_j\Phi^j + \sum_{j=1}^{n-1} \overline{S}_j(\Psi - \mu_j Q_{55}^{(j+1)}\Theta_j^{(2)}) +$$

$$\left[\sum_{j=1}^{n-1}\left[\left(\frac{Q_{55}^{(j+1)}}{Q_{55}^{(j)}} - 1\right)\Theta_j^{(1)} - \mu_j Q_{55}^{(j+1)}\Theta_j^{(2)}\right] - \phi_3^{(1)} - \phi_4^{(n)}\right]\frac{\partial w}{\partial x}, \tag{14}$$

$$w(x, z, t) = w(x, t).$$

where

$$\mu_j = 0, \text{ if } j \in \Pi.$$

$$\Theta_j^{(2)} = \begin{cases} \phi_2^{(i)} & j = i \quad \text{layer } (i) \\ 0 & \text{others} \end{cases} \tag{15}$$

With the above displacement field, it is possible to obtain the equilibrium equations. Since the finite element solutions, instead of the closed-form solutions, are of major concern in this study, the equilibrium equations are omitted here.

Finite Element Formulation

Static Analysis—The total potential energy of a composite beam of interest can be written as

$$PE = \int_{\Omega}\left\{\int_{-h/2}^{h/2}\left[\frac{1}{2}\sigma_x\epsilon_x + \frac{1}{2}\tau_{xz}\gamma_{xz} - P_z w\right]dz + \sum_{i=1}^{n}\frac{1}{2}T_i\Delta U_i\right\}dA \tag{16}$$

$$T_i = \tau_{xz}^{(i)} = \tau_{xz}^{(i+1)} \quad \text{at} \quad z = z_i$$

After considering the continuity requirements on the composite interfaces and assuming constant μ for each element, the strains in layer (i) can be expressed in terms of the reduced variables.

$$\epsilon_x^{(i)} = \left\{ \frac{dU_{i-1}}{dx} \phi_1^{(i)} + \frac{dU_i}{dx} \phi_2^{(i)} + \frac{d\overline{S}_{i-1}}{dx} \phi_3^{(i)} + (A_1^{(i)}\phi_4^{(i)} - \mu_i Q_{55}^{(i+1)}\phi_2^{(i)}) \frac{d\overline{S}_i}{dx} \right\}$$
$$+ (B_1^{(i)}\phi_4^{(i)} - \mu_i Q_{55}^{(i+1)}\phi_2^{(i)}) \frac{d^2w}{dx^2}$$

$$= [N_n^{(i)}]\{X_1^{(i)}\}$$

$$\gamma_{xz}^{(i)} = \left\{ U_{i-1}\frac{d\phi_1^{(i)}}{dz} + U_i\frac{d\phi_2^{(i)}}{dz} + \overline{S}_{i-1}\frac{d\phi_3^{(i)}}{dz} + \overline{S}_i\left(A_1^{(i)}\frac{d\phi_4^{(i)}}{dz} - \mu_i Q_{55}^{(i+1)}\frac{d\phi_2^{(i)}}{dz} \right) \right\}$$
$$+ \left(A_1^{(i)}\frac{d\phi_4^{(i)}}{dz} - \mu_i Q_{55}^{(i+1)}\frac{d\phi_2^{(i)}}{dz} + 1 \right)\frac{dw}{dx} \tag{17}$$

$$= [N_s^{(i)}]\{X_s^{(i)}\}$$

where

$$A_1^{(i)} = \begin{cases} \dfrac{Q_{55}^{(i+1)}}{Q_{55}^{(i)}} & i = 1, 2, \ldots, n-1 \\ 1 & i = n \end{cases}$$

$$B_1^{(i)} = \begin{cases} \dfrac{Q_{55}^{(i+1)}}{Q_{55}^{(i)}} - 1 & i = 1, 2, \ldots, n-1 \\ 0 & i = n \end{cases}$$

$$[N_s^{(i)}] = [\tilde{\phi}_{1,z}^{(i)}, \tilde{\phi}_{2,z}^{(i)}, \tilde{\phi}_{3,z}^{(i)}, \tilde{\phi}_{4,z}^{(i)}, \tilde{\phi}_{5,z}^{(i)}]$$

$$\{X_2^{(i)}\} = \left\{ U_{i-1}, \overline{S}_{i-1}, U_i, \overline{S}_i, \frac{dw}{dx} \right\}^T = \mathbf{x}^{(i)T} \tag{18}$$

$$[N_n^{(i)}] = [\tilde{\phi}_1^{(i)}, \tilde{\phi}_2^{(i)}, \tilde{\phi}_3^{(i)}, \tilde{\phi}_4^{(i)}, \tilde{\phi}_5^{(i)}]$$

$$\{X_1^{(i)}\} = \left\{ \frac{dU_{i-1}}{dx}, \frac{d\overline{S}_{i-1}}{dx}, \frac{dU_i}{dx}, \frac{d\overline{S}_i}{dx}, \frac{d^2w}{dx^2} \right\}^T = \mathbf{x}'^{(i)T}$$

In Eq 18, the following variables are defined

$$\tilde{\phi}_1^{(i)} = \phi_2^{(i)}$$

$$\tilde{\phi}_2^{(i)} = \phi_3^{(i)} \tag{19}$$

$$\tilde{\phi}_3^{(i)} = \phi_2^{(i)}$$

$$\tilde{\phi}_4^{(i)} = (A_1^{(i)}\phi_4^{(i)} - \mu_i Q_{55}^{(i+1)}\phi_2^{(i)})$$

$$\tilde{\phi}_5^{(i)} = (B_1^{(i)}\phi_4^{(i)} - \mu_i Q_{55}^{(i+1)}\phi_2^{(i)})$$

Substituting strains and stresses into the Eq 16 and then integrating through the thickness, the total potential energy can be concluded as follows:

$$PE = \int_\Omega \left\{ \frac{1}{2} \sum_{i=1}^{n} (x^{'(i)T}A^{(i)}x^{'(i)} + x^{(i)T}(B^{(i)} + C^{(i)})x^{(i)}) - P_z w \right\} dA \tag{20}$$

The matrices $A^{(i)}$, $B^{(i)}$, and $C^{(i)}$ in the above equation have the following definitions

$$A_{lm}^{(i)} = \int_{z_{i-1}}^{z_i} Q_{11}^{(i)}\tilde{\phi}_l^{(i)}\tilde{\phi}_m^{(i)} \, dz$$

$$B_{lm}^{(i)} = \int_{z_{i-1}}^{z_i} Q_{55}^{(i)}\tilde{\phi}_{l,z}^{(i)}\tilde{\phi}_{m,z}^{(i)} \, dz \tag{21}$$

$$C_{lm}^{(i)} = \mu_i (Q_{55}^{(i+1)})^2 \{N_p\}_l^T \{N_p\}_m \quad l, m = 1, 2, \ldots, 5$$

where

$$\{N_p\} = \{0, 0, 0, 1, 1\} \tag{22}$$

Hence, the nodal variables in the thickness direction can be assembled first. The assembly of the finite elements in the x-direction can then be performed. In this study, Hermit cubic shape functions is also used for x-direction assembly. Accordingly, the potential energy can be expressed as

$$PE = \frac{1}{2}\hat{X}^T K \hat{X} - \hat{X}^T F \tag{23}$$

where $\hat{X}$ is the vector for nodal variables, K is the total stiffness matrix, and F is the associated external loading vector.

Vibration Analysis—In vibration analysis, the kinetic energy associated with the assumed displacement field can be written as

$$KE = \frac{1}{2} \int_\Omega \int_{-h/2}^{h/2} (\dot{u}^2 + \dot{w}^2) \, dz dA \tag{24}$$

Substituting the time derivatives of u and w into the above equation yields

$$KE = \frac{1}{2} \int_\Omega \sum_{i=1}^{n} (\dot{x}^{(i)T}K^{(i)}\dot{x}^{(i)} + \rho_i h_i w^2) \, dA \tag{25}$$

where ρ_i is the density of layer (i), $\dot{x}^{(i)} = \partial x^{(i)}/\partial t$ and

$$K_{lm}^{(i)} = \int_{z_{i-1}}^{z_i} \rho_i \tilde{\phi}_l^{(i)} \tilde{\phi}_m^{(i)} \, dz \tag{26}$$

With the same fashion as used in deriving the total potential energy, the kinetic energy can be expressed in terms of the mass matrix M and the vector for nodal variables, $\hat{X}$, i.e.,

$$KE = \frac{1}{2}(\hat{X}^T M \hat{X}) \tag{27}$$

Experimental Studies and Comparisons

In order to justify the feasibility of using the interlayer shear slip theory for delamination analysis and to explore the effect of delamination on the composite performance, composite beams with different kinds of delamination were investigated. In addition to the finite element analysis, some experimental studies which included central deflection and vibration frequency measurements were also performed.

Central Deflection

Composite beams with stacking sequence of [0] and fabricated from 3M's glass/epoxy tapes were examined. The specimens had effective dimensions of 80 by 25.4 by 3 mm, while the artificial central delamination had a length either 25.4 or 50.8 mm. The delamination was achieved by embedding a piece of TFE-fluorocarbon film at the midplane of the composite beams. The beams were then subjected to three-point bending with a maximum load of 0.3 kN as depicted in Fig. 3a. The central deflections of the composite beams were measured by a dial gage. The experimental results are normalized by the central deflection of a composite beam without delamination and are shown in the corresponding diagrams.

Delamination Length—The relation between the central deflection and the delamination length can be found in Fig. 4. In this study, the delaminations are located on the midplane of the composite beams and are symmetric with respect to the beam centers. The normalized delamination length vanishes when there is no delamination in the composite beam which equals to unity when the beam is completely delaminated. Reasonable agreements between the analytical solutions and the experimental results can be concluded. Also shown in Fig. 4, the central deflection remains insensitive to the delamination length when the delamination does not extend to the specimen end. However, if the delamination is very close to the beam end, a significant increase in central deflection takes place.

Delamination Location—It was interesting to evaluate the effect of delamination location on the beam deflection. In this study, the delamination size remained 25.4 mm, while the center of the delamination was moved along the midplane between the beam center and the beam end. As indicated in Fig. 5, the change of the central deflection is moderate except when the delamination approaches the beam end. The same conclusion is also obtained in the previous section.

The primary goals in this study were to examine two special cases of delamination, so-called central delamination and end delamination. The former refers to a delamination whose center coincides with the beam center, while the later is for a delamination which

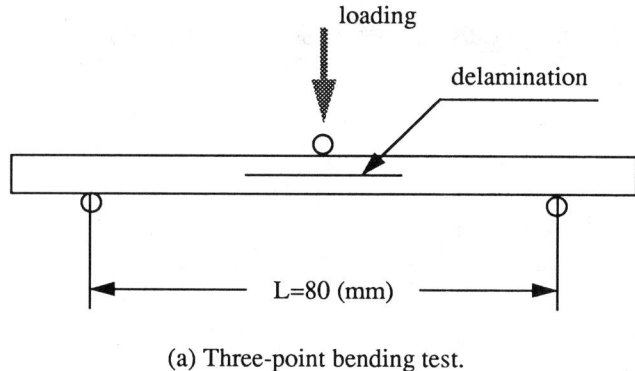

(a) Three-point bending test.

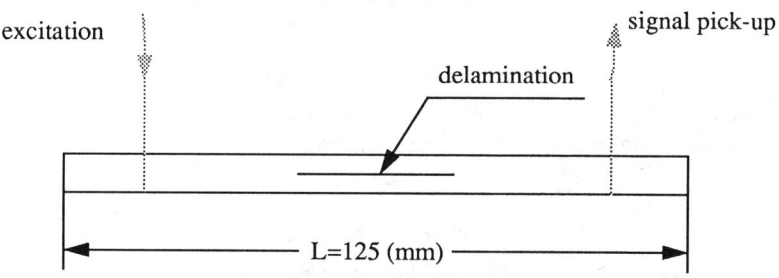

(b) Free-free vibration test.

FIG. 3—*Experimental setup for both three-point bend and free-free vibration tests.*

extends to the beam end. The central delamination is used in the center-notch flexure (CNF) test while the end delamination is used for the end-notch flexure (ENF) test. These two tests are important techniques for interlaminar strength characterization. As can be seen from Figs. 4 and 5, the central deflection is not sensitive to central delamination but is very sensitive to end delamination.

Delamination Position—In general, delaminations in composite laminates are not necessary to be located in the laminate midplane. The sensitivity of delamination at different thickness positions should also be considered. In this study, both 24.5 and 50.8-mm central delamination were investigated. The delaminations were positioned at the center of the composite beams while being moved from the midplane to the top surface. Shown in Fig. 6 are the results from finite element analysis as well as the experiments. Apparently, delamination of 25.4 mm causes only very little change to central deflection while that of 50.8 mm shows some influence when the central delamination is very close to the mid-plane, which is designated as the normalized delamination position of 0.5. The experimental results and the analytical solutions seem to have a discrepancy. It is believed that this may be due to the embedded TFE-fluorocarbon layer, which is neglected in the analytical solution. Similar doubt has also been expressed by other investigators [18].

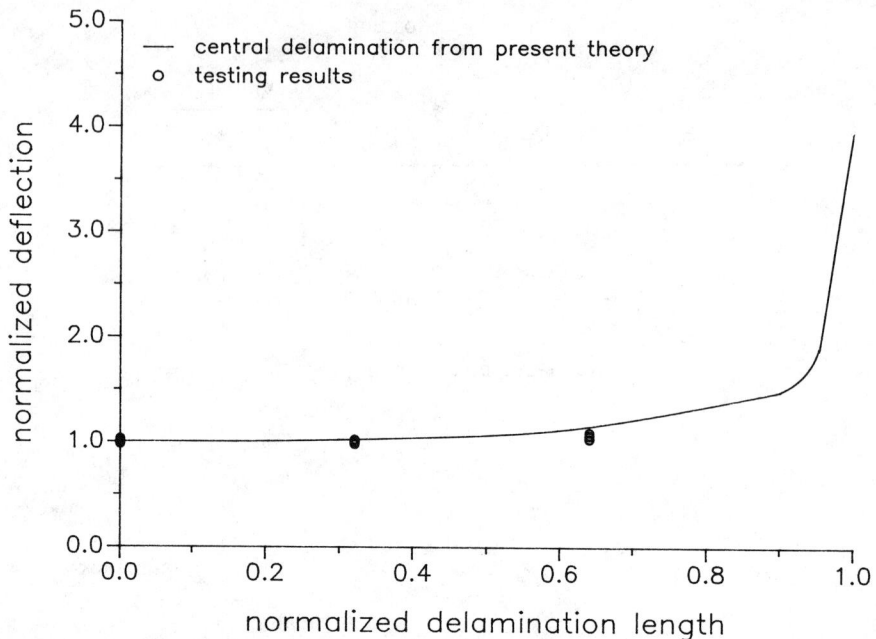

FIG. 4—*Normalized central deflection of glass/epoxy beams with central midplane delamination as a function of delamination length.*

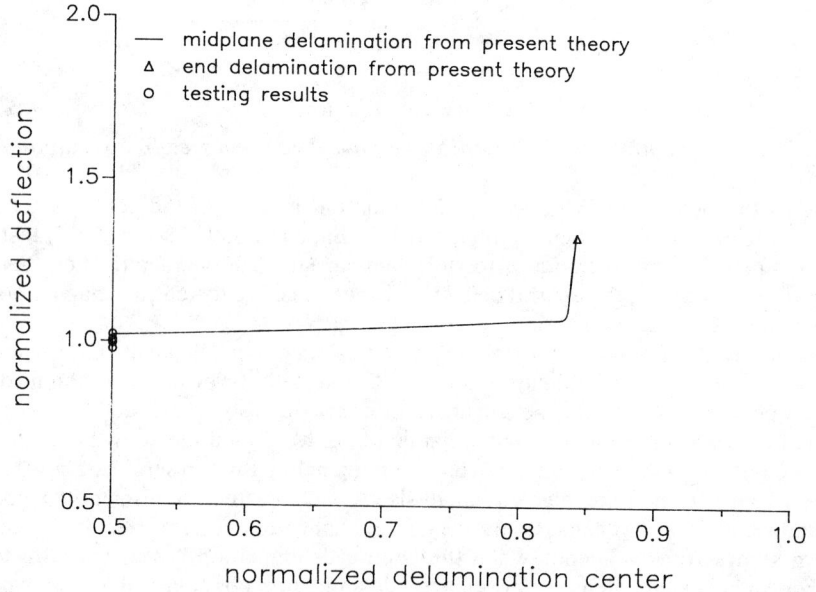

FIG. 5—*Normalized central deflection of glass/epoxy beams with 25.4 mm midplane delamination as a function of location of delamination center.*

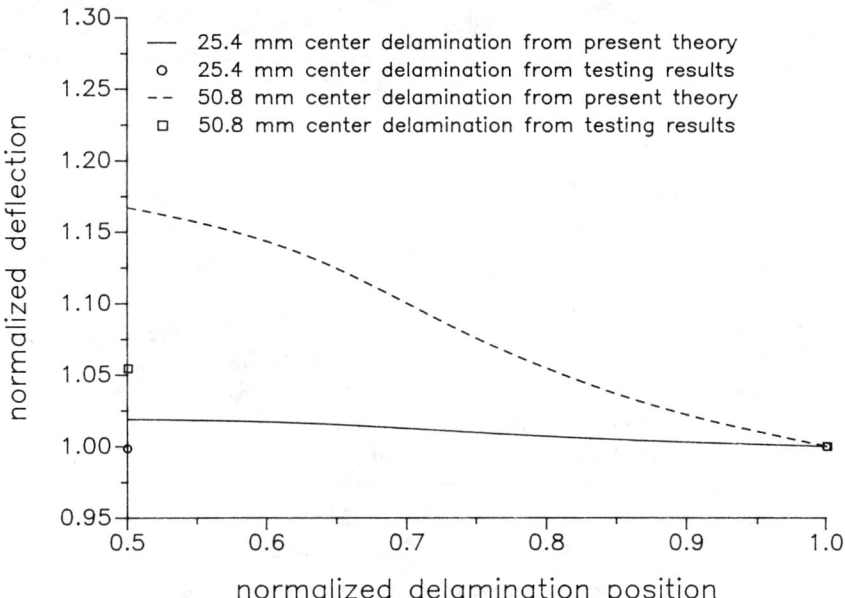

FIG. 6—*Normalized central deflections as a function of delamination position for glass/epoxy beams with 25.4 and 50.8 mm central delamination, respectively.*

Vibration Frequency

In studying the effect of delamination on vibration frequency, the ASTM Test Method for Young's Modulus, Shear Modulus, and Poisson's Ratio for Ceramic Whitewares by Resonance (C 848–78) was performed. The specimen configuration is depicted in Fig. 3b. The length of the testing specimens was 125 mm, while the width and thickness remained the same as those used in the previous testing for central deflection. However, in order to avoid the controversy due to the TFE-fluorocarbon layer, the specimens were made of two pieces of pure glass and joined by epoxy glue on the interface. Depending on the bonding area, different kinds of delamination were examined. The effects of delamination length on the composite performance are discussed as follows.

Delamination Length—The vibration frequency of a composite beam with central delamination was investigated. Shown in Fig. 7 are the first and the second modes of vibration from both experimental measurements and finite element analysis. Similar to central deflection analysis, the vibration frequency of the first mode does not change significantly until the delamination length extends to the beam end. The analytical solutions seem to reasonably agree with the experimental results.

The measurements for the second mode also agree very well with the analytical solutions. It can be concluded from Fig. 7 that the major change of frequency takes place when the delamination is of intermediate length. There is no abrupt change in frequency as the delamination extends to the beam end as occurs in the case of the first mode.

A further study of the relation between delamination and vibration frequency was performed for simply supported graphite/epoxy beams, which had received much attention by some other investigators [18]. The first four modes are depicted in Fig. 8. Although the boundary condition is different from the investigations for pure glass, the first two modes

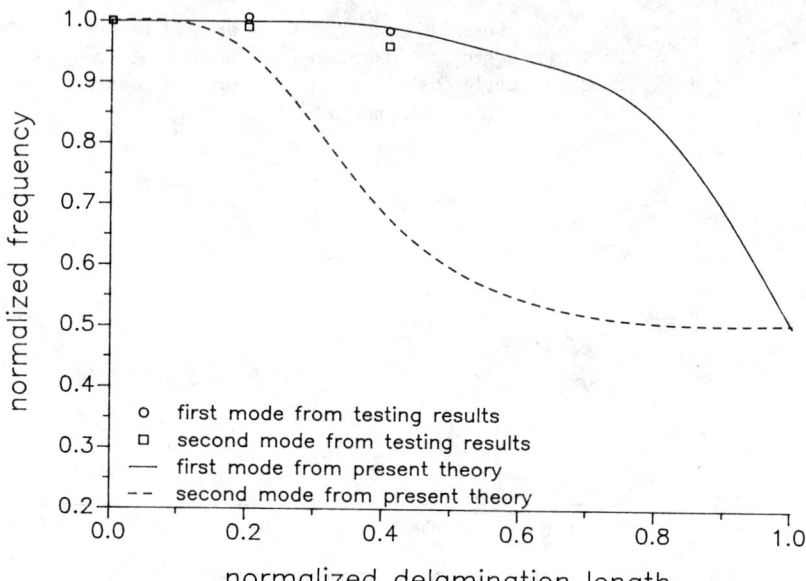

FIG. 7—*Normalized natural frequencies of glass/epoxy beams with central midplane delamination of different lengths.*

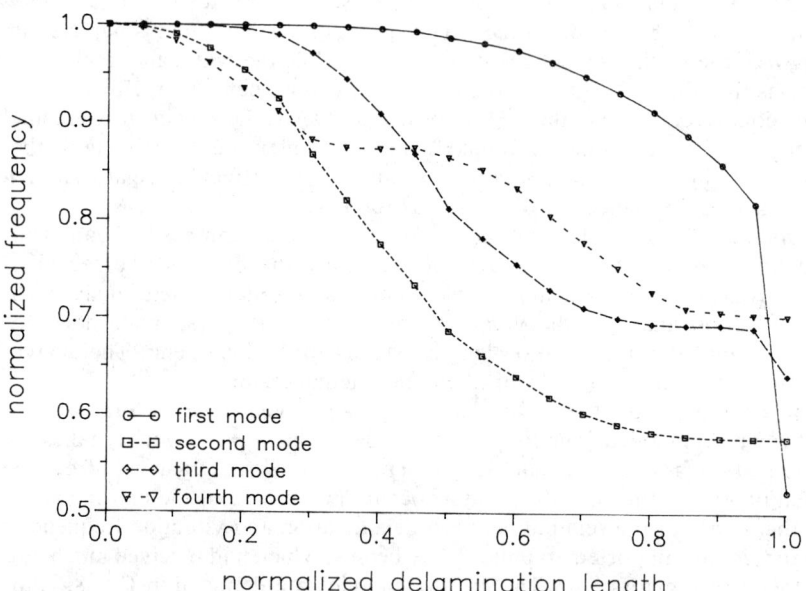

FIG. 8—*Normalized natural frequencies of graphite/epoxy beams with central midplane delamination as a function of delamination length.*

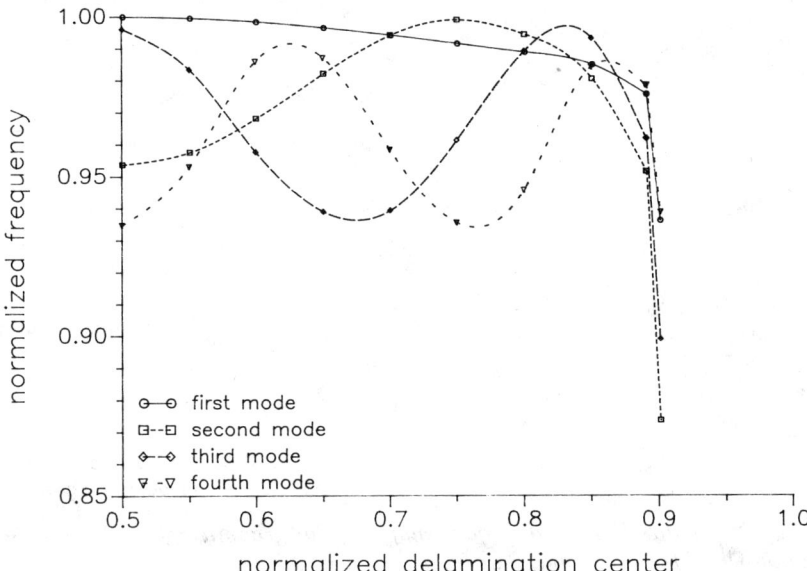

FIG. 9—*Normalized natural frequencies of graphite/epoxy beams with 20% midplane delamination as a function of location of delamination center.*

appear to have the same trends as those shown in Fig. 7. It is also noted that the odd modes are very sensitive to the end delamination while the even modes are not.

Delamination Location—Shown in Fig. 9 is the normalized frequency as a function of the location of the delamination center. The delamination length is this study was 20% of the beam length, while its center was moved along the midplane from the beam center to

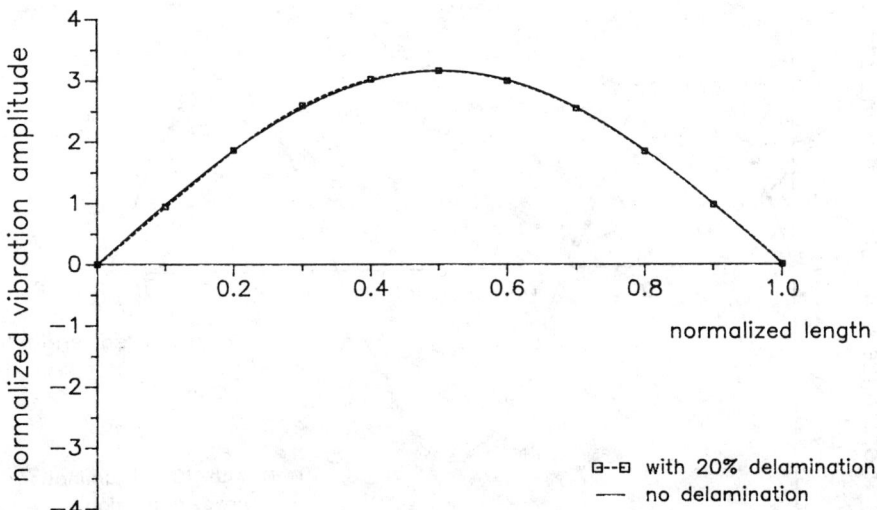

FIG. 10—*Normalized first mode shape of a graphite/epoxy beam with 20% delamination when the delamination center is at x = 0.2.*

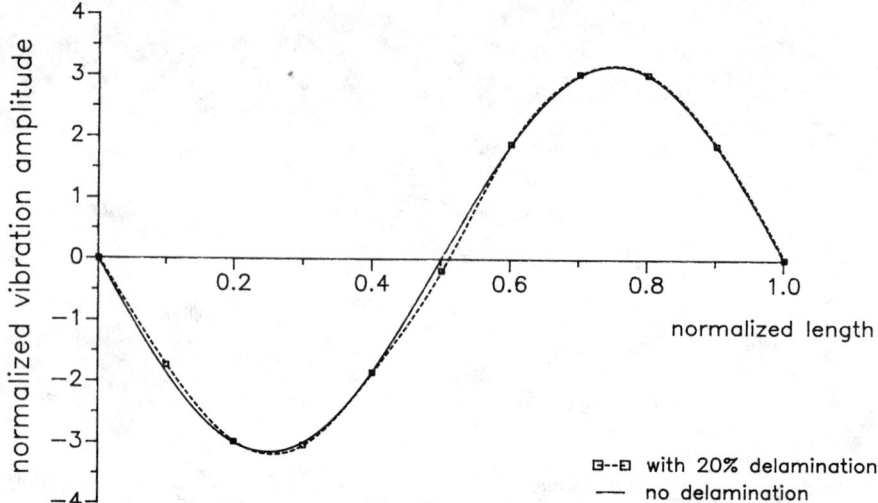

FIG. 11—*Normalized second mode shape of a graphite/epoxy beam with 20% delamination when the delamination center is at* x = 0.2.

the beam end. For all four modes, the frequencies experience significant reduction as they change from central delamination to end delamination. In addition, the patterns of the change seem to match the vibration modes. The largest changes in each mode seem to coincide with the nodal points of the individual mode. This implies that the vibration modes may have significant change if the delamination is located at the nodal point.

Shown in Figs. 10 through 13 are the changes of amplitudes of the four vibration modes when the center of a 20% central delamination is located around a nodal point of the fourth

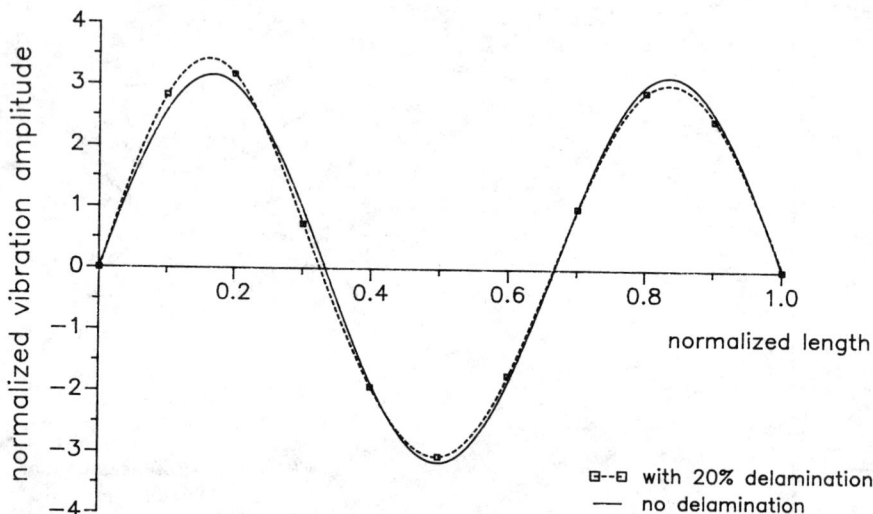

FIG. 12—*Normalized third mode shape of a graphite/epoxy beam with 20% delamination when the delamination center is at* x = 0.2.

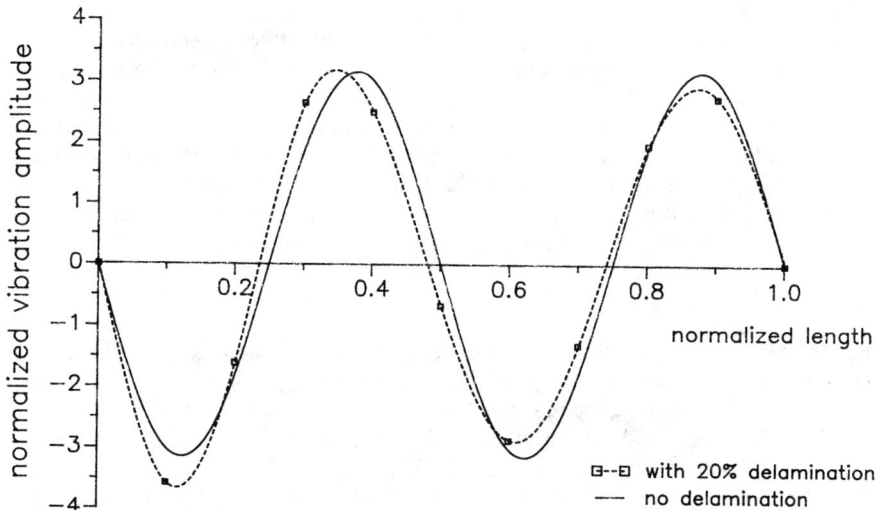

FIG. 13—*Normalized fourth mode shape of a graphite/epoxy beam with 20% delamination when the delamination center is at* x = *0.2.*

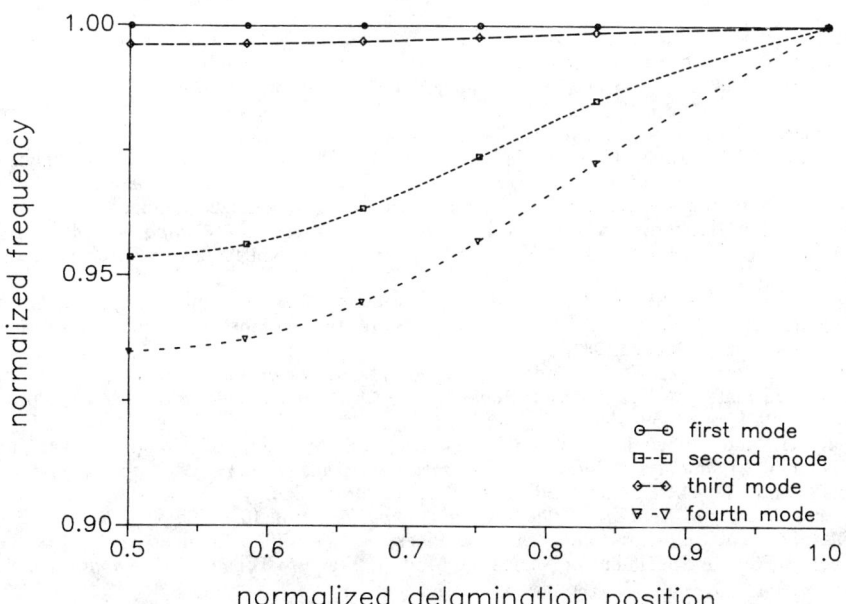

FIG. 14—*Normalized natural frequencies of graphite/epoxy beams with 20% central delamination as a function of delamination position.*

mode, i.e., 25% from one end. Hence, among the four vibration modes, the fourth mode experiences the most significant change in vibration amplitude, as shown in Fig. 13.

Delamination Position—The change of vibration frequency due to the movement of a 20% central delamination in the thickness direction is not significant, although the even modes seem to experience a little higher change than the odd ones. The value of 0.5 of the normalized delamination position in Fig. 14 corresponds to the midplane, while 1.0 corresponds to the top surface of the composite beam. Since the transverse shear stress is maximum at midplane but vanishes at the surface, the delamination has more significant influence on the vibration frequency if positioned on the midplane.

Conclusions

Both the experimental results and the analytical solutions indicate that the changes of central deflection and vibration frequency due to the changes of delamination length and location are not significant except when the delamination approaches the beam end. In addition, from the study of delamination position, it is concluded that the change of central deflection and vibration frequency can be distinguishing if the delamination is located at midplane. Since both beam end and midplane are locations of maximum transverse shear stress, the transverse shear effect seems to play a very important role in the deflection and vibration of composite beams with delamination. Moreover, the analytical results indicate that if delamination coincides with a nodal point of vibration, the vibration frequency can undergo significant change. Accordingly, it is concluded from these studies that it is important to include the interlaminar shear effect in the delamination analysis.

References

[1] Carlsson, L. A., Gillespie, J. W. Jr., and Pipes, R. B., "On the Analysis and Design of End Notch Flexure (ENF) for Mode II Testing," *Journal of Composite Materials,* Vol. 20, 1986, pp. 594–604.
[2] Maikuma, H., Gillespie, J. W. Jr., and Whitney, J. M., "Analysis and Experimental Characterization of the Center Notch Flexure Test Specimen for Mode II Interlaminar Fracture," *Journal of Composite Materials,* Vol. 23, 1989, pp. 757–786.
[3] Tracy, J. J. and Pardoen, G. C., "Effect of Delamination on the Flexural Stiffness of Composite Laminates," *Thin-Walled Structures,* Vol. 6, 1988, pp. 371–383.
[4] Whitney, J. M., "Analysis of Interlaminar Mode II Bending Specimens Using a Higher Order Beam Theory," *Journal of Reinforced Plastics and Composites,* Vol. 9, 1990, pp. 523–536.
[5] Barbero, E. J. and Reddy, J. N., "An Application of the Generalized Laminate Plate Theory to Delamination Buckling," *Proceedings of American Society for Composites,* 4th Technical Conference, Blacksburg, VA, 3–5 Oct. 1989, Technomic Publishing Co., Lancaster, PA, pp. 244–253.
[6] Russell, A. J. and Street, K. N., "Factors Affecting the Interlaminar Fracture Energy of Graphite/Epoxy Laminates," *Proceedings,* Fourth International Conference on Composite Materials, Tokyo, Japan, 1982, pp. 279–286.
[7] Chai, H., Babcock, C. D., and Knauss, G., "One-Dimensional Modeling of Failure in Laminated Plates by Delamination Buckling," *International Journal of Solids and Structures,* Vol. 17, 1981, pp. 1069–1083.
[8] Chai, H. and Babcock, C. D., 'Two-Dimensional Modeling of Compressive Failure in Delaminated Laminates," *Journal of Composite Materials,* Vol. 19, 1985, pp. 67–98.
[9] Lu, X. and Liu, D., "Finite Element Analysis of Strain Energy Release Rate at Delamination Front," *Journal of Reinforced Plastics and Composites,* Vol. 10, 1991, pp. 279–292.
[10] Gillespie, J. W. Jr., Carlsson, L. A., and Pipes, R. B., "Finite Element Analysis of the End Notched Flexure Specimen for Measuring Mode II Fracture Toughness," *Composites Science and Technology,* Vol. 27, 1986, pp. 177–197.
[11] Liu, D. and Lu, X., "An Interlayer Bonding Theory for Delamination and Imperfect Interface Analysis," *Proceedings of American Society for Composites,* 6th Technical Conference, Albany, NY, 7–9 Oct. 1991, Technomic Publishing Co., Lancaster, PA, pp. 89–96.

[*12*] Becker, J., Carey, G. F., and Oden, J. T., *Finite Elements—Special Problems in Solid Mechanics,* Vol. V, Prentice-Hall Inc., Englewood Cliffs, NJ, 1983.
[*13*] Ambartsumyan, S. A., *Theory of Anisotropic Plates,* J. E. Ashton, Ed., Technomic Publishing Co., Stanford, CT, 1970.
[*14*] Vinson, J. R. and Sierakowski, R. L., *The Behavior of Structures Composed of Composite Materials,* Martinus Nijhoff Publishers, Dordrecht, 1986.
[*15*] Pagano, N. J. and Soni, S. R., *Interlaminar Response of Composite Materials,* N. J. Pagano, Ed., Elsevier Science Publishing Co. Inc., New York, 1989, pp. 1–86.
[*16*] Lee, C. Y. and Liu, D., "An Interlaminar Stresses Continuous Theory for Laminated Composite Beams Analysis," *AIAA Journal,* Vol. 29, 1991, pp. 2010–2012.
[*17*] Newmark, N. M., Seiss, C. P., and Viest, I. M., "Tests and Analysis of Composite Beams with Incomplete Interaction," *Proceedings,* Society for Experimental Stress Analysis, Vol. 9, Society for Experimental Mechanics, Bethel, CT, 1951, pp. 73–79.
[*18*] Tracy, J. J. and Pardoen, G. C., "Effect of Delamination on the Natural Frequencies of Composite Laminates," *Journal of Composite Materials,* Vol. 23, 1989, pp. 1201–1215.

Intralaminar and Interlaminar Fracture

Gretchen B. Murri[1] and Roderick H. Martin[2]

Effect of Initial Delamination on Mode I and Mode II Interlaminar Fracture Toughness and Fatigue Fracture Threshold

REFERENCE: Murri, G. B. and Martin, R. H., "**Effect of Initial Delamination on Mode I and Mode II Interlaminar Fracture Toughness and Fatigue Fracture Threshold,**" *Composite Materials: Fatigue and Fracture, Fourth Volume, ASTM STP 1156,* W. W. Stinchcomb and N. E. Ashbaugh, Eds., American Society for Testing and Materials, Philadelphia, 1993, pp. 239–256.

ABSTRACT: Static and fatigue double-cantilever beam (DCB) and end-notch flexure (ENF) tests were conducted to determine the effect of the simulated initial delamination on interlaminar fracture toughness, G_c, and fatigue fracture threshold, G_{th}. Unidirectional, 24-ply specimens of S2/SP250 glass/epoxy were tested with Kapton inserts at the midplane at one end to simulate an initial delamination. Four insert thicknesses were used: 13, 25, 75, and 130 μm. Some specimens were also tested with either tension or shear precracks as the initial delamination. To determine G_{th}, fatigue tests were conducted by cyclically loading specimens until delamination growth was detected. The fatigue fracture threshold was defined to be the maximum cyclic strain energy release rate, G_{max}, below which no delamination growth would occur in less than 1×10^6 cycles.

For the DCB specimens, consistent values of Mode I fracture toughness, G_{Ic}, were measured for specimens with inserts of thickness 75 μm or thinner, or with shear precracks. The fatigue DCB tests gave similar values of G_{Ith} for the 13, 25, and 75-μm specimens. Results for the shear precracked specimens were significantly lower than for specimens without precracks.

Results for both the static and fatigue ENF tests showed that measured G_{IIc} and G_{IIth} values decreased with decreasing insert thickness, so that no limiting thickness could be determined. Results for specimens with inserts of 75 μm or thicker were significantly higher than the results for precracked specimens or specimens with the 13 or 25-μm inserts.

KEY WORDS: delamination, ENF, DCB, strain energy release rate, fracture toughness, Mode I, Mode II, fatigue, fatigue fracture threshold

Nomenclature

A Constant in delamination growth law
a Delamination length
a_0 Initial delamination length
B Exponent in delamination growth law
b Specimen width
C Specimen compliance, δ/P

[1]Research engineer, U.S. Army Aerostructures Directorate, NASA Langley Research Center, Hampton, VA 23681-0001.
[2]Research scientist, Analytical Services and Materials, Inc., Hampton, VA 23681.

$\dfrac{da}{dN}$ Delamination growth per loading cycle

E_{11} Axial modulus of specimen

G Strain energy release rate

G_{I} Mode I strain energy release rate

G_{II} Mode II strain energy release rate

G_{c} Interlaminar fracture toughness

G_{Ic} Mode I interlaminar fracture toughness

G_{IIc} Mode II interlaminar fracture toughness

G_{max} Maximum cyclic strain energy release rate

G_{Imax} Maximum cyclic strain energy release rate for Mode I loading

G_{IImax} Maximum cyclic strain energy release rate for Mode II loading

G_{th} Fatigue fracture threshold for delamination growth onset

G_{Ith} Mode I fatigue fracture threshold

G_{IIth} Mode II fatigue fracture threshold

G_{13} Transverse shear modulus for a unidirectional laminate

h Specimen half-thickness

L ENF specimen half-span

m Constant for DCB compliance expression

n Exponent for DCB compliance expression

P Out-of-plane load

P_{c} Maximum load in static ENF test

P_{max} Maximum load in cyclic DCB test

P_{NL} Load at which load-displacement curve becomes nonlinear

R $\delta_{\mathrm{min}}/\delta_{\mathrm{max}}$ in fatigue tests

δ Load-point displacement

δ_{max} Maximum load-point displacement under cyclic loading

δ_{min} Minimum load-point displacement under cyclic loading

Introduction

In order to optimize the use of fiber-reinforced composite materials in primary aircraft structures, the damage tolerance of such materials under both static and fatigue loading must be established. The most common failure mechanism in laminated composites is delamination. Therefore, characterization of a composite material must include measurement of its interlaminar fracture toughness. However, there are still important unresolved issues concerning the measurement of interlaminar fracture toughness which this paper will address. The opening (Mode I) and sliding shear (Mode II) fracture toughnesses can be measured as a delamination grows in a specimen with a simulated initial delamination. This study considered the effect of several different types of initial delamination on interlaminar fracture toughness measured using the double-cantilever beam (DCB) and end-notched flexure (ENF) specimens under static and fatigue loading.

The DCB and ENF specimens are typically used to measure the delamination strain energy release rates due to Mode I (tension) and Mode II (shear) loading, respectively. Both of these tests may be performed statically to determine interlaminar fracture toughness, G_{c}, or cyclically to determine a fatigue fracture threshold, G_{th}, below which no delamination growth will occur. The DCB and ENF specimens are unidirectional laminates with an insert at the midplane at one end to simulate an initial delamination. During the manufacture of the specimen, a resin pocket may form at the tip of the insert. The size of the resin pocket depends on the thickness of the insert. A delamination starting at the insert tip must first extend beyond this resin pocket before it can grow along the interface between two plies. Values of G_c and G_{th} measured for a delamination growing through the resin pocket may be unconservative [1,2].

One method that has been used to avoid the problem of propagating the delamination through the resin pocket is to precrack the specimen. That is, before testing, the initial delamination is extended through the resin pocket and a short distance along the midplane, creating a sharp delamination front.

A Mode I, or tension, precrack may be created by wedging open the delamination faces until a delamination grows through the resin pocket. The disadvantage of this method is that it can cause fiber bridging. In the DCB specimen, fiber bridging increases the apparent opening mode strain energy release rate, G_I, as the delamination continues to grow [3–6]. However, in structural configurations, delamination usually occurs between plies of different orientation where fiber bridging would not be present [7]. Therefore, only the values of G_I measured before fiber bridging begins can be considered a material property. Reference 8 indicated that tension precracking may be unsuitable for the ENF specimen as well. If fiber bridging exists in the ENF specimen after tension precracking, the bridged fibers must be broken before the delamination can grow under the shear loading. This may result in higher values of G_{II} than if no fiber bridging was present.

Alternatively, a shear precrack may be created by using the ENF test fixture to apply a shear load to either the DCB or ENF specimen. The shear precracking does not cause fiber bridging, but it may create damage ahead of the delamination tip [9] that would not be present in a delamination not formed in shear, nor in a structure in which delaminations initiate at a structural discontinuity. This damage may result in measuring overly conservative values of G.

For composite materials under fatigue loading, the problems of fiber bridging and damage ahead of the delamination front must also be considered. Fatigue crack growth in metals has typically been characterized by relating crack growth per loading cycle to a cyclic stress intensity factor [10]. For composites, this method was modified to relate delamination growth to cyclic strain energy release rate using a power law of the form $da/dN = AG_{max}^B$, where da/dN is the delamination growth rate per fatigue cycle. The constants A and B are determined experimentally [11–13]. However, Refs 11 through 13 showed that for both the DCB and ENF specimens, the exponent B is so large (as high as 10 for the DCB specimen) that small errors in the applied load can result in large errors (up to an order of magnitude) in the predicted delamination growth rate. For example, for the DCB specimen with a material for which $B = 9$, a 10% error in the applied load value reduces the predicted da/dN by 61%. Therefore, the delamination growth approach is not suitable as a damage tolerance analysis for composites.

An alternative method for determining a fatigue fracture threshold for fatigue loading was proposed in Ref 13 and demonstrated for the AS4/PEEK material. Each specimen was cyclically loaded until delamination growth could be detected in the specimen. Specimens were tested over a range of G_{max} values, and the log of the number of cycles to delamination growth onset (log N) was plotted against G_{max}. Specimens that reached approximately 1 to 3 million cycles without delaminating were considered runouts. In this way a fatigue fracture threshold, G_{th}, below which delamination will not occur at less than 1 million cycles can be determined. However, the specimens used in Ref 13 contained inserts of 130-μm thickness, and it is not certain how these thick inserts may have affected the results.

In this study the effects of the insert thickness and precracking techniques on G_c and G_{th}

are investigated for both DCB and ENF specimens. The DCB test results are from Ref 5. Tests were conducted on glass/epoxy specimens with Kapton inserts of either 13, 25, 75, or 130-μm thickness, without precracks, growing the delamination directly from the insert tip. Tests were also conducted on specimens with precracks as the initial delamination. For the DCB specimens, shear precracks were created in specimens with 13-μm inserts. For the ENF specimens, both tension and shear precracks were grown from various thickness inserts. Results are compared for each type of initial delamination for both specimens.

Materials

Schematic diagrams of the DCB and ENF specimens are shown in Figs. 1 and 2, respectively. The specimens used in this study were 24-ply unidirectional specimens of S2/SP250 glass/epoxy made from a 122°C cure prepreg manufactured by the 3M Corp. Panels were laid up at the NASA Langley Research Center using four different thicknesses of Kapton film for the inserts: 13, 25, 75, and 130 μm. To prevent the inserts from adhering to the laminate surfaces, the insert material was sprayed with a release agent before the panels were cured. In this study, calculations were performed assuming material properties of E_{11} = 45.5 GPa, and G_{13} = 6.07 GPa.

The DCB and ENF specimens were cut from the same panels. All specimens were approximately 25 mm wide and had a nominal thickness, $2h$, of 4.7 mm. The DCB and ENF specimens were approximately 146 and 159 mm long, respectively, with inserts approximately 57 mm long. The specimens were determined to have an average fiber volume fraction of 62.8%.

Load was applied to the DCB specimens through piano hinges, indicated in Fig. 1, which were bonded to the specimen with Hysol EA9309, a two-part, room-temperature cure adhesive. Prior to testing, the DCB and ENF specimens were dried for 19 h. The

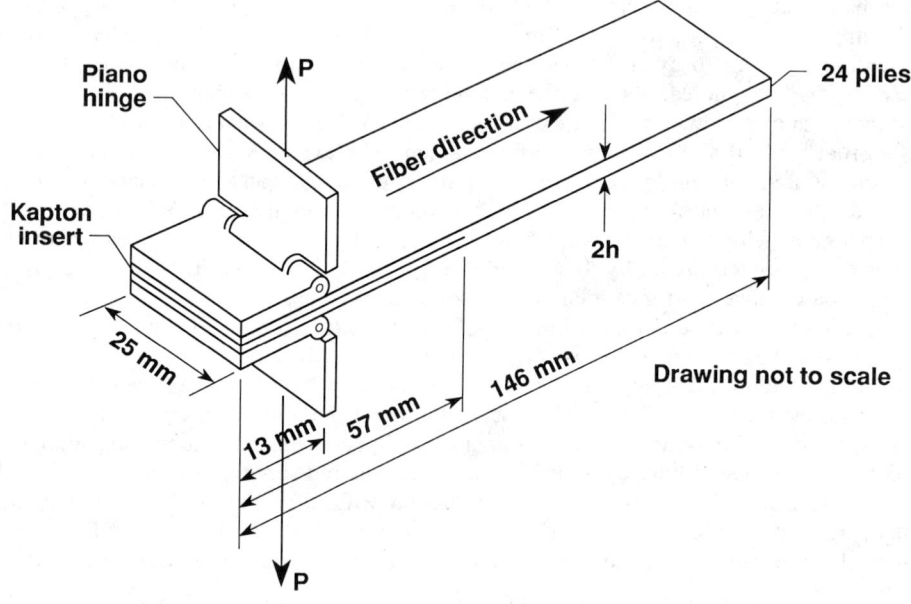

FIG. 1—*Diagram of DCB specimen.*

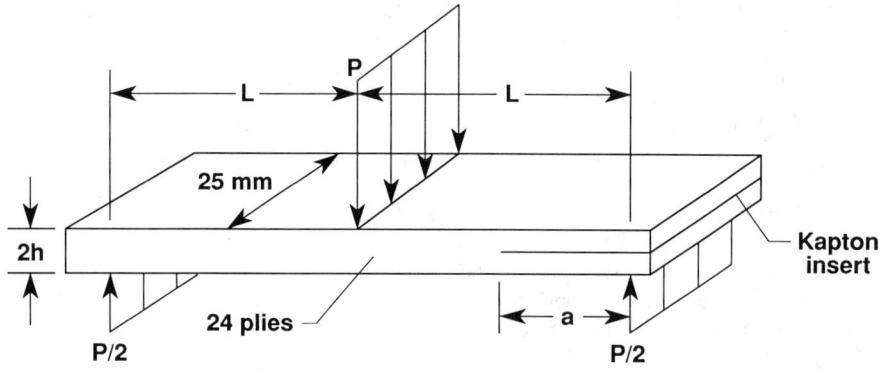

FIG. 2—*Diagram of ENF specimen.*

drying process consisted of heating for 1 h at 95°C, 1 h at 110°C, 16 h at 125°C, and 1 h at 150°C. After the specimens cooled to room temperature, they were stored in a dessicator until tested.

Experimental Procedure

Double Cantilevered Beam

The DCB test apparatus is shown in Fig. 3. Static DCB tests were conducted under displacement control at a crosshead displacement rate of 0.5 mm/min. The beam-opening displacement, δ, was measured using the displacement of the crosshead. Load was applied

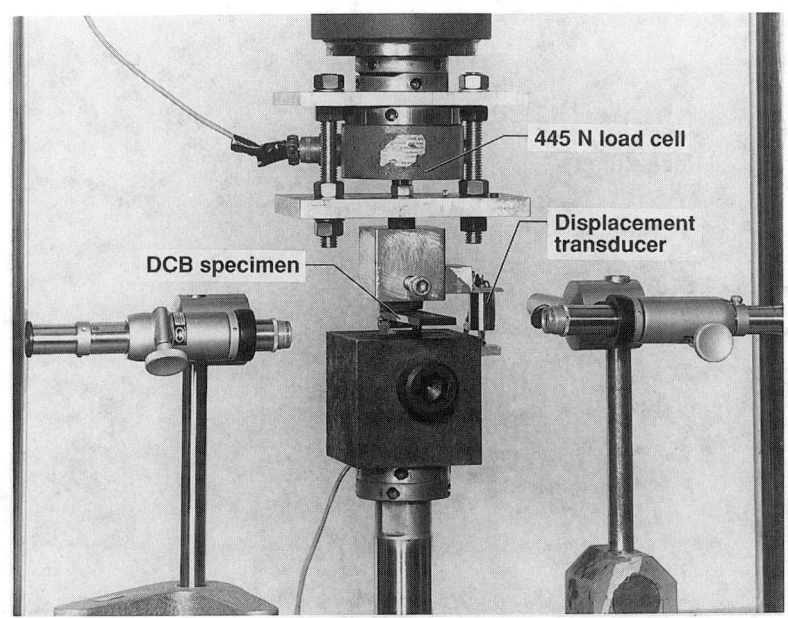

FIG. 3—*DCB test apparatus.*

to the specimen until the delamination had extended approximately 12 mm from the insert tip. An X-Y chart recorder was used to record the load and displacement of the specimen.

The fatigue DCB tests were also run under displacement control in a servohydraulic loading machine. The specimens were assumed to be linear elastic and were tested using a frequency of 10 Hz and an R-ratio of 0.5, where R is equal to $\delta_{min}/\delta_{max}$. Delamination growth onset was determined by visually inspecting the specimen edge with an optical microscope of $\times 60$ magnification and recording the number of loading cycles at which delamination growth onset was observed. Testing was continued until the delamination had grown approximately 0.25 mm. To obtain a curve relating the maximum cyclic G at delamination growth onset, G_{Imax}, and the corresponding number of loading cycles to delamination growth onset, N, specimens with different insert thicknesses or with shear precracks were cycled using a range of different maximum displacements.

End-Notched Flexure

The ENF test fixture, shown in Fig. 4, was mounted in a servohydraulic load frame. The specimen rested on the two outer rollers, and load was applied by the center roller. These rollers were mounted on ball bearings and were free to rotate. Because the specimen was delaminated on only one end, it deflected unsymmetrically, resulting in small side forces which tended to shift the specimen on the rollers as load was applied. The restraining bar shown in Fig. 4 prevented shifting of the specimen and was free to rotate as the specimen deflected during the test. The total span between the outer rollers, $2L$, was 101.6 mm. The specimen was placed on the rollers so that the distance from the outer roller to the tip of the insert, a_0, was approximately $L/2$ (Fig. 2). The specimen displacement was measured by means of a direct-current differential transducer (DCDT) mounted under the center of the specimen, with the rod supported by a spring. Prior to loading, the location of the outer

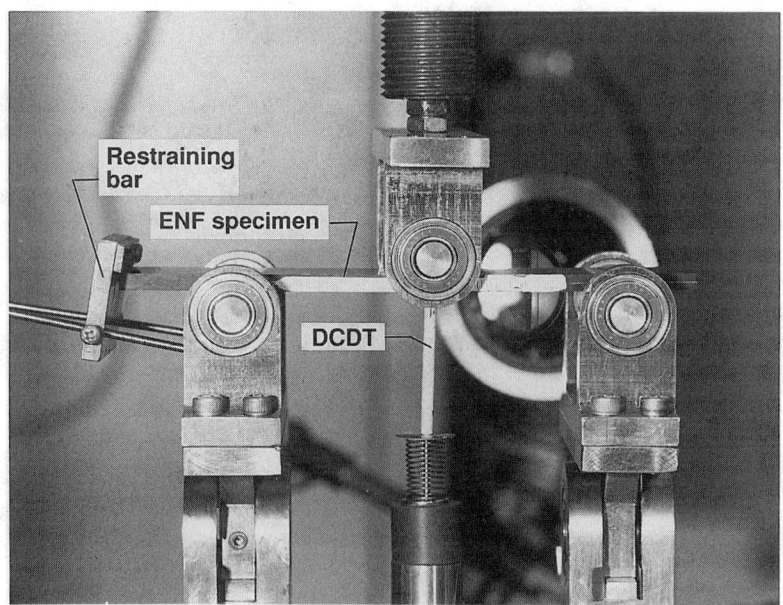

FIG. 4—*ENF test apparatus.*

load point was marked on the specimen for use later in determining the initial delamination length. Tests were conducted under displacement control, both statically and in fatigue. To determine G_{IIc}, specimens were loaded at a crosshead displacement rate of 2.5 mm/min until they delaminated. The load and center point displacement were recorded with an X-Y chart recorder. The specimens failed unstably in every case, with the delamination extending rapidly from the insert to the point under the central roller or slightly past it.

Cyclic ENF tests were run using the same apparatus as the static ENF tests. Specimens were loaded to the mean load and were then cycled sinusoidally at a frequency of 10 Hz and an R-ratio of 0.5. Each specimen was cycled until delamination growth onset could be detected. By testing at a range of cyclic G_{IImax} values, a threshold was determined below which delamination growth would not occur. An optical microscope with magnification of $\times 32$ was attached to the test stand to aid in visually determining the onset of delamination growth. Additionally, delamination growth was detected using the load output of the test stand load cell. Under displacement control, as the delamination grows, the maximum load decreases. Therefore, the load output was monitored with a digital voltmeter. Delamination growth of about 0.2 mm corresponded to an approximate 2% drop in the indicated load value. If a specimen reached approximately 1 to 3 million loading cycles with no indication of delamination growth onset, it was considered a runout and the test was stopped.

Precracking

Specimens of both the DCB and ENF laminates were tested with precracks. Both tension and shear loading were used to precrack the ENF specimens; only shear precracks were used for the DCB specimens because tension precracks initiate fiber bridging which increases the initial G_{Ic} value, and the results are therefore meaningless. Shear precracks were created for both the DCB and ENF specimens using the ENF test fixture. Specimens were positioned in the fixture so that the initial delamination length was slightly less than the half-span, L. The specimen was then loaded statically to delamination growth onset. The delamination extended to a point just under the central loading roller. For the ENF specimens, an optical microscope with magnification of $\times 32$ was used to locate the tip of the new delamination front, which was then marked on the specimen edge. After testing, the laminate was completely split by hand along the midplane to expose the fracture surfaces, and the exact location of the delamination tip prior to testing was determined by observing the delaminated surfaces. The previously mentioned load point mark was then used to determine the actual value of initial delamination length, a_0, to use in data reduction. If the precracking had not produced a straight delamination front, a_0 was determined to be the average of the initial delamination lengths at the edges and in the center of the specimen.

Tension precracks were created in the ENF specimens by clamping the specimens across the width, just ahead of the insert tip, and inserting a thin wedge (such as a putty knife blade) at the insert end of the laminate. The surfaces were pried apart until the delamination grew to the clamp. The new delamination front was located using the same technique that was used for the shear precracked specimens. In Ref 9, it was shown that surfaces that delaminated under tension loading look markedly different from surfaces that delaminated due to shear. Also, surfaces that delaminated under the same type of loading have a different appearance depending on whether the load was applied statically or cyclically. Therefore, for all cases except the ENF specimens with shear precracks that were tested statically, it was easy to verify the assumed delamination tip location in the precracked specimens after testing, by splitting the specimen and examining the delamination surfaces. The method of locating the new delamination tip with a microscope after precracking proved to be very accurate for the cases that could be checked after testing and therefore was assumed to be equally good for the static ENF tests with shear precracks.

Data Reduction

The equations used in this study to calculate the various strain energy release rates are given in the following section. They are based on classical linear beam theory expressions. Further explanation and derivations can be found in Refs *14* and *15* for the DCB and ENF tests, respectively.

Double Cantilevered Beam

The static fracture toughness, G_{Ic}, was calculated using Eq 1

$$G_{Ic} = \frac{nP\delta}{2ba} \tag{1}$$

where n, the exponent in the equation relating specimen compliance to delamination length ($\delta/P = ma^n$), was determined by testing and ranged from 2.37 to 2.56, with an average value of 2.52. The effect of fiber bridging on the value of n was not taken into account. As fiber bridging occurs, the measured compliance decreases and hence values of n will decrease. Because the effect on n is difficult to quantify, however, the experimentally determined value of n was used. This was considered a conservative approach.

The visually observed onset of delamination growth from an insert corresponded to a deviation from linearity in the load-displacement plot, as shown in Fig. 5. The loads and displacements corresponding to this deviation from linearity were used in Eq 1. However, for the specimens that were precracked in shear, the P–δ curve prior to delamination growth onset was nonlinear, as in Fig. 6, possibly indicating a change in the shape of the

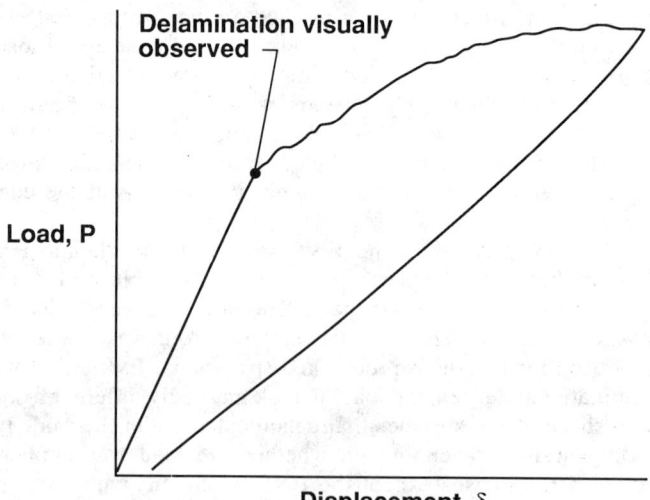

FIG. 5—*Load-displacement plot for DCB specimen with 13-μm insert and no precrack.*

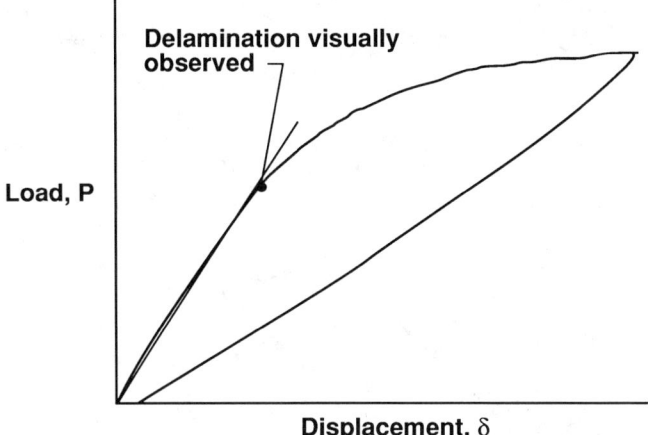

Delamination visually observed

Load, P

Displacement, δ

FIG. 6—*Load-displacement plot for DCB specimen with shear precrack.*

damage zone. Because there is no well-defined linear region for those cases, G_{Ic} was determined using the load and displacement values corresponding to delamination growth visible at the edges. The maximum Mode I strain energy release rate for cyclic loading was calculated from Eq 2 using the average value of n ($n = 2.52$) from the static tests and the maximum cyclic load and displacement at the first cycle.

$$G_{Imax} = \frac{nP_{max}\delta_{max}}{2ba} \qquad (2)$$

End-Notched Flexure

In Ref 9 it was shown that the beam theory equation for G_{IIc} with a correction for transverse shear agreed very well with predicted values from a 2-D finite element analysis for specimens with delamination lengths in the range used in this study. This equation is of the form

$$G_{II} = \frac{9P^2a^2C}{2b(2L^3 + 3a^3)}[1 + 0.2(E_{11}/G_{13})(h/a)^2] \qquad (3)$$

where C is the compliance of the ENF specimen, and the E_{11}/G_{13} term in Eq 3 is the shear correction term. A typical ENF load-displacement diagram for a specimen with a shear precrack is shown in Fig. 7. For the shear precracked and 13 and 25-μm insert specimens, the curve was relatively linear from the beginning of loading until the specimen was close to the final failure load. Just before the specimen delaminated unstably, the P–δ curve became nonlinear. The final failure load and the load at the deviation from linearity are indicated on the figure as P_c and P_{NL}, respectively. A greater degree of load-displacement nonlinearity was evident for the specimens with inserts thicker than 25 μm and in specimens with tension precracks. The onset of nonlinearity in the curve may be an indication of the beginning of subcritical delamination growth in the laminate. However, the delamination growth occurred very rapidly, and no visual observation of the beginning of

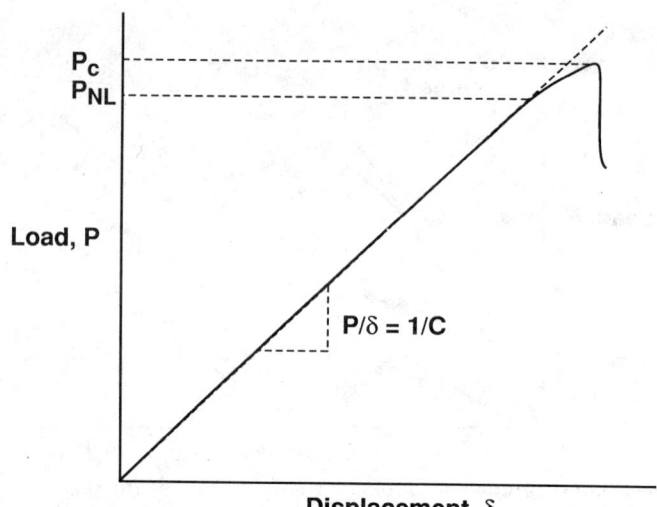

FIG. 7—*Load-displacement diagram for ENF specimen with shear precrack.*

delamination growth was possible. Therefore, Mode II fracture toughness values were calculated using both P_c and P_{NL} for the value P in Eq 3.

The maximum Mode II strain energy release rate for cyclic loading was calculated using Eq 4

$$G_{IImax} = \frac{9P_{max}^2 a^2 C}{2b(2L^3 + 3a^3)} [1 + 0.2(E_{11}/G_{13})(h/a)^2] \qquad (4)$$

where P_{max} is the maximum cyclic load, and a is the initial delamination length. The compliance, C, is the slope of the initial linear portion of the load-deflection curve for both Eqs 3 and 4.

Results and Discussion

Resin Pockets

Figures 8 through 10 show photos of the region around the insert tip for specimens with 13, 25, and 130-μm inserts, respectively. There is no obvious resin pocket for the 13-μm insert. However, a resin pocket is visible at the tip of the 25-μm insert, and there is a very obvious resin pocket associated with the 130-μm inserts.

Static Tests

Figure 11 shows G_{Ic} results for the DCB tests. As the delamination was allowed to grow in the static tests, fiber bridging was evident and resulted in a large (factor of 6) increase in G_{Ic}. Therefore, the values of G_{Ic} in Fig. 11 represent the initial delamination growth, that is, before any fiber bridging. The values shown are averages, with the number of specimens of each type tested indicated on the figure, along with the range of the data. The most scatter was observed for the 130-μm insert specimens, where the measured G_{Ic} values were within ±10% of the mean value. The measured fracture toughnesses are very similar

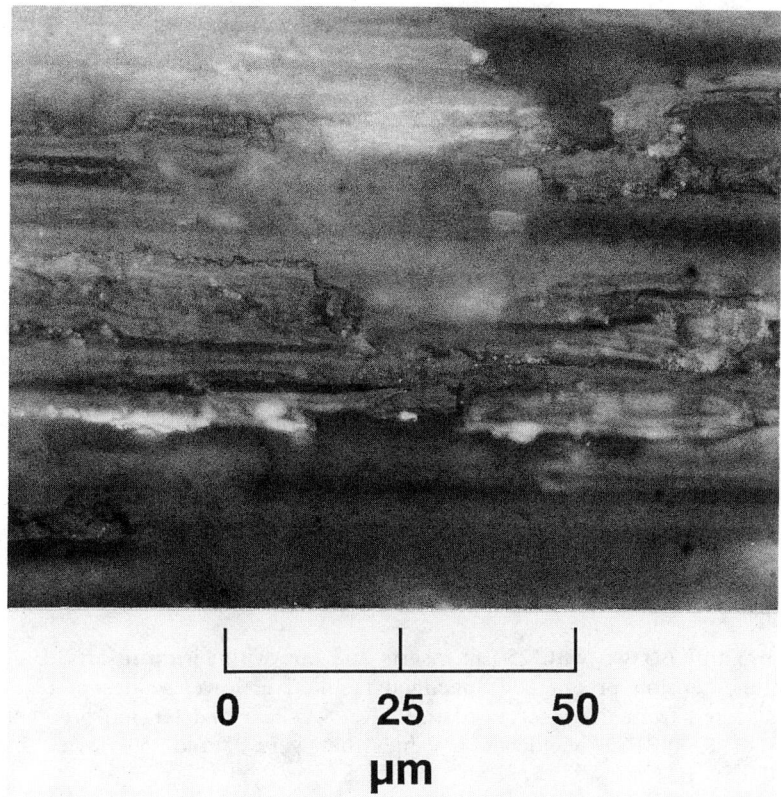

FIG. 8—*Specimen-edge photo showing 13-μm Kapton insert.*

for specimens with insert thicknesses of 13, 25, and 75 μm and for those with shear precracks. These results are consistent with the results of Ref *16* on adhesive bond thickness effects of an epoxy resin on Mode I fracture toughness. Reference *16* showed that an increase in fracture toughness occurred for bond thicknesses greater than 63.5 μm for a toughened epoxy resin (BP907).

Figure 12 shows the results for G_{IIc}, calculated using P_c and P_{NL}. For each case, the bars represent the average value of the listed number of specimens, and the range of the data is indicated. The calculated values for G_{IIc} were within ±13% of the mean values for all except the P_{NL} calculations for the 75-μm insert specimens and both calculations for the shear precracked specimens. Those cases showed significantly more scatter, with the range of the data within ±20 and ±30% of the mean G_{IIc} for the 75-μm insert specimens and the shear precracked specimens, respectively. Also, as Fig. 12 shows, there was more variation in G_{IIc} between the different initial delamination types than there was for the DCB tests. The results calculated using P_c yielded the highest values of G_{IIc} for the two thickest inserts, 75 and 130 μm, and the specimens with 13 and 25-μm inserts yielded the lowest values, with the 13-μm specimens being the lowest of the group. The two types of precracked specimens yielded very similar results that fell between the results for the 25 and 130-μm specimens. In other recent studies, however, the lowest values of G_{IIc} were measured from precracked specimens. For example, Ref *8* shows results for three different

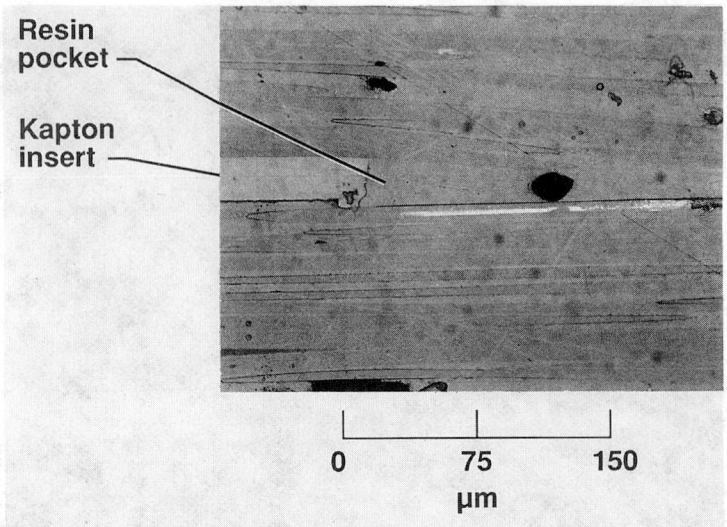

FIG. 9—*Specimen-edge photo showing 25-μm Kapton insert and resin pocket.*

composite materials, two with 25-μm inserts and one with 13-μm inserts. For all three materials, the tension precracked specimen resulted in lower values of G_{IIc} than the specimens tested from the inserts. Also, in Ref 9, shear and tension precracked ENF specimens of T300/BP907 yielded G_{IIc} values that were 25 and 35% lower than values

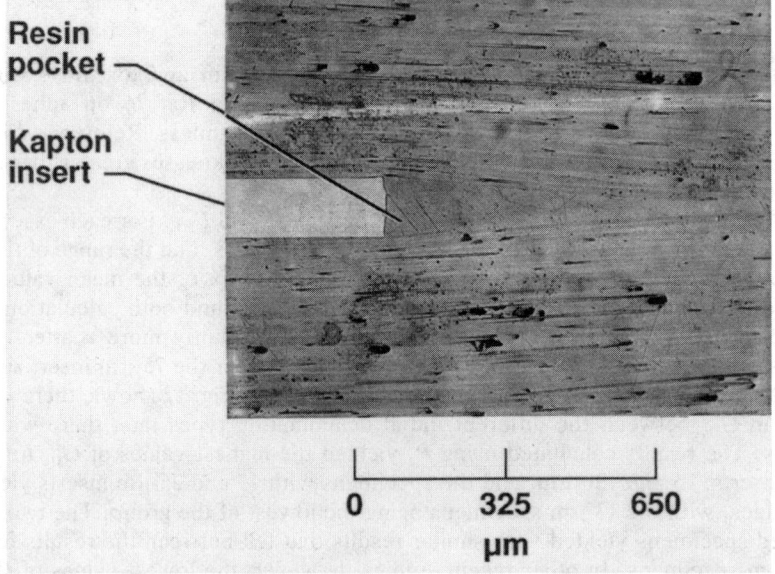

FIG. 10—*Specimen-edge photo showing 130-μm Kapton insert and resin pocket.*

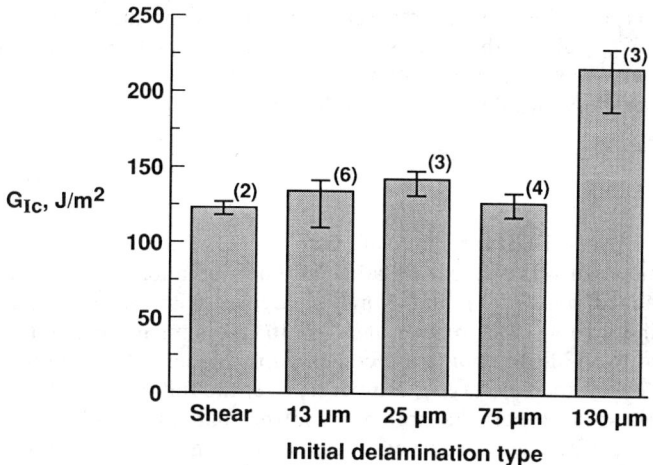

FIG. 11—*Mode I interlaminar fracture toughness for S2/SP250 DCB specimen.*

measured from inserts of double layers of 13-μm Kapton. The difference in results between Refs *8* and *9* and the current study may be a result of the fact that the glass/epoxy used in this study is very brittle and very prone to fiber bridging.

Results from the same ENF specimens, calculated using P_{NL} in Eq 3, are also shown in Fig. 12. In this case the lowest value corresponds to the 130-μm inserts, because those specimens exhibited a large amount of nonlinearity in the loading curve, possibly due to plasticity in the large resin pocket. There was also more load-displacement nonlinearity for the tension precracked specimens than for the other types, as shown by the larger differences in G_{IIc} between the P_c and P_{NL} calculations. This may be due to bridged fibers deforming and breaking [*8*]. The results for the 75-μm inserts, using P_{NL}, were again higher

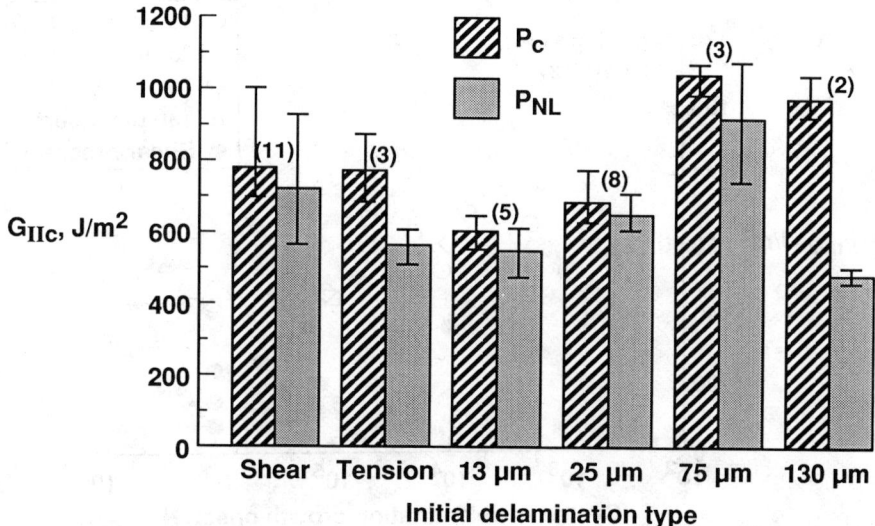

FIG. 12—*Mode II interlaminar fracture toughness for S2/SP250 ENF specimen.*

than the other types. The tension precracked specimens and the 13 and 25-μm specimens all had similar G_{IIc} values; the results for the shear precracked specimens were slightly higher. However, excluding the 130-μm insert specimens, whether the results are calculated using P_c or P_{NL}, G_{IIc} decreases with decreasing insert thickness without reaching a constant level.

Fatigue Tests

Results of the cyclic DCB tests are shown in Fig. 13. Each data point shown represents a single specimen that was cyclically loaded until delamination growth onset was detected, or until it reached between 1 and 3 million cycles with no delamination growth, as indicated by the arrows. The results at $N = 10^6$ are similar for the 13, 25, and 75-μm specimens, and the 130-μm results are considerably higher. There is considerable scatter in the data for every specimen type. However, over the full range of N, the results for the specimens with shear precracks are much lower than the results measured from the inserts. The same effect was observed in Ref *13* for a graphite/thermoplastic composite. The shear precracking process creates damage ahead of the delamination front in the form of microcracks [9]. As these microcracks coalesce, the delamination grows sooner than it would in an undamaged laminate. Therefore, under fatigue loading, delamination growth onset occurs much sooner in specimens that were precracked in shear.

For each of the specimen types, a visual best-fit curve was drawn as a lower bound to the data set, and the fatigue fracture threshold, G_{Ith}, was defined as the value of the curve at $N = 1 \times 10^6$ cycles. The various G_{Ith} values are given in Fig. 14. The results are consistent with the static results except for the very low value for the shear precracked case. The G_{Ith} results for the 13, 25, and 75-μm cases are again very similar to each other, whereas values for the 130-μm case are about 40% higher.

Fatigue test results for the ENF specimens are shown in Fig. 15. Because the G_{IIc} values for the 75 and 130-μm insert specimens were similar and were much higher than for the thinner inserts, specimens with 130-μm inserts were not included in the cyclic tests.

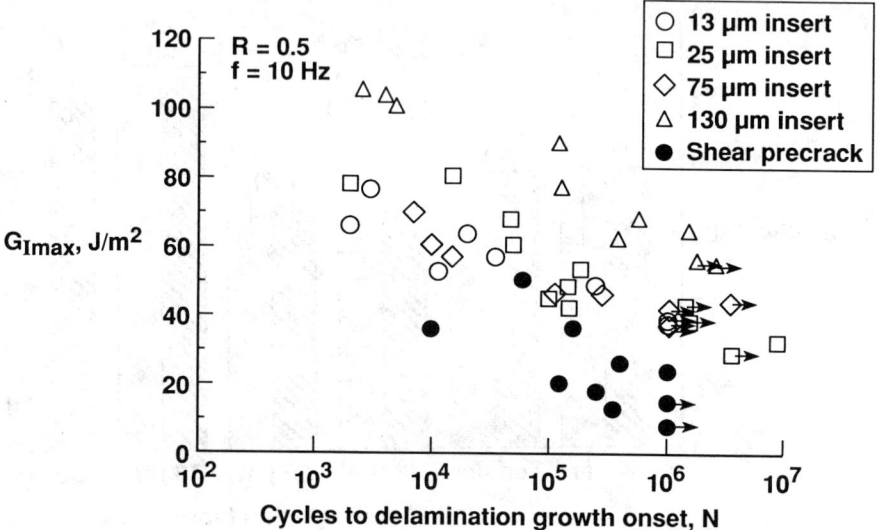

FIG. 13—*Fatigue delamination onset in S2/SP250 DCB specimens.*

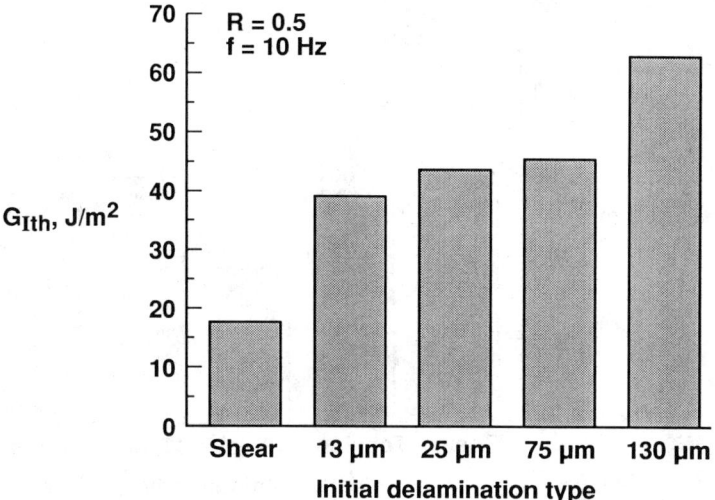

FIG. 14—*Effect of initial delamination on G_{Ith} in S2/SP250.*

Results at $N = 1 \times 10^6$ were similar for the two precracked specimen types and the 25-μm insert specimens. Results for the specimens with 75-μm inserts are noticeably higher over the range of N, whereas results for those specimens with 13-μm inserts are considerably lower. Visual best fit curves were drawn through the lower bound of each data set, and G_{IIth} values were chosen from the curve at $N = 1 \times 10^6$. Those G_{IIth} results are compared in Fig. 16. The results are similar to the static results, in that G_{IIth} decreases with decreasing insert thickness for the three thicknesses tested. The threshold value for the 13-μm

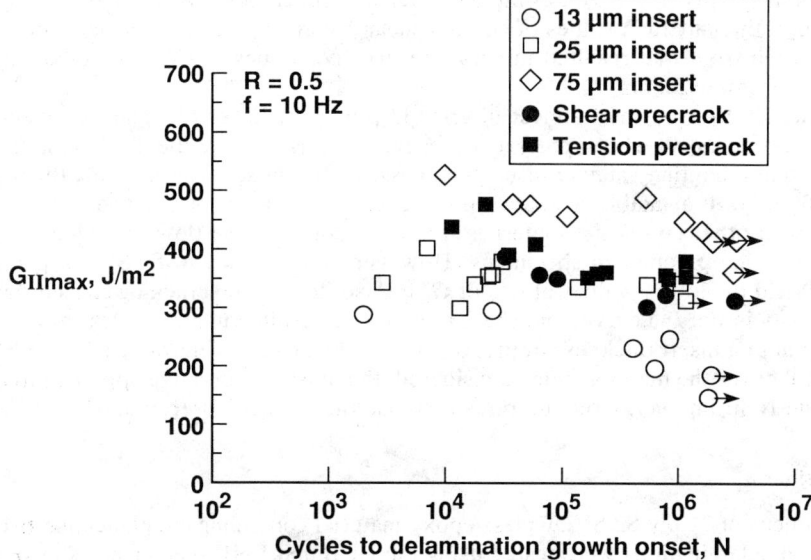

FIG. 15—*Fatigue delamination onset in S2/SP250 ENF specimens.*

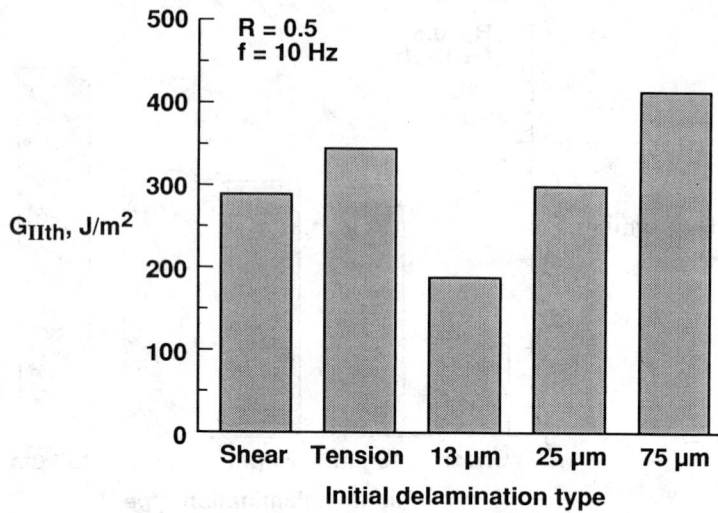

FIG. 16—*Effect of initial delamination on* G_{IIth} *in S2/SP250.*

insert specimens is much lower than any of the other cases and is only half the threshold value for the 75-μm insert specimens. The G_{IIth} values for the precracked specimens were in the same range as the results for the 25-μm insert specimens.

Summary of Results

For the glass/epoxy material tested, only the thinnest (13-μm) inserts yielded specimens with no visible resin pocket at the insert tip. However, results for the DCB tests under both static and fatigue loading were similar for insert thicknesses up to 75 μm. Precracking should not be used for DCB testing since tension precracks create fiber bridging which results in artificially high values of G_I, and shear precracking creates damage ahead of the delamination front that resulted in very conservative values of G_{Ith}, as well as nonlinear load-displacement plots.

For the ENF tests, the thinnest inserts, 13 μm, gave lower G_{IIc} and G_{IIth} values than either the tension or shear precracked specimens. However, the ENF results do not indicate that a limiting value of insert thickness can be chosen. It is possible that a thinner insert material, if available, would result in even lower values of G_{IIc} and G_{IIth}. In similar studies with other composite materials, the most conservative (lowest) values of G were measured using precracked specimens. However, in this study of S2/SP250, precracking did not yield the lowest values of G_{IIc} or G_{IIth}. Also, tension precracking caused significant nonlinearity in the P-δ curve, and shear precracking resulted in greater scatter in the data. Until a limiting insert thickness or precracking method can be determined for the ENF test in general or for the material being considered, the most conservative approach to measuring Mode II strain energy release rates is to use the thinnest insert material available.

Conclusions

Specimens of 24-ply S2/SP250 glass/epoxy material containing midplane inserts to simulate initial delaminations were tested using the DCB and ENF specimens. The specimens were tested statically to determine the interlaminar fracture toughness, G_c, and in fatigue to determine the fatigue fracture threshold, G_{th}. To study the effect of the simulated initial delamination on G_c and G_{th}, four different insert thicknesses—13, 25, 75, and 130 μm—as well as either tension or shear precracks were used. Results of this study showed that for S2/SP250 laminates:

1. A resin pocket will form at the tip of the insert used to simulate the delamination. Thicker inserts will result in thicker resin pockets. However, for the thinnest inserts used in this study (13 μm) no resin pocket was visible.

2. It is best to use the thinnest insert material possible for DCB testing to eliminate the resin pocket or keep it as small as possible. However, for the material tested in this study, the DCB fracture toughness and fatigue fracture threshold were very similar for insert thicknesses up to 75 μm. Inserts that were 130-μm thick gave significantly higher test results in both cases and should not be used.

3. The static and cyclic ENF tests showed that G_{IIc} and G_{IIth} decrease with decreasing insert thickness without reaching an apparent limiting level. Therefore, a conservative approach requires that the insert material used for the ENF test should be as thin as possible and not greater than 13 μm.

4. Shear precracking creates damage ahead of the delamination front. For the DCB specimens in fatigue, this damage resulted in significantly lower values of fatigue fracture threshold, G_{Ith}.

5. For the ENF specimens, tension and shear precracking yielded similar values of G_{IIc} and G_{IIth}. Results obtained from the precracked ENF specimens were higher than results for the thinnest (13 μm) insert specimens.

References

[1] Murri, G. B. and O'Brien, T. K., "Interlaminar G_{IIc} Evaluation of Toughened Resin Matrix Composites Using the End-Notched Flexure Test," AIAA-85-0647, *Proceedings,* 26th AIAA/ASME/ASCE/AHS Conference on Structures, Structural Dynamics and Materials, Orlando, FL, April 1985, ASME, New York, pp. 197–202.

[2] O'Brien, T. K., Johnston, N. J., Raju, I. S., Morris, D. H., and Simonds, R. A., "Comparisons of Various Configurations of the Edge Delamination Test for Interlaminar Fracture Toughness," *Toughened Composites, ASTM STP 937,* N. J. Johnston, Ed., American Society for Testing and Materials, Philadelphia, 1987, pp. 275–294.

[3] Russell, A. J., "Factors Affecting the Opening Mode Delamination of Graphite/Epoxy Laminates," Defence Research Establishment Pacific (DREP), Victoria, British Columbia, Canada, Materials Report 82-Q, December 1982.

[4] Johnson, W. S. and Mangalgiri, P. D., "Investigation of Fiber Bridging in Double Cantilever Beam Specimens," *Journal of Composites Technology and Research,* Vol. 9, No. 1, Spring 1987, pp. 10–13.

[5] Martin, R. H., "Effect of Initial Delamination on G_{Ic} and G_{Ith} Values from Glass/Epoxy Double Cantilever Beam Tests," *Proceedings,* American Society for Composites, Third Technical Conference, Seattle, WA, September 1988, pp. 688–701.

[6] Davies, P., Cantwell, W., and Kausch, H. H., "Measurement of Initiation Values of G_{Ic} in IM6/PEEK Composites," *Composites Science and Technology,* Vol. 35, 1989, pp. 301–313.

[7] O'Brien, T. K., "Generic Aspects of Delamination in Fatigue of Composite Materials," *Journal of the American Helicopter Society,* Vol. 32, No. 1, January 1987, pp. 13–18.

[8] Russell, A. J., "Initiation of Mode II Delamination in Toughened Composites," *Composite Materials: Fatigue and Fracture, Third Volume, ASTM STP 1110,* T. K. O'Brien, Ed., American Society for Testing and Materials, Philadelphia, 1991.

[9] O'Brien, T. K., Murri, G. B., and Salpekar, S. A., "Interlaminar Shear Fracture Toughness and Fatigue Thresholds for Composite Materials," *Composite Materials: Fatigue and Fracture, Second Volume, ASTM STP 1012,* P. A. Lagace, Ed., American Society for Testing and Materials, Philadelphia, 1989, pp. 222–250.

[10] Clark, W. G. Jr. and Hudak, S. J. Jr., "Variability in Fatigue Crack Growth Rate Testing," *Journal of Testing and Evaluation,* Vol. 3, No. 6, 1975, pp. 454–476.

[11] Bathias, C. and Laksimi, A., "Delamination Threshold and Loading Effect in Fiber Glass Epoxy

Composite," *Delamination and Debonding of Materials, ASTM STP 876,* W. S. Johnson, Ed., American Society for Testing and Materials, Philadelphia, 1985, p. 217.

[12] Ramkumar, R. L. and Whitcomb, J. D., "Characterization of Mode I and Mixed Mode Delamination Growth in T300/5208 Graphite/Epoxy," *Delamination and Debonding of Materials, ASTM STP 876,* W. S. Johnson, Ed., American Society for Testing and Materials, Philadelphia, 1985, pp. 315–335.

[13] Martin, R. H. and Murri, G. B., "Characterization of Mode I and Mode II Delamination Growth and Thresholds in Graphite/PEEK Composites," *Composite Materials: Testing and Design, Ninth Volume, ASTM STP 1059,* S. P. Garbo, Ed., American Society for Testing and Materials, Philadelphia, 1990, pp. 251–270.

[14] Whitney, J. M., Browning, C. E., and Hoogsteden, W., "A Double Cantilever Beam Test for Characterizing Mode I Delamination of Composite Materials," *Journal of Reinforced Plastics and Composites,* Vol. 1, October 1982, pp. 297–313.

[15] Russell, A. J., "On the Measurement of Mode II Interlaminar Fracture Energies," Defence Research Establishment Pacific (DREP), Victoria, B.C., Canada, Materials Report 82-O, December 1982.

[16] Chai, H., "Bond Thickness Effect in Adhesive Joints and its Significance for Mode I Interlaminar Fracture of Composites," *Composite Materials: Testing and Design (Seventh Conference), ASTM STP 893,* J. M. Whitney, Ed., American Society for Testing and Materials, Philadelphia, 1986, pp. 209–231.

Sotiris Kellas,[1] John Morton,[2] and Karen E. Jackson[3]

Damage and Failure Mechanisms in Scaled Angle-Ply Laminates

REFERENCE: Kellas, S., Morton, J., and Jackson, K. E., **"Damage and Failure Mechanisms in Scaled Angle-Ply Laminates,"** *Composite Materials: Fatigue and Fracture, Fourth Volume, ASTM STP 1156,* W. W. Stinchcomb and N. E. Ashbaugh, Eds., American Society for Testing and Materials, Philadelphia, 1993, pp. 257–280.

ABSTRACT: The effect of specimen size upon the response and strength of $\pm 45°$ angle-ply laminates has been investigated for two graphite fiber-reinforced plastic systems and several stacking sequences. The first material system was an epoxy-based system, AS4 fibers in 3502 epoxy, and the second was a thermoplastic-based system, AS4 fibers in PEEK matrix. For the epoxy-based system, two generic $\pm 45°$ layups were studied; $(+45°_n/-45°_n)_{2S}$ (blocked plies), and $(+45°/-45°)_{2nS}$ (distributed plies), where $n = 1, 2, 3,$ and 4. In the case of the thermoplastic system, only the layup with distributed plies was investigated, $(+45°/-45°)_{2nS}$, for $n = 1$ and 2. The in-plane dimensions of the specimens were varied such that the width/length relationship was $12.7 \times n/127 \times n$ mm, for $n = 1, 2, 3,$ or 4.

It is shown that the stress/strain response and the ultimate strength of these angle-ply laminates depends on the laminate thickness and the type of generic layup used. The ultimate strength of the epoxy matrix material was found to be much more sensitive to specimen size when compared to the thermoplastic matrix system. Scaling effects defined with respect to the first ply failure, strain at ultimate failure, and ultimate strength are isolated and discussed. Furthermore, it is shown that first ply failure occurs in the surface plies as a result of normal rather than shear stresses. The implications of the experimental findings upon the validity of the $\pm 45°$ tension test, which is used to determine the in-plane shear response of unidirectional composites, are discussed.

KEY WORDS: scaling, composites, laminates, angle-ply, tension, shear, standard test, damage mechanisms

It is well known that some engineering materials exhibit strength scaling effects; that is, the strength of the material is some function of its absolute size, as well as a function of its geometric features. In particular it has been observed that specimens of brittle isotropic materials which are subjected to tensile loads exhibit strength scaling effects which can be described by one of two scaling relationships. The first relationship is based on Weibull statistics and states that the strength ratio, S_1/S_2, of two geometrically similar specimens is related to the ratio of the material volume, V_2/V_1, as

$$\frac{S_1}{S_2} = \left(\frac{V_2}{V_1} \right)^{\frac{1}{m}}$$

where $V_2 > V_1$ and m is known as the shape parameter, thought to be constant for a given

[1]Lockheed Engineering and Sciences Co., NASA Langley Research Center, Hampton, VA 23681-0001.

[2]Virginia Polytechnic Institute and State University, Blacksburg, VA 24061-0219.

[3]U.S. Army Vehicle Structures Directorate, Army Research Laboratory, NASA Langley Research Center, Hampton, VA 23681-0001.

material [1]. The second relationship [2] is based on a fracture mechanics approach and states that the ratio of critical stresses, σ_1^c/σ_2^c, is related to a scaling factor λ as

$$\frac{\sigma_1^c}{\sigma_1^c} = \frac{1}{\sqrt{\lambda}}$$

where the critical stress, σ^c, is defined within the framework of linear elastic fracture mechanics as the stress required for unstable crack propagation, and λ is generally defined as the ratio of a geometric size in the model to that of the full-scale structure.

In the case of unidirectional composites under tensile loads, it has been shown that the Weibull statistics based model can be used to describe the changes in apparent strength with specimen size within reasonable limits [3]. However, composites of practical importance are usually laminated with off-axis ply orientations, and simple scaling laws, or models, become inadequate in the description of the strength scaling effects. In their laminated form composites are complex structures, and their structural response bears no simple relationship to the size of the individual unidirectional plies. Moreover, in the case of laminated composites, the definition of a scaling effect is rather more complex since first ply failure is as important as stiffness, strain at failure, and strength. In other words, two scaled specimens of the same layup may exhibit similar strengths while first ply failure may occur at different applied stresses. In this simple example, the scaling effect from the ultimate strength point of view is negligible or zero. However, a scaling effect does exists from the first ply failure point of view, and under certain design requirements this may be the most important parameter.

The in-plane geometry of laminated composite materials can be scaled by simply varying the in-plane dimensions according to some known ratio, λ, called the geometric scale factor. However, since the fiber diameter cannot be scaled, there are two practical methods of scaling the laminate thickness. The first method is known as "ply level scaling" in which a full-scale laminate is constructed by increasing the individual ply thickness by blocking several layers of identical fiber orientation [4–6]. The second method is known as "sublaminate level scaling," where a full-scale laminate is constructed by repeating a basic sublaminate about the plane of symmetry. For $\pm 45°$ laminates where the $(+45°/-45°)_{2S}$ laminate represents the baseline (model size), $(+45°_n/-45°_n)_{2S}$ and $(+45°/-45°)_{2nS}$ represent the ply and sublaminate level scale model laminates. For values of n greater than 1, the two basic methods of scaling are also represented schematically in Fig. 1. If the in-plane dimensions of these scale model laminates are also sized in proportion to the model, the scaling procedure is known as three dimensional. On the other hand, the in-plane dimensions can remain fixed for the model and the full-scale laminates. In this case, the scaling procedure is referred to as thickness scaling. Likewise, if the thicknesses of the model and full-scale laminates are the same and the in-plane dimensions are scaled, the procedure is referred to as in-plane scaling. All three methods of dimensional scaling are shown schematically in Fig. 2. The thickness and in-plane scaling are referred to as one- and two- dimensional scaling techniques, respectively, and are, therefore, incomplete so far as the three-dimensional composite laminate is concerned.

Even though some three-dimensional scaling studies have been performed [4–6] due to economic and experimental constraints, the bulk of the published research has dealt with one- or two-dimensional scaling problems [7–11]. In the present experimental study, the influence of specimen size upon the stress/strain response of $\pm 45°$ laminates is investigated from all scaling points of view, including ply and sublaminate thickness scaling levels. While the main portion of this experimental study is centered around the behavior

Baseline (model) Size

$(\pm 45°/\pm 45°)$ s

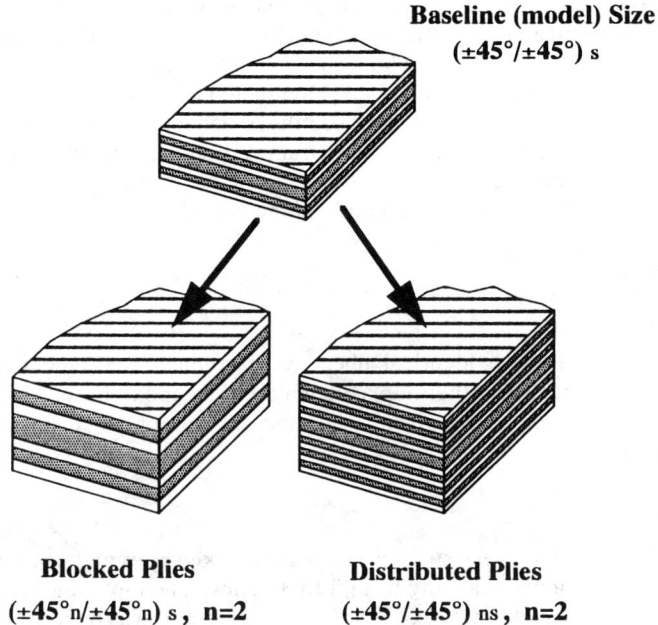

Blocked Plies

$(\pm 45°n/\pm 45°n)$ s, n=2

Distributed Plies

$(\pm 45°/\pm 45°)$ ns, n=2

FIG. 1—*Schematic of thickness build-up procedures for composite laminates. Note that while in both techniques the extensional stiffness remains constant with increasing laminate thickness, the bending stiffness will remain constant only in the case of ply level scaled laminates.*

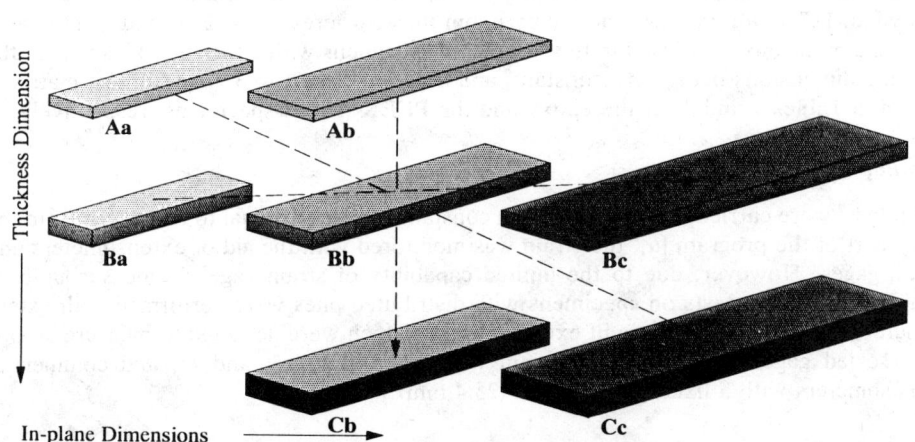

FIG. 2—*Schematic of dimensional scaling procedures for composite laminates. The thickness dimension is increased according to one of the two methods shown in Fig. 1. Note that specimens which lie on the diagonal line (Aa, Bb, and Cc) represent three-dimensional scaling.*

of a graphite/epoxy material system, a second system, graphite/PEEK, was introduced, and studied briefly to provide a better understanding of the problem of scaling in matrix-dominated layups. The purpose of choosing the ±45° layup for this investigation is twofold. First, this represents the only matrix-dominated layup where the plies are thought to be loaded primarily in shear, and, second, the ±45° tension test is one of the most popular tests for determining the shear stress/strain response of unidirectional composites. Consequently, any dependencies in the shear response based on specimen size and stacking sequence should be well understood and characterized.

Experimental Details

Material Systems

Two material systems with identical fibers were studied, AS4 fibers in 3502 epoxy matrix and AS4 fibers in PEEK thermoplastic matrix. The epoxy-based system was layed up and cured in-house, and the PEEK-based material was obtained in its molded form directly from the manufacturer.

Stacking Sequence

Two generic ±45° layups were studied, one with blocked plies and one with distributed plies with stacking sequences ranging from 8 to 32 plies. The 8-ply laminate consisted of unidirectional plies arranged in a $(+45°/-45°)_{2S}$ sequence and was denoted the baseline or model stacking sequence. In the case of the epoxy-based system, six additional "scaled-up" laminates were tested with the following stacking sequences, $(+45°_n/-45°_n)_{2S}$ (blocked plies) and $(+45°/-45°)_{2nS}$ (distributed plies), where $n = 2$, 3, and 4. For the PEEK matrix system, one "scaled-up" laminate was tested with stacking sequence, $(+45°/-45°)_{4S}$ (distributed plies), in addition to the baseline stacking sequence.

Specimen Geometry

Specimens scaled in three dimensions had the general specimen dimensions, $12.7 \times n$ mm wide, $127 \times n$ mm long, and 1.0 by n mm thick, where $n = 1$, 2, 3, and 4. The same general dimensions were used in the rest of the specimens with either the thickness or the in-plane dimensions being held constant, and n varied from 1 to 3. The full test matrix is shown in Tables 1 and 2 for the epoxy and the PEEK matrix specimens, respectively.

Loading Mode

All tests were carried out in tension at a constant strain rate equal to 0.028 min^{-1}. In the early part of the program [6], the strain was monitored with the aid of extensometers and strain gages. However, due to the limited capability of strain gages in measuring large strains, subsequent tests on specimens with distributed plies were performed with extensometers only. Both custom-built extensometers, which were designed to measure strains over scaled gage sections ($38 \times n$ mm where $n = 1$, 2, 3, and 4), and commercial extensometers with a fixed gage section (25.4 mm) were used.

Nondestructive Damage Evaluation

Nondestructive damage examination was carried out on specimens preloaded to a given stress level. The given stress levels for each specimen type were selected according to the predetermined stress/strain curves for a given family of specimens. Following the first

TABLE 1—*Test matrix for AS4/3502 indicating the specimen dimensions and the type of layups tested in each case. Note that specimens* Aa, Bb, Cc, *and* Dd *are scaled in three dimensions.*

Nominal Thickness, mm	Nominal Width by Nominal Length, mm by mm			
	12.7 by 127 (Size *a*)	25.4 by 254 (Size *b*)	38.1 by 381 (Size *c*)	50.8 by 508 (Size *d*)
1.0 (8 plies)		. . .	. . .	. . .
Layup *A*	Baseline	Distributed	. . .	. . .
2.0 (16 plies)	. . .	Blocked	. . .	. . .
Layup *B*	Distributed	Distributed	. . .	. . .
3.0 (24 plies)	. . .	. . .	Blocked	. . .
Layup *C*	Distributed	. . .	Distributed	. . .
4.0 (32 plies)	. . .	Blocked	. . .	Blocked
Layup *D*	. . .	Distributed	. . .	Distributed

loading/unloading cycle, specimens were soaked in a zinc iodide solution before being X-rayed. Subsequently, both free edges of loaded specimens were carefully polished and then examined under an optical microscope. In most cases the whole procedure was repeated again with a load step slightly higher than the preceding one. At least one specimen of each specimen type was examined in this way.

Experimental Results

Stress/Strain Response

Typical stress/strain plots of specimens scaled in three dimensions are shown in Figs. 3 and 4 for the epoxy and the PEEK matrix systems, respectively. The plots shown in these figures are typical of a given specimen type from a number of replicate tests (seven being the least number of replicate tests).

In the case of the epoxy matrix system, typical stress/strain plots are presented in Fig. 3 for the 8-ply baseline case and for each generic layup (distributed or blocked plies), with $n = 2, 3$, and 4. The stress/strain response of the $\pm 45°$ laminates depends on the specimen size and the stacking sequence. Specimens with blocked plies exhibit a brittle-like behavior, whereas specimens with distributed plies exhibit a ductile-like behavior. Moreover, specimens with blocked plies show a consistent reduction in strength with increasing specimen size, as opposed to an increase in strength with increasing specimen size exhibited by the specimens with distributed plies. Note that while the strain at failure for the 32-ply specimen with blocked plies is approximately 1%, the strain at failure of the

TABLE 2—*Test matrix for AS4/PEEK indicating the specimen dimensions and the type of layup tested in each case. Note that specimens* Aa, *and* Bb *are scaled in three dimensions.*

Nominal Thickness, mm	Nominal Width by Nominal Length mm by mm	
	12.7 by 127 (Size *a*)	25.4 by 254 (Size *b*)
1.0 (8-plies)	Baseline	Distributed
Layup *A*		
2.0 (16-plies)	Distributed	Distributed
Layup *B*		

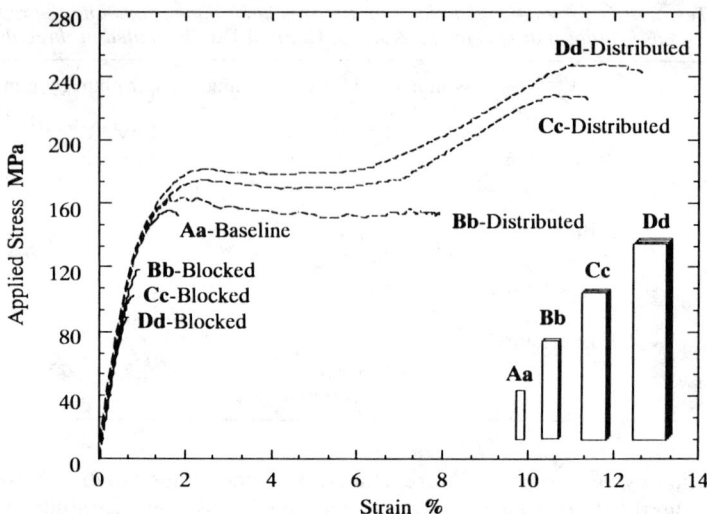

FIG. 3—*Stress/strain response of epoxy matrix × specimens scaled in three dimensions. Aa, Bb, Cc, and Dd are equal to 12.7 × n mm wide, 127 × n mm long and 1 × n mm thick where* n = 1, 2, 3, and 4) respectively.

corresponding 32-ply specimen with distributed plies has exceeded 12%. Furthermore, for these two specimens the difference in the ultimate strength is approximately 170%. Clearly, so far as the ultimate strength and strain at failure are concerned, there are significant scaling effects. The magnitude and direction of the scaling effect, whether increasing or decreasing relative to the baseline, depends upon the scaling level used, ply or sublaminate level.

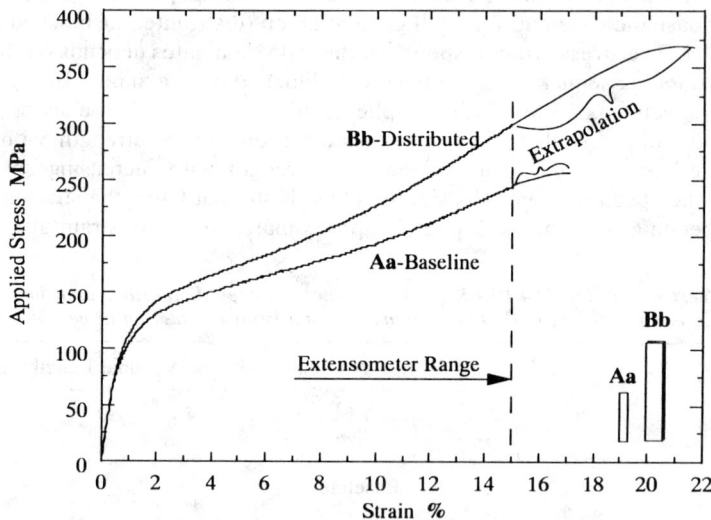

FIG. 4—*Stress/strain response of PEEK matrix specimens scaled in three dimensions. Note that the extensometer range of 15% was exceeded. Beyond this point, both curves were extrapolated to the maximum measured stress.*

The stress/strain plots for AS4/PEEK, shown in Fig. 4, also indicate the existence of a scaling effect for the two specimen sizes, *Aa* and *Bb*. As in the case of the epoxy matrix material for specimens with distributed plies, both the strength and the strain at failure increase with specimen size. However, when corresponding specimens are compared, there exists a basic difference between the stress/strain response of the two material systems. That is, unlike the epoxy matrix specimens, the PEEK matrix specimens do not exhibit a negative gradient in their stress/strain responses. Moreover, the PEEK matrix specimens appear to be more ductile when compared to the corresponding epoxy matrix specimens with a difference in strain at failure of approximately 15% strain, as depicted in Fig. 5, which shows a comparison of the stress/strain response of a baseline epoxy and a baseline PEEK matrix specimen. In the magnified portion of Fig. 5, it is also shown that the initial extensional stiffness of the PEEK matrix specimens is slightly lower than that of the epoxy matrix specimens.

The effect of laminate thickness for epoxy matrix specimens of the same in-plane

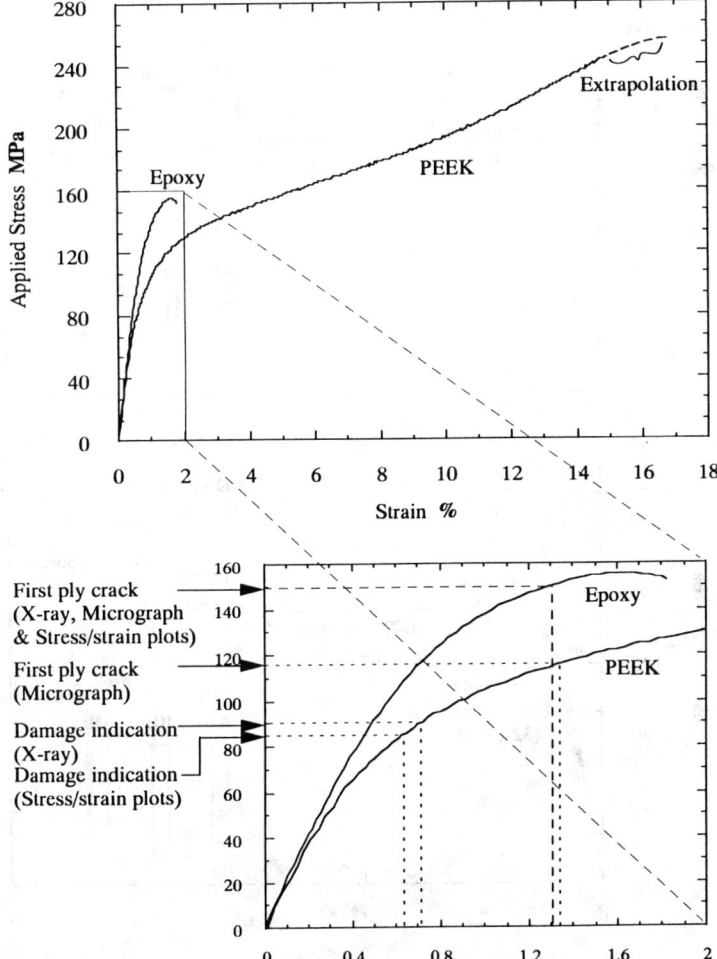

FIG. 5—*Comparison of the stress/strain response of epoxy and PEEK matrix, baseline, specimens.*

dimensions is shown in Fig. 6. Clearly, the laminate thickness has a significant influence upon the stress/strain behavior. Specimens with distributed plies show an increase in strength and strain at failure with increasing laminate thickness, while the contrary is true for specimens with blocked plies. An almost identical effect was observed for the three-dimensional scaled specimens, if the results of Figs. 3 and 6 are compared. This observation indicates that in-plane scaling will have a lesser effect on strength and strain at failure than thickness scaling. Indeed, the stress/strain plots in Fig. 7 show that, within a large portion of their stress/strain curve, epoxy matrix specimens with scaled in-plane dimensions exhibit a very similar response.

The fact that narrow specimens (with width/thickness ratios less than 12.7) fail at somewhat lower strains than the corresponding wider specimens has more to do with the higher gripping constraint rather than the stress field being different in the gage section of the

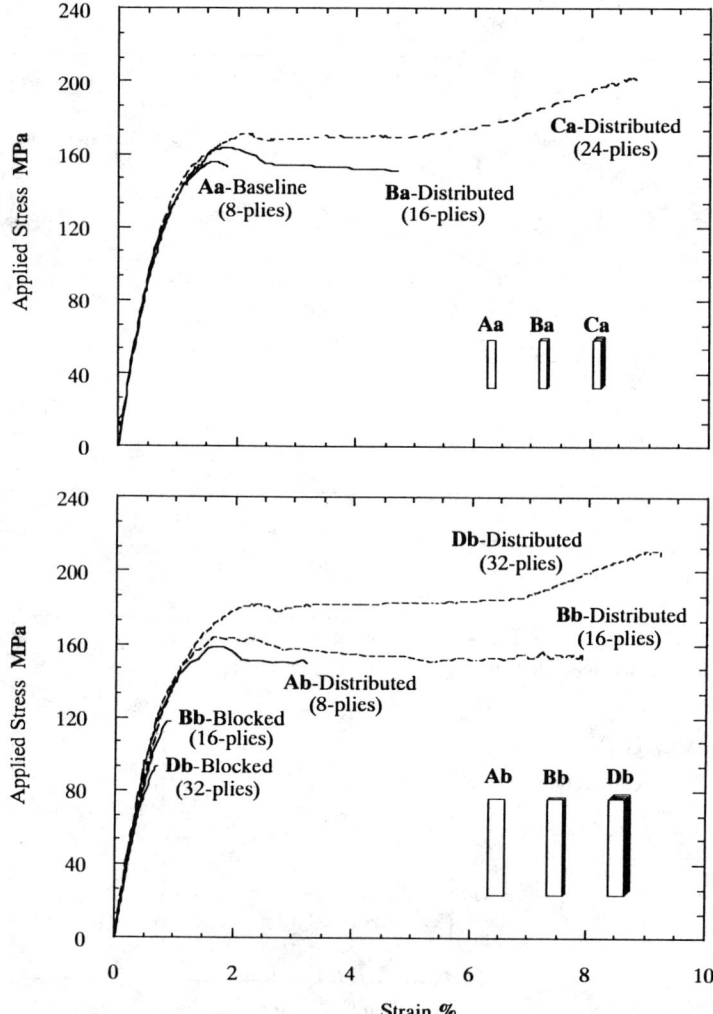

FIG. 6—*Stress/strain response of epoxy specimens scaled in one dimension. The in-plane dimensions are kept fixed: (top) 12.7 × 127 mm, and (bottom) 25.4 × 254 mm.*

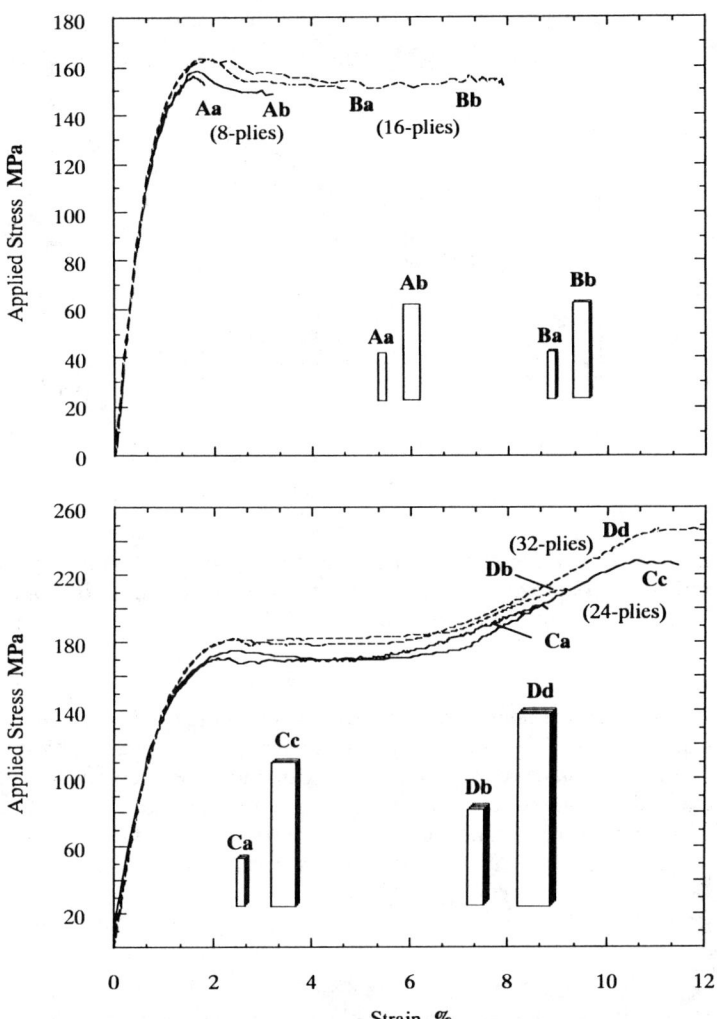

FIG. 7—*Stress/strain response of epoxy matrix specimens scaled in two dimensions. The thickness is kept fixed:* (top) *1 or 2 mm and* (bottom) *3 or 4 mm.*

specimens. The higher gripping constraint in the narrow specimens can be explained with the introduction of a grip constraint factor equal to the maximum applied load divided by the gripped area. Clearly, each time the width of the specimen is halved, the grip constraint factor is doubled since the specimen is expected to carry one-half of the original load with only one-quarter of the original area gripped. Note that, in practice, the actual gripping constraint becomes even more severe by the fact that a local high stress concentration exists at the specimen free edge region closest to the jaws. Thus the relative constraint effect increases further as the specimen width decreases. The effect of the higher gripping constraint in the narrow specimens is verified experimentally by the fact that the ultimate failure in the narrow specimens was either associated or occurred close to the gripping region.

Comparisons of two-dimensional scaled PEEK matrix specimens are shown in Fig. 8.

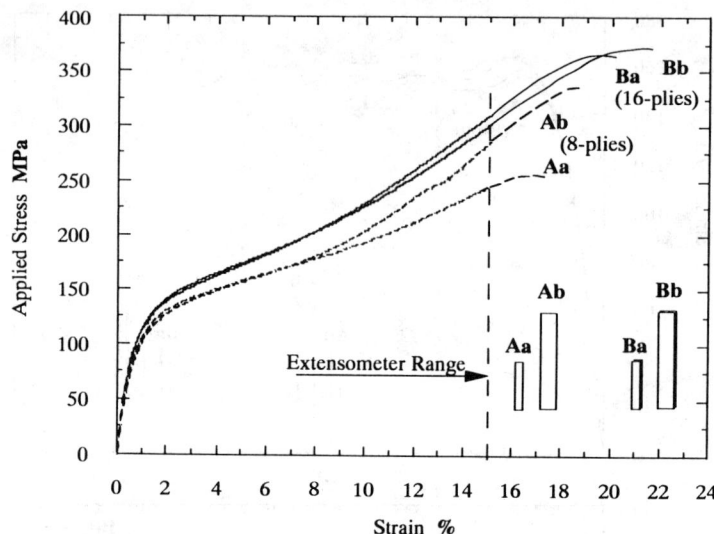

FIG. 8—*Stress/strain response of PEEK matrix specimens scaled in one and two dimensions. Note that the extensometer range of 15% was exceeded. Beyond this point all curves were extrapolated to the maximum measured stress.*

While evidence of a definite dependency of strength upon the thickness dimension is not as strong as in the case of the epoxy matrix specimens, it is still clear that the laminate thickness is the predominant scaling dimension. Compared with epoxy, the PEEK matrix specimens showed a larger scatter at large strains (typically above 8%). Such scatter is thought to be, at least partly, an artifact of the extensometer output, since some knife edge slipping could occur as a result of the smooth surface texture of the PEEK matrix specimens.

Damage Evaluation

So far, scaling effects have been approached from the point of view of stress/strain responses, ultimate strengths, and strains at failure. An additional important factor for laminated composites is the initiation of damage and first ply failure. Evidence of scaling effects related to first ply failure for the two material systems is presented in Figs. 9 and 10 for epoxy and PEEK matrix specimens, respectively. These figures show a combined view of damage as documented by enhanced X-ray radiographs of the gage section and micrographs of the polished specimen free edges. There is a clear correlation between damage initiation and stiffness changes observed in the stress/strain curve of a given specimen type.

In general, all ±45° laminates (with the exception of 24 and 32 blocked ply laminates) presented in Fig. 3 share approximately the same initial extensional stiffness. Changes in extensional stiffness are usually expected to occur when the material enters a nonlinear state or when damage occurs. Clearly then, when a group of ±45° laminates follow an identical path up to a point where one deviates from the rest, then that point could coincide with failure initiation. For the 24 and 32 blocked ply laminates it has been shown [6] that damage exists in virgin (nonloaded) specimens. Therefore, the stiffness deviations for these two specimens from the rest, even though they are small at low strains, can be

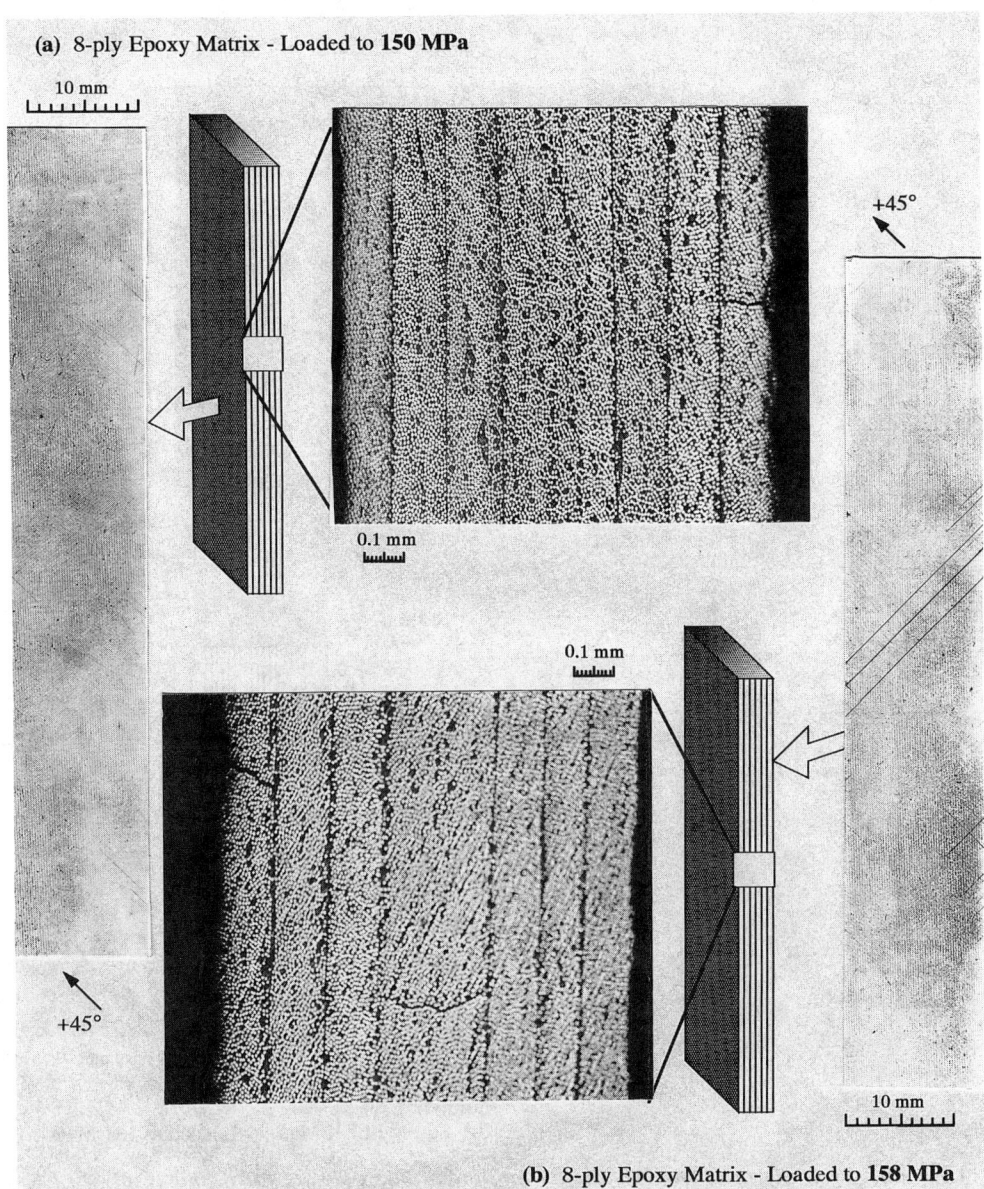

FIG. 9—*First and second ply failures for 8-ply epoxy matrix shown in* (a) *and* (b), *respectively.*

attributed to the observed transply cracks. In the case of distributed ply specimens, where no damage was observed in virgin specimens, noticeable stiffness deviations occur at approximately 150 MPa. The first specimen to show such a deviation is the baseline (8-ply) specimen. Damage examination of these specimens, preloaded to approximately 150 MPa, showed that damage has initiated (Fig. 9a) and that it is confined to the surface plies. Further damage occurred in the middle plies at a stress (158 MPa) corresponding approximately to the peak value of stress for this specimen (Fig. 9b).

(a) 8-ply PEEK Matrix - Loaded to **115 MPa**

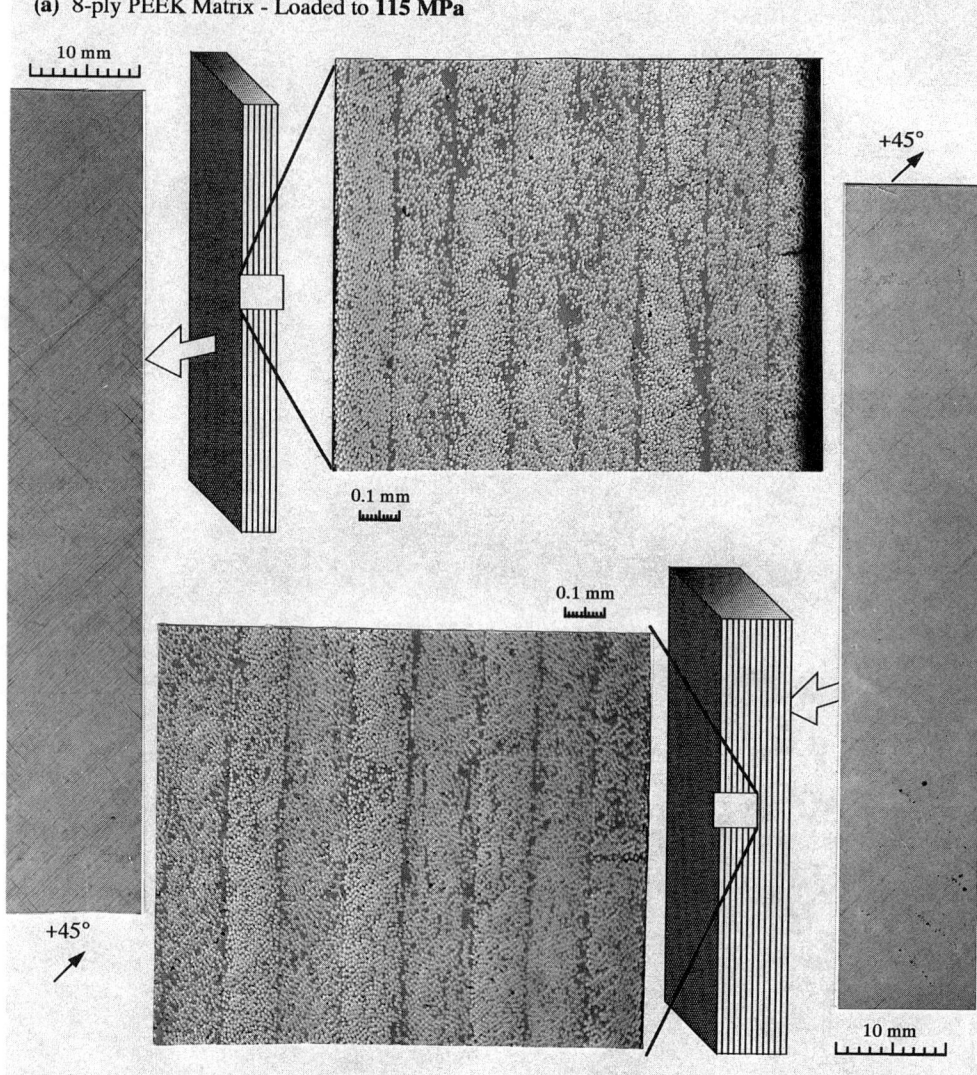

(b) 16-ply PEEK Matrix - Loaded to **131 MPa**

FIG. 10—*First ply cracking (visible by an optical microscope) for 8 and 16-ply PEEK matrix specimens, shown in (a) and (b), respectively, Note that, while transply cracks are only visible on the surface (+45°) plies, the X-ray radiographs indicate that there is damage in both +45° and −45° directions.*

Following the same argument, failure in the 8-ply PEEK matrix specimens should occur at approximately 85 MPa where the stress/strain curve of the 8-ply specimen deviates from that of the 16-ply specimen (Fig. 4). However, microscopic examination of polished edges of specimens loaded to 85 MPa and beyond did not produce evidence of transply cracks. On the other hand, X-ray radiographs did contain dark lines at 45°, which are normally associated with transply cracks. Damage, as detected in the X-ray radiographs, became clear in PEEK matrix specimens which were loaded to a stress of approximately 90 MPa and beyond. At this point there is, however, no conclusive experimental evidence as to the nature of this damage, but one possible explanation is fiber debonding. Once several neighboring fibers debond from the supporting matrix, their effect on the stress/strain response of the laminate could be similar to that of transply cracking, and such damage is X-ray detectable. Clear evidence of transply cracking in the 8-ply PEEK matrix specimens was observed at stresses of 117 MPa and beyond, as shown in Fig. 10a. Such cracks were first observed to occur in the surface plies (the +45° direction) and, as the X-ray radiograph of Fig. 10a indicates, there co-exists damage in the −45° direction. The damage in the −45° direction is, however, not seen on the photomicrograph of the specimen's edge, that is, while some of the dark lines at +45° correspond to transply cracks, a large number of lines at +45° and −45° correspond to, possibly, fiber debonding. Moreover, comparing Fig. 10a and 10b, it can be seen that equivalent damage in the 16-ply specimens occurs at approximately 131 MPa, that is, an increase of approximately 14%. Likewise, from the data in Fig. 3, it can also be deduced that first ply failure in the 16-ply epoxy specimens occurs at approximately 159 MPa, compared to 150 MPa for the 8-ply specimens, which represents a 6% increase in first ply stress. The delay in damage development from one specimen size to the next is a scaling effect, and this type of scaling effect appears to depend on the material system. For a direct comparison, the stress values at which damage initiates in the baseline epoxy and PEEK matrix specimens are indicated in Fig. 5.

Another important difference in the damage development in different size specimens is shown in Fig. 11a and 11b for two epoxy matrix specimens loaded to a stress corresponding to the point on the stress/strain curve where the gradient is zero (Fig. 3). While damage in the 8-ply specimen develops around the initial damage, damage in the 16-ply specimen is more dispersed throughout the gage section, as seen in both the X-ray radiograph and the micrograph of such a specimen.

Discussion

Scaling effects in laminated composites can be defined with regard to several parameters, the most important of which are strength, strain at failure, stress/strain response, and first-ply-failure stress. If, for example, two geometrically scaled specimens exhibit different stress/strain responses, this would be a scaling effect. Likewise, if the two specimens fail at different ultimate stresses, this would also be a scaling effect. All four factors are important and will be discussed separately, but, since the initiation of damage is the parameter that, in many cases, controls the strength and stress/strain response of a laminated composite, it will be discussed first. Finally, the effect of specimen size will be examined with reference to the standard ±45° tension test used to obtain the in-plane shear stress/strain response of unidirectional composite materials.

First-Ply Failure

The understanding of the processes that control the first-ply failure in angle-ply laminates has implications which extend beyond the demands of the present study, since most practical composite laminates contain angle plies on their surfaces for improved damage tolerance. Therefore, any factors that can influence the failure of these outer plies, such as ply thickness and constraint of the neighboring plies, should be identified and well characterized.

From the experimental evidence it has been deduced that first ply failure occurs in the

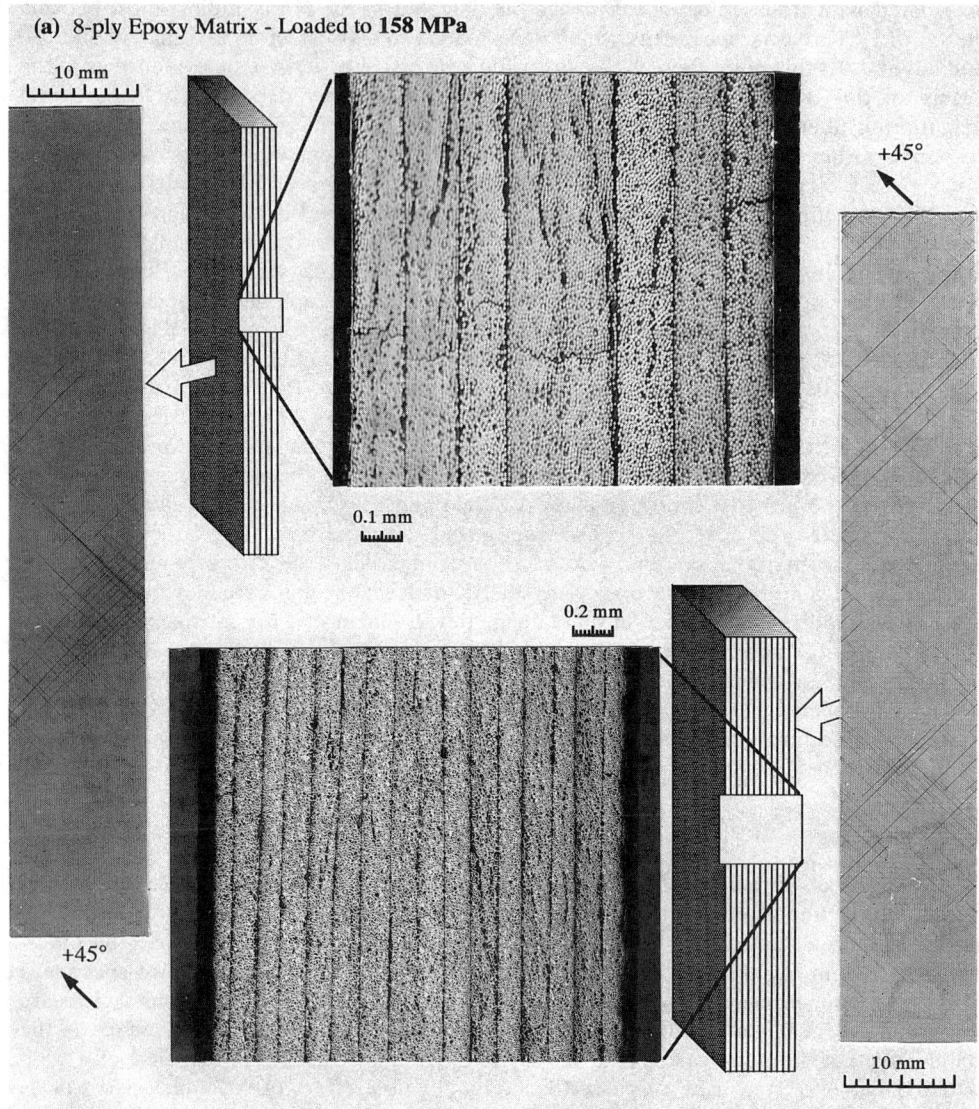

(a) 8-ply Epoxy Matrix - Loaded to **158 MPa**

10 mm

0.1 mm

+45°

0.2 mm

(b) 16-ply Epoxy Matrix - Loaded to **164 MPa**

+45°

10 mm

FIG. 11—*Developed damage states in 8 and 16-ply epoxy matrix specimens shown in* (a) *and* (b), *respectively, Note that damage in the 8-ply specimen is very localized, whereas in the 16-ply specimen it is well distributed.*

surface plies, followed by failure of the mid-plane blocked plies. This was found to be true for all specimens sizes and laminate thicknesses. Therefore, it appears that a ±45° laminate has two "weak links," the surface and the middle plies. The question is, what is the cause of weakness in these plies? If the classical lamination theory assumptions are applied, it would be expected that all plies in a symmetric ±45° laminate be strained in an identical manner, and, therefore, first ply failure would occur arbitrarily in any one of the plies, not necessarily the surface plies. However, in practice, the top (free) surface of the

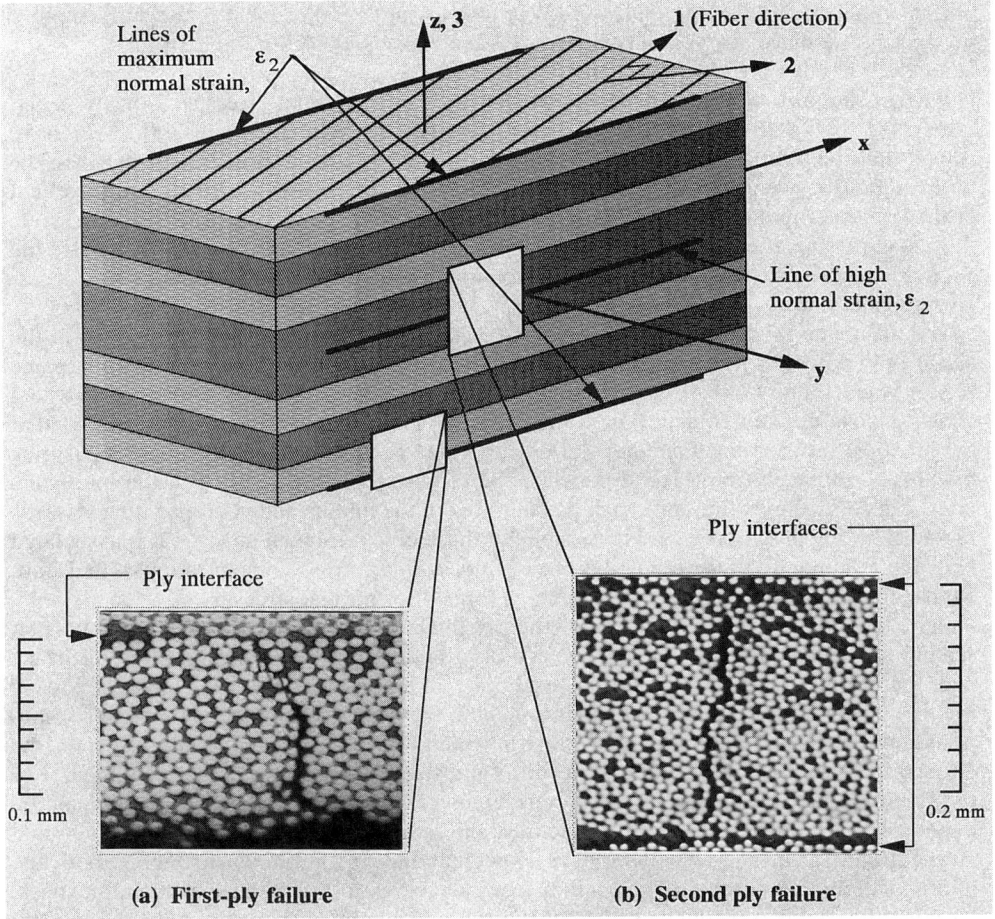

(a) **First-ply failure**   (b) **Second ply failure**

FIG. 12—*Close-up view of first and second ply failures in epoxy matrix specimens. Laminate and ply axes: x-y-z are the laminate axes with x being along the loading direction; 1-2-3 are the ply axes with 1 along the fiber direction. Note that the ϵ_2 is tensile for tensile applied loads and compressive for compressive applied loads.*

outermost plies in a symmetric $\pm 45°$ laminate is not constrained in the same way as the bottom surface, which is bonded to the adjacent $-45°$ ply. As a result, in-plane shear strains, γ_{xy}, develop on these top surfaces close to the free edges of the laminate when an axial load is applied in the x-direction (see Fig. 12 for axes notation). It can be shown, using a simple strain transformation approach [*12*], that such shear strains can be transformed to significant normal strains transverse to the fiber direction, ϵ_2. Since the free surface material is unrestrained in the direction transverse to the fibers, cracking occurs. Therefore, first ply failure in $\pm 45°$ laminates is primarily a normal fracture (or fracture occurring primarily in the opening mode) and not a shear mode fracture. This argument is supported by the opening nature and starting position of first ply cracks. An example of such a crack is shown in Fig. 12a. Likewise, the mid-region of the center . . . $-45°/-45°$ . . . plies is subjected to a reduced constraint when compared to the interply region, where the constraint from the neighboring $+45°$ plies is at a maximum. As a result, large normal strains can develop in a similar way (but lesser degree) to that in the surface plies, producing one more site for normal failure. The regions of high normal strain, ϵ_2, on the laminate free edge are shown schematically in Fig. 12.

Further supporting evidence of the presence of large normal stress, σ_2, or strain, ϵ_2, transverse to the fiber direction has been presented by Whitney et al. [13]. They have shown that the ultimate compressive strength of ($\pm 45°$)$_\text{s}$ laminates is 28% higher than the ultimate tensile strength. The increase in compressive strength was attributed to the effect of the stress component, σ_2, which under a compressive applied load is also compressive, thus suppressing the initiation of (transverse cracking) first ply failure. In effect, the "weak laminate sites" shown in Fig. 12 become "strong laminate sites" under a compressive applied load.

The sequence of ply failures is shown in Fig. 9a and 9b where first ply failure in the surface ply occurs at a stress of 150 MPa, followed by second ply failures in the mid-plane region when the stress is equal to approximately 158 MPa. Moreover, Fig. 12a (which shows a close-up view of a surface transply crack) indicates that the crack has initiated at the junction between the specimen surface and the specimen free edge and is propagating towards the first ply interface. Likewise, the close-up view of a crack in the mid-plane plies, Fig. 12b, shows that the crack has initiated in the middle and is propagating towards the neighboring ply interfaces. Neither one of these cracks, shown in Fig. 12, is associated with interlaminar fracture (and, therefore, interlaminar stress concentrations), and both appear to propagate in an opening mode as a result of normal stresses.

Since ply constraint is a function of the ply thickness, this appears to be an important scaling parameter. The thicker the surface ply, the less the constraint on the top surface and, therefore, the greater the normal strain, ϵ_2. Likewise, the thicker the block of plies at the mid-plane of the laminate, the greater ϵ_2, and, therefore, the lower the first and second-ply-failure stresses. Following the manner in which first and second ply failures occur, the effect of these ply failures upon the different specimen types can be examined. For example, the first-ply-failure stress of specimens scaled at the ply level (Fig. 1) would be expected to be reduced as the ply thickness increases, and, therefore, a lower ultimate strength and strain at failure would be expected. Indeed, as shown in Fig. 3, both the ultimate strength and the strain at failure are reduced with increasing specimen thickness. On the other hand, specimens scaled at the sublaminate level share the same surface ply thickness and the same mid-plane ply thickness, a single ply on the surface and two blocked plies in the middle. In this case the relative severity of the weak links is reduced with increasing specimen thickness, and, therefore, the strength and strain at failure increase with increasing specimen thickness. Note that the ($+45°/-45°$)$_{2n\text{S}}$ laminate is equivalent to a ($+45°\{[-45°/+45°]_{2n-1}/-45°_2/[+45°/-45°]_{2n-1}\}+45°$) laminate which, basically, is the same as ($+45°\{$unbalanced core$\}+45°$) where the shear/extension coupling of the unbalanced core will depend on n; the greater n, the smaller the shear/extension coupling. That, in effect, means that the constraint at the interface between the unbalanced core and the surface plies is reduced as n increases, resulting in some degree of relaxation of the in-plane shear strain, γ_{xy}, on the free surfaces, and hence a higher first-ply-failure stress. The effect of the failed plies on the rest of the laminate also depends on the total number of plies, and this will be discussed below.

An issue concerning the importance of the specimen dimensions is raised in this investigation. Which is the most important specimen dimension? Is it the specimen thickness, or is it the specimen width, or is it the specimen width/thickness ratio? Figure 3 shows that the specimen size, for a given scale-up procedure, affects the first-ply-failure stress. For example, considering the specimens scaled at the sublaminate level and looking at the points where the stress/strain curve of a given specimen size deviates from the rest, it can be deduced that the first-ply-failure stress for the 8- and 16-ply specimens is approximately 150 and 159 MPa, respectively. Figure 6a shows that even though the 16-ply specimen, Ba, has a narrower width (12.7 mm versus 25.4 mm in Fig. 3) than the 16-ply specimen, Bb, the first-ply-failure stress is still approximately 159 MPa. That is, the change in specimen width did not influence the first ply strength. Likewise, the 8-ply specimen, Ab, in Fig. 6b has a width twice as large as the Aa specimen of Fig. 3 and, yet, first ply failure still occurs at approximately 150 MPa. Again, the specimen width did not affect the first ply strength. Clearly, the laminate thickness appears to be the most important specimen dimension. A remaining question is the importance of the specimen width/thickness ratio.

The issue of the specimen width/thickness ratio has been approached by many researchers who have anticipated that interlaminar stresses can influence the failure mechanisms of angle-ply laminates, e.g. Ref 14. Interlaminar stresses have also been thought to have been responsible for the different stress/strain responses in distributed and blocked ply ±45° laminates [15]. In some cases, interlaminar stresses were also thought to be important enough to influence the specimen design for the ±45° tension specimen [16]. If such edge stresses were significant in the mechanical response of the ±45° laminates, specimens of different width/thickness ratios would exhibit different first-ply-failure stresses. However, the present experimental findings indicate that, within the range of stresses that first ply failures occur, there is no significant difference in the stress/strain response of such laminates (Fig. 7). Clearly, specimens with the same number of plies exhibit a similar stress/strain response within a large portion of the stress/strain curve. In particular, the 24-ply specimens, Ca and Cc in Fig. 7b, which have width/thickness ratios of 4.2 and 12.7, respectively, exhibit a more or less identical stress/strain response up to about 6% strain, well beyond the point of first ply failure. A similar behavior has also been observed in the PEEK matrix specimens (Fig. 8). The fact that the narrow specimens fail at lower strains than the corresponding wider specimens is thought to be due to the relatively higher grip constraint that the narrow specimens have to endure. For this reason, most narrow specimens failed either very close or within the gripping region.

When it was apparent that the first ply failure in the epoxy specimens occurred in a normal rather than shear mode, it was postulated that a tough matrix system with a high opening mode crack resistance would sustain higher loads prior to first ply failure. For this reason, the PEEK matrix specimens were introduced. However, as the experimental results indicate, the first-ply-failure stress for the PEEK matrix specimens appears to be substantially lower than the corresponding epoxy specimens (Fig. 5). Even if the first-ply-failure stress in the PEEK matrix specimen is considered to occur at the point where a visible (by optical microscope) ply crack initiates, that stress is still about 22% lower than the equivalent stress in the corresponding epoxy matrix specimen. Clearly, the toughness of the matrix alone is not a sufficient parameter for ply toughness. Additional factors that may contribute to the rather low value of first-ply-failure stress are the relative ply stiffnesses, which control the surface deformations of the laminate and the fiber/matrix interfacial strength. This subject requires further investigation.

Stress/Strain Response

Classical lamination theory predicts that undamaged ±45° laminates share the same initial stress/strain response irrespective of the stacking sequence. This is clearly shown in Fig. 3 for the five laminates, Aa, Cc, Dd with distributed plies and Bb with blocked plies. The two laminates with blocked plies, Cc and Dd, appear to have a slightly lower initial stiffness as a result of residual stress-induced damage in these specimens [6]. Clearly, the total stress/strain response of angle-ply laminates depends not only upon the generic layup (blocked or distributed plies), but also upon the laminate thickness.

The stress/strain response of epoxy matrix specimens changes from brittle to ductile as the thickness of the blocked plies is reduced or the thickness of the distributed ply laminates is increased. These large changes in the stress/strain response can be attributed to transply cracking (or fiber/matrix debonding). When such cracks are formed, the laminate stiffness appears to be decreasing initially, and, in the case of the epoxy specimens, the tangent stiffness even becomes negative. When the laminate stiffness reaches a minimum value, the specimen either fails due to the fact that severe localized straining takes place in a small region of the gage section, where the crack density is greater, or the specimen stiffness increases as the crack density increases due to fiber alignment with the loading direction (this effect is known as fiber scissoring). In other words, if a $(+45°/-45°)_{2nS}$ laminate can survive a given deformation, before ultimate failure occurs the stiffness will increase as a result of fiber scissoring. Specimens that fall into this category are expected to be those with brittle matrices and a large number of ply interfaces like the epoxy matrix specimens with 24 or greater number of distributed plies, or specimens with a compliant matrix and a relatively weak fiber matrix interface like the PEEK matrix specimens. Note that once a significant amount of transply cracking occurs, the integrity of the laminate depends solely upon the ply interface region which holds the damaged plies together. Therefore, the larger the number of ply interfaces in a laminate, the larger the axial strain that a specimen can endure. In this case, the uniform distribution of the transply cracks within the volume of the specimen is also important. The more evenly distributed the cracks, the higher the axial strain, which results in large fiber rotations which, in turn, result in a higher laminate stiffness and ultimate strength.

The absence of a negative gradient in the stress/strain response of PEEK matrix specimens constitutes a characteristic difference between corresponding PEEK and epoxy matrix specimens. The absence of a negative (or even zero) gradient in the stress/strain response of PEEK matrix specimens can be attributed to the larger strains that can be accommodated by the PEEK matrix leading to larger fiber scissoring at a given applied stress (see Fig. 5).

Strength and Strain at Failure

It has been shown that the first ply failure for a given generic $\pm 45°$ layup (with blocked or distributed plies) depends primarily upon the laminate thickness. Moreover, it has been shown that, within a large range of applied loads, the stress/strain response depends primarily upon the laminate thickness. From the point of view of the ultimate strength and strain at failure, it appears that the laminate thickness is also the most important factor. Wide specimens, in general, exhibited a slightly higher ultimate strength and strain at failure; however, this was attributed to a loading artifact rather than an actual material scaling effect. This represents a typical practical constraint, where a loading condition as simple as a tensile load cannot be reproduced in an equivalent manner for any set of scaled specimens.

While, in the case of distributed ply laminates, the increase in first-ply-failure stress with specimen thickness is partly responsible for the observed increases in ultimate strength, the end result should depend on additional factors such as the number of ply interfaces within a laminate, the distribution of damage within the laminate, and the overall effect of the failed plies on the remainder of the laminate. For example, in an 8-ply laminate, failure of the surface plies followed by failure in the mid-plane plies is roughly equivalent to a 50% ply loss. However, failure of the same four plies in a 32-ply thick laminate corresponds to only 12.5% ply loss. Clearly, the influence of the four failed plies will be detrimental in an 8-ply laminate. Likewise, in a 32-ply laminate, constructed out of blocked plies, the first and second ply failures correspond to 50% ply loss (eight plies in the two +45° blocks on the surface and eight plies in the −45° block in the middle) with a detrimental effect on strength. Moreover, specimens with blocked plies exhibit a lower first-ply-failure stress. In some cases, first ply failure occurs even before any mechanical load is applied to the specimens due to the influence of curing stresses on the weaker, blocked plies [17]. As a result, the strength of the 32-ply laminate with blocked plies is much lower than the strength of the 8-ply baseline specimen.

For the purpose of a direct comparison, the normalized strength and strain at failure of the seven different types of ±45° epoxy matrix specimens, which were geometrically scaled in three dimensions, are plotted in Fig. 13. Note that for specimens scaled on the sublaminate level, very large strains were accommodated, which resulted in fiber scissoring. Consequently, due to these large deformations, the fibers were no longer oriented

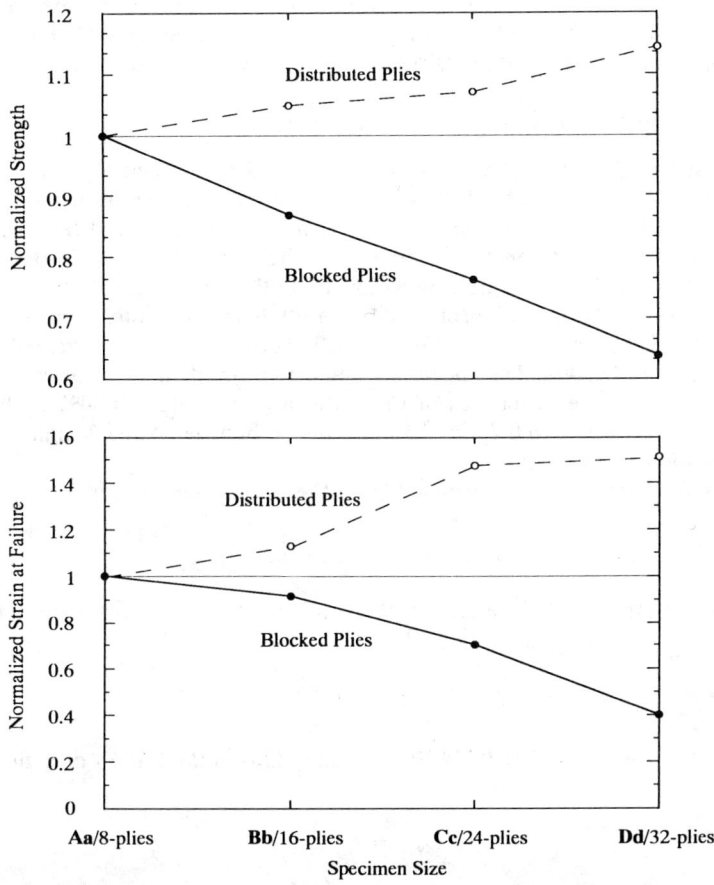

FIG. 13—*Normalized strength* (top) *and normalized strain at failure* (bottom) *versus specimen size for geometrically scaled epoxy specimens. For specimens with distributed plies an apparent strength (or strain at failure) value is used, corresponding to the point where the gradient of the stress/strain curve is zero, (see Fig. 3).*

at $\pm 45°$. In fact, it can be shown that for a Poisson's ratio, $\nu_{xy} = 0.7$ (a typical value for a $\pm 45°$ graphite/epoxy laminate) and an applied strain, $\epsilon_x = 2\%$, the fiber rotation is approximately $1.0°$ [12]. Therefore, an apparent ultimate strength of the $\pm 45°$ laminates, scaled at the sublaminate level, was chosen to represent an average $\pm 45°$ in-plane shear strength rather than the laminate ultimate strength. Note that the in-plane shear stress is equal to one-half of the applied stress [18]. The stress chosen to describe the apparent strength of the $\pm 45°$ epoxy matrix laminates corresponds to the point where the gradient of the stress/strain curves first becomes zero (Fig. 3). Also, an apparent strain at failure corresponds to the same point on the stress/strain curves. This point for the 32-ply laminates corresponds to approximately 2.4% axial strain, a small enough value to neglect any fiber scissoring. As shown in Fig. 13, the normalized values of the apparent strength and strain at failure show the same trend with increasing specimen size: an increase with size for specimens with distributed plies and a decrease with size for specimens with blocked plies. Note that the strength and strain at failure of the specimens with distributed plies cannot show an indefinite increase in strength with specimen size. When a given large number of plies is reached, volume effects due to the variability of the material will become important, and, therefore, a plateau in the curves of Fig. 13 is to be expected. In fact, Fig. 13b indicates that the 32-ply laminates, with distributed plies, are very close to such a plateau.

The $\pm 45°$ Standard Shear Test

The ASTM Standard Practice for In-plane Shear Stress-Strain Response of Unidirectional Reinforced Plastics (ASTM D 3518) is one of the simplest shear tests available for the determination of the in-plane shear stress/strain response of fiber-reinforced unidirectional composites. In addition to its simplicity, the $\pm 45°$ shear test is one of few tests available that is used for the determination of both the shear stiffness and the shear strength. The test was originally proposed by Petit [19] and was later improved (simplified) by Rosen [18]. The test procedure involves a balanced and symmetric $\pm 45°$ composite specimen loaded in tension. The specimen dimensions are defined according to the ASTM Test Method for Tensile Properties of Fiber-Resin Composites (D 3039), which allows a specimen thickness between 0.5 and 2.5 mm, that is, between 4 and 20 standard thickness (0.125 mm) unidirectional plies.

The shear stiffness can be determined from the simple relationship

$$G_{12} = \Delta\tau_{12}/\Delta\gamma_{12}$$

where the shear stress, τ_{12}, is related to the applied load, P, and the cross sectional area, A, as

$$\tau_{12} = P/2A$$

and the shear strain, γ_{12}, is related to the normal strains in the x and y directions, ϵ_x and ϵ_y, as

$$\gamma_{12} = (\epsilon_x - \epsilon_y)$$

Consequently, the shear strength, S_{12}, is given by

$$S_{12} = P_{ult}/2A$$

where P_{ult} is the ultimate applied load.

Since the ratio $\Delta\tau_{12}/\Delta\gamma_{12}$ is defined in ASTM D 3518 as being the slope of the shear stress/strain curve within the linear portion of the curve, the value of the modulus appears to be independent of the specimen size. However, the shear strength, as defined by the standard, depends strongly upon the specimen thickness. Moreover, the choice of stacking sequence is left to the user, which according to the present experimental evidence, will also have a strong influence on the measured shear strength. The validity of the $\pm45°$ tensile specimen for the determination of the shear strength has been the subject of an earlier publication by the authors [12], where it was argued that, in the case of a brittle matrix system like the 3502 epoxy, there are two choices for the ultimate load, P_{ult}. The first choice is the one suggested by ASTM D 3518, that is, the load at which the specimen breaks apart. The second choice is the load at which ply failures occur throughout the laminate thickness. Since the above equations, which define the shear strength, are only valid within a small range of applied strains and prior to the onset of damage, the second choice of load is more appropriate. This value of ultimate load (or stress) was defined as the point where the gradient of the stress/strain curve is first equal to zero. It was argued further that, since the first and second ply failures are normal rather than shear failures, the shear strength is defined by

$$S_{12} = (P_{ult}/2A) = \sigma_o/2$$

where σ_0 is the stress value at which the gradient of the stress/strain curve is zero and will always be underestimated. However, since the effect of the first and second ply failures on the rest of the laminate is diminished, when the number of the total distributed plies is increased, the result of the $\pm45°$ tension test will be more accurate. It was, therefore, recommended [12] that a stacking sequence with distributed plies be clearly specified in ASTM D 3518, with a minimum specimen thickness of 3 mm (24 plies).

While the above recommendations are valid for material systems with brittle matrices and/or strong fiber/matrix interfacial bond, materials with tough or compliant matrices like the AS4/PEEK system cannot be treated. In this case, there is no obvious point on the stress/strain curve that can be used in the definition of an in-plane shear strength. Therefore, the only remaining alternative is the use of an off-set construction based on some design parameter such as, for example, the 0.4% strain. In this case, since the stress/strain response of PEEK matrix specimens was also sensitive to the specimen thickness, it is recommended that the minimum thickness of 3 mm (24 plies) becomes the standard minimum thickness for all material systems.

Conclusions

First Ply Failure/Epoxy System

First ply failure in the epoxy matrix specimens with distributed plies occurred in the form of transply cracks in the surface plies as a result of normal and *not* shear fracture. Likewise, second ply failure which occurred in the middle block of two $-45°$ plies was also the result of normal rather than shear fracture. It was shown that the first-ply-failure stress depends primarily on the specimen thickness and not the specimen width. The larger the number of distributed plies the greater the first-ply-failure stress. In the case of specimens with blocked plies the first-ply-failure stress decreased as the thickness of the blocked plies was increased.

First Ply Failure/PEEK System

In the case of specimens with PEEK matrix, first ply damage initiated at relatively low applied stresses. Such damage, which was visible in the enhanced X-ray radiographs but could not be verified by optical microscopy, was much lower than the first-ply-failure stress in the corresponding epoxy specimens. Evidence of transply cracks were first observed to occur in the surface plies followed by cracks in the middle plies. These transply cracks occurred at stresses which were still lower than the first-ply-failure stress in the corresponding epoxy matrix specimens. As in the case of the epoxy matrix specimens, the stress value at which transply cracking initiated was primarily a function of the specimen thickness and not the specimen width.

Stress/Strain Response/Epoxy System

As a result of the sensitivity of the first-ply-failure stress to the specimen thickness, a comparison of the stress/strain response of different thickness specimens could be used as an indicator of damage initiation. The deviation of a given stress/strain curve from the curve of a second geometrically scaled specimen corresponded to the point at which damage could be detected by nondestructive damage examination.

In general, the stress/strain response of the epoxy matrix specimen depended primarily upon two factors: the generic layup type and the laminate thickness. Specimens with blocked plies exhibited a brittle-like stress/strain response, and their ultimate strength and strain at failure increased with decreasing blocked ply thickness. Specimens with distributed plies, however, exhibited a ductile-like stress/strain response, and the ultimate strength and strain at failure increased with increasing specimen thickness. The specimen width had no significant influence on the stress/strain response.

The stress/strain response of the epoxy specimens with distributed plies was characterized by a region in which the gradient of the stress/strain curves became zero and then negative before it became positive again at slightly higher strains. This behavior was more obvious in specimens with 24 and 32 plies.

Stress/Strain Response/PEEK System

As in the case of epoxy matrix specimens, the stress/strain response of the PEEK matrix specimens proved to be a useful tool in determining the point of damage initiation. For this particular material system the technique of comparing the stress/strain responses of two or more scaled specimens as a means of determining the point on the stress/strain curve at which damage initiated was more useful than the two nondestructive techniques used.

In comparison with the corresponding epoxy matrix specimens, the PEEK matrix specimens exhibited a more ductile stress/strain response, having a lower initial longitudinal stiffness and failing at higher strains. In the case of the baseline (8-ply) specimens, the difference in the strain at failure between the epoxy and the PEEK matrix specimens was approximately 15%. Another basic difference in the stress/strain response of the two material systems took place at a strain of approximately 2%, where the gradient of the stress/strain response of the epoxy matrix specimens became zero and then negative. This behavior was not observed in the PEEK specimens.

Strength and Strain at Failure/Epoxy Matrix

Both the ultimate strength and the strain at failure were a function of the generic layup (blocked or distributed plies) and the specimen thickness. Specimens with distributed plies showed an increase in ultimate strength and strain at failure with increasing specimen thickness. In specimens with blocked plies, as the ply thickness increased, the ultimate strength and the strain at failure were reduced.

Strength and Strain at Failure/PEEK Matrix

Similar behavior to that of the epoxy matrix specimens was also observed in the PEEK matrix specimens. The ultimate strength and strain at failure increased with increasing specimen thickness. The ultimate strength and strain at failure of PEEK matrix specimens was always greater than the ultimate strength and strain at failure of corresponding epoxy matrix specimens.

±45° Standard Shear Test

This work has highlighted the need to redefine the specimen configuration in the ASTM D 3518 test if a representative value of the shear strength, as a material property, is to be determined. It is recommended that the minimum specimen thickness be changed from 0.5 to 3.0 mm (or from 4 to 24 plies) and the stacking sequence be clearly specified ($\pm 45°/\pm 45°$)$_{ns}$ where the value of n is at least 3.

Acknowledgments

This study was supported by NASA Langley Research Center under grant NAS1-18471. Thanks are due to the contract monitor Huey D. Carden. The discussions with Robert M. Jones are also gratefully acknowledged.

References

[1] Zweben, C. "The Effect of Stress Nonuniformity and Size on the Strength of Composite Materials," *Composites Technology Review,* Vol. 3, No. 1, 1981.

[2] Atkins, A. G. and Caddell, R. M., "The Laws of Similitude and Crack Propagation," *International Journal of Mechanics Science,* Pergamon Press, Elmsford, NY, Vol. 16, 1974.

[3] Batdorf, S. B., "Note on Composite Size Effects," *Journal of Composites Technology and Research,* Vol. 11, No. 1, 1989.

[4] Morton, J., "Scaling of Impact-Loaded Carbon-Fiber Composites," *AIAA Journal,* Vol. 26, No. 8, 1988, pp. 989–994.

[5] Jackson, K. E., "Scaling Effects in the Static and Dynamic Response of Graphite-Epoxy Beam-Columns," Ph.D. thesis, Engineering Science and Mechanics Dept., Virginia Polytechnic Institute and State University, Blacksburg, VA, 1990.

[6] Kellas, S. and Morton, J., "Strength Scaling in Fiber Composites," NASA Contractor Report 4335, November 1990.

[7] Rodini, B. T. and Eisenmann, J. R., "An Analytical and Experimental Investigation of the Edge Delamination in Composite Laminates," *Fibrous Composites in Structural Design,* E. M. Lenoe, Ed., Plenum Press, New York, 1978, pp. 441–457.

[8] Lagace, P., Brewer, J., and Kassapoglou, C., "The Effect of Thickness on Interlaminar Stresses and Delamination in Straight-Edged Laminates," *Journal of Composites Technology and Research,* Vol. 9, No. 3, 1987, pp. 81–87.

[9] Camponeschi, E. T. Jr., "Compression Testing of Thick-Section Composite Materials," Research and Development Report, DTRC-SME-89/73, David Taylor Research Center, Bethesda, MD, Ship Materials Engineering Dept., October 1989.

[10] Wisnom, M. R., "Relationship Between Strength Variability and Size Effect in Unidirectional Carbon Fibre/Epoxy," *Composites,* Vol. 22, No. 1, January 1991, pp. 47–52.

[11] Crosman, F. W. and Wang, A. S. D., "The Dependence of Transverse Cracking and Delamination of Ply Thickness in Graphite/Epoxy Laminates," *Damage in Composite Materials, ASTM STP 775,* K. Reifsnider, Ed., American Society for Testing and Materials, Philadelphia, 1982, pp. 118–139.

[12] Kellas, S., Morton, J., and Jackson, K. E., "An Evaluation of the ±45° Tensile Test for the Determination of the In-plane Shear Strength of Composite Materials," *Proceedings, ICCM VIII*, Honolulu, Hawaii, July 1991, S. W. Tsai and S. Springer, Eds., SAMPE, Ovina, CA.

[13] Whitney, J. M., Daniel, I. M., and Pipes, R. B., "Experimental Mechanics of Fiber Reinforced Composite Materials—Revised Edition," Brookfield Center, CN, *SEM Monograph No. 4*, Section 4.3.6, Society for Experimental Mechanics, 1984, pp. 187–191.

[14] Murthy, P. L. N. and Chamis, C. C., "Free-Edge Delamination: Laminate Width and Loading Conditions Effects," *Journal of Composites Technology and Research*, Vol. 11, No. 1, Spring 1989, pp. 15–22.

[15] Terry, G., "A Comparative Investigation of Some Methods of Unidirectional, In-plane Shear Characterization of Composite Materials," *Composites*, October 1979, pp. 233–237.

[16] Chatterjee, S. N., Wung, E. C. J., Ramnath, V., and Yen, C. F., "Composite Specimen Design Analysis-Volume I: Analytical Studies," MTL TR 91-5, January 1991.

[17] Flaggs, D. L. and Kural, M. H., "Experimental Determination of the In Situ Transverse Lamina Strength in Graphite/epoxy Laminates," *Journal of Composite Materials*, Vol. 16, March 1982, pp. 103–116.

[18] Rosen, B. W., "A Simple Procedure for Experimental Determination of the Longitudinal Shear Modulus of Unidirectional Composites," *Journal of Composite Materials*, Vol. 6, October 1972, pp. 552–554.

[19] Petit, P. H., "A Simplified Method of Determining the In-plane Shear Stress/Strain Response of Unidirectional Composites," *Composite Materials: Testing and Design, ASTM STP 460*, American Society for Testing and Materials, Philadelphia, 1969, pp. 83–93.

Masaki Hojo[1] and Takahira Aoki[2]

Thickness Effect of Double Cantilever Beam Specimen on Interlaminar Fracture Toughness of AS4/PEEK and T800/Epoxy Laminates

REFERENCE: Hojo, M. and Aoki, T., "**Thickness Effect of Double Cantilever Beam Specimen on Interlaminar Fracture Toughness of AS4/PEEK and T800/Epoxy Laminates,**" *Composite Materials: Fatigue and Fracture, Fourth Volume, ASTM STP 1156,* W. W. Stinchcomb and N. E. Ashbaugh, Eds., American Society for Testing and Materials, Philadelphia, 1993, pp. 281–298.

ABSTRACT: The influence of specimen thickness on Mode I interlaminar fracture behavior was investigated with laminates made from ICI APC-2 prepregs (AS4/PEEK) and from Toray P2212 prepregs (T800/epoxy). The mechanisms of the influence of the thickness effect were discussed on the basis of the direct measurement of the fiber bridging. Double cantilever beam (DCB) specimens 3, 4, 5, and 8 mm thick were prepared from the same unidirectional laminate of 8 mm in thickness in order to avoid molding effects. Fracture toughness, G_{IC}, at the initiation of crack growth was carefully measured in order to clarify the increment of G_{IC} from the initiation values with crack propagation. The initiation values of G_{IC} were independent of the specimen thickness both for AS4/PEEK and T800/3631 laminates. The G_{IC} values for AS4/PEEK laminates increased quickly from the initiation values and leveled off. The thicker specimen indicated slightly higher propagation values of G_{IC}. The thickness effect obtained here was much smaller than that obtained from specimens of originally different thickness. For the case of T800/epoxy laminates, two series of tests from different panels indicated different propagation values of G_{IC}. For one panel, the values of G_{IC} increased continuously with crack length, whereas for the other panel, the G_{IC} values were almost constant, and agreed with the initial values. The influence of different panels and the scatter of data was much larger than the influence of specimen thickness. This suggests that the propagation values of G_{IC} can be used as an indication of fiber alignment of laminates which cannot be detected by the initiation values of G_{IC}. Fiber-bridged zone length was measured in order to discuss the mechanism of the *R*-curve behavior. A weak correlation between the fiber-bridged zone length and the increase of G_{IC} from the initiation value was observed.

KEY WORDS: fracture, composite materials, delamination, interlaminar fracture toughness, thickness effect, fiber bridging

Composite materials used in primary structures require performance reliability based on damage tolerance. The improvement of the interlaminar fracture toughness of advanced polymer matrix composite is of current interest, particularly with regard to their post-impact compressive strength, which is currently the limiting factor in many design applica-

[1]Associate professor, Mesoscopic Materials Research Center, Faculty of Engineering, Kyoto University, Kyoto 606-01, Japan.
[2]Associate professor, Department of Aeronautics and Astronautics, the University of Tokyo, Tokyo 113, Japan.

tions [1]. Mode I interlaminar fracture toughness, G_{IC}, with toughened resin systems has been extensively investigated by using double cantilever beam (DCB) specimens [2,3]. Many experimental results of carbon fiber-reinforced laminates with toughened resin system and glass fiber-reinforced laminates show that G_{IC} values depend on delamination crack length and laminate thickness [4–7]. For the case of the crack length dependency, the initiation values of G_{IC}, G_{ICi}, were lower than values of G_{IC} measured during propagation. Thus, the importance of the initiation values, G_{ICi}, and the R-curve behavior (an increase of G_{IC} with crack length) in advanced composites has been recognized by several researchers [4,8,9]. Moreover, these points have been extensively discussed in the standardization activities by ASTM Subcommittee D30.06 on Interlaminar Properties, the European Structural Integrity Group, VAMAS, and the Japanese group for Japan Industrial Standards (JIS) [10–12].

Research on the toughening mechanisms is very important for developing tougher composites. Although many authors attributed the R-curve behavior to fiber bridging, the mechanisms and the contribution of fiber bridging are not yet well understood [13]. The fiber bridging causes the crack closure force on crack surfaces, which is a function of the crack opening displacement. Since this fiber-bridged zone is much larger than the region where the stress and strain field is governed by the singularity at the crack tip, use of specimens with different crack opening displacement may provide a key to separate the contribution of the damage process zone ahead of the crack tip and that of the fiber-bridged zone to the interlaminar fracture toughness.

One way to verify this hypothesis is to carry out fracture toughness tests by using DCB specimens of different thickness. Previous experimental results of Wang et al. [5], Russell and Street [6], and Prel et al. [7] show that the propagation values of fracture toughness, G_{IC}, increase with increasing laminate thickness. However, molding laminates of different thickness might bring about different mechanical properties. The difference in degree of crystallinity is one of the most possible indications for the case of thermoplastic matrix composites. The study by Phillips and Wells [14] on intralaminar fracture toughness shows that increasing laminate thickness results in decrease of fracture toughness.

Although many authors pointed out the importance of fiber bridging, only qualitative observation has been reported. Quantitative information on the exact nature of fiber bridging is missing. Recently, Davies et al. [15] and Hojo et al. [16] tried to find the influence of specimen thickness by machining down the specimen in thickness direction. Their experimental results suggested that the influence of the specimen thickness on the propagation values was much reduced by machining down the specimens. They mainly discussed the propagation values of fracture toughness. Direct measurement of the fiber bridging was not carried out in their research.

In the present study, the influence of specimen thickness on interlaminar fracture behavior has been investigated under Mode I opening loading by using DCB specimens. Thinner specimens were obtained by machining down thicker specimens in order to investigate the effect of pure thickness. Special care was taken to measure the initiation values of fracture toughness. Furthermore, direct measurement of the bridged zone length was carried out to find the relation to the increase of fracture toughness from its initiation value. The present study is an extension of Ref 16, with additional experimental results.

Experimental Procedure

Materials and Specimens

The 64-plies of unidirectional laminates with the nominal thickness of 8 mm were made from ICI prepregs APC-2 (AS4/PEEK) and from Toray prepregs P2261 (T800H/3631). An

AS4/PEEK panel was fabricated by ICI Wilton. Two T800/3631 panels were fabricated by Toray Shiga, which were both laid up from the same roll of prepreg and molded in an autoclave at the same time. For the CF/PEEK laminate, a DuPont DSC System 9900 was used at a heating rate of 20°C/min to measure the degree of crystallinity [17]. The degree of crystallinity was 33% for the laminate tested here. The exact thickness of the laminates was 8.3 mm for AS4/PEEK and 8.9 mm for T800/3631.

DCB specimens (20 mm wide) were used for the interlaminar fracture toughness tests. The fiber direction of the specimen was parallel to the crack growth direction. Figure 1 shows the DCB specimen and the loading apparatus [18]. The distance between the center of the pin and the specimen surface was designed to be 3 mm in order to minimize the geometrical nonlinearity. For the case of AS4/PEEK specimens, a starter slit of about 40 mm in length was introduced into the specimen by inserting folded aluminum foil (total thickness = 30 μm) at mid-thickness. Mode I precracks of lengths from 6 to 8 mm were introduced into many of the specimens tested here from the folded Al starter slit. This was done by clamping the specimen across the entire width approximately at the end of the starter slit and then manually wedging open the specimen. For the case of T800/3631 specimens, a starter slit of about 20 mm in length was introduced into the specimen by inserting PTFE film (thickness 12 μm). No additional precracks were introduced before starting the tests. DCB specimens of 3, 4, and 5 mm thick were prepared from the same unidirectional laminates of the nominal thickness of 8 mm by milling down in the thickness direction in order to avoid other possible effects, such as the molding effect. The 8-mm-thick DCB specimens were prepared from the original 8-mm-thick laminates without machining in the thickness direction.

None of the specimens, including the milled-down ones, exhibited crack opening behavior without loading or crack closure behavior. This indicated that no significant residual stress was developed during the curing process. Thus, the series of milled-down specimens promised the effect of pure thickness on the fracture behavior.

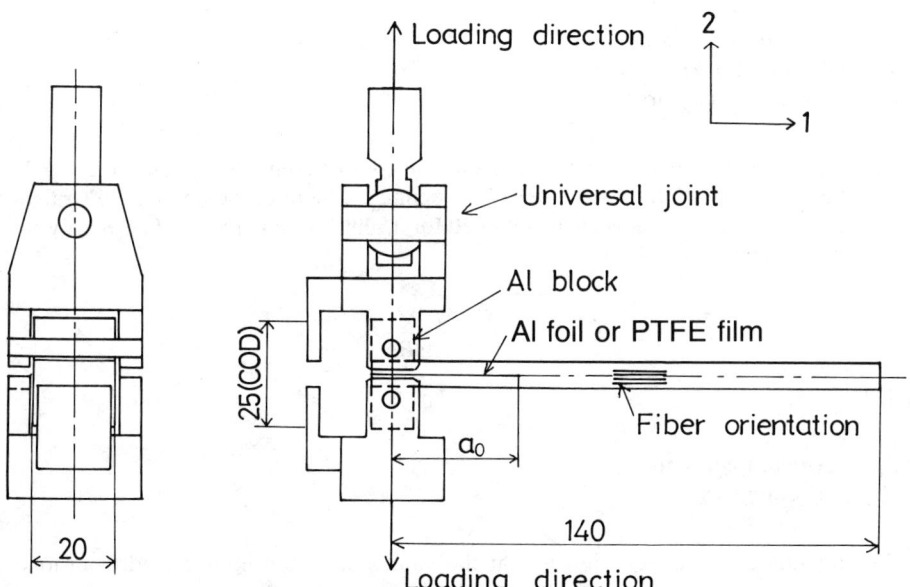

FIG. 1—*DCB specimen with loading apparatus (dimensions in mm).*

Interlaminar Fracture Tests

Interlaminar fracture toughness tests were carried out in a computer-controlled 10-kN servohydraulic testing machine with a load cell capacity of 490 N. The loading rate was 0.25 to 0.5 mm/min for AS4/PEEK laminate and 1 mm/min for T800/3631 laminates. The lower rates for the AS4/PEEK laminate were chosen because the stability of propagation could be increased [19]. The effect of the loading rate on the fracture toughness was considered to be negligible for the range of rate used in this study [20]. The specimen was loaded until the crack extended about 10 mm, and then the specimen was unloaded. This procedure was repeated until the crack had propagated about 80 mm. Since the permanent deformation was very small, some specimens were unloaded down to about 90% of the fracture load in order to measure crack length. The crack length was measured on both sides of the specimen with traveling microscopes at a magnification of $\times 100$ when specimens were unloaded down to about 90% of the fracture load. For the case of the thinner specimens (3 and 4 mm), the crack length was also measured every 5 mm by using the reference lines on both sides of the specimen because the speed of crack propagation was slow enough to measure crack length during loading. Displacement of the testing machine was used as the crack opening displacement (COD) on the load line. For the case of the thicker specimens (5 and 8 mm), COD was also measured by an extensometer attached to the loading apparatus, as shown in Fig. 1. The testing environment was 50% RH at 23°C.

For some specimens, photographs were taken at the crack length measurement, and these photographs were used for the measurement of the length of the fiber-bridged zone.

Values of the interlaminar fracture toughness G_{IC}, were calculated using Kageyama's analytical compliance method [12]

$$a/h = A_1\lambda^{1/3} + A_0 \qquad (1)$$

where

$$\lambda = \text{compliance,}$$
$$a = \text{measured crack length,}$$
$$h = \text{half thickness, and}$$
$$A_0, A_1 = \text{empirical parameters.}$$

Figure 2 shows the relation between the cubic root of compliance and the crack length obtained experimentally for an AS4/PEEK specimen. The fit of the data to the straight line is quite good. A similar relation was obtained for T800/3631 specimens. The energy release rate, G, is expressed as follows

$$G = \frac{3P_c^2\lambda^{2/3}}{2A_1Bh} \qquad (2)$$

where

$$P_c = \text{critical load, and}$$
$$B = \text{specimen width.}$$

The advantage of this equation is that the values of G can be obtained from load and compliance. Since the permanent deformation after unloading was small, compliance can

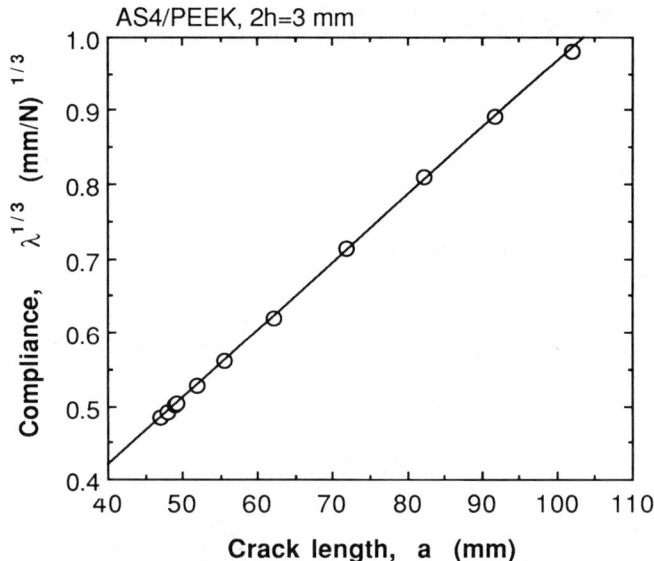

FIG. 2—*Relation between cubic root of compliance and crack length for AS4/PEEK laminate (2h = 3 mm).*

be calculated by (COD)/P_c on the load-COD curve. Thus, the values of G corresponding to any point on the load-COD curve can be obtained by using Eq 2. In addition, the crack length at this point can be calculated by using Eq 1. In the present study, A_1 values in Eq 1 were first calculated for each specimen. Then the values for G were calculated by using Eq 2. The points on load-COD curves where the crack length was measured were usually used for the calculation of the G values. For the case of AS4/PEEK laminates, several points between the initiation of nonlinearity at the initial loading and the maximum load were used without measuring crack length in order to detect the change of the fracture toughness in detail. Since the values of (COD)/a are less than 0.4 in the present experimental conditions, geometrical nonlinearity was not taken into account in the calculation of G [21].

Experimental Results and Discussion

Crack Growth Behavior

Figure 3a shows load-COD(δ) curves for a 3-mm-thick specimen of AS4/PEEK laminate with a Mode I precrack. Overall crack propagation was stable. Unstable crack growth occurred in about one out of two specimens tested in each case only once during total propagation. At the initial loading, slight nonlinearity was observed before reaching the maximum load [22]. Davies [10] attributed this nonlinearity to the initiation of delamination in the center of the specimen width. It is also reported that this point corresponds closely to the first acoustic emission [4]. This point has been suggested in the ASTM round-robin tests [11] as the initiating point. Thus, the initiation (G_{ICi}) was taken as the deviation of the load-displacement curve from linearity for all AS4/PEEK specimens with Mode I precracks. For the case of the specimens without Mode I precracks, crack propagation from the starter slit was unstable. Since the total thickness of the starter slit

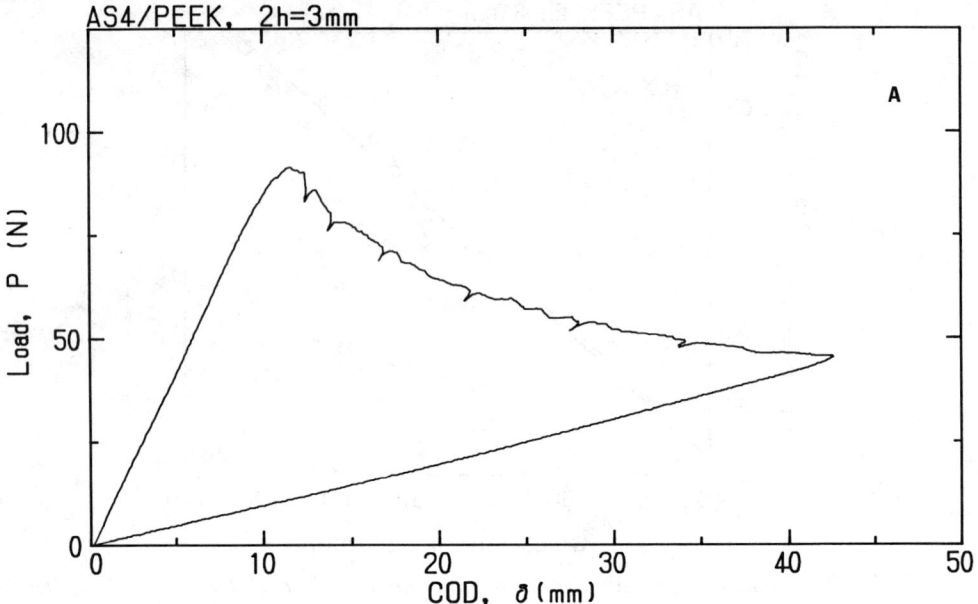

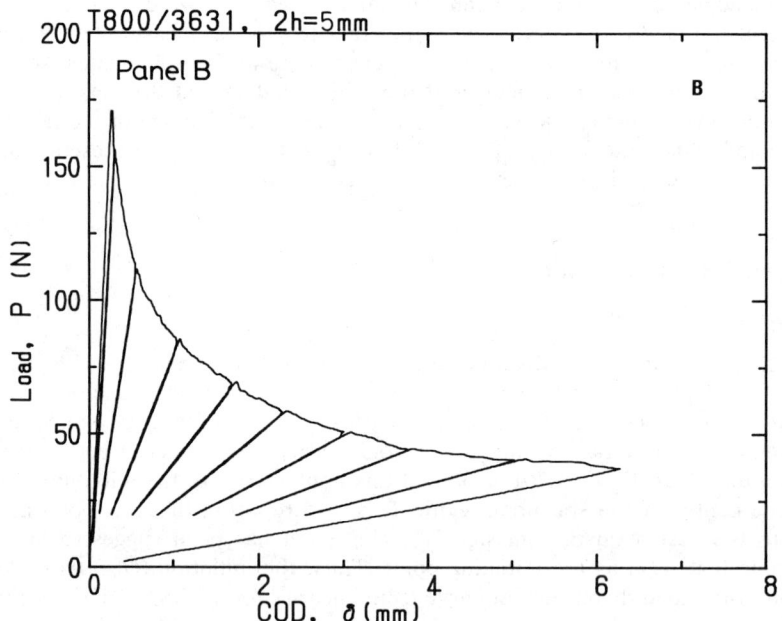

FIG. 3—*Relation between load and displacement:* (a) *AS4/PEEK laminate,* 2h *= 3 mm. Mode I precrack;* (b) *T800/3631 laminate, Panel B,* 2h *= 5 mm.*

was 30 μm, this instability might be caused by the blunting of the starter slit. Promotion of instability by the thicker starter slit was reported by other researchers [8]. The following cracks grew in a manner similar to the specimens with Mode I precracks. In this case, only the propagation values of the fracture toughness, G_{IC}, were calculated. In contrast to the complicated crack growth in AS4/PEEK laminates, the crack growth in T800/3631 was stable, as shown in Fig. 3b. Since the initiation defined by the deviation of the load-displacement curves from linearity almost agreed with the maximum load at the initial loading, the maximum load was used for the calculation of the initial values, G_{ICi}.

Influence of Specimen Thickness for AS4/PEEK Laminates

Changes of G_{IC} with crack growth (R-curves) are shown in Figs. 4A and 4B. In each figure, the first point indicated by the arrow shows the initiation value, G_{ICi}. For all cases tested here, G_{IC} values increased quickly from the initiation and then leveled off at values much higher than the initiation values. Figure 5 shows the relation between G_{ICi} and the specimen thickness, $2h$. In this figure, each data point represents a different specimen in which the initiation value was measured. The initiation values are almost independent of the specimen thickness. Recently, Davies et al. reported that the very thin insert (less than 13 μm) without precracks gives the lower bound of the G_{ICi} values [8]. They also pointed out that the Mode I precrack will give much higher G_{ICi} values, which are almost the same as propagation values. However, the present results shown in Fig. 5 ($G_{ICi} = 1.1$ to 1.3 kN/m) agree fairly well with the G_{ICi} values obtained by using a very thin insert without precracks in Ref 8 ($G_{ICi} = 1.1$ to 1.2 kN/m). Thus, we think that the present G_{ICi} values are reasonable.

Figure 6 shows the relation between the propagation values, G_{IC}, and the specimen thickness at $a = 80$ mm. In contrast to the initiation values, the propagation values show a clear effect of specimen thickness. The results of Davies et al. [15] for IM6/PEEK laminates and their review of the published results for AS4/PEEK laminates are also reproduced in this figure. Open symbols indicate the results of AS4/PEEK laminates, and solid symbols indicate IM6/PEEK laminates. Square marks indicate that the series of specimens were milled down from thicker specimens, and triangle marks indicate that the specimens were cut directly from original laminates. The solid square symbol at $2h = 3$ mm indicates the results of the laminate which was milled down from the laminate of the original thickness of 5 mm. The result of this 5-mm-thick laminate is indicated by the solid triangle at $2h = 5$ mm. The effect of laminate thickness in the present study (solid line) is much smaller than that obtained by Davies et al. with milling down the thickness (dashed line). They found a similar large effect of the thickness for the results of IM6/PEEK laminates and for their review of AS4/PEEK laminates without milling down the thickness [15]. Present experimental results show that the molding effect is larger than the genuine thickness effect on the propagation values of G_{IC}.

Influence of Specimen Thickness for T800/3631 Laminates

For the case of T800/3631 laminates, two panels of the nominal thickness of 8 mm were used for testing. These two panels were designated as Panel A and Panel B. Figure 7 shows the results obtained from Panel A, and Fig. 8 shows those obtained from Panel B.

R-curves obtained from Panel A differed greatly from those obtained from Panel B. For Panel A, G_{IC} values increased continuously with crack length, and they did not reach plateau values. G_{IC} values at $a = 100$ mm were twice as much as those at initiation. For Panel B, the increase of G_{IC} values with crack length was almost negligible. Figure 9 shows the relation between the initiation values, G_{ICi}, and the specimen thickness for both panels.

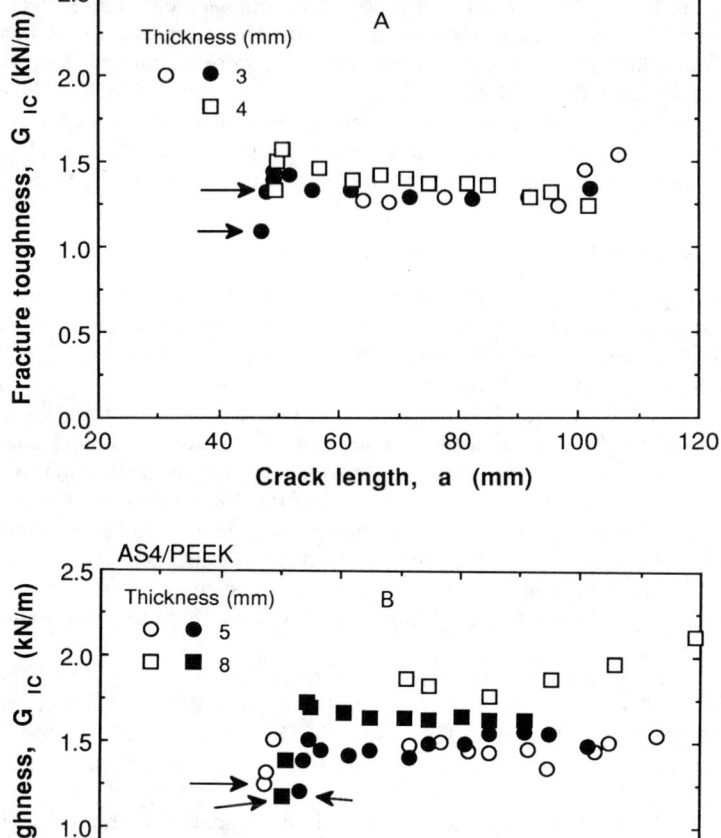

FIG. 4—*Relation between fracture toughness and crack length for AS4/PEEK laminate:* (A) *thickness = 3 and 4 mm (arrows indicate the initiation values);* (B) *thickness = 5 and 8 mm.*

The initiation values for two panels are almost the same and are independent of the specimen thickness.

The relation between G_{IC} values at $a = 80$ mm and specimen thickness is summarized in Fig. 10. Although the slight increase of G_{IC} with thickness is shown for both panels, the change is within the scatter range of the data points. As will be mentioned later, we observed extensive fiber bridging in the specimens from both panels. It is noteworthy that the effect of the specimen thickness is minimal for specimens exhibiting extensive fiber bridging.

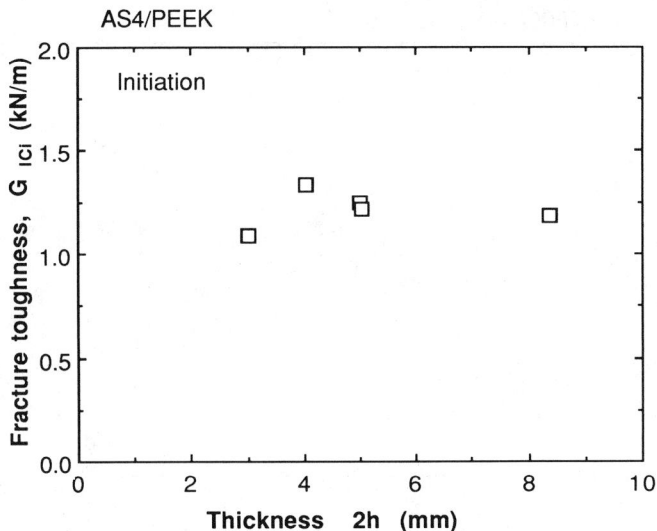

FIG. 5—*Effect of the thickness of initiation value of fracture toughness,* G_{ICi} *for AS4/PEEK laminate.*

Observation of Fiber Bridging

Figure 11 shows the formation of fiber bridging with crack propagation for the AS4/PEEK specimen of the thickness of 5 mm. Fiber bridges were continuously formed and broken. They were more prominent for the thicker specimen than for the thinner specimen in AS4/PEEK laminates.

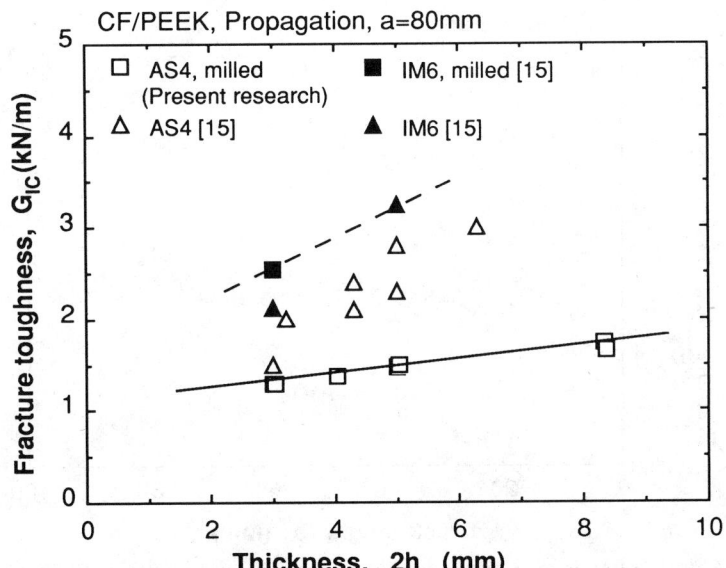

FIG. 6—*Effect of the thickness on propagation value of fracture toughness,* G_{IC} *for AS4/PEEK laminate.*

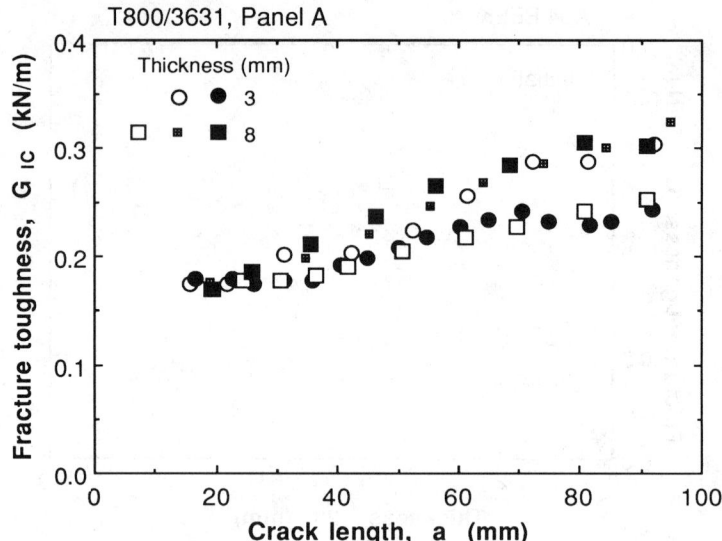

FIG. 7—*Relation between fracture toughness and crack length for T800/3631 laminate, Panel A.*

The length of the bridged zone was measured in some specimens according to the definition shown in Fig. 12. This definition was tentatively chosen for this study because other parameters, such as the distribution of the cohesive force and the density of the bridged fibers in the width direction, were difficult to evaluate at the present moment.

The bridged zone length of AS4/PEEK laminates is shown against the crack length in

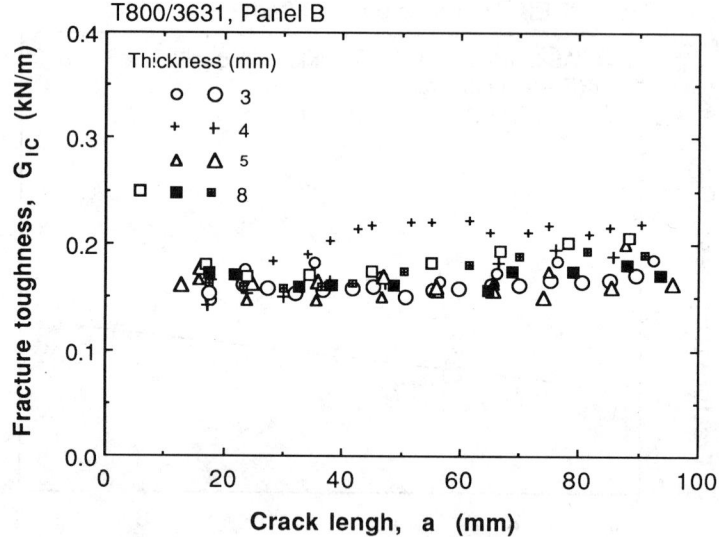

FIG. 8—*Relation between fracture toughness and crack length for T800/3631 laminate, Panel B.*

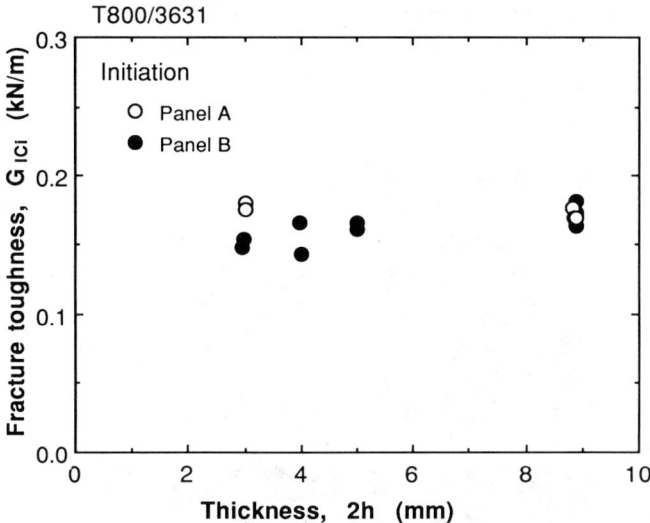

FIG. 9—*Effect of the thickness on initiation value of fracture toughness, G_{ICi} for T800/ 3631 laminate.*

Fig. 13. For the case of T800/3631 laminates, the bridged zone length was measured for only 3-, 4-, and 8-mm-thick specimens. The results are shown in Fig. 14. It is interesting that the bridged zone length increases with the increase of the specimen thickness in these figures.

The bridged zone length at $a = 80$ mm was plotted against the thickness of the specimen in Fig. 15. Since the formation of fiber bridging and corresponding R-curve behavior

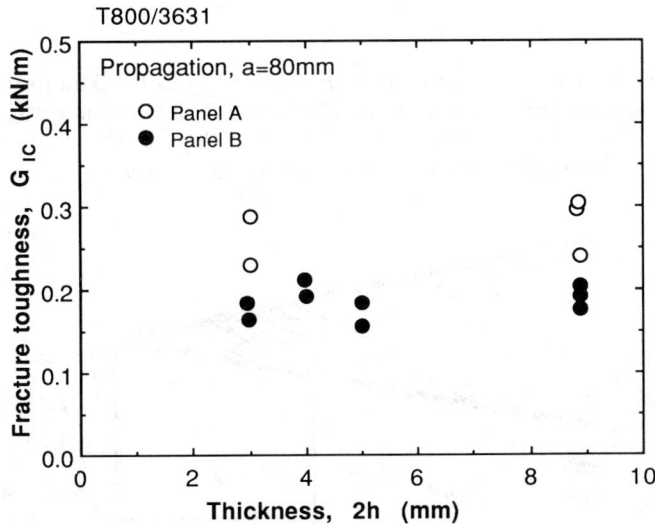

FIG. 10—*Effect of the thickness on propagation value of fracture toughness, G_{IC} for T800/3631 laminate.*

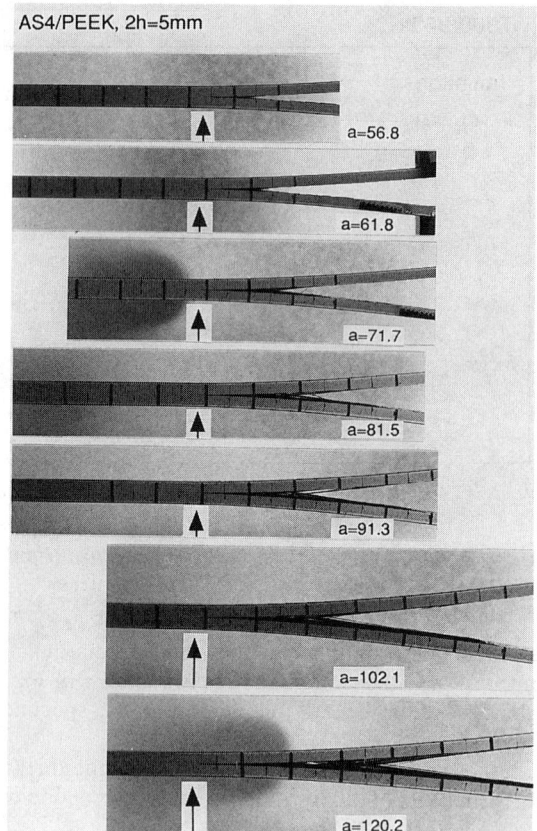

FIG. 11—*Formation of fiber bridging with crack growth for 5-mm-thick AS4/PEEK laminate.*

depend on each specimen, a single point in Fig. 15 indicates the data for each specimen whose bridged zone length was measured. The fit of the data to the straight line is very good for AS4/PEEK laminates. This fact suggests that there exists a critical value of the crack opening displacement at which the bridged fibers were broken. For the case of T800/

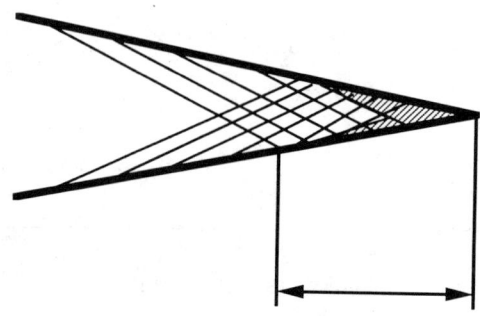

Bridged zone length, D

FIG. 12—*Definition of the measurement of bridged zone length.*

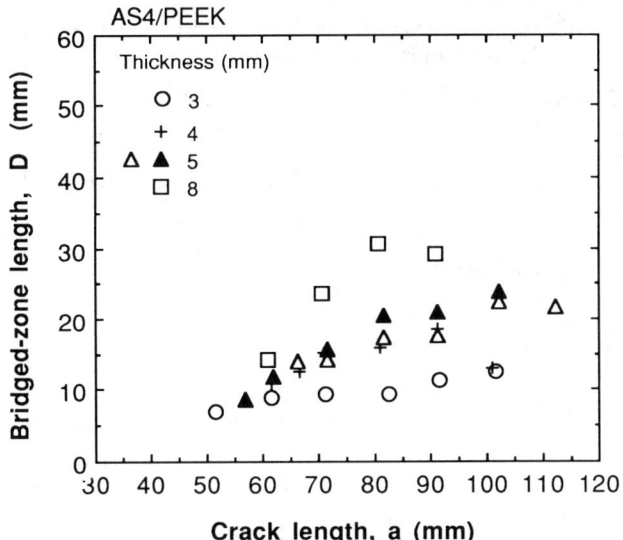

FIG. 13—*Relation between bridged-zone length and crack length for AS4/PEEK lami-nate.*

3631 laminates, the scatter is large, and the slope of the straight line is smaller than that for AS4/PEEK laminates.

In order to find out the relation between the bridged zone length and the fracture toughness values, the bridged zone length was plotted against the normalized increment of the fracture toughness from the initiation value, $(G_{IC} - G_{ICi})/G_{ICi}$, at the crack length of 80 mm in Fig. 16. This figure shows that the increase of fracture toughness can be roughly

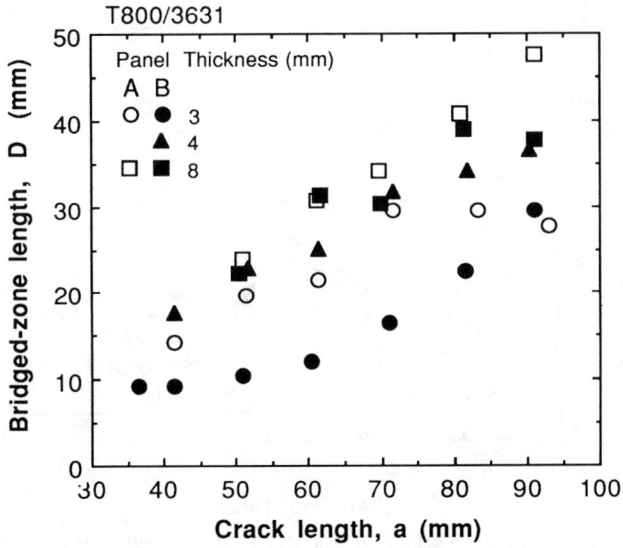

FIG. 14—*Relation between bridged-zone length and crack length for T800/3631 lami-nate.*

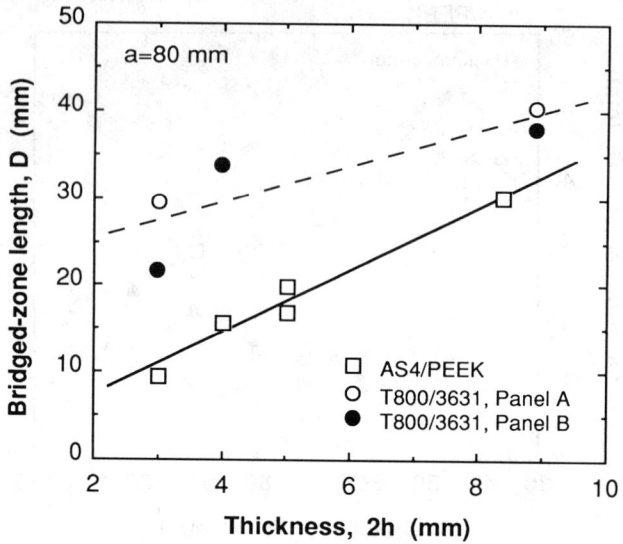

FIG. 15—*Relation between bridged zone length at a = 80 mm and thickness.*

estimated by the bridged zone length. A correlation seems to exist between the increase of the fracture toughness and the bridged zone length. For the case of AS4/PEEK laminates, the higher propagation values, G_{IC}, for thicker specimens shown in Fig. 6 were due to the increased contribution of fiber bridging and breakage [5–7,15].

For T800/3631 laminates, the thickness effect on fracture toughness was almost negligible (Fig. 10). We also found weak correlations between bridged zone length and specimen thickness (Fig. 15) and positive correlations between bridged zone length and the nor-

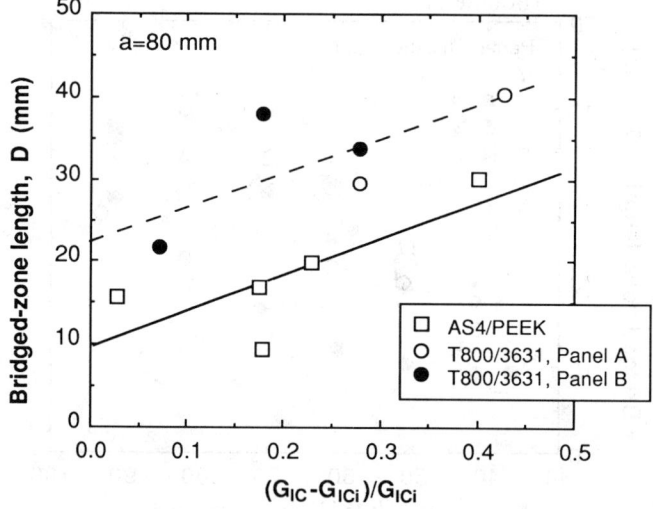

FIG. 16—*Relation between bridged-zone length and normalized increment of fracture toughness, $(G_{iC} - G_{ICi})/G_{ICi}$ at a = 80 mm.*

malized increment of the fracture toughness (Fig. 16). These results look contradictory. One of the possible explanations for T800/3631 laminates is that, if the closure force of the fiber bridging is very small, then only the highly bridged region near the crack tip contributes to the fracture toughness values. Since the crack opening displacement near the crack tip is determined by the energy release rate, the effect of thickness is minimal. This might be the reason why there was no effect of the thickness on the propagation values, G_{IC}, for T800/3631 laminates. Another possible explanation for the T800/3631 laminate is the large scatter of the contribution of fiber bridging, which is suggested in Figs. 10 and 15. Figure 16 indicates that, for a given bridged zone length, T800/3631 laminates have much less contribution to the increase of fracture toughness than AS4/PEEK laminates. The threshold bridged zone length of the increase of the fracture toughness for T800/3631 laminates is larger than that for AS4/PEEK laminates due to the smaller COD corresponding to the smaller G_{IC}. Because of this relatively smaller contribution of fiber bridging, the bridged zone length measurement did not present a clear answer to the effect of the thickness for the case of T800/3631 laminates. Further discussion is given with microscopic observation.

Fiber bridging also gives the reason for the scatter of the R-curves observed in Figs. 7 and 8. For example, measurement of the width of bridged fiber bundles was carried out for 3-mm-thick specimens of T800/3631, Panel A. The total width of the bridged fiber bundles in the width direction was about 23% for one specimen and 12% for the other specimen. This difference corresponds well to the difference of the R-curves in Fig. 7. For the case of 3-mm-thick specimens of Panel B, the total width of the bridged fiber bundles was only 3% for one specimen and 2% for the other. Moreover, this large difference in the density of the fiber bridging indicates that the measurement of the bridged zone length is not enough for the quantitative evaluation of the contribution of the fiber bridging, especially for T800/3631 laminates. This fact also supports the former discussion on the contradictory results between Figs. 10 and 16.

As mentioned before, many authors attributed the thickness effect to the fiber bridging [5,15]. The relation between crack opening displacement and the bridged zone was suggested [15]. However, our bridged zone length observation showed that the contribution of the fiber bridging is not directly transferred to the increase of the fracture toughness for T800/3631 laminates. The measurement of the distribution of closure force will provide a clearer understanding. The mechanisms of fiber bridging are very complicated, and further systematic research is necessary in the future.

Microscopic Observation

Figure 17 is a micrograph of the transverse section of the 3-mm-thick T800/3631 specimen. In this figure, the upper surface indicates the fracture surface. The direction of crack propagation is vertical to this figure, traveling from front side to back side. We carefully observed whether the crack path was on the prepreg interface or not in these micrographs. For the case of the specimen from Panel A, the crack path was mainly between the prepreg interfaces. Only about 20% of the crack path was on the prepreg interface for the specimen from Panel A. On the other hand, more than 80% of the crack path was on a single prepreg interface for the specimen from Panel B. When the crack path is inside one prepreg layer, fiber bridging is expected since fibers are entangled within layers. It is easy to understand that the difference of the crack path corresponds to the difference of the density of the bridged fibers. In the present study, the density of bridged fibers was not measured. This difference of the density is one of the reason why the bridged zone length did not give a clear understanding of the mechanism of relative increase of the fracture toughness from the initiation values.

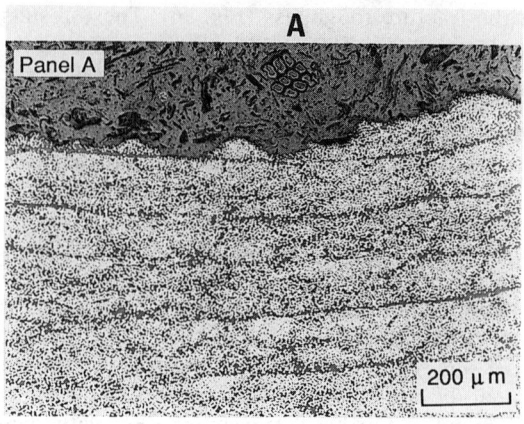

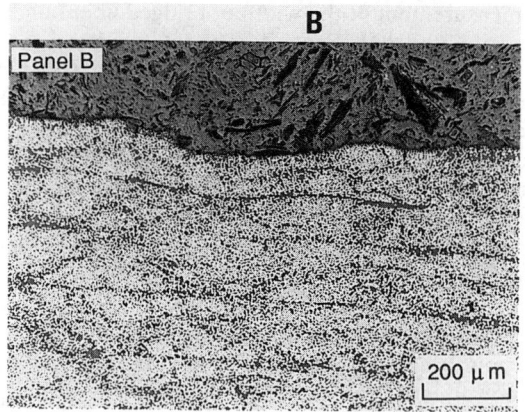

FIG. 17—*Transverse section of specimen for T800/3631 laminates (Thickness = 3 mm).* (A) *Panel A,* (B) *Panel B.*

In our previous study, the difference of the waviness was measured on the surface of the starter slits for T800/3631 laminates. Surface waviness values (WCA) were 31 μm for Panel A and 12 μm for Panel B. It is easy to understand that higher roughness promotes formation of fiber bridging. This fact is another explanation for the difference of the fiber bridging between two panels. However, other possible factors, such as difference of the thickness of the prepreg interface and microscopic distribution of fibers, may also differ between two panels. Further research is necessary.

As mentioned, Figs. 10, 15, and 16 seem to give contradictory results for T800/3631 laminates. This can be explained by the scatter of the contribution of fiber bridging which was caused by the difference of the density of bridged fiber bundles in each specimen. The small slope of the dashed line in Fig. 15 indicates that the change of the bridged zone length is comparable to the scatter at each specimen thickness. The same slope for AS4/PEEK laminates in Fig. 16 indicates that the bridged zone length can be correlated to the normalized increase of the fracture toughness for individual specimen. However, the scatter of the formation of fiber bridging for T800/3631 laminates concealed the effect of thickness on fracture toughness (Fig. 10).

Finally, the experimental results obtained for T800/3631 laminates suggest that the difference of the waviness of the prepreg interface and the density of the fiber bridging can be detected by the propagation values of fracture toughness. This difference does not affect the initiation values. Thus, the evaluation of the propagation values of G_{IC}, as well as the initiation values, is very important for characterization of the quality of CFRP laminates.

Conclusions

The relation between the interlaminar fracture toughness and the crack length was investigated by using DCB specimens of different thickness. The thinner specimens used in this study were milled down from the same thicker laminates in order to investigate the pure effect of the thickness of the specimen. While a single panel was used for AS4/PEEK laminate, two series of specimens from different panels were used for T800/3631 laminates.

The initiation values of the fracture toughness were independent of the specimen thickness both for AS4/PEEK laminates and T800/3631 laminates. For the case of T800/3631 laminates, the initiation value did not depend on two different panels.

For the case of AS4/PEEK laminates, the propagation values of fracture toughness increased sharply from the initiation values and then leveled off. The plateau values obtained from thicker specimens were slightly higher than those from thinner specimens. The genuine thickness effect on the propagation values obtained here was much smaller than the effect resulting from molding laminates of different thickness.

For the case of T800/3631 laminates, the propagation values from one panel increased continuously with crack length, whereas those from the other panel were almost constant with respect to crack length. The effect of specimen thickness on the propagation values at a certain crack length was smaller than the scatter of the data points for the two panels tested here.

The bridged zone length was measured for specimens of both laminates. For the case of AS4/PEEK laminates, the bridged zone length was related to the increase of the fracture toughness from the initiation values. Fiber bridging was more prominent for thicker specimens and was directly related to the thickness effect of the fracture toughness. On the other hand, for the case of T800/3631 laminates, contribution of the fiber bridging was not so clear as that of AS4/PEEK laminates. For the individual specimen, there was a weak relation between the bridged zone length and the increase of the fracture toughness from the initiation values. The scatter of the bridged zone length was large enough to conceal the thickness effect.

Acknowledgments

We wish to acknowledge the valuable discussion with Professor Kazuro Kageyama of the University of Tokyo and Professor Tatsuzo Koga of the University of Tsukuba. We also thank Yasutoshi Tateishi, a graduate student at the University of Tsukuba, for conducting the experiment.

References

[1] Bradley, W. L. and Cohen, R. N., "Matrix Deformation and Fracture in Graphite-Reinforced Epoxies,"*Delamination and Debonding of Materials, ASTM STP 876,* W. S. Johnson, Ed., American Society for Testing and Materials, Philadelphia, 1985, pp. 389–410.
[2] Davies, P. and Benzeggagh, M. L., "Interlaminar Mode-I Fracture Testing," *Application of Fracture Mechanics to Composite Materials,* K. Friedrich, Ed., Elsevier, Amsterdam, 1989, pp. 81–112.

[3] Sela, N. and Ishai, O., "Interlaminar Fracture Toughness and Toughening of Laminated Composite Materials: a Review," *Composites,* Vol. 20, No. 5, 1989, pp. 423–435.

[4] de Charentenay, F. X., Harry, J. M., Prel, Y. J., and Benzeggagh, M. L., "Characterizing the Effect of Delamination Defect by Mode I Delamination Test," *Effect of Defects in Composite Materials, ASTM STP 836,* American Society for Testing and Materials, Philadelphia, 1984, pp. 84–103.

[5] Wang, S. S., Suemasu, H., and Zahlan, N. M., "Interlaminar Fracture of Random Short-Fiber SMC Composite,"*Journal of Composite Materials,* Vol. 18, 1984, pp. 574–594.

[6] Russell, A. J. and Street, K. N., "Factors Affecting the Interlaminar Fracture Energy of Graphite/Epoxy Laminates," *Proceedings of the 4th International Conference on Composite Materials (ICCM4),* the Japan Society for Composite Materials, Tokyo, 1982, pp. 279–286.

[7] Prel, Y. J., Davies, P., Benzeggagh, M. L., and de Charentenay, F. X., "Mode I and Mode II Delamination of Thermosetting and Thermoplastic Composites," *Composite Materials: Fatigue and Fracture, Second Volume, ASTM STP 1012,* American Society for Testing and Materials, Philadelphia, 1989, pp. 251–269.

[8] Davies, P., Cantwell, W., and Kausch, H. H., "Measurement of Initiation Values of G_{IC} in IM6/PEEK Composites," *Composite Science and Technology,* Vol. 35, 1989, pp. 301–313.

[9] Hashemi, S., Kinloch, A. J., and Williams, J. G., "Mechanisms of Delamination in a Poly(ether sulphone)-Fibre Composite," *Composite Science and Technology,* Vol. 37, 1990, pp. 429–462.

[10] Minutes of Meeting Between ASTM Task Group D30.02.02, European Group on Fracture, and Japan Industrial Standards Group, Orlando, 8 November 1989.

[11] Minutes of ASTM Task Group D30.02.02 on Interlaminar Fracture Toughness, San Antonio, 13 November 1990.

[12] Kageyama, K. and Hojo, M., "Proposed Methods for Interlaminar Fracture Toughness Tests of Composite Laminates,"*Proceedings of the 5th Japan–U.S. Conference on Composite Materials,* Japan Society for Composite Materials, Tokyo, 1990, pp. 227–234.

[13] Huang, X. N. and Hull, D., "Effects of Fiber Bridging on G_{IC} of a Unidirectional Glass/Epoxy Composite," *Composite Science and Technology,* Vol. 35, 1989, pp. 283–299.

[14] Phillips, D. C. and Wells, G. M., "The Stability of Transverse Cracks in Fiber Composites," *Journal of Materials Science Letters,* Vol. 1, 1982, pp. 321–324.

[15] Davies, P., Cantwell, W., Moulin, C., and Kausch, H. H., "A Study of the Delamination Resistance of IM6/PEEK Composites," *Composite Science and Technology,* Vol. 36, 1989, pp. 153–166.

[16] Hojo, M., Tateishi, Y., and Aoki, T., "Influence of Specimen Thickness on Mode I Interlaminar Fracture Toughness of AS4/PEEK and T800/Epoxy Laminates," *Proceedings of the 5th Japan–US Conference on Composite Materials,* Japan Society for Composite Materials, 1990, pp. 193–200.

[17] Talbott, M. F., Springer, G. S., and Berglund, L. A., "The Effects of Crystallinity on the Mechanical Properties of PEEK Polymer and Graphite Fiber Reinforced PEEK," *Journal of Composite Materials,* Vol. 21, 1987, pp. 1056–1081.

[18] Hojo, M., Tanaka, K., Gustafson, C. G., and Hayashi, R., *Proceedings of the 7th International Conference on Composite Materials (ICCM7),* Vol. 2, Pergamon Press, Elmsford, NY, 1989, pp. 511–516.

[19] Davies, P. and de Charentenay, F. X., "The Effect of Temperature on the Interlaminar Fracture of Tough Composites," *Proceedings of the 6th International Conference on Composite Materials and 2nd European Conference on Composite Materials (ICCM6 & ECCM2),* Elsevier, London, Vol. 3, 1987, pp. 284–294.

[20] Smiley, A. J. and Pipes, R. B., "Rate Effects on Mode I Interlaminar Fracture Toughness in Composite Materials," *Journal of Composite Materials,* Vol. 21, 1987, pp. 670–687.

[21] Devitt, D. F., Schapery, R. A., and Bradley, W. L., "A Method for Determining the Mode I Delamination Fracture Toughness of Elastic and Viscoelastic Composite Materials," *Journal of Composite Materials,* Vol. 14, 1980, pp. 270–285.

[22] Gillespie, J. W., Carsson, L. A., and Smiley, A. J., "Rate-Dependent Mode I Interlaminar Crack Growth Mechanisms in Graphite/Epoxy and Graphite/PEEK," *Composite Science and Technology,* Vol. 28, 1987, pp. 1–15.

Eileen Armstrong-Carroll,[1] Bassel Iskandarani,[2] Ihab Kamel,[2] and Thomas M. Donnellan[1]

The Influence of Interleaf Deformation Behavior and Film-Resin Adhesion on the Fracture Toughness of Interleaved Composites

REFERENCE: Armstrong-Carroll, E., Iskandarani, B., Kamel, I., and Donnellan, T. M., **"The Influence of Interleaf Deformation Behavior and Film-Resin Adhesion on the Fracture Toughness of Interleaved Composites,"** *Composite Materials: Fatigue and Fracture, Fourth Volume, ASTM STP 1156,* W. W. Stinchcomb and N. E. Ashbaugh, Eds., American Society for Testing and Materials, Philadelphia, 1993, pp. 299–317.

ABSTRACT: Two-dimensional laminated composite materials are susceptible to damage by interlaminar crack growth. Delaminations severely degrade composite load-carrying capability. Interleaved composites are fabricated with thin films inserted between composite prepreg plies. Previous work has shown that interleaved composites have significantly higher delamination and impact resistance than composites without interleaves.

This paper examines interleaving as a technique for the improvement of delamination resistance in composites. In particular, the influence of the deformation behavior of the interleaf film and film adhesion on composite fracture toughness is studied. Teflon and Kapton interleaf film materials are compared. Kapton displays predominantly brittle deformation behavior. Teflon exhibits substantial deformation before failure. Both films are plasma treated to vary film-resin bond strengths. The fracture toughness of composites interleaved with these films is determined. Plasma treatments are shown to increase film-resin adhesion and fracture toughness. The effectiveness of Teflon as an interleaf material may be limited by the toughness of this film. Fracture toughness specimens are found to fail at the film-plasma grafted polymer interface. Additional optimization of the film-grafted polymer bond is necessary for improved performance.

KEY WORDS: polymer matrix composites, interleaved composites, plasma treatment

The increased thermal and structural requirements of emerging aircraft designs necessitate the use of composites which operate in the 350 to 400°F (177 to 204°C) temperature range. Bismaleimide (BMI) composites afford the required properties and have received a great deal of attention for composite structures. BMI materials systems developed initially are extremely brittle and display microcracking after processing or repeated thermal exposure. In composites there is a correlation between resin brittleness and reduced material compressive strength after impact (CAI) [1]. Brittle BMI composites have CAI values of approximately 124 to 138 MPa (18 to 20 ksi) based on the Boeing version of the test. Conventional epoxy-based composites, by comparison, have CAI strengths in the range of

[1]Materials engineers, Naval Air Development Center, Code 6064, Warminster, PA 18974-5000.
[2]Materials engineering graduate student and department head, Materials Engineering Dept., respectively, Drexel University, Philadelphia, PA 19104.

172 to 276 MPa (25 to 40 ksi). The damage sensitivity of these materials affects both design efficiency and materials selection for composite components in aircraft designs.

There are a number of approaches being investigated for the improvement of the impact resistance of brittle matrix composites. Formulation approaches center on the addition of ductile rubber or thermoplastic constituents to the resin, which act to increase the fracture energy of the material [2]. One variant of this approach is the design of multiple phase resins which separate spatially in the composite during processing [3]. Physical reinforcement techniques, such as stitching and braiding, improve impact resistance via out-of-plane fiber orientations [4]. These fiber orientations allow for more effective stress translation through the fibrous reinforcements after an impact event. With resin formulation and physical reinforcement techniques, some compromise in thermal stability or in in-plane structural properties is accepted in order to improve the damage tolerance of the composite.

Another approach toward toughness improvement involves the insertion of films of adhesive or thermoplastic materials at the interply interfaces in composite laminates [5–10]. The material used as the interleaf must have high strain to failure for energy absorption during fracture and must possess a high modulus for retention of composite properties.

Previous work [6] on interleaved composites is directed at epoxy-based systems. Interleaf materials used are tough epoxy adhesives. This past work concludes that tailoring of the film thickness and film-resin interdiffusion provides a good combination of toughness and in-plane properties.

Another previous work [5] examines interleaving of BMI composites. Although film adhesives are studied, better results are obtained with thin thermoplastic films. One in particular, the "E" film, provides CAI values of 214 MPa (31 ksi) in an Cycom 3100/IM6 composite system. Correlations between the CAI test and material parameters of the composite are studied in this past work. The fracture energy in shear, G_{IIc}, is found to be a sensitive indicator of material response to postimpact compression-type tests. One conclusion that the authors of this past work made is that the degree of improvement in BMI composites is limited by the adhesion between the film and the BMI resin. This conclusion is based on a comparison of the surface deformation characteristics observed on the interleaf and resin in epoxy and BMI composite samples.

To date, no fundamental study has reported on the importance of the film characteristics of high-temperature interleaf materials or on the effect of interleaf/resin interaction on interleaved composite performance.

In this work the importance of the material properties of Kapton and Teflon interleaf films in BMI interleaved composites is examined. For this work, radio frequency (RF) plasma treatment is used to provide the potential for systematic variation of surface chemistry and adhesion. The use of a wide variety of plasmas for surface modification of films and fibers is common in the literature [11–13].

Plasma technology provides a valuable means of modifying surface structure and composition of materials without altering bulk properties. Plasma treatments typically produce altered regions which range from angstroms to microns in thickness.

In the present research, plasma has been used to etch the polymer film surface and to graft a thin layer of polymeric material covalently to the interleaf film. This coating is specially designed to contain reactive chemical groups that are capable of bonding to the composite resin system during the curing cycle.

In this study allylamine is selected as a monomer for plasma polymerization. The choice of allylamine is based on past successful experience in using this monomer for plasma surface modification [13] and on consideration of the possible curing reactions of the grafted layer with the BMI resin used in the composite. Plasma polymerized allylamine is a good source of primary and secondary amine groups. The plasma-grafted polymer layer can play the role of coupling agent between the interleaf and the bismaleimide resin. A chemical reaction between the amine and imine groups provided by the grafted layer and the bismaleimide resin is expected to occur at the interface during the curing cycle. This reaction results in a direct covalent bonding between the interleaf and the resin as presented in Fig. 1.

The goal of this effort is the identification of the key materials characteristics of interleaf films which control film-resin adhesion and the energy absorption behavior of interleaved composite laminates. Film adhesion is enhanced through plasma treatments of interleaf films. The fracture behavior of interleaved composites is characterized to determine the effect of film deformation behavior on fracture toughness.

This paper is divided into four sections: materials, procedures, results/discussion, and conclusions. In the results and discussions sections, treated film surface chemistry, plasma treatment reproducibility, film surface topography, film tensile properties, and interleaved composites characterization are discussed. Findings based on results from this section are summarized in the Conclusions section.

Materials

The films studied were a 12.7-μm Kapton polyimide film from DuPont and a 12.7-μm Teflon film from ChemFab. The films were incorporated into composites with American Cyanamid's Cycom 3100/IM6 BMI/graphite prepreg material.

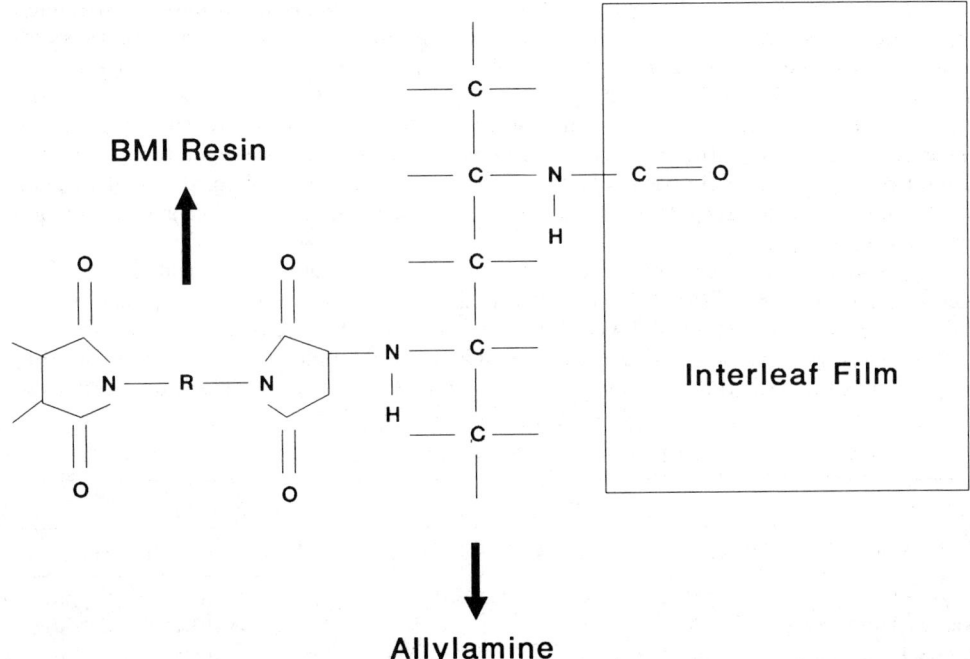

FIG. 1—*Possible interleaf film-resin bonding mechanisms. Schematic depicts how grafted-allylamine polymer acts as a coupling agent between the interleaf film and BMI resin.*

Procedures

Plasma Treatments—All films were solvent cleaned with toluene or acetone prior to treatment. Plasma treatments were performed in a Branson IPC 3000 Series 13.5 MHZ Rf plasma reactor. In this study, films were plasma treated in a two-step process which involved an argon etch followed by an allylamine polymerization. The argon etch removed surface impurities. The allylamine monomer was polymerized in the plasma atmosphere and then condensed on the interleaf film in the plasma chamber. The allylamine polymer acts as a coupling layer between the film and the resin. The process variables examined were plasma power, pressure, and time. The experimental conditions used control the chemistry and thickness of the coating. Higher plasma power levels led to increases in crosslink density. Higher pressure levels increased the allylamine condensation rate. The plasma-polymerized layer thickness increased with plasma exposure time. Initial treatment levels were selected based on earlier work [13]. The plasma treatments examined are summarized in Table 1.

Fourier Transform Infrared Spectroscopy (FTIR)—Fourier transform infrared spectroscopy was performed with a Perkin Elmer 1800 FTIR with a 4 cm^{-1} accuracy. With FTIR, resonance of chemical bonds at infrared frequencies cause infrared radiation to be absorbed. Film coating thickness and chemistry were monitored with sodium chloride salt tablets exposed with the films in the plasma chamber. Isolation of peaks due to the extremely thin coatings from the bulk film material was not possible with conventional FTIR spectroscopy. FTIR spectroscopy with an attenuated total reflection (ATR) sample mount was performed on plasma-treated films and end-notch flexure failure surfaces. With ATR, infrared radiation propagates through the germanium crystal and enters the specimen at an angle close to the critical angle. The depth of penetration into the specimen ranges from less than a micron to a few microns as a function of wavelength. ATR spectra were obtained with a trapezoidal germanium crystal possessing an incident angle of 60°. Spectra are plotted in transmission as a function of wavelength. No units are given for the vertical axes when the FTIR plots are staggered for comparison. When the plots are staggered, the shape of the plots can be compared, but the relative magnitude of valleys cannot be compared. Relative thickness measurements are made from the relative peak intensities of absorption spectra. The carbon dioxide peak at 2360 cm^{-1} is an experimental artifact.

Microscopy—Surfaces of fractured specimens were examined with optical and scanning electron microscopy (SEM). A Nikon optical microscope was used for the optical microscopy. An Amray Model AMR 1000 microscope was used for the SEM work.

Processing Procedures—All composite panels (including panels with no interleaf film) were processed with the recommended cure and post-cure procedures for interleaved

TABLE 1—*Interleaf film plasma treatments.*

	Argon			Allylamine		
Sample	Power, W	Pressure, N/m^2	Time, min	Power, W	Pressure, N/m^2	Time, min
Kapton Treatment 1	50	66.6	10	70	106.6	30
Kapton Treatment 2	150	93.1	15	100	93.1	50
Teflon Treatment 1	50	66.6	10	70	106.6	20
Teflon Treatment 2	50	66.6	10	70	106.6	30
Teflon Treatment 3	150	93.1	15	100	93.1	50

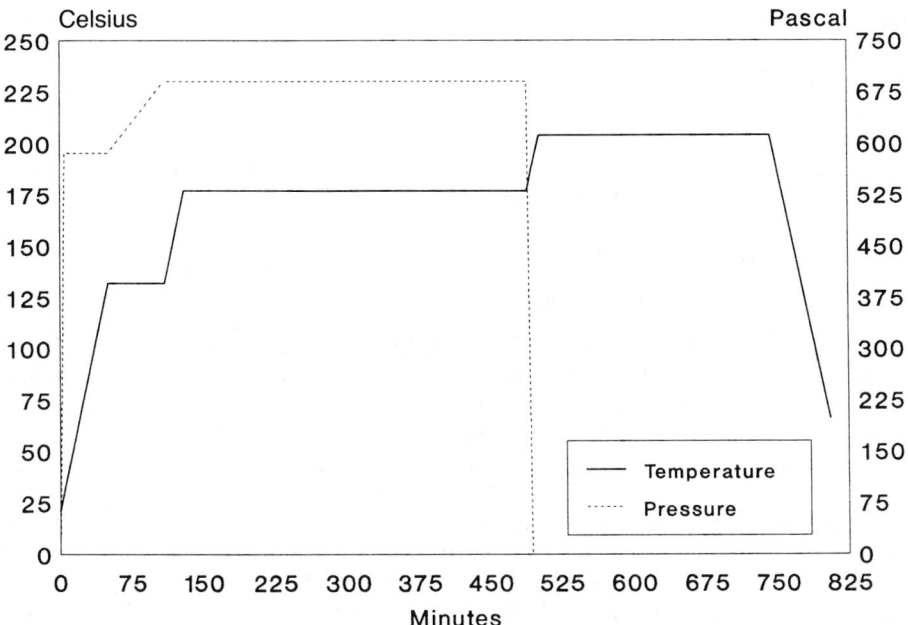

FIG. 2—*Processing cycle for interleaved Cycom 3100/IM6.*

Cycom 3100/IM6 developed by American Cyanamid. The processing cycle is diagrammed in Fig. 2.

Mechanical Tests

Film Tension Test—Tension tests were performed according to ASTM Test Methods for Tensile Properties of Thin Plastic Sheeting (D 882–81). The tests were performed with a Model 1122 table top Instron machine. The gage length used was 5.08 cm. A crosshead speed of 2.54 cm/min was used for the Kapton film. A crosshead speed of 50.8 cm/min was used for the Teflon film.

End Notch Flexure Test (ENF)—The end notch flexure test performed was based on ASTM Subcommittee D30.02 ENF test round-robin instructions. ENF tests were performed on a Model 1122 table top Instron machine. Crack starter film inserts were 3.2 cm long and are composed of 12.7-μm Teflon. Precracks approximately 4 cm long were created using a razor blade. Coupons were placed in the three-point bend loading fixture so that the end of the precrack region was located halfway between the upper and a lower loading nose. A crosshead speed of 2.54 mm/min was used. The beam theory calculation method for determining Mode II critical crack propagation energy was used. The compliance value was measured directly from the load deflection curve for each sample. The equation used for determination of mode two critical crack propagation energy is

$$G_{\text{IIC}} = \frac{9P_c^2 a^2 C}{2b(2L^3 + 3a^3)}$$ (1)

where

P_c = critical load,
a = crack length,
C = compliance,
b = sample width, and
L = distance between upper and lower loading noses.

Results and Discussion

Chemistry and Reproducibility of Plasma Treatments

The surface chemistry of the plasma-treated films is characterized with FTIR spectroscopy using an ATR crystal. The chemistry of the plasma polymerized allylamine is also documented with FTIR spectroscopy on salt tablets. Previous work [13] assigns the following functional groups to plasma polymerized allylamine: 3000 to 2800 cm^{-1}, aliphatic carbon single-bond hydrogen stretching; 2190 cm^{-1}, carbon triple-bond nitrogen stretching; and 1630 cm^{-1}, carbon double-bond nitrogen stretching. The spectra of the plasma polymerized grafts show the conversion of primary amine groups in the monomer to either a nitrile or an imine group. These functional groups are advantageous for the coupling reaction between the modified interface and the resin of the lamina.

Figure 3 shows spectra of untreated Teflon and Teflon Treatment No. 1. Both salt tablet and ATR crystal FTIR spectrum collection techniques are shown in this figure. The spectra of the treated samples exhibit the three allylamine signature valleys. Thus both ATR and salt tablet collection techniques characterize plasma treatment chemistry. Note the absence of these peaks with the untreated film spectrum. Similar results are observed with treated versus untreated Kapton FTIR spectra.

FTIR spectra provide information on the reproducibility of the plasma treatments.

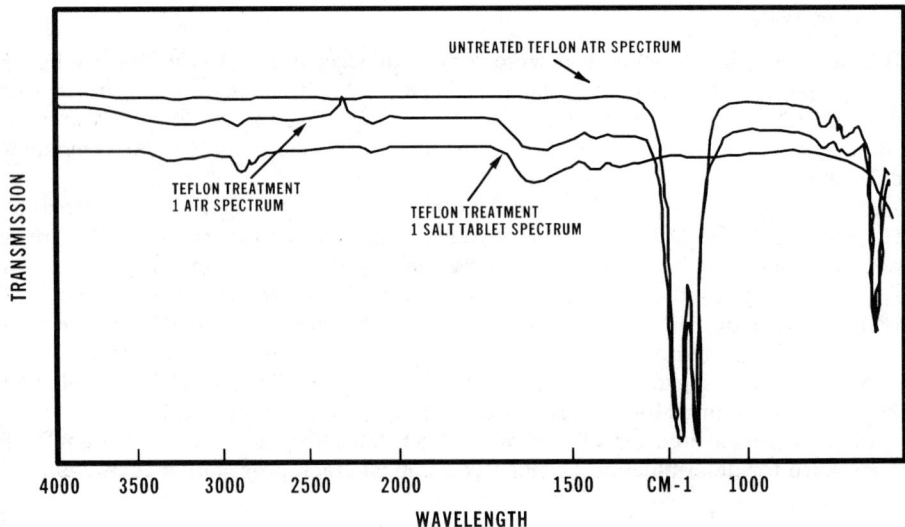

FIG. 3—*Transmission spectra of untreated and plasma Treatment 1 Teflon. Three allylamine valleys are visible on spectra of plasma-treated Teflon collected from a salt tablet and an ATR crystal.*

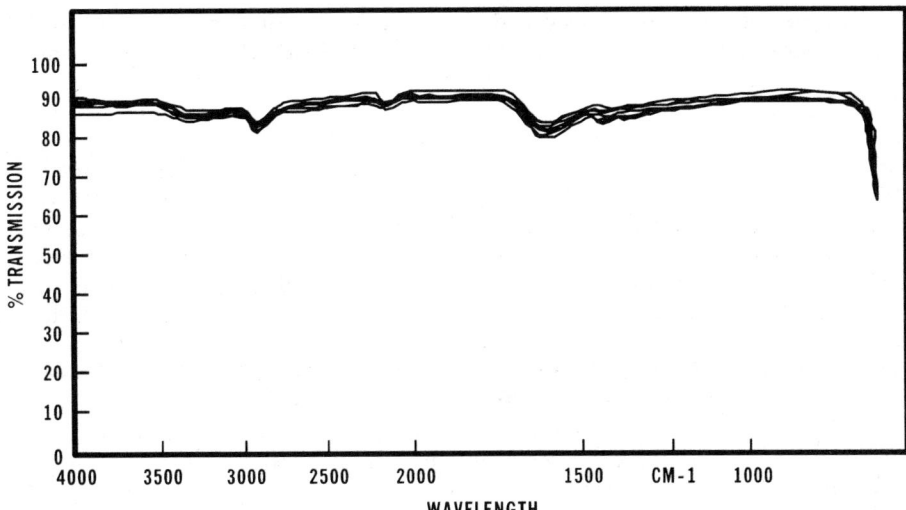

FIG. 4—*Reproducibility of plasma treatments as collected with salt tablets placed in the plasma reactor during film treatment. Seven different batches of Teflon Treatment 1 are plotted.*

Figure 4 shows the reproducibility of the Teflon No. 1 plasma treatment process for seven different plasma runs. In general the treatments were chemically reproducible. Some variability in spectra occurs. Most of this variability can be attributed to baseline differences in the salt tablets.

Reproducibility within five areas from the same Teflon film is depicted in Fig. 5. Variability in coating thickness is more apparent in Fig. 5. Some of the variability is due to

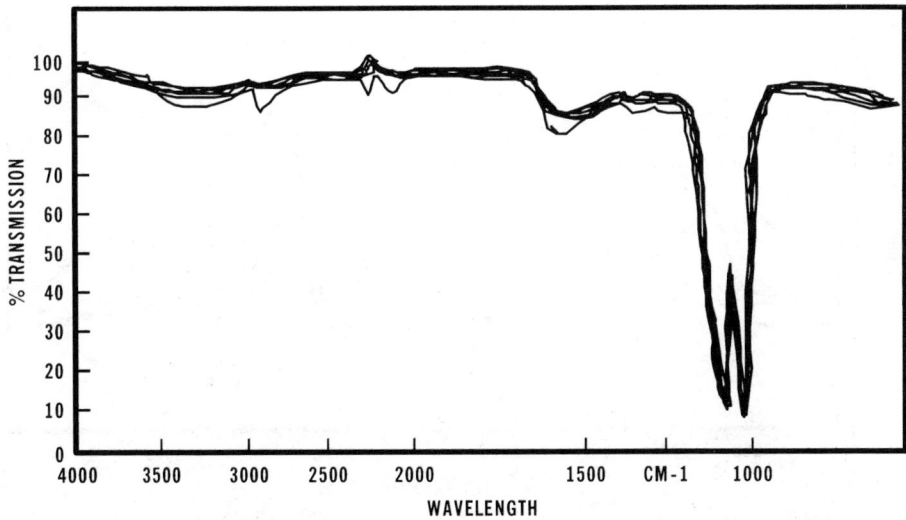

FIG. 5—*Reproducibility of plasma treatments as collected with ATR spectra of treated films. Five different regions of a Teflon Treatment 1 film are plotted.*

ATR experimental variables. However, there is additional variability not accounted for by ATR experimental conditions. This variability is due to heterogeneity in the plasma flow field over the large surface area of the film being treated.

FTIR spectra show that the differences between spectra due to different plasma treatments is significantly greater than the scatter within a given treatment. Figure 6 depicts spectra obtained with Teflon plasma Treatments 1 and 3. The differences in Treatment 1 and Treatment 2 spectra are significantly greater than those between the two Treatment 1 spectra shown. These differences are quantified by measuring the intensity of the 1630 cm^{-1} absorbance peak. There is a 30% scatter in peak intensity within the batch runs of Teflon Treatment 1. The 1630 cm^{-1} peak intensity of the Teflon Treatment 3 spectrum is 160% greater than the most intense Teflon Treatment 1 peak. The increases in peak intensity correspond with increases in the number of allylamine functional groups. The peak intensities are indicative of the relative thickness of the grafted allylamine layer. The ratio between the carbon triple-bond nitrogen and the carbon double-bond nitrogen peak increases with plasma treatment severity. This increase denotes changes in the chemistry of the allylamine layer. The increases of plasma treatment power, time, and pressure in Treatment 3 are responsible for these differences.

FTIR spectra of untreated and treated Kapton show results similar to the Teflon results documented in this section.

Film Surface Topography

SEM microscopy shows evidence of surface topographical changes induced by the plasma treatments. Figures 7 through 9 show the surfaces of Teflon in an untreated state and after two different treatments. An effect of the treatments is to increase the coating thickness of the allylamine, which eventually covers Teflon surface characteristics.

The treated Kapton film surface is smooth and featureless (Fig. 10). The topography of

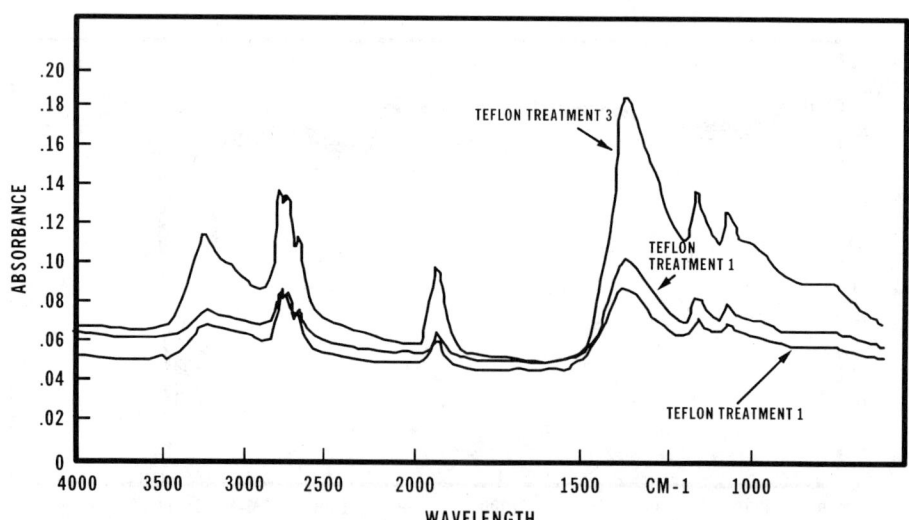

FIG. 6—*Absorption spectra of plasma-treated Teflon. The differences in peak intensity due to the more severe plasma treatment are greater than those due to variations in reproducibility.*

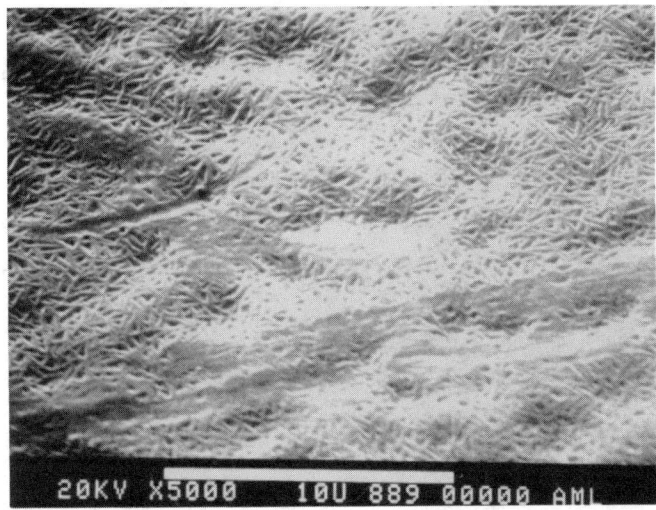

FIG. 7—*Untreated Teflon film. Topography is marked by a dendritic structure.*

untreated Kapton is also smooth and featureless. Thus SEM techniques cannot distinguish smoothing due to grafted allylamine on the surface of Kapton films.

Tensile Properties of Interleaf Films

The results of the tension tests performed on the film materials are shown in Table 2. An examination of the untreated film properties shows that there is a significant difference between the deformation behavior of the Kapton film and the Teflon film. The Kapton film has a high modulus and fails in a brittle fashion with very little plastic deformation. The

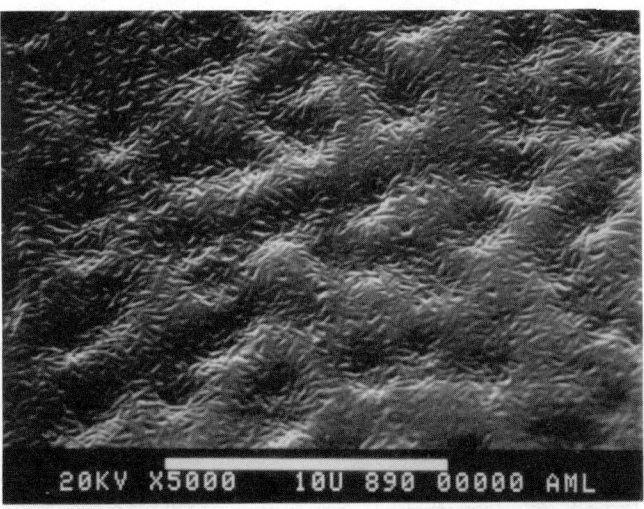

FIG. 8—*Teflon film treated with Treatment 2. Valleys in film topography are partially obscured by grafted allylamine.*

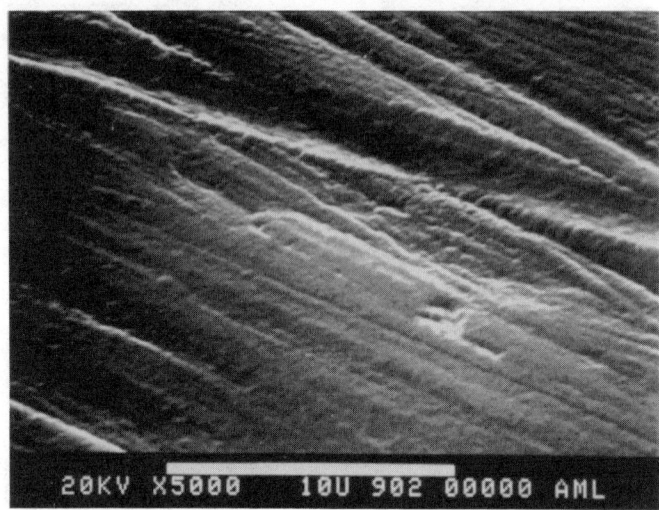

FIG. 9—*Teflon film treated with Treatment 3. Original Teflon topography is totally obscured by plasma treatment.*

Teflon film has a lower modulus and deforms extensively prior to failure. Estimates of material toughness from the area under the stress-strain curve indicate that the Kapton film is tougher (138 MPa) than the Teflon film (97 MPa).

The modulus of the Teflon films increases with plasma treatment. The effect of plasma treatment on the Kapton and Teflon film properties is to reduce the film tensile strength and elongation. There is a number of possible explanations for these reductions. The reductions could be the result of surface embrittlement due to chain scission, a film surface roughening effect, or modification of the mechanical properties of the film's surface due to the plasma coating.

FIG. 10—*Kapton film treated with Treatment 2. Similar to untreated Kapton film topography.*

TABLE 2—*Tensile properties of interleaf films.*

Sample	Elastic Limit			Failure	
	Stress, MPa	Strain, cm/cm	Modulus, MPa	Stress, MPa	Strain, cm/cm
Kapton untreated	56	0.02	2830	202	0.44
Kapton Treatment 2	38	0.01	3377	176	0.54
Teflon untreated	13	0.90	14	31	5.90
Teflon Treatment 2	15	0.03	462	29	2.86
Teflon Treatment 3	17	0.03	681	22	0.82

A comparison of the surfaces of tested tension specimens from the treated and untreated films (Figs. 11 through 14) indicates that the grafted-allylamine layer embrittles the film surface. The allylamine layer fails by cleavage. The vertical stria in Figs. 12 and 14 indicate this brittle failure mode. The amount of deformation at failure is greater in the Teflon film, and the surface stria are more widely spaced. The increased ability of Teflon to plastically deform and tear as well as the high strain rate used in testing the Teflon films contribute to the effect observed.

The results indicate that the coating deposited by the argon/allylamine treatment embrittles the film surface and lowers the failure strain of the treated films. It would be expected that the effectiveness of these film materials as interleaves is adversely affected by the embrittlement of the surface. The diminished failure strain and strength of the treated materials may indicate that the notch sensitivity of the film materials is increased with the argon/allylamine plasma treatment. Clearly, plasma treatments can significantly alter the material properties of interleaf films.

Interleaved Composite Characterization

Plasma treatments significantly alter ENF test results. End notch flexure test data are shown in Table 3. The untreated Kapton film possesses a G_{IIc} similar to that of the uninterleaved material. The similar G_{IIc} values indicate that the energy absorbed in film

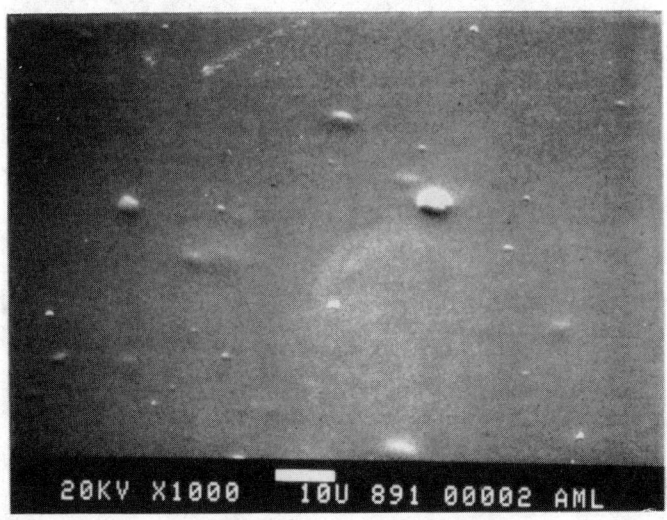

FIG. 11—*Untreated Kapton film failed in tension. Failure surface is featureless.*

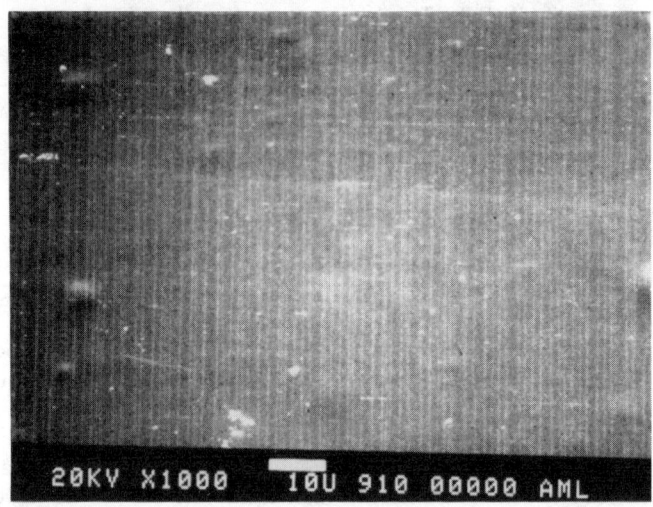

FIG. 12—*Kapton film Treatment 2 failed in tension. Failure surface possesses several vertical cleavage striae.*

deformation and fracture must be equivalent to that absorbed during resin fracture in the uninterleaved material. Plasma treatment of the Kapton provides an approximately 100% improvement in G_{IIc}. The more severe Kapton film treatment possesses a slightly lower G_{IIc} value than Kapton Treatment 1. The G_{IIc} values of Kapton Treatments 1 and 2 are within a standard deviation of each other.

Plasma treatment also increases G_{IIc} for Teflon film laminates. It is not possible to measure the G_{IIc} of the untreated Teflon interleaved composite because the bond between

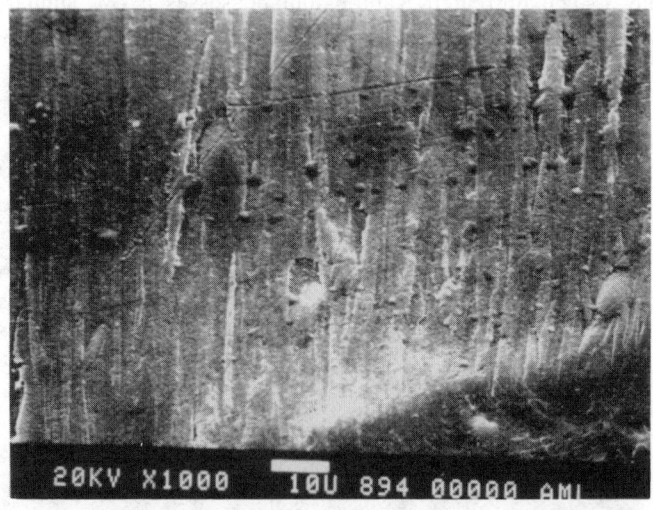

FIG. 13—*Untreated Teflon film failed in tension. Outer film layers deform and tear away from film. Material behaves in this manner since it is composed of several thin cast-Teflon layers.*

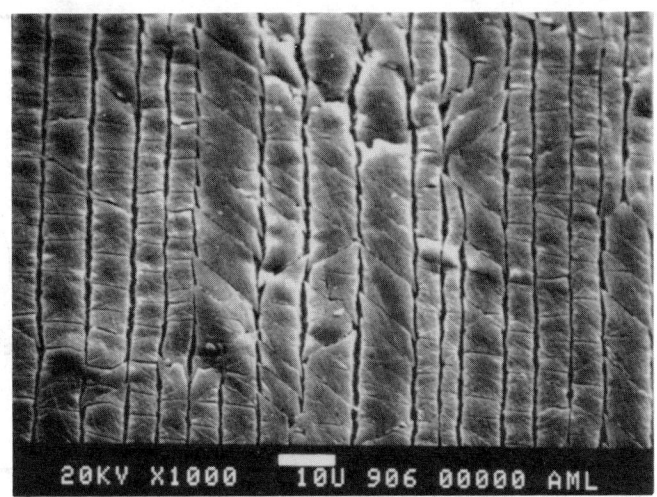

FIG. 14—*Teflon film Treatment 3 failed in tension. Failure surface marked by several vertical cleavage striae. Within each stria the failure surface is cleaved at several locations.*

the resin and the film is too weak, causing the specimens to fail prematurely. The G_{IIc} values of Teflon Treatments 1 and 2 are within a standard deviation of each other. Teflon Treatment 2 is similar to Treatment 1 except that the allylamine exposure time is increased 50%. The G_{IIc} value for Teflon Treatment 3 is approximately double the Treatment 1 and 2 values.

Teflon plasma Treatment 3 and 2 conditions are similar to Kapton plasma Treatment 2 and 1 conditions, respectively. However, while the more severe plasma treatment doubles G_{IIc} in the Teflon case, the G_{IIc} of Kapton is essentially unchanged. This result is not surprising since the effect of different plasma treatments on the film surfaces will depend on the chemical structure of the polymer film. Thus optimization of treatment conditions is sensitive to film surface chemistry. Thus the G_{IIc} of Kapton interleaved composites may increase with plasma treatments in which either allylamine monomers possess a higher concentration of reactive groups or another monomer is used.

The locus of failure for ENF samples is determined with FTIR. ATR spectra of the interleaf side of ENF failure samples were examined. Figures 15 and 16 show ATR spectra

TABLE 3—*Edge notch flexure test results.*

Sample	G_{IIC} J/m^2
No interleaf film	468
Kapton untreated	476
Kapton Treatment 1	949
Kapton Treatment 2	755
Teflon untreated	⋯
Teflon Treatment 1	189
Teflon Treatment 2	208
Teflon Treatment 3	392

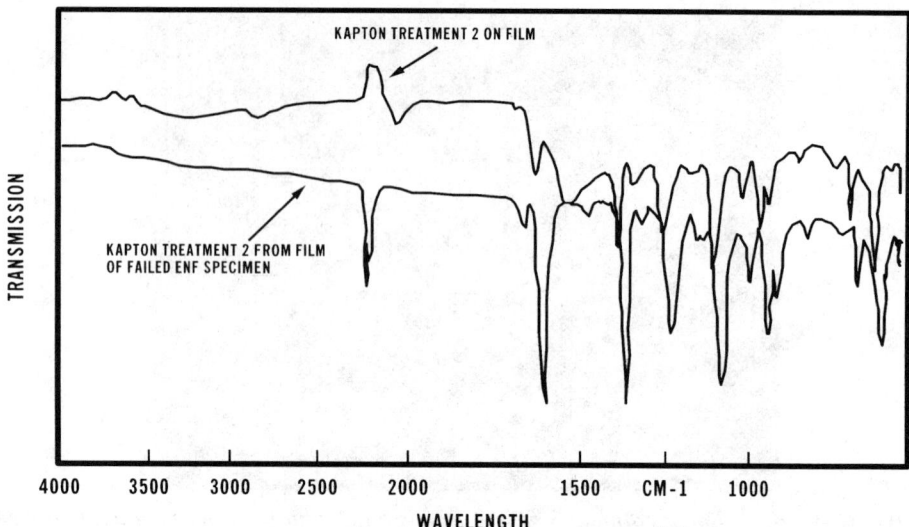

FIG. 15—*Spectrum of an ENF failure surface compared with spectrum of a plasma-treated film for Kapton Treatment 2. Allylamine valleys are absent on the ENF failed film surface.*

of plasma-treated films and the film side of failed ENF specimens with Kapton and Teflon interleaf films. These spectra indicate that the transmission valleys due to allylamine do not appear on the fractured samples. With both film materials, failure occurs at the film-grafted polymer interface. These results demonstrate the importance of film adhesion in the fracture process with interleaved composites.

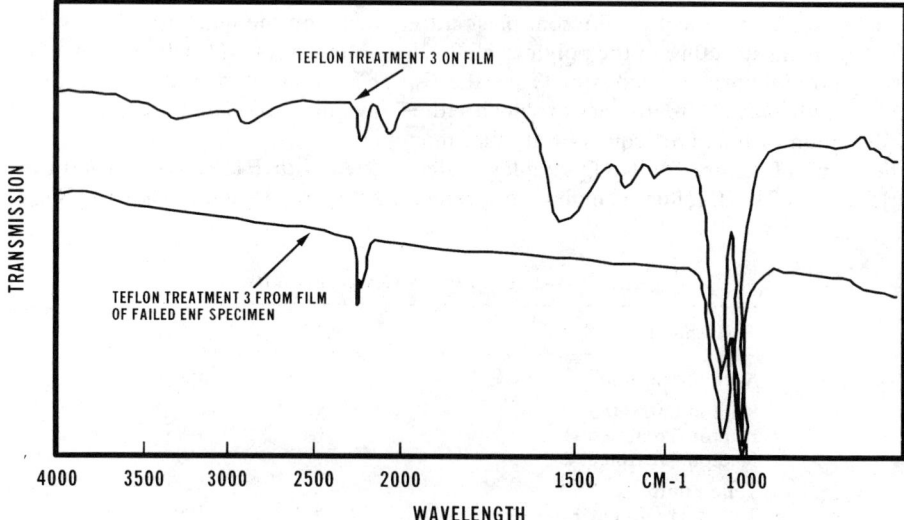

FIG. 16—*Spectrum of ENF failure surface compared with spectrum of plasma-treated film for Teflon Treatment 3. Allylamine valleys are absent on the ENF failed film surface.*

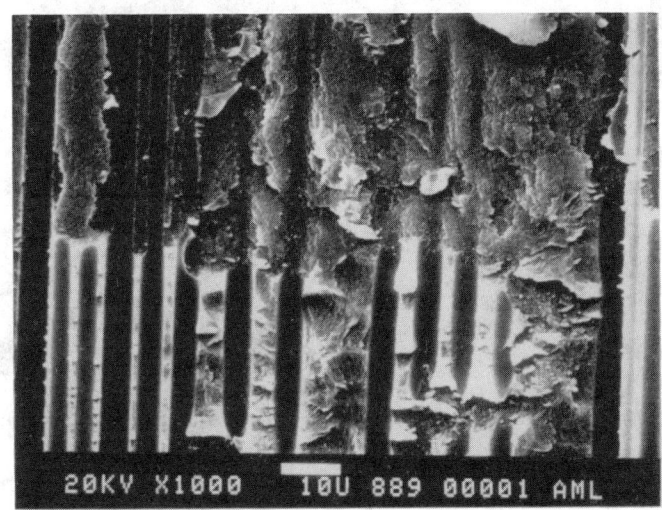

FIG. 17—*Cycom 3100/IM6 composite failed through ENF testing. Fracture energy absorbed through resin fracture.*

The ENF failure surfaces were characterized with SEM. Figure 17 shows the ENF failure surface of a composite without an interleaf. With this material, crack propagation energy is absorbed via resin fracture and fiber bridging. The G_{IIc} value of this composite is comparable to that of composites interleaved with untreated Kapton. Figure 18 shows the failure surface of a composite interleaved with untreated Kapton. With this material, crack propagation energy is absorbed via film tearing. The fracture surface of plasma-treated Kapton interleaved composites (Fig. 19) contains fewer sites of film tearing. However, at these sites, film deformation is more extensive. A limited number of hackles (Figs. 19 and

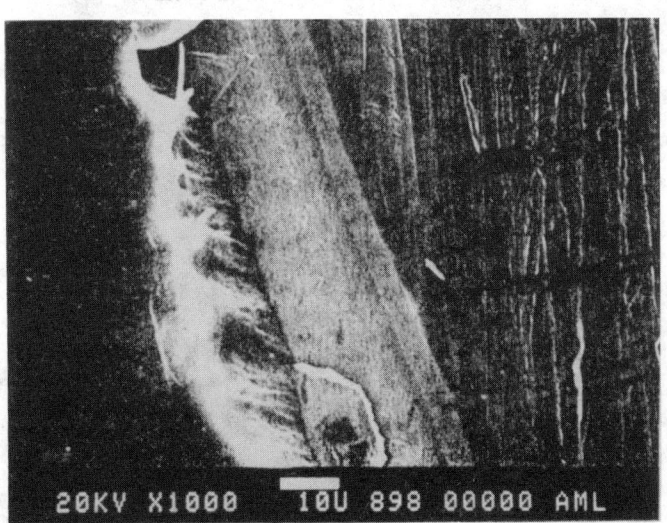

FIG. 18—*Cycom 3100/IM6 composite interleaved with Kapton, failed through ENF testing. Additional fracture energy absorbed through film tearing.*

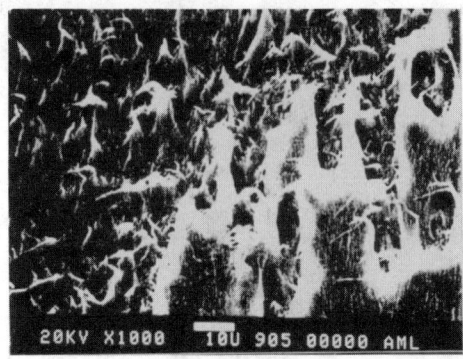 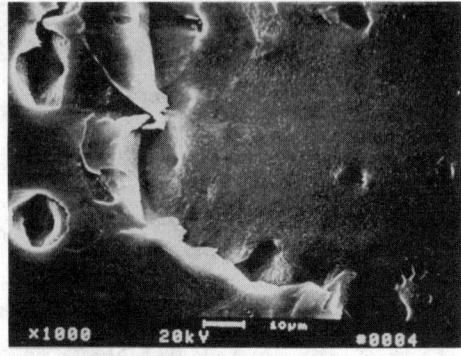

FIG. 19—*Cycom 3100/IM6 composite, interleaved with Kapton Treatment 2, failed through ENF testing. More deformation is evident at film-tearing sites. Some hackles are evident.*

20) are evident on the interleaf surface. Even the addition of a brittle film as an interleaf material markedly improves the crack propagation energy absorbed.

Since tension test results show that Teflon extensively plastically deforms before failure, it was expected that the ENF failure surfaces of Teflon interleaved composites would be characterized by extensive plastic deformation features. However, for Teflon Treatment 2, Fig. 21 shows that there is limited deformation on the ENF failure surface. Viewing this region at a higher magnification, Fig. 22 shows that on a local level the Teflon definitely deforms during fracture. The effect of the more severe No. 3 plasma treatment on plastic deformation of the Teflon interleaf during Mode II crack propagation is viewed in Fig. 23. In this case the Teflon interleaf deforms extensively. The contrast between the failure surface with Treatment 2 (Fig. 21) and Treatment 3 (Fig. 23) is marked. This increase in energy absorption may be due to increased adhesion between the film and the resin.

It is interesting to note that film deformation by itself is not a sufficient determinant of

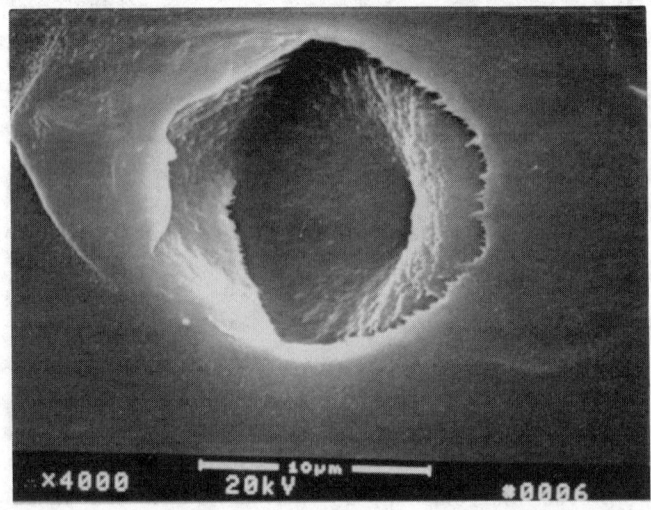

FIG. 20—*Region of Fig. 19 viewed at higher magnification. A closeup of a hackle.*

FIG. 21—*Cycom 3100/IM6 composite, interleaved with Teflon Treatment 2, failed through ENF testing. Plastic deformation is subtle.*

interleaf effectiveness. Comparison of Figs. 19 and 23 show that the Teflon film deforms much more than the Kapton film, even though the Kapton provides approximately twice the critical crack propagation energy. An explanation may be that the G_{IIc} of the interleaved composite is dependent on the toughness of the film material. If this explanation is valid, it is probable that improvements in the level of adhesion would produce limited additional increases in the composite fracture toughness with Teflon interleafs.

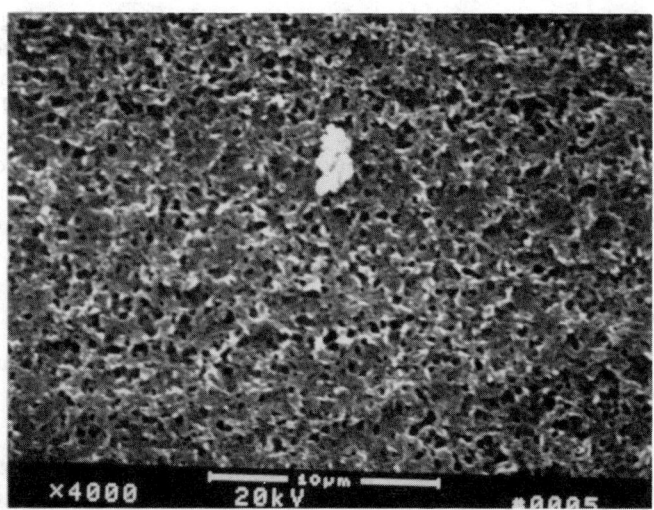

FIG. 22—*Cycom 3100/IM6 composite, interleaved with Teflon Treatment 2, failed through ENF testing. Cratered surface denotes extensive local deformation of Teflon film.*

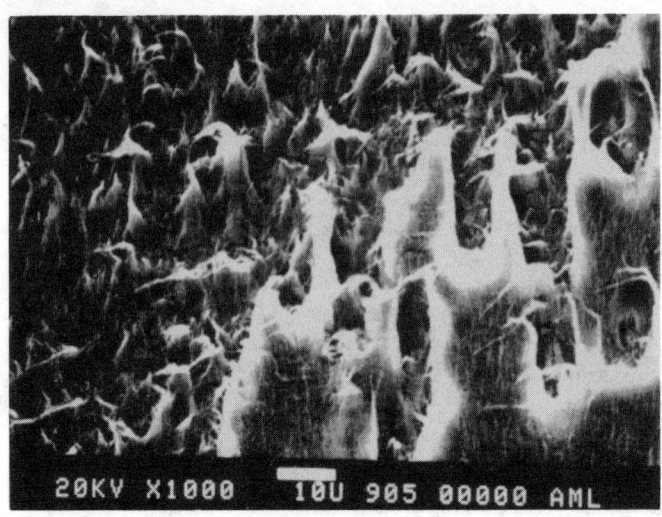

FIG. 23—*Cycom 3100/IM6 composite, interleaved with Teflon Treatment 3, failed through ENF testing. Extensive tearing and stretching of Teflon film.*

Conclusions

The fracture characteristics of brittle composites can be significantly improved through interleaving. This study demonstrates the usefulness of RF plasma treatment techniques for modification of the surface characteristics of interleaf film materials. Even though plasma-treated Kapton fails in a brittle manner when interleaved into a composite, this film provides a marked improvement in critical crack propagation energy. The G_{IIc} improvements with the soft Teflon film material are not as significant and may be limited by the film toughness.

Although the treatments presented here increase the critical crack propagation energy of Cycom 3100/IM6 composites, evidence that the locus of failure in ENF testing is at the film-grafted polymer interface shows that plasma treatments should be improved to maximize film-resin adhesion. A high degree of adhesion between the film and the composite resin is a requirement to maximize fracture toughness in interleaved composites.

A general approach for the development of novel interleaved composites would entail selection of films with high toughness followed by plasma tailoring of surface characteristics for adhesion to the matrix of interest. Surface treatments which will provide optimum interleaf performance should focus on film-resin adhesion characteristics.

Acknowledgments

The authors would like to thank D. Alley, B. Duffy, S. Kwong, J. Katilaus, B. Pregger, and K. Hutchins for assistance in specimen preparation and testing.

References

[1] Williams, J. G. and Rhodes, M. D., in *Composite Materials: Testing and Design (6th Conference), ASTM STP 787,* I. M. Daniel, Ed., American Society for Testing and Materials, Philadelphia, 1983.
[2] Riew, C. K. and Gillham, J. K., "Rubber-Modified Thermoset Resins," *Advances in Chemistry,* Series 208, American Chemical Society, Washington, DC, 1984.

[3] Turpin, R. L. and Green, A. L., in *Proceedings, 35th SAMPE Symposium and Exhibition,* Society for the Advancement of Process and Materials Engineering, Covina, CA, 1990, pp. 1079–1088.

[4] Ko, F. K., *Textile Structural Composites,* T. W. Chow and F. K. Ko, Eds., Elsevier Science, Netherlands, 1989, pp. 129–171.

[5] Dolan, G. L. and Masters, J. E., in *Proceedings,* 20th International SAMPE Technical Conference, Society for the Advancement of Process and Materials Engineering, Covina, CA, 1988, pp. 34–45.

[6] Kreiger, R. B., in *Proceedings,* 29th SAMPE Symposium and Exhibition, Society for the Advancement of Process and Materials Engineering, Covina, CA, 1984, pp. 1570–1584.

[7] Evans, J. E. and Masters, J. E., in *Toughened Composites, ASTM STP 937,* N. Johnston, Ed., American Society for Testing and Materials, Philadelphia, 1987, pp. 413–436.

[8] Masters, J. E., Courter, J. L., and Evans, R. E., in *Proceedings,* 31st International SAMPE Symposium and Exhibition, Society for the Advancement of Process and Materials Engineering, Covina, CA, 1986, pp. 844–858.

[9] Armstrong-Carroll, E. and Donnellan, T. M., in *Proceedings,* 1990 Materials Congress, American Society of Civil Engineers, New York, 1990.

[10] Masters, J. E., *Interlaminar Fracture of Composites,* E. A. Armanios, Ed., Trans Tech Publications, Switzerland, 1989, pp. 317–347.

[11] Wertheimer, M. R. and Schreiber, H. P., *Journal of Applied Polymer Science,* Vol. 26, 1981, pp. 2087–2096.

[12] Petrie, E. M. and Chottiner, J. C., in *Proceedings,* 40th Annual SPE Technical Conference and Exhibition: ANTEC 82, Society of Plastics Engineers, St. Louis, MO, 1982, pp. 777–779.

[13] Krishnamurthy, V., Kamel, I. L., and Wei, Y., *Journal of Polymer Science,* Part A, Vol. 27, 1989, pp. 1211–1224.

Steven J. Hooper[1] and Ramaswamy Subramanian[1]

Effects of Water and Jet Fuel Absorption on Mode I and Mode II Delamination of Graphite/Epoxy

REFERENCE: Hooper, S. J., and Subramanian, R., "**Effects of Water and Jet Fuel Absorption on Mode I and Mode II Delamination of Graphite/Epoxy,**" *Composite Materials: Fatigue and Fracture, Fourth Volume, ASTM STP 1156,* W. W. Stinchcomb and N. E. Ashbaugh, Eds., American Society for Testing and Materials, Philadelphia, 1993, pp. 318–340.

ABSTRACT: The environmental effects of water and jet fuel absorption on AS4/3501-6 graphite/epoxy were measured in terms of the changes in the interlaminar fracture toughness. Saturated double cantilever beam (DCB) and end notched flexure (ENF) specimens were tested to determine the environmental effects of moisture sorptions on critical strain energy release rates, G_{Ic} and G_{IIc}, respectively. These data are compared with baseline data obtained for specimens tested in the room temperature dry (RTD) condition. This phase of the study revealed that moisture absorption of either water or jet fuel tends to toughen this material and that this change is more significant for Mode I than for Mode II.

This paper also presents the results of a fractographic analysis of the fracture surfaces developed in these unidirectional specimens. These results document the changes in the surface morphology due to the environmental conditions studied. Significant differences were observed for Mode I fracture surfaces. SEM fractographs of delamination surfaces for DCB specimens reveal that the moisture absorption condition results in fracture surfaces where the fibers are covered with resin. This corresponds to the Mode I toughening, which is quantified in terms of an increase in G_{Ic}. The differences in the fractographs of saturated and room temperature dry ENF specimens are less profound, with the hackle formation in the former case being more deformed than the latter.

KEY WORDS: composite materials, environmental effects, moisture absorption, jet fuel absorption, Mode I delamination, Mode II delamination, crack propagation, fractography, graphite/epoxy, failure mechanisms, strain energy release rate

Composite materials have found widespread use in a variety of products where their high strength-to-weight and stiffness-to-weight ratios can be used to advantage. The first generation resins developed for use in advanced composites were graphite/epoxy systems which were highly cross-linked to produce both high stiffness and good hot/wet strength [1,2]. The principal disadvantage of these systems is that they are very brittle and thus exhibit very poor damage tolerance. Damage in these systems is often observed in the form of delaminations, which are caused by a variety of sources including residual thermal stresses generated during processing, damage resulting from manufacturing processes such as drilling, low-energy impact damage, and stresses at discontinuities such as ply drop-offs or free edges. Delamination is an important failure mechanism of these structures since it reduces their stiffness, strength, and fatigue life [3].

Hertzberg [4] pointed out that fracture surface markings of a material generally reveal

[1]Assistant professor and graduate student, respectively, Department of Aerospace Engineering, Wichita State University, Wichita, KS 67208.

the crack origin, the direction of crack propagation, and the state of stress that prevailed during the fracture event. He also stated that though certain micromechanisms are common to all materials, other fracture patterns are unique to a particular material as a result of different microstructures and deformation processes. Similarly, Smith and Grove [5] indicated that examinations of graphite/epoxy fracture surfaces facilitate identification of the origin, direction of crack growth, and load conditions associated with premature component fracture. These authors also discussed the usefulness of fractographic studies in identifying the causes of component failure and maintained that understanding the cause of a premature failure is critically important feedback to design engineers.

Several authors [5–11] have employed fractography to characterize damage developed in interlaminar fractures. These investigations have generally evaluated a material's performance by conducting parameter studies in which the resin system was varied, from brittle to tough, while the fiber type was held constant. These studies were limited to consideration of room temperature dry (RTD) conditions since environmental conditions were not considered. Advanced composite materials are commonly exposed to various forms of moisture in typical service conditions. Examples include water, in the form of rain or high humidity, jet fuel, lubricants, and solvents.

A number of studies [12–15] have reported that the interlaminar fracture toughness of graphite/epoxy composites increases as these materials absorb moisture. Most considered only the environmental effects of exposure to water or water vapor.

The objectives of this study were to (1) to identify the effects of moisture absorption on the morphology of the delamination surface and (2) to relate these differences to the changes in fracture toughness for the particular fluid/material combination. The fluids selected for this study included water, JP-4 military jet fuel, and Jet-A commercial jet fuel. The fractographic studies were performed using double cantilever beam (DCB) and end-notched flexure (ENF) specimens to measure Mode I and Mode II critical strain energy release rates (SERR), respectively. While failures of typical aerospace structures are unlikely to occur in either pure Mode I or pure Mode II, these loading conditions represent fundamental building blocks in an effort to understand the more common mixed Mode loading conditions.

Materials

DCB and ENF specimens were fabricated at Wichita State University from 24 unidirectional plies of ICI Fiberite HYE1337AU graphite/epoxy. This prepreg is equivalent to Hercules AS4/3501-6 graphite/epoxy and was manufactured by Fiberite under license from Hercules. The average cured ply thickness for both DCB and ENF specimens was 0.02 cm (0.008 in.), and the average fiber volume ratio was 60%. The specimens were cured using a modified Fiberite C-9 [16] autoclave cycle. The modification consisted of immediately applying pressure when the autoclave temperature reached 394 K (250°F). The specified cure temperature was 450 K (350°F). The specimens contained starter cracks which were formed using a single 0.0076 mm (0.3 mil) layer of Kapton film which had been sprayed with a release agent.

The DCB specimens were 17.78 cm (7.00 in.) long and 2.54 cm (1.00 in.) wide, with a 6.35-cm (2.5-in.)-long Kapton insert at the midplane of one end for delamination initiation. Aluminum hinges 2.54 cm (1.00 in.) wide were bonded to the specimen with Hysol EA9320 adhesive prior to vacuum drying. A bonding fixture was utilized to ensure proper alignment of the hinges on both sides of the specimens. The position of the hinge was varied relative to the end of the 0.0076-mm (0.3-mil)-thick Kapton insert so as to define nine different initial crack lengths from 3.81 cm (1.5 in.) to 6.35 cm (2.5 in.). Thus, at least nine different specimens were fabricated for each of the four environmental conditions. The ENF specimens were 15.24 cm (6 in.) long and 2.54 cm (1.00 in.) wide, with a 3.05-cm (1.20-in.) midplane 0.0076-mm (0.3-mil) Kapton insert at one end.

In order to establish uniform initial conditions for the moisture sorption process, all specimens were preconditioned in a vacuum drying cycle wherein the specimens were vacuum bagged, placed in an oven, and heated as described in Ref 12. A chart recording was produced in order to monitor the oven controller for each batch of specimens. Immediately after the vacuum drying cycle, the specimens were weighed to the nearest 0.0001 g. A total of 40 DCB specimens and 40 ENF specimens were prepared. Each of these were divided into four groups, one for each environmental condition to be tested. One group was tested in the room temperature dry condition while the other three groups were respectively soaked in water, Jet A commercial jet fuel, and JP-4 military jet fuel. The latter three groups were soaked at room temperature. These specimens were removed from the fluid, wiped dry, and weighed once a month. A specimen was assumed to be saturated when it exhibited no weight gain for three successive weighings. This required approximately 200 days. They were then stored in an immersed state until they were tested. The average percent weight gains at the end of these periods are indicated in Table 1. The data for the edge delamination tensile test (EDT) specimens represents the average values for the four laminates previously reported in Ref 12.

The data for the DCB specimens revealed a lower weight gain than for the ENF or EDT specimens. This difference can be attributed to the presence of the hinge tabs on the DCB specimens when they were soaked in the fluids. The hinge tabs reduced the surface area exposed to the fluids for the DCB specimens and thus affected the data.

Experimental Procedure

Delamination fracture toughness tests for opening mode (G_{Ic}) and shear mode G_{IIc} were performed using a material testing system (MTS 810) servo-hydraulic test stand in stroke control at rates of 8.5 $(10)^{0.7}$ m/s (0.2 in./min) and 4.23 $(10)^{0.7}$ m/s (0.1 in./min), respectively. Tests were controlled and data collected using a personal computer-based data acquisition system.

Mode I Double Cantilever Beam Test

The double cantilever beam test was conducted using the special fixture shown in Fig. 1. This fixture consisted of a 2224-N (500-lb) load cell, a linear variable differential transducer (LVDT), and a gripping device for the DCB specimen. The edge of the coupon was observed through a microscope of 10 to 60× magnification power. Load, ram displacement, and LVDT displacement were continuously monitored and recorded using the data acquisition system. Real-time analog plots of the load versus crack-opening displacement were also recorded using an X-Y plotter. The onset of delamination was identified by the

TABLE 1—*Percent weight gain due to moisture absorption.*

	Specimen		
Environment	DCB	ENF	EDT
Water	0.930	1.390	2.150
Jet-A	0.360	0.520	0.728
JP-4	0.380	0.580	0.816

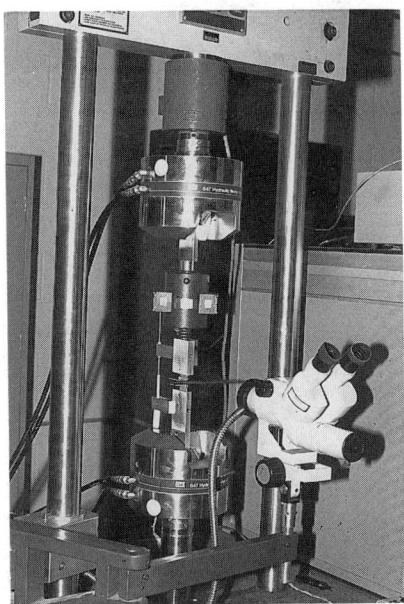

FIG. 1—*Double cantilever beam test.*

drop in the load-displacement curve. The onset was simultaneously verified by visual observations using the microscope. After the delamination had grown a few millimetres, the MTS machine was placed in a hold state, the crack tip was marked, and the specimen was unloaded. This mark was taken to be the new crack length for the subsequent loading. The data reduction technique used to obtain the strain energy release rate values was performed as follows: the compliance corresponding to each crack length was obtained using a linear regression analysis, and the compliance calibration method was employed to calculate G_{Ic}.

Mode II End-Notched Flexure Test

A three-point loading fixture shown in Fig 2 was used in the end-notched flexure tests. This fixture was designed such that spans of 5.08 cm (2.0 in.), 7.62 cm (3.00 in.), and 10.16 cm (4.00 in.) were obtainable between the outside supports. The shorter span lengths were used in shear precracking, and the 10.16-cm setting was employed during the delamination tests. Specimens were precracked and then wedged open in order to provide a sharp crack tip with an identifiable region between the shear precracked region and the delaminated region.

A LVDT was used to measure the center point displacement during these tests, and a 2224-N (500-lb) load cell was used to measure the applied load. The load, LVDT displacement, and the MTS crosshead displacement data were acquired and stored in a personal computer. A real-time analog plot was also generated on a X-Y plotter. One specimen from each group was used to determine the load-displacement curves for crack lengths of 1.91, 2.54, 3.175, 3.81, 4.44, and 5.08 cm by locating the specimen in the appropriate positions in the ENF fixture. Care was exercised during these measurements to avoid applying a load which would extend the delamination. G_{IIc} was calculated from these data using the compliance calibration method.

FIG. 2—*End-notched flexure test.*

Results and Discussion

The results for delamination onset are presented in Table 2 and Figs. 3 through 7. Figures 3 and 4 show the pure Mode I and Mode II fracture toughness data distributions for the environmental conditions considered. Both Mode I and Mode II fracture toughness have been found to increase with saturation in the three fluids. The percent weight gains for the Mode I and Mode II specimens are presented in Figs. 5 and 6. The percent weight gain observed for specimens soaked in water was nearly twice that observed for specimens soaked in jet fuel. This same trend was also evident in the data presented in Ref *12*. Another notable observation was that the percent weight gain for 24-ply unidirectional specimens (DCB and ENF) was approximately half of the percent weight gain measured for EDT specimens with 12 or fewer plies. The percentage increase in the Mode I fracture toughness is found to be much more than that found in Mode II.

The plot of G_{Ic} versus G_{IIc} as shown in Fig. 7 illustrates the shift in the fracture toughness curves. All of the mixed-mode data presented in this plot was obtained from EDT specimens as reported in Ref *12*. This figure clearly shows the increase in the fracture toughness due to moisture absorption. A careful examination of this figure reveals that the

TABLE 2—*Environmental effects on* G_{Ic} *and* G_{IIc}.

Environment	Mode I, kJ/m³ (in. × lb/in.³)	Mode II, kJ/m³ (in. × lb/in.³)
RTD	0.111 (0.634)	0.814 (4.654)
Water saturated	0.170 (0.971)	0.955 (5.459)
Jet-A saturated	0.170 (0.973)	0.962 (5.498)
JP-4 saturated	0.159 (0.911)	1.021 (5.834)

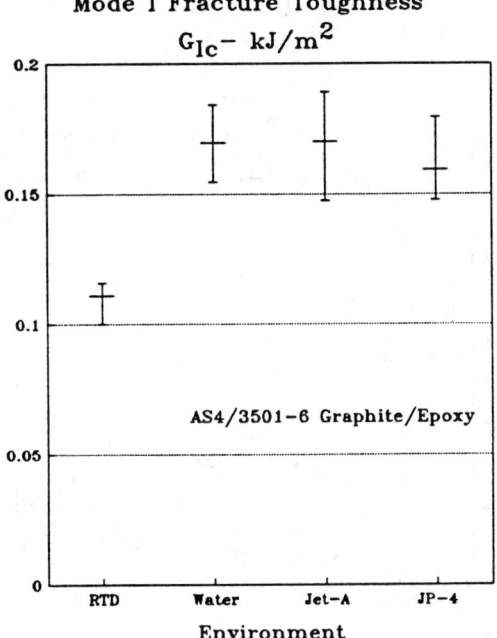

FIG. 3—*Environmental effects on Mode I fracture toughness.*

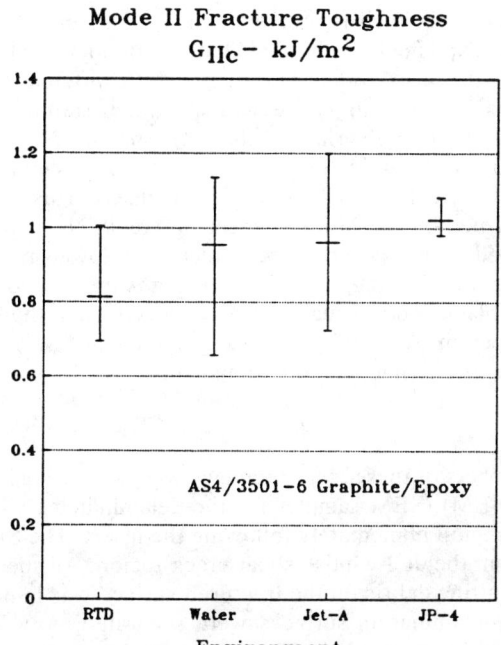

FIG. 4—*Environmental effects on Mode II fracture toughness.*

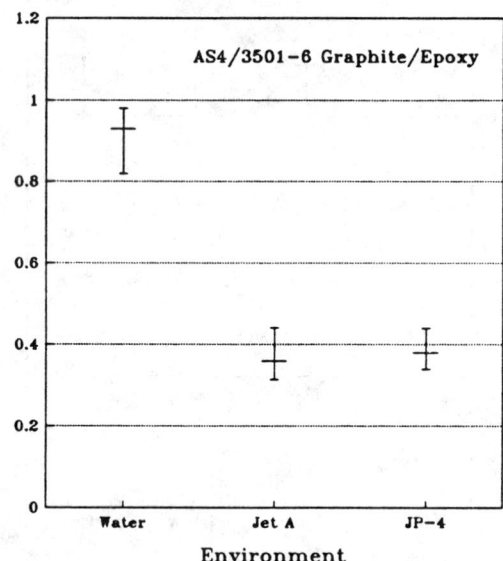

FIG. 5—*Percent weight gain for DCB specimens.*

G_{Ic} for a saturated EDT specimen is greater than that obtained from a saturated DCB specimen. This can be explained by the fact that the thinner EDT specimens absorbed significantly more fluid. It is also evident that there is little difference between the fracture toughness of specimens saturated in jet fuel and specimens saturated in water.

The photographs of the fracture surface of the delaminated ENF specimens, Figs. 8a to 8d reveals several regions of delamination. In each figure, proceeding from left to right, the initial region is the starter crack formed by the Kapton insert. This is followed by a narrow dull whitish region, formed by the Mode II shear precrack. Next, the shiny black region corresponds to the Mode I wedge precrack, which is followed by another dull whitish Mode II shear crack region. The left side of this region is the location of the natural crack tip. Finally, the shiny black Mode I region was produced when the specimens were split open for post-test inspection. As is evident in these figures, no change in these characteristics is observed for the environmental conditions considered.

Fractographic Results

The fractography aspects of these test specimens were studied using a Phillips scanning electron microscope (SEM). SEM samples for the delaminated DCB specimen were cut from the pure Mode I region immediately following the insert. The SEM samples for ENF specimens were cut from the dull-whitish shear crack region. All specimens were approximately 2 cm square. Cutting debris on the fractured surface was minimized by cutting the specimens before the delamination surfaces were separated. Any loose debris that did occur was removed with compressed air. The delaminated surface was then sputter coated with an approximately 20-nm-thick layer of gold-palladium to avoid charging.

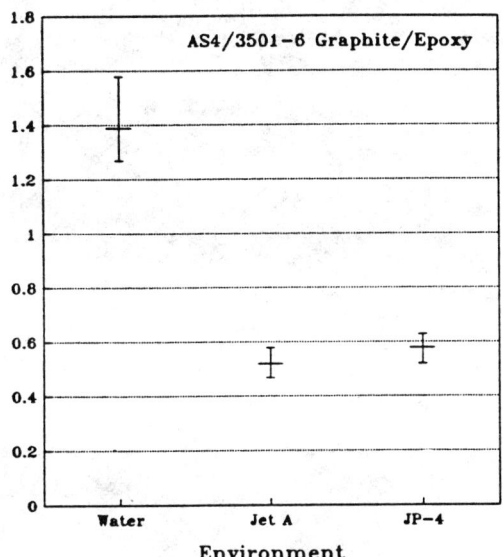

FIG. 6—*Percent weight gain for ENF specimens.*

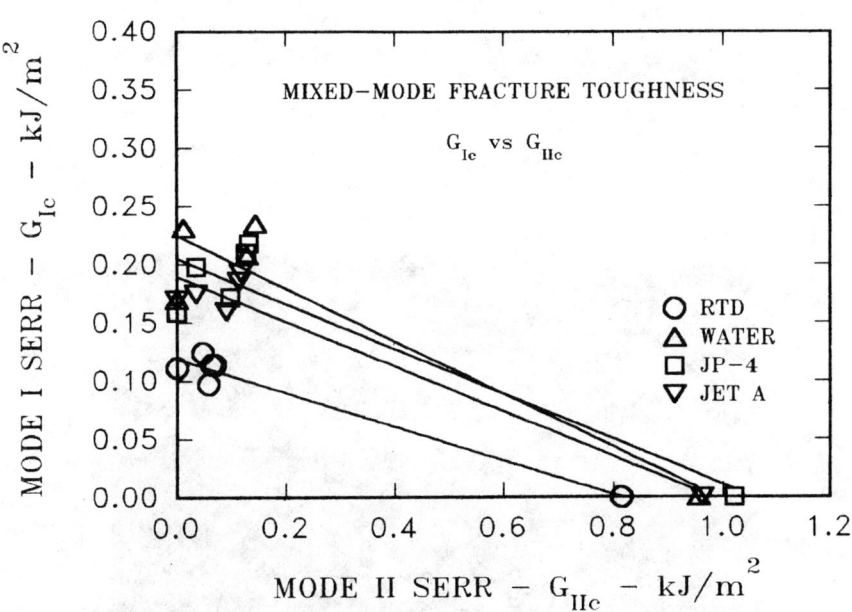

FIG. 7—*Environmental effects of moisture absorption on fracture toughness.*

FIG. 8a—*Delamination surface of an ENF specimen, RTD condition.*

FIG. 8b—*Delamination surface of an ENF specimen, water saturated condition.*

FIG. 8c—*Delamination surface of an ENF specimen, JP-4 saturated condition.*

FIG. 8d—*Delamination surface of an ENF specimen, Jet-A saturated condition.*

Mode I Delamination Fracture: Room Temperature Dry

Typical fractographs of AS4/3501-6 graphite/epoxy material are shown in Figs. 9 and 10. It is found that the matrix generally failed in tension near the fiber matrix interface in agreement with the work of Arcan et al. [6]. Large numbers of fibers appear clean with occasional areas of matrix covering the fiber. The surface also contains a number of fiber ends which are due to fiber breakages. This is due to the fiber-bridging phenomenon exhibited by the DCB specimen (Fig. 11). The matrix directly under the fiber shows deep matrix cracks inclined towards the left (Fig. 10). These hackles are inclined in the opposite direction on the opposite fracture surface. The crack propagation in this figure was from the left to right. As discussed in Ref 6, failure occurs at those points where the tensile stress is maximum. At the crack tip, this occurs on an interface normal to the crack. Away from the crack front, it is in a direction inclined towards the direction of crack propagation. This results in the formation of the inclined hackles.

FIG. 9—*Mode I fractograph (×203) RTD condition.*

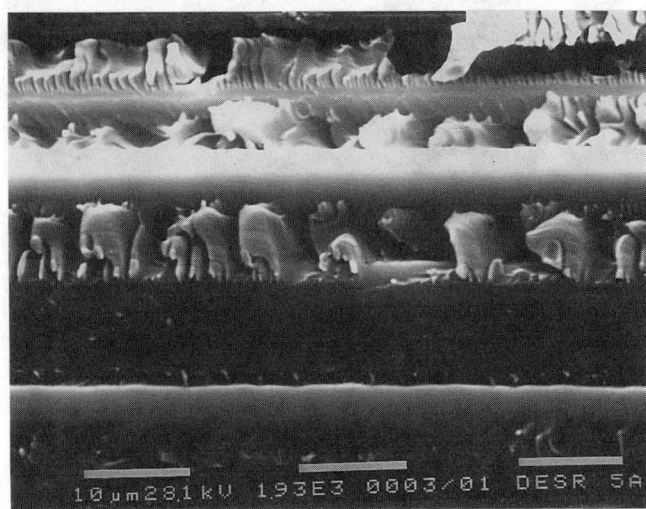

FIG. 10—*Mode I fractograph (×1930) RTD condition.*

Mode I Delamination Fracture: Water-Saturated Condition

Figures 12 to 16 show typical fractographs for graphite/epoxy AS4/3501-6 material tested in a water-saturated condition. In all three figures, crack propagation is from right to left. The broken fibers are also evident in Figs. 12 to 16, but now we find that most fibers are completely covered with resin. This is indicative of a greater amount of energy absorption due to extensive matrix deformation. This is indicative of a ductile fracture of the type seen in AS4/Dow-Q6 system [2]. Apparently the fiber/resin adhesion seems to have improved, allowing the load to be effectively transferred to the resin. This results in the kind of extensive resin deformation seen here. We also notice the absence of any hackle formation other than in regions directly under the pulled out fibers. The vertical line on the right side of Fig. 12 is the mark due to the Kapton insert, used to start the delamination. The other noticeable feature is the presence of river markings. Figures 13 and 14 show the

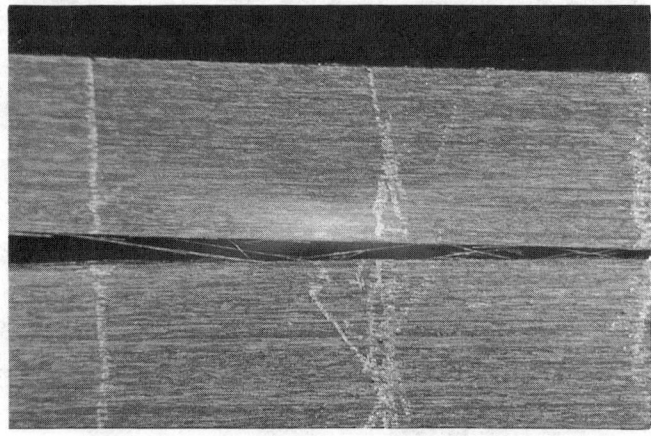

FIG. 11—*Fiber bridging in a DCB specimen.*

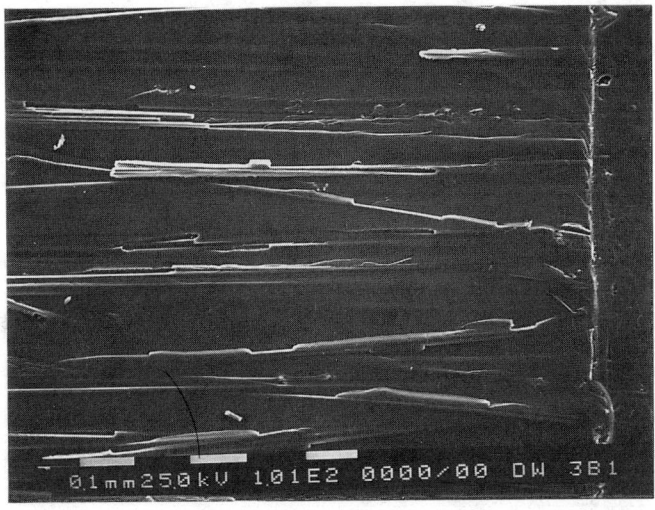

FIG. 12—*Mode I fractograph (×101) water saturated condition.*

opposite faces of the delaminated surface containing corresponding river markings. The crack propagation direction is from the right to the left. This general directional sense can also be concluded from the direction of the confluence of river markings. The river markings are formed by the joining together of the various microcracks. These river markings almost always terminate at a fiber matrix interface. It must be pointed out that, if only one fracture surface is studied, it is possible at times to mistake extraneous elements (impurities) as part of the fractographic features. These are easily detected by examining both surfaces of the delaminated region of interest. Figures 15 and 16 show the opposite faces of the region of fiber pullout. Note that the matrix material is still adhering to the

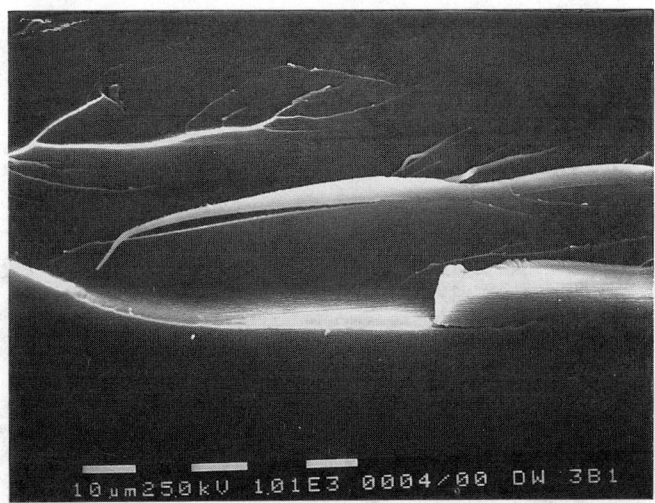

FIG. 13—*Mode I fractograph (×1010) water saturated condition.*

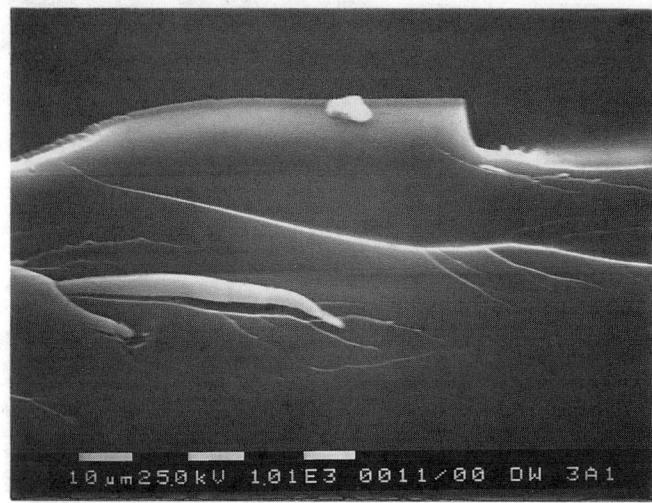

FIG. 14—*Mode I fractograph (×1010) water saturated condition.*

fiber ends. There are also very tiny hackle-like features present in the grooves from which the fibers have been pulled out.

Mode I Delamination Fracture: Jet A and JP-4 Conditions

Figures 17 to 20 show the features seen in a JP-4-saturated delaminated DCB specimen. The features seen in Jet A-saturated specimens are nearly identical to these features. Hence, only fractographs of specimens saturated in JP-4 are presented. As in the water-saturated case, there is a noticeable absence of hackles. The predominant features present are the river markings. Figures 18 and 19 show the opposite surfaces of the delaminated

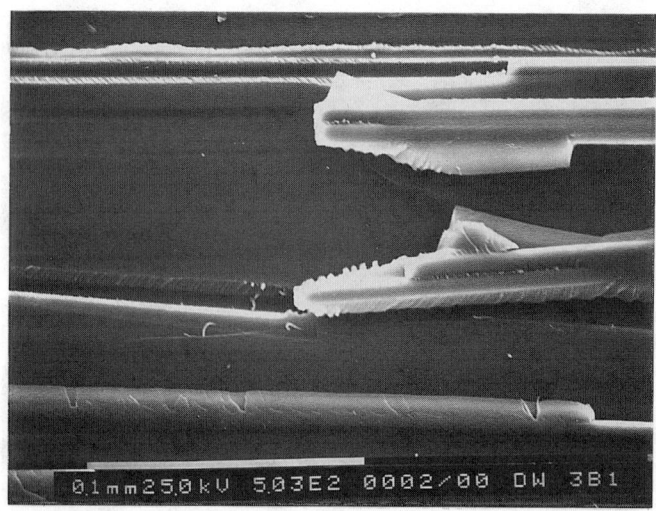

FIG. 15—*Mode I fractograph (×503) water saturated condition.*

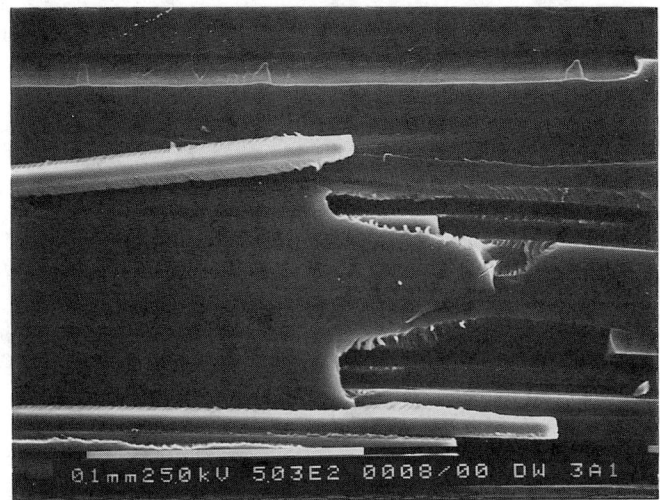

FIG. 16—*Mode I fractograph (×503) water saturated condition.*

region with river markings. As was pointed out earlier, the river markings terminate at the fiber matrix interface. On closer examination, Figs. 17 and 20 reveal the presence of broken fiber ends. The fibers are almost completely covered by the matrix material. There are far fewer regions of fiber matrix separation than in the case of the specimen tested at the RTD condition. Smith and Grove [5,7] pointed out that the extent to which these occur depends on the volume percent of reinforcing fibers and the proximity of the fracture plane to these fibers. Since all specimens were cut from panels with a 60% fiber volume ratio, the differences observed were attributed to the sorption of the fluids. The extensive regions of cohesive fracture of matrix material accompanied by a significant amount of fiber breakages tend to increase the Mode I fracture toughness considerably. There is also a greater

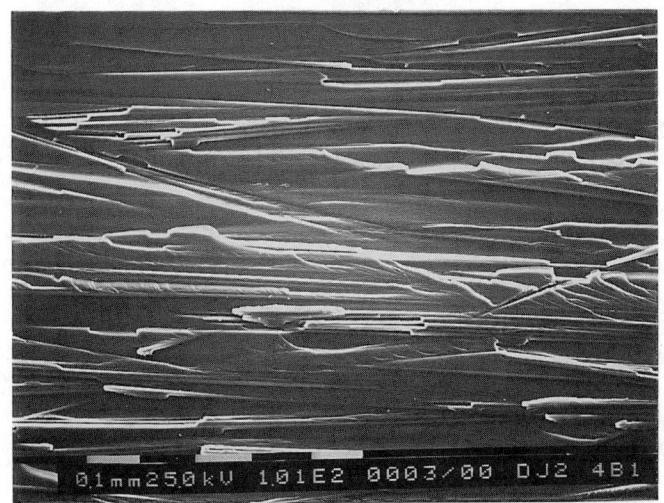

FIG. 17—*Mode I fractograph (×101) JP-4 saturated condition.*

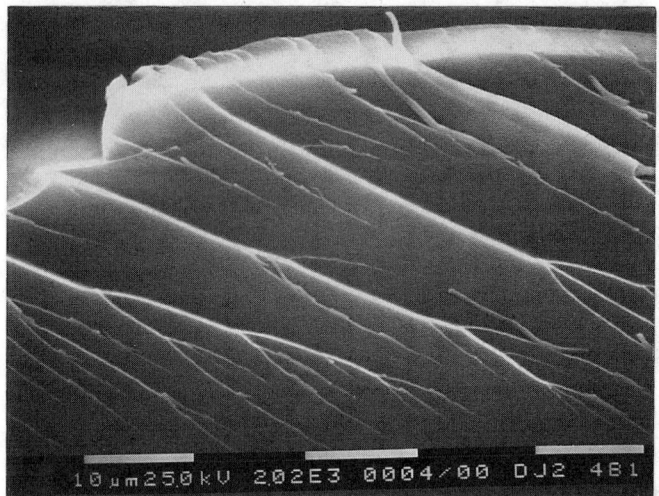

FIG. 18—*Mode I fractograph (×2020) JP-4 saturated condition.*

amount of fiber-resin adhesion. It must be mentioned that there is not much difference between the water-saturated and the jet fuel-saturated cases.

Mode II Delamination Fracture: Room Temperature Dry Condition

Typical fractographs of AS4/3501-6 tested at RTD are presented in Figs. 21–25. As established by many researchers [2,5–7], for RTD condition, a much larger number of hackle formations are found in Mode II regions as compared with Mode I regions. Microcrack nucleation in advance of the crack tip result in hackles in brittle systems such as graphite/epoxy. The appearance of the resulting fracture surface is flat and unlike Mode I

FIG. 19—*Mode I fractograph (×2020) JP-4 saturated condition.*

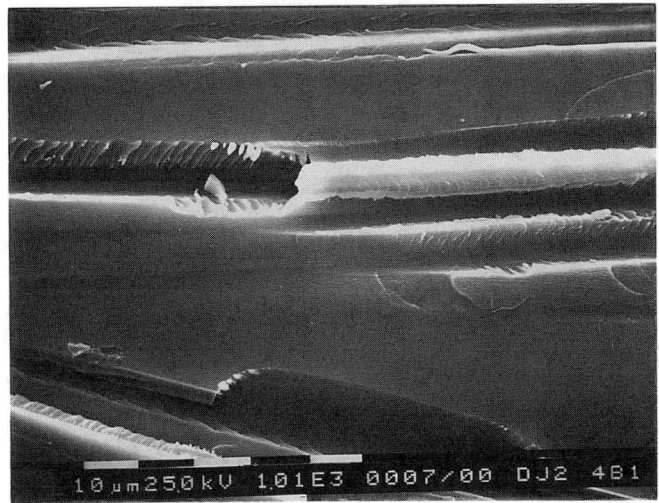

FIG. 20—*Mode I fractograph (×1010) JP-4 saturated condition.*

delamination, containing no deep grooves. Figures 22 and 23 show the opposing surfaces of a delaminated region from the upper and lower halves of an ENF specimen, respectively. A comparison of these two figures shows the direction of inclination of the hackles to be opposite each other. The direction of crack propagation is from right to left in all of these figures. On the delaminated surface of the lower half, the direction of the shear stress is from right to left and the crack propagation direction is also from right to left. Whereas on the delaminated surface of the upper half, the direction of the shear stress is from the left to the right but the crack propagation direction is from the right to the left. This observation is in agreement with the conclusion of Smith and Grove [5,7], in that the direction of the interlaminar shear can be determined by examining the directional tilt of

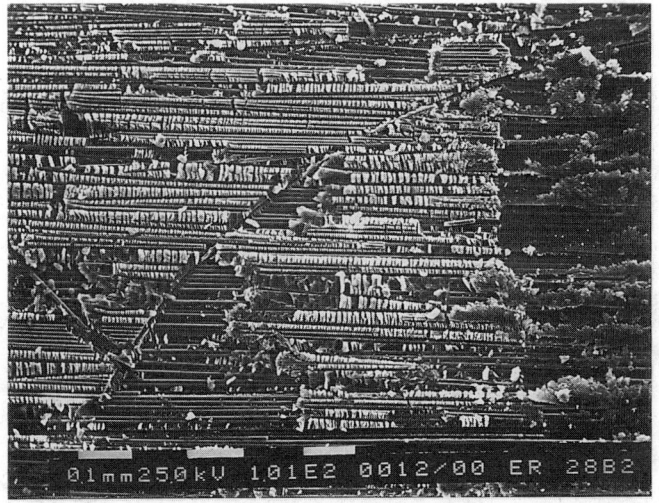

FIG. 21—*Mode II fractograph (×101) RTD condition.*

FIG. 22—*Mode II fractograph (×503) RTD condition.*

the hackles. Figures 24 and 25 show the opposite surfaces of the same small delaminated region, with Fig. 24 being that of the delaminated surface on the upper half of the ENF specimen and Fig. 25 the delaminated surface of the lower half of the specimen. This conclusively proves the formation of areas of scalloped or concave-shaped resin fracture resulting from the separation of the hackles from the fracture surface as was stated by Smith and Grove [5,7]. The hackles seen at high magnification (Fig. 25) are more closely packed in this case than the Mode I (Fig. 10) RTD case. These appear to be vertical, in agreement with the work of Hibbs and Bradley [1].

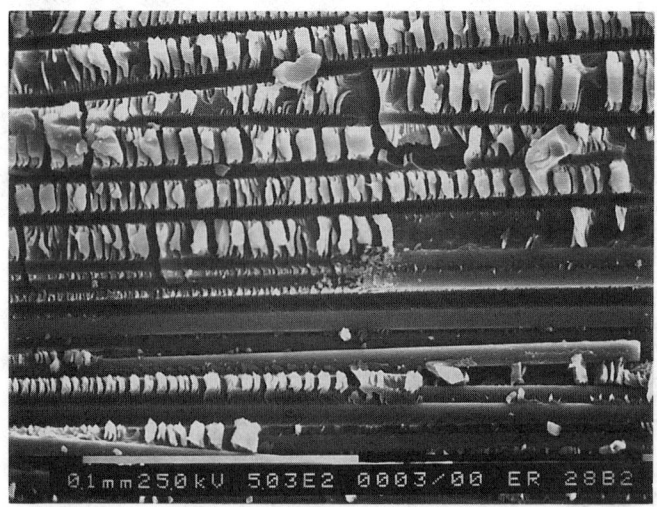

FIG. 23—*Mode II fractograph (×503) RTD condition.*

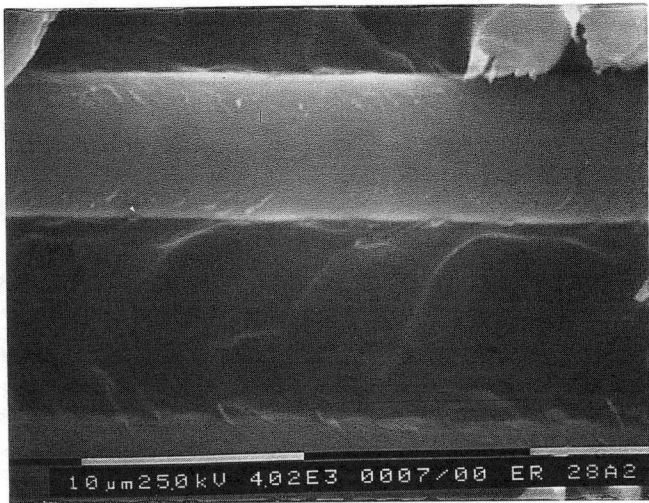

FIG. 24—*Mode II fractograph (×4020) RTD condition.*

Mode II Delamination Fracture: Water-Saturated Condition

Figures 26 and 27 are typical fractographs of AS4/3501-6 tested after being saturated with water. Figure 26 shows clearly the abrupt transition from a Mode I opening mode precrack to a Mode II shear crack. The features seen in the right half of Fig. 26 are identical to those seen in Fig. 12, the Mode I water-saturated case. The left half shows a large number of hackle formation accompanied by some matrix deformation. The fractograph in Fig. 27 shows, in part, a regular formation of hackles along with some region of cohesive fracture in the resin and some deformed hackles. Though the presence of deformed hackles was not extensive, the few present point to slightly more matrix

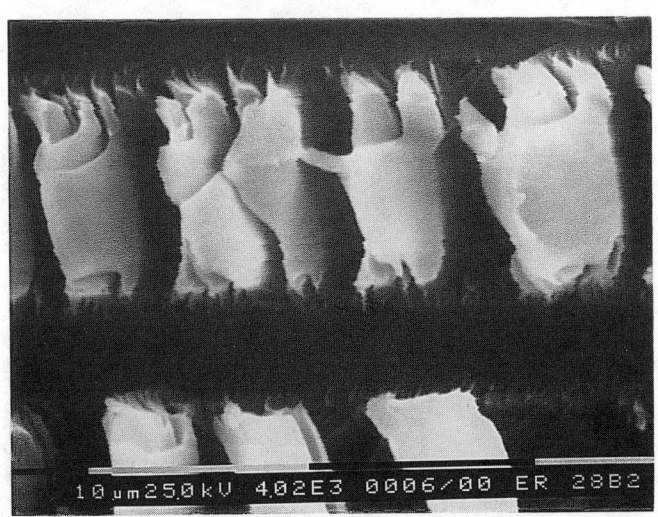

FIG. 25—*Mode II fractograph (×4020) RTD condition.*

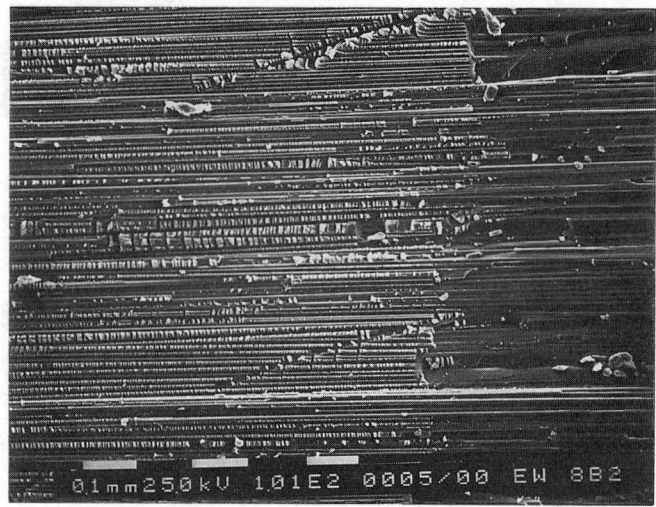

FIG. 26—*Mode II fractograph (×101) water saturated condition.*

deformation for this specimen than for the RTD specimen. A comparison of Figs. 27 and 23 reveals, with the exception of matrix deformation and deformed hackles there, little difference in their features.

Mode II Delamination Fracture: Jet Fuel-Saturated Condition

Figures 28–31 show the fractographs of an ENF specimen saturated in Jet A. There was little difference between the specimens saturated in Jet A and those saturated in JP-4. A comparison of Fig. 28 with Fig. 21 shows the presence of a greater amount of deformed hackles in the jet fuel-saturated case than in the RTD case. Figures 29 and 30 are the

FIG. 27—*Mode II fractograph (×503) water saturated condition.*

FIG. 28—*Mode II fractograph (×101) Jet-A saturated condition.*

opposite surfaces of the same delaminated region in the specimen. The hackles that are formed exhibit more deformation than was evident for either the RTD or water-saturated cases.

Conclusions

The environmental effects of sorption of water and jet fuel on the interlaminar fracture toughness of AS4/3501-6 graphite/epoxy were investigated. DCB and ENF specimens were saturated with either commercial jet fuel, JP-4 military jet fuel, or water and then tested. Fractographic analyses of the delamination surfaces of these specimens were later

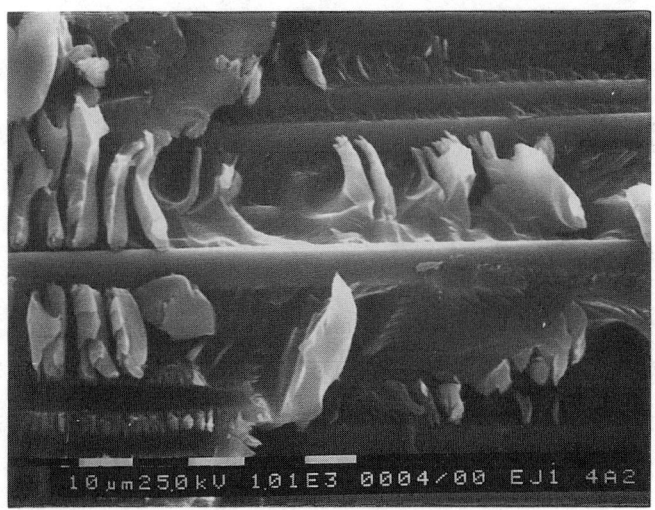

FIG. 29—*Mode II fractograph (×1010) Jet-A saturated condition.*

FIG. 30—*Mode II fractograph (×1010) Jet-A saturated condition.*

performed using a scanning electron microscope. Based on the results of this investigation, it is concluded that:

1. Both G_{Ic} and G_{IIc} increase due to environmental exposure. The Mode I toughness exhibits a greater percentage increase than does the Mode II toughness.
2. The environmental effects produce significant changes in the surface morphology of Mode I fracture surfaces.
3. The environmental effects of water absorption do not significantly alter the surface morphology of Mode II fracture surfaces from those developed in RTD specimens.

FIG. 31—*Mode II fractograph (×503) Jet-A saturated condition.*

However, the effects of jet fuel absorption do significantly modify the Mode II fracture surface morphology.

Acknowledgments

This work was partially supported by the Army Research Office, Contract Number DAAL 03-87-G-003, and also by the National Institute of Aviation Research at Wichita State University. The authors would like to thank Boeing Military Airplane Company for supplying the JP-4 military jet fuel used in this study. The authors would like to thank T. Kevin O'Brien and Rod Martin of NASA Langley Research Center for their many helpful suggestions during the course of this project.

References

[1] Hibbs, M. F. and Bradley, W. L., "Correlations Between Micromechanical Failure Processes and the Delamination Toughness of Graphite/Epoxy Systems," *Fractography of Modern Engineering Materials: Composites and Metals, ASTM STP 948*, J. E. Masters and J. J. Au, Eds., American Society for Testing and Materials, Philadelphia, 1987, pp. 68–97.

[2] Masters, J. E., "Characterization of Impact Damage Development in Graphite/Epoxy Laminates," *Fractography of Modern Engineering Materials: Composites and Metals, ASTM STP 948*, J. E. Masters and J. J. Au, Eds., American Society for Testing and Materials, Philadelphia, 1987, pp. 238–258.

[3] O'Brien, T. K., "Mixed-Mode Strain Energy Release Rate Effects on Edge Delamination of Composites," *Effects of Defects in Composite Materials, ASTM STP 836*, American Society for Testing and Materials, Philadelphia, 1984, pp. 125–142.

[4] Hertzberg, R. W., "Fracture Surface Micromorphology in Engineering Solids," *Fractography of Modern Engineering Materials: Composites and Metals, ASTM STP 948*, J. E. Masters and J. J. Au, Eds., American Society for Testing and Materials, Philadelphia, 1987, pp. 5–36.

[5] Smith, B. W. and Grove, R. A., "Determination of Crack Propagation Directions in Graphite/ Epoxy Systems," *Fractography of Modern Engineering Materials: Composites and Metals, ASTM STP 948*, J. E. Masters and J. J. Au, Eds., American Society for Testing and Materials, Philadelphia, 1987, pp. 154–173.

[6] Arcan, L., Arcan, M., and Daniel, I. M., "SEM Fractography of Pure and Mixed-Mode Interlaminar Fractures in Graphite/Epoxy Composites," *Fractography of Modern Engineering Materials: Composites and Metals, ASTM STP 948*, J. E. Masters and J. J. Au, Eds., American Society for Testing and Materials, Philadelphia, 1987, pp. 41–67.

[7] Grove, R. and Smith, B., "Compendium of Post-Failure Analysis Techniques for Composite Materials," Air Force Wright Aeronautical/Laboratories, Report No. AFWAL-TR-86-4137, January 1987.

[8] Hunston, D. L., Moulton, R. J., Johnston, N. J., and Bascom, W. D., "Matrix Resin Effects in Composite Delamination: Mode I Fracture Aspects," *Toughened Composites, ASTM STP 937*, N. J. Johnston, Ed., American Society for Testing and Materials, Philadelphia, 1987, pp. 74–94.

[9] Jordan, W. M. and Bradley, W. L., "Micromechanisms of Fracture in Toughened Graphite-Epoxy Laminates," *Toughened Composites, ASTM STP 937*, N. J. Johnston, Ed., American Society for Testing and Materials, Philadelphia, 1987, pp. 95–114.

[10] Hibbs, M. F., Tse, M. K., and Bradley, W. L., "Interlaminar Fracture Toughness and Real-Time Fracture Mechanisms of Some Toughened Graphite/Epoxy Composites," *Toughened Composites, ASTM STP 937*, N. J. Johnston, Ed., American Society for Testing and Materials, Philadelphia, 1987, pp. 115–130.

[11] Bascom, W. D., Boll, D. J., Hunston, D. L., Fuller, B., and Phillips, P. J., "Fractographic Analysis of Interlaminar Fracture," *Toughened Composites, ASTM STP 937*, N. J. Johnston, Ed., American Society for Testing and Materials, Philadelphia, 1987, pp. 131–149.

[12] Hooper, S. J., Subramanian, R., and Toubia, R. F., "The Effects of Moisture Absorption and Edge Delamination Part II. An Experimental Study of Jet Fuel Absorption on Graphite/Epoxy," *Composite Materials: Fatigue and Fracture, Third Volume, ASTM STP 1110*, T. K. O'Brien, Ed., American Society for Testing and Materials, Philadelphia, 1991, pp. 89–106.

[13] O'Brien, T. K., Raju, I. S., and Garber, D. P., "Residual Thermal and Moisture Influences on the Strain Energy Release Rate Analysis of Edge Delamination," *Journal of Composites Technology and Research*, Vol. 8, No. 2, 1986, pp. 37–47.

[*14*] Kriz, R. D. and Stinchcomb, W. W., "Effects of Moisture, Residual Thermal Curing Stresses, and Mechanical Load on Damage Development in Quasi-Isotropic Laminates," *Damage in Composite Materials, ASTM STP 775*, K. L. Reifsnider, Ed., American Society for Testing and Materials, 1982, pp. 63–80.

[*15*] Russell, A. J. and Street, K. N., "Moisture and Temperature Effects on the Mixed-Mode Delamination Fracture of Unidirectional Graphite/Epoxy," *Delamination and Debonding of Materials, ASTM STP 876*, W. S. Johnson, Ed., American Society for Testing and Materials, 1985, pp. 349–370.

[*16*] *ICI Fiberite Materials Handbook*, ICI Fiberite Corp., Tempe, AZ, 15 March, 1989.

Erian A. Armanios[1] and Jian Li[1]

Interlaminar Fracture Analysis of Unsymmetrical Laminates

REFERENCE: Armanios, E. A. and Li, J., "**Interlaminar Fracture Analysis of Unsymmetrical Laminates,**" *Composite Materials: Fatigue and Fracture, Fourth Volume, ASTM STP 1156*, W. W. Stinchcomb and N. E. Ashbaugh, Eds., American Society for Testing and Materials, Philadelphia, 1993, pp. 341–360.

ABSTRACT: A simple analytical model is developed to predict the interlaminar stresses and the total strain energy release rate for delaminated unsymmetrical laminates. The method is based on a shear deformation theory and a sublaminate approach. The solution is obtained in closed form, and the controlling parameters are isolated. Results for a class of laminates with $[-\theta, (90-\theta)_2, -\theta, \theta, (\theta-90)_2, \theta]_T$ layup exhibiting extension-twist coupling are provided. Potential delamination sites are predicted by applying the analysis to laminates with successive delaminated interfaces. Critical interfaces are determined based on the peel stress distribution and the total strain energy release rate.

KEY WORDS: composite, interlaminar stress, strain energy release rate, free edge delamination, fracture

Interlaminar stress and delamination analyses of laminated composites have been studied extensively for laminates under extension. However, the majority of the work is confined to symmetric laminates. While symmetric configurations are simple to analyze and possess unique design and manufacturing characteristics, their damage modes often alter their initial symmetry. An example is the case of a laminate subjected to unsymmetrical loading such as bending. Free edge delamination is expected to initiate on one side of the symmetry plane, creating an unsymmetrical effect. This lack of symmetry may be neglected at crack initiation [1]. Its effects, however, on the laminate behavior in the crack growth stage can be significant.

Unsymmetrical laminate designs are also used for elastic tailoring in order to create a coupling between deformation modes such as extension-twist. Understanding and predicting the interlaminar fracture is of crucial importance in designing against such failures.

The objective of this work is to develop a simple analytical method to predict the interlaminar stresses and total strain energy release rate in unsymmetrical laminated composites.

Mathematical Model

For a symmetric laminate under anti-symmetric loading such as bending or torsion or an unsymmetrical laminate under extension, free edge delaminations may develop in the upper or lower half of the laminate as shown in Fig. 1. One half of the laminate cross section can be modeled using four sublaminates as shown in Fig. 2. The sublaminate

[1]Associate professor and graduate research assistant, respectively, School of Aerospace Engineering, Georgia Institute of Technology, Atlanta, GA 30332.

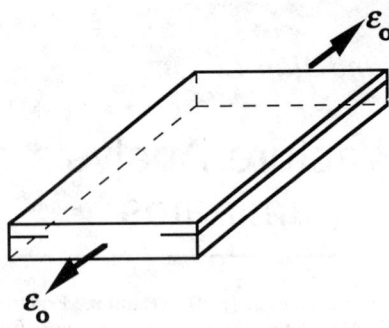

FIG. 1—*Delaminated unsymmetrical laminate under extension.*

numbering scheme and the coordinate system appear in Fig. 2. Sublaminates 2 and 3 represent the group of plies above and below the delamination, respectively, while Sublaminates 1 and 0 represent the same group of plies in the uncracked portion.

The laminate is assumed to be initially flat. Unsymmetrical layups induce warping, bending, or twisting as a result of residual curing stresses. Appropriate changes in design can lead to hygrothermally stable laminates. Such a design strategy is outlined in Ref *2* in order to produce a family of unsymmetrical laminates that do not produce changes in curvature with changes in temperature or moisture content.

Governing Equations

Each sublaminate can be treated as a homogeneous anisotropic elastic body bounded by a cylindrical surface. A generalized plane deformation [*3*] exists in the laminate shown in Fig. 1 when it is subjected to a remote extension such that the stress tensor is independent of the x-coordinate. The displacement field within each sublaminate may be written as

$$u(x, y, z) = \epsilon_o \cdot x + \kappa \cdot x \cdot (z + \delta) + U(y) + z \cdot \beta_x(y)$$

$$v(x, y, z) = V(y) + z \cdot \beta_y(y) + C \cdot x \cdot (z + \delta) \tag{1}$$

$$w(x, y, z) = -\tfrac{1}{2}\kappa \cdot x^2 - C \cdot x \cdot (y + \rho) + W(y)$$

where u, v, and w denote displacements relative to the x, y, and z axes, respectively. The axial extension strain is ϵ_o. The arbitrary constants, δ and ρ, associated with each sublaminate are to be determined from continuity of displacements between sublaminates and

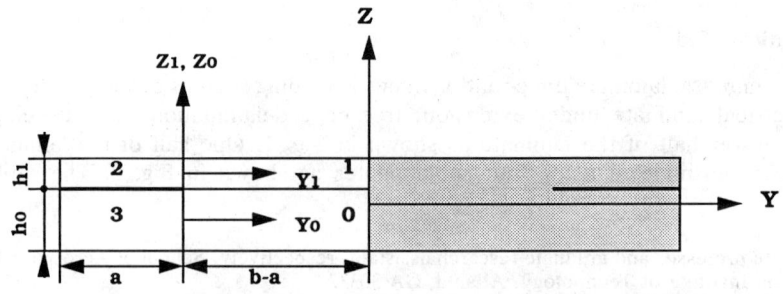

FIG. 2—*Sublaminate model and coordinate system.*

from the overall boundary conditions. The relative angle of rotation and bending curvature are denoted by C and κ, respectively. These result from the coupling effects associated with unsymmetrical layups. Shear deformation is recognized through the rotations, β_x and β_y. The bending about the z-axis is neglected since the sublaminate thickness is small compared to the width.

The corresponding strains are defined as

$$\epsilon_{xx} = \epsilon_{xx}^o + z\kappa_x \quad \epsilon_{yy} = \epsilon_{yy}^o + z\kappa_y \quad \epsilon_{zz} = 0$$

$$\gamma_{xy} = \gamma_{xy}^o + z\kappa_{xy} \quad \gamma_{yz} = \gamma_{yz}^o \quad \gamma_{xz} = \gamma_{xz}^o \tag{2}$$

The strain components associated with the reference surface are denoted by superscript "o". These and the associated curvatures are defined as

$$\epsilon_{xx}^o = \epsilon_o + \kappa \cdot \delta \quad \epsilon_{yy}^o = V_{,y} \quad \gamma_{xy}^o = U_{,y} + C \cdot \delta$$

$$\kappa_x = \kappa \qquad \kappa_y = \beta_{y,y} \quad \kappa_{xy} = \beta_{x,y} + C \tag{3}$$

$$\gamma_{yz}^o = \beta_y + W_{,y} \qquad \gamma_{zx}^o = \beta_x - C \cdot (y + \rho)$$

where partial differentiation is denoted by a comma. The constitutive relationship can be written as resultant forces and moments in terms of strains and curvatures as follows

$$\begin{Bmatrix} N_x \\ N_y \\ N_{xy} \\ M_x \\ M_y \\ M_{xy} \end{Bmatrix} = \begin{bmatrix} A_{11} & A_{12} & A_{16} & B_{11} & B_{12} & B_{16} \\ A_{12} & A_{22} & A_{26} & B_{12} & B_{22} & B_{26} \\ A_{16} & A_{26} & A_{66} & B_{16} & B_{26} & B_{66} \\ B_{11} & B_{12} & B_{16} & D_{11} & D_{12} & D_{16} \\ B_{12} & B_{22} & B_{26} & D_{12} & D_{22} & D_{26} \\ B_{16} & B_{26} & B_{66} & D_{16} & D_{26} & D_{66} \end{bmatrix} \begin{Bmatrix} \epsilon_{xx}^o \\ \epsilon_{yy}^o \\ \gamma_{xy}^o \\ \kappa_x \\ \kappa_y \\ \kappa_{xy} \end{Bmatrix} \tag{4}$$

$$\begin{Bmatrix} Q_y \\ Q_x \end{Bmatrix} = \begin{bmatrix} A_{44} & A_{45} \\ A_{45} & A_{55} \end{bmatrix} \begin{Bmatrix} \gamma_{yz}^o \\ \gamma_{xz}^o \end{Bmatrix} \tag{5}$$

For a sublaminate of thickness h, the stiffness coefficients are defined as

$$(A_{ij}, B_{ij}, D_{ij}) = \int_{-h/2}^{h/2} \overline{Q}_{ij}(l, z, z^2) \cdot dz \tag{6}$$

where $\overline{Q}_{ij}$ are the transformed reduced stiffnesses as defined in Ref 4.

The equilibrium equations can be written as

$$N_{xy,y} + t_{ux} - t_{lx} = 0$$

$$N_{y,y} + t_{uy} - t_{ly} = 0$$

$$Q_{y,y} + p_u - p_l = 0$$

$$M_{xy,y} - Q_x + \frac{h}{2} \cdot (t_{ux} + t_{lx}) = 0 \tag{7}$$

$$M_{y,y} - Q_y + \frac{h}{2} \cdot (t_{uy} + t_{ly}) = 0$$

where the interlaminar shear and peel stresses at the sublaminate upper and lower surfaces are denoted by t_{ux}, t_{uy}, p_u, and t_{lx}, t_{ly}, p_l, respectively, as shown in Fig. 3.

Solution Procedure

The above collection of equations will be applied to each sublaminate. To suppress the rigid body motion of the laminate, it is assumed that the center of the laminate is fixed. That is, for $X = Y = Z = 0$, we have

$$u = v = w = 0$$

$$\frac{\partial v}{\partial X} = \frac{\partial w}{\partial X} = \frac{\partial v}{\partial Z} - \frac{\partial w}{\partial Y} = 0 \tag{8}$$

Continuity Conditions

The continuity of displacements at the interface between sublaminates 0 and 1 are

$$u_0\left(x, y, \frac{h_0}{2}\right) = u_1\left(x, y, -\frac{h_1}{2}\right)$$

$$v_0\left(x, y, \frac{h_0}{2}\right) = v_1\left(x, y, -\frac{h_1}{2}\right) \tag{9}$$

$$w_0\left(x, y, \frac{h_0}{2}\right) = w_1\left(x, y, -\frac{h_1}{2}\right)$$

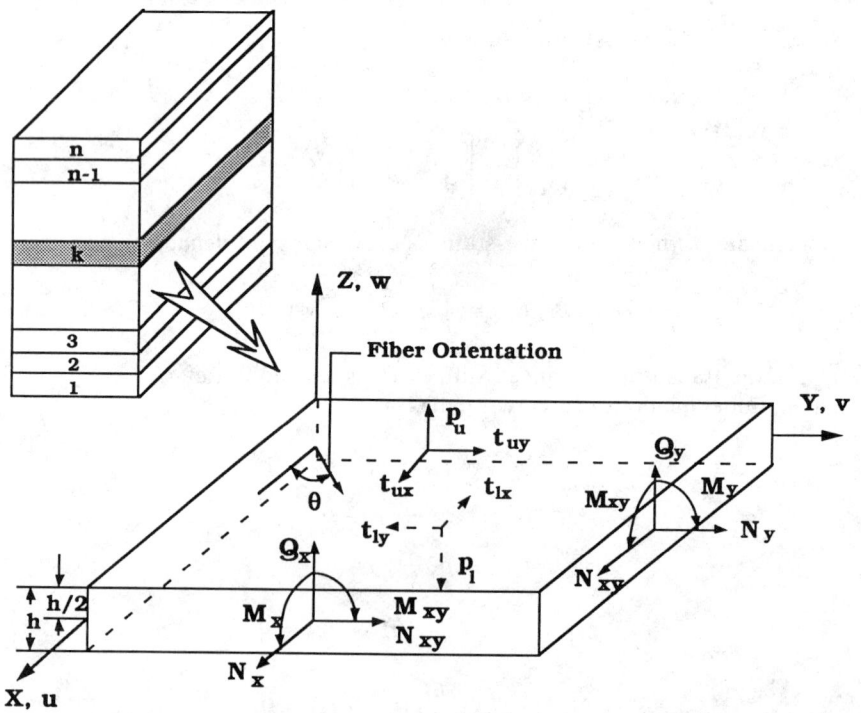

FIG. 3—*Sublaminate notation and sign convention.*

where the subscripts refer to the respective sublaminate. This convention is adopted in the subsequent equations. The thicknesses of Sublaminates 1 and 0 are denoted by h_1 and h_0, respectively. By substituting Eq 1 into Eqs 8 and 9, the constants δ and ρ for each sublaminate can be determined as

$$\delta_1 = \delta_2 = \frac{h_0}{2} \quad \delta_0 = \delta_3 = -\frac{h_1}{2} \quad \rho_0 = \rho_1 = a - b \tag{10}$$

where b is the semi-width of the laminate, and a is the delamination length.
Substitute Eq 1 into Eq 9 and use Eq 10 to obtain

$$U_0 = U_1 - \frac{h_0}{2}\beta_{0x} - \frac{h_1}{2}\beta_{1x}$$

$$\tag{11}$$

$$V_0 = V_1 - \frac{h_0}{2}\beta_{0y} - \frac{h_1}{2}\beta_{1y}$$

Sublaminate Equilibrium Equations

The responses associated with Sublaminates 0 and 1 are coupled through the interface continuity conditions. The upper surface of Sublaminate 1 and the bottom surface of Sublaminate 0 are stress free. The shear and peel stresses at the interface will be denoted by t_x, t_y, and p, respectively. Apply the equilibrium Eq 7 to Sublaminates 0 and 1 to get

$$N_{xy1} + N_{xy0} = 0$$

$$N_{y1} + N_{y0} = 0$$

$$Q_{y1} + Q_{y0} = 0$$

$$M_{xy1,y} + \frac{h_1}{2}N_{xy1,y} - Q_{x1} = 0$$

$$\tag{12}$$

$$M_{xy0,y} - \frac{h_0}{2}N_{xy0,y} - Q_{x0} = 0$$

$$M_{y1,y} + \frac{h_1}{2}N_{y1,y} - Q_{y1} = 0$$

$$M_{y0,y} - \frac{h_0}{2}N_{y0,y} - Q_{y0} = 0$$

For Sublaminate 2, the upper and lower surfaces are stress free. The equilibrium equations reduce to

$$N_{y2} = N_{xy2} = M_{y2} = Q_{y2} = 0$$

$$\tag{13}$$

$$M_{xy2,y} - Q_{x2} = 0$$

Similarly, the equilibrium equations for Sublaminate 3 can be written as

$$N_{y3} = N_{xy3} = M_{y3} = Q_{y3} = 0$$
$$M_{xy3,y} - Q_{x3} = 0 \tag{14}$$

Solution for the Rotations

The continuity conditions between Sublaminates 1 and 2, and 0 and 3, and the boundary conditions at the free ends of Sublaminates 2 and 3 can be written as

$$N_{y1}|_{y_1=0} = 0 \qquad N_{xy1}|_{y_1=0} = 0 \qquad M_{y1}|_{y_1=0} = 0$$

$$M_{y0}|_{y_0=0} = 0 \qquad M_{xy1}|_{y_1=0} = M_{xy2}|_{y_1=0}$$

$$M_{xy0}|_{y_0=0} = M_{xy3}|_{y_0=0} \tag{15}$$

$$M_{xy2}|_{y_1=-a} = 0 \qquad M_{xy3}|_{y_0=-a} = 0$$

Substitute Eqs 1 through 3 into Eqs 4 and 5 for each sublaminate and substitute the results into Eqs 12 through 14 to obtain a set of ordinary differential equations in terms of the sublaminate rotations. Solving these equations and using Eq 15 to determine the integration constants yields

$$\{\beta\} = [\Phi][e^{-sy}]\{G\} + \{\hat{\beta}\} \cdot C \cdot (y + a - b) \tag{16}$$

where $\{\beta\} = \{\beta_{1x}, \beta_{0x}, \beta_{1y}, \beta_{0y}\}^T$ is a vector containing the four rotations for Sublaminates 1 and 0 and $[\Phi] = [\{\phi_1\}, \{\phi_2\}, \{\phi_3\}, \{\phi_4\}]$ is a four-by-four matrix whose columns are the four eigenvectors $\{\phi_1\}$ through $\{\phi_4\}$. The exponential matrix $[e^{-sy}]$ is a four-by-four diagonal matrix of eigenvalues. The ith diagonal element is e^{-s_iy} corresponding to the ith eigenvalue, s_i. When s_i equals zero, e^{-s_iy} is replaced by $-(y - b + a)$. The eigenvalues and eigenvectors are determined from Eq A-1 of the Appendix.

The vector $\{G\} = \{G_1, G_2, G_3, G_4\}^T$ in Eq 16 represents four integration constants and can be expressed as

$$\{G\} = [\bar{G}] \begin{Bmatrix} \epsilon_o \\ \kappa \\ C \end{Bmatrix} \tag{17}$$

where $[\bar{G}]$ is given by

$$[\bar{G}] = \begin{bmatrix} \dfrac{1}{S} \end{bmatrix} [\Phi]^{-1}[M]^{-1}\langle[L] + [M]\{\hat{\beta}\}(0,0,1)\rangle \tag{18}$$

The diagonal matrix $[1/s]$ in Eq 18 is a matrix whose ith element is the inverse of the ith eigenvalue. For a zero eigenvalue the diagonal element equals 1. The matrices $[M]$ and $[L]$ are given in Eqs A-11 and A-12 of the Appendix. The vector $\{\hat{\beta}\}$ in Eqs 16 and 18 can be expressed as

$$\{\hat{\beta}\} = [\Phi][\bar{K}][\Phi]^T\{q\} \tag{19}$$

where $[\bar{K}]$ and $\{q\}$ are given in Eqs A-13 and A-14 of the Appendix.

The mid-plane displacement functions of Sublaminate 1 are related to the rotations of Sublaminates 1 and 0 through the following relations

$$\begin{Bmatrix} V_{1,y} \\ U_{1,y} \end{Bmatrix} = [\tilde{T}] \begin{Bmatrix} \epsilon_o \\ \kappa \\ C \end{Bmatrix} - [T]\{\beta_{,y}\} \tag{20}$$

where matrices $[\tilde{T}]$ and $[T]$ are given in Eqs A-9 and A-10 of the Appendix. The rotation, β_{2x}, of Sublaminate 2 can be expressed as

$$\beta_{2x} = [\sinh (s_c y) \quad \cosh (s_c y) \quad y + a - b][H] \begin{Bmatrix} \epsilon_o \\ \kappa \\ C \end{Bmatrix} \tag{21}$$

where

$$s_c = \sqrt{\dfrac{A_{55}^1 - \dfrac{(A_{45}^1)^2}{A_{44}}}{\overline{M}_3^1}} \tag{22}$$

The third element of vector $\{\overline{M}^1\}$ given in Eq A-15, is denoted by $\overline{M}_3^1$. The remaining displacement functions associated with Sublaminate 2 can be related to β_{2x} by

$$\begin{Bmatrix} V_{2,y} \\ U_{2,y} \\ \beta_{2y,y} \end{Bmatrix} = -[\tilde{R}^1] \begin{Bmatrix} \epsilon_o \\ \kappa \\ C \end{Bmatrix} - \{R^1\}\beta_{2x,y} \tag{23}$$

Similarly,

$$\beta_{3x} = [\sinh (s_d y) \quad \cosh (s_d y) \quad y + a - b][I] \begin{Bmatrix} \epsilon_o \\ \kappa \\ C \end{Bmatrix} \tag{24}$$

and

$$\begin{Bmatrix} V_{3,y} \\ U_{3,y} \\ \beta_{3y,y} \end{Bmatrix} = -[\tilde{R}^0] \begin{Bmatrix} \epsilon_o \\ \kappa \\ C \end{Bmatrix} - \{R^0\}\beta_{3x,y} \tag{25}$$

where s_d can be obtained by replacing superscript 1 with 0 in Eq 22. Matrices $[\tilde{R}^1]$, $[\tilde{R}^0]$, $\{R^1\}$, $\{R^0\}$, $[H]$, and $[I]$ are defined in Eqs A-17 through A-19 of the Appendix.

Resultant Forces and Moments

The axial resultant forces and moments in Sublaminates 1 and 0 are given by

$$\{N^1\} = [L^1] \begin{Bmatrix} \epsilon_o \\ \kappa \\ C \end{Bmatrix} + [M^1]\{\beta_{,y}\} \tag{26a}$$

where

$$\{N^1\} = (N_{x1} \quad N_{y1} \quad N_{xy1} \quad M_{x1} \quad M_{y1} \quad M_{xy1})^T \tag{26b}$$

$$\{N^0\} = [L^0] \left\{ \begin{matrix} \epsilon_o \\ \kappa \\ C \end{matrix} \right\} + [M^0]\{\beta_{,y}\} \tag{27a}$$

where

$$\{N^0\} = (N_{x0} \quad N_{y0} \quad N_{xy0} \quad M_{x0} \quad M_{y0} \quad M_{xy0})^T \tag{27b}$$

$$\left\{ \begin{matrix} Q_{x1} \\ Q_{x0} \\ Q_{y1} \\ Q_{y0} \end{matrix} \right\} = [K]\{\beta\} - \{q\} \cdot C \cdot (y + a - b) \tag{28}$$

Matrices $[K]$, $[M^1]$, $[M^0]$, $[L^1]$, and $[L^0]$ are given in the Appendix as Eqs A-3 through A-8.

The resultant forces and moments associated with Sublaminates 2 and 3 can be expressed as

$$\{\overline{N}^1\} = [\overline{L}^1] \left\{ \begin{matrix} \epsilon_o \\ \kappa \\ C \end{matrix} \right\} + \{\overline{M}^1\}\beta_{2x,y} \tag{29a}$$

where

$$\{\overline{N}^1\} = (N_{x2} \quad M_{x2} \quad M_{xy2})^T \tag{29b}$$

$$Q_{x2} = \left(A_{55}^1 - \frac{(A_{45}^1)^2}{A_{44}^1} \right) \{\beta_{2x} - C(y + a - b)\} \tag{30}$$

$$\{\overline{N}^0\} = [\overline{L}^0] \left\{ \begin{matrix} \epsilon_o \\ \kappa \\ C \end{matrix} \right\} + \{\overline{M}^0\}\beta_{3x,y} \tag{31a}$$

where

$$\{\overline{N}^0\} = (N_{x3} \quad M_{x3} \quad M_{xy3})^T \tag{31b}$$

$$Q_{x3} = \left(A_{55}^0 - \frac{(A_{45}^0)^2}{A_{44}^0} \right) \{\beta_{3x} - C(y + a - b)\} \tag{32}$$

Matrices $[\overline{L}^1]$, $[\overline{L}^0]$, $\{\overline{M}^1\}$, and $\{\overline{M}^0\}$ are defined in Eqs A-15 and A-16 of the Appendix.

Interlaminar Stresses

The interlaminar stresses between Sublaminates 1 and 0 are found from the equilibrium equations

$$t_x = N_{xy1,y}$$

$$t_y = N_{y1,y} \tag{33}$$

$$p = Q_{y1,y}$$

The peel stress in Eq 33 needs to be modified as outlined in Ref 5. The modified peel stress can be written as

$$\hat{p} = Q_{y1,y} - c_1 s_3 e^{-s_3 y} - c_2 s_4 e^{-s_4 y} \tag{34}$$

where

$$c_1 = \frac{Q_{y1}(0) - 2s_4 \left(M_{y1}(b-a) + \dfrac{h_1}{2} N_{y1}(b-a) \right)}{\dfrac{s_4}{s_3} - 1}$$

$$\tag{35}$$

$$c_2 = -\frac{s_4}{s_3} \frac{Q_{y1}(0) - 2s_3 \left(M_{y1}(b-a) + \dfrac{h_1}{2} N_{y1}(b-a) \right)}{\dfrac{s_4}{s_3} - 1}$$

and the eigenvalues in Eq 34 are arranged in increasing order of magnitude: $s_1 \leq s_2 \leq s_3 \leq s_4$.

Loading Conditions

The laminate resultant force and moments can be written as

$$F_x = 2 \int_0^{b-a} (N_{x1} + N_{x0})dy + 2 \int_{-a}^0 (N_{x2} + N_{x3})dy$$

$$M_x = 2 \int_0^{b-a} (M_{x1} + M_{x0} + \delta_1 N_{x1} + \delta_0 N_{x0})dy$$

$$+ 2 \int_{-a}^0 (M_{x2} + M_{x3} + \delta_1 N_{x2} + \delta_0 N_{x3})dy \tag{36a}$$

$$M_{xy} = 2 \int_0^{b-a} [M_{xy1} + M_{xy0} + \delta_1 N_{xy1} + \delta_0 N_{xy0} - (Q_{x1} + Q_{x0})(y + a - b)]dy$$

$$+ 2 \int_{-a}^0 [M_{xy2} + M_{xy3} - (Q_{x1} + Q_{x0})(y + a - b)]dy$$

The last equation in Eq 36a can be simplified by using Eqs 12 through 15 to get

$$M_{xy} = 4 \int_0^{b-a} [M_{xy1} + M_{xy0} + \delta_1 N_{xy1} + \delta_0 N_{xy0}]dy + 4 \int_{-a}^0 [M_{xy2} + M_{xy3}]dy \quad (36b)$$

Substitute Eqs 26 through 32 into Eq 36 and set $M_x = M_{xy} = 0$ for unidirectional extension to get

$$\begin{bmatrix} \alpha_{11} & \alpha_{12} & \alpha_{13} \\ \alpha_{21} & \alpha_{22} & \alpha_{23} \\ \alpha_{31} & \alpha_{32} & \alpha_{33} \end{bmatrix} \begin{Bmatrix} \epsilon_o \\ \kappa \\ C \end{Bmatrix} = \begin{Bmatrix} \dfrac{F_x}{2} \\ 0 \\ 0 \end{Bmatrix} \quad (37)$$

where α_{ij} are given in Eq A-20 of the Appendix, and F_x is the applied axial force. Solve Eq 37 for the induced bending curvature, κ, and twist C in terms of the extensional strain to obtain

$$\begin{Bmatrix} \kappa \\ C \end{Bmatrix} = - \begin{bmatrix} \alpha_{22} & \alpha_{23} \\ \alpha_{32} & \alpha_{33} \end{bmatrix}^{-1} \begin{Bmatrix} \alpha_{21} \\ \alpha_{31} \end{Bmatrix} \epsilon_o \quad (38)$$

Strain Energy Release Rate

The total strain energy release rate of the laminate can be calculated as

$$G_T = - \frac{\partial U}{\partial a} = - \frac{\epsilon_o}{4} \frac{\partial F_x}{\partial a} \quad (39)$$

where U is one half of the total strain energy of the laminate per unit length. The strain energy is equal to one half of the work done by the external force per unit length. Substitute from Eq 37 into Eq 39, the total strain energy release rate can be expressed as

$$G_T = - \frac{\epsilon_o}{2} (\alpha_{11,a}\epsilon_o + \alpha_{12,a}\kappa + \alpha_{13,a}C + \alpha_{12}\kappa_{,a} + \alpha_{13}C_{,a}) \quad (40)$$

A closed-form solution for a delaminated unsymmetrical laminate under extension has been developed. The induced bending curvature, κ, and twist C are estimated from Eq 38. The interlaminar stresses can be obtained from Eqs 33 and 34, and the strain energy release rate can be obtained from Eq 40.

Application

The present solution is applied to the analysis of a class of laminates with $[-\theta, (90-\theta)_2, -\theta, \theta, (\theta-90)_2, \theta]_T$ layup under uniaxial extension. Values of $\theta = 10°, 20°, 30°, 40°, 45°, 50°, 60°,$ 70°, and 80° are considered in this study. This class of laminates is designed to exhibit extension-twist coupling with no initial warping due to curing stress. The applied axial strain is ϵ_o. The laminate geometry and material properties are given in Table 1.

The analysis involves two steps. The first step is to determine the potential delamination sites or critical interfaces. This is done by assuming delaminations at the free edges of one interface at a time and predicting the interlaminar peel stress in the neighborhood of the crack-tip. The closed-form expressions provided by the present analysis make it possible

TABLE 1—*Material properties of graphite/epoxy.*

Graphite/Epoxy
$E_{11} = 16.44$ Msi (113.31 GPa)
$E_{22} = E_{33} = 1.41$ Msi (9.73 GPa)
$G_{12} = G_{13} = 0.82$ Msi (5.66 GPa)
$G_{23} = 0.50$ Msi (3.40 GPa)
$v_{12} = v_{13} = 0.25$
$v_{23} = 0.42$
Ply thickness $H = 0.0055$ in. $(0.14 \times 10^{-3}$ m)
Semi-width $b = 68H$

to perform this successive analysis procedure at minimum computational effort. A positive peel stress represents a potential critical interface. This is demonstrated in Table 2, where the sign of the peel stress at the delamination tip associated with each interface for $\theta = 20°$ is provided. The critical interface lies between 20 and -20 layers, where the peel stress is tensile. The remaining interfaces show a compressive peel stress at the delamination tip which retards delamination growth in the graphite/epoxy material system.

The influence of the layup angle θ on the critical interface locations appears in Fig. 4. The nondimensionalized stress parameter $\sigma_z/(E_{11}\epsilon_o)$ value at the crack tip is plotted against θ. Laminates with $\theta = 20°, 30°, 40°,$ and $45°$ have tensile interfacial peel stress values, while the remaining layups exhibit compressive interfacial stress at all interfaces. The numbers shown in Fig. 4 above each bar denote the critical interfaces for the laminates with tensile peel stress. Interface 1 lies between the first and second plies starting from the bottom of the laminate. Potential critical interfaces are shown in Table 2 for the case of $\theta = 20°$. While laminates with $\theta = 20°$ show one critical interface, the remaining laminates show multiple potential delamination sites. These occur at symmetric locations since the peel stress through-the-thickness distribution is symmetric about the mid-plane of the laminate for $\theta = 30°, 40°,$ and $45°$. This can be explained as follows: since the family of laminates considered is antisymmetric, Interfaces 1 and 2 are the same as Interfaces 5 and 4, respectively, if the laminate is turned upside down.

The second step is to perform a complete analysis of the laminate with a delamination at the free edges of the critical interface. Figures 5 and 6 show the interlaminar peel stress, σ_z, and shear stress, τ_{xz}, distribution ahead of the delamination, respectively, for the case of $\theta = 20°$ with delaminations located at the laminate mid-plane. The initial delamination size is ten ply thicknesses. The shear stress, τ_{yz}, and induced bending curvature are found to be zero for this case. Since the induced bending curvature is zero, there are only three terms which contribute to the total strain energy release rate (G_T). The strain energy release rate terms associated with ϵ_o, C, and $C_{,a}$ are denoted G_ϵ, G_C, and $G_{C,a}$, respectively, and are shown in Fig. 7 along with the G_T distribution. The strain energy release rate has been nondimensionalized by $E_{11}H(\epsilon_o)^2$ in the figure. The largest contribution to G_T is associated with the extension behavior (G_ϵ). The contributions associated with the

TABLE 2—*Potential critical interfaces* (a = *10*H).

Laminate layup	20	-70_2	20	-20	70_2	-20
Interface	1	2	3	4	5	
Peel stress	Negative	Negative	Positive	Negative	Negative	

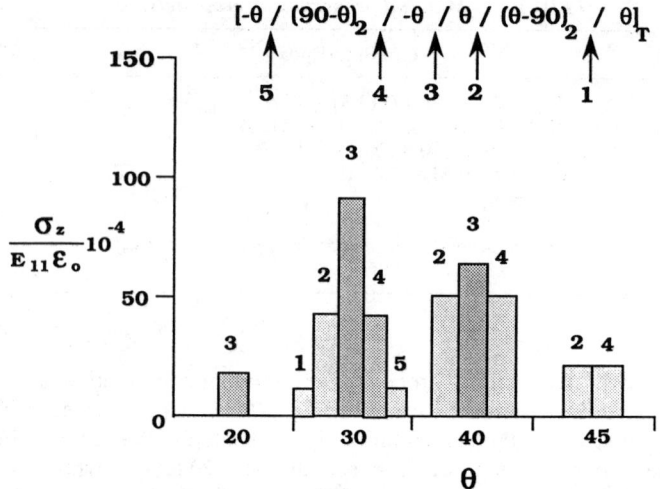

FIG. 4—*Interlaminar peel stress parameters at the critical interfaces.*

induced twist (G_C and $G_{C,a}$), however, result in a reduction of G_ϵ and are dependent on the delamination size.

The effect of the induced curvature on the total strain energy release rate is further investigated for $\theta = 30°$ and $45°$. Although the uncracked laminate does not exhibit

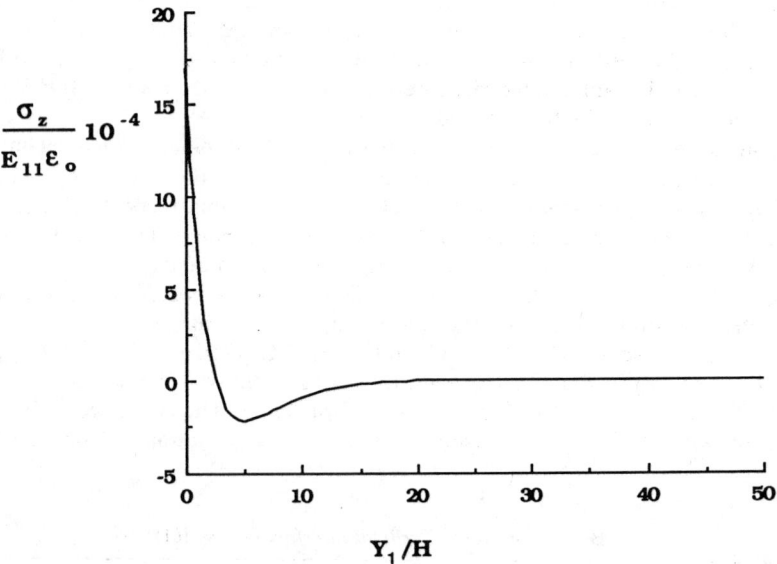

FIG. 5—*Interlaminar peel stress distribution ahead of a delamination* a $= 10H$ *at 20,-20 interface.*

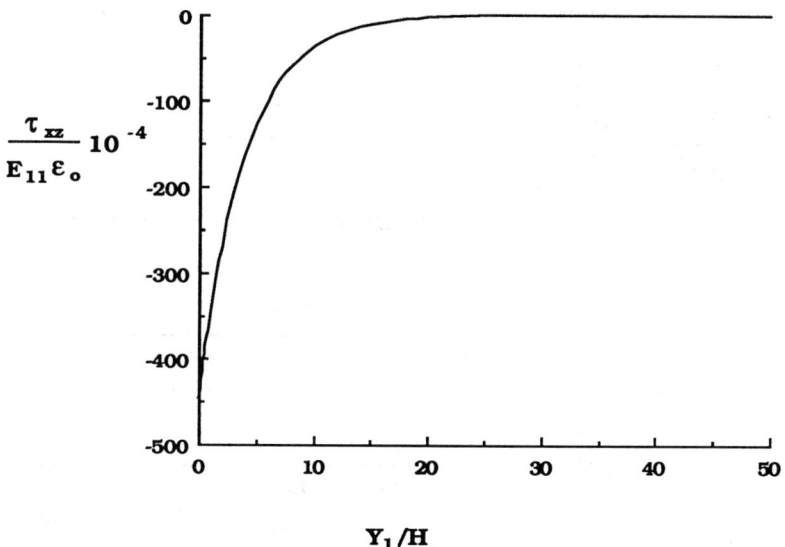

$$\frac{\tau_{xz}}{E_{11}\varepsilon_o} 10^{-4}$$

Y_1/H

FIG. 6—*Interlaminar shear stress distribution ahead of a delamination* a $=$ *10H at 20,-20 interface.*

extension bending coupling ($B_{11} = B_{12} = B_{22} = 0$), the presence of an off-midplane delamination induces bending curvature. Figures 8 through 10 show the variation of the strain energy release rate with delamination size at the critical interfaces for $\theta = 30°$ and $45°$. It is worth noting that for off-midplane delaminations, bending curvature is induced.

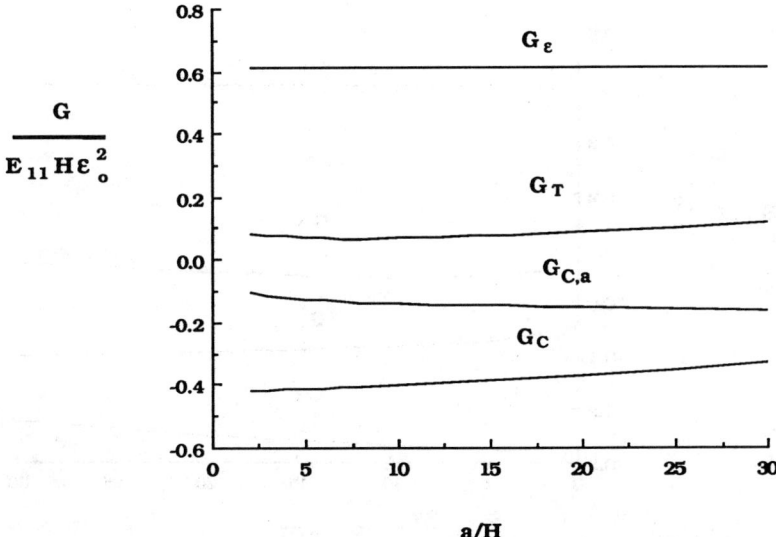

$$\frac{G}{E_{11}H\varepsilon_o^2}$$

a/H

FIG. 7—*Total strain energy release rate variation with delamination length for* $\theta = 20$ *at Interface 3.*

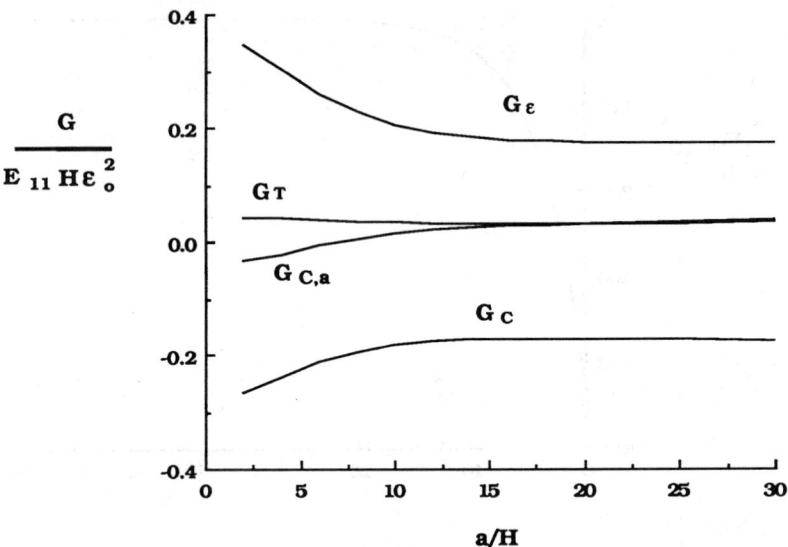

FIG. 8—*Total strain energy release rate variation with delamination length for* $\theta = 30$ *at Interface 2.*

However, its contributions to the total strain energy release rate is negligible compared to the extension and induced twist contributions for the cases studied. For a delamination size approaching zero, the strain energy release vanishes. A value of G_T corresponding to ten ply thicknesses was used to predict delamination onset.

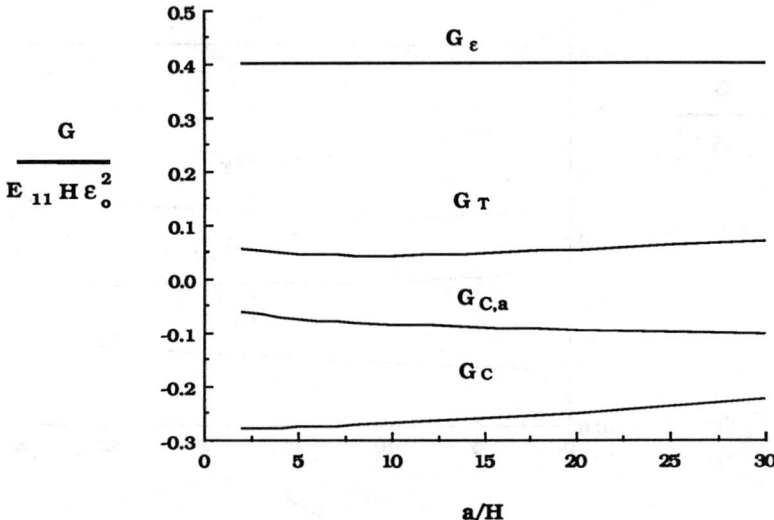

FIG. 9—*Total strain energy release rate variation with delamination length for* $\theta = 30$ *at Interface 3.*

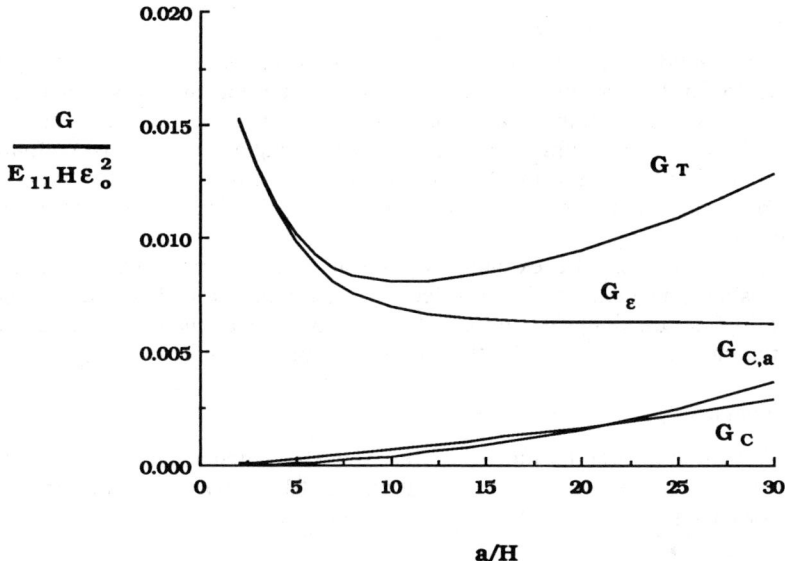

FIG. 10—*Total strain energy release rate variation with delamination length for* θ = 45 *at Interface 2.*

The total strain energy release rate parameter, $G_T/[E_{11}H(\epsilon_o)^2]$, associated with the critical interfaces for θ = 20°, 30°, 40°, and 45° appears in Fig. 11. The laminate with θ = 20° shows the highest total strain energy release rate parameter at the −20°/20° interface. Laminates with θ = 20° and 30° are most likely to delaminate at the mid-plane, while delaminations in laminates with θ = 40° and 45° are likely to occur along the second and fourth interfaces.

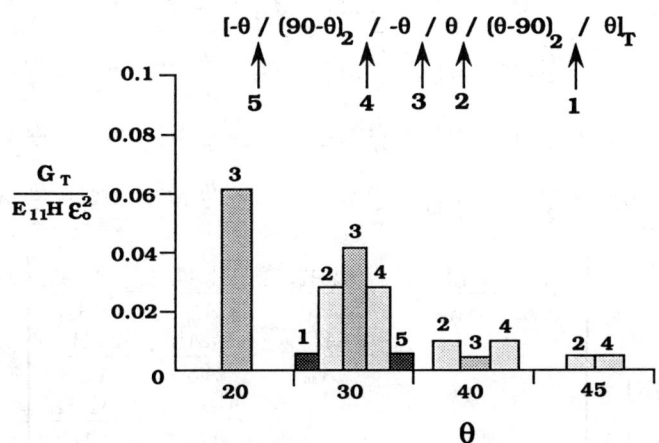

FIG. 11—*Total strain energy release rate associated with the critical interfaces.*

Conclusion

An analytical model for the analysis of free edge delamination in unsymmetrical laminates is developed. Closed-form expressions for the interlaminar stresses and total strain energy release rate are obtained. The analysis is applied to the prediction of the critical interfaces in a class of unsymmetrical laminates exhibiting extension-twist coupling based on the peel stress distribution through the thickness and the total strain energy release rate. It is found that the induced twist has a significant influence on the total strain energy release rate.

The present model can be extended to the analysis of symmetric laminates under bending, torsion, and their combined effect by modifying the end boundary conditions. Moreover, the fracture analysis of initially bent and twisted laminates can be performed by adding initial curvature and twist to the assumed displacement field.

Acknowledgments

This work was sponsored by the Army Research Office under the Center of Excellence for Rotary Wing Aircraft Technology Grant DAAL 03-88-C-0003. This support is gratefully acknowledged.

APPENDIX

Definition of Parameters

The eigenvalues and eigenvectors in Eq 16 are determined from the following eigen-equations

$$s^2[F]\{\phi\} = [K]\{\phi\} \tag{A-1}$$

where $[F]$ and $[K]$ are 4 by 4 matrices defined as

$$[F] = \begin{bmatrix} M_{61}^1 + \dfrac{h_1}{2}M_{31}^1 & M_{62}^1 + \dfrac{h_1}{2}M_{32}^1 & M_{63}^1 + \dfrac{h_1}{2}M_{33}^1 & M_{64}^1 + \dfrac{h_1}{2}M_{34}^1 \\[2mm] M_{61}^0 - \dfrac{h_0}{2}M_{31}^0 & M_{62}^0 - \dfrac{h_0}{2}M_{32}^0 & M_{63}^0 - \dfrac{h_0}{2}M_{33}^0 & M_{64}^0 - \dfrac{h_0}{2}M_{34}^0 \\[2mm] M_{51}^1 + \dfrac{h_1}{2}M_{21}^1 & M_{52}^1 + \dfrac{h_1}{2}M_{22}^1 & M_{53}^1 + \dfrac{h_1}{2}M_{23}^1 & M_{54}^1 + \dfrac{h_1}{2}M_{24}^1 \\[2mm] M_{51}^0 - \dfrac{h_0}{2}M_{21}^0 & M_{52}^0 - \dfrac{h_0}{2}M_{22}^0 & M_{53}^0 - \dfrac{h_0}{2}M_{23}^0 & M_{54}^0 - \dfrac{h_0}{2}M_{24}^0 \end{bmatrix} \tag{A-2}$$

$$[K] = \begin{bmatrix} A_{55}^1 - \dfrac{(A_{45}^1)^2}{A_{44}} & -\dfrac{A_{45}^1 A_{45}^0}{A_{44}} & \dfrac{A_{45}^1 A_{44}^0}{A_{44}} & -\dfrac{A_{45}^1 A_{44}^0}{A_{44}} \\[2mm] & A_{55}^0 - \dfrac{(A_{45}^0)^2}{A_{44}} & -\dfrac{A_{45}^0 A_{44}^1}{A_{44}} & \dfrac{A_{45}^0 A_{44}^1}{A_{44}} \\[2mm] & & \dfrac{A_{44}^1 A_{44}^0}{A_{44}} & -\dfrac{A_{44}^1 A_{44}^0}{A_{44}} \\[2mm] \text{SYM} & & & \dfrac{A_{44}^1 A_{44}^0}{A_{44}} \end{bmatrix} \tag{A-3}$$

where the superscripts 1 and 0 refer to the respective sublaminates. The coefficient A_{44} in Eq A-3 is defined as

$$A_{44} = A_{44}^1 + A_{44}^0 \tag{A-4}$$

The parameters M_{ij}^1 and M_{ij}^0 in $[F]$ are elements of matrices $[M^1]$ and $[M^0]$ given by

$$[M^1] = \begin{bmatrix} B_{16}^1 & 0 & B_{12}^1 & 0 \\ B_{26}^1 & 0 & B_{22}^1 & 0 \\ B_{66}^1 & 0 & B_{26}^1 & 0 \\ D_{16}^1 & 0 & D_{12}^1 & 0 \\ D_{26}^1 & 0 & D_{22}^1 & 0 \\ D_{66}^1 & 0 & D_{26}^1 & 0 \end{bmatrix} - \begin{bmatrix} A_{12}^1 & A_{16}^1 \\ A_{22}^1 & A_{26}^1 \\ A_{26}^1 & A_{66}^1 \\ B_{12}^1 & B_{16}^1 \\ B_{22}^1 & B_{26}^1 \\ B_{26}^1 & B_{66}^1 \end{bmatrix} [T] \tag{A-5}$$

$$[M^0] = \begin{bmatrix} -\dfrac{h_1}{2}A_{16}^0 & B_{16}^0 - \dfrac{h_0}{2}A_{16}^0 & -\dfrac{h_1}{2}A_{12}^0 & B_{12}^0 - \dfrac{h_0}{2}A_{12}^0 \\ -\dfrac{h_1}{2}A_{26}^0 & B_{26}^0 - \dfrac{h_0}{2}A_{26}^0 & -\dfrac{h_1}{2}A_{22}^0 & B_{22}^0 - \dfrac{h_0}{2}A_{22}^0 \\ -\dfrac{h_1}{2}A_{66}^0 & B_{66}^0 - \dfrac{h_0}{2}A_{66}^0 & -\dfrac{h_1}{2}A_{26}^0 & B_{26}^0 - \dfrac{h_0}{2}A_{26}^0 \\ -\dfrac{h_1}{2}B_{16}^0 & D_{16}^0 - \dfrac{h_0}{2}B_{16}^0 & -\dfrac{h_1}{2}B_{12}^0 & D_{12}^0 - \dfrac{h_0}{2}B_{12}^0 \\ -\dfrac{h_1}{2}B_{26}^0 & D_{26}^0 - \dfrac{h_0}{2}B_{26}^0 & -\dfrac{h_1}{2}B_{22}^0 & D_{22}^0 - \dfrac{h_0}{2}B_{22}^0 \\ -\dfrac{h_1}{2}B_{66}^0 & D_{66}^0 - \dfrac{h_0}{2}B_{66}^0 & -\dfrac{h_1}{2}B_{26}^0 & D_{26}^0 - \dfrac{h_0}{2}B_{26}^0 \end{bmatrix} - \begin{bmatrix} A_{12}^0 & A_{16}^0 \\ A_{22}^0 & A_{26}^0 \\ A_{26}^0 & A_{66}^0 \\ B_{12}^0 & B_{16}^0 \\ B_{22}^0 & B_{26}^0 \\ B_{26}^0 & B_{66}^0 \end{bmatrix} [T] \tag{A-6}$$

$$[L^1] = \begin{bmatrix} A_{11}^1 & A_{11}^1\delta_1 + B_{11}^1 & A_{16}^1\delta_1 + B_{16}^1 \\ A_{12}^1 & A_{12}^1\delta_1 + B_{12}^1 & A_{26}^1\delta_1 + B_{26}^1 \\ A_{16}^1 & A_{16}^1\delta_1 + B_{16}^1 & A_{66}^1\delta_1 + B_{66}^1 \\ B_{11}^1 & B_{11}^1\delta_1 + D_{11}^1 & B_{16}^1\delta_1 + D_{16}^1 \\ B_{12}^1 & B_{12}^1\delta_1 + D_{12}^1 & B_{26}^1\delta_1 + D_{26}^1 \\ B_{16}^1 & B_{16}^1\delta_1 + D_{16}^1 & B_{66}^1\delta_1 + D_{66}^1 \end{bmatrix} - \begin{bmatrix} A_{12}^1 & A_{16}^1 \\ A_{22}^1 & A_{26}^1 \\ A_{26}^1 & A_{66}^1 \\ B_{12}^1 & B_{16}^1 \\ B_{22}^1 & B_{26}^1 \\ B_{26}^1 & B_{66}^1 \end{bmatrix} [\tilde{T}] \tag{A-7}$$

$$[L^0] = \begin{bmatrix} A_{11}^0 & A_{11}^0\delta_0 + B_{11}^0 & A_{16}^0\delta_0 + B_{16}^0 \\ A_{12}^0 & A_{12}^0\delta_0 + B_{12}^0 & A_{26}^0\delta_0 + B_{26}^0 \\ A_{16}^0 & A_{16}^0\delta_0 + B_{16}^0 & A_{66}^0\delta_0 + B_{66}^0 \\ B_{11}^0 & B_{11}^0\delta_0 + D_{11}^0 & B_{16}^0\delta_0 + D_{16}^0 \\ B_{12}^0 & B_{12}^0\delta_0 + D_{12}^0 & B_{26}^0\delta_0 + D_{26}^0 \\ B_{16}^0 & B_{16}^0\delta_0 + D_{16}^0 & B_{66}^0\delta_0 + D_{66}^0 \end{bmatrix} - \begin{bmatrix} A_{12}^0 & A_{16}^0 \\ A_{22}^0 & A_{26}^0 \\ A_{26}^0 & A_{66}^0 \\ B_{12}^0 & B_{16}^0 \\ B_{22}^0 & B_{26}^0 \\ B_{26}^0 & B_{66}^0 \end{bmatrix} [\tilde{T}] \tag{A-8}$$

$$[\tilde{T}] = \begin{bmatrix} A_{22} & A_{26} \\ A_{26} & A_{66} \end{bmatrix}^{-1} \begin{bmatrix} A_{12} & B_{12}^1 + B_{12}^0 + A_{12}^1\delta_1 + A_{12}^0\delta_0 & B_{26}^1 + B_{26}^0 + A_{26}^1\delta_1 + A_{26}^0\delta_0 \\ A_{16} & B_{16}^1 + B_{16}^0 + A_{16}^1\delta_1 + A_{16}^0\delta_0 & B_{66}^1 + B_{66}^0 + A_{66}^1\delta_1 + A_{66}^0\delta_0 \end{bmatrix} \tag{A-9}$$

$$[T] \begin{bmatrix} A_{22} & A_{26} \\ A_{26} & A_{66} \end{bmatrix}^{-1} \begin{bmatrix} B_{26}^1 - \dfrac{h_1}{2}A_{26}^0 & B_{26}^0 - \dfrac{h_0}{2}A_{26}^0 & B_{22}^1 - \dfrac{h_1}{2}A_{22}^0 & B_{22}^0 - \dfrac{h_0}{2}A_{22}^0 \\ B_{66}^1 - \dfrac{h_1}{2}A_{66}^0 & B_{66}^0 - \dfrac{h_0}{2}A_{66}^0 & B_{26}^1 - \dfrac{h_1}{2}A_{26}^0 & B_{26}^0 - \dfrac{h_0}{2}A_{26}^0 \end{bmatrix} \tag{A-10}$$

The elements of matrices $[L]$ and $[M]$ in Eq 18 are given by

$$[L] = \begin{bmatrix} L_{21}^1 & L_{22}^1 & L_{23}^1 \\ L_{31}^1 & L_{32}^1 & L_{32}^1 \\ L_{51}^1 & L_{52}^1 & L_{53}^1 \\ L_{51}^0 & L_{52}^0 & L_{53}^0 \end{bmatrix} \tag{A-11}$$

$$[M] = \begin{bmatrix} M_{21}^1 & M_{22}^1 & M_{23}^1 & M_{24}^1 \\ M_{31}^1 & M_{32}^1 & M_{33}^1 & M_{34}^1 \\ M_{51}^1 & M_{52}^1 & M_{53}^1 & M_{54}^1 \\ M_{51}^0 & M_{52}^0 & M_{53}^0 & M_{54}^0 \end{bmatrix} \tag{A-12}$$

The matrix $[\overline{K}]$ is a diagonal matrix expressed as

$$[\overline{K}] = \langle [\Phi]^T [K][\Phi] \rangle^{-1} \tag{A-13}$$

the diagonal element $\overline{K}_{ii}$ corresponding to a zero eigenvalue is zero.

The vector $\{q\}$ in Eq 20 is expressed as

$$\{q\} = \left(A_{55}^1 - \frac{A_{45}}{A_{44}} A_{45}^1 \quad A_{55}^0 - \frac{A_{45}}{A_{44}} A_{45}^0 \quad A_{45}^1 - \frac{A_{45}}{A_{44}} A_{44}^1 \quad A_{45}^0 - \frac{A_{45}}{A_{44}} A_{44}^0 \right)^T \tag{A-14}$$

where A_{45} is defined in the same way as Eq A-4.

Vector $\{\overline{M}^1\}$ and matrix $[\overline{L}^1]$ in Eq 20 are defined as

$$\{\overline{M}^1\} = \begin{Bmatrix} B_{16}^1 \\ D_{16}^1 \\ D_{66}^1 \end{Bmatrix} - \begin{bmatrix} A_{12}^1 & A_{16}^1 & B_{12}^1 \\ B_{12}^1 & B_{16}^1 & D_{12}^1 \\ B_{26}^1 & B_{66}^1 & D_{26}^1 \end{bmatrix} \{R^1\} \tag{A-15}$$

$$[\overline{L}^1] = \begin{bmatrix} A_{11}^1 & B_{11}^1 + A_{11}^1\delta_1 & B_{16}^1 + A_{16}^1\delta_1 \\ B_{11}^1 & D_{11}^1 + B_{11}^1\delta_1 & D_{16}^1 + B_{16}^1\delta_1 \\ B_{16}^1 & D_{16}^1 + B_{16}^1\delta_1 & D_{66}^1 + B_{66}^1\delta_1 \end{bmatrix} - \begin{bmatrix} A_{12}^1 & A_{16}^1 & B_{12}^1 \\ B_{12}^1 & B_{16}^1 & D_{12}^1 \\ B_{26}^1 & B_{66}^1 & D_{26}^1 \end{bmatrix} [\tilde{R}^1] \tag{A-16}$$

where

$$[\tilde{R}^1] = \begin{bmatrix} A_{22}^1 & A_{26}^1 & B_{22}^1 \\ A_{26}^1 & A_{66}^1 & B_{66}^1 \\ B_{22}^1 & B_{66}^1 & D_{22}^1 \end{bmatrix}^{-1} \begin{bmatrix} A_{12}^1 & B_{12}^1 + A_{12}^1\delta_1 & B_{26}^1 + A_{26}^1\delta_1 \\ A_{16}^1 & B_{16}^1 + A_{16}^1\delta_1 & B_{66}^1 + A_{66}^1\delta_1 \\ B_{12}^1 & D_{12}^1 + B_{12}^1\delta_1 & D_{26}^1 + B_{26}^1\delta_1 \end{bmatrix} \tag{A-17}$$

$$\{R^1\} = \begin{bmatrix} A_{22}^1 & A_{26}^1 & B_{22}^1 \\ A_{26}^1 & A_{66}^1 & B_{66}^1 \\ B_{22}^1 & B_{66}^1 & D_{22}^1 \end{bmatrix}^{-1} \begin{Bmatrix} B_{26}^1 \\ B_{66}^1 \\ D_{26}^1 \end{Bmatrix} \tag{A-18}$$

In Eqs A-15 through A-18, replace δ_1 with δ_0, and superscript 1 with 0 to obtain the respective matrices and vectors with superscript 0.

Matrix $[H]$ in Eq 21 is given by

$$[H] = \begin{bmatrix} H_{11} & H_{12} & H_{13} \\ H_{21} & H_{22} & H_{23} \\ 0 & 0 & 1 \end{bmatrix} \tag{A-19a}$$

where

$$H_{11} = (L_{61}^1 - \overline{L}_{31}^1 - \Psi_{61})/(s_c \overline{M}_3^1)$$

$$H_{12} = (L_{62}^1 - \overline{L}_{32}^1 - \Psi_{62})/(s_c \overline{M}_3^1)$$

$$H_{13} = (L_{63}^1 - \overline{L}_{33}^1 - \Psi_{63} - \overline{M}_3^1 + M_{6i}^1 \overline{\beta}_i)/(s_c \overline{M}_3^1)$$

$$H_{21} = H_{11} \cot(s_c a) + \frac{\overline{L}_{31}^1}{s_c \overline{M}_3^1} \, csch(s_c a) \tag{A-19b}$$

$$H_{22} = H_{12} \cot(s_c a) + \frac{\overline{L}_{32}^1}{s_c \overline{M}_3^1} \, csch(s_c a)$$

$$H_{23} = H_{13} \cot(s_c a) + \frac{1}{s_c} \left(\frac{\overline{L}_{33}^1}{\overline{M}_3^1} + 1 \right) csch(s_c a)$$

$$[\Psi] = [M^1][M]^{-1} \langle [L] + [M]\{\overline{\beta}\}(0,0,1) \rangle \tag{A-19c}$$

The elements of matrix $[I]$ in Eq 24 can be derived from $[H]$ by substituting superscript 0 for 1 and s_d for s_c. Matrix $[\alpha]$ in Eq 37 is expressed as

$$[\alpha] = \begin{bmatrix} F_{11}^1 + F_{11}^0 + \overline{F}_{11}^1 + \overline{F}_{11}^0 & F_{12}^1 + F_{12}^0 + \overline{F}_{12}^1 + \overline{F}_{12}^0 & F_{13}^1 + F_{13}^0 + \overline{F}_{13}^1 + \overline{F}_{13}^0 \\ F_{41}^1 + F_{41}^0 + \overline{F}_{21}^1 + \overline{F}_{21}^0 & F_{42}^1 + F_{42}^0 + \overline{F}_{22}^1 + \overline{F}_{22}^0 & F_{43}^1 + F_{43}^0 + \overline{F}_{23}^1 + \overline{F}_{23}^0 \\ F_{61}^1 + F_{61}^0 + \overline{F}_{31}^1 + \overline{F}_{31}^0 & F_{62}^1 + F_{62}^0 + \overline{F}_{32}^1 + \overline{F}_{32}^0 & F_{63}^1 + F_{63}^0 + \overline{F}_{33}^1 + \overline{F}_{33}^0 \end{bmatrix} +$$

$$\begin{bmatrix} 0 & 0 & 0 \\ \delta_1(F_{11}^1 + \overline{F}_{11}^1) + \delta_0(F_{11}^0 + \overline{F}_{11}^0) & \delta_1(F_{12}^1 + \overline{F}_{12}^1) + \delta_0(F_{12}^0 + \overline{F}_{12}^0) & \delta_1(F_{13}^1 + \overline{F}_{13}^1) + \delta_0(F_{13}^0 + \overline{F}_{13}^0) \\ \delta_1 F_{31}^1 + \delta_0 F_{31_1}^0 & \delta_1 F_{32}^1 + \delta_0 F_{32}^0 & \delta_1 F_{33}^1 + \delta_0 F_{33}^0 \end{bmatrix} \tag{A-20}$$

The parameters in Eq A-20 are elements from the following matrices,

$$[F^1] = \langle [L^1] + [M^1]\{\overline{\beta}\}(0,0,1) \rangle (b - a) - [M^1][\Phi][\zeta][\overline{G}] \tag{A-21a}$$

where $[\zeta]$ is a diagonal matrix whose elements are defined as

$$\zeta_{ii} = 1 \qquad \text{for} \quad s_i \neq 0 \tag{A-21b}$$

$$\zeta_{ii} = a - b \qquad \text{for} \quad s_i = 0 \tag{A-21c}$$

and

$$[\overline{F}^1] = [\overline{L}^1]a + \{\overline{M}^1\}(\xi(a))[H] \tag{A-22a}$$

where

$$(\xi(a)) = (\sinh (s_c a) \quad 1 - \cosh (s_c a) \quad a) \tag{A-22b}$$

The elements of $[F^0]$ and $[\overline{F}^0]$ can be derived from Eqs A-21 through A-22 by substituting superscript 0 for 1, s_d for s_c, and I_{ij} for H_{ij}.

References

[1] Armanios, E. A. and Rehfield, L. W., "Interlaminar Fracture Analysis of Composite Laminates Under Bending and Combined Bending and Extensions," *Composite Materials: Testing and Design (Eighth Conference), ASTM STP 972*, J. D. Whitcomb, Ed., American Society for Testing and Materials, Philadelphia, 1988, pp. 81–94.

[2] Winckler, S. I., "Hygrothermally Curvature Stable Laminates with Tension-Torsion Coupling," *Journal of the American Helicopter Society*, Vol. 31, No. 7, July 1985, pp. 56–58.

[3] Lekhnitskii, S. G., *Theory of Elasticity of an Anisotropic Elastic Body*, Holden-Day, San Francisco, 1963, pp. 107–108.

[4] Vinson, J. R. and Sierakowski, R. L., *The Behavior of Structures Composed of Composite Materials*, Martinus Nijhoff Publishers, 1986, p. 47.

[5] Armanios, E. A. and Badir, A. M., "Hygrothermal Influence on Mode I Edge Delamination in Composites," *Composites Structures*, Vol. 15, 1990, pp. 323–342.

Micromechanics and Interfaces

L. J. Hart-Smith[1]

Some Observations on the Analysis of In-Plane Matrix Failures in Fibrous Composite Laminates

REFERENCE: Hart-Smith, L. J., "**Some Observations on the Analysis of In-Plane Matrix Failures in Fibrous Composite Laminates,**" *Composite Materials: Fatigue and Fracture, Fourth Volume, ASTM STP 1156,* W. W. Stinchcomb and N. E. Ashbaugh, Eds., American Society for Testing and Materials, Philadelphia, 1993, pp. 363–380.

ABSTRACT: The prediction of matrix failures of fibrous composite laminates under in-plane loading is discussed. It is shown that a complete micromechanical analysis is needed to analyze such failures, even when attention is focused on significant structural damage as opposed to discontinuous microcracking. It is shown that matrix-dominated composite structures could not be certified even if they could be analyzed reliably. All that is needed is an ability to distinguish between those laminates which fail in the matrix and those which fail in the fibers since structurally significant matrix failures preceding the breaking of the fibers automatically signify inferior laminates. This permits the reliable use of failure theories very much simpler than conventional micromechanics to predict fiber-dominated strengths with no loss of accuracy. There is actually no need to calculate matrix-dominated strengths. Indeed, since fewer material properties are needed and are easier to measure than the much larger number needed for a full micromechanical assessment, physically realistic analyses confining attention to failures of only the fibers are actually more accurate than any micromechanical prediction of laminate strength under in-plane loads.

KEY WORDS: composite materials, fibrous composites, matrix failures, certification

During the past few years, the author has published several papers on predicting the strength of fibrous composite laminates under in-plane loads by progressively more scientific orthotropic adaptations of the classical maximum-shear-stress (Tresca) yield criterion for ductile isotropic materials. The earliest derivations covered the application of mental-arithmetic techniques to tape and fabric laminates, respectively [1,2]. These were followed by the first computerized version in which the BLACKART computer code prepared by a colleague was shown to be capable of demonstrating excellent agreement with experiment provided that the transverse properties of the lamina were characterized by someone with a thorough understanding of the mechanics of composites [3]. Otherwise, this program was shown to be as unreliable as any of the other computer codes.

Subsequently, the same failure criterion was expressed in the strain plane rather than in the stress plane to yield a formulation that would consistently generate reliable estimates of laminate strength for novices and expert alike [4–7].

However, one caveat still remains; the methods are meaningful only for fiber-dominated failures of well-designed cross-plied laminates. That may not seem to be a severe restriction since no competent designer would knowingly waste strong expensive fibers in a laminate so poorly designed that the matrix would fail first. Unfortunately, none of the

---

[1]Douglas Aircraft Co., McDonnell Douglas Corp., Long Beach, CA 90846.

innumerable pseudoscientific abstract mathematical failure "theories" for composite materials published during the last quarter century or so contain any such limitations on their applicability, so current audiences seem uncomfortable with such a constraint. Moreover, the new theories do not, of themselves, differentiate between good and bad laminates.

Accordingly, it is appropriate to address the issue of why this author consciously excluded matrix failures from his failure theory and to explain how the reader can use other means to distinguish between the good laminate patterns, which fail in the fibers, and the bad ones, which fail prematurely in the matrix.

It is appropriate at this point to clearly differentiate between structurally insignificant microcracks in a resin matrix and matrix failures. Microcracks are disconnected and small enough not to cause any appreciable reduction in the ability of the matrix to support the load-carrying fibers. On the other hand, matrix failures, being large and widespread, leave the fibers unable to resist the applied loads. Microcracks are known to occur within high-temperature-cured laminates during the cooldown after cure and clearly precede fiber-dominated failures of the laminates. However, they build up exponentially as mechanical loads are increased or as the result of fatigue loading. There is usually no strain level at which there is a quantum jump in the density of such cracks (except for all 90° test coupons) so, in a well-designed fiber-dominated laminate, microcracks cannot be construed as triggering fiber failures. The best of carbon-epoxy laminates, for example, suffer only slight matrix cracking before the fibers fail. However, even the best-designed fiber-glass-polyester laminates will usually suffer massive cracking and crazing in the matrix long before the high-strain fibers fail. In this context, the matrix may reasonably be considered to have "failed" first. The words "matrix failure" in this paper refer exclusively to widespread damage resulting in the loss of ability to support the fibers and specifically exclude structurally insignificant disconnected microcracks.

It should be noted that a comprehensive micromechanics analysis must be conducted to calculate accurate matrix-dominated strengths. While some of these analyses have been performed for both structurally insignificant microcracking and major damage, they are not often used in industry because they are considered to be overly complicated—at least by those who have failed to grasp that no meaningful simpler alternatives are available. This assertion of the need for micromechanical analyses for matrix-dominated failures is so contrary to current practice that an explanation is warranted. For fiber-dominated failures in advanced composites such as carbon-epoxy, the longitudinal stress in the fiber is almost directly proportional to the corresponding stress in the lamina. It is *only* because of this that the failures can be predicted by a single analysis formulated at the lamina level rather than by separate analyses in terms of fiber and matrix stresses. This relationship, shown in Fig. 1, exists whenever strong, stiff fibers are embedded in soft matrices. However, because resin matrices are traditionally soft and have very high coefficients of thermal expansion, relatively intense stresses develop in the matrix within and between the plies during cooldown after typical high-temperature cures. The residual thermal stresses are often an order of magnitude greater than the additional stresses that can be withstood under purely mechanical loads. Consequently, the matrix stresses are very definitely *not* proportional to applied mechanical loads, as shown in Fig. 1. Thus, while fiber-dominated failures might be characterized at the lamina level, matrix-dominated failures cannot be: analyses accounting for *discrete* fiber and matrix constituents in the composite are then necessary.

Some researchers erroneously claim that, since the matrix stresses are linearly related to the applied loads, the lack of proportionality can be compensated for by eliminating all the residual stress terms from the theory and reducing the theoretical matrix stress at failure to a measured transverse-tension strength of a unidirectional lamina. This approach could be

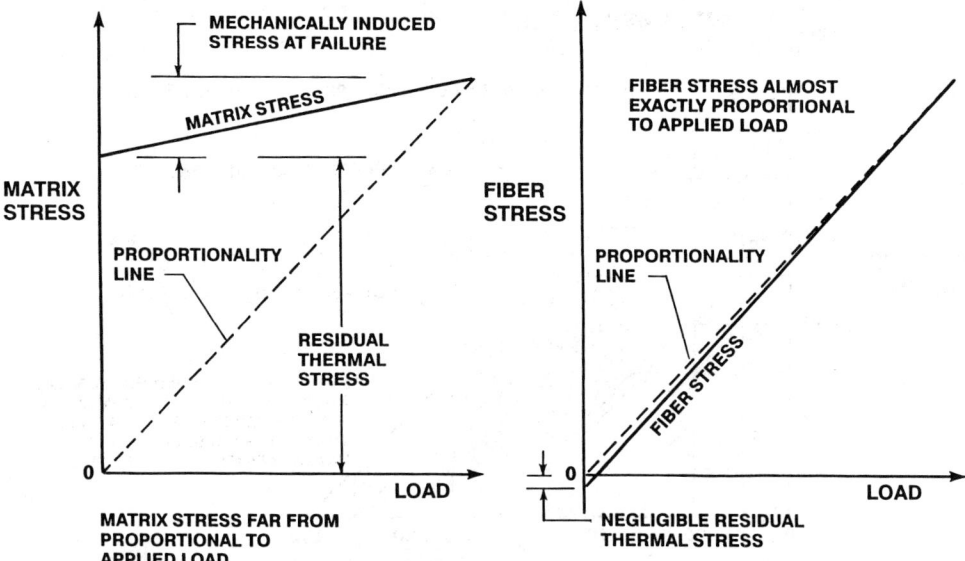

FIG. 1—*Improper characterization of matrix stresses in homogenized models of fiber-polymer composites even when fiber stresses are adequately approximated.*

made to work if one were to acknowledge that the intralaminar residual stresses vary with the fiber pattern, leaving a different apparent matrix strength for each fiber pattern. This necessary refinement is customarily disregarded by those who consider micromechanics too difficult for day-to-day use. Moreover, this approach could never characterize matrix failures in woven fabric laminates because the longitudinal and "transverse" measured strengths are both fiber dominated. One would therefore predict grossly different matrix-dominated strengths for one laminate made from unidirectional preimpregnated tape and another made from resin-transfer molding of a dry preform containing exactly the same fibers and matrix as the first laminate.

Although matrix-dominated failures have been predicted by theories condemned by the author without any of the refinements he believes are needed, the computed answers are not necessarily correct. Since the theories used contain many approximations, their validity can be assured only by deriving the same answer from a more complex analysis without the approximations. This has not been done, so the matrix failures predicted by tensor-polynomial interaction methods have not been validated and should be rejected since they are so clearly contradicted by a physical assessment of the situation.

Rather than complicate a simple theory containing all the features necessary to predict the strength of well-designed fiber-dominated laminates by extending it to cover inferior matrix-dominated laminates as well, the author advocates separate analyses to identify the fiber patterns that should not be used. To this end, he has employed micromechanics to establish limits on clustering parallel plies of fibers [8]. A small fraction of the resin was analytically extracted from each ply and concentrated as a very thin interlayer between the fiber-dominated remainders of the plies. This avoids the so-called free-edge singularity. The objective of this analysis was to identify the circumstances under which edge delaminations would precede failure of the fibers. The conclusions reached for the case of typical carbon-epoxy laminates confirmed the empirical findings of others who had studied cases in which composite panels delaminated before they could be removed from the

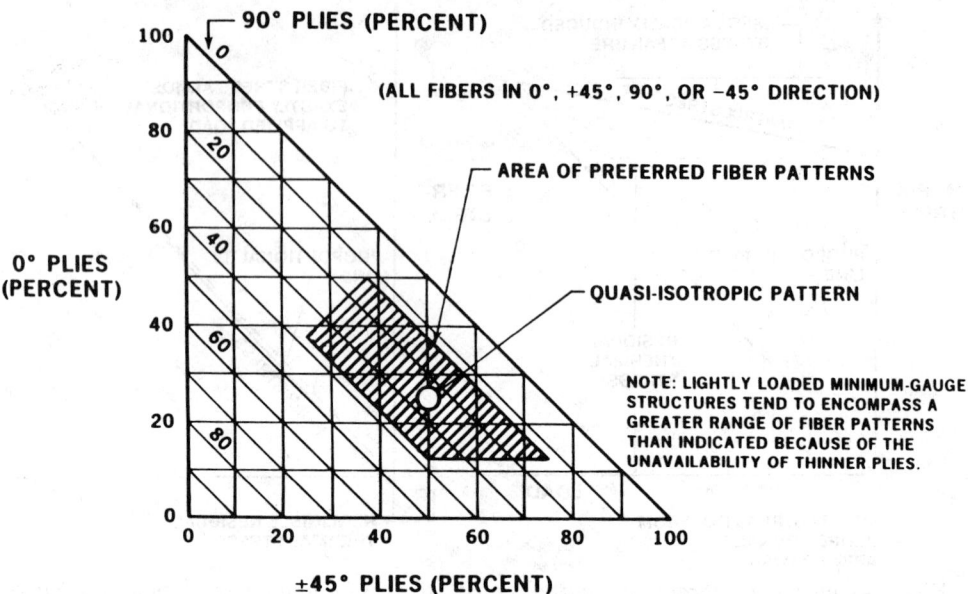

FIG. 2—*Selection of layup pattern for fibrous composite laminate.*

autoclave. Either source of information can be used to restrict the fiber pattern selection, thereby excluding the possibility of matrix failures without having to actually calculate the loads at which premature failures would otherwise occur.

Figure 2 shows recommended fiber patterns for typical carbon-epoxy laminates, calling for thorough interspersion of the plies in the various orientations and placing both upper and lower limits on the fraction of fibers in any one direction. This particular diagram was prepared for bolted structures. More flexibility is available for unrepairable or throw-away bonded or co-cured structures which can be designed to higher strain levels.

The concept of using refined analyses to validate the domain of applicability of a simpler theory should be used far more extensively than it is. It was used to establish that the simple ''ten-percent rule''[2] was never unconservative with respect to the predictions of more accurate—and more complex—analyses, and that nor was it ever unduly conservative [1]. The refined analysis in Ref 1 was never intended to be used in its own right; it was derived for the express purpose of generating confidence in the use of the much simpler theory.

Micromechanics should likewise never be the preferred method to calculate the strength of fiber-dominated laminates. Uncertainty about the accuracy of the much more numerous material input properties would always render such an analysis less accurate than a generalization of the Tresca criterion provided that the fibers really did fail before the matrix. Even if the analysis were analytically and physically perfect, low allowable material properties because of scatter would lead one to avoid such predictions. Apart from the

---

[2]The ten-percent rule is a simple rule-of-mixtures procedure for predicting the strengths and stiffnesses of cross-plied laminates for preliminary design purposes. Each 90° or ±45° ply is characterized as having 10% of the strength and stiffness of an equivalent 0° ply for a 0° load. The reference lamina strengths are measured in the specific environment of interest. Fiber-dominated in-plane shear strengths and stiffnesses are established similarly, using the ±45° plies as the reference.

usual elastic constants for the unidirectional lamina, the only reference strengths needed for the new analysis methods are those for uniaxial tension and compression [4–7]. The more complicated micromechanics theory would be needed only to ensure that the simpler theory was not misapplied. The transverse strengths measured for other failure "theories" belong to a totally unrelated failure criterion.

If, however, the matrix really did fail first for a typical fiber-polymer composite, it would be impossible to certify any structure made from such a laminate. So, again, there is no need to calculate the actual (inferior) strength of matrix-dominated laminates except to ensure that such fiber patterns are not used for structural applications.

Critical Experiment to Quantify Effects of Matrix Microcracking

The awareness of matrix cracking in composite laminates has been raised by O'Brien [9]. However, this work has emphasized the use of nonstructural test coupons in which excessive numbers of parallel tape layers were clustered together to ensure that extensive matrix cracking and delaminations would precede any fiber failures (see Fig. 3). Otherwise, the matrix failures would not occur at all, or they would be so interwoven with fiber failures that interpretation of the data would be questionable at best. O'Brien's own tests have shown that the incidence of microcracking in well-interspersed structural laminates is far less than with his specialized coupons. However, the question remains as to whether or not this whole effect may reasonably be ignored for what the author would call well-designed laminates.

The following proposal is made to resolve this issue. If a large 0°/90° laminate is fatigued gently in each of the fiber directions in turn, it should be possible to establish a saturation density of microcracks related to the applied stress levels. These would be expected to have little, if any, effect on subsequently measured 0° or 90° tensile strengths. There might be a measurable loss of compressive strength in each of those directions, but compression testing is now known to be so variable that small differences between conditioned and virgin coupons would not offer positive proof of a structural change.

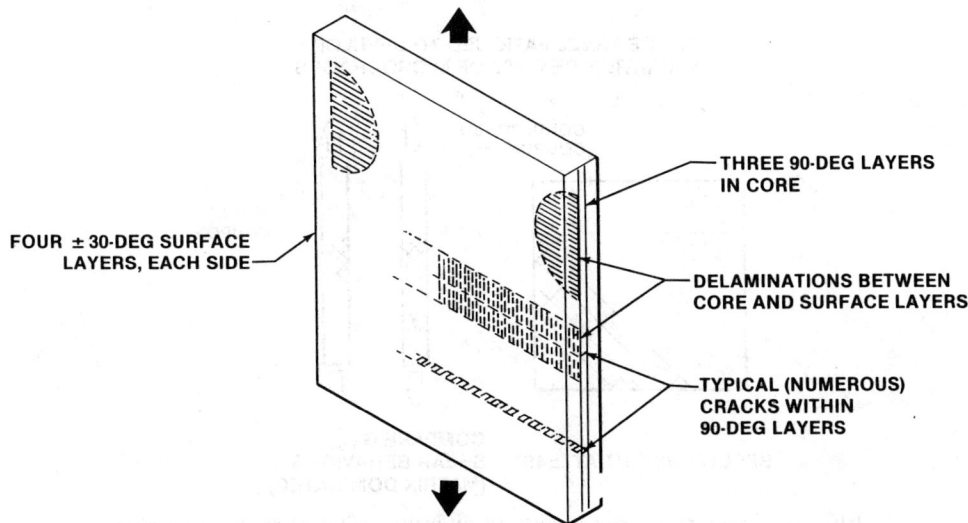

FIG. 3—*Delaminations due to clustering parallel plies.*

However, tension tests on ±45° laminates cut from the fatigue-conditioned panels above would definitely be expected to show an appreciable reduction in modulus if the saturation crack density were high enough. So, the experiment described in Fig. 4 should be capable of determining the level of microcrack density needed to cause varying degrees of loss of in-plane ($G_{0/90}$) shear stiffness, which is a truly matrix-dominated property of the composite laminate. This method of generating the microcracks would, in turn, establish the operating stress levels in structural laminates needed to induce such densities of microcracks. The effects on all fiber-dominated properties would be appreciably less. So, if the comparative tests on conditioned and virgin ±45° tensile coupons failed to show an appreciable loss of stiffness, except for unrealistically high fatigue stresses to induce the microcracks, one could assume with confidence that microcracking was not a structurally significant problem for well-interspersed cross-plied laminates made from conventional fiber-polymer composites.

The opposite finding from such tests would create a need for further testing to quantify the loss in fiber-dominated properties, particularly in compression. But existing tests on structurally impractical matrix-dominated laminates, such as in Fig. 3, in no way imply that microcracking is necessarily of serious concern for structural laminates merely because customary composite failure theories fail to account properly for stacking-sequence and ply-thickness effects.

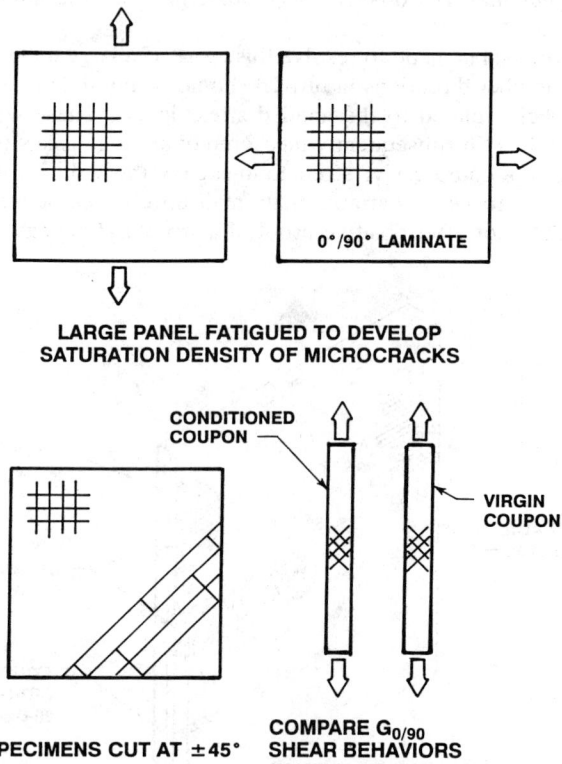

FIG. 4—Suggested experiments to quantify effect of matrix cracking.

Need for Micromechanical Analyses of Matrix Failures

In characterizing the strength of fiber-polymer composites, there must be separate failure criteria for the fiber, the matrix, and possibly also the interface. Further, if one of those ingredients can fail in more than one mode, additional failure criteria are needed as truncations of the original ones. The notion that one single mathematical formula can be made to apply to an entire lamina, as if it were truly a homogeneous material, is simply a physically unrealistic oversimplification. Attempts to do so, many of which are recorded in Ref 10, have led to innumerable abstract mathematical failure "theories" for composites.

Taking as a typical example the quadratic criterion in stress space as defined by Tsai on p. 11-5 of Ref 11,

$$F_{ij}\sigma_{ij} + F_i\sigma_i = 1, \quad (i,j = 1,2,3,4,5,6)$$

where σ's are the stresses and the F's are the strength parameters.

$$F_{xx} = 1/XX', F_{yy} = 1/YY', F_{SS} = 1/S^2$$

$$F_x = 1/X - 1/X', F_y = 1/Y - 1/Y', F_S = 0$$

and the interaction term is given as

$$F_{xy} = F_{xy}^*[F_{xx}F_{yy}]^{1/2}.$$

Here, X and X' are the longitudinal tensile and compressive strengths of the lamina, while Y and Y' are the corresponding transverse strengths. S is the lamina in-plane shear strength, and F_{xy} or its nondimensionalized term F_{xy}^* can be determined only by biaxial tests. (It is therefore usually either set equal to $-1/2$ in accordance with the generalized von Mises' criterion or set equal to zero.)

Thus, within this single failure criterion, one term X is governed by the tensile strength of the fiber. Another, X', may be governed by compressive instability of that fiber. Yet another term, Y, is clearly matrix dominated, although the transverse tensile strength is diminished considerably by the presence of the fibers. Y' can be established at almost any number up to the limiting hydraulic capacity of the test machine by varying the size or shape of the composite block being compressed. S is clearly matrix dominated, and the remaining term, F_{xy}^*, is undefined. Yet the original criterion loses all scientific validity the moment that even one of the reference strengths X, X', Y, Y', or S does not refer to the same composite constituent as the others or whenever different states of combined stress cause failure by different modes.

In the case of the author's failure model, Refs 4 through 7, the dominant criterion is failing the fibers by shear at 45° to the fiber axes, but there is often a need to apply a further limit to the compressive strength estimate alone so as not to exceed some experimentally established stability limit, particularly for the newer (small-diameter) high-strain carbon fibers. The matrix-failure envelope is assumed to lie entirely *outside* the corresponding envelope for the fibers, implying a restriction to strong, stiff fibers embedded in a soft matrix.

Unlike the preceding quadratic failure criterion which requires at least five experimentally determined reference strengths before it can be implemented, the author's generalization of the Tresca condition can be completely defined by knowing only one reference strength. The remaining points on the failure envelope can all be established on the basis of the first one—because they all refer to the failure of a homogeneous material, the fiber, under a single mechanism, shear. This is illustrated in Fig. 5. The step-by-step construction of the envelope, starting from the measured uniaxial tension strain-to-failure, is explained in Fig. 6. Actually, a little thought will show that the envelope can be completed from any single starting point.

Figure 7 shows how an additional failure mode, in compression, can be superimposed on the first without altering any prediction for the first failure mode throughout the states of stress for which the second mode does not govern. In contrast with this, any change in even one reference strength for the quadratic failure criterion and similar criteria will relocate every point around the failure envelope with the exception of the other reference strengths. On what basis should one be asked to believe that a reduction in unidirectional compression strength due to microbuckling, for example, should be expected to increase the biaxial tension strength, as shown in Fig. 8. Or why should the decrease in transverse-tension strength shown in Fig. 9 enhance the biaxial compression strength of the laminate?

Any scientifically valid application of a formula like the preceding quadratic expression would repeat it as many times as necessary to confine it to one mode of failure at a time within only one constituent of the composite laminate. That would obviously dramatically increase the number of reference strengths needed and would, in effect, be tantamount to using micromechanics rather than the neoclassical macromechanics. While the author's

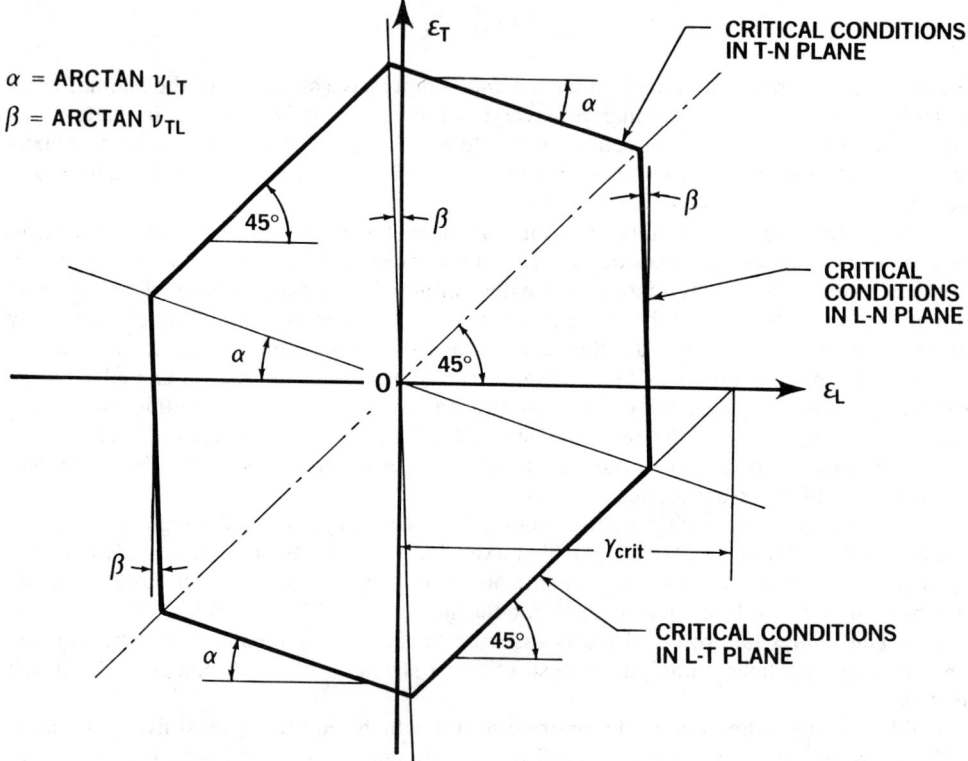

FIG. 5—*Shear failure envelope for orthotropic materials having equal tensile and compressive strengths.*

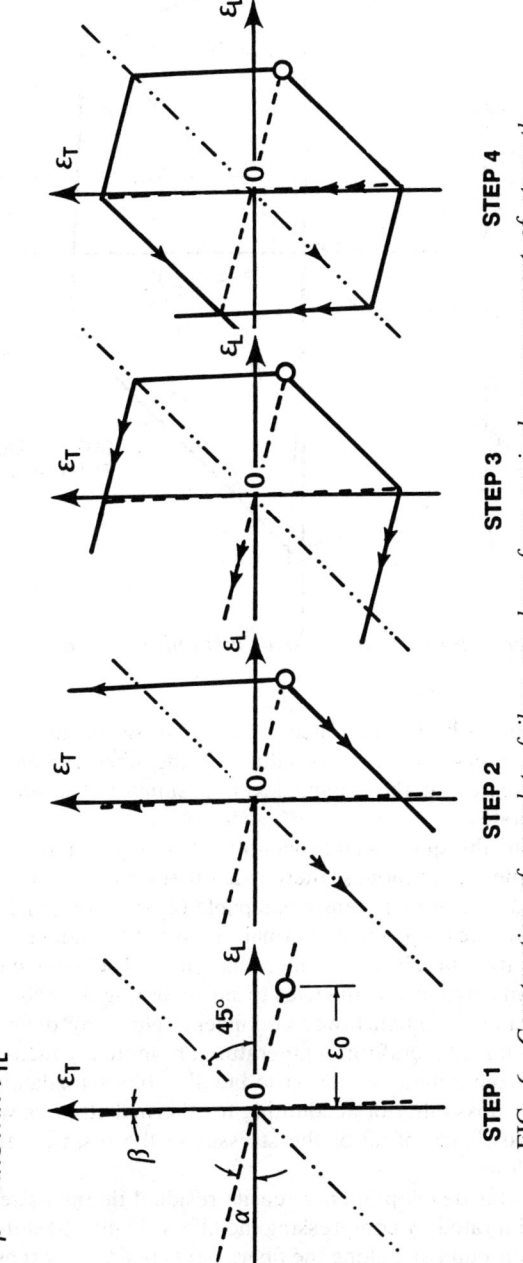

FIG. 6—*Construction of complete failure envelope from a single measurement of strength.*

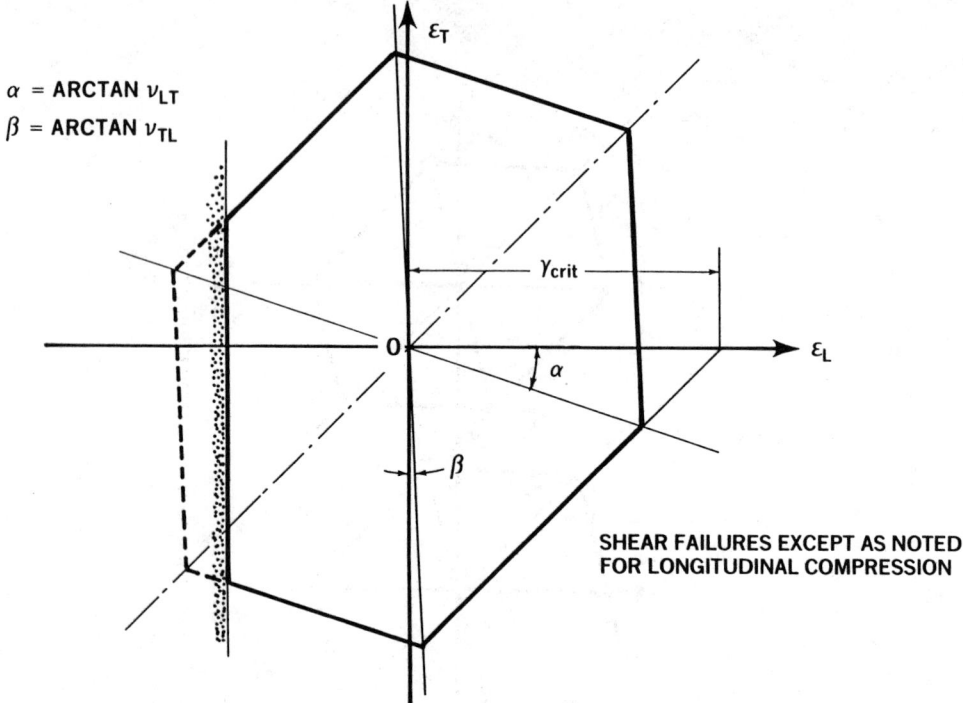

FIG. 7—*Superposition of compressive instability limit on shear-failure envelope.*

analysis may seem to be a macromechanical analysis of an homogenized lamina, it is definitely not; it refers only to the failure of the fibers, even though the strains are conveniently expressed at the lamina level to simplify the superposition of additional possible failure modes.

Quite apart from the questionable validity of a single smooth failure criterion in the presence of multiple failure modes, there is another crucial reason why analyses based on an homogenized lamina are inherently incapable of predicting matrix-dominated strengths for composite laminates. Residual thermal stresses are induced in the resin matrix by curing at a temperature far above the operating range. The resin matrix typically has a high coefficient of thermal expansion and tries to shrink during the cool-down, but is resisted by the fibers, which have a much lower coefficient. Now, by definition, homogeneity and thermal stresses while at a uniform temperature are mutually incompatible. Therefore, the use of a homogenized lamina as the basic building block for laminate analysis automatically excludes any possibility of accounting for thermal stresses within a lamina. And, in the absence of knowledge of all of the stresses in the resin matrix, it is not possible to predict matrix failures.

Fibrous composites develop intense tensile residual thermal stresses in the resin which are internally equilibrated by compressing the fibers. Figure 10 shows how the resin matrix attempts to shrink around and along the fibers but is resisted by those fibers which are very much stiffer and have much lower (and in the case of carbon and aramid fibers almost zero) coefficients of thermal expansion.

Some authors have maintained that residual thermal stresses are automatically ac-

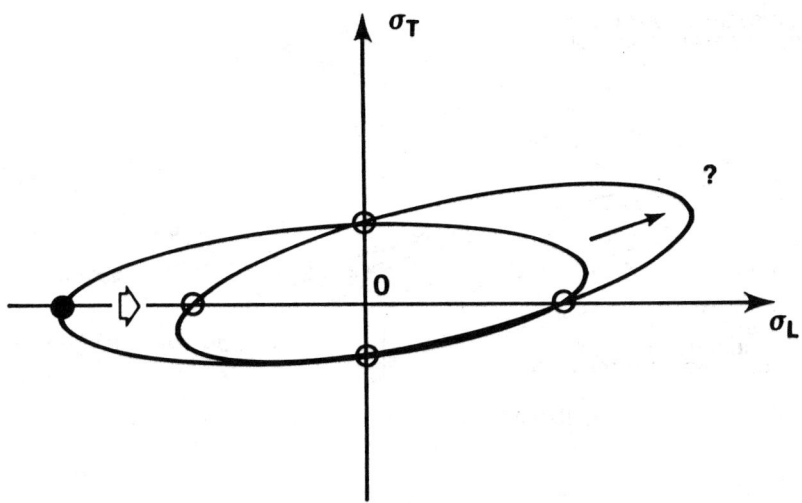

**REDUCED COMPRESSION STRENGTH DUE TO MICROBUCKLING OR
SIMILAR EFFECT IS PREDICTED TO *INCREASE* BIAXIAL TENSILE STRENGTH**

FIG. 8—*Common fallacy with abstract mathematical characterization of fibrous composite laminates.*

counted for because they are present in the test coupons and influence the measured transverse tensile strain to failure. These stresses are indeed present, but not to the same extent as in structural cross-plied laminates. So there can be no reliance on compensating errors, as a comparative assessment of 0° and 0°/90° laminates shows. The residual stresses

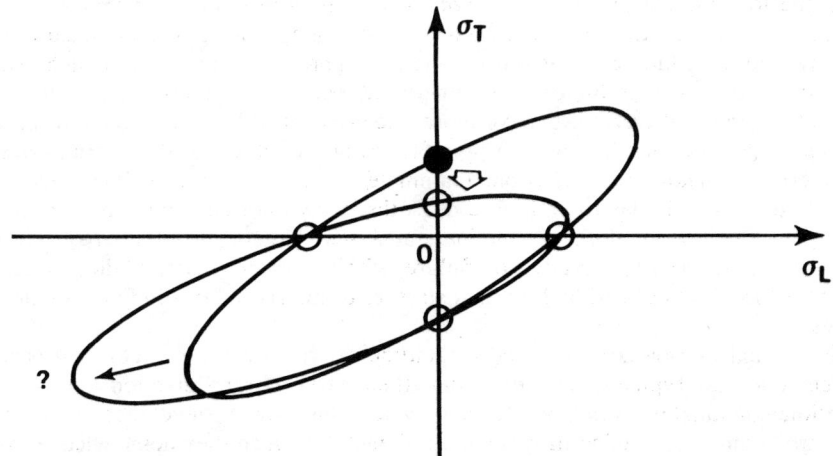

**WOULD ANYONE BELIEVING THIS ANALYSIS BE WILLING
TO PUT TO SEA FOR THE DIVING TRIALS?**

FIG. 9—*"Improved" composite material for submarine hulls by decreasing transverse-tension strength of unidirectional lamina.*

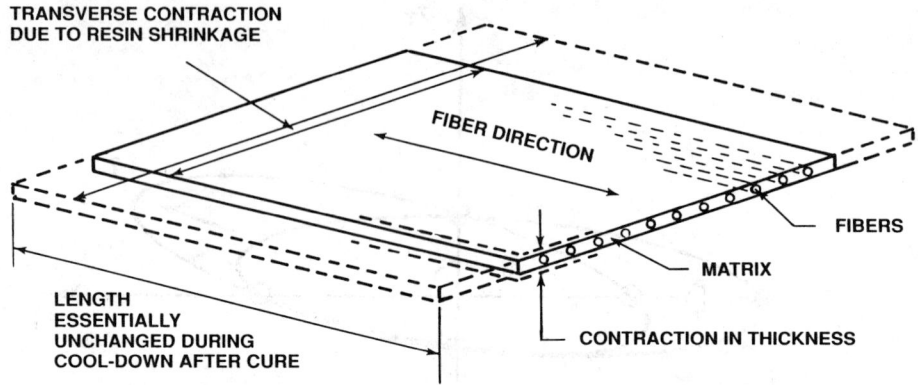

FIG. 10—*Shrinkage of resin matrix around fibers.*

in the matrix parallel to any of these fibers will be roughly equal for carbon-epoxy laminates, but, for the cross-plied laminate, the transverse matrix stresses will be far higher than for the unidirectional laminate. Any orthogonal fibers prevent the matrix from contracting in that direction as much as is possible in the absence of such fibers. So the transverse tensile strain to failure in the 0°/90° laminate (predicted for first-ply failure) should be at a far lower mechanical strain than for the all-0° laminate. Yet this distinction is only rarely made by structural analysts. And, in any event, test data show that the fibers fail completely before there is any major cracking in the matrix, implying a higher mechanical strain to failure.

Further, if the 0°/90° laminate happens to be in the form of a woven fabric rather than a cross-plied tape, it is customary to assign a transverse strain to first-ply failure essentially equal to the longitudinal failure strain, not to something less than half that strain. This inappropriate differentiation between transverse strains to failure is maintained even for satin-weave cloths, which contain seven bundles of parallel fibers between each orthogonal bundle, in contrast with the one-over one-under pattern for plain-weave cloths. Satin-weave cloths should, therefore, behave more like cross-plied laminates made from layers of unidirectional tape than like plain-weave fabrics in regard to residual thermal stresses. Or is the correct analogy that cross-plied laminates made from unidirectional tape should be considered to be similar to satin-weave cloths, which are conventionally treated the same way as plain-weave cloths for laminate analyses? Homogenized failure criteria that do not distinguish between modes of failure within a lamina simply do not account properly for the effects of residual thermal stresses or matrix failures in fibrous composite laminates.

These residual thermal stresses, the magnitude of which can be established easily by micromechanics, are typically far greater than the measured transverse-tension strength of unidirectional laminates. Some researchers would have us believe that such residual stresses creep out by changing the thickness of each ply since they acknowledge that the in-plane deformations necessary to relieve these stresses would be impossibly large.

The most telling argument against this hypothesis is that, if it were correct, the same creep of the resin matrix to relieve residual thermal stresses would also occur due to sustained loads in flight, or on the ground, destroying the stabilization of the fibers under compressive loads. Consider, for example, the upper surface of a wing during flights of many days duration for unmanned reconnaissance aircraft, or even the 15 h or more per leg on some passenger flights. While the cold at high altitude would retard any creep, that mechanism would not be available for an aircraft parked in the desert and flown infrequently. It is not proper to postulate that creep will always occur when needed to relieve troublesome stresses but will never occur at other times when doing so would create rather than eliminate a problem.

This thinking is akin to the common belief that there can be no residual thermal stresses if a composite panel is flat, while such stresses do exist whenever a panel is visibly warped. Ironically, flatness is no guarantee that residual thermal stresses are not present, and warping is an indication of the relief of thermal stresses rather than of their generation. Proof that these intralaminar thermal stresses do not disappear with time is easily provided by experiments on "bimetallic" strips of unbalanced laminates. These remain bowed and do not flatten out. The most common example of a thermally unbalanced composite laminate is a single layer of satin-weave cloth which has most of the 0° fibers on one surface and most of the 90° fibers on the other. A plain-weave cloth, on the other hand, would be essentially flat after curing, but would not be stress free. The stress-free temperature of an unbalanced laminate, as deduced by calculations based on measured curvatures or as measured by seeking the temperature at which the strips flatten out in an oven, has been consistently found to be only some 14°C (25°F) below the 180 or 120°C (350 or 250°F) cure temperature. Obviously, this technique cannot be used to establish the stress-free temperature of thermally balanced panels which do not warp. Only micromechanics can account for this major source of stress in the matrix.

An alternative excuse sometimes offered for neglecting these intralaminar residual thermal stresses and confining attention to the smaller increment of interlaminar stress associated with the laminating process is to postulate that the thermal stresses will be relieved by the swelling stresses associated with moisture absorption. It might be possible to counteract these two effects for some laboratory coupons if one waits long enough, but, given the extremely small diffusion rate for water absorption by either resin or fiber, it is difficult to accept the idea that this mechanism can remain effective while a jet aircraft rapidly changes its temperature between ground level and the stratosphere. Actually, far from enhancing the case of those who would ignore the intralaminar thermal stresses to justify continued use of homogenized laminae in their strength analyses, this very argument reinforces the position that a composite structure governed by structurally significant matrix failures is inherently uncertifiable. This is explained in the following section.

The features of a meaningful analysis that include the effects of thermal stresses are presented in Fig. 11. Separate failure envelopes are shown for the fibers and the matrix, and the origins of each are offset to reflect the residual thermal stresses both along and around the fibers. Fiber failures would dominate for some states of biaxial stress, while matrix failures would govern for others. This kind of characterization can never be generated once the model has been oversimplified by homogenizing the unidirectional lamina.

The Uncertifiability of Matrix-Dominated Composite Structures

If the matrix is soft in comparison with the strong, stiff fibers embedded in it, the fiber-dominated strengths of composites can be established without also determining the state of stress in the matrix. Some empirical allowance has to be made for a reduction in compressive strength of the fibers under hot, wet conditions compared with their room-temperature properties, but the internal fiber stresses corresponding with given external loads can

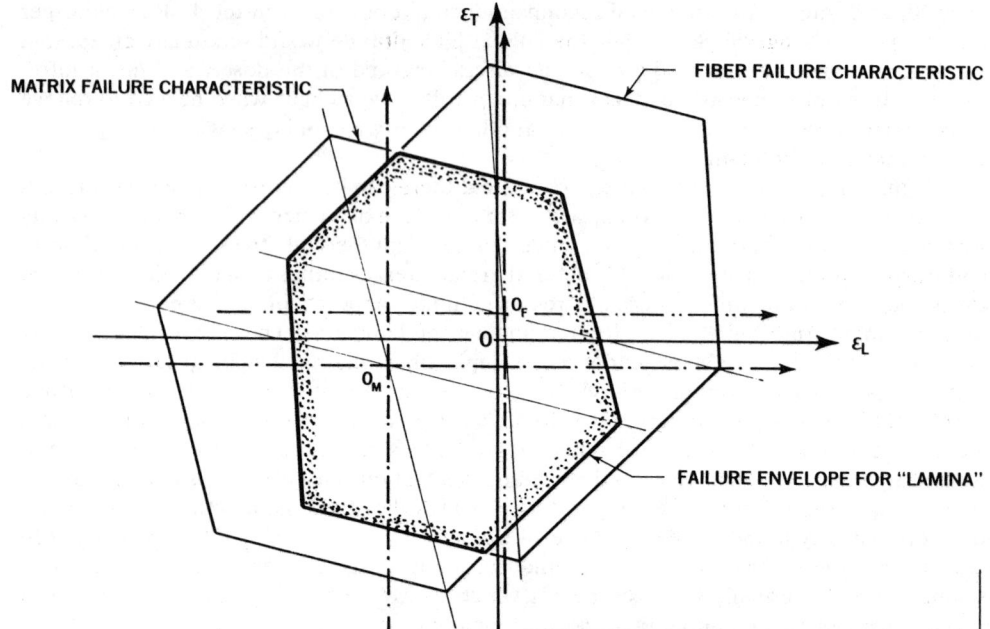

FIG. 11—*Separate failure characteristics for fibers and matrix.*

be determined with a high degree of precision. If that were not so, predicting fiber-dominated strengths could be as unreliable as predicting those dominated by the matrix.

With matrix-dominated failures, there are many unknowns or uncertainties. As mentioned earlier, moisture absorption can relieve some of the residual thermal stresses. But how much? Also, this implies for matrix-dominated failures that the strength of an aircraft would change with the location of the airport and the time of the year, or even within a day. And, according to structural analysis in terms of "progressive failures" of composite laminates, the laminate strength should change with the intensity of microcracks in the resin. Worse, the compressive strength is then predicted to be dramatically different depending on whether the loading was monotonic or if it included subcritical tensile loads earlier in the life of the structure. There is simply no way to guarantee any specified matrix-dominated strength for a composite structure because neither the state of stress in the matrix nor its strength can be established reliably beyond the first load cycle, even with micromechanics.

Nevertheless, there are many fibrous composite structures in use today in which matrix failures do precede fiber failures. And it is possible to perform rational, if not strictly scientific, analyses of their strength. A case in point is pressure vessels made from brittle resins reinforced by glass fibers, for which the ultimate strain-to-failure is typically twice that of the resin. Such low-cost materials make sense because of the very high tensile strength of glass fibers. Matrix cracking in this case does not represent a structural problem, but it would allow the contents to escape, so it is customary to wind the vessel around a high-strain liner.

In much the same way that both longitudinal and transverse strains were incorporated in the original formulation of the maximum-strain failure model for composite laminates, an arbitrary transverse-tension cutoff can be superimposed on the basic failure envelope in

Fig. 5. This has been done in Fig. 12. As shown here, the cutoff is more properly represented as a constant transverse *stress* line rather than as the limit on strain customarily accepted in implementing the maximum-strain failure model.

This "lamina" failure envelope can be generalized to arbitrary cross-plied laminates by superimposing it on its own mirror image about the biaxial stress line, leading to the family of lines in Fig. 13 for different operating temperatures. As shown, the effect of a limit on the transverse stress rather than on the strain is a more severe curtailment of the biaxial than the uniaxial tension strength. But, with this diagram, it becomes easy to see how ridiculous it would be to try to estimate a matrix-dominated strength. As expected, Fig. 13 shows that the extent of the matrix cutoff is diminished as the operating temperature is increased toward the stress-free (cure) temperature. But suppose that, before raising the temperature from a presumed room temperature state, a structure had first been chilled to subzero temperatures, inducing many microcracks in the process. The strength at elevated temperature would than have been reduced. And it would be reduced even more if, while the structure was at the subzero temperature, the density of microcracks had been increased still further by applying subcritical mechanical tensile loads. It is not difficult to prescribe some rational matrix cracking design cutoff in the manner of Fig. 13. But accurately predicting the load level at which catastrophic matrix failure would occur on some arbitrary cycle would not be practical, even if it were possible to define every preceding load cycle.

One is forced to accept the conclusion that matrix-dominated strengths are so variable

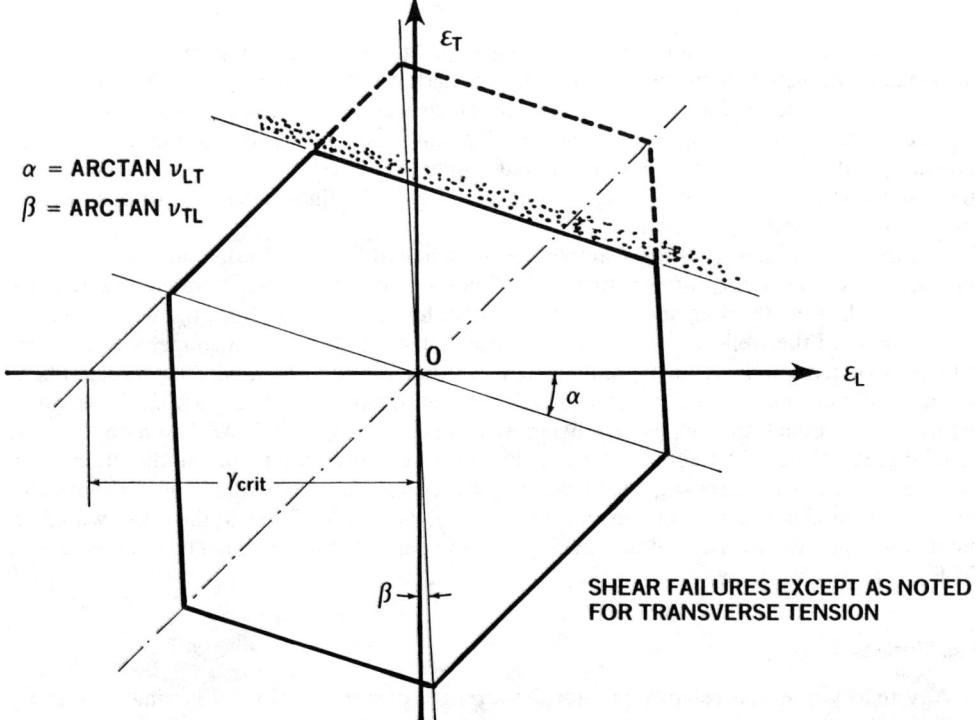

FIG. 12—*Transverse-tension matrix failure superimposed on shear failure envelope for fibers.*

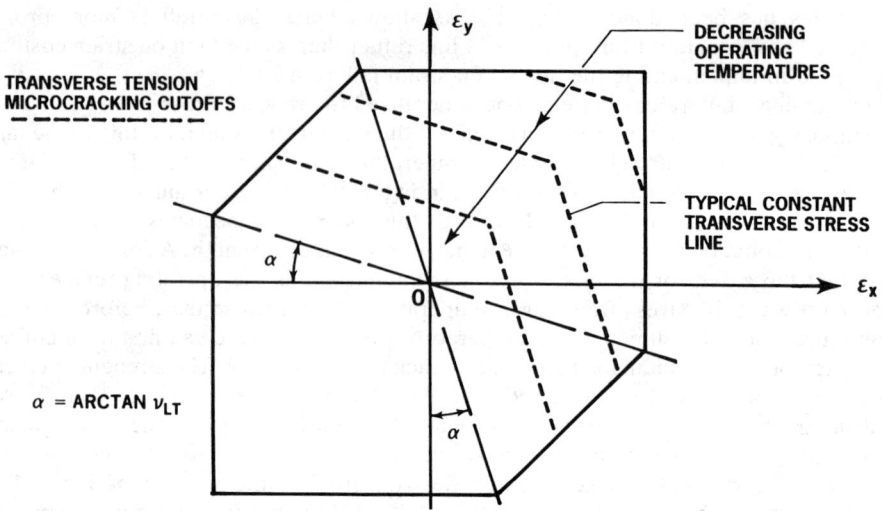

NOTE: NO CHANGE TO PREDICTED FAILURES FOR <u>OTHER</u> STATES OF STRESS

FIG. 13—*Physically realistic treatment of matrix cracking occurring before fiber failures (e.g., high-strain glass fibers in a brittle epoxy matrix).*

that one could not be assured of achieving or maintaining such a strength. Only well-designed composite structures in which the strength is determined by fiber failures which are themselves not a function of matrix properties could be relied on. So there is no apparent need to accurately predict matrix-dominated strengths. Accordingly, the author concentrated on predicting fiber-dominated laminate strengths and on identifying those fiber sequences which were prone to result in matrix-dominated failures to ensure that they were not used.

One may question the author's simple fiber-dominated failure criterion depicted in Fig. 5 in relation to the ability of a matrix full of microcracks to transmit transverse tension loads, which is the mechanism by which the upper left quadrant of the figure was truncated with respect to the well-known maximum-strain failure model. Such a concern is probably true for isolated unidirectional laminae. However, most structural laminates contain fibers in many directions. Consider the case of in-plane shear on a ±45° laminate, which is equivalent to equal and opposite tension and compression on a 0°/90° laminate. Even a cracked matrix would be able to transfer transverse compression from one direction to an orthogonal fiber with a tensile axial load. So, while such matrix cracking might expand the failure envelope of the fibers in one direction, the failure of the fibers in the other would be unaffected, and the overall failure envelope for the entire laminate would be unchanged, as in Fig. 14.

Conclusions

Any theory that can reliably predict the strength of fiber-dominated laminates is in no way deficient because it cannot predict matrix-dominated failures as well. In fact, the only interest of structural design engineers in calculating matrix-dominated strengths is to identify those laminates that should not be used.

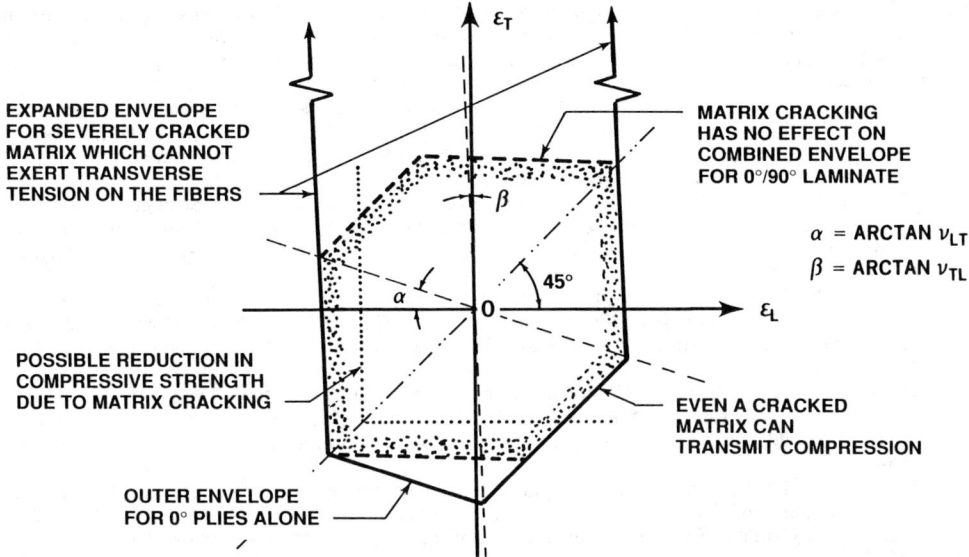

HIGH-DENSITY MICROCRACKING DOES NOT EXPAND EFFECTIVE FAILURE ENVELOPE

FIG. 14—*Apparent expansion of failure envelope for unidirectional lamina in presence of extensive microcracking.*

The need for calculating matrix stresses in composite laminates is to identify thresholds below which matrix failures will not occur. The actual loads at which failures would occur in poorly designed laminates are unimportant for conventional fiber-polymer composites used in aircraft structures.

Omitting consideration of matrix failures leads to simpler theories for fiber-dominated strengths alone. This simplification is also likely to yield more accurate estimates of laminate strengths because of the greatly reduced number of material properties needing to be characterized to implement the theory. A full micromechanics model involves many more material properties, some of which are extremely difficult to measure. Most structures engineers see this need for greater input data as a significant impediment in using micromechanics for other than research.

The customary homogenization of the fibers and resin in the innumerable previous failure "theories" for composite materials has made it impossible to characterize the state of stress in the resin matrix, thereby precluding the possibility of ever calculating matrix-dominated strengths with such theories.

Finally, it is noted that matrix-dominated strengths are so variable that they could not be relied on if they were the primary mode of failure of a composite structure. Under the circumstances, such a structure could never be certified, even by test. Consequently, there is no need to accurately predict the strength of matrix-dominated composite structures.

References

[1] Hart-Smith, L. J., "Simplified Estimation of Stiffness and Biaxial Strengths for Design of Carbon-Epoxy Composite Structures," Douglas Paper 7548, presented to Seventh Conference on Fibrous Composites in Structural Design, Denver, CO, 17–20 June 1985, published in *Proceedings*, AFWAL-TR-85-3094, pp. V(a)–17 to V(a)–52.

[2] Hart-Smith, L. J., "Simplified Estimation of Stiffness and Biaxial Strengths of Woven Carbon-Epoxy Composites," Douglas Paper 7632, *Closed Session Proceedings,* 31st International SAMPE Symposium and Exhibition, Las Vegas, Nevada, 7–10 April 1986, Society for the Advancement of Material and Process Engineering, Covina, CA. pp. 83–102.

[3] Peterson, D. A. and Hart-Smith, L. J., "A Rational Development of Lamina-to-Laminate Analysis Methods for Fibrous Composites," Douglas Paper 7928, *Composite Materials: Testing and Design, (9th Volume), ASTM STP 1059,* S. P. Garbo, Ed., American Society for Testing and Materials, Philadelphia, 1990, pp. 121–164.

[4] Hart-Smith, L. J., "A New Approach to Fibrous Composite Laminate Strength Prediction," Douglas Paper 8366, *Proceedings,* CP-3087, Part 2, 8th DoD/NASA/FAA Conference on Fibrous Composites in Structural Design, Norfolk, VA, 28–30 Nov. 1989, NASA, Washington, DC, pp. 663–693.

[5] Hart-Smith, L. J., "A Strain-Based Maximum-Shear-Stress Failure Criterion for Fibrous Composites," Douglas Paper 8376, *Proceedings,* CP-902, Part 2, 31st AIAA/ASME/ASCE/AHS/ASC Structures, Structural Dynamics and Materials Conference, Long Beach, CA, 2–4 April 1990, pp. 714–722.

[6] Hart-Smith, L. J., "Fibrous Composite Laminate Strength Predictions Demystified," Douglas Paper 8430, *Proceedings,* 11th European SAMPE meeting, Basel, Switzerland, 29–31 May 1990, H. L. Hornfeld, Ed., Society for the Advancement of Material and Process Engineering, pp. 365–380.

[7] Hart-Smith, L. J., "A Scientific Approach to Composite Laminate Strength Prediction," Douglas Paper 8467, *Composite Materials: Testing and Design, (Tenth Conference), ASTM STP 1120,* G. C. Grimes, Ed., American Society for Testing and Materials, Philadelphia, 1992, pp. 142–169.

[8] Hart-Smith, L. J., "A Simple Two-Phase Theory for Thermal Stresses in Cross-Plied Composite Laminates," IRAD Report MDC-K0338, Douglas Aircraft Co. Long Beach, CA, November 1986.

[9] O'Brien, T. K., "Characterization of Delamination Onset and Growth in a Composite Laminate," NASA Langley Technical Memorandum 81940, January 1981.

[10] Rowlands, R. E., "Strength (Failure) Theories and their Experimental Correlation," *Handbook of Composites,* Vol. 3, *Failure Mechanics of Composites,* G. C. Sih and A. M. Skudra, Eds., North Holland Publishing Co., Amsterdam, 1985, pp. 71–125.

[11] Tsai, S. W., "*Composites Design,* 4th ed.," Think Composites, Dayton, Paris, and Tokyo, 1988.

Gregory P. Carman,[1] John J. Lesko,[1] Ahmad Razvan,[1] and Kenneth L. Reifsnider[1]

Model Composites: A Novel Approach for the Evaluation of Micromechanical Behavior

REFERENCE: Carman, G. P., Lesko, J. J., Razvan, A., and Reifsnider, K. L., "**Model Composites: A Novel Approach for the Evaluation of Micromechanical Behavior,**" *Composite Materials: Fatigue and Fracture, Fourth Volume, ASTM STP 1156,* W. W. Stinchcomb and N. E. Ashbaugh, Eds., American Society for Testing and Materials, Philadelphia, 1993, pp. 381-400.

ABSTRACT: Microlevel damage events in composite materials are extremely important issues when addressing the remaining strength and life of the system. These events include fiber fractures, matrix cracks, and fiber end effects. Development of accurate representations of these phenomena at the local level is difficult and until this time unverifiable. An experimental procedure is suggested which involves a macromodel composite to generate data that are utilized to validate (or invalidate) current and proposed micromechanical analysis of composite materials containing damage. Quantitative experimental data on the perturbed strain field are measured with embedded strain gages and internal Fabry Perot fiber optic strain sensors, while qualitative measurements are accomplished with data obtained from a birefringent matrix. The majority of the tests described herein are representative of continuous fiber composites containing a fiber fracture. A controlled fiber fracture is achieved in the model composite at a predetermined location. Stress concentration and ineffective length (i.e. size of the perturbed stress field) measurements are experimentally determined for different fiber volume fractions and interphase coatings (i.e. on fiber). Initial results suggest that crack propagation plays a significant role in the stress redistribution which occurs in the vicinity of a fiber fracture. We suggest that the interphase region affects the formation and propagation of cracks in the composite. The methodology described in this paper allows a researcher to systematically study the effect of various physical parameters, such as fiber volume fraction, constituent properties, and interphases, on the local stress state in a material system.

KEY WORDS: fiber fracture, sensors, ineffective length, micromechanical interphase

The thrust of current research is aimed at presenting a direct approach to understanding and validating analytical micromechanical models under development and others already present in the literature. The fact that local physical parameters at the microlevel significantly influence the macroperformance of composite materials signifies that the micromechanic models should be accurate in every detail. To ensure that these models are precise, a test technique must be developed to provide quantitative experimental data on microphenomena. A test methodology such as this one should also assist in identifying new directions for subsequent modeling efforts, particularly for optimizing the selection of fiber, matrix, and interphase for specific applications. Traditionally, the greatest obstacles to obtaining these local data are the inability to closely examine internal damage events as

[1]Graduate research assistant, graduate research assistant, graduate project assistant, and Alexander Giacco professor, respectively, Materials Response Group, Dept. of Engineering Science and Materials, Virginia Polytechnic Institute and State University, Blacksburg, VA 24061-0219.

they occur and the relatively small dimensions being examined. However, we do point out that recent success utilizing laser Raman spectroscopy [1] to quantitatively measure fiber axial strains in a model composite has yielded promising results.

Previous attempts to validate theoretical micromechanical models may be subdivided into two categories: volume-averaging techniques [2–10], and qualitative studies [11–14] on modeled situations. While both approaches yield useful information, they fall short of providing the quantitative data correlation needed to assess the validity of the micromechanical predictions. The volume averaging techniques smear out interaction effects and other microdetails which may (or may not) be addressed in the analysis. Utilizing these results to substantiate a model depiction of microstress redistribution may lead to erroneous conclusions about the microparameters which are essential to global-level strength predictions. Likewise, the techniques that employ a modeled situation rely on theoretical micromechanics, which have not been validated, to generate any quantitative results. These models encompass a large number of broad assumptions which, if incorrect, may drastically alter the basic understanding of damage progression purported by the microanalysis.

The macromodel composite described in this paper does not suffer from the inconsistencies stated above and thus offers the unique opportunity to quantitatively assess the microdetails which affect stress redistribution near damaged regions. This unique investigation employs a scaled-up representation of a composite lamina (i.e. one ply). Scaling up the fiber-matrix interaction allows for the quantitative and qualitative studies of typical damage states without loss of generality. At this level, the damage is examined by visual inspection and monitored with accepted measurement techniques. Continuum mechanics [15] accurately describe the constituent interaction at the microlevel and the macrolevel, such that the scaled-up version provides representative data depicting the stress redistribution in the actual laminate. Within this macromodel composite, internal quantitative measurements of strain are obtainable in all constituents (i.e. fiber, matrix, and interphase) in the presence of particular damage events (i.e. in this paper, fiber fracture) which are simulated in a controlled and known manner.

When utilizing a scaled-up version of a composite lamina to study the stress redistribution in the vicinity of a fiber fracture, the question of scale effects immediately surfaces. Fracture mechanics predict that stress redistribution in the vicinity of a crack [16] varies as $(a/r)^{1/2}$ (at least for the case considered here), where a is the crack size and r is the radial location of the point in question. By scaling up the crack size and the radial position simultaneously, one finds that scale effects on the stress/strain state are not an issue. That is, the stress/strain at similar radial locations in either the macromodel composite or the actual laminate are identical. However, we do point out that the stress intensity factors (these vary as $a^{1/2}$) for the two systems are not the same. Therefore, while the stress intensity factors are different (we do not intend to address these values), the stress/strain values are the same. The measurement of these latter quantities, that is, the strain state near a fiber fracture, are the principal focus of the present paper. Quantitative measurements of strain in the macromodel composite are accomplished with embedded fiber optic strain sensors and resistance strain gages, while qualitative measurements are made with visual inspection of the photoelastic resin. We point out that the use of embedded sensors to interrogate the state of the material has direct application in the theoretical and experimental development of smart material structures [17].

The tests described in this paper provide experimental evidence concerning the interaction between fiber, interphase, and matrix in the presence of a fiber fracture. A fiber fracture in the model composite is shown to be achieved in a controlled manner at a predetermined location in the composite. The stress concentration and ineffective length (i.e. size of perturbed stress field) resulting from a fiber fracture are quantitatively and qualitatively measured for different fiber volume fractions and interphases. Experimentally determined values for these quantities are the primary interest in the present analysis due to their importance in making uniaxial tensile strength predictions [9]. The principal focus of developing the current methodology is to provide a technique to examine the effects of fiber volume fraction, irregular fiber spacing, constituent properties, and interphases on stress redistribution. Results presented in this paper also include data on short fiber composites. A methodology is thus presented to study directly the effects of various physical parameters and their relation to stress redistribution due to internal microdamage.

Material

The macromodel composite described in this paper represents a unidirectional E-glass/ epoxy composite. The fibers in the macromodel are scaled up 150:1 when compared to conventional E-glass fibers. As illustrated in Fig. 1 and photographically depicted in Fig. 2, the model represents a typical composite lamina in which a single row of fibers are arranged in a plane. The macromodel composite, as well as the actual laminate, can ideally be separated into three distinct constituent material regions: the fiber, matrix, and interphase. These three material regions and the interfacial quality (i.e. between the fiber and matrix) within the macromodel are the topic of this section.

The macromodel composite contains an epoxy resin system (PLM-9) which is represen-

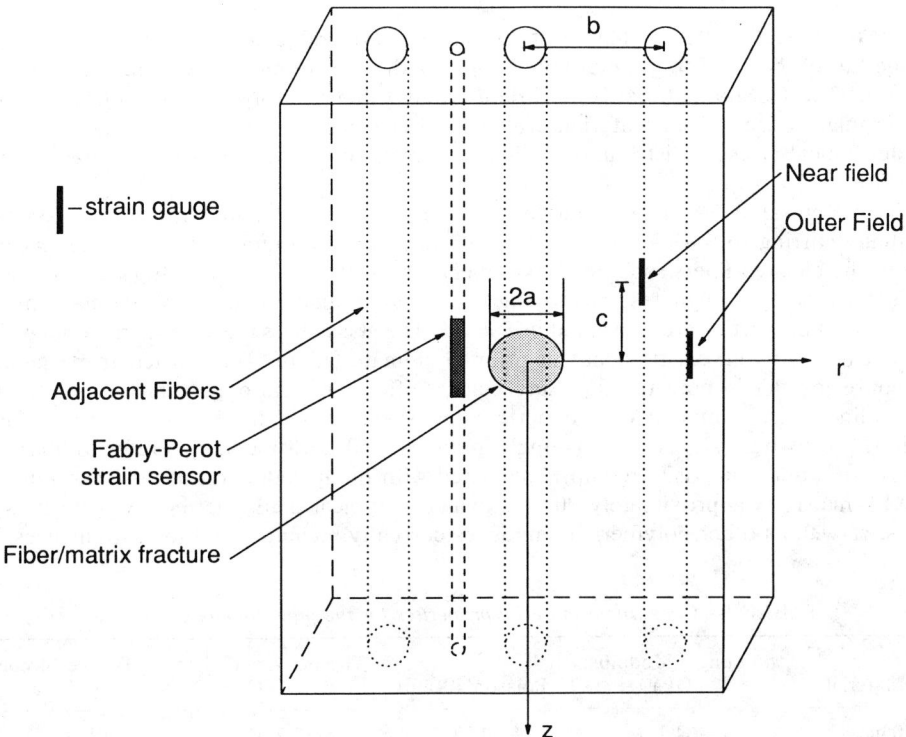

FIG. 1—*Illustration (not drawn to scale) of the model composite with embedded strain gages and their respective locations.*

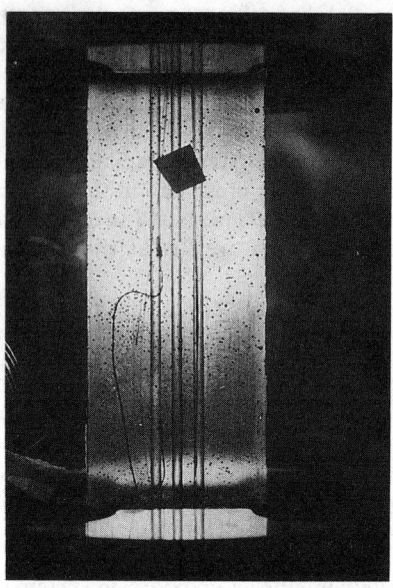

FIG. 2—*Typical photograph of a manufactured model composite containing internal strain sensors mounted in a servohydraulic test frame.*

tative of typical matrices that exist in actual polymeric composites. The mechanical properties of the PLM-9, as determined with standard experimental tests, are presented in Table 1. The PLM-9 material is a brittle thermoset that is easily cast into relatively large rectangular shapes. This material, which is birefringent, represents an ideal choice to model typical stress redistributions which occur in actual laminates that contain a fiber fracture.

The structural fibers embedded in the macromodel composite are borosilicate glass rods (actually stirring rods which will be referred to as fibers) 3 mm in diameter (i.e. $d_f = 3$ mm). The cleaned fibers/rods bond extremely well with the PLM-9 matrix as determined by a fiber push-through test employing a continuous ball indenter. A summary of the mechanical properties are presented in Table 1. The glass fibers are a brittle material which exhibit purely linear elastic behavior prior to ultimate failure. This particular mechanical response (occurs in both the fiber and matrix) offers an ideal opportunity to study proposed linear elastic models of polymeric composites. In fact, the fiber-to-matrix stiffness ratio [E_{zz}^f/E_{zz}^m where E_{ZZ} is axial Young's modulus, and the superscripts refer to fiber (f), matrix (m), and composite (comp)] calculated with the introduction of this fiber into the PLM-9 matrix is approximately 20. This value is typical of E-glass/epoxy material systems, as well as other polymeric composites currently manufactured for various uses.

TABLE 1—*Constituent material properties for the macromodel composite.*

Material	Young's Modulus, GPa	Poisson's Ratio	Thermal Coefficient, 1/°C	Tensile Strength, MPa
7740 glass	62.7	0.20	3.25E-6	. . .
PLM-9 epoxy	3.3	0.36	70.0E-6	50.0
PES	3.65	. . .	46.6E-6	91.7

Again, this paper mainly addresses the stress redistribution that occurs in a macromodel composite containing a fiber fracture. To study these phenomena, a controlled fiber failure must occur at a predetermined location in the composite. This failure is accomplished by notching the fiber with a diamond saw prior to its placement in the composite. The flaw introduced on the fiber surface provides a sufficient stress concentration to fracture the embedded fiber during tensile loading of the macromodel. It has been demonstrated through tests of ten individual glass fiber samples that the monolithic notched fibers fail at a load level of 200 ± 30 N. The notch in some of the more recent model composite tests has been replaced with a mechanical score. This introduces a comparatively larger stress concentration on the fiber, which causes it to fail at loads approximately 25% less than a notched one.

Coatings may be applied to the surface of the glass fibers prior to their placement in the mold. These coatings represent a finite interphase region which exists between the fiber and matrix in actual composite systems. Typical examples of the ones used in these tests include: bismaleimide (BMI), which provides a finite region of a thermoset material; polyethersolfone (PES) (see Table 1 for mechanical properties), which provides a finite region of a thermoplastic material; and epoxide, which provides a finite region of a thermoset material. The bismaleimide material represents a relatively brittle interphase which bonds well with the matrix material (i.e. both are thermosets). The epoxide should also provide excellent adhesion between the interphase and matrix region; however, this material is relatively more ductile than the bismaleimide and thus offers a distinctly different type of response to study. The PES is a thermoplastic material which may not chemically bond to epoxy; however, mechanical interlocking occurs due to surface texture and thermal contraction effects (i.e. residual stress due to manufacturing caused by the thermal mismatch between fiber and matrix) as evidenced with a fiber push-through test.

The interface between the fiber and matrix may be altered with the application of comparatively thin films whose thicknesses are considered inconsequential in the macromodel. These materials include silane coupling agents, which promote adhesion [18], and vacuum grease, which prevents adhesion.

Measurement Techniques

In this study, quantitative and qualitative methods are utilized to study the stress redistribution around fiber end effects. The birefringent behavior of the matrix material provides a qualitative understanding of the stress redistribution around a fiber fracture. By viewing the composite in the desired state (i.e. with an applied load following fiber fracture), information about the localized stress state is obtained. The stress gradients produce different hues of color which correspond to isostress regions produced by the internal strain concentration. Photographs taken at applied load levels are subsequently analyzed to provide qualitative information about the effect local fiber volume fraction, interphases, and crack size have on the size of the perturbed stress field (i.e. ineffective length). A micrometer is employed to measure the discolored region for a determination of the ineffective length. Although this technique provides a good indication of the size of the perturbed stress field, precise quantitative information is difficult to obtain from this methodology. This obstacle is attributed to the three-dimensional stress field generated by the interaction between fibers, which results in superposition and interference of the photoelastic data.

Two types of resistance strain gages, internal and external, are utilized to study the mechanical response of the macromodel composite. External resistance strain gages (350 Ω) are applied on the surface of the composite and oriented in the axial direction to obtain quantitative information about the global stiffness values (i.e. E_{zz}^{comp}). These gages are applied to a defect-free composite to ensure that no interaction occurs between the internal disturbance and the external strain patch. The external strain patches have a gage length of 6.35 mm. The purpose of this strain patch is two-fold. First, it provides measurement of global composite strain to determine the elastic modulus. Second, in a defect-free continuous macromodel composite, the external strain patch should measure the same strain as an embedded sensor oriented in the direction of the fiber (i.e. generalized plane strain assumption). These results can be used to validate (or invalidate) the internal measurement technique.

The other type of resistance strain gage, an internal one, is attached directly to the glass fiber surface. This gage monitors the strain state exhibited by an embedded fiber over a relatively small dimension. The internal gages are adhered (see Fig. 1) with epoxy adhesive (AE-10 micromeasurements group) and oriented so as to monitor the fiber's axial strain. These relatively high-resistance (4500 Ω) gages provide extremely sensitive measurements of strains exhibited by the embedded glass fibers. In fact, the sensing length of the internal strain gages are smaller than a fiber diameter (i.e. 2.54 mm). This allows measurements of fiber strains at the fiber diameter d_f level with a resolution of 3 microstrain. The lead wires, soldered to the strain gages, are an order of magnitude smaller than the glass fibers. These wires are carefully positioned during the manufacturing process so that they do not perturb the strain field being measured.

Fabry-Perot fiber optic strain sensors (FP-FOSS) are another quantitative method employed to study the internal stress redistribution in the macromodel. FP-FOSS's interferometers are embedded in the macromodel to monitor the axial strains (see Lesko et al. [17] for a detailed description) present at specific locations in the matrix around a fiber fracture (see Fig. 1). The sensing length of these transducers vary from sensor to sensor due to the unique nature of the FP-FOSS (i.e. each interfermeter is individually made by hand). Nominally the length varies from 2 to 4 mm, which permits internal matrix strain measurements at the d_f level. The diameter of the fiber optic sensors is approximately 150 μm, which is an order of magnitude smaller than a fiber diameter. This size is sufficiently small, so that the FOSSs are essentially an unobtrusive measuring device (discussed in Results section also).

Fabrication

The macromodel composite is cast inside an aluminum mold whose internal dimensions are 50.8 by 7.4 by 304.8 mm (see Fig. 1). The mold is thoroughly cleaned with acetone prior to manufacturing the macromodel. TFE release agent is applied to the mold to permit easy extraction of the fabricated product.

Following mold preparation, a choise is made on the number and the spacing of the fibers embedded during the manufacturing process. The present macromodel composite represents a 2-D model capable of containing six fibers aligned in a row. For a composite containing one fiber, the global fiber volume fraction is 1.87% ($v_f^g = 0.0187$), while for a six-fiber composite it is 11.2%. This value is utilized for the theoretical calculations of global stiffness values. However, when studying microeffects around a fiber fracture, the local fiber volume fraction (i.e. v_f^l is the pertinent quantity. This value is determined by the spacing between the fibers placed in the macromodel. In this paper, two uniform fiber spacings are addressed, that is, center-to-center fiber spacings (defined by the distance "b" shown in Fig. 1) of 4.6 and 6.2 mm. These spacings are representative of 20% ($v_f^l = 0.20$) and 15% ($v_f^l = 0.15$) local fiber volume fractions, respectively (larger fiber spacing correlates to smaller fiber volume fraction). Variable fiber spacing is obtained by skewing the central fiber in the composite towards one of its neighbors. This represents a composite containing unevenly spaced fibers on which a large number of tests have been run (see Carman et al. [*19*] and Carman [*20*]). By embedding fiber lengths smaller than the length of the composite, short fiber systems can be manufactured.

Having chosen fiber spacing, the fibers are cleaned with acetone to remove any oils which may have been deposited on them due to handling. A choice is made on whether to alter the interface and/or include an interphase coating. Usually the interphase/interface coatings (if any) are applied to the fibers prior to positioning them in the mold, but following the application of the resistance strain gages. However, in a number of tests, the resistant strain gages are adhered to the deposited interphase region on the glass fiber. The two coatings applied to the fiber to alter the level of fiber/matrix adhesion are silane coupling agent (z-6040) and vacuum grease. These materials are not regarded as depositing a finite interphase region on the fiber. The silane coupling agent (45 parts by volume) is thoroughly mixed with isopropyl alcohol (5 parts) and distilled water (50 parts) and allowed to age for 6 h. The fibers are subsequently dipped into the solution once and allowed to dry. This theoretically deposits one atomic layer of the silane coupling agent on the glass rod. In contrast, the vacuum grease is applied to the fibers by hand and subsequently wiped off to leave a "thin" film.

Interphase regions are applied to the fiber with either a pouring or a dipping procedure to deposit a layer whose thickness is 0.1 to 0.2 mm thick. The thermoset BMI comes in a liquid form which is somewhat viscous at room temperature. The BMI is preheated to decrease the viscosity of the polymer and is subsequently poured over the fibers. The manufacturer recommends curing the BMI at 250°C; however, the strain gage adhesive is rated at 100°C. Therefore, the fibers are baked at 100°C for a minimum of 1 h to cure the deposited BMI which is adhered to the fibers. At this temperature the BMI does solidify; however, it is not fully cured. The epoxide interphase is mixed following the manufacture's recommended instructions. Once prepared, the liquid mixture is poured over the fibers and then allowed to dry at room temperature overnight. The manufacture of the thermoplastic interphase is accomplished by dissolving a powder form of PES in methylene chloride. The fibers are dipped into this solution and allowed to dry at room temperature for 24 h.

Following these procedures, the fibers are placed in their appropriate locations and secured in the model. This includes the fibers which have strain gages adhered to them and the fibers which have been notched. The positioning of these fibers in the mold is extremely important since this determines the location of the fiber fracture relative to the location that strain data are obtained. After the glass fibers are in place, the fiber optic strain sensors are positioned in the composite. Once again, the orientation of the sensor's gage length is extremely important since this is the location that data will be recorded.

All gage positions are referred to the z coordinate (see Fig. 1) whose origin is defined by the strain concentration (e.g. short fiber and fiber crack) present in the composite. This axial distance from the crack to the gage is referred as "c" depicted in Fig. 1. The resistance strain gages are placed on adjacent fibers at essentially two radial locations, either facing the crack (near field) or facing away from it (outer field), as shown in Fig. 1. These terms are employed in the discussion of test results in subsequent sections.

With the fibers and sensors in the desired location, the mold is sealed and put in an oven to soak for a minimum of 1 h at 50°C. While the mold is being preheated, the PLM-9 matrix material is prepared for pouring. Following the manufacturer's recommended instructions, the appropriate amounts of hardener and curing agent are heated to the appropriate temperatures and thoroughly mixed. The liquid matrix is slowly poured into the mold and subsequently placed in a controlled oven at the desired cure temperature for an allotted time period.

The temperature and time period to cure the composite are extremely crucial. If the composite is cured at an elevated temperature, significant residual stresses are present in the fiber and matrix. In fact, for composites cured at 100°C, the residual compressive stresses in the fiber are so large that internal fiber fracture cannot be achieved without failing the entire composite. Therefore, to circumvent this problem, a cure cycle of 24 h at 37°C (this differs from the manufacturer's instructions) is utilized. This cure cycle minimizes the compressive stresses in the fibers, but serves to expedite the fabrication of the composite and appears to provide a sufficiently cured composite. A photograph of the fabricated composite mounted in a servohydraulic test frame is presented in Fig. 2.

Results and Discussion

A variety of results are presented in this section which describe the model composite's ability to represent (and the embedded sensor's ability to measure) specific physical situations which occur in actual composite systems. To ensure that the internal measurements are accurate, a subsection details a validation procedure for the sensors. Following the validation of internal measurements, a brief description is presented on the experimental results pertaining to a short fiber composite. However, the bulk data presented in this paper detail stress redistribution which occurs near a fiber fracture in composites containing single-fiber, three-fiber (this subsection describes two distinct types of crack propagation sequences), and six-fiber systems. The final subsection presents results describing interphase effects in composites and their relation to fiber fracture and crack propagation. Specific problems which arose in the fabrication of composites containing an interphase region are also discussed in this section.

The tests performed on the model composite system are accomplished by either thermal loading (only in defect-free composites) in a controlled oven or by axial tension (i.e. fiber direction) with a servohydraulic test machine.

Validation

Before embedded sensors can be used to corroborate theoretical micromechanical predictions, the data obtained from the sensors must be validated with standard testing and analytical techniques. This is accomplished by applying a tensile load (load in the z-direction) to the composite, which is void of strain concentrators (i.e. fiber fractures, matrix cracks). For this state of stress it is well accepted by the scientific community that a unidirectional continuous composite experiences a state of generalized plane strain in the axial direction. This implies that the axial strain (i.e. strain is ϵ) experienced by the macromodel is equal to that experienced by the fiber and matrix. This classic theoretical argument for continuous fiber composites is mathematically expressed as

$$\epsilon_{zz}^{m} = \epsilon_{zz}^{f} = \epsilon_{zz}^{comp} \tag{1}$$

The generalized plane strain assumption stated above is generally accepted and provides an avenue to test the strain-sensing ability of the FP-FOSS and the embedded resistance strain gages. This is accomplished by comparing the strain measurements made by the external strain patch ϵ_{zz}^{comp} to measurements made by the internal sensors (i.e. the resistance gages measure ϵ_{zz}^{f}, while FP-FOSS measure ϵ_{zz}^{m}). Upon making this comparison, the strain response from all three sensors should be indistinguishable (see Eq 1) if the sensors are unobtrusive. If they are obtrusive, the embedded gages will be the initiation of stress risers in the composite. Therefore, the data obtained from the embedded sensors at these locations will be different from the data obtained from the external sensor. When compar-

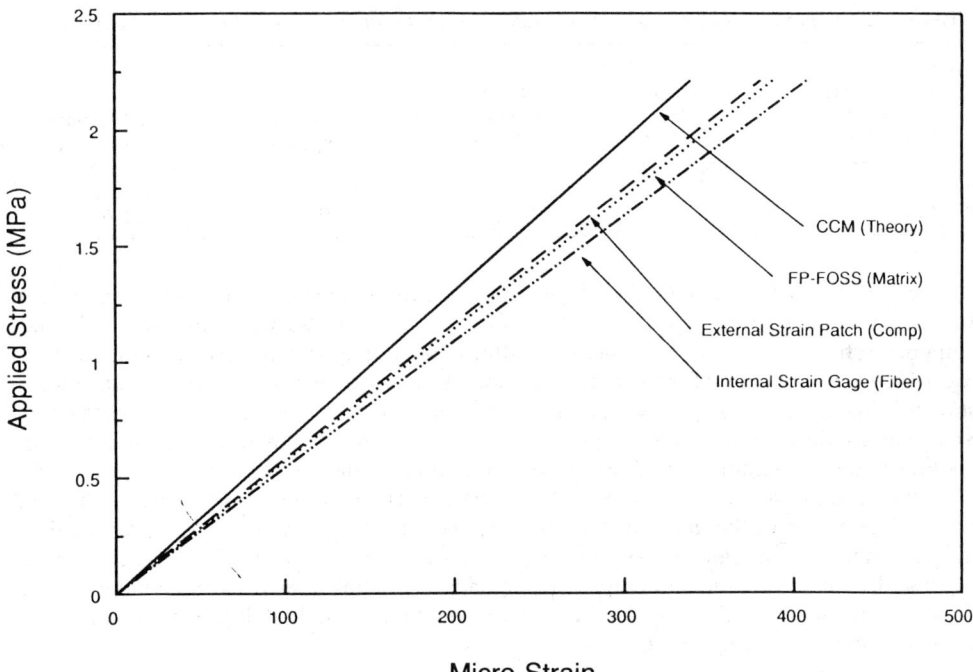

FIG. 3—*Comparison of strain measurements obtained from an external strain patch, an internal strain gage, and a fiber optic strain sensor to a theoretical prediction from a CCM model.*

ing the strain record for the three sensors as a function of applied stress (Fig. 3) for a three-fiber system (global $v_f^g = 5.67\%$), a small deviation of about 5% is found. This small deviation is well within expected experimental error. When comparing the strain response of these sensors to an accepted micromechanical prediction for an upper bound on composite stiffness (i.e. Concentric Cylinders Model [21]), reasonable agreement is observed. Note that the CCM model was formulated assuming the model composite could be represented with a cylindrical element such that the ratio of fiber to matrix diameters is equivalent to the global fiber volume fraction (i.e. $d_f^2/d_m^2 = v_f^g$). Therefore from these results we conclude that the embedded sensors are a nonobtrusive technique for the measurement of strain in the macromodel composite. We also suggest that the modeling methodologies which are commonly employed to predict stiffness values of actual composites adequately represent the macromodel. This is true, even though the actual laminate and the macromodel are not axisymmetric as is assumed in the CCM formulation. These facts provide additional supportive evidence suggesting that the macromodel composite adequately represents an actual lamina.

A summary of test data obtained on defect-free model composites for both single- and three-fiber systems are presented in Table 2. The mechanical properties shown in this table are determined with data obtained with the internal resistant strain gages. Theoretical results calculated from a displacement formulation with a CCM model (upper bounds) are also presented in the table.

In tension tests performed on the model composite, the stress versus strain plots are essentially linear curves (see representative Fig. 3). The experimentally determined

TABLE 2—*Experimental and theoretical results for material properties of the model composite.*

7740-Glass/PML-9, Fiber Volume Fraction	Longitudinal Modulus, GPa Exponent	Longitudinal Modulus, GPa CCM	Thermal Expansion, 1/°C Exponent	Thermal Expansion, 1/°C CCM
0.0187	3.5	4.4	44.5E-6	52.7E-6
0.0561	5.6	6.6	25.4E-6	35.6E-6

Young's modulus E_{zz} (presented in Table 2) for the three-fiber system is representative of 18 tests with a scatter in data of $+0.5$ GPa or -0.3 GPa, while in the single-fiber composite, three specimens exhibited a scatter in data of approximately ± 0.41 GPa. With regard to thermal loading for the determination of the coefficient of thermal expansion (α_{zz} shown in Table 2), the strain versus temperature plots (see Fig. 4) are linear curves below 87°C and nonlinear above this temperature. It appears that the matrix thermal expansion coefficient and longitudinal modulus become functionally dependent on temperature as the glass transition temperature is approached. The experimental results presented in Table 2 are representative of the linear portion of the curves analyzed. The α_{zz} data presented for these composites are an average of two tests with a variance of $\pm 2\%$. These results suggest that the macromodel composite responds to loading similar to an actual laminate. This is evidenced by the applicability of the predictive methodologies used for actual laminate coupons to the macromodel.

Short Fiber Systems

The short fiber composite results assist in demonstrating the ability of this technique to model various physical situations which occur in composites. A transient ($c = 4.8$ mm) and a far-field ($c \gg d_f$ or essentially outside the perturbed stress field) resistance strain gage are employed to study the stress redistribution in a short single-fiber system loaded in uniaxial tension and compression. A plot of applied composite stress versus the local strain at each gage location is displayed in Fig. 5. The curves generated from the strain data obtained from the gages are linear for either loading condition. The stress/strain slope at the far-field gage is 3.6 GPa in tension and 3.5 GPa in compression. These values are typical of a continuous fiber composite system, that is, the strain gage is outside the ineffective length [10]. The slope at the transient gage is 6.9 and 6.4 GPa in tension and compression, respectively. By taking a ratio of the transient slope to the far-field slope, an effective decrease in strain at the fiber end is computed to be 0.53 (i.e. *transient/far-field*). By analyzing photographs of the stress redistribution, as depicted by the photoelastic data, an approximate ineffective length of 15 mm is measured.

A comparison between the experimental results described above to various micromechanical models present in the literature is presented as follows. Calculations accomplished with the Whitney and Drzal [12] model of a single broken fiber in an infinite matrix suggests an ineffective length of 51 mm. Performing similar calculations with Rosen's [22] classical shear lag solution, an ineffective length of 62 mm is computed (the efficiency parameter is taken to be 0.95 and the global fiber volume fraction used is 0.0187). However, we do note that Rosen's model may not be applicable for a single-fiber system. Both theoretical results appear to significantly overestimate the experimental data presented above. On the other hand, by back-calculating the exponential parameter λ, which dictates ineffective length, from the experimental data (i.e. $\lambda = 0.316$ mm^{-1}) with

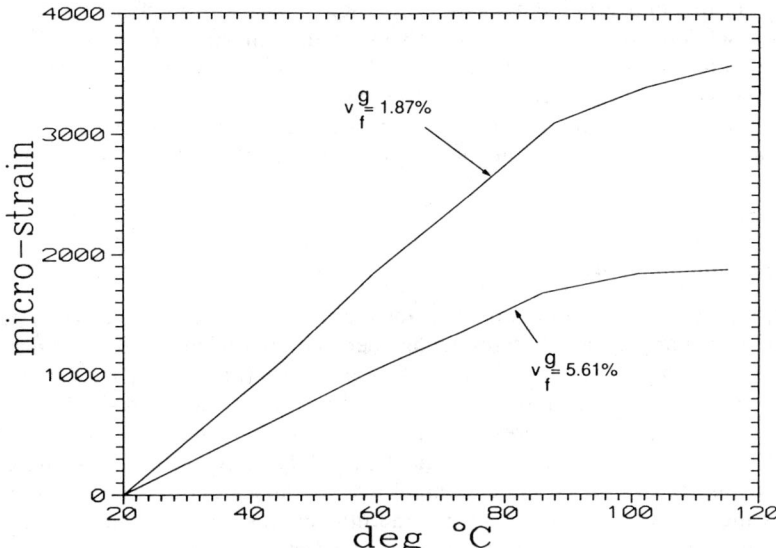

FIG. 4—*Strain temperature plot using embedded resistance gages for a single-* (v_f^g = *1.87) and three-fiber* (v_f^g = *5.61) model composite.*

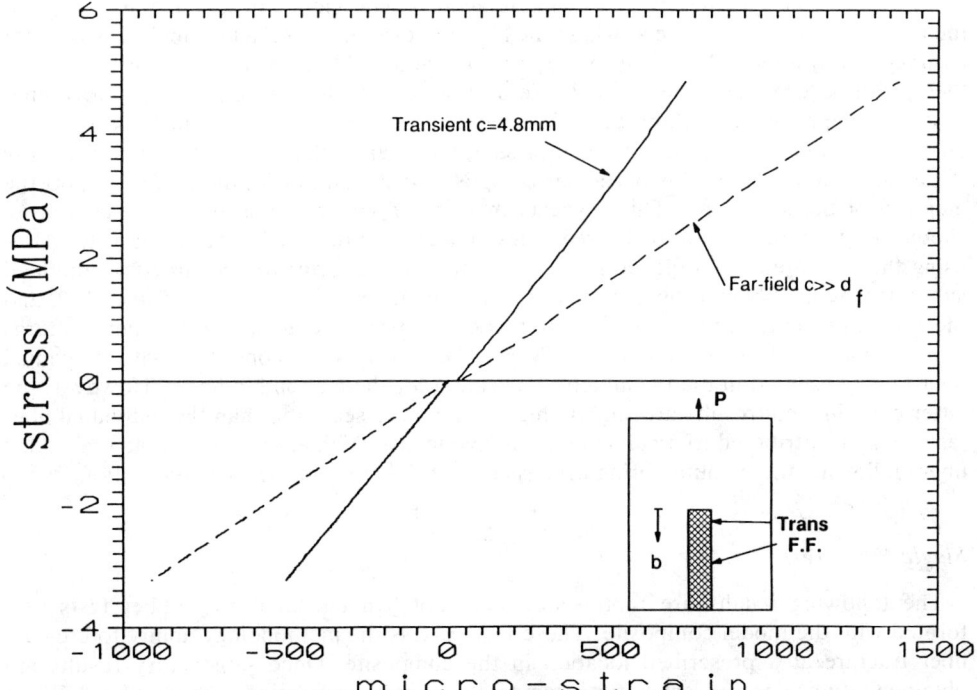

FIG. 5—*Stress strain plot using embedded resistance gages for a short fiber composite with the transient gage at* c = *4.8 mm.*

the short fiber model presented by Carman and Reifsnider [10] and using this result in their stress analysis, one calculates an effective decrease in strain of 0.50. This is nearly identical (within 6%) to the measured value (0.53) presented above. Therefore, it appears that the functional dependence (i.e. exponential) of stress and thus strain on z is being adequately addressed in this and the other models. However, the experimental findings suggest that the derivation of the parameter λ in the theoretical models needs to be reevaluated.

Continuous Fiber Systems

Herein, strain concentration is defined for a continuous fiber composite system as the strain at a particular gage location following fiber fracture divided by the strain at the same location prior to fiber fracture. Thus, a far-field gage (i.e. $c \gg d_f$) exhibits no strain concentration or SC = 1.0. In this study, single-fiber, three-fiber, and six-fiber composites are employed. All test results presented in this paper involving fiber fracture are accomplished by loading the composite in uniaxial tension. The stress-strain plots are essentially linear curves up to composite failure. Initial tests on fiber fracture were accomplished by applying a thermal load to the composite and utilizing the constituents thermal mismatch to start a fiber fracture. However, this led to additional curing of the composite and increased the residual compressive stresses on the fiber, which negated the phenomena being studied (fiber fracture). Therefore, only test results on macromodel composites loaded in uniaxial tension are discussed in the following sections.

When a model composite which contains a notched fiber is loaded in uniaxial tension, the notched fiber fractures with an audible acoustic emission. For a single-fiber system, the traction on the composite at which the fiber fractures is 1.2 MPa (2 samples), while, for a three-fiber system with no interphase, it occurs at 3.55 MPa (18 samples). By comparing these results, one finds the applied stress at which the fiber fractures in a three-fiber system is approximately three times greater than the single-fiber system (this is for the notched fibers). This should not be surprising, considering that the fibers are significantly stiffer than the matrix in this particular composite system and will, therefore, support the majority of the axial load. This suggests that fiber fracture in the three-fiber composite should occur at an applied load three times greater than the single fiber composite. Also, using the assumption that generalized plane strain exists prior to fiber fracture, the load supported by the fibers at fiber fracture can be calculated with the use of Tables 1, 2, and the results presented above. Performing this calculation, one finds that the fiber fractures at an applied load between 160 and 270 N. These results are consistent with individual fiber tests presented in the manufacturing section for the monolithic fibers. However, the latter experimental results are slightly higher and more scattered than the individual fiber tests. This is attributed to residual compressive loads which are present in the embedded fibers following the manufacturing process (i.e. shrinkage and thermal contraction).

Single Fiber Test

The following results are representative of notched uncoated single-fiber tests performed with the model composite. These tests assess the methodology ability to create a fiber fracture at a prescribed location in the composite. Once satisfactory results are obtained, studies are undertaken on composite systems containing multiple fibers. When the fiber fractures in the composite, energy is released in a manner similar to a spring which has been instantaneously released. The energy released is dissipated in the form of a shock wave, matrix cracking, and fiber slipping. In the single-fiber model composites, the fiber fractured and a crack propagated a finite distance into the matrix.

The crack size is usually $a < 2.5$ mm for a single-fiber composite. By viewing the composite with an applied load under polarized plates, a representative stress redistribution pattern in the vicinity of the fiber fracture is observed. The ineffective length for a single fiber composite appears to be approximately 13 mm. This length measurement is comparable (as it should be) to the results presented in the short fiber section. As noted in that section, the analytical models currently in the literature appear to overestimate this length. This strongly suggests that a revaluation of this quantity may be necessary in these analytical developments (at least for small fiber volume fractions).

Three-Fiber System (Small Crack)

For the uncoated three-fiber (one fiber notched/scored) model composite systems, two distinct fracture patterns are discovered: a large crack ($a = b - d_f/2 \equiv r_a$), and a small crack ($a = d_f/2 \equiv r_f$). The results obtained on composites containing the small crack are first presented, while the results typical of a large crack are described in a subsequent section. In both of these sections a local fiber volume fraction of 15 and 20% is addressed. A small crack occurs in a model composite when a fiber fractures and the crack propagates a small distance into the matrix. This length is typically smaller than 0.5 mm (i.e. $a \approx r_f$), and crack arrest occurs at or near the fractured fiber/matrix interface. A photograph depicting this typical fiber fracture pattern is presented in Fig. 6.

In the $v_f^l = 0.15$ composite, the small crack occurred in 8 out of 15 specimens tested. The center fiber fractures at an applied traction of approximately 3.3 MPa. From strain gages placed on adjacent fibers in the near-field region at $c = 0$, a strain concentration measurement of approximately 1.28 is determined. While for strain gages placed on adjacent fibers in the outer field region at $c = 0$, a strain concentration measurement of 1.14 is found. With the use of photographs depicting the birefringent behavior, an ineffective length is measured to be on the order of 6 to 10 mm (see Fig. 7). Employing Rosen's [22] shear lag model to provide a theoretical prediction for comparison purposes, a 30-mm length is calculated. In this calculation, the local fiber volume fraction is utilized. As alluded to in previous sections of this paper, the theoretical value determined for the ineffective length from Rosen's [22] model is larger than the measured value.

In model composite tests performed on comparatively larger fiber volume fraction composites (20%), 5 out of 15 specimens exhibited small cracks ($a \approx r_f$). The fiber fractured in these composites at an applied traction similar to that quoted above (i.e. 3.3 MPa). The measured strain concentration on an adjacent fiber in the near-field region at $c = 0$ for this fiber volume fraction is determined to be 1.4. This strain concentration value is larger than that measured (i.e. 1.28) in the smaller fiber volume fraction composite described above. In composites with relatively larger fiber volume fraction, the fiber spacing is smaller and thus the strain concentration should be larger (that is, the adjacent fiber is closer to the crack). From strain data obtained in the outer field region at $c = 0$ on an adjacent fiber, a strain concentration of 1.11 is found. This value is actually lower than the previous value for a larger fiber volume fraction. That is, the maximum strain concentration (near field $c = 0$) increases, but the minimum strain concentration (outer field $c = 0$) is measured to decrease as fiber volume fraction increases.

For the 20% fiber volume fraction composite, the ineffective length measured from the photoelastic data (see Fig. 8) suggested values between 4 and 8 mm. These values are less than those measured (i.e. 6 to 10 mm) for the smaller fiber volume fraction quoted above (also compare Fig. 7 to Fig. 8). Employing the shear lag model, an ineffective length of 26

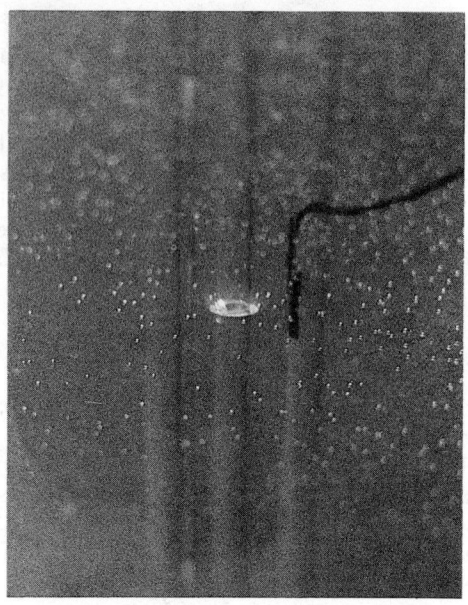

FIG. 6—*Photo depicting crack arrest (i.e. small crack* a = r_f*) at the fractured fiber matrix interphase.*

FIG. 7—*Representative photograph of the stress redistribution in a model composite due to a fiber fracture (*a = r_f *and* v_f^1 = 0.15*).*

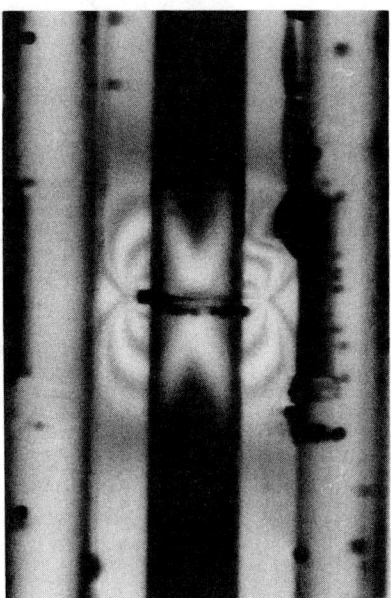

FIG. 8—*Representative photograph of the stress redistribution in a model composite due to a fiber fracture* (a = r_f *and* v_f^l = 0.20).

mm is calculated for comparison purposes. This once again appears to be an overestimate of the measured values determined from the photoelastic results. Shear lag solutions assume that the matrix does not support an axial load. This assumption should cause these models to overpredict ineffective length values, as presented in this paper. Nonetheless, both the theoretical models and the experimental results suggest a decreasing ineffective length for increasing fiber volume fraction. As the fiber volume fraction increases, the local stiffness of the composite increases and thus decreases the length over which the stress is redistributed.

Three-Fiber System (Large Crack)

The following section describes the tests results obtained on model composite systems in which the fiber fractured and propagated a relatively large crack into the matrix ($a = r_a$). A larger crack occurs in a model composite when the fiber fractures and the crack propagates up to the adjacent fibers. As in the preceding section, two distinct fiber spacings are addressed which are representative of local fiber volume fractions of 15 and 20%. The reader should be aware that the crack size in these larger crack composites decreases with increasing fiber volume fraction. That is, in a 15% v_f^l model composite, $a = 4.9$ mm, while in the 20% v_f^l model composite, $a = 2.7$ mm. In the previous section the crack size was not a function of local fiber volume fraction as it is here.

For the 15% fiber volume fraction composite, a large crack occurred in 7 of the 15 specimens tested. Strain data obtained at the near-field gage location at $c = 0$ suggest a strain concentration of 1.76. This is in sharp contrast to the 1.28 value measured (presented in previous section) in a composite containing the same fiber volume fraction with a small crack (see Carman et al. [*19*] for a mechanics treatment). For the outer field region at $c = 2.5$ mm, a strain concentration of 1.25 is measured on the adjacent fiber. This suggests that a large strain gradient exists across the diameter of the fiber. The ineffective length measured from pictures taken of the photoelastic resin did not seem to be consistent with the test performed on the larger crack size. Values as low as 16 mm and as large as 26 mm are observed. The inability to distinguish a definitive demarcation in the fringe patterns along the boundary of the ineffective length caused a high degree of skepticism in these measurements. Nonetheless, this length is considerably larger than the ineffective length experimentally determined for the same fiber volume fraction composite with a small crack (i.e. 6 to 10 mm). Referring once again to Rosen's [22] shear lag model, the ineffective length is independent of crack size and thus predicts the same value quoted in the previous section (i.e. 30 mm) for this fiber volume fraction. This theoretical value appears to be a reasonable estimate of the ineffective length for this fiber volume fraction and crack size.

In the 20% fiber volume fraction composite containing a large crack, some interesting discoveries are presented. Considering the strain data obtained on an adjacent fiber in the near-field region at $c = 0$, the strain concentration values measured are virtually indistinguishable (i.e. within the accuracy of the measurement devices) from the measurements made on the 15% fiber volume fraction composite (i.e. 1.76). This is attributed to the relatively compliant matrix not contributing significantly to the load borne by the unbroken fiber. This is assumed in shear lag type of solutions where v'_f does not effect strain concentration [29]. Once again, this is in sharp contrast to the measurement made on the same fiber volume fraction composite with a smaller crack size (i.e. 1.40). The ineffective length measurements accomplished with this composite system and crack size is also comparable to that described above (14 to 24 mm compared to 4 to 8 mm for similar fiber volume fractions with smaller crack size). The length measured in these tests is also subject to some skepticism due to the lack of a definitive demarcation. As noted in the previous section, Rosen's [22] shear lag model predicts an ineffective length for this fiber volume fraction of 26 mm. Therefore, Rosen's model appears to provide reasonable estimates for ineffective lengths in macromodel composites containing larger cracks. This is attributed to the lack of axial load transfer in the matrix material (i.e. it is actually cracked). This assumption (i.e. matrix supports no axial load) forms the basis for shear lag models and would therefore be expected to provide better correlation with a macromodel composite containing a large crack.

Six-Fiber System

The following subsection details the results obtained on a model composite containing six fibers. The fiber spacing in this composite system is $b = 4.2$ mm and represents a composite whose local fiber volume fraction is 20%. This fiber volume fraction is identical to the three-fiber system with a comparable fiber spacing described in the two previous sections. The results presented in this subsection depict the model composite's ability to investigate multiple fiber fracture which occurs in composite systems. In the current study, the two central fibers are notched in the same axial plane (i.e. $z = 0$), with one notched more severely than the other. The more severely notched fiber should theoretically fail first, followed by the less severely notched fiber. Strain gages in this composite system are attached to fibers in the near-field region closest to the more severely notched fiber at $c = 0$ and in the near-field region of the fiber closest to the less severely notched fiber at $c = 21$ mm. That is, the strain gages bracket the notched fibers at the specified axial locations.

Upon loading the six-fiber composite, the more severely notched fiber fails at an applied traction of 3.15 MPa. The fiber fracture did not propagate a crack into the matrix and is adequately represented by a small crack ($a = 1.5$ mm $= r_f$). Following this fiber failure, data are collected from the two strain gages to determine the strain concentration which exists in the neighboring fibers. The strain concentration determined from the results of the gage located at $c = 21$ mm is found to be 1; in other words, no strain concentration exists. The strain concentration on the near-field gage at $c = 0$ is found to be 1.40. This value is nearly identical to the strain concentration value determined for the three-fiber composite with a similar crack size and fiber spacing described previously [see section on three-fiber system (small crack)]. These results strongly suggest that the stress field is unperturbed outside the first set of adjacent fibers to a fiber fracture.

After collecting these data, the composite is again loaded until the second fiber fails. This occurs at an applied traction of 2.43 MPa. When the second fiber fails in the composite, a crack did not form in the matrix material. Thus, the two fractured fibers are adequately represented by two independent small crack sizes. The data collected from the strain gages on the fibers reveal a strain concentration at $c = 21$ mm of 1.1, while at $c = 0$, the strain concentration is 1.7.

After collecting these data, the composite is again loaded, at which time two more adjacent fibers fail (a total of four have failed at this time). These fibers fail at an applied traction of 4.54 MPa, and the crack does extend into the matrix material. Only the two end fibers and the matrix material outside this region remain intact. The strain concentration calculated from the strain gages located at $c = 21$ mm is found to be 0.6 (net decrease due to fiber fracture), while the gage at $c = 0$ became inoperable due to damage.

Interface/Interphase

As noted in the material section, both silane coupling agent and vacuum grease are applied to the structural fibers to alter the adhesion between the fibers and the surrounding matrix. The silane coupling agent applied to the glass rods generates a composite with sporadic voids or air pockets along the interface between the fiber and matrix. These voids predominantly occur around the resistance strain gages attached to the fibers. Nonetheless, in regions which are devoid of these air pockets, the adhesion between the glass rods and the matrix is strong. However, the difference in the fiber/matrix interface strength in silane-treated fibers as compared to untreated fibers appears to be insignificant. This statement is based on a fiber push-through test accomplished on a coated and an uncoated specimen. The fact that an uncoated fiber exhibits adequate adhesion and the silane coupling agent actually causes sporadic voids along the interface leads one to the conclusion that an uncoated fiber is an excellent choice for the study of composites in which the fibers are strongly bonded to the matrix. The vacuum grease applied to the glass fibers produces a composite in which virtually no adhesion occurs between the fiber and matrix, that is, no chemical bond and only minimal mechanical interlocking exists. The fact that the glass rods can actually be removed from the composite by hand attests to this fact.

In model composites containing embedded glass fibers coated with epoxide, a number of small air bubbles are present along the surface of the interphase (i.e. interface between the interphase and matrix). However, in several composites manufactured, the air bubbles are not present. The differences noted between the two batches is that the amount of hardener added to one batch was different from the other. In the majority of the composites manufactured with the epoxide interphase, the air bubbles present on the interphase surface inhibited accurate strain measurements at the gage locations. This is mainly due to fiber slipping in the matrix, which causes a slip stick–phenomena to be recorded. However, in these tests it is found that when the composite fails, the fibers pull out of the matrix. The exposed fiber length remains coated with the epoxide, such that failure occurred between the interphase and matrix regions.

In composite materials containing fibers coated with BMI, there appears to be regions of relatively large voids along the interface between the interphase and matrix. This may be due to the relatively low cure temperature cycle (i.e. 100°C instead of 200°C) performed on the interphase material prior to manufacturing the composite. This cure cycle is used because of the thermal constraints on the gage adhesive used to secure the strain gages to the fibers (see manufacturing section). The large voids present in this composite are in sharp contrast to the small bubbles present in the epoxide-coated composite system. Once again, due to the presence of these voids, no accurate strain measurements can be accomplished regarding fiber fracture. Upon loading the composite the strain readings jump, thus signifying a fiber/interphase slip. As noted for the composite described above, when the composite fails the fibers pull out of the matrix region. However, in this case the fiber is clean (as could be seen by eye). The failure occurs between the interphase and the fiber in this composite system.

The only quantitative results regarding the effect of interphases on strain redistribution presented in this paper are with data obtained from a PES-coated fiber system. The presence of voids along the interphase surface is minimal in the macromodel. Also, when the composite catastrophically fails, little to no fiber pullout is present. This further signifies that the interface between fiber/matrix/interphase is sufficiently strong to transfer load between the fiber and matrix. Therefore, this material system represents an adequate model to initiate studies on interphase effects and their relationship to stress redistribution in the presence of fiber fracture.

In tests performed on the PES-coated three-fiber model composite system ($v'_f = 15$), the center fiber fails at an applied traction of approximately 3.35 MPa. This value is comparable to the stress level achieved in a large portion of the uncoated three-fiber systems. When the fiber fractures, the crack propagates half the distance between the fiber and matrix ($a = b/2 + 1.5$ mm). In comparison to composites containing uncoated fibers made under the same circumstances, the crack propagated up to the adjacent fiber ($a = b - 1.5$ mm $\equiv r_a$). The latter crack is a larger crack. This result suggests that the interphase coating is capable of altering the subsequent crack propagation into the matrix. The energy released when the fiber fractures is distributed between shock wave formation, matrix cracking, and fiber slip. It is believed that in an uncoated fiber system no slip occurs. However, in a coated system, especially for PES coatings where only mechanical interlocking may be present (see Material section), fiber slip may occur. This would tend to decrease the size of the crack formed in the matrix. In fact, a smaller crack is observed in composites coated with PES. No definitive results are presented, which indicate that local fiber slip is actually occurring in this composite even though it is postulated in this paper.

The data obtained on a near-field gage placed on an adjacent fiber at $c = 0$ reveals a strain concentration of 1.46, while the strain concentration on an outer field gage at $c = 0$ reveals a strain concentration of 1.14. These strain concentration values are not believed to be significantly altered by the interphase coating but are attributed to crack size. This statement is based on the facts that the interphase is capable of transferring load between the fiber and matrix, the interphase stiffness is not appreciably different than that of the matrix, and the interphase region is not extraordinarily large. This statement is further substantiated by results obtained by adhering strain gages to the interphase region. Comparisons performed between the data obtained from these sensors to strain measurements performed at comparable locations on the fiber revealed indistinguishable results.

The final results obtained with the PES-coated fiber system involve a regular array of fiber breaks in the composite. The composite is initially loaded until the notched fiber fractures, at which time data are gathered on strain concentration effects. An additional load is applied to the composite, at which time an adjacent fiber fractures at $z = 18$ mm.

Upon further loading of the composite, additional fiber fractures occur sporadically in the composite. The fractures are arranged in a staircase type of distribution, such that no two exist in the same z plane. This phenomenon of staircasing is not noted in any of the uncoated specimens. These results suggest that the strain concentration in the PES-coated fiber composites may be redistributed over a larger region (i.e. larger ineffective length), such that multiple fiber fracture occurs in the composite as opposed to catastrophic failure.

Conclusion

A test methodology employing a macromodel composite utilizing embedded strain gages and a birefringent matrix was presented. Results obtained with the embedded resistance gages and the embedded fiber optic strain sensors (FP-FOSS) were validated with classical test and analytical techniques. These techniques included model composites subjected to thermal effects and mechanical loading sequences. The ability to vary specific physical parameters in the experimental model, such as fiber aspect ratio, fiber volume fraction, and interphase/interface, in a systematic fashion enables the current technique to study various physical aspects present in actual composite systems. This model was shown to be useful for the validation (or invalidation) of micromechanical models (e.g. fiber fracture analysis) presently being used or currently being developed by the scientific community.

The capability to initiate a fiber fracture at a specified location and load level was demonstrated. It was revealed that significantly different strain concentrations and ineffective lengths occur in PMC composites which contain different fiber volume fractions and crack sizes. These results raised serious doubts on using shear lag type of solutions to analyze the stress redistribution around a fiber fracture under certain circumstances. The fact that multiple fiber fracture could be achieved and interphase coatings could be applied in a methodical fashion demonstrated the versatility of the model. These tests suggested that the perturbed stress field around a fiber fracture does not significantly extend past the first set of adjacent fibers. Furthermore, interphase coatings may alter the subsequent crack propagation into the matrix material. These studies demonstrate that this experimental technique can model various physical phenomena which occur in actual composite systems.

Acknowledgments

This research has been supported by grants from the Virginia Institute for Material Systems at Virginia Tech and the National Science Foundation under grant No. MSS-9115380 for which we are grateful. We would also like to thank the Fiber Electro Optic Research Center at Virginia Tech, headed by Richard O. Claus, for their technical support with the fiber optic sensors.

References

[1] Galiotis, C., "Interfacial Studies on Model Composites by Laser Raman Spectroscopy," *Composite Science & Technology,* Vol. 42, 1991, pp. 125–150.

[2] Cox, H. L., "The Elasticity and Strength of Paper and Other Fibrous Materials,"*British Journal of Applied Physics,* Vol. 3, 1952, pp. 72–79.

[3] Choon, T. C. and Sun, C. T., "Stress Distributions Along a Short Fiber in Fiber Reinforced Plastics," *Journal of Material Sciences,* Vol. 15, 1980, pp. 931–938.

[4] Fukuda, H. and Chou, T., "An Advanced Shear-Lag Model Applicable to Discontinuous Fiber Composites," *Journal of Composite Materials,* Vol. 15, 1981, pp. 79–91.

[5] Chou, T., Seiichi, N., and Minoru, T., "A Self-Consistent Approach to the Elastic Stiffness of Short-Fiber Composites," *Journal of Composite Materials,* Vol. 14, 1980, pp. 178–187.

[6] Laws, N. and McLaughlin, R., "The Effect of Fibre Length on the Overall Moduli of Composite Materials," *Journal of Mech. Phys. Solids,* Vol. 27, 1979, pp. 1–13.

[7] Aboudi, J., "Elastoplasticity Theory for Composite Materials," *Solid Mechanics Archives,* Vol. 11, 1986, pp. 141–183.

[8] Carman, G. P., "Modeling the Elastic Behavior of Current and Proposed Metal Matrix Composites," masters thesis, University of Alabama, Tuscalosa, 1988.

[9] Batdorf, S. B., "Tensile Strength of Unidirectionally Reinforced Composites—I" *Journal of Reinforced Plastics & Composites,* Vol. 1, 1981, pp. 153–177.

[10] Carman, G. P. and Reifsnider, K. L., "Micro-Mechanics of Short Fiber Composites," *Composites Science & Technology,* Vol. 43, January 1992.

[11] Verpoest, I., Desaeger, M., and Keunings, R., "Critical Review of Direct Micromechanical Test Methods for Interfacial Strength Measurements in Composites," *Controlled Interphases in Composites,* Proceedings of the Third International Conference on Composite Interface, H. Ishida, Ed., 1990, pp. 653–665.

[12] Whitney, J. M. and Drzal, L. T., "Axisymmetric Stress Distribution Around an Isolated Fiber Fragment,"*Toughened Composites, ASTM STP 937,* N. J. Johnson, Ed., American Society for Testing and Materials, Philadelphia, 1987, pp. 179–196.

[13] Ashbee, K. H. G. and Ashbee, E., "Photoelastic Study of Epoxy Resin/Graphite Fiber Load Transfer," *Journal of Composite Materials,* Vol. 22, 1988, pp. 602–615.

[14] James, M. R., Morris, W. L., and Cox, B. N., "A High Automated Strain-Field Mapper," *Experimental Mechanics,* 1990, pp. 60–67.

[15] Fredrick, D. and Chang, T. S., *Continuum Mechanics,* Scientific Publishers Inc., Boston, MA, 1972, pp. 34–37.

[16] Broek, D., *Elementary Engineering Fracture Mechanics,* Martinus Nijhoff Publishers, Boston, MA, 1987, pp. 3–29 and 75–91.

[17] Lesko, J. J., Carman, G. P., Fogg, B. R., Miller, W. V., Vengsarkar, A. M., Reifsnider, K. L., and Claus, R. O., "Embedded Fabry-Perot Fiber Optic Strain Sensors in the Macro-Model Composite," *Optical Engineering SPIE,* January 1992.

[18] Walker, P., "The Use of Silane Adhesion Promoters in Polymer Industries," *Materials and Processing,* Birmingham, United Kingdom, 1989, pp. 227–235.

[19] Carman, G. P., Lesko, J. J., and Reifsnider, K. L., "Micromechanical Analysis of Fiber Fracture," *Composite Materials: Fatigue and Fracture,* this publication.

[20] Carman, G. P., "Micromechanics of Finite Length Fibers in Composite Materials," Ph.D. dissertation, Virginia Tech, Blacksburg, VA, December 1991.

[21] Hashin, Z. and Rosen, R. W., "The Elastic Moduli of Fiber-Reinforced Materials," *Journal of Applied Mechanics,* 1964, pp. 223–232.

[22] Rosen, B. W., "Fiber Composite Materials," *Proceedings of American Society of Metals,* Chap. 3, 1964, pp. 37–75.

John J. Lesko,[1] *Gregory P. Carman,*[1] *David A. Dillard,*[1] *and*
Kenneth L. Reifsnider[1]

Meso-Indentation Testing of Composite Materials as a Tool for Measuring Interfacial Quality

REFERENCE: Lesko, J. J., Carman, G. P., Dillard, D. A., and Reifsnider, K. L., "Meso-Indentation Testing of Composite Materials as a Tool for Measuring Interfacial Quality," *Composite Materials: Fatigue and Fracture, Fourth Volume, ASTM STP 1156*, W. W. Stinchcomb and N. E. Ashbaugh, Eds., American Society for Testing and Materials, Philadelphia, 1993, pp. 401–418.

ABSTRACT: A unique application of indentation testing to composite materials is presented. The primary goal is to evaluate the sensitivity of conventional penetration techniques to fiber-matrix bond strength in laminated composites. The Vickers fixed-load hardness test and the continuous ball indentation test (CBIT) are employed for the penetration of continuous fiber-reinforced epoxy systems. These indentations are performed in the axial (fiber) direction of laminated unidirectional composites. Both techniques demonstrate significant sensitivity to fiber-matrix adhesion in graphite/epoxy systems which possess a systematic variation to their interfacial characteristics. The CBIT displayed the best opportunity for making quantitative measurements of interfacial strength. An innovative micromechanics model of the composite contact by a spherical indenter predicts trends and failure phenomena observed in the tests. Experimental and analytical results suggest that the ball penetration test is sensitive to the shearing characteristics of the composite. Preliminary results also reveal interrelationships between composite strength properties and hardness values.

KEY WORDS: Vickers hardness, continuous ball indentation testing, composite hardness testing, meso-indentation of composites, interfacial shear strength, interfacial adhesion assessment

The interface/interphase which exists between fiber and matrix strongly influences the properties and overall performance of its composite [1–3]. The strength of this region has recently received a great deal of attention in the literature, as it has been shown to have profound effects on laminate strength [4–7]. However, no one test for measuring interfacial shear strength (ISS) stands out as an accurate and precise means of achieving such measurements in a laminated continuous fiber composite. It is believed that this deficiency is associated with the inability to produce and measure the appropriate phenomena. Both macro and micro techniques possess inadequacies in developing and understanding the state of stress at the interface/interphase when investigating the strength of this zone. Moreover, interpretation of these results provides additional uncertainty in the values obtained.

In an effort to overcome the detracting features of the present technology, meso-

[1]Graduate research assistant, research scientist, associate professor, and Alexander Giacco professor, respectively, Material Response Group, Dept. of Engineering Science and Mechanics, Virginia Polytechnic Institute and State University, Blacksburg, VA 24061-0219.

indentation testing of composite materials is investigated as a potential means of measuring interfacial quality. In particular, two very different indentation techniques will be considered: the Vickers fixed-load microhardness test and the continuous ball indentation test (CBIT). These methods are performed with the load applied along the fiber axis of laminated composite sections as depicted in Fig. 1. Such indentations/penetrations contact a region of both fibers and matrix, unlike the single fiber micro-indentation test (SFMI), which measures ISS by pushing a single fiber from its surrounding matrix.

In general, this study is an introduction to the concept of composite meso-indentation testing (a level of investigation between micro and macro techniques). The observations which are made serve to highlight the potential for future application of these and similar techniques. The primary objectives of this study are stated as follows: (1) to determine if penetration testing of composites is sensitive to interfacial quality; (2) to investigate the mechanics which contribute to the observed indentation responses; and (3) to assess the viability of employing composite penetration testing to make quantitative measures of ISS.

Materials

Three graphite/epoxy systems which possess as much as a 219% difference in ISS were provided by the Composite Materials and Structures Center at Michigan State University. The AS-4C, AS-4, and AU-4 fibers in an Epon 828 mPDA epoxy were reported to possess an ISS of 81.4, 68.3, and 37.2 MPa, respectively, as determined by the single fiber fragmentation test (SFFT) [4]. The AU-4 fiber composite was shown to exhibit frictional sliding as an interfacial failure mechanism, while the AS-4 fiber SFFT specimen revealed interfacial crack propagation. In addition, the AS-4 fiber composite was stated to have an ISS "near" that of the neat matrix yield strength, as observed in the SSFT coupon. The AS-4C fiber composite was reported to possess an "interphase of less plasticity" resulting in matrix crack propagation. Madhukar and Drzal [4–7] fully characterized these systems, showing that with a decrease in ISS the strength of the laminates decreased as well when loaded in tension, compression, shear, and flexure. These three systems provide an excellent opportunity to explore the sensitivity of composite indentation testing to interfacial quality.

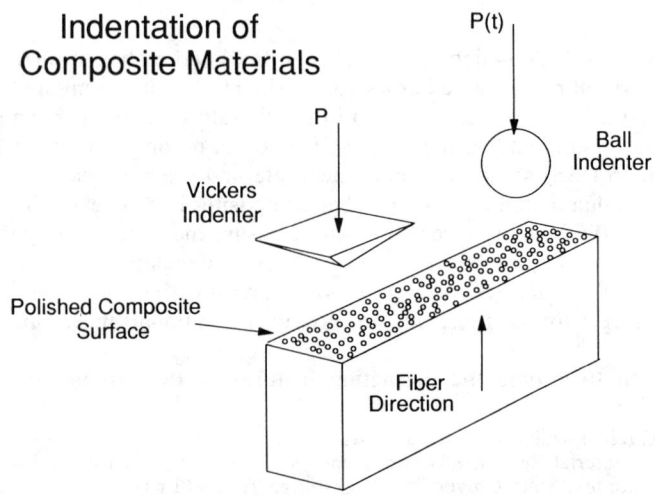

FIG. 1—*Indentation testing of unidirectional continuous fiber composites.*

Sample Preparation

Unidirectional composite sections 25 mm long by 12 mm wide, and nominally twelve plies thick or greater, were cut from laminated panels by means of a water-cooled diamond saw. The specimens were then mounted in a standard low-shrinkage, room-temperature-cure, metallographic mounting epoxy. The mounted sections were polished in a two-step process while ensuring that the fiber direction remained perpendicular to the polished surface. Both stages of polish occurred in the presence of water at room temperature, with each specimen experiencing no more than 30 min at 100% humidity during the entire process. Immediately following polishing, the specimens were conditioned in a desiccator (at roughly 10% humidity) at room temperature for a minimum of one week prior to testing. Scanning electron microscopy (SEM) of the specimen surfaces revealed a uniform and even surface with little to no preferential polishing of the fiber to the matrix. The fiber ends appeared cleanly in the surrounding matrix and possessed a smooth finish similar to that of the epoxy.

Vickers Fixed Load Microhardness Testing

A standard microhardness tester fitted with a Vickers diamond square pyramid is employed for fixed-load testing. All penetrations were performed with the standard instrument settings as calibrated by the manufacturer. A single predetermined load, P, is applied for indentation and withdrawn after the set duration of ~ 15 s. A typical residual indentation in a polymeric matrix composite is shown in Fig. 2. For graphite/polymer systems, 2 to as many as 500 fibers are contacted for loads ranging from 25 to 3000 g. The diagonals of

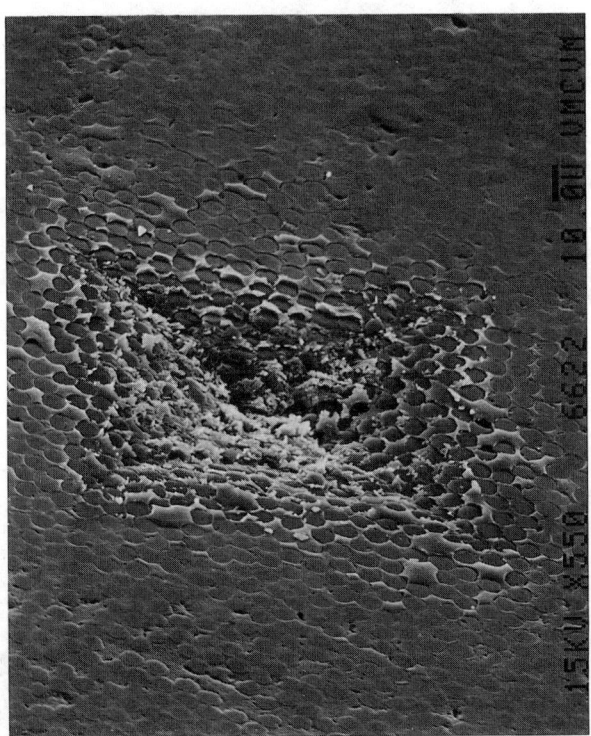

FIG. 2—*SEM of a typical Vickers residual impression in a graphite/epoxy composite.*

the square residual impression are carefully measured, averaged, and used to determine a contact area, A_c. The hardness, or applied indenter pressure, is computed by dividing P by A_c, resulting in a value with units of stress or pressure. Sources of error arise from uncertainties in determining the length of the diagonals. Identifying the outer bounds of the indentation in this inhomogeneous surface further complicates the task. These difficulties may have contributed to the scatter in data, which is at worst two to four times greater than the scatter observed on isotropic materials under similar conditions. It is, however, important to point out that with a uniform material (i.e., a standard hardness test block), hardness testing of this kind is generally very precise. Thus, Vickers data obtained on composites is still useful as was demonstrated by Lesko [8].

Five indentations were made at each of five loads and averaged to determine the Vickers hardness (HV) as a function of load level. These trials, performed on the three varied interface composites, reveal some sensitivity of Vickers indentation test to the interfacial quality, as shown in Fig. 3. The error bars are the actual scatter in data and do not represent standard deviation. Error bars were excluded for the AS-4 composite so as to avoid cluttering the plot; however, the scatter for this system is similar to its counterparts. The respective residual impressions were all very similar in degree and type of damage produced, with the only difference being the relative sizes of the impressions. The characteristic damage observed is typical of that shown in Fig. 2. All three systems exhibit an analogous trend in hardness as a function of load, differing only in their relative level of hardness. Previous studies suggest that the increase in hardness at the lower loads results from the manner in which damage is produced by the indenter. This is believed to be related to the number of fibers contacted and the properties of the matrix material [8].

Thus, the Vickers technique has demonstrated the ability to distinguish between dis-

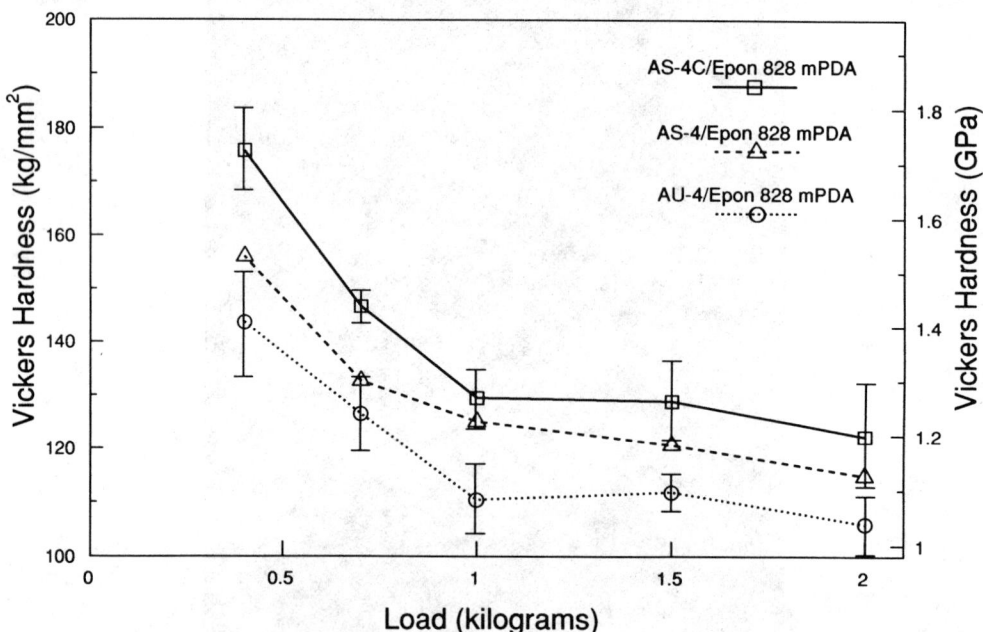

FIG. 3—*Vickers hardness as a function of load for the graphite/Epon 828 mPDA composite with three variations in interfacial strengths.*

tinctly contrasting interfaces displayed in the AU-4 and AS-4C composites. The variations observed between the Vickers hardness results and the reported ISS [7] are not in similar proportions and reveals only limited sensitivity. However, it is important to point out that this sensitivity was observed using a relatively small section of actual composite, completed in a short period of time, and performed with equipment common to a materials laboratory. This provided the incentive for further investigation of other penetration techniques for testing composites.

Continuous Ball Indentation Testing

Load and Indenter Penetration from the CBIT

In contrast with the fixed-load pyramid indenter test which provides a fully plastic indentation, the CBIT is capable of producing fully elastic and/or elastic-plastic indentations. The incorporation of a load and displacement transducer within the indentation system also allows for the real-time measurement of applied indenter pressure throughout the process. Load is measured through a load cell (with a resolution of ~10 g) placed in line with the applied indenter load. The indenter penetration depth is monitored with an LVDT possessing a resolution of 0.1 μm. The displacement is measured relative to the indenter head and the specimen surface. Indentations are made quasi-statically in displacement control at a rate of ~4 μm of penetration per second with a 1.58-mm hardened steel ball. This system is an experimental test frame developed by the lead author and Dr. Ronald Armstrong of the Department of Mechanical Engineering at the University of Maryland.

The result is a continuous record of load, P, and indenter penetration depth, h, as shown in Fig. 4 for indentations made in the varied interface composites. The recorded P versus h response consists of a loading portion (origin to Point C) and an unloading portion (Point C to Point D), resulting in a residual penetration depth at Point D. The points marked A are believe to represent an elastic limit (subsurface interfacial failure) for the respective materials. Indentations which do not exceed Point A reveal minimal hysteresis in the loading and unloading cycle. The hysteresis results from two mechanisms: frictional sliding at the fiber-matrix interface, and permanent deformation of the constituents. The energy dissipated in frictional sliding is much less than that absorbed to deform the material for the same amount of displacement. In addition, penetrations made below this limit reveal little to no evidence of residual depth or surface damage. The response beyond Point A is predominantly plastic deformation that includes interface failure as well as matrix deformation.

The curves of Fig. 4 provide some qualitative information about the level of adhesion between the fiber and matrix. This is best seen by comparing the hysteresis loop for the AU-4 composite to the other two stronger interface composites. As the ISS increases, the width of this loop appears to become larger. When comparing the residual penetration depths, the AU-4 is the most shallow of the three systems. This suggests that the weak interface is not able to remain adhered to the fiber, resulting in a frictional sliding situation. Having broken the bond between fiber and matrix, the constituents are able to slide past one another, which allows them to return to a nearly undeformed position (i.e. shallow residual depth) when unloaded. Conversely, the stronger interface is capable of transferring a sufficient amount of stress so as to permanently deform the surrounding matrix. With the fibers still adhered to the deformed epoxy, elastic recovery is minimized, resulting in a larger hysteresis loop and a deeper residual impression.

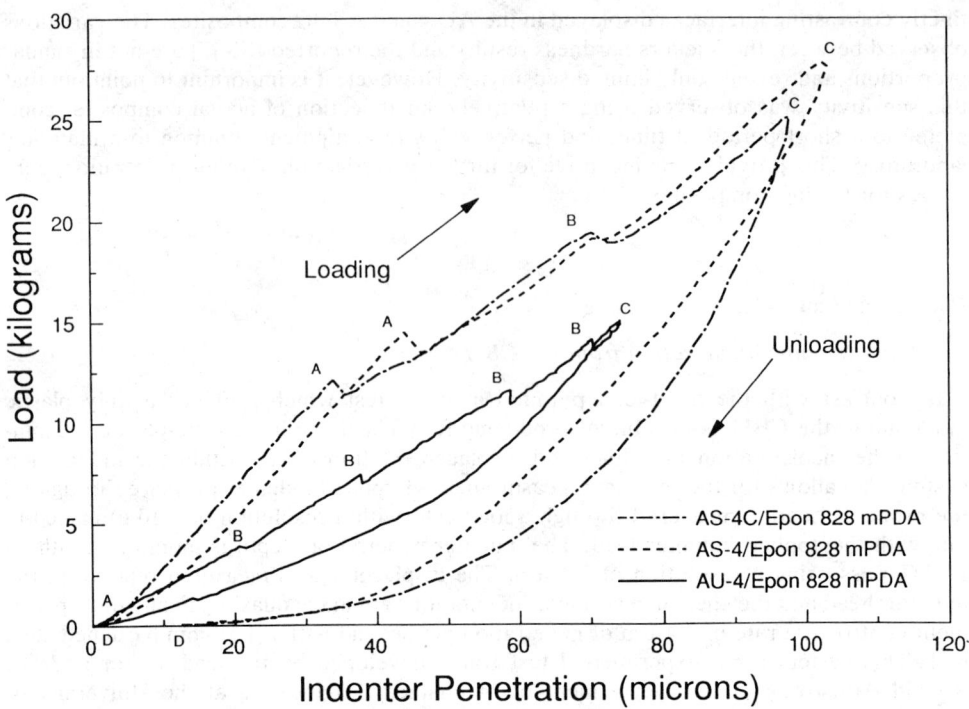

FIG. 4—*The load versus indenter penetration response from the CBIT of graphite/Epon 828 mPDA for three variations in interfacial strength.*

Review of the Residual Impressions

Scanning electron micrographs (SEMs) of the residual ball impressions reveal damage information consistent with the CBIT data. Figures 5 through 7 are representative of ball penetration damage in AS-4C, AS-4, and AU-4 fiber composites, respectively. It is important to point out that, for the AS-4C and AS-4 fiber composites, this type of damage was only observed after exceeding Point A. For the AU-4 fiber composite this damage is believed to appear on the surface at the first Point B. Clearly the AU-4 composite responds quite differently than its two stronger interface counterparts. Concentric rings of fiber-matrix interface failure are distinguishable, with little plastic deformation occurring in the matrix. There appears to be four rings of this kind, which correspond to the four secondary breaks, Point B, of Fig. 5. In contrast, micrographs of the AS-4C and AS-4 fiber composites are very similar, however differing greatly from that of the AU-4 composite. Both possess a central region that is depressed but appears to be otherwise undisturbed. Also, two distinct and characteristic deformation modes are observed on opposing halves of the impression. One zone appears "abraded" with the fibers bent over towards the center of the indentation (right portion of Fig. 6 and upper left of Fig. 7). On the opposite side is a region of more deeply depressed material with the fiber normal to the surface, the "depressed" zone. In both cases the matrix is grossly deformed and appears to remain adhered to the fibers.

As will be shown in the analytical study, the subsurface response is crucial to the understanding of this behavior. Preliminary studies reveal the presence of characteristic

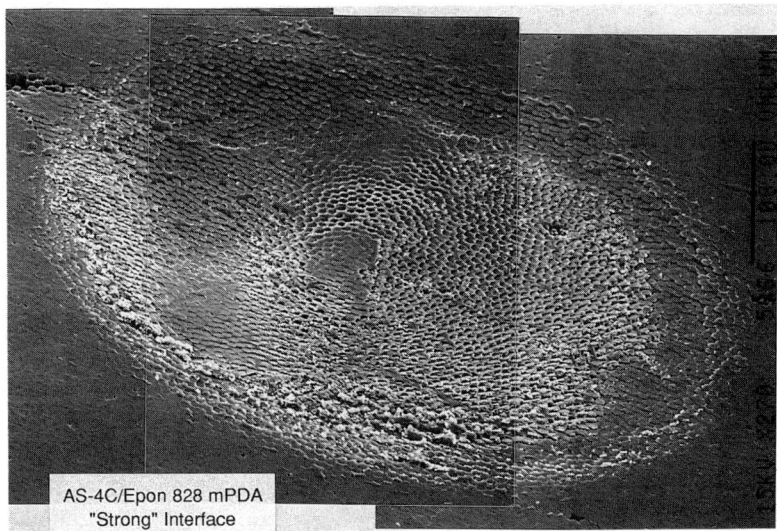

FIG. 5—*SEM of a residual ball impression in AS-4C/Epon 828 mPDA.*

subsurface damage in all three systems, which is the result of interfacial failure. These subsurface features are repeatable, consistent, and correspond to the particular surface damage mentioned above. This characteristic deformation state causes the transition at Point A, i.e. interfacial failure. Further investigation using various destructive and nondestructive techniques will be used to describe in detail the initiation and propagation of damage.

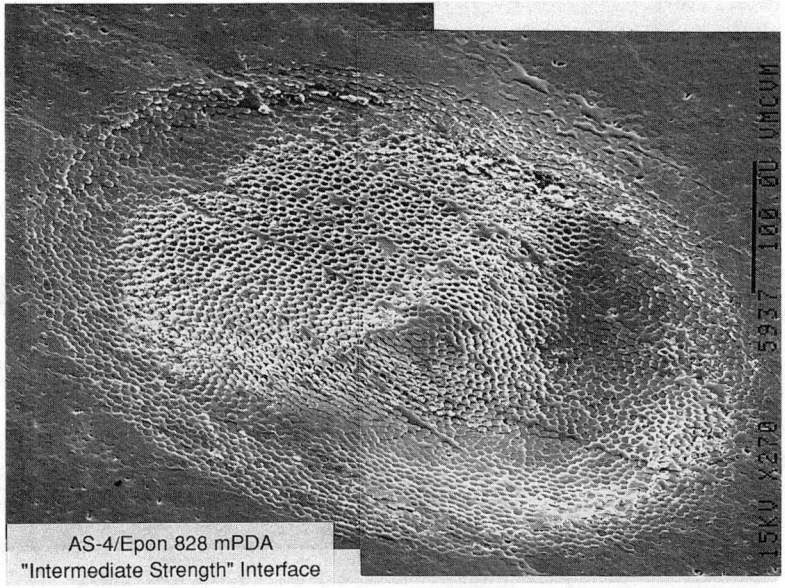

FIG. 6—*SEM of a residual ball impression in AS-4/Epon 828 mPDA.*

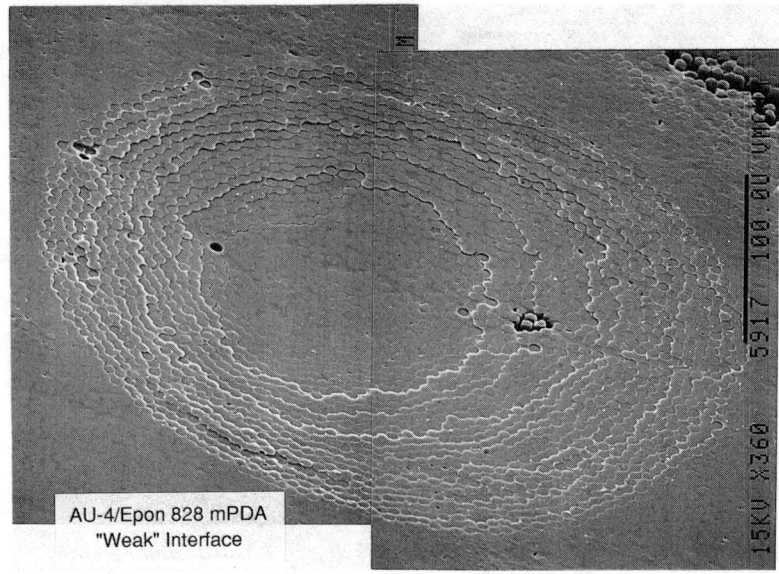

FIG. 7—*SEM of a residual ball impression in AU-4/Epon 828 mPDA.*

Representative Stress-Strain Response from the CBIT

A more quantitative assessment of interfacial adhesion can be made by reducing the data in Fig. 4 to a record of average applied indenter pressure as a function of a representative strain quantity. First, the contact diameter, $d = 2a$, is computed from the indenter penetration, h, by,

$$d = 2\sqrt{Dh - h^2} \qquad (1)$$

where D is the diameter of the ball penetrator. The elastic and plastic mean hardness pressure (MHP) produced by the indenter is simply calculated from

$$\text{MHP} = \frac{P}{A_c} = \frac{4P}{\pi d^2} \qquad (2)$$

where A_c is the projected contact area of the ball on the surface. With the MHP as the stress, the strain is represented as a ratio of the contact diameter to the ball diameter, d/D [8]. These physical quantities, which result from the mechanics of contact, have been shown to precisely describe the stress-strain behavior of a material under penetration by a spherical penetrator [9]. For a fully elastic contact situation described by Hertz [10], the slope of the MHP versus d/D line is given by the following constant

$$\frac{\text{MHP}}{d/D} = \frac{4\bar{E}}{3\pi} \qquad (3)$$

where the contact radius is related to the load through the Hertzian relation

$$\frac{d}{2} = \sqrt[3]{\frac{3PD}{4}\bar{E}}; \quad \bar{E} = \text{contact stiffness} \qquad (4)$$

Thus, any inelastic response will cause a deviation from this linear relationship.

The resulting representative stress-strain responses for the loading portion only are shown in Fig. 8. The computed curves are noisy at the lower range, and some information has been lost due to the low resolution of the data collection system, particularly in the measurement of load. This is particularly true for the AU-4 system which deviated from linearity at a very low load, ~2 kg. Refinement of the technique is currently underway to improve the sensitivity of measurements made at this level.

Figure 8 explicitly provides quantitative stress and strain information from the composite's response to a ball indentation. This representation of the data exposes nonlinearities, marked *NL*, in the response of the two stronger interface composites at an MHP just under 600 MPa. Again, the transitions marked at Point A signify a change from elastic to plastic dominated deformation. The subsequent secondary breaks, Point B, are easily distinguishable; however, they do not appear to affect the general constant trend in the hardness. Following Point A, the AS-4C and AS-4 composites respond at a constant hardness of about ~550 MPa while the AU-4 composite maintains a hardness of ~380 MPa. What is believed to the most important information gained from Fig. 8 is the maximum mean hardness pressure (MMHP) (i.e. the MHP observed at Point A) which is believed to provides a measure of interfacial quality. This assertion is quantified more precisely through the following discussion and a summary presented at the end of this section.

Table 1 summarizes the average representative stress-strain information based on eleven, ten, and seven indentations in AS-4C, AS-4, and AU-4 fiber composites, respectively. The averages and standard deviations reported are computed from Weibull analy-

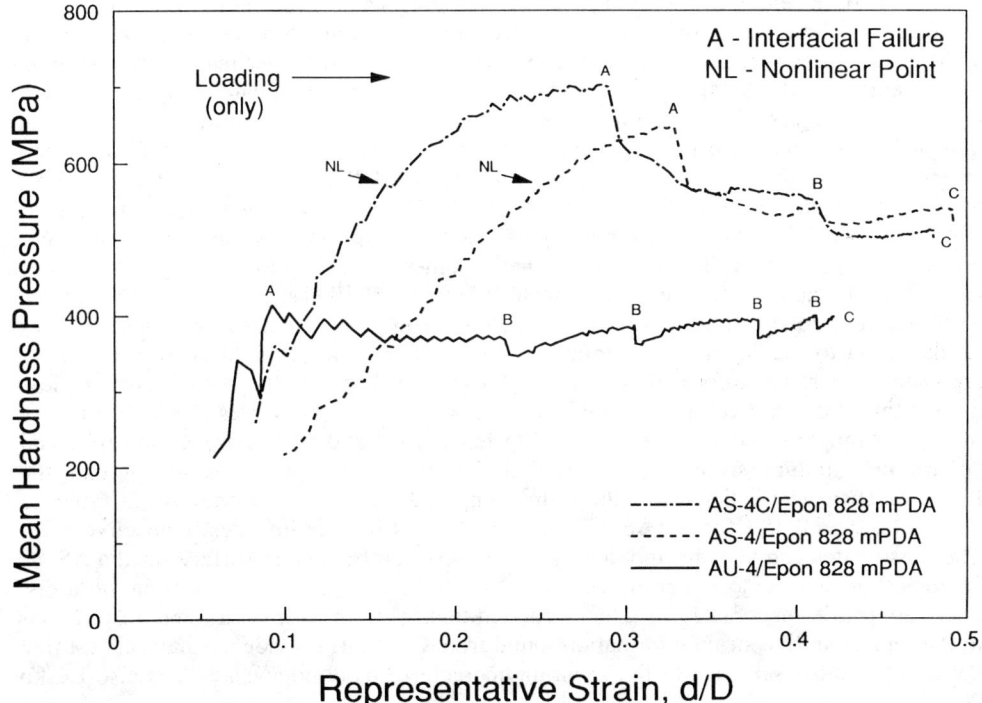

FIG. 8—*Representative stress versus strain response from the CBIT of three graphite/ Epon 828 mPDA systems of varied interfacial strength.*

TABLE 1—*Summary of the CBIT results on graphite/Epon 828 mPDA systems with correlations to composite laminate strength properties.*

	AS-4C/Epon 828 mPDA "Strong" Interface	AS-4/Epon 828 mPDA "Intermediate Strength" Interface	AU-4/Epon 828 mPDA "Weak" Interface
MMHP, MPa	698 ± 49	659 ± 38	460 ± 88
d/D at MMHP	0.274 ± 0.040	0.318 ± 0.026	0.098 ± 0.020
MHP at nonlinear, MPa	604 ± 42	593 ± 46	...
d/D at nonlinear	0.192 ± 0.060	0.199 ± 0.050	...
ISS/MMHP	0.12	0.10	0.08
Shear strength/MMHP	0.14	0.13	0.10
Compression strength/ MMHP	1.7	1.4	1.5

sis, which provided the best representation of the distribution of data. The MMHP for the AS-4C composite is roughly 50% greater than that for the AU-4 system. There is a 6% difference in MMHP between the AS-4C and AS-4 composites. In relative comparisons with SFFT results, the trends in MMHP are as expected with some similarities in proportion. However, the authors do not intend to make direct comparisons with the SFFT since fundamental differences exist between the two techniques (e.g., specimen type, loading, data interpretation). The average data for the nonlinear points, NL, of Fig. 8, for the AS-4C and AS-4 fiber composites are very close in both MHP and d/D, possibly suggesting that they are related to a similar phenomena (i.e. matrix plasticity).

Since definite relationships exist between the interfacial quality and the laminate performance [4–7], there may also exist correlation of indentation response data to the laminate performance. Thus, ratios of ISS and composite properties (shear and compression strength) to the MMHP are made to study trends and investigate interrelations. We investigated compression strength due to the nature of the contact situation and with the knowledge that the shearing responses at the fiber-matrix interface is crucial to the micromechanics of compression strength. The ISS/MMHP ratios for the three systems average, 0.10, that is, the ISS is 10% of the MMHP. Comparably, the laminate shear strength to MMHP ratios are roughly 12%. In contrast, the compression strength averages about 150% of the MMHP. In recent studies of other systems whose laminate properties have been characterized, similar relationships have been observed.

Besides the correlations to strength data, the CBIT strain information reveals a very similar trend to maximum engineering shear strain (i.e. strain to failure). Reviewing the representative strain information (d/D at MMHP) in Table 1, the AS-4C appears less ductile than the AS-4 composite while the AU-4 composite possesses the least ductility overall. Madhukar and Drzal reported that tension (0° and 90°), compression (0°), and flexure (90°) failure strains increased with increasing ISS [4–6]. Assuming that the Iosipescu shear test is linear to failure, the computed maximum shear strain was found to be 1.2, 1.5, and 0.75% for the AS-4C, AS-4, and AU-4 fiber composites, respectively [5]. These estimates can be substantiated by the reports that the failure surface for the AS-4C composite Iosipescu specimen possessed a "brittle interphase layer," resulting in a distinctively brittle response [5,7]. The calculated failure strains display a trend very similar to the representative strain information found from CBIT. This evidence may suggest that the CBIT is most sensitive to the shearing characteristics of composites (see also Lesko [8]).

In summary, the authors believe that the MMHP can provide a meaningful quantitative representation of the interfacial condition. This assertion is based on the following obser-vations:

1. The representative stress-strain response from the CBIT data is a quantitative repre-sentation of the material response under penetration possessing the ability to discern nonlinearities.
2. Characteristic deformation is observed as a result of interfacial failure initiated at the MMHP (i.e. Point A), which is found at the surface and within the contacted mate-rial.
3. There is evidence that the representative strain information at the MMHP correlates to the shear response characteristics reported for the three systems [5,7]. This suggests a sensitivity to shearing mechanisms which may explain the correlation of MMHP to the interface shear properties.

Thus, with the ability to quantitatively measure the applied stresses and displacements, with the knowledge when failures/nonlinearities are occurring, and the knowledge of what type of failures occur, quantitative assessments of strength may be made through this technique given an accurate analytical analysis. The analysis in this case (i.e. composite interface failure) must understand how the constituents interact. Thus, a micromechanical analysis is employed to investigate the interface stress state due to ball penetration.

Micromechanical Analysis

The elastic ball contact of a semi-infinite transversely isotropic half space (TIHS) [11] exposes a subsurface shear that is atypical of indentation in isotropic bodies. All the normal stresses (σ_{ZZ}, σ_{RR}, and $\sigma_{\Theta\Theta}$) are compressive in and below the region of contact. The shearing stress, σ_{RZ}, is maximum at $R/a = 0.865$ and $Z/a = 0.65$, as shown in Fig. 9, differing greatly from the location of maximum normal stresses which occur at $R = Z = 0$ [8]. At this location of maximum shear, the shearing stress reaches ~15% of the MHP. Comparisons of the contact diameter from the CBIT response and the diameter of inter-face failure for the AU-4 composite (central ring of Fig. 5) suggests that the failure is initiated just inside the diameter of contact [8]. Thus, this region of maximum shearing stresses is suspect as a possible failure location.

An elastic micromechanical analysis was developed to explore the pointwise stresses in the constituents [12]. The unique analysis is based on a "cellular" modeling approach where a representative element consisting of fiber, matrix, and interphase(s) is selected from the TIHS, as shown in Fig. 10. Notice also in Fig. 10 the relationship between the MHP and the peak hardness pressure (PHP), i.e. PHP = 1.5 MHP, which is the result of the distributed ball contact pressure over the contact area. Having selected the desired cellular element of the $xr\theta$ system, boundary conditions are placed on the cylindrical element consistent with the state of stress at that region within the continuum as shown in Fig. 11. The traction applied at the top surface, due to the parabolic-like distribution of pressure produced by the ball indenter, is proportional according to the relative stiffness of each constituent. The tractions on the perimeter of the element are those prescribed by the stress state of the elastic solution employed above [11]. All stresses are forced to zero as $z \rightarrow \infty$. In addition, stress continuity exists across the interface(s) within the element. The present formulation does not address compatibility of the displacements at the element boundaries and therefore does not constitute an exact solution. Thus, only the boundary conditions and the equilibrium equations are satisfied for the present model.

The interphase region is excluded for the results computed here due to the lack of elastic

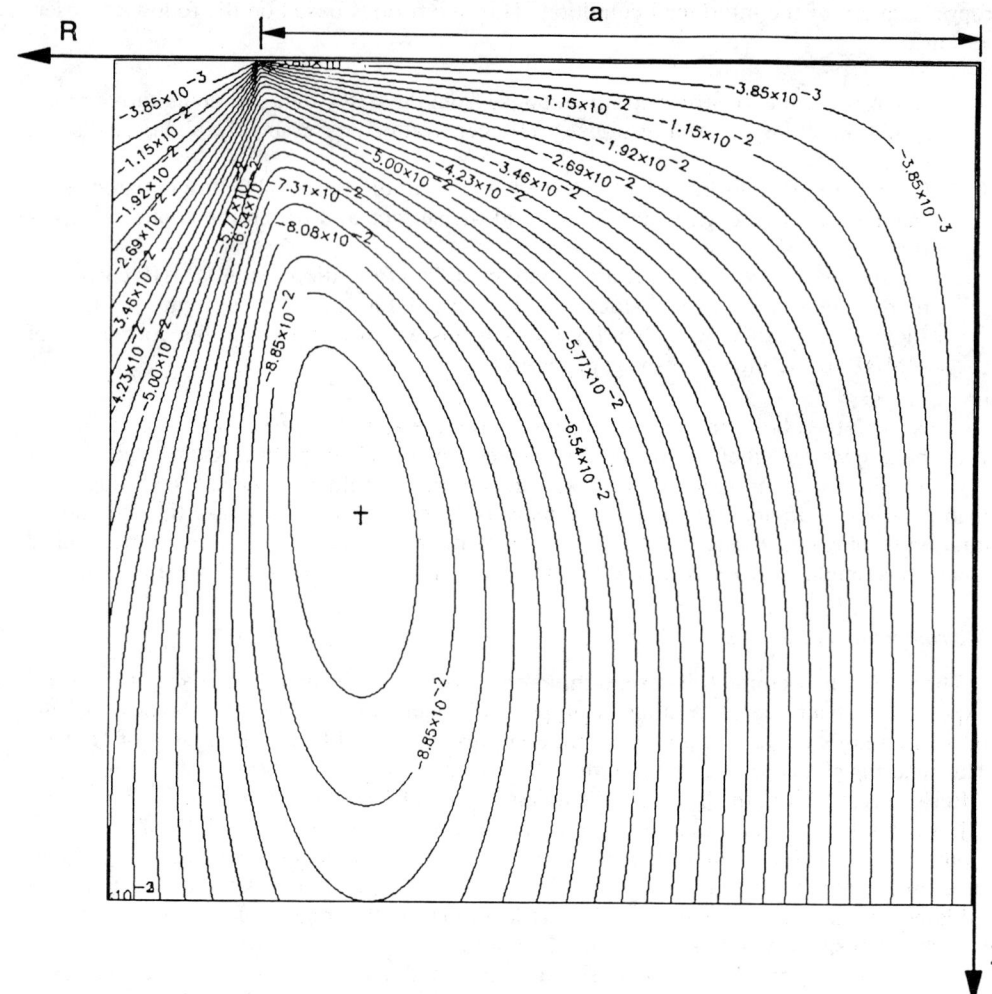

FIG. 9—*The shearing stress, σ_{RZ}, normalized to the PHP, for the elastic ball contact of the transversely isotropic half space of graphite/Epon 828 mPDA. The contact radius, a, denotes the boundary of contact.*

properties for this finite region. The maximum interface shear stress was found to occur at the location of maximum σ_{RZ} as previously depicted in Fig. 9. The interface stress, σ_{RZ}^i, is transformed from the cellular system, normalized to PHP, and plotted in Fig. 12 against the normalized radius of the element. The highest shearing stress occurs at the interface furthest from the center of indentation ($\theta = 180°$) at a value of about -0.104 of PHP. The prediction of this failure locus is validated by the micrograph of Fig. 13. The break in the interface is clearly observed at the outer interface, furthest from the center of contact.

Applying the relationship predicted by the micromechanics analysis to the first nonlinear point (i.e. MMHP) for the AU-4 system, the maximum interface shear stress value is computed to be 71.8 MPa/10.4 ksi. We have assumed that the composite responds elastically up to this point. This implies that the fiber-matrix bond remains intact and no other damage (e.g., plasticity) occurs in the fiber or matrix. We realize that an additional

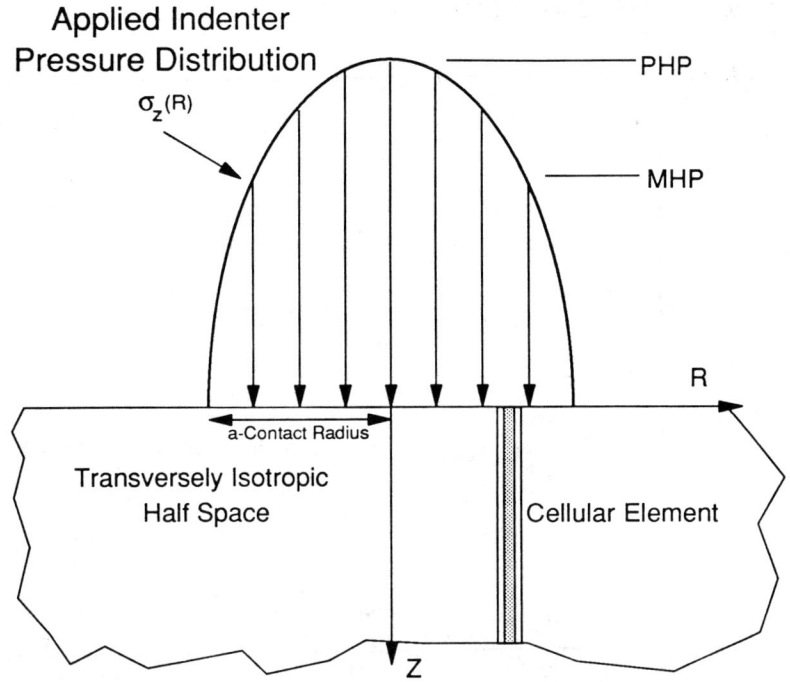

FIG. 10—*The conceptual illustration of the cellular element modeling approach for the micromechanics analysis of composite indentation by a ball penetrator.*

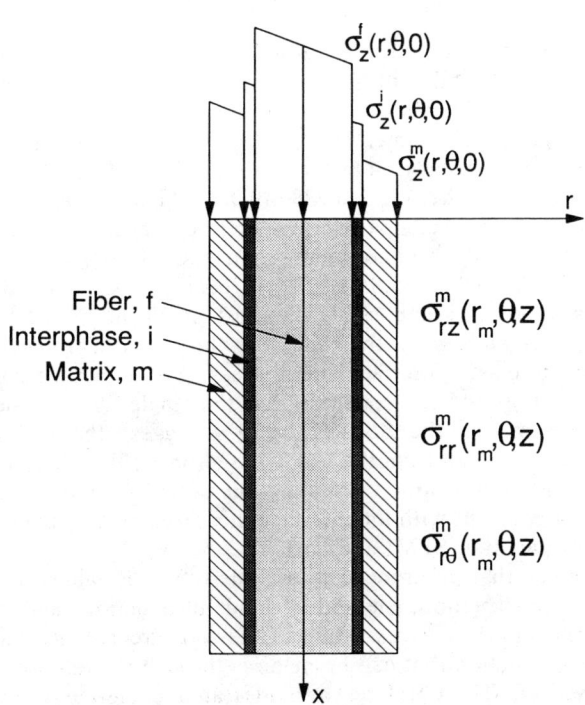

FIG. 11—*The arrangement of boundary conditions on the selected cellular element.*

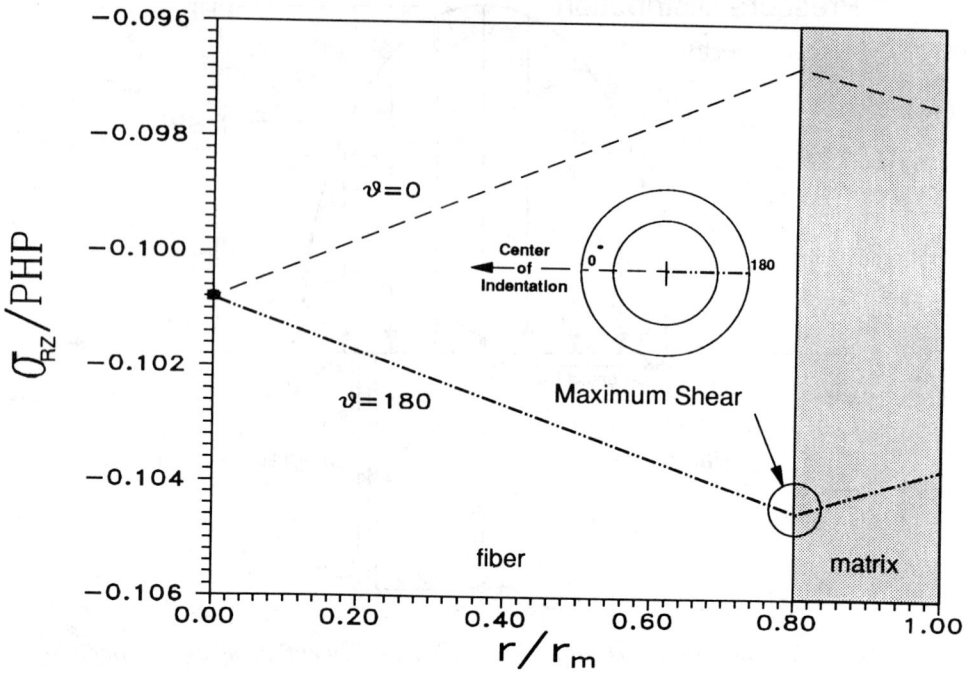

FIG. 12—*The maximum predicted interface shear stress, σ_{RZ}, normalized to PHP (element position at* R/a = 0.865, *maximum shear stress at* Z/a = 0.65*).*

nonlinear analysis may be required to accurately determine the maximum interface shear stress in the AS-4C and AS-4 due to the apparent nonlinearities present in these systems.

This predicted maximum interfacial shear stress is ~90% higher than the reported value of ISS from the single fiber fragmentation test (SFFT) of 37.2 MPa (5.4 ksi). Herrera-Franco et al. [*13*] also reported a value for ISS of 23.44 MPa/3.4 ksi for a micro-bead debond test (MBDT) test on AU-4/Epon 828 mPDA, which is roughly a third of the CBIT assessment. However, when comparing the interface shear stress from CBIT to that obtained from the single fiber micro-indentation test (SFMI) on an actual composite (AU-4/Epon 828 mPDA: ISS = 55.8 MPa/8.1 ksi) [*13*], the CBIT prediction is only 28% greater. The latter comparison is possibly the most significant in that both are performed on an actual composite and obtain their measure through a micromechanical analysis of the point of maximum elastic interface shear stress. A summarized comparison of these data is presented in Fig. 14. We conclude that the single fiber (modeled) system tests provide a lower assessment of the ISS. Verpoest [*14*] suggests that with an improved shear lag analysis, the ISS measured from the SFFT based on Kelly's shear lag model (used for the values reported here [*7*]) is low. Correcting this value based on this assertion would in fact provide better agreement with the interface measurements obtained by the composite interface test techniques (i.e. SFMI and CBIT).

Nonetheless, we see that different tests provide different values of ISS due to differences in specimen construction, method of load introduction, and understanding and interpretation of the interface stress state. Thus, no direct comparison can be made. However, the one distinction that can be made is through the interpretation of the value and how it can be used. The CBIT test provides an average measurement of numerous

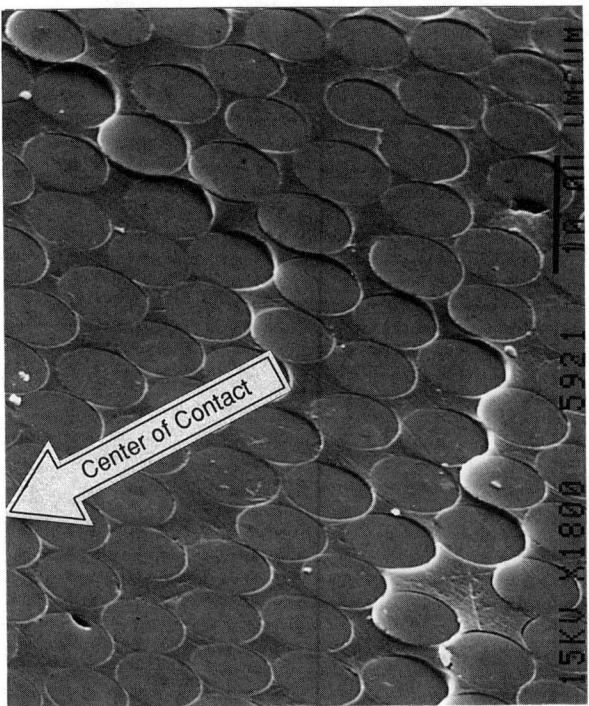

FIG. 13—*Interfacial failure in AU-4/Epon 828 mPDA from ball penetration, revealing the failure at the fiber face furthest from the center of contact.*

interfaces within an as-manufactured composite containing the representative fiber arrangement, morphology, etc. Therefore, we suggest that this may reflect a more representative measure of the interface properties in an actual composite. Any inaccuracies in the maximum interface shear stress obtained from the CBIT analysis are possibly due to the absence of an exact elastic solution and/or through errors associated with experimentally determining the applied contact pressure. Further study is underway to explore those errors associated with the technique and the elastic analysis.

With knowledge of the matrix stresses calculated from the micromechanics model, a matrix failure analysis can be performed to assist our understanding of the nonlinear responses observed in Fig. 8. This idealized study assumes that the fiber does not fail, the fiber-matrix bond is not broken during the process, and the matrix material close to the fiber is representative of the bulk matrix properties. The stress state in the matrix close to the fiber was determined for five R/a values from the center of contact to the point of maximum shear stress, $R/a = 0.865$. Although the model is not valid outside this region, the stresses become less severe beyond this point and are less likely to be a location of failure. The octahedral shear stress, τ_{oct}, normalized to the PHP, was computed and plotted as a function of Z/a, as shown in Fig. 15. The maximum τ_{oct} occurs at the point of maximum σ_{RZ} located at $R/a = 0.865$ and $Z/a = 0.65$. This again may suggest that the ball indentation test is sensitive to the shearing characteristics of the composite. Another region exhibiting large octahedral shear stress occurs at the surface near the center of contact. The high octahedral stress at the surface is the result of the applied normal

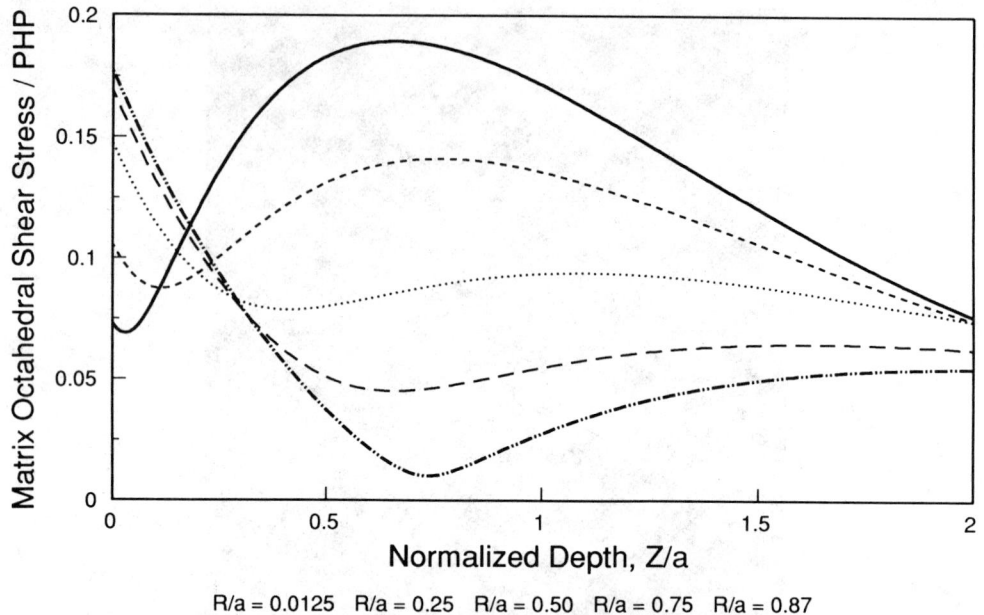

FIG. 14—*Comparison of ISS measured on AU-4/Epon 828 mPDA composite from four techniques: micro-bead debond test, single fiber fragmentation test, single micro-indentation test, and continuous ball indentation test.*

traction. On the other hand, the larger τ_{oct} below the surface is generated (mainly) by the shearing stress, σ_{RZ}. Thus, the matrix yield, if that is in fact what is taking place, could be due to compressive stresses at the surface and/or shearing stresses below the surface.

Conclusions and Remarks

This introductory investigation into composite indentation testing reveals results which are very promising or future applications of these and similar techniques. Five main points were demonstrated in this study:

1. Vickers pyramid and ball indenter geometries are sensitive to the interfacial quality of continuous fiber graphite/epoxy composites penetrated in the fiber direction.
2. Both the fixed-load and the continuous loading/measurement techniques were capable of sensing variations in the fiber-matrix interface strength inherent in different composite systems.
3. CBIT could potentially provide an inexpensive and efficient means of performing, at the very least, comparative studies of interfacial quality. With further research of different materials, empirical correlations may provide a means of estimating composite properties.
4. Of the two techniques, the CBIT possess the greatest potential for making quantitative measures of interfacial bond strength.
5. Experimental and analytical work suggests that the ball indentation testing is most sensitive to the shearing characteristics of composites.

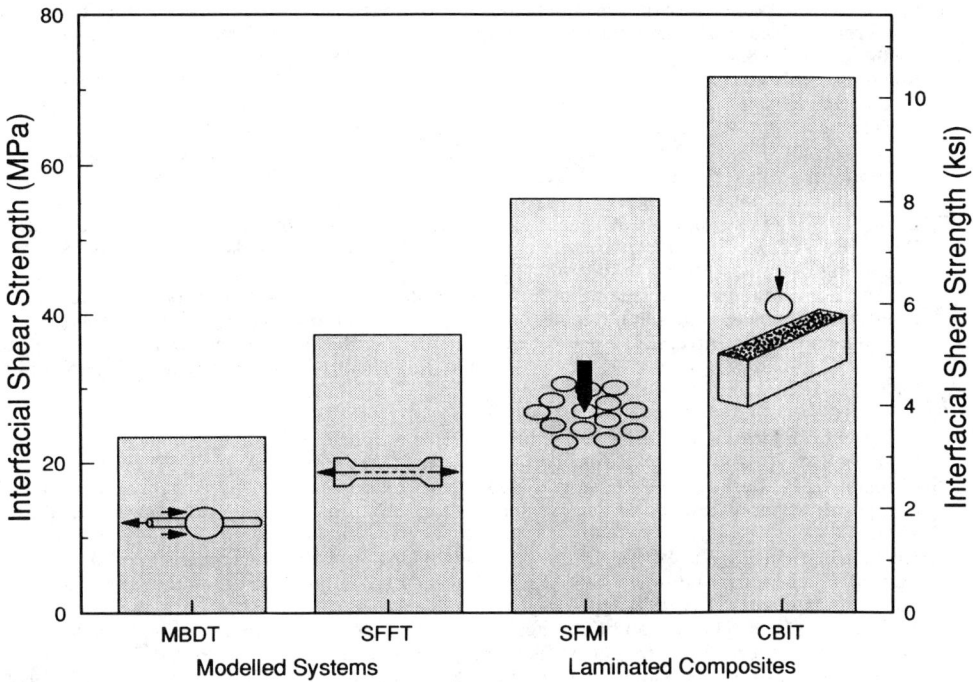

FIG. 15—*Matrix octahedral shear stress, normalized to PHP, shown as a function of normalized depth, Z/a, for five selected R/a values.*

Acknowledgments

This research is supported by the Air Force Office of Scientific Research through grant number AFOSR-89-0216, Evolution Mechanics. John Lesko would like to recognize the Adhesive and Sealant Council, Inc. for their support through the Educational Foundation. The authors acknowledge Scott Case for his assistance and continued work on the CBIT test, which has improved our understanding of the technique. The authors also thank the University of Maryland, Dept. of Mechanical Engineering, for the use of the continuous indentation system (TEX-CIS™). The donation of the well-characterized composite materials from the Composite Materials and Structure Center, Michigan State University, is greatly appreciated. These systems allowed the authors an extraordinary opportunity to explore the capabilities of composite penetration testing. The authors also acknowledge the support, assistance, and constructive discussion from the other members of the Materials Response Group, who added to the success of this work.

References

[1] Sharpe, L. H., "Some Thoughts About the Mechanical Response of Composites," *Journal of Adhesion,* Vol. 6, 1974, pp. 15–21.
[2] Drzal, L. T., "The Interphase in Epoxy Composites," *Advances in Polymer Science,* Vol. 75, 1986, pp. 1–32.
[3] Swain, R. E., Reifsnider, K. L., Jayaraman, K., and El-Zein, M., "Interface/Interphase Concepts in Composite Material Systems," *Journal of Thermoplastic Composite Materials,* Vol. 3, 1990, pp. 13–23.
[4] Madhukar, M. S. and Drzal, L. T., "Effect of Fiber-Matrix Adhesion on the Longitudinal

Compressive Properties of Graphite/Epoxy Composites," *Proceedings,* Fifth Technical Conference of the American Society for Composites, Technomic Publishing Co., Lancaster, PA, 1990, pp. 849–858.

[5] Madhukar, M. S. and Drzal, L. T., "Fiber-Matrix Adhesion and its Effect on Composite Mechanical Properties. I. In-plane and Interlaminar Shear Behavior of Graphite/Epoxy Composites," *Journal of Composite Materials,* August 1991, pp. 932–957.

[6] Madhukar, M. S. and Drzal, L. T., "Fiber-Matrix Adhesion and its Effects on Composite Material Properties. II. Tensile and Flexural Behavior of Graphite/Epoxy Composites," *Journal of Composite Materials,* August 1991, pp. 958–991.

[7] Drzal, L. T., "Fiber-Matrix Interphase and its Effect on Adhesion and Composite Mechanical Properties," *Controlled Interphases in Composite Materials,* H. Ishida, Ed., Elsevier Science Publishing Co., Inc., New York, 1990.

[8] Lesko, J. J., "Indentation Testing of Composite Materials: A Novel Approach to Measuring Interfacial Characteristics and Engineering Properties," masters thesis, Virginia Polytechnic Institute and State University, Engineering Sciences and Mechanics Dept., March 1991.

[9] Lesko, J. J. and Armstrong, R. W., "Thermal Expansion Driven Indentation Stress-Strain System," U.S. Patent 5,133,210, July 1992.

[10] Hertz, H., *Journal of Mathematics (Crelle's Journal),* Vol. 92, 1881.

[11] Dahan, M. and Zarka, J., "Elastic Contact between a Sphere and a Semi Infinite Transversely Isotropic Body," *International Journal of Solids and Structures,* Vol. 13, 1977, pp. 229–238.

[12] Carman, G. P., Lesko, J. J., Reifsnider, K. L., and Dillard, D. A., "Micromechanical Model of Composite Materials Subjected to Ball Indentation," *Journal of Composite Materials,* March 1993.

[13] Herrera-Franco, P., Wu, W. L., Drzal, L. T., and Hunston, D. L., "Comparison of Methods to Access Fiber-Matrix Interface Strength," *Proceedings,* 13th Meeting of the The Adhesion Society, Clearwater, FL, 1991, pp. 21–23.

[14] Verpoest, I., Desaegkr, M., and Keunings, R., "Critical Review of Direct Micromechanical Test Methods for Interfacial Strength Measurements in Composites," *Controlled Interphases in Composite Materials,* H. Ishida, Ed., Elsevier Sciences Publishing Co., Inc., New York, 1990, pp. 653–666.

D. D. Edie,[1] *J. M. Kennedy,*[1] *R. J. Cano,*[2] *and R. A. Ross*[3]

Evaluating Surface Treatment Effects on Interfacial Bond Strength Using Dynamic Mechanical Analysis

REFERENCE: Edie, D. D., Kennedy, J. M., Cano, R. J., and Ross, R. A., "**Evaluating Surface Treatment Effects on Interfacial Bond Strength Using Dynamic Mechanical Analysis,**" *Composite Materials: Fatigue and Fracture, Fourth Volume, ASTM STP 1156,* W. W. Stinchcomb and N. E. Ashbaugh, Eds., American Society for Testing and Materials, Philadelphia, 1993, pp. 419–429.

ABSTRACT: Three series of torsional dynamic mechanical analysis (DMA) tests were conducted on unidirectional carbon/epoxy composites to determine differences in interfacial bond strength. For the first series, composites were manufactured with PAN-based and pitch-based carbon fibers. The fibers had a proprietary surface treatment. For the second series of tests, the composites were made with four grades of pitch-based carbon fibers. Two different methods of surface treatment were used on the fibers. As a comparison, short beam shear tests were conducted on the composites made with fibers which had different surface treatments. The final series of tests were conducted on composite samples made from one grade of pitch-based fibers with four different fiber volume fractions. All samples were made using the same matrix material. Dynamic mechanical analysis was able to clearly differentiate between the fiber/matrix bond strength of pitch-based and PAN-based fibers. The interlaminar shear strength correlated well with the DMA results, indicating that DMA can measure the effectiveness of various surface treatments. The dynamic mechanical loss modulus increased with increasing fiber volume fraction or total surface area.

KEY WORDS: dynamic mechanical analysis, composites, fiber/matrix interface strength, pitch-based fibers, PAN-based fibers

The bonding at the fiber/matrix interface plays a major role in the mechanical behavior of a composite material. In polymer matrix composites, perhaps the most important function of the matrix is to distribute the applied stress among the fibers. Because the applied stress must be transferred across the fiber/matrix interface, transverse, longitudinal, and shear strength of a polymer matrix composite depend heavily on the interfacial bond strength. Thus, bonding must be maximized if the full strength of the reinforcing fiber is to be realized, making the accurate characterization of interfacial bonding in composite materials critical.

Numerous tests have been devised to directly or indirectly characterize the fiber/matrix interface properties of composites. Commercially, the short beam shear test is used to determine interlaminar shear strength, and the double cantilever beam (DCB) test is used

[1]Dow professor of chemical engineering and associate professor of mechanical engineering, respectively, Center for Advanced Engineering Fibers and Their Composites, Clemson University, Clemson, SC 29634-0909.

[2]Research engineer, Materials Division, NASA Langley Research Center, Hampton, VA 23665-5225.

[3]E. I. DuPont de Nemours and Company, Inc., 4501 North Access Road, Chattanooga, TN 37415.

to determine interlaminar fracture properties due to Mode I loading. However, neither of these tests provides an accurate characterization of the fiber-matrix interface. Also, they are not able to isolate the contribution of the interface from the contribution of the matrix and fiber to the property measured without resorting to a detailed micromechanics analysis of the composite. Other tests, such as the microdebond, the fiber pullout, or fiber indentation tests, are used to measure the interface strength, but these tests are performed on single fibers and stress concentrations greatly influence the results.

Recently, Edie et al. [1] demonstrated, using torsional dynamic mechanical analysis (DMA), that the interface contribution can be experimentally separated from the matrix and fiber contributions. Thus, they were able to apply dynamic analysis to characterize the interfacial bonding of composites. In their investigation, composite specimens fabricated using surface-treated fibers were subjected to SBS, DCB, and DMA testing. The results were compared to those obtained from composite specimens fabricated with fibers which had not been exposed to any surface treatment. The results indicated that DMA testing detected the effect of surface treatment with much greater accuracy than the standard mechanical tests.

In this investigation, three series of DMA tests were performed on unidirectional composite samples made from pitch-based and PAN-based fibers as well as the neat resin used to make the composites. In the first series, DMA tests were conducted on composites made from PAN-based and pitch-based fibers with proprietary surface treatments. For the second series of tests, the effects of different methods of surface treatment were studied. Two sets of samples were fabricated using fibers which were exposed to two different surface treatments, and one set was fabricated from fibers with no surface treatment. Four pitch-based fiber types were evaluated. Also, the interlaminar strength of the samples in this group was measured using the short beam shear test. The influence of total interfacial area on the dynamic properties of the composite was investigated in a third study. For this series, samples were tested which had different fiber volume fractions.

Background

Dynamic mechanical analysis has traditionally been used to study viscous response of fluids and viscoelastic response of solids. The theoretical development supporting the use of DMA assumes that the material being evaluated is isotropic and homogeneous. DMA measurements on composite materials reflect contributions from each constituent in the system, as well as effects of fiber volume fraction and laminate and fiber orientation. Isolating the effect of the numerous parameters associated with a composite is difficult. However, DMA can be a valuable empirical technique when carefully applied to composite systems.

When a purely elastic material is deformed, it stores all of the strain energy, while a purely viscous material dissipates all of the deformation energy as heat. A viscoelastic material, lying between an elastic and a viscous material, will dissipate some fraction of the deformation energy. In a composite material, consisting of fibers (essentially elastic), a polymeric matrix (viscoelastic), and a fiber-matrix interface, the deformation energy will be dissipated mainly in the matrix and at the interface. Figure 1 shows a typical sample geometry and loading for a torsional DMA test.

Relation Between Stress and Strain

In the torsional DMA test used for this investigation, a sinusoidal rotation

$$\phi = \phi_0 \sin \omega t \tag{1}$$

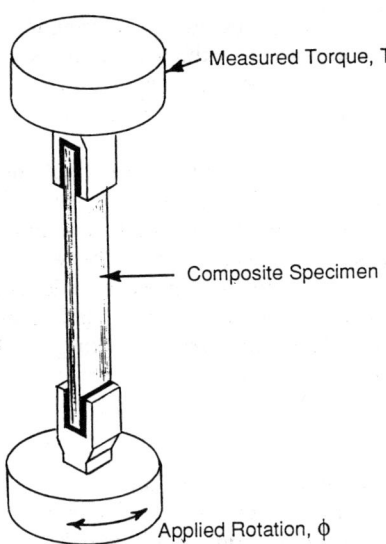

FIG. 1—*Test geometry for dynamic mechanical test.*

was applied to a unidirectional composite specimen. The resulting shear strain has the form

$$\gamma(x, y) = \gamma_0(x, y) \sin \omega t \tag{2}$$

where $\gamma_0(x, y)$ is the maximum strain amplitude which is dependent on the position on the sample cross section, and ω is the angular frequency. The specimen responds with a stress given by

$$\tau(x, y) = \tau_0(x, y) \sin (\omega t + \delta) \tag{3}$$

where $\tau_0(x, y)$ is the amplitude of the stress. The phase lag between the stress and the applied strain is given by δ. For a purely viscous material, δ equals 90°, while for a purely elastic material, δ equals 0°. However, for a viscoelastic material, such as a polymer-matrix composite, δ lies between 0° and 90°. In a dynamic mechanical test, stress, τ, and strain, γ, are related by

$$\tau = \gamma_0[G' \sin (\omega t) + G'' \cos (\omega t)] \tag{4}$$

where G' and G'' are defined as the storage modulus and the loss modulus, respectively [2]. As the phase angle, δ, approaches zero, the storage modulus approaches the elastic modulus and the loss modulus goes to zero.

A common alternate approach is to define the complex modulus as

$$G^* = G' + iG'' \tag{5}$$

where $i = \sqrt{-1}$. This results in the stress-strain relation

$$\tau = G^*\gamma \tag{6}$$

where τ is now a complex number. The elastic component is given by the storage modulus, G', and the viscous component by the loss modulus, G''. Because the phase shift between stress and strain is denoted as δ, it can be shown that the energy dissipation caused by the viscous nature of a viscoelastic material can be completely characterized by the loss modulus, G''.

As indicated, both the shear strain and stress are functions of position on the cross section of the specimen. The stress distribution was determined from the measured torque using the elasticity solution for a rectangular section loaded in torsion [3]. The shear strain was related to the angle of twist in a similar manner. These assumptions on the stress and strain distributions imply that the viscous properties are independent of the stress or strain level in the specimen. Further, the elasticity solution used herein assumes the material to be isotropic and homogeneous. Thus, the stresses acting on the cross section must be viewed as equivalent stresses.

Energy Dissipation

Edie et al. [1] indicated that if the complex modulus, G^*, varies significantly from specimen to specimen (because of differences in fiber alignment, for example) the energy dissipation normalized by the complex modulus, ΔD, may better characterize the effects of the fiber/matrix interface, where

$$\Delta D = \pi \gamma_0^2 \tan \delta \qquad (7)$$

In other words, for small δ, a variation in the tan δ of the samples also may be expected to reflect the magnitude of the energy dissipated during the test. Therefore, when composite materials are subjected to dynamic mechanical testing, any decrease in the strength of the fiber/matrix bonding should be reflected by an increase in the measured values of tan δ and G''.

Experimental

Sample Preparation

In the initial phase of the present study, E. I. DuPont de Nemours and Co., Inc. supplied a total of twelve unidirectional composite samples with a nominal fiber content of 60% by volume for DMA analysis. The samples were prepared by compression molding carbon fiber bundles which had been impregnated with epoxy resin (EPON 828 + 360-L methylene dianiline curing agent). Four specimens contained high modulus E-105 pitch-based carbon fibers and four contained lower modulus E-55 pitch-based fibers. Both varieties of pitch-based fibers had received the standard surface treatment to improve interfacial bonding. The exact process used is proprietary, but presumably an oxidizing medium was employed. The remaining four samples were fabricated using AS4 PAN-based carbon fibers. The composite samples (1.3 cm wide and 0.2 cm thick) were cut to lengths of 6.3 cm for DMA testing.

In the second phase of the study, the influence of two methods of surface treatment on the fiber/matrix bond strength was investigated. For this evaluation, eleven 13-cm-long unidirectional composite samples were fabricated using pitch-based carbon fibers with a range of moduli: three samples were fabricated using DuPont E-35, E-75, and E-125 fibers which had been surface treated by a proprietary ozone treatment; four specimens were made using DuPont E-35, E-75, E-105, and E-125 fibers which had been surface treated by a proprietary plasma etching technique; and, as a control, four of the composite samples contained E-35, E-75, E-105, and E-125 fibers which had not received any surface treatment.

In the third phase of the present study, the effect of fiber volume fraction and fiber/matrix interfacial area on the dynamic mechanical properties of composite samples was determined using five types of specimens which were provided by DuPont. For this series of tests, E-105 pitch-based fibers exposed to a standard surface treatment were used to fabricate unidirectional composite specimens with fiber volume fractions of 12, 17, 38, and 60%. Again, a common epoxy matrix (EPON 828 + 360L) was used for all specimens. In addition to these four types of composite specimens, neat epoxy specimens were fabricated to serve as a control.

Procedure for the DMA Testing

The dynamic properties of the 6.3-cm-long composite specimens were measured in torsion using a Rheometrics RDS-7700 dynamic mechanical spectrometer. Standard rectangular fixtures supplied by Rheometrics were used to hold the specimen during the test. The top fixture was held stationary, and a cyclic torsional twist was applied to the bottom fixture during the test. The torque and angle of twist were recorded. Knowing the relationship between torque and shear stress and angle of twist and shear strain, the complex shear modulus of the specimen was determined by Eq 6. The instrument also measured the phase lag, δ, between the applied strain and the resulting stress, allowing G'' to be calculated. Measurements were made at room temperature over a frequency range of 0.1 to 25 rad/s with a maximum angle of twist of 0.0005 rad.

Results and Discussion

Recall that, at a constant rotation, the loss shear modulus is a direct measurement of the energy dissipated in the sample during testing. Thus, for samples formed from the same volume percent of purely elastic carbon fibers and an identical viscoelastic matrix material, a variation in G'' should reflect a difference in bonding at the fiber/matrix interface. The sensitivity of the dynamic mechanical test to differences in interfacial bond strength of composites made from PAN-based and pitch-based fibers was evaluated in the initial series of tests. As previously mentioned, the same matrix polymer was used for all specimens. One group was reinforced with Hercules AS-4 PAN-based carbon fibers, a second group with DuPont E-55 pitch-based carbon fibers, and the third group with DuPont E-105 pitch-based carbon fibers. Even though the results of this series of tests, listed in Table 1, were obtained at a frequency of 0.1585 rad/s, they are typical of the trends observed over the entire frequency range.

The loss shear moduli, G'', of the composite specimens ranged from three to six times that of the control specimens made from pure epoxy, even though the carbon fibers, which do not generate viscous losses, comprised 60% of these composites. The larger loss moduli in the composite can be due to the fiber/matrix interface, defects in the composite due to manufacturing, and residual stresses in the matrix. Samples of the composite were sectioned to visually determine fiber volume fraction and to inspect for voids and delaminations. From these inspections it was found that the composites were essentially void and delamination free, which implies that the contribution due to macroscopic defects in the matrix is small. The change in the apparent DMA properties of the matrix due to residual stresses is potentially significant. Superposing the dynamic loading with the residual stresses can increase the dynamic properties; however, limited experience with polymer matrix composites indicates that this effect is small [4]. Consequently, these test results indicate

TABLE 1—*Storage shear modulus, G', loss shear modulus, G", tan δ for carbon fiber/epoxy composite specimens from various types of fiber and neat epoxy specimens at 25°C and 0.1585 rad/s.*

Fiber Type	G', GPa	G'', MPa	tan δ ($\times 10^2$)
Hercules AS-4 PAN-based	6.1 ± 0.7[a]	89.0 ± 16.0[a]	1.46 ± 0.2[a]
DuPont E-55 pitch-based	6.0 ± 0.2	160.0 ± 28.0	2.66 ± 0.3
DuPont E-105 pitch-based	6.1 ± 0.2	180.0 ± 31.0	2.90 ± 0.2
Neat epoxy	1.5 ± 0.2	33.4 ± 5.0	2.28 ± 0.2

[a]Standard deviation.

that at ambient temperature the fiber/matrix interface contributes very significantly to the loss shear modulus of composite specimens.

In fact, only twelve tests were required to determine that the average loss shear modulus and the average tan δ of the specimens fabricated from the pitch-based fibers was significantly higher (at a 95% confidence level using a *t* distribution test) than those of the specimens made from PAN-based fibers. Because pitch-based carbon fibers are known to form a weaker interfacial bond with polymer matrices than PAN-based fibers, these higher G'' and tan δ values measured for the pitch fiber specimens would be expected. Note that the average G'' and tan δ values measured for the specimens made from the high modulus variety of pitch-based fibers were higher than those measured for the specimens containing low modulus pitch-based fibers (again, significant at a 95% confidence level). This seems reasonable since high modulus fibers are known to have less reactive surfaces, yielding a weaker fiber/matrix interfacial bond.

The second series of tests evaluated the effect of surface treatment and fiber modulus on the DMA properties. The loss shear moduli and tan δ's of specimens containing low modulus E-35 fibers were lower than those containing higher moduli fibers, indicative of stronger fiber/matrix bonding, as shown in Fig. 2 (also Table 2). However, the dynamic analyses suggest that neither surface treatment significantly improved interfacial bonding for the E-35 fibers. For higher moduli fibers, the DMA results showed that, even though both methods of surface treatment reduced the viscous losses during dynamic testing, the loss moduli and tan δ's of the composite specimens containing plasma surface-treated fibers were lower than those with ozone-treated fibers. This was observed over the entire range of frequencies tested. Thus DMA analysis indicates that plasma etching is superior to ozone treatment for improving the interfacial bond strength of the E-75, E-105, and E-125 fibers.

To corroborate this, SBS tests were conducted by DuPont on an identical set of samples. Figure 3 shows the interlaminar shear strengths measured in this series of tests. Again, as expected, specimens made using E-35 fibers exhibited the highest interfacial shear strength. However, SBS testing did indicate a slight improvement in fiber/matrix bonding with surface treatment. No difference in failure mode was observed for the two surface treatments; the catastrophic failure mode was a shear failure between plies in the laminate. Where failure initiated was not determined. For the E-75, E-105, and E-125 fibers, the higher viscous losses observed in dynamic analysis corresponded to lower interlaminar shear strengths, as measured by SBS testing. Thus, the SBS testing confirmed that, for E-75, E-105, and E-125 fibers, the plasma etching technique is superior to ozone treatment.

In the third series of tests the influence of fiber volume fraction was studied by testing

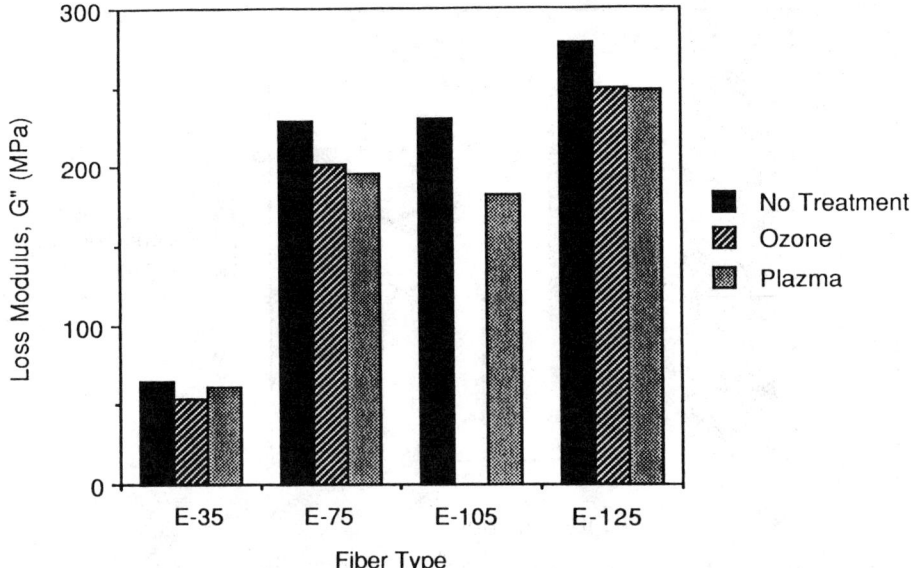

FIG. 2—*Effect of surface treatment on loss modulus, G".*

epoxy matrix samples containing 0, 12, 17, 38, and 60% carbon fibers by volume. Approximately 18 DMA tests were conducted on each type of specimen, and the dynamic properties are plotted versus frequency in Figs. 4, 5, and 6. As Fig. 4 shows, the storage modulus increases with increasing fiber volume fraction, and it is virtually independent of fre-

TABLE 2—*A comparison of ozone and plasma surface treatment on G", tan δ (at 25°C and 0.1585 rad/s), and the interlaminar shear strength of epoxy-matrix composites containing 60 volume percent fibers.*

Fiber Type	tan δ ($\times 10^2$)	G", MPa	Interlaminar Shear Strength, MPa
	No Surface Treatment		
DuPont E-35[a]	1.19 ± 0.17[b]	65 ± 9[b]	79.3
DuPont E-75	4.05 ± 0.70	229 ± 43	59.3
DuPont E-105	4.15 ± 1.23	231 ± 75	35.8
DuPont E-125	4.53 ± 0.60	278 ± 43	33.8
	Ozone Surface Treatment		
DuPont E-35[a]	0.98 ± 0.11	54 ± 7	86.2
DuPont E-75	3.73 ± 0.58	202 ± 15	70.3
DuPont E-125	4.48 ± 0.72	250 ± 47	69.7
	Plasma Surface Treatment		
DuPont E-35[a]	1.16 ± 0.16	61 ± 10	91.1
DuPont E-75	3.38 ± 0.58	196 ± 34	78.0
DuPont E-105	3.16 ± 0.54	183 ± 31	77.3
DuPont E-125	4.47 ± 0.96	248 ± 64	69.7

[a]Different DMA fixture used for E-35 specimens.
[b]Standard deviation.

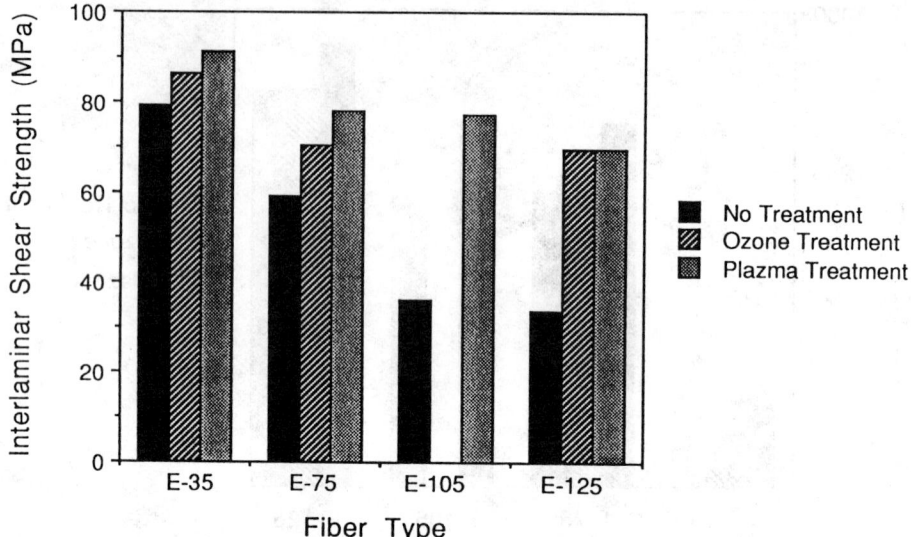

FIG. 3—*Effect of surface treatment on interlaminar shear strength.*

quency. Since G' is a measure of elastic contributions, one would expect it to increase with fiber volume fraction and be relatively independent of frequency. In Fig. 5, the loss modulus shows a similar result. Although being more frequency dependent than G', the values of G'' are larger for the higher fiber volume fractions. One would also expect this trend since increasing the fiber fraction will increase the interfacial area, which tends to contribute to the composite's viscous nature and thus increases G''. In both figures one can observe that, as the fiber fraction increases from 38 to 60%, the increase in G' and G'' is

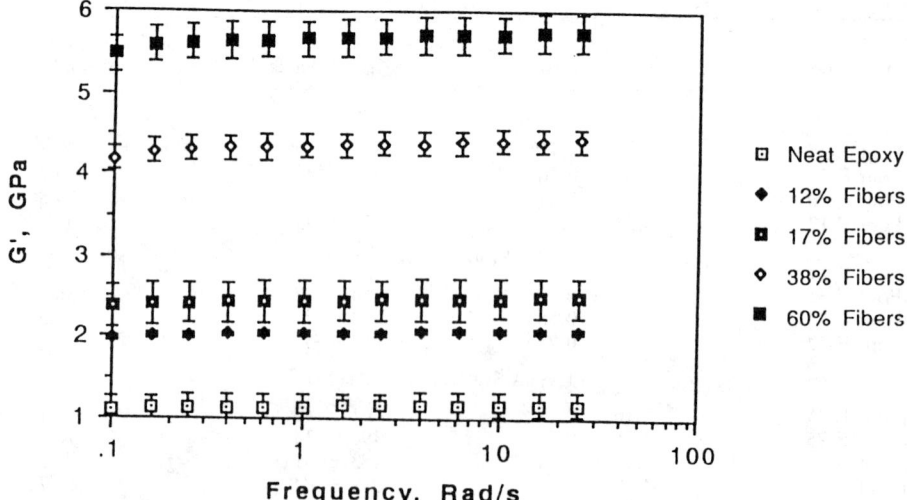

FIG. 4—*Storage modulus for various fiber volume fractions of composites made with E-105 pitch-based fibers.*

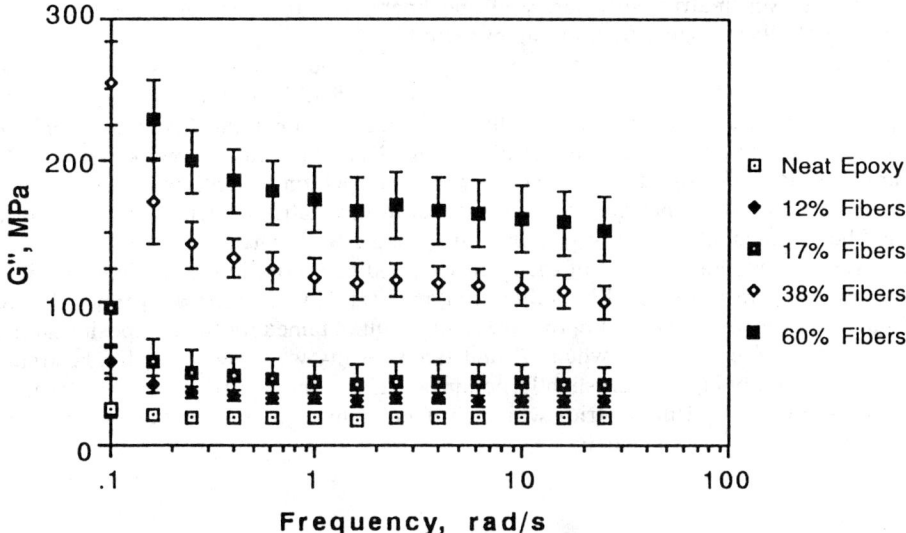

FIG. 5—*Loss modulus for various fiber volume fractions of composites made of E-105 pitch-based fibers.*

much larger than when the fiber volume fraction increases from 12 to 38%. This seems to indicate a nonlinear relationship.

Figure 6 presents the results obtained for tan δ. The tan δ values of the neat epoxy are very similar or larger than the values obtained for the 12% fiber volume samples. Because tan δ is a measure of the viscous nature of the sample, this indicates that the viscous loss created by the interface of the 12% sample is nearly identical to that lost by replacing 12%

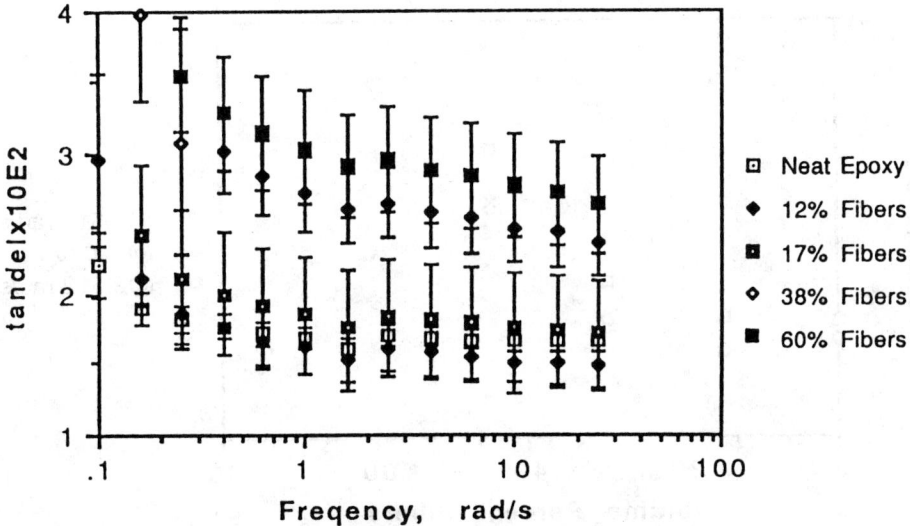

FIG. 6—*Tan δ for various fiber volume fractions.*

of the viscoelastic matrix with purely elastic fibers. For the composite samples, tan δ increases with fiber volume fraction, as expected.

When the dynamic properties (at a given frequency) were plotted versus volume fraction, the results showed that elastic properties (represented by G') increase steadily with increasing fiber fraction. However, as indicated in Fig. 7, tan δ initially decreases and then increases at fiber volume fractions greater than 12%. Thus, the response between the dynamic mechanical properties and the fiber volume fraction is nonlinear.

The results were then normalized with respect to the matrix contributions and plotted versus fiber surface area. By normalizing with respect to the matrix effects, one may be able to view the results in terms of the interfacial and fiber contributions alone. In Fig. 8, the G'' results were normalized by subtracting the matrix volume fraction times the loss modulus determined for the neat epoxy from the result obtained for the composite sample. Similar trends were observed when G' and tan δ results were normalized. The trends, while more monotonic, are still slightly nonlinear. This nonlinearity may be the result of the test motion of the Rheometrics which, in fact, induces a combination of shear and tensile stress within the specimen.

Conclusions

In this investigation torsional dynamic mechanical testing, a nondestructive technique, was used to study the bonding at the fiber/matrix interface in unidirectional composites made from PAN-based and pitch-based carbon fibers. The effects of fiber surface treatment and fiber volume fraction on DMA were studied. From the results obtained in this investigation, the following can be concluded.

1. During torsional DMA, a composite material with poor interfacial bonding will dissipate more energy than the same composite with good interfacial bonding.
2. Dynamic mechanical analysis is a reliable technique for detecting differences in interfacial bond strength in polymer matrix composites.

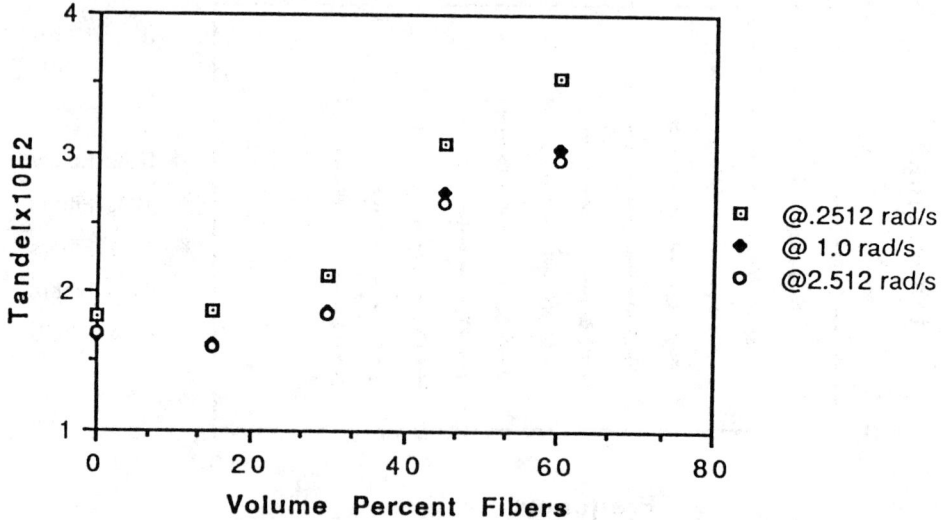

FIG. 7—*Tan δ as a function of fiber volume percent.*

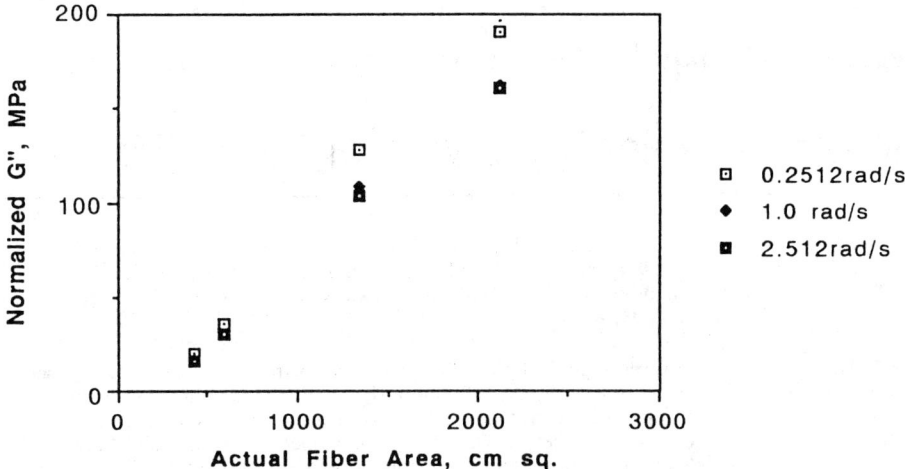

FIG. 8—*Normalized loss modulus versus interfacial area.*

3. Dynamic mechanical analysis clearly differentiates between the fiber/matrix bond strength of PAN-based and pitch-based carbon fibers.
4. The effectiveness of different surface treatments can be measured with DMA.
5. Higher viscous losses, G'', as determined by DMA, corresponded to lower interlaminar strength, as measured by the short beam shear test.
6. The dynamic mechanical properties, G', G'', and tan δ were determined to increase with increasing fiber volume fraction. While not exactly linear, the viscous characteristics represented by G'' and tan δ appear to correlate well with total interfacial area.

Acknowledgments

The authors wish to thank the E. I. DuPont de Nemours and Co., Inc. for supporting this investigation.

References

[1] Edie, D. D., Banerjee, A., Kennedy, J. M., and Cano, R. J., "Characterization of Interfacial Bond Strength by Dynamic Analysis," *Proceedings,* Fiber-Tex 89, NASA Conference Publication 3082, 1989, pp. 349–361.
[2] Ferry, J. D., *Viscoelastic Properties of Polymers,* 3rd ed., John Wiley, New York, 1980.
[3] Timoshenko, S. P. and Goodier, J. N., *Theory of Elasticity,* 3rd ed., McGraw-Hill, New York, 1970.
[4] Wang, S. F. and Oagle, A. A., "Influence of Aging on Transient and Dynamic Mechanical Properties of Carbon Fiber/Epoxy Composites," *SAMPE Quarterly,* Vol. 20, No. 2, January 1989.

Gregory P. Carman,[1] John J. Lesko,[1] and Kenneth L. Reifsnider[1]

Micromechanical Analysis of Fiber Fracture

REFERENCE: Carman, G. P., Lesko, J. J., and Reifsnider, K. L., "**Micromechanical Analysis of Fiber Fracture,**" *Composite Materials: Fatigue and Fracture, Fourth Volume, ASTM STP 1156,* W. W. Stinchcomb and N. E. Ashbaugh, Eds., American Society for Testing and Materials, Philadelphia, 1993, pp. 430–452.

ABSTRACT: A closed-form analytical model is developed in this paper to study the perturbed stress field in a general unidirectional composite system containing fiber fractures. The model includes analysis of eccentrically oriented fibers and multiple fiber fractures with the use of an approximating geometry. The theory developed incorporates a functional dependence of stress/strain concentration on constituent material properties, fiber volume fraction, crack size, multiple fiber fracture, and fiber eccentricity. Utilizing a mechanics of materials approach with classic elasticity concepts, the approximate stress field is generated for each constituent region (e.g. fiber, matrix, and interphase). The analysis employs an annular ring of fibers to model the adjacent unbroken fibers and a novel fiber discount methodology to examine multiple fiber fractures. Analytical results of strain concentration are compared with direct experimental measurements to corroborate the model's depiction of the effect specific physical parameters (crack size and axial position) have on this quantity. Parametric studies are performed with variables such as fiber volume fraction, constituent stiffness properties, crack size, and eccentrically located fibers to investigate their significance on general trends.

KEY WORDS: fiber fracture, crack size, variable fiber spacing, interphase, strain concentration, ineffective length, tensile strength

The influence of microparameters (e.g., interphase between fiber and matrix) on the behavior of composite materials is now being realized by the scientific community. The fact that there is a direct relationship between these parameters and macromechanical strength properties (Madhukar and Drzal [1]) has initiated a research thrust for the development of accurate micromechanical representations of such phenomena. The literature currently abounds with microsolutions for a number of various problems which are formulated to derive stiffness and strength information. Up until now the validation of these micromechanical models has been accomplished with the use of volume averaging concepts and their comparison with macromechanical predictions. This, however, leads to "smearing out" of important details in the analysis, which results in questionable validation of the models.

It is well understood that the tensile strength of most unidirectional composites (and fiber-dominated multiaxial laminates) is controlled by fiber failure. A fiber fracture generally causes a strain concentration to exist in the surrounding fiber, matrix, and interphase regions. The stress redistribution problem involving a fiber fracture was formulated by Rosen [2] with classic "shear lag" assumptions. Hedgepeth and Van Dyke [3] extrapolated

---

[1]Graduate research assistant, graduate research assistant, and Alexander Giacco professor, respectively, Materials Response Group, Dept. of Engineering Science and Mechanics, Virginia Polytechnic Institute and State University, Blacksburg, VA 24061-0219.

this methodology to include stress concentration as a function of fiber fracture arrangement in the composite. A detailed analysis of fiber fracture was also accomplished by Harlow and Phoenix [4] in a 2-D tape. There have been attempts at modeling variable fiber spacing using the shear lag approach such as Fukuda's [5]. Various finite element models have been proposed to develop a precise solution to this problem such as Adams' [6]. Models like these can subsequently be used in strength formulations, such as that described by Batdorf [7] to "bridge" the gap between micromechanics and desired laminate level strength predictions. While each model, in it's own right, provides useful information, they, however, do not incorporate all the required parameters to adequately address a general class of problems. These include a lack of stress concentration dependence on material properties, fiber volume fraction, crack size, multiple fiber fracture, interphase effects, and variable fiber spacing.

In this paper, we present a model which is capable of incorporating these parameters in a concise and straightforward formulation. The closed-form solution is developed based on a mechanics of materials approach and classical elasticity concepts. We assume that the fibers can be modeled as annular rings to eliminate the local stress variation with angle. Through experimental evidence, we suggest a functional dependence of axial strain on radial position. By force balance, constitutive relations, equilibrium equations, and boundary conditions, a solution is arrived at which describes the approximate stress redistribution in each material region. By comparing the theoretical predictions of strain with direct experimental measurements, a corroboration of the model's depiction of strain concentration on crack size and axial position is obtained. This provides a foundation not only for the results provided by the model but for the assumptions employed in its formulation. Parametric studies are undertaken to suggest general trends in strain concentration values as a function of specific physical parameters.

Theoretical Development

To develop a solution to the stress redistribution around a fiber fracture, a classic superposition technique is employed. This involves separating the problem into a far-field analysis and a near-field analysis (e.g. Carman and Reifsnider [8]). The following equation is representative of this approach.

$$\sigma'^n_{ij} = \sigma^n_{ij} + \tilde{\sigma}^n_{ij} \tag{1}$$

The stresses with primes (σ') are the complete stress solution, the stresses with tildes ($\tilde{\sigma}$) are the near-field solution, and the stresses without any markings are the far-field solution. The superscript n refers to either the matrix, fiber, or composite (m, f, or c). While the near-field derivation presented in this paper does not explicitly contain an interphase material, the basic form of the analysis will not change with its addition. We present only the derivation of the near-field solution and assume the reader is familiar with the far-field solution. The latter analysis has been presented in a variety of forms and is not essential to the present derivation. The reader is referred to the methodology presented by Pagano and Tandon [9].

In posing the near-field problem, we assume that a fiber fracture has occurred in a composite with a hexagonal array of fibers, as shown in Fig. 1. The value r_f is the radius of the fiber, r_c is the crack size, and r_a is the closest adjacent fiber dimension. The present analysis, while case specific, does not require the assumption of a hexagonal array. The size of the fracture, r_c, may (or may not) extend a finite distance into the matrix, up to the

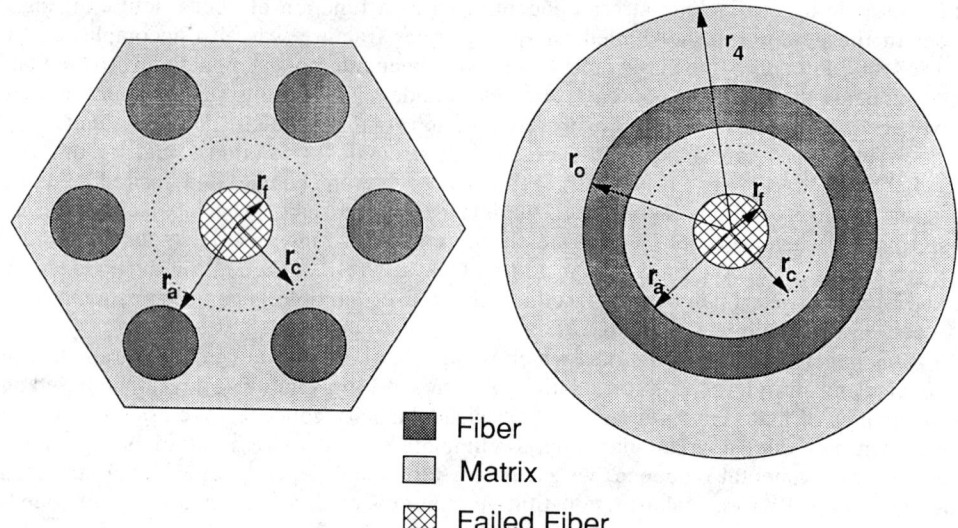

Fiber

Matrix

Failed Fiber

FIG. 1—*Illustration of a composite containing a fiber fracture which extends into the matrix modeled as an annular ring problem.*

adjacent fibers (i.e. $r_c = r_a$). Prior to the formation of the crack, the load was transferred across this material region; however, following crack propagation, the surface of the crack is traction free. To simulate this mathematically, a traction is applied to the crack face. To calculate the net force applied to the crack face, the following equation is employed.

$$P_0 = \int_0^{2\pi} \int_0^{r_f} \sigma_{zz}^f r \, dr \, d\theta + \int_0^{2\pi} \int_{r_f}^{r_c} \sigma_{zz}^m r \, dr \, d\theta \tag{2}$$

Note that the far-field stress terms are used in this calculation and the subscript z refers to the axial direction. This equation is representative of the net force, P_0, resulting from the applied traction in the near-field solution. This can be expanded out to the following.

$$P_0 = [C_{11}^f + 2C_{12}^f A]\epsilon_{zz}^c \pi r_f^2 + \left[C_{11}^m - 2C_{12}^m \frac{V_f}{(1 - V_f)} A \right] \epsilon_{zz}^c \pi (r_c^2 - r_f^2) \tag{3}$$

where

$$A = (C_{12}^m - C_{12}^f) \left[C_{22}^f + C_{23}^f + \frac{(C_{22}^m + C_{23}^m)v_f + C_{22}^m - C_{23}^m}{1 - v_f} \right]^{-1}$$

where C_{ij}^n is defined to be the contracted stiffness tensor for the nth constituent ($1 - z$, $2 - r$, $3 - \theta$), ϵ_{zz}^c is the global strain, and v_f is the fiber volume fraction in the composite. Equation 3 can be closely approximated by the following simplified formula with the assumption that transverse strains do not significantly affect axial stresses.

$$P_0 \simeq C_{11}^f \epsilon_{zz}^c \pi r_f^2 + C_{11}^m \epsilon_{zz}^c \pi (r_c^2 - r_f^2) \tag{4}$$

If the crack does not extend into the matrix, the second term in Eqs 2, 3, and 4 is exactly zero. Also, including an interphase region would require one additional integral term in Eq 2.

We now make the assumption that the fibers, which are immediately adjacent to the fractured fiber, can be represented by an annular ring of material (see Fig. 1). This reduces the problem to an axisymmetric case. The point stresses thus calculated are not expected to be the exact values for a composite containing distinct fibers; however, they should depict appropriate trends in stress variations of interest. The inner radius of the fibrous annular ring is represented by the inner dimension of the closest fiber, r_a. The outer radius of the fibrous annular ring, r_o, is chosen such that the area of the ring is equivalent to the surrounding fibers (for a hexagonal array the ring area is equal to $6\pi r_f^2$). The outer radius of the matrix annular ring (i.e. r_4) is characterized by the global fiber volume fraction, that is, for a hexagonal array $r_4^2 = 7r_f^2/v_f$. By taking a representative slice of the composite in the plane of the crack, a typical force element can be constructed, as shown in Fig. 2. By summing the forces on this element, the following equation can be obtained.

$$P_0 - F_1 - F_2 - F_3 = 0 \tag{5}$$

where

$$F_n = \int_0^{2\pi} \int_{r_{inner}}^{r_{outer}} \bar{\sigma}_{zz}^n r \, dr \, d\theta$$

F represents the average force present in each of the finite regions depicted in Fig. 2, which is generated by the near-field traction. These F's are calculated by area averaging

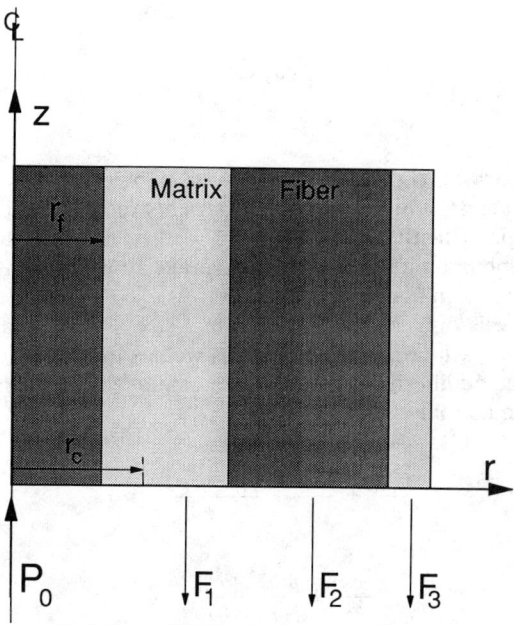

FIG. 2—*Drawing depicting the force balance on a composite element near a fiber fracture subjected to near field tractions.*

the axial stress $\bar{\sigma}_{zz}^n$ on the nth face, as presented in Eq 5 and depicted in Fig. 2. However, these stress terms are unknown at this time. To solve this problem, knowledge of the functional dependence of each axial stress term on the r-coordinate must be established. We realize that the axial strain ($\bar{\epsilon}_{zz}^n$) is a continuous function in r (if slip does not occur) without point discontinuities, except at the crack tip. This is a direct result of the displacement continuity condition which exists at the interface between constituents. We assume that this function can adequately be represented by a smooth continuous function, even though in actual composite systems the slope (at most) may be discontinuous at the interface between constituents. Based on experimental evidence (Carman [10]), the following radial variation of the axial strain in r is proposed.

$$\bar{\epsilon}_{zz}^n = \frac{B}{r^{1.5}} f(z) \tag{6}$$

The constant B in Eq 6 is an unknown, and the function $f(z)$ represents the dependence of the axial strain on the z variable, which is yet to be determined. We impose the constraint $f(0) = 1$, which is purely arbitrary and merely reflects the interaction between the constant B and the function itself. Noting that the axial stress for each constituent is dominated by the axial strain for this problem ($\bar{\epsilon}_{zz} \gg \epsilon_{rr}$ or $\epsilon_{\theta\theta}$) and the fact that $C_{11}^n \gg C_{12}^n$, the following relationship is written.

$$\bar{\sigma}_{zz}^n \simeq C_{11}^n \bar{\epsilon}_{zz} \tag{7}$$

By using Eqs 2, 5, 6, and 7, an explicit relation (implicit in fiber volume fraction) for B in terms of the applied load, the fiber's position, crack radius, and the constituent material properties is obtained.

$$B = \frac{1}{4\pi} P_0 [C_{11}^m (r_a^{1/2} - r_c^{1/2} + r_4^{1/2} - r_0^{1/2}) + C_{11}^f (r_0^{1/2} - r_a^{1/2})]^{-1} \tag{8}$$

All of the quantities on the right side of Eq 8 are known, and therefore a closed-form solution (Eqs 6 and 7) is presented for the axial strain/stress redistribution on neighboring fibers and matrix regions in the crack plane ($z = 0$). From this formulation (Eq 6), a maximum ($r = r_a$), a minimum ($r = r_0$), and an average strain concentration (area average) can be calculated for the annular ring of fibers.

The preceding analysis only provides the normal stress in the plane of the crack. It is useful to have the axial variation in this stress, as well as the other stress components. To solve this problem, the equilibrium equations are employed. These may be written for an axisymmetric problem as follows.

$$\frac{\partial \sigma_{rr}}{\partial r} + \frac{\partial \sigma_{rz}}{\partial z} + \frac{\sigma_{rr} - \sigma_{\theta\theta}}{r} = 0$$

$$\tag{9}$$

$$\frac{\partial \sigma_{zz}}{\partial z} + \frac{\partial \sigma_{rz}}{\partial r} + \frac{\sigma_{rz}}{r} = 0$$

By using Eqs 6 and 7 in Eq 9, the following stresses are suggested.

$$\bar{\sigma}_{zr}^n = \frac{\partial f(z)}{\partial z} \left[\frac{D_1^n}{r} - \frac{2BC_{11}^n}{r^{1/2}} \right]$$

(10)

$$\bar{\sigma}_{\theta\theta}^n = \bar{\sigma}_{rr}^n = \frac{\partial^2 f(z)}{\partial z^2} [-D_1^n \ln (r) + 4BC_{11}^n r^{1/2} + D_2^n]$$

The D_i^n terms in Eq 10 are constants determined from boundary conditions yet to be posed. The fiber fracture problem has historically been treated as a shear transfer problem. In that light, it is interesting to point out that the coupling term (B) in $\bar{\sigma}_{xr}$ decays as $1/r^{0.5}$. While the shear stress dependence on the radial coordinate is similar to that obtained from fracture mechanics, we do point out that our crack tip is not located at $R = 0$.

To evaluate the constants in Eq 10, an internal traction is applied at $r = r_c$ [$\bar{\sigma}_{rz}(r_c, z)$ and $\bar{\sigma}_{rr}(r_c, z)$]. To formulate these boundary conditions, we suggest that the fractured fiber and matrix region ($r \leq r_c$) can be represented by a single fractured fiber system (fiber and matrix homogenized into one element). This homogeneous fiber system has material stiffness properties which are equivalent to the fractured fiber and failed matrix which it is representing. The mechanical properties for this local element are easily determined from a concentric cylinder model approach (see Pagano and Tandon [9]). Furthermore, the surrounding undamaged composite region ($r > r_c$) is also assumed to be represented by a homogeneous material whose properties are that of the global composite. The problem, thus posed, can be construed as a single fractured fiber embedded in an infinite matrix. We assume that the tractions at $r = r_c$ for either the single fractured fiber in an infinite matrix or the problem of a fractured fiber and matrix embedded in a composite are approximately equivalent. Thus, the response of each system in the linear elastic region, and the distribution of stress along the axial direction near the crack face, is similar. This rational is based on equivalency arguments classically made and similar to those published by Christensen [11] in his description of three phase composite models. For example, Christensen's model of a composite subjected to axial loading where the composite is assumed to be homogeneously distributed around the periphery of the matrix as opposed to discrete elements that give rise to local stress concentrations even in this damage-free composite.

A solution to the problem of a single fractured fiber in an infinite matrix has been formulated by Whitney and Drizal [12]. Their results are utilized in a fundamentally different manner than originally derived. The stress terms calculated at the cracked fiber matrix interface in their derivation are employed to pose the traction boundary conditions at $r = r_c$. By enforcing continuous tractions at each interface between constituents, the approximate stress field for each region is found. [Note: displacement continuity is not satisfied.] The boundary conditions and continuity conditions are represented by the following sets of equations.

$$\bar{\sigma}_{xr}(r_c, z) = \frac{-P_0}{2\pi r_c} \lambda^2 z e^{-\lambda z}$$

$$\bar{\sigma}_{rr}(r_c, z) = \frac{P_0}{4\pi} (-\lambda^2 + \lambda^3 z) e^{-\lambda z}$$

(11)

$$\bar{\sigma}_{ir}^n = \bar{\sigma}_{ir}^{(n+1)}$$

and

$$\lambda = \frac{2}{r_c} \left[\frac{C_{55}^{\text{comp}} S_{11}^{fm}}{1 + 4S_{12}^{fm} C_{55}^{\text{comp}}} \right]^{1/2}$$

where S_{ij}^{fm} is the compliance tensor for the fractured fiber and cracked matrix element. The reader is referred to Christensen [11] for a detailed and concise representation of these stiffness quantities. By applying the boundary conditions in Eq 11 to the stresses in Eq 10, the following relations are developed.

$$f(z) = (1 + \lambda z)e^{-\lambda z}$$

$$D_1^m = \frac{P_0}{2\pi} + 2BC_{11}^m r_c^{1/2}$$

$$D_1^n = D_1^{n-1} + 2Br_n^{1/2}(C_{11}^n - C_{11}^{n-1}) \tag{12}$$

$$D_2^n = \frac{P_0}{4\pi} + D_1^n \ln (r_c) - 4BC_{11}^n r_c^{1/2}$$

$$D_2^n = D_2^{n-1} + \ln r_n(D_1^n - D_1^{n-1}) + 4Br_n^{1/2}(C_{11}^{n-1} - C_{11}^n)$$

In the above equations, λ is a parameter which affects the extent to which the predicted perturbed stress field extends in the axial direction, normally defined as a region of *low* stress, not *high* stress.

Some brief results will be presented to provide corroboration that the boundary conditions posed in Eq 11 are appropriate. Carman et al. [13] has suggested that "ineffective" lengths predicted by shear lag models overestimate this region (for small fiber volume fraction $v_f < 0.25$). In addition, these observations suggest that a crack size parameter should be incorporated into the analysis. By utilizing the approach stated above, we have actually included a dependence of ineffective length on crack size, as well as constitutive properties. When computing an ineffective length for a 15% fiber volume fraction glass/epoxy composite (see Table 1 for mechanical properties) where $r_c = r_f$, the augmented model gives 14.7 r_f and Rosen's [2] shear lag model estimates 19.2 r_f. When calculating the ineffective length for a $v_f = 0.65$, the current model predicts 8.4 r_f, while Rosen's [2] model predicts 7.5 r_f. Therefore, it appears as though the proposed approach for modeling the axial stress redistribution is appropriate. This is based on the theoretical comparison with Rosen's [2] accepted model. We also note that the theoretical model proposed for the calculation of λ, presented in Eq 11, is valid only for polymeric and metal matrix composites.

TABLE 1—*Constituent material properties.*

Material	Young's Modulus, E, GPa	Poisson's Ratio, ν	Thermal Coefficient, α, 1/°C	Tensile Strength, σ, MPa
7740 glass	62.7	0.20	3.25E-6	...
PLM-9 epoxy	3.3	0.36	70.0E-6	50.0

Multiple Fiber Fracture

A formulation is presented to calculate the stress/strain concentrations which occur in composites containing multiple fiber fractures. The proposed methodology utilizes the analysis presented in the previous section along with a fiber discount scheme and a superposition technique.

The formulation for a composite which contains two adjacent fiber fractures (see Fig. 3) is described as follows. Each of the fiber fractures are assumed to exist in the same z ($z =$ 0) plane of the composite. Therefore we postulate the second fractured fiber will not support any of the axial load redistributed from the first fractured fiber. This seems reasonable when considering the plane defined by $z = 0$ (crack plane). Here both fibers are fractured, and thus neither is capable of supporting any load. Therefore, the second fractured fiber is discounted from the initial part of the analysis. This mathematically consists of reducing the area of the fibrous annular ring surrounding the first fiber fracture (see Fig. 1) by an amount equal to second fiber fracture πr_f^2 ($r_o^2 - r_a^2 = 5r_f^2$). This effectively increases the load which must be supported by the unbroken fibers adjacent to the first fractured fiber when the second fiber fails. By using Eqs 2 through 8 as described in the preceding section, the strain concentration in neighboring regions can be calculated.

To address the stress redistribution caused from the second fractured fiber, the process described above is repeated. However, in this case, the first fractured fiber is discounted and the stress redistribution around the second is analyzed. By superimposing this result on the one described above, the stress concentrations in a composite containing two fiber fractures is obtained. We point out that the first and second fibers have different sets of neighbors (see Fig. 3); thus, they distribute the stress to different fibers. This produces a strain concentration distribution which is not symmetric (following superposition) around either fiber fracture.

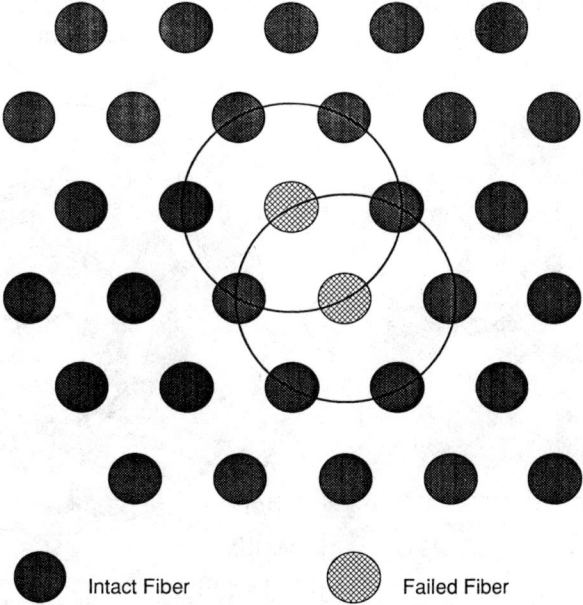

Intact Fiber   Failed Fiber

FIG. 3—*Typical illustration of two adjacent fiber fractures used in the computation of stress concentration as a result of multiple fiber fractures.*

To analyze a composite containing multiple fiber fractures (greater than two) the same sequence of steps is employed. Each fractured fiber is analyzed as a separate problem. This requires that all other fractured fibers be discounted from the analysis. After analyzing the stress redistribution which results from each fiber fracture separately, the solutions are superimposed to develop stress concentration values in a composite containing multiple fiber fracture. By utilizing the homogeneous element concept described in the preceding section (smearing all broken fibers into a single broken fiber), a dependence of ineffective length on fiber fracture can actually be postulated with this approach. As fiber fracture increases, r_c increases and λ decreases (see Eq 11), which results in an increase in ineffective length as one would intuitively expect.

Variable Fiber Spacing

A similar methodology to that outlined in the initial section is employed to study the effect of variable fiber spacing on the stress redistribution near a fiber fracture. This issue is addressed in the context of a hexagonal array of fibers for simplicity. The derivation in this paper describes one fiber which has been eccentrically oriented in the array during the manufacturing process (see Fig. 4). The methodology presented here can easily be extrapolated to numerous eccentric fibers. We propose that this problem can be approximated geometrically by an annular ring of fibers as presented in the initial formulation. However, unlike the initial analysis, multiple rings will be required to adequately address this problem.

For a composite containing one fiber eccentrically oriented, two rings are constructed (see Fig. 4). The outer fibrous ring is representative of the five regularly spaced fibers. This ring is located at the inner radius corresponding to r_a, and the outer radius is adjusted such that the area is equal to $5\pi r_f^2$. The inner fiber ring, which represents the single eccentric fiber, is located at an inner radius of r_a'. The outer radius of this ring corresponds to an equivalent area equal to πr_f^2. By utilizing equations similar to Eqs 2 through 8, general trends in stress concentration values for each constituent region can be obtained.

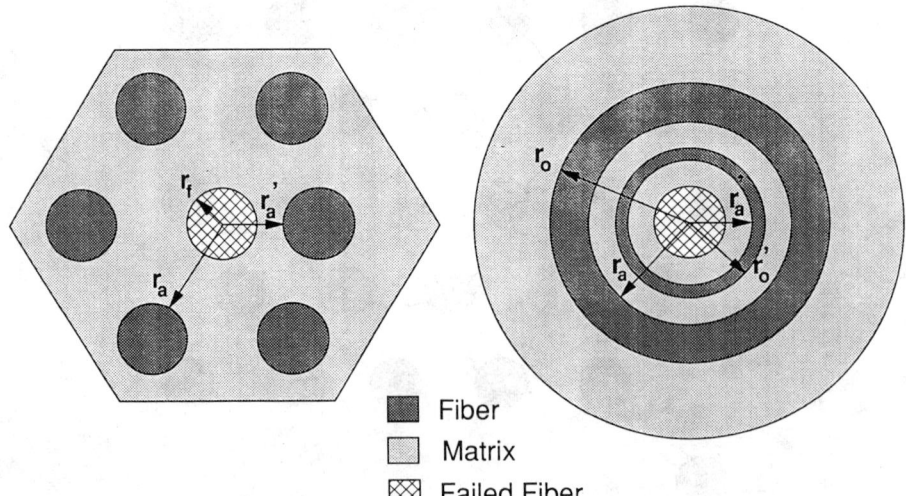

■ Fiber
▢ Matrix
▨ Failed Fiber

FIG. 4—*Illustration of a composite containing an eccentrically located fiber and a fractured fiber modeled as an annular ring problem.*

In describing the results of this analysis, it is convenient to define an eccentricity parameter η. We propose (see Fig. 4)

$$\eta = \frac{r'_a - r_f}{r_a - r_f} \qquad (13)$$

When η equals unity, the composite contains regularly spaced fibers, and when it is zero the eccentrically oriented fiber touches the fractured fiber. As noted above, the model is not limited to one eccentrically oriented fiber. By the addition of multiple rings, any number of variably spaced fibers can be represented.

Results and Discussion

A number of parametric studies are presented in this section to depict the influence various physical parameters have on the stress/strain concentration in composites containing fractured fibers. The constitutive material properties for a typical polymeric composite are presented in Table 1. These mechanical properties are used in all theoretical calculations unless otherwise noted. Results from Hedgepeth and Van Dyke's [3] model are included in figures where applicable.

In Fig. 5, the experimental results developed by Carman et al. [13] for a polymeric model composite $v_f = 0.15$ with internal sensors (resistance gages and fiber optic sensors) are plotted along with theoretical predictions from the current model and that of Hedgepeth and Van Dyke's model [3]. The y-ordinate corresponds to strain concentration, defined as the full-field axial strain normalized to the far-field axial strain ($\epsilon'_{zz}/\epsilon^c_{zz} = 1 + \bar{\epsilon}_{zz}/\epsilon^c_{zz}$). The x-abscissa represents the normalized distance from the crack tip (fractured fiber is located at $r/r_f = 1$) at the plane of the crack, that is, $z = 0$ in both the theoretical

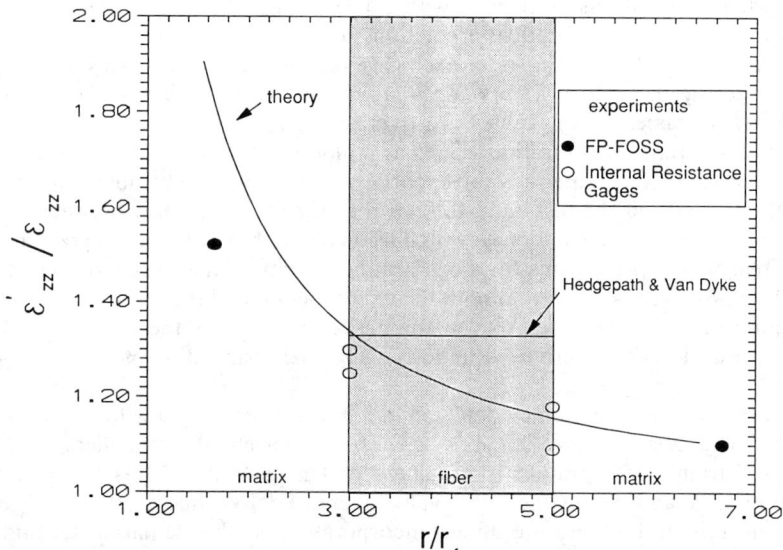

FIG. 5—*Comparison of strain concentration values determined by experimental techniques and theoretical predictions as a function of radial distance ($z = 0$) from the fractured fiber. Results depict a 15% fiber volume fraction model composite where $r_c = r_f$.*

calculations and test data. Experimental data are presented for both the fiber (determined by resistance strain gages) and the matrix regions (determined by Fabry-Perot fiber optic strain sensors FP-FOSS–Lesko et al. [14]). The fact that the experimental data points appear to be arranged in a smooth curve provides a fundamental understanding for the assumption made in Eq 6. The overall agreement between the theoretical and experimental results presented in Fig. 5 appears to be quite good. The experimental data point at $r/r_f =$ 1.6 exhibits the largest discrepancy with the current theory. It was suggested by Lesko et al. [14] that the experimental data point is actually lower than the point strain at this location due to the inordinately large sensing length of this particular fiber optic strain sensor. That is, the measured strain represents an average strain over the entire gage length which should be smaller than the value in the plane of the crack. Nonetheless, it appears for this fiber volume fraction and crack size that reasonable predictions for strain concentration are obtained with the current theory.

To investigate further the assumptions made in Eq 6 and Eq 11 and the model's prediction of strain concentration values, a comparison is made between the theoretical predictions (current model and Hedgepeth and Van Dyke [3]) and experimental results for a composite with a similar fiber volume fraction as described above. However, in this composite, the fiber fractured and a crack propagated into the matrix up to the adjacent fiber. The crack size is thus equal to the adjacent inner fiber dimension ($r_c = r_a$), which is considerably larger (over twice as large) than in the previous composite modeled. This represents a fundamentally different problem than presented in the previous paragraph. These results can be employed to investigate the appropriateness of the assumed radial stress variation made in Eq 6. The sensors used to monitor the internal strain concentrations in this composite were placed at different axial distances from the crack plane. This permits an evaluation of the predicted axial stress redistribution developed from the boundary conditions posed in Eq 11. λ has been experimentally suggested for this material system by Carman et al. [13] with the use of photoelastic data to be approximately $0.7125/r_f$, while calculations performed with Eq 11 predict that λ is $0.30/r_f$.

Plots corresponding to theoretical predictions utilizing both the experimental λ and the theoretical λ are presented in Fig. 6 for each gage location ($z = 0$, $z = 1.66r_f$, and $z = 2r_f$). With regard to the theoretical λ curves $z = 1.66r_f$ and $z = 2r_f$, it appears as though the predicted strain concentration values are overestimates of the test data. However, the general trend in strain concentration values as a function of axial coordinate are appropriate. With regard to the experimental λ curves, one immediately notes the agreement between the theory and the test data. In fact, the largest discrepancy between the theory and the experimental data is only 6%, which leads one to believe that the present model is providing reasonable predictions for axial strain concentration as a function of crack size and axial location. The results from both the experimental and theoretical λ curves aide in corroborating the assumption employed for the formulation of the boundary conditions described in Eq 11. These were used to derive the axial distribution of stress presented in Eqs 10 and 12.

The results from the larger crack depicted in Fig. 6 are contrasted with the smaller crack presented in Fig. 5 ($r_c = r_f$ versus $r_c = r_a$). The reader should immediately note that a much larger strain concentration is predicted and measured for this larger crack size (compare Figs. 5 and 6); in fact, the value increased by almost 50% (1.25 to 1.76). Presently, models in the literature do not incorporate a crack size parameter and suggest no dependence of strain concentration on crack size (e.g., Hedgepeth and Van Dyke predict 1.33 [3] and do not provide matrix strain concentrations) even though experimental results suggest this. The results presented in these two figures demonstrate the model's ability to predict strain concentration as a function of crack size, axial location, and radial

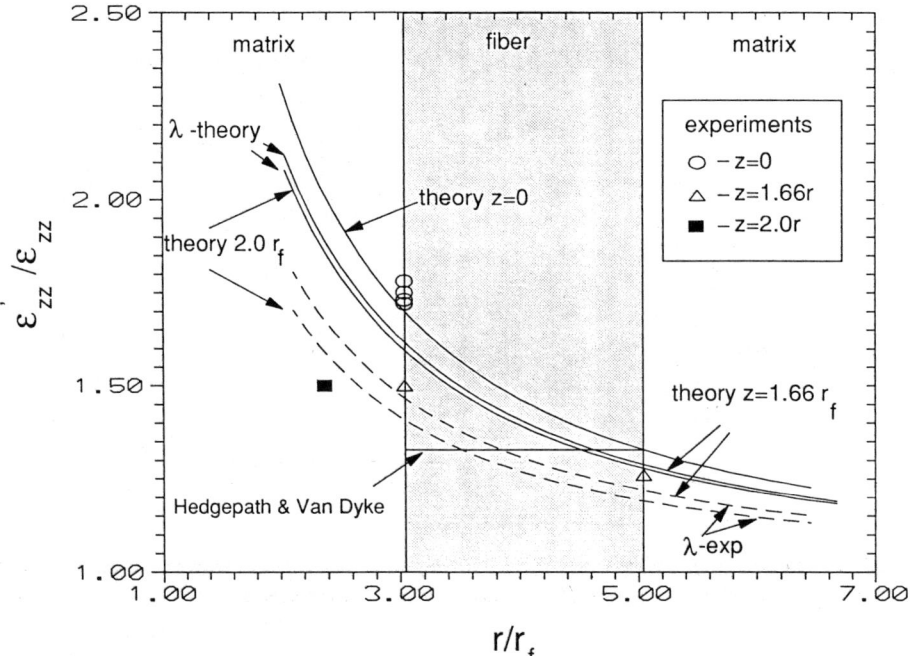

FIG. 6—*Comparison of strain concentration values determined by experimental techniques and theoretical predictions as a function of radial distance from the fractured fiber. Results depict a 15% fiber volume fraction model composite where* $r_c = r_a$.

position. For a corroboration of the model's dependence of strain concentration on fiber volume fraction, the reader is referred to Carman [*10*].

In Fig. 7, a plot of the average strain concentration in adjacent fibers versus constituent properties for the present model and Hedgepeth and Van Dyke's model [*3*] is shown for three fiber volume fractions and a small crack, $r_c = r_f$. Hedgepeth and Van Dyke's model does not incorporate a dependence of strain concentration on fiber volume fraction and thus only one curve is presented. One immediately recognizes that the present model predicts strain concentration increases with increasing C_{11}^f/C_{11}^m, as is expected. That is, as the matrix stiffness decreases relative to the fiber stiffness, the fiber supports a larger portion of the load. Note that for larger values of C_{11}^f/C_{11}^m, the strain concentration values appear to become independent of material properties (constant). However, the value at which this occurs varies as a function of v_f. For $v_f = 0.65$ it happens at approximately $C_{11}^f/C_{11}^m = 15$, for $v_f = 0.40$; $C_{11}^f/C_{11}^m = 20$, and for $v_f = 0.20$; $C_{11}^f/C_{11}^m = 30$. This suggests, for composites which fall to the right of these values (typically polymeric systems), matrix properties have little influence on strain concentration. For composites which fall to the left of these values (typically metal and ceramic systems) the load-carrying capabilities of the matrix significantly affects the strain concentration.

A parametric study is presented in Fig. 8 depicting the effect larger cracks ($r_c = r_a$) have on the average strain concentration in adjacent fibers in the plane of the crack as a function of constituent properties. The strain concentrations shown in Fig. 8 decay for increasing values of the stiffness ratio. However, this is the exact opposite of the trend described in Fig. 7 for the smaller crack. When considering a composite with a larger crack, P_0

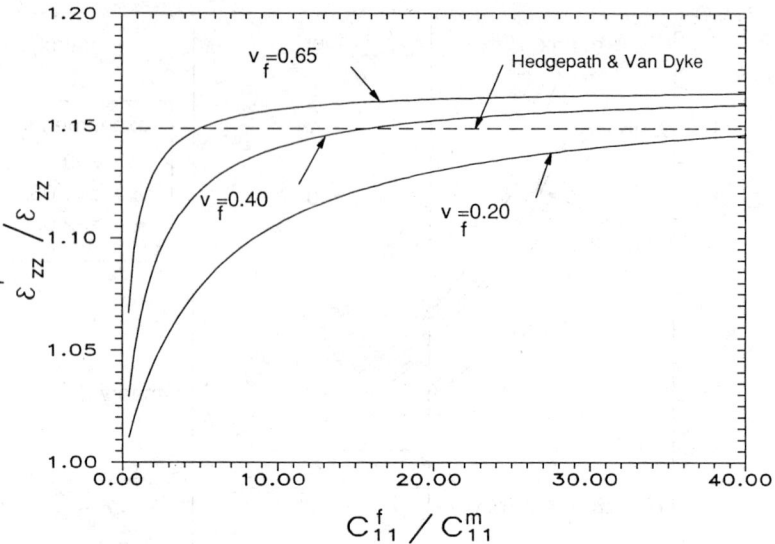

FIG. 7—*Figure displaying the functional dependence of average strain concentration on material properties for three fiber volume fractions and small crack size,* $r_c = r_f$.

represents a combination of the forces originally supported by the fractured fiber and the fractured matrix (see Eqs 3 and 4); while in the composite with the smaller crack, P_0 represents only the force originally supported by the fractured fiber. The composite with the larger crack has a larger P_0 term which results in larger strain concentrations as

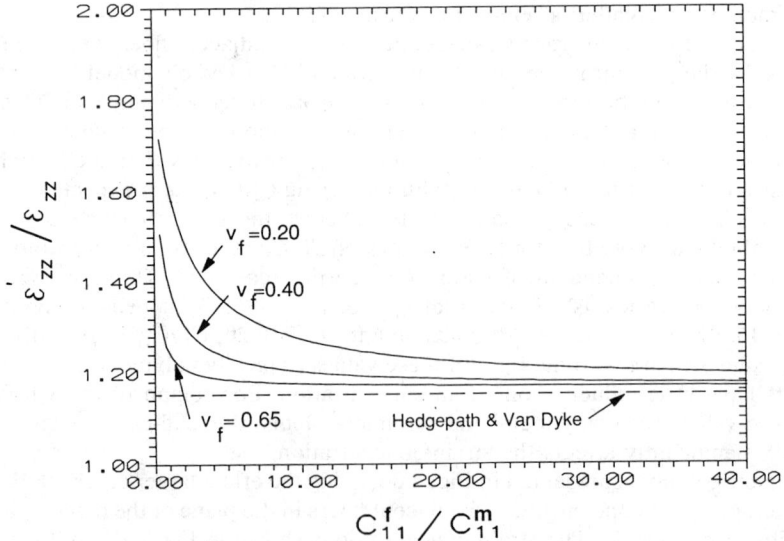

FIG. 8—*Figure displaying the functional dependence of average strain concentration on material properties for three fiber volume fractions and large crack size,* $r_c = r_a$.

compared to the smaller crack sizes. However, as the stiffness of the matrix decreases (C_{11}^f/C_{11}^m increases), the load originally supported by the cracked matrix decreases and results in a decrease in P_0. This causes a decrease in strain concentration as a function of increasing stiffness ratio. One can also see in Fig. 8 that the ordering of the strain concentration as a function of fiber volume fraction is exactly the opposite of that presented in Fig. 7. Composites with smaller fiber volume fractions contain fibers which are spaced further apart. Thus, when a fiber fracture propagates up to the adjacent fibers, the crack in the smaller fiber volume fraction composites is actually larger. Furthermore, since the crack is larger, the term P_0 is larger due to the additional matrix region cracked in the composite (see Eqs 3 and 4). This in turn causes smaller fiber volume fraction composites to exhibit larger average strain concentration for larger crack sizes.

Figure 9 depicts the variation of strain concentration values in the adjacent fibers as a function of v_f for a typical polymeric composite (see Table 1). The maximum, minimum, and average strain concentrations, described in the previous section, are presented in this plot as well as Hedgepeth and Van Dyke's [3] prediction (which gives only the average value). As v_f increases, the maximum and average strain concentrations increase, as intuition would suggest. However, the minimum strain concentration exhibits a maximum value at approximately $v_f = 0.20$. We believe that as the fibers are moved closer together (larger v_f), the fiber region nearest the fractured fiber is actually supporting a larger portion of the load (local stiffness increases), which results in a decrease in the load supported by the outer fiber region. This hypothesis is substantiated by the steady increase exhibited in the maximum strain concentration for increasing v_f. The reader should also notice that for larger fiber volume fraction, the average strain concentration becomes relatively constant. Models present in the literature (e.g., Hedgepeth and Van Dyke [3] predict a strain concentration of 1.15) actually suggest this independence of stress/strain concentration on

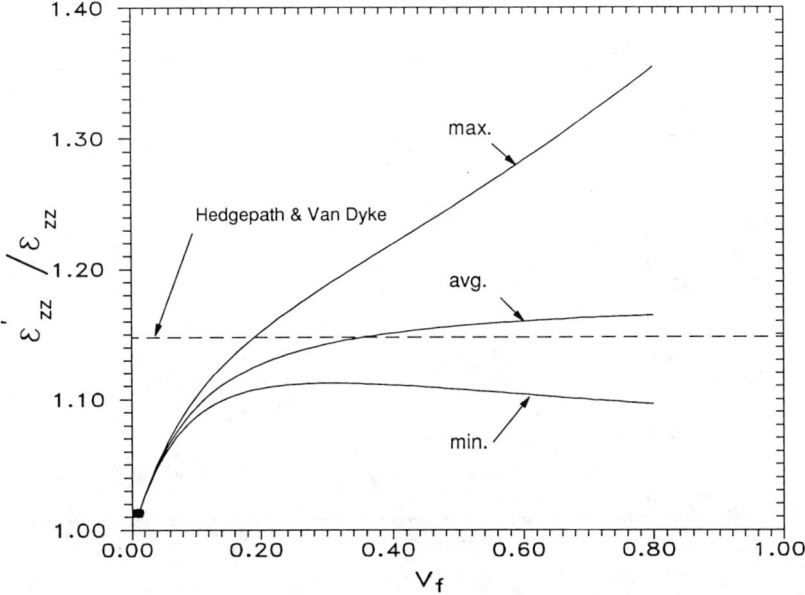

FIG. 9—*Figure displaying the functional dependence of the maximum, minimum, and average strain concentration on fiber volume fraction for an Glass/epoxy system.*

fiber volume fraction. However, these latter models do not depict the extreme stress/strain variation which actually exists across the fiber dimension predicted by the current theory, as presented in Fig. 9.

Figure 10 depicts the maximum strain concentration on an adjacent fiber in a composite with a large crack, $r_c = r_a$, as a function of v_f for four different stiffness ratios [ranging from a typical metal matrix composite (MMC) to a typical polymer matrix composite (PMC)]. By increasing the fiber volume fraction, there is an initial decrease in strain concentration followed by an increase. The curves in this figure suggest a minimum in the maximum strain concentration for a composite material which is a function of both v_f and the stiffness ratio. As fiber volume fraction increases, the crack size decreases (at least in $r_c = r_a$ composites), which causes a decrease in strain concentration. However, as v_f increases, the local stiffness increases, which causes an increase in strain concentration. These competing physical mechanisms cause a minimum in the maximum strain concentration. It has been suggested that the maximum strain concentration dominates the predicted strength of unidirectional composites (e.g., Gao et al. [14] and Harlow and Phoenix [4]) whose Weibull shape parameter is large.

In Fig. 11, the variation of average strain concentration as a function of normalized crack size $(r_c - r_f)/(r_c - r_f)$ is presented for three fiber volume fractions. At a normalized value equal to zero, the crack does not extend into the matrix $r_c = r_f$, and when it is unity, the crack extends up to the adjacent fiber, $r_c = r_a$. For cracks which do not extend into the matrix, the smaller v_f composites exhibit a smaller strain concentration, as is expected. However, for cracks which propagate up to the neighboring fibers (normalized value of 1), there is a reversal in the ordering of the strain concentration. For large fiber spacing (smaller v_f), a crack which propagates to the adjacent fibers is actually a larger crack as compared to the small fiber spacing (larger v_f). This was noted and discussed in previous paragraphs. Therefore, we conclude that smaller v_f composites with larger matrix cracks should exhibit larger average strain concentrations.

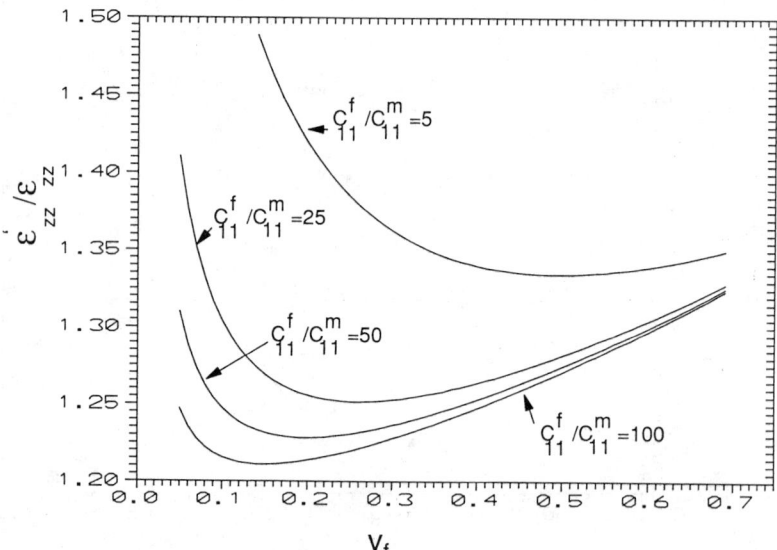

FIG. 10—*Figure depicting a minimum in the maximum strain concentration on adjacent fibers for different stiffness ratios.*

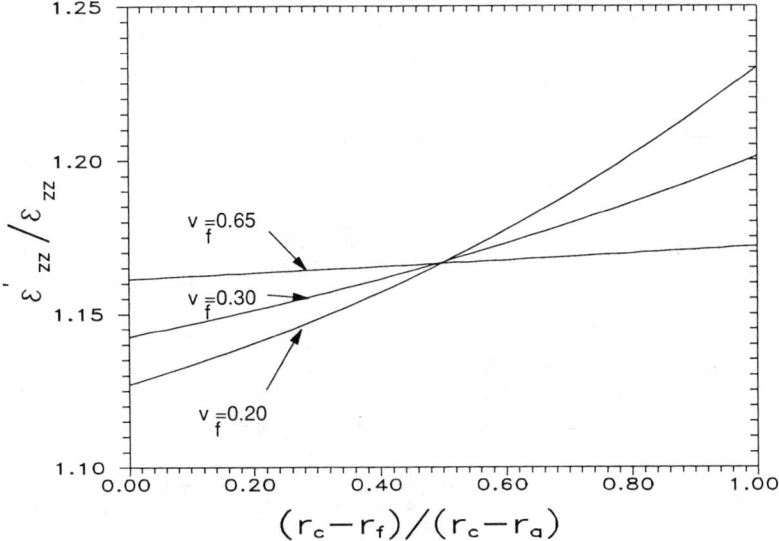

FIG. 11—*Figure depicting the variation of average strain concentration with normalized crack radius for five fiber volume fractions in a glass/epoxy system.*

Extrapolating the significance of this result to actual composites can assist in the design and processing of these materials to improve the tensile strength. For instance, it is desirable to control the crack dimensions. This can be achieved with the application of an interphase coating (Carman et al. [*13*]) and/or by adding toughening mechanisms to the matrix material. However, if this is not feasible, Fig. 7 suggests that a decrease in matrix stiffness or an increase in fiber volume fraction will cause a decrease in the strain concentration.

The model developed herein provides more than just axial fiber strain concentrations in the crack plane at discrete values of r. To demonstrate this, the methodology described in Eqs 10 through 12 is used to provide stress results for continuous values of r at given z locations in Fig. 12. The normalized stresses (Note: near field stress versus far-field fiber stress $\bar{\sigma}_{ij}^n(\sigma_{zz}^f)$ plotted correspond to σ_{zz} at $z = 0$ and σ_{rz} at $z = 1/\lambda$. The absolute value of the shearing stress is plotted to provide results on the same graph. For this material system, λ was calculated with Eq 11 and found to be $0.86/r_f$. The reader should immediately notice that a variation in axial stress exists in both the fiber and matrix materials. This plot suggests, for this polymeric matrix system, that the fiber supports the majority of the axial load. However, unlike shear lag analysis, the matrix supports a normal stress equivalent to approximately 10% of the fiber axial stress. For the variation of shear stress, it appears as though the transfer of normal load to the fiber is through a matrix shearing mechanism, as expected. However, the shear is predominately dissipated in the fiber region and not the matrix. Thus, unlike the shear lag models, a substantial shear stress is also present in the fiber.

The final plot (Fig. 13) presented in this section depicts the effect of multiple fiber fractures on average strain concentration. We assume in these calculations that fiber fractures occur in fibers which are adjacent to one another (see Fig. 3). The strain concentration values presented in Fig. 13 corresponds to the fiber which exhibits the largest average strain concentration in the composite for the given number of fibers fractured.

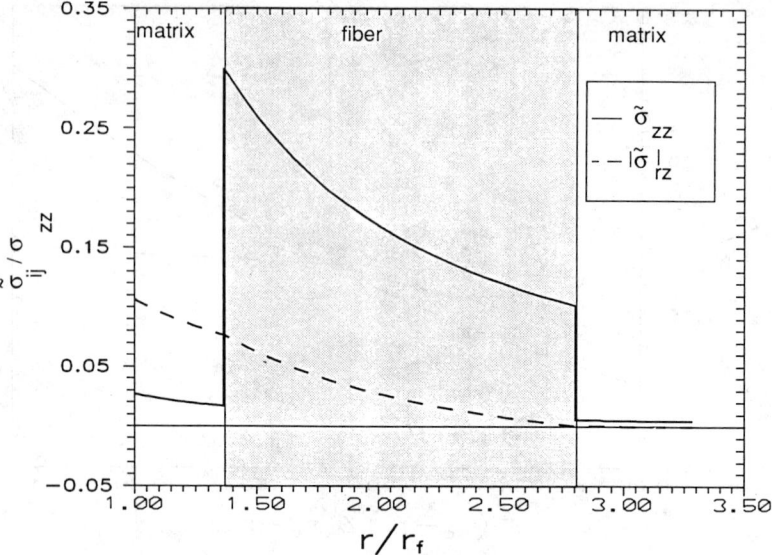

FIG. 12—*Plot of normalized stress versus normalized radial distance for axial stress (z = 0) and absolute value of shear stress (z = 1/λ) in glass/epoxy v_f = 0.65.*

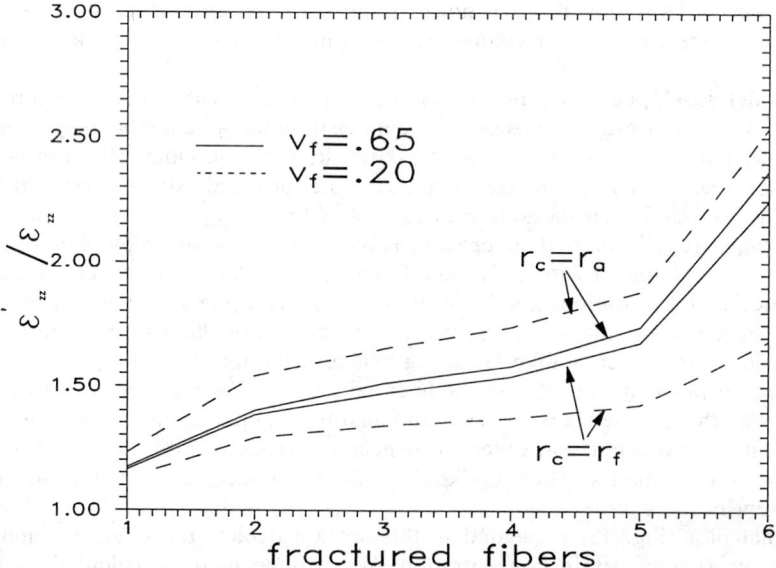

FIG. 13—*Figure representing the influence of fractured fibers on the largest average strain in a glass/epoxy system for two different fiber volume fractions and crack radii.*

Predictions are made for four combinations of v_f and r_c in Fig. 14. As is expected, with increasing fiber failures, the strain concentration increases. Note that after five fibers fracture, there is a drastic increase in strain concentration. This suggests that the composite should be incapable of sustaining a large number of fractured fibers in a given region prior to composite failure. This is substantiated by the experimental results presented by Jamison [16] and the theoretical predictions of Batdorf [7]. Furthermore, the $v_f = 0.20$ curve actually brackets the larger $v_f = 0.65$. This phenomenon was noted and explained in a previous paragraph of this section. Preliminary experimental results on multiple fiber fracture in a 2-D model composite by Carman et al. [13] have found for two adjacent fiber fractures the maximum strain concentration ($r = r_a$) is 1.7, while the current theory predicts a maximum value of 1.66 at this precise location. Models presently in the literature, such as Hedgepeth and Van Dyke [3], predict an average value of 1.60, while the current theory predicts an average value of 1.61. These initial results provide an indication of the model's capability to predict strain concentration values for multiple fiber fractures. In conclusion, the ability to use this model in formulations such as Batdorf's [7] to provide quantitative trends in strength as a function of specific parameters is a natural extension of this work.

Variable Fiber Spacing

The experimental data for a model composite $v_f = 0.15$ described by Carman et al. [13] are presented for two distinct crack geometries: $r_c = r_f$ for small cracks and $r_c = r'_a$ for large cracks. Two theoretical curves are presented for each crack geometry, which corresponds to the strain concentration on either an eccentric fiber, $r = r'_a$, or an aligned fiber, $r = r_a$ (Fig. 4). Experimental results are shown for three different η values of 0.16, 0.31, and 1.0 (Eq 13), with $\eta = 1.0$ representative of uniformly spaced fibers.

The experimental data shown in Fig. 14 are represented by the unfilled symbols which correspond to the curves with the solid symbols. The reader should note that for $\eta = 1$ the experimental data for the eccentric fiber are also representative of an aligned fiber. The comparison between the present theory and experimental results appears to be fairly reasonable. In regard to the strain readings obtained on the eccentric fiber $r = r'_a$ for the composite with the large crack size $r_c = r'_a$, the theoretical predictions are well within 10% error of the experimental data for both values of η (0.16 and 1.0). For the strain concentration on an aligned fiber $r = r_a$ with this crack size, the theoretical data are close to the experimental results (remember that for $\eta = 1$, the strain concentration at $r = r'_a = r_a$). The predicted trend in strain concentration as a function of eccentricity parameter for either the aligned fiber or the eccentric fiber appears to be in reasonably good agreement with the experimental results.

For the strain measurements in a composite with a small crack $r_c = r_f$ for an eccentric fiber $r = r'_a$ shown in Fig. 14, the error is less than 20%. Here, it appears as though strain concentration decreases as η increases for both the theoretical and experimental data. However, the theoretical predictions suggest a more severe decline than the experimental data depict. In reviewing the results of strain concentration on an aligned fiber $r = r_a$ for this crack size, it appears that an outlying data point at $\eta = 0.36$ may be present. If we disregard these data, good agreement is obtained between the theoretical predictions and experimental results. The general trend in the predicted strain concentration as a function of η also appears to be appropriate when comparing it to experimental data. In conclusion, it appears as though reasonable quantitative and qualitative strain concentration values are obtainable with the present theory for various eccentricity parameters and crack sizes.

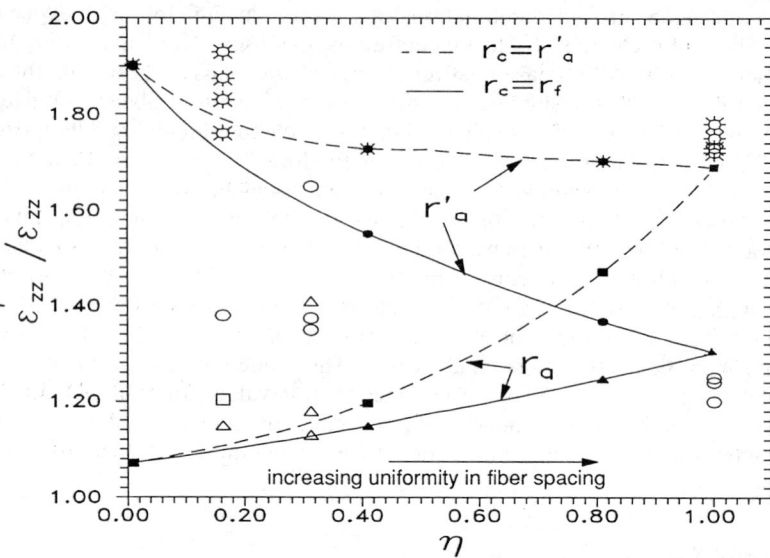

FIG. 14—*Comparison of the theoretical predictions to the experimental results of a model composite (Carman et al. [14]). Note: the experimental data in the figure are represented by the hollow symbols.*

A parametric study is now presented to demonstrate the effect variable fiber spacing has on maximum strain concentration trends in a typical polymeric composite, $v_f = 0.65$ (review Fig. 4). In Fig. 15, a plot is presented of the maximum strain concentration versus the eccentricity parameter η for the eccentrically located fiber $r = r'_a$ and for the adjacent fiber $r = r_a$. The curves are representative of two crack sizes a fiber fracture ($r_c = r_f$), and a full matrix crack ($r_c = r'_a$). Note the differences in strain concentration for a fiber fracture as compared to a full matrix crack described in the preceding section. For increasing alignment (η increases), the eccentrically located fiber ($r = r'_a$) exhibits a decrease in strain concentration for either crack size, as expected. However, the aligned fibers ($r = r_a$) display an increase in strain for increasing alignment. The aligned fibers are required to support a larger portion of the load as the eccentrically located fiber becomes more uniformly spaced. Additionally, the higher fiber volume fractions (Fig. 15) do not appear to have as large a dependence on crack size or η as does the smaller fiber volume fractions (Fig. 14). Therefore, it is suggested that eccentricity parameters affect smaller v_f more severely than do the larger v_f composites. This appears reasonable, noting that for small v_f composites, the measured distance between the eccentrically located fiber and the aligned fibers is greater than in large v_f. The larger distance correlates to larger strain concentration effects. This does not imply that the larger fiber volume fractions are not affected. One can see from Fig. 15 that this is not the case.

In Fig. 16, a plot of maximum strain concentration versus v_f is presented for an eccentrically located fiber for three different η. First, for fibers uniformly spaced ($\eta = 1$), the strain concentration increases for increasing fiber volume fraction. However, for $\eta = 0$, the exact opposite trend is noted. That is, as v_f increases, strain concentration decreases. A value of $\eta = 0.5$ is also presented to demonstrate the apparent trend from an increasing strain concentration to a decreasing strain concentration as a function of v_f.

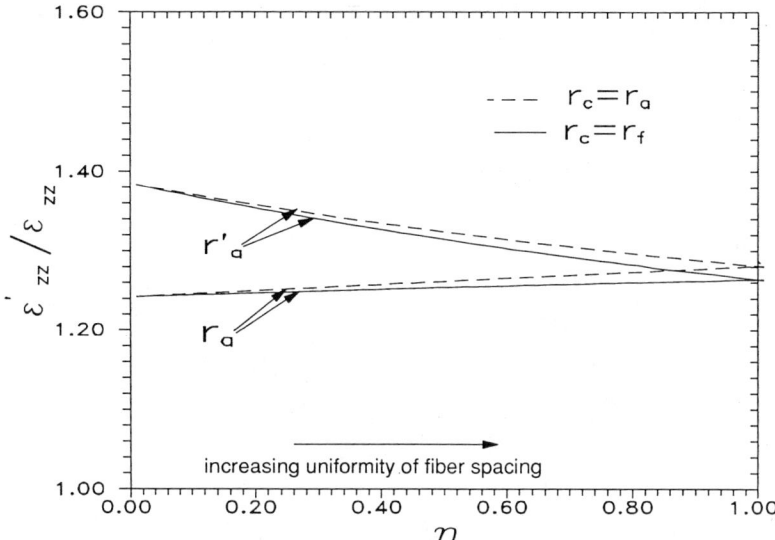

FIG. 15—*Figure displaying the functional dependence of maximum strain concentration on eccentricity parameter η for a glass/epoxy system $v_f = 0.65$ for two different crack sizes and locations.*

This drastic change in the variation of strain concentration for different values of η is explained as follows. We described in the previous section that, for uniformly spaced fibers, the maximum strain concentration increases as the fiber volume fraction increases (see the $\eta = 1$ curve). For a composite containing an eccentrically located fiber, the maximum strain concentration may decrease (or increase), depending on the degree of eccentricity. Assume that the eccentrically located fiber touches the fractured fiber as depicted in the figure ($r'_a = r_f$ and $\eta = 1$). For small v_f, the aligned fibers are further away from the fractured fiber than in large v_f. Therefore, the eccentrically located fiber supports more of the load in small v_f. As v_f increases, the aligned fibers move closer to the fractured fiber and thus support a larger portion of the load. This reduces the strain concentration on the eccentrically located fiber. Therefore, strain concentration trends may be severely affected in composites containing varying degrees in the uniformity of fiber spacing.

Conclusions

A theoretical micromechanics model has been presented which provides an approximate solution for the stress redistribution which occurs in composites containing fiber fractures. This theoretical model is functionally dependent on physical parameters such as constituent material properties, fiber volume fraction, and crack size. Its use is not limited to polymeric composites but is applicable to a general class of composite systems. The model includes a quantitative analysis of variable fiber spacing which occurs in virtually every composite manufactured. A novel fiber discount methodology was proposed to study multiple fiber fractures, which are of extreme importance when attempting to predict tensile strength of fiber-dominated composite laminates.

The strain concentration values predicted by the model were compared to direct experi-

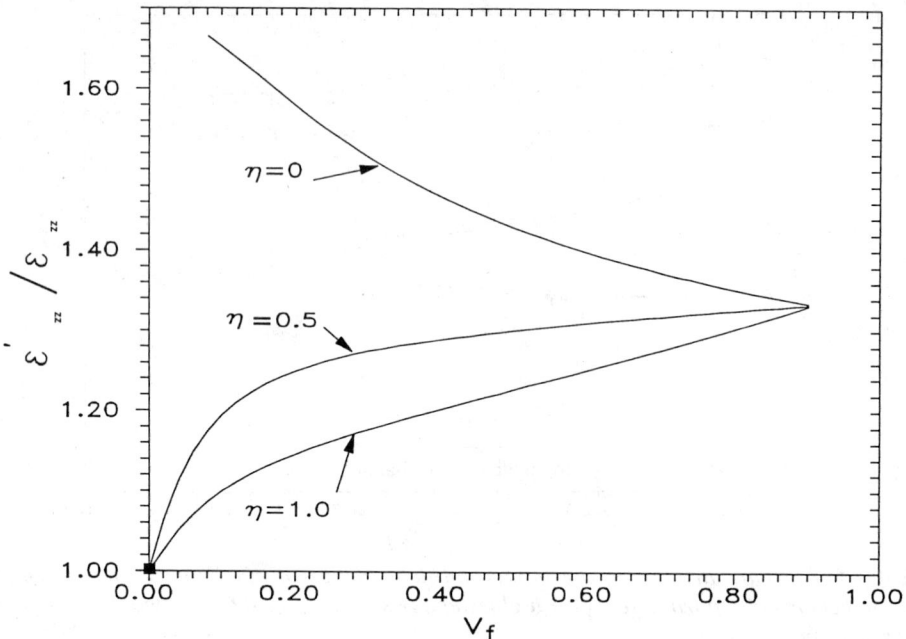

FIG. 16—*Plot depicting the the functional dependence of maximum strain concentration on fiber volume fraction for a glass/epoxy system for three different eccentricity parameters.*

mental data as well as accepted theoretical models. The test data consisted of composites containing two different crack sizes with strain measurements performed at various axial locations. These results provide a justification of the model's ability to depict the dependence of strain concentration on crack size, radial position, and axial position. Corroboration of the model's prediction of strain concentration as a function of fiber fractures was also suggested with experimental and theoretical results. Reasonable correlation was obtained between strain concentration values predicted by the model and direct experimental measurements performed on eccentrically oriented fibrous composites. Specifically, in this comparison the predicted trends in strain concentration as a function of eccentricity parameter were insinuated to be correct.

A number of parametric studies were performed utilizing the present theoretical approach. The following general conclusions regarding strain concentration on adjacent fibers were noted.

1. A minimum occurs in the maximum strain concentration as a function of fiber volume fraction and stiffness ratio for composites with large cracks ($r_c = r_a$).
2. Average strain concentration values increase, as the stiffness ratios C_{11}^f/C_{11}^m increase in composites with small cracks ($r_c = r_f$). However, in composites which contained a relatively larger crack ($r_c = r_a$), the average strain concentration decreases.
3. In composites containing small crack sizes ($r_c = r_f$), the average and the maximum strain concentration increases as fiber volume fraction increases. However, in com-

posites with large cracks ($r_c = r_a$), the average strain concentration decreases as fiber volume fraction increases.

4. Crack size appears to significantly alter the stress redistribution in composites.

5. Variable fiber spacing seems to substantially modify the affect fiber volume fraction has on the maximum strain concentration in composites. Depending upon the degree of eccentricity, strain concentration may increase or decrease as v_f increases.

The model did not explicitly incorporate an interphase region; however, the addition of this region is easily achieved. Work is presently being performed on developing a quantitative relation between physical parameters in the model and composite tensile strength.

Acknowledgments

This research has been supported by grants from the Virginia Institute for Material Systems at Virginia Tech, the National Science Foundation-Science and Technology Center for High Performance Polymeric Adhesives and Composites at Virginia Tech, and the National Science Foundation under grant no. MSS-9115380.

References

[1] Madhukar, M. S. and Drzal, L. T., "Effect of Fiber-Matrix Adhesion on Longitudinal (0°) Compressive Properties of Graphite/Epoxy Composites," *Proceedings,* 5th Technical Conference of the American Society of Composites, 1990, pp. 849–858.

[2] Rosen, B. W., "Fiber Composite Materials," Chap. 3, *American Society of Metals,* Metals Park, OH, 1964, pp. 37–75.

[3] Hedgepeth, J. M. and Van Dyke, P., "Local Stress Concentration in Imperfect Filamentary Composite Materials," *Journal of Composite Materials,* 1968.

[4] Harlow, G. and Phoenix, S. L., "Probability Distributions for the Strength of Composite Materials I: Two-Level Bounds," *International Journal of Fracture,* Vol. 17, No. 4, 1981, pp. 347–372.

[5] Fukuda, H., "Statistical Strength of Unidirectional Composites with Random Fiber Spacing," *Composites 86: Recent Advances in Japan and the United States,* Proceedings of the Third Japan-U.S. Conference on Composite Materials, K. Kawata, S. Umekawa, A. Kobayashi, Eds., Japan Society for Composite Materials, Tokyo, 1986, pp. 307–314.

[6] Adams, D. F., "Micromechanical Predictions of Crack Propagation and Fracture Energy in a Single-Fiber Boron/Aluminum Composite," NASA-Lewis Technical Report, UWME-DR-201-101-1, Nsg-3217.

[7] Batdorf, S. B., "Tensile Strength of Unidirectionally Reinforced Composites—I," *Journal of Reinforced Plastics and Composites,* Vol. 1, 1981, pp. 153–177.

[8] Carman, G. P. and Reifsnider, K. L., "Micromechanics of Short Fiber Composites," *Composites Science and Technology,* accepted for publication, 1990.

[9] Pagano, N. J. and Tandon, G. P., "Elastic Response of a Multidirectional Coated-Fiber Composites," *Composites Science and Technology,* Vol. 31, 1988, pp. 273–293. Carman, G. P., "Micromechanics of Finite Length Fiber in Composite Materials," Ph.D. dissertation, Virginia Tech, September 1991.

[10] Whitney, J. M. and Drzal, L. T., "Axisymmetric Stress Distribution Around an Isolated Fiber Fragment," *Toughened Composites, ASTM STP 937,* N. J. Johnston, Ed., American Society for Testing and Materials, Philadelphia, 1987, pp. 179–196.

[11] Christensen, R. M., *Mechanics of Composite Materials,* Wiley Interscience Publication, New York, 1979.

[12] Carman, G. P., Lesko, J. J., Razvan, K., and Reifsnider, K. L., "Model Composites: A Novel Approach for the Evaluation of Micromechanical Behavior," this publication.

[13] Lesko, J. J., Carman, G. P., Reifsnider, K. L., Vengsarkar, A., Miller, W., Fogg, B., and Claus, R., "Embedded Fabry-Perot Fiber Optic Strain Sensors in the Macro-Model Composite," *Optical Engineering,* January 1992.

[*14*] Gao, Z., Reifsnider, K. L., and Carman, G. P., "Strength Prediction and Optimization of Composites with Statistical Fiber Flaw Distributions," *Journal of Composite Materials*, Vol. 26, No. 11, 1991, pp. 1678–1705.

[*15*] Jamison, R. D., "Advanced Fatigue Damage Development in Graphite Epoxy Laminates," Ph.D. dissertation, Engineering Science and Mechanics, Virginia Tech, August 1982, pp. 146–180.

Zhanjun Gao[1] and Kenneth L. Reifsnider[2]

Micromechanics of Tensile Strength in Composite Systems

REFERENCE: Gao, Z. and Reifsnider, K. L., **"Micromechanics of Tensile Strength in Composite Systems,"** *Composite Materials: Fatigue and Fracture, Fourth Volume, ASTM STP 1156,* W. W. Stinchcomb and N. E. Ashbaugh, Eds., American Society for Testing and Materials, Philadelphia, 1993, pp. 453–470.

ABSTRACT: This paper deals with the tensile strength of composites. The fibers of the composites are considered to have a statistical distribution of flaws of the Weibull type. The tensile strength formulation developed explicitly contains the Weibull shape factor, m, stress concentration factors due to fiber fractures, C_i, and the ineffective length, δ. Here, m describes the uniformity of the fiber strength. The stress concentration factors, C_i, characterize the disturbance of stress in neighboring unbroken fibers when i numbers of fiber have broken together. They are related directly to the subsequent event of fracture propagation perpendicular to the fiber direction, which is an important failure mechanism in ceramic composites. The ineffective length, δ, hinges on the concept of "the weakest link" along the fiber direction and is the dominant factor in the bundle strength representation of the strength of polymer-based composites. There is a lack of systematic analysis addressing the competition of these two important failure mechanisms and their influence on the tensile strength of the composites. In this paper, we combine the theory of bundle strength with the mechanics of local stress concentration to develop a micromechanical model for tensile strength, and study the influence of micromechanical properties on the tensile strength of composites and optimal design for tensile-controlled loading applications.

Using the strength formulas developed in this paper, we study how tensile strength is influenced by the changes of the micromechanical properties, m, C_i, and δ, and also by the properties of the fiber and matrix, E_f and G_m. It is shown that the existence of stress concentrations due to fiber fractures increases the chance of local crack propagation and, therefore, decreases the strength values predicted by the bundle strength theory. Stress concentrations play a dominate role in determining the magnitudes of tensile strength, especially for large values of m. However, since the stress concentrations remain virtually constant, for composites with different E_f and G_m and regularly spaced fibers, the variation of the strength curve as a function E_f and G_m is still controlled by the changes of the ineffective length, δ. The predictions of strength from the present model are compared to experimental measurements to validate the accuracy of the formulation.

The strength consideration is more complicated when the existence of irregular fiber spacing and the lack of quantitative descriptions of the spacing are taken into account. We show that, even for polymeric composites, optimal strength (when material properties E_f and G_m are taken as design variables) can be achieved due to certain fiber spacing variations. Therefore, the fiber spacing of composites, although difficult to control in the manufacturing process and hard to determine for a specimen, must be considered carefully to eliminate, isolate, or correctly represent its effects on strength and other macromechanical properties of engineering importance.

KEY WORDS: tensile strength, fiber fractures, stress concentrations, irregular fiber spacing

[1]Assistant professor, Dept. of Mechanical and Aeronautical Engineering, Clarkson University, Potsdam, NY 13699.

[2]Alexander Giacco professor, Dept. of Engineering Science and Mechanics, Virginia Polytechnic Institute and State University, Blacksburg, VA 24061.

Micromechanical approaches have been suggested for strength considerations, especially for the calculation of the strength of fiber-reinforced composites wherein the final failure mode is controlled by local stresses and strength considerations. For continuous fiber-reinforced composite materials, bundle strength consideration is utilized in the calculation of tensile strength [1], where fibers are treated as having a statistical distribution of flaws. Each fiber fracture creates a segment of the length δ on each side of the break. The fiber fracture does not cause any changes of stresses outside the section bounded by the length δ, called the ineffective length. Composite failure occurs when the remaining unbroken fibers at the weakest cross section are unable to resist the applied load. The composite strength is evaluated herein as a function of the statistical strength characteristics of the fiber population and of parameters, such as ineffective length, that define the composite materials properties. Bundle strength concept formulas are well established and generally take the form suggested by Tsai and Hahn [2]:

$$X_b(\delta) = \frac{E_f \epsilon}{(me\delta)^{1/m}} \tag{1}$$

where

δ = the ineffective length,
E_f = Young's modulus of the fiber,
ϵ = the average strain to the failure of fibers (with length δ),
m = the shape factor of the Weibull distribution of the fibers with length δ, and
e = the Naperian logarithm base.

The ineffective length, δ, in general, depends on the properties of fibers, matrix, as well as their interface.

The above analyses have the drawback of neglecting the effect of stress concentrations on the neighboring fibers caused by fiber fractures, which results in overestimates of strength, especially for small values of the ineffective length, δ. The micromechanical strength formulas which consider such an effect on tensile strength are typified by the relationship suggested by Batdorf [3]. Batdorf abandons the chain of the bundle model and concentrates on the formation and growth of multiple fiber fractures. This permits treating interaction of fiber fractures in different bundles and assignment of different ineffective lengths to fiber breaks of different multiplicity. Batdorf uses the weakest link theory to determine the number of isolated fiber fracture (singlets), double fracture (doublets), and multiplets of arbitrary order as a function of stress. It turns out that if the fracture of individual fibers obeys Weibull's distribution, a plot of ln (Q_i) versus ln (σ) is a straight line of the slope, im. Here σ is applied stress, Q_i is the number of i-plets formed during loading to stress σ, and m is the Weibull modulus. The envelope of the Q_i curves serves as a failure line. Use of the failure line leads directly to a rational failure criterion based on a Griffith-type instability.

The objectives of this paper are: to develop a representation of tensile strength which combines the theory of bundle strength with the mechanics of local stress concentration, to study the influence of micromechanical properties and irregular fiber spacing on the tensile strength of composites, and to shed light on the optimal design of composite systems for tensile-controlled composites.

Probabilities Associated with Fiber Fractures

We will assume that when a stress, σ, is applied uniformly over a fiber of length L, the cumulative probability of failure, P_f, obeys the Weibull distribution

$$P_f(\sigma) = 1 - \exp\left[-L\left(\frac{\sigma}{\sigma_0}\right)^m\right] \tag{2}$$

where m and σ_0 are parameters of the distribution. In the above expression, the length, L, is a dimensionless quantity, i.e., the actual length of the fibers in microns normalized by a characteristic length taken as 400 μm. In this paper, all the other quantities with dimension of length are normalized by the same length.

If there are totally N fibers in the composite, the number of isolated broken fibers, Q_1, (called singlets after Batdorf [3]) that will have been created by the time the stress rises to σ is

$$Q_1 = NP_f(\sigma) = N\left\{1 - \exp\left[-L\left(\frac{\sigma}{\sigma_0}\right)^m\right]\right\} \tag{3}$$

A singlet becomes a doublet (two adjacent fibers broken) when one of the neighboring fibers breaks in its over-stressed region. The probability that a neighboring fiber failure can be found from the general form given by Weibull [4] as

$$P_1 = 1 - \exp\left[-\lambda_1\left(\frac{\sigma}{\sigma_0}\right)^m\right] \tag{4}$$

where

$$\lambda_1 = 2\int_0^{\delta_1}\left[\frac{\sigma_f^*(x)}{\sigma}\right]^m dx \tag{5}$$

and $\sigma_f^*(x)$ is the tensile stress of the neighboring fiber. If n_1 is the number of nearest neighbors of a singlet, the probability that a given singlet becomes a doublet is then

$$n_1\left\{1 - \exp\left[-\lambda_1\left(\frac{\sigma}{\sigma_0}\right)^m\right]\right\} \tag{6}$$

Therefore, the number of doublets, Q_2, created in loading to stress, σ, thus becomes

$$Q_2 = Q_1 n_1\left\{1 - \exp\left[-\lambda_1\left(\frac{\sigma}{\sigma_0}\right)^m\right]\right\} \tag{7}$$

In general, if Q_i is the number of i-plets (i adjacent fibers broken) at stress level σ, we have the following iterative relation

$$Q_{i+1} = Q_i n_i\left\{1 - \exp\left[-\lambda_i\left(\frac{\sigma}{\sigma_0}\right)^m\right]\right\} \tag{8}$$

where

$$\lambda_i = 2 \int_0^{\delta_i} \left[\frac{\sigma_f^*(x)}{\sigma} \right]^m dx \tag{9}$$

and n_i is the number of nearest neighbors an i-plet has. When the stress distribution in the neighboring fibers, $\sigma_f^*(x)$, is assumed to vary linearly over the distance δ_i, i.e.,

$$\sigma_f^*(x) = \sigma \left[C_1 + \frac{x}{\delta_1} (1 - C_1) \right] \tag{10}$$

then Eq 9 is rewritten as

$$\lambda_i = 2\delta_i \frac{C_i^{m+1} - 1}{(C_i - 1)(m + 1)} \tag{11}$$

If we expand the exponential and keep only the first order term, we obtain Batdorf's formulation

$$Q_{i+1} = Q_i n_i \lambda_i \left(\frac{\sigma}{\sigma_0} \right)^m \tag{12}$$

A diagram of σ versus Q_i based upon Eq 8 can be used to determine the strength of the composite.

A schematic diagram of the failure envelope created by Q_i is shown in Fig. 1. Let the stress, σ_i, be defined as the stress value at the intersection of Q_i and the horizontal line $Q_i = 1$. We look for the first σ_i such that $\sigma_i > \sigma_{i+1}$. This σ_i represents the ultimate failure stress. For the case shown in Fig. 1, the first σ_i satisfying $\sigma_i > \sigma_{i+1}$ is σ_3. Since σ_4 is less than σ_3, as soon as the stress level reaches the value σ_3, at which first triplet is formed, it exceeds σ_4, and we begin to see quadruplets and so forth, and the material fails. In other words, σ_3 is the onset of instability. The failure stress is actually the intersection of the failure envelope and the horizontal line, $Q_i = 1$.

The formulation shown in Eq 8 hinges on the micromechanics analysis which provides the dependence of the ineffective length, δ, and stress concentration factor, C, of the unbroken fibers on the material properties of the fibers and matrix. This is the subject of the next section.

Micromechanics Analysis

We start the micromechanics analysis of ineffective length, δ, and the stress concentration factor, C, by examining and modifying the classic shear lag analysis.

The equilibrium element used in the formulation of the shear lag model is shown in Fig. 2. The tensile stress outside the region of the broken fiber and surrounding matrix is considered as a constant, equal to the average value in the composite, i.e., σ_c. Therefore, the stress concentration near the broken fiber is actually averaged over the whole composite and is represented by σ_c in the equilibrium equation in the fiber direction

$$\pi r_f^2 \sigma_f(x) + (r_c^2 - r_m^2) \pi \sigma_c(x) = \pi r_c^2 \sigma_a \tag{13}$$

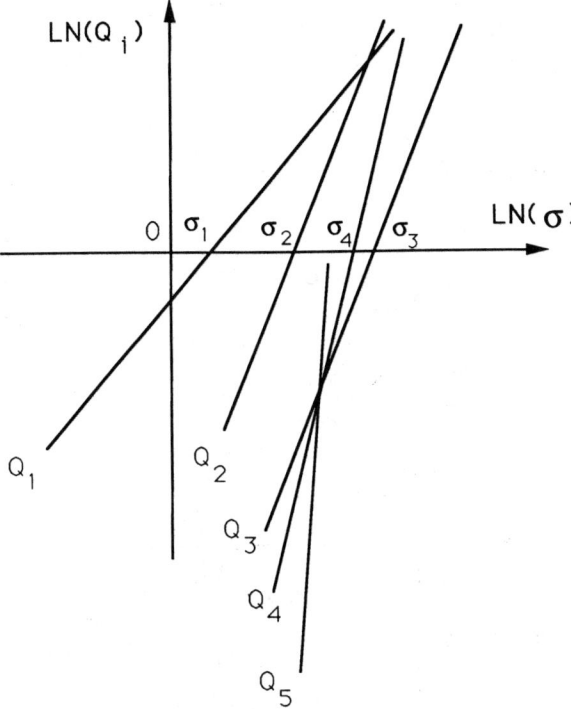

FIG. 1—*Failure envelope based on fiber fractures.*

where r_f, r_c, and r_m are the radius of fiber, matrix, and composite, respectively. σ_f and σ_c are stresses of the broken fiber and the composite. σ_a is the applied stress.

The average shear strain, $\gamma_m(x)$, in the matrix region can be defined as

$$\gamma_m(x) = \frac{u_c(x) - u_f(x)}{r_m - r_f} \tag{14}$$

where u_c and u_f are displacements of the composite and the broken fiber, respectively.

The equation for the equilibrium of the broken fiber over the distance dx is written as follows (see the diagram on the bottom of Fig. 2)

$$\pi r_f^2 \frac{d\sigma_f(x)}{dx} + 2\pi r_f \tau_m(x) = 0 \tag{15}$$

The shear stress obtained by solving the shear lag equations, Eqs 13, 14, and 15, is written as [5]

$$\tau_m(x) = \frac{-r_f \eta \sigma_a E_f}{2E_c} \exp(-\eta x) \tag{16}$$

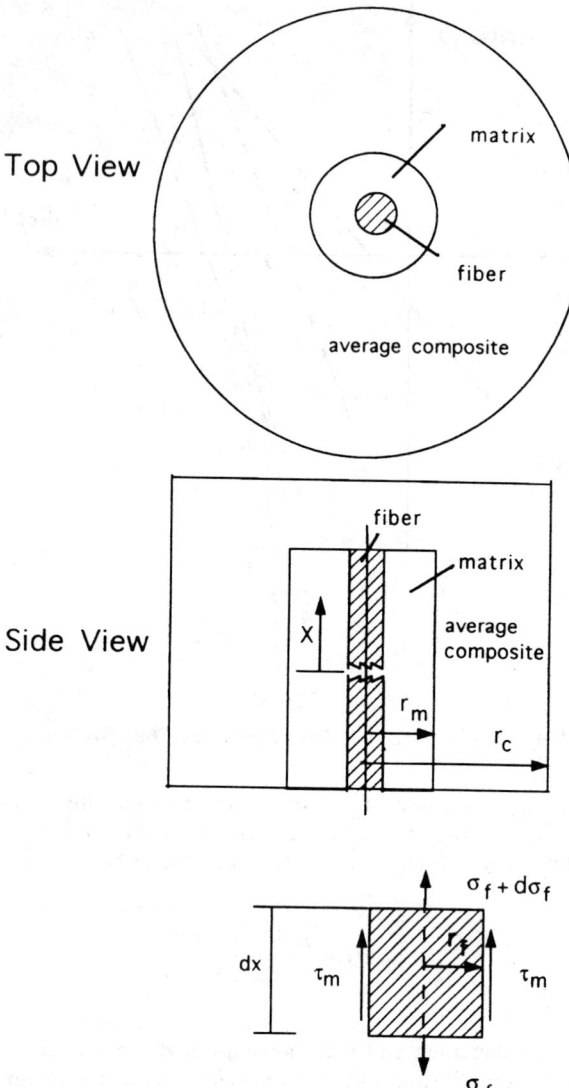

FIG. 2—*A broken fiber, matrix region, and average composite: the equilibrium element used in formation of shear lag model.*

and

$$\sigma_f(x) = \frac{\sigma_a E_f}{E_c} [1 - \exp(-\eta x)] \tag{17}$$

where

$$\eta = \left[\frac{2G_m}{E_f(r_m - r_f)r_f} \right]^{0.5} \tag{18}$$

where E_f is Young's modulus of the fibers, and G_m is the shear modulus of the matrix. Please note that Eq 18 is a simplified expression for η, in which only the leading term is kept.

In the shear lag analysis as shown in Eqs 14 and 15, only the broken fiber is considered and the rest of the composite is treated as a homogeneous material with uniform displacement, $u_c(x)$. However, in terms of fiber fracture (crack) propagation, it is the neighboring fibers that are over-stressed and may be caused to fail. Therefore, with regard to tensile failure, it is necessary to include the neighboring unbroken fibers in the local stress analysis.

The equilibrium element for the shear lag analysis shown in Fig. 2 is modified to include the neighboring unbroken fibers (Fig. 3). Here, the neighboring fibers are no longer considered as a part of the average composite. Their displacements and stresses are distinct from those of the broken fiber or the average composite. Let σ_f^* and u_f^* be the stress and displacement in one of the neighboring fiber (for example, unbroken fiber FABH). The quantity

$$\frac{u_f^*(x) - u_f(x)}{2(r_m - r_f)} \tag{19}$$

defines the average shear deformation in matrix region EFGH, which is different from the one in Eq 14. The shear strain in the matrix region EFGH varies from Point E to Point F in the matrix. In order to establish the equilibrium equation of the broken fiber (Eq 15), the shear stress, $\tau_m(x)$, of the matrix adjacent to the broken fiber (along Line E) was used. Therefore, we introduce a correction factor, α, to the average shear strain expression of Eq 14 and write

$$\gamma_m(x) = \alpha \frac{u_f^*(x) - u_f(x)}{2(r_m - r_f)} \tag{20}$$

Please note that, although we still use the same notation $\gamma_m(x)$, it is different from the one defined in Eq 14.

Since we have modified the shear strain, $\gamma_m(x)$, of the shear lag analysis into the one shown in Eq 20, the shear stress in the matrix $\tau_m(x)$ of the shear lag analysis (Eq 16) also needs to be modified. We imposed

$$\tau_m(x) = \frac{-r_f \eta \sigma_a E_f}{2E_c} \exp(-\eta x) f(x) \tag{21}$$

where $f(x)$ is an influence function with regards to stress concentration effects. As a first approximation, we choose

$$f(x) = Ax + B \tag{22}$$

in which A and B are constants to be determined later. By using different forms of approximation for the function $f(x)$, we can modify the analysis to consider other forms of damage influence present in composites.

Now we determine the constants A and B in Eq 22 and α in Eq 20 and then obtain the shear stress in matrix $\tau_m(x)$. Differentiating Eq 20 and using the definition of Hooke's law produce

$$\frac{2(r_m - r_f)}{\alpha G_m} \frac{d\tau_m(x)}{dx} = \frac{1}{E_f} [\sigma_f^*(x) - \sigma_f(x)] \tag{23}$$

The stress concentration factor, C, in the unbroken fiber is defined as

$$C = \frac{\sigma_f^*(0)}{\sigma_f^*(\infty)} \tag{24}$$

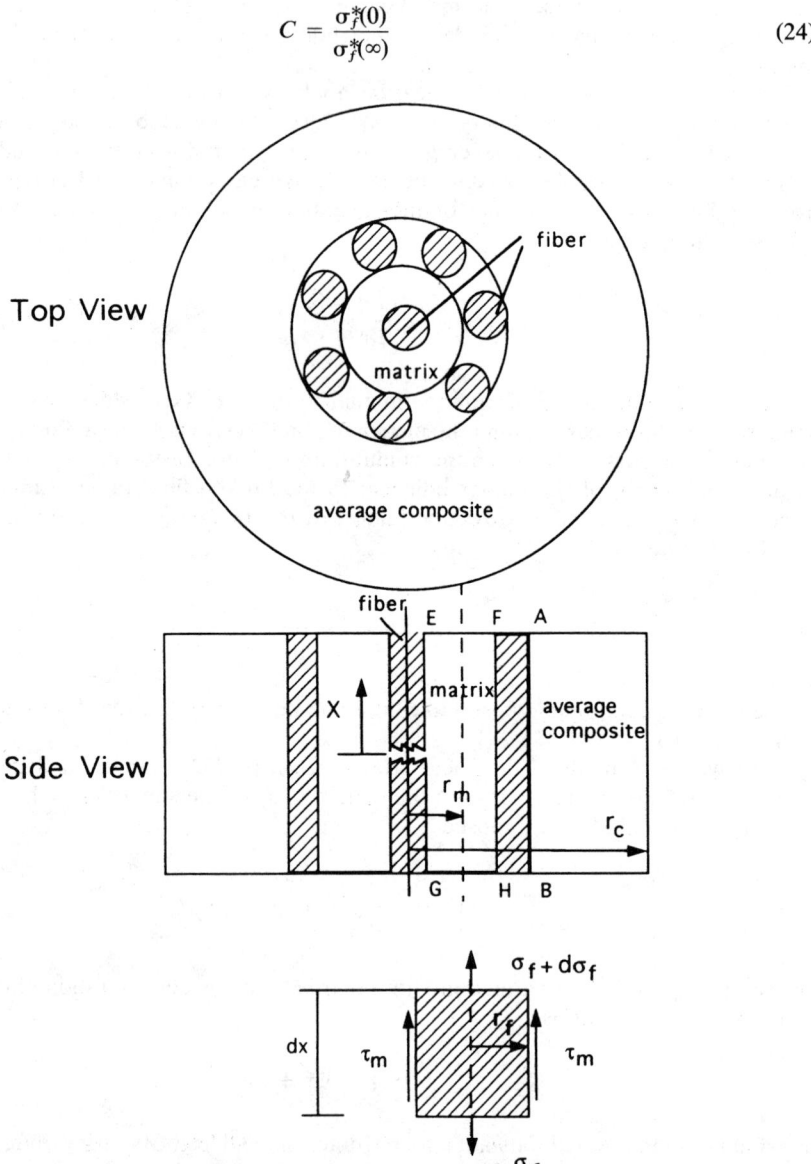

FIG. 3—*A broken fiber, neighboring unbroken fibers, matrix region, and average composite: the equilibrium element used in the present analysis to study the local stress concentration.*

therefore, from Eq 23:

$$\frac{2(r_m - r_f)}{\alpha G_m} \frac{d\tau_m(x)}{dx}\bigg|_{x=0} = \frac{1}{E_f} \sigma_f^*(0) = \frac{1}{E_f} C\sigma_f^*(\infty) = \frac{1}{E_f}\left(C\frac{\sigma_a E_f}{E_c}\right) = C\frac{\sigma_a}{E_c} \qquad (25)$$

where the far-field fiber stress $\sigma_f^*(\infty)$ is assumed to be equivalent to $\sigma_a E_f/E_c$.

At large axial distances from the fracture position, $\sigma_f^*(x)$ and $\sigma_f(x)$ approach the far-field fiber stress value. Therefore, in the limit, the right hand side of Eq 23 is identically zero. Noting that the shear stress, $\tau_m(x)$, in the left hand side of Eq 23 decays exponentially (see Eq 21), we impose

$$\frac{1}{E_f}(\sigma_f^*(x) - \sigma_f(x))\big|_{x=\delta} = \frac{\sigma_a}{E_c}\exp(-\eta\delta) \qquad (26)$$

which yields

$$\frac{2(r_m - r_f)}{\alpha G_m}\frac{d\tau_m(\delta)}{dx} = \frac{\sigma_a}{E_c}\exp(-\eta\delta) \qquad (27)$$

where δ is the ineffective length.

Using Eq 21, we can rewrite Eqs 25 and 27 as

$$-\frac{2}{\alpha\eta}(A - \eta B) = C \qquad (28)$$

$$-\frac{2}{\alpha\eta}(A - \eta\delta A - \eta B) = 1 \qquad (29)$$

When Eqs 28 and 29 are solved, we obtain

$$A = \frac{\alpha(1 - C)}{2\delta}, B = \frac{\alpha C}{2} + \frac{\alpha(1 - C)}{2\delta\eta} \qquad (30)$$

Therefore, the shear stress defined in Eq 21 is finally written as

$$\tau_m(x) = \frac{-r_f\eta\sigma_a E_f}{2E_c}\exp(-\eta x)\frac{\alpha}{2}\left[\frac{(1 - C)}{\delta}x + C + \frac{(1 - C)}{\delta\eta}\right] \qquad (31)$$

The shear lag stress expressions, Eqs 16 and 17, are independent of quantities such as interfacial bonding of composites. The objective of this micromechanical analysis is to correctly represent the dependence of the local stresses on such quantities through the stress concentration factor, C, and the ineffective length, δ. As we know, both the stress concentration factor and the ineffective length are dependent of interfacial bonding. Therefore, Eq 31 is suitable to be used to represent the influence of interfacial bonding. However, the shear stress expression shown in Eq 31, as well the normal stress expression to be seen later, is the result of modifying the shear lag analysis for x near the fracture, i.e., for x between zero and δ. Therefore, it is valid only for x between zero and δ.

Now we solve for the normal stress in the broken fiber. Combining Eqs 15 and 31 and solving for $\sigma_f(x)$, we obtain

$$
\sigma_f(x) = -\int_0^x \frac{2\tau_m(t)}{r_f} dt
$$

$$
= -\frac{\alpha \sigma_a E_f}{2E_c} \left\{ \left[C + \frac{(1 - C)}{\delta \eta} \right] \exp(-\eta x) + \frac{(1 - C)}{\delta \eta} (\eta x + 1) \exp(-\eta x) + K \right\}
$$

(32)

where K is a constant. From the traction-free condition at the fiber broken end, $\sigma_f(0) = 0$, we have

$$
K = -C - 2\frac{(1 - C)}{\delta \eta}
$$

(33)

Therefore,

$$
\sigma_f(x) = \frac{\alpha \sigma_a E_f}{2E_c} \left\{ \left[C + \frac{(1 - C)}{\delta \eta} \right] [1 - \exp(-\eta x)] \right.
$$

$$
\left. - \frac{(1 - C)}{\delta \eta} [(\eta x + 1) \exp(-\eta x) - 1] \right\}
$$

(34)

Note that the above expression for $\sigma_f(x)$ is different from the one obtained from the shear lag analysis presented in Eq 17. The latter equation does not contain the stress concentration factor, C, explicitly.

The ratio of the fiber stress at $x = \delta$ to the far-field fiber stress is defined as the efficiency parameter, ϕ (after Rosen [1])

$$
\phi = \frac{\sigma_f(\delta)}{\sigma_f(\infty)}
$$

(35)

Therefore, by using Eq 34 and the fact that $\sigma_f(\infty) = \sigma_a E_f / E_c$, we can rewrite as

$$
\phi = \frac{\alpha}{2} \left\{ \left[C + \frac{(1 - C)}{\delta \eta} \right] [1 - \exp(-\eta \delta)] - \frac{(1 - C)}{\delta \eta} [(\eta \delta + 1) \exp(-\eta \delta) - 1] \right\}
$$

(36)

The efficiency parameter, ϕ, is usually chosen approximately equal to unity. Equation 36 provides a unique relationship between the ineffective length, δ, and the stress concentration factor, C, in the unbroken fibers. For sufficiently large values of δ, a stress concentration should not exist (i.e., $C = 1$). Therefore by Eq 36, α is equal to 2. Equation 36 can thus be rewritten as

$$
\phi = \left[C + \frac{(1 - C)}{\delta \eta} \right] [1 - \exp(-\eta \delta)] - \frac{(1 - C)}{\delta \eta} [(\eta \delta + 1) \exp(-\eta \delta) - 1]
$$

(37)

A plot of C versus $\delta \eta$ is presented in Fig. 4. It is conceivable to change the values of δ and keep η constant by altering the interfacial bond of a composite. Note that as C

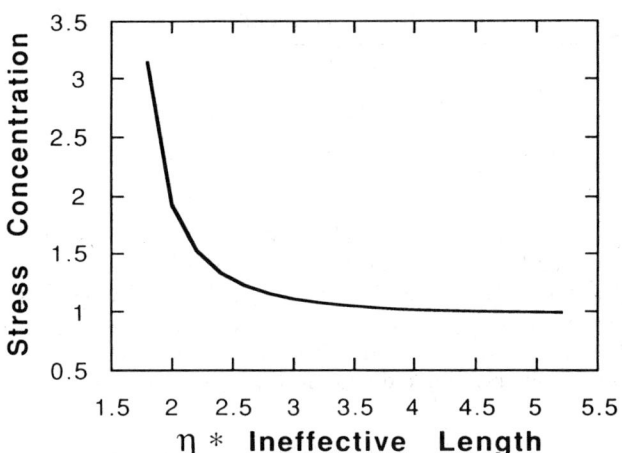

FIG. 4—*Relationship between stress concentration factor, C, and ineffective length, δ. Here η is defined in Eq 18.*

increases δ decreases, which suggests that when the ineffective length becomes smaller, the resulting stress concentration in the unbroken fiber becomes larger. This implies that as the stress concentration effects are confined to a smaller region (i.e., δ decreases), the value of the stress concentration for a particular x must increase to maintain equilibrium.

Local stress concentrations around broken fibers are needed together with Eq 37 to apply the strength formulation shown in Eq 8. While there are no closed-form three-dimensional elasticity solutions available for the stress concentration factors, different types of two-dimensional models have been developed; linear analysis [6,7], elastic plastic analysis [8], and bilinear stress-strain curves [9] have been included in finite element calculations. Several analytical solutions have been obtained by using the shear lag analysis [10,11]. These solutions are typified by Hedgepeth and Van Dyke's formulation [11]. In their analysis, they present formulations for the static stress concentration factors in the

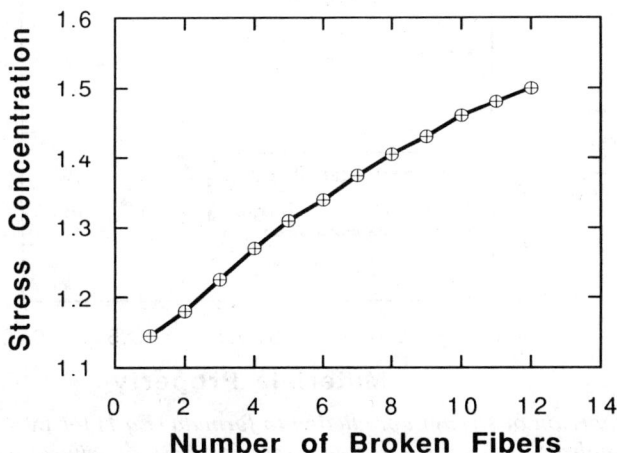

FIG. 5—*Stress concentration factor as a function of number of fibers broken together, after Hedgepeth and Van Dyke [11].*

TABLE 1—*Values of* i *and* n_i.

i^a	1	2	3	4	5	6	7	8	9	10	11	12
$n_i{}^b$	6	8	9	10	11	12	12	14	14	15	15	16

[a]Number of fibers broken together.
[b]Number of closest neighbors.

unbroken fibers of a square or hexagonal array where specified fibers are broken. The results of Hedgepeth and Van Dyke's [11] are shown in Fig. 5.

With values of C_i from Fig. 5, Eq 37 is solved for the ineffective length, δ_i, for a given composite material. Once δ_i and C_i are determined, we can obtain Q_i from Eq 8. Hence, the failure envelope is constructed, and failure stress is predicted.

The total number of fibers, denoted as N in Eq 3, is computed from the formula

$$N = \frac{\text{Area of cross section of composite}}{\text{Area of one fiber}} * v_f \tag{38}$$

Hence, for the given volume fraction, v_f, and m (shape factor of Weibull's distribution), λ_i can be computed from Eq 11. With the value of N from Eq 38 and the values of n_i from Table 1, Q_i can be obtained from Eq 8 as a function of σ.

Strength Predictions and Experimental Validation

We consider tensile strength variations as a function of the material properties, E_f and G_m, at a fixed volume fraction of 60%. The dimension of the composite specimen is 127 mm long, 12.7 mm wide, and 2.54 mm thick.

In Fig. 6, we present predictions of strength as a function of the material property, E_f/G_m, for the shape factor of Weibull distribution, m, at the value of 20. Comparisons are

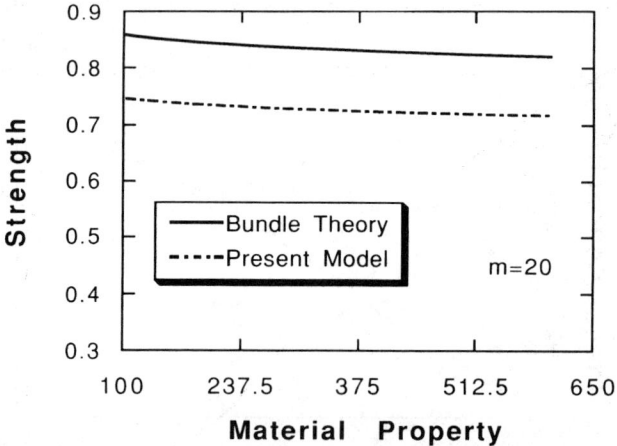

FIG. 6—*Comparison of strength prediction to formula (Eq 1) for* m = 20. *The vertical axis is the (normalized) strength, i.e., the stress of fibers at failure divided by* σ_0. *The abscissa, material property, implies the ratio* E_f/G_m. *Similar coordinates are used in the rest of the figures.*

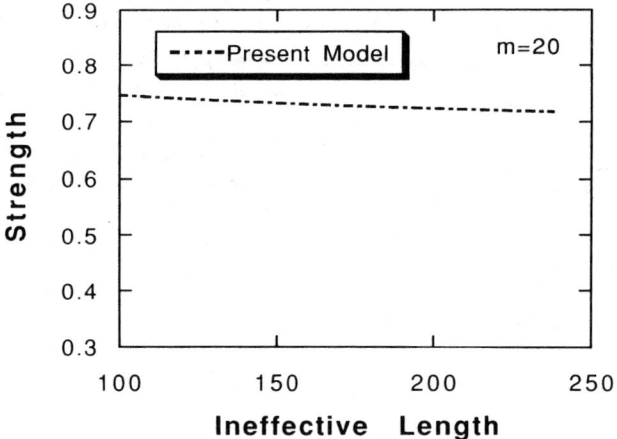

FIG. 7—*Strength prediction as function of the ineffective length,* δ, *in the units of* μm.

made with the bundle strength formula (Eq 1). The vertical axis in Fig. 6 is the normalized strength of the composite, i.e., fiber stress at failure divided by the Weibull parameter σ_0. Note that the results presented for the bundle theory are larger than those in the present analysis. Due to the neglect of stress concentration factor of the neighboring fiber, one obtains an overestimate of the strength values from Eq 1. In Fig. 7, we plot the strength predictions as a function of the ineffective length.

The shape factor of the Weibull distribution, m, plays an important role in changing the values of strength [12,13]. Figure 8 shows strength predictions of the present model for different values of the shape factor, m. If we first fix a value of material property E_f/G_m, it is seen from this figure that as the shape factor m decreases, the performance of the composite decreases by comparison to the average fiber strength. For example, for a shape factor of 20 and material property E_f/G_m at the value of 375, the strength (normalized by σ_0, which is related to the average strength of the fibers) is about 0.72. For a shape factor of 4 and the same value of E_f/G_m, the strength is about 0.46. Furthermore, the

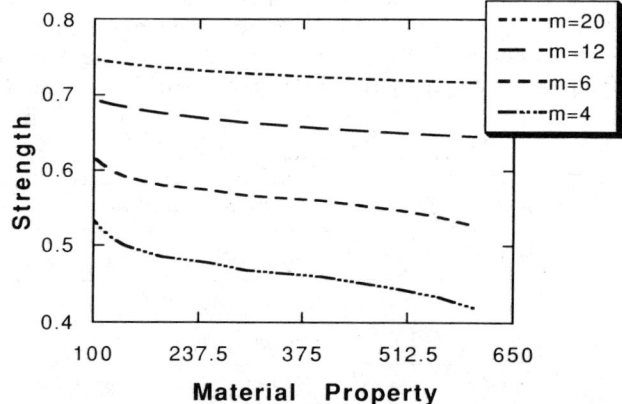

FIG. 8—*Strength predictions for different values of* m.

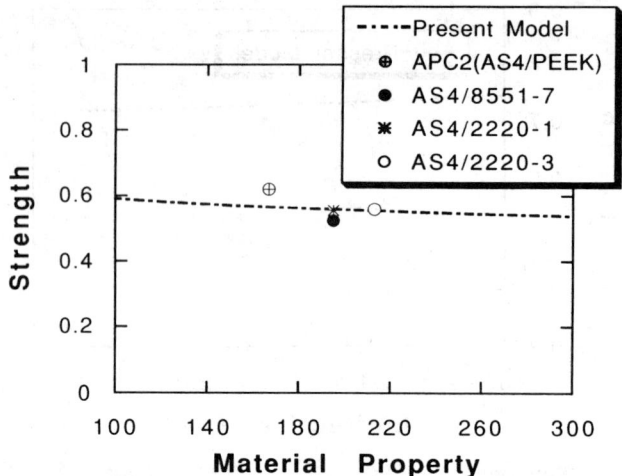

FIG. 9—*Comparison of strength prediction to experimental measurements* [14].

strength curve is more flat for a larger value of m. This suggests that the material property E_f/G_m has lesser influence on the variations of strength for a larger m.

The comparison of our predictions to experimental measurements of strength is provided in Fig. 9. The experimental data of carbon/epoxy composite systems have been reported by Coquill and Adams [*14*]. The matrix and fiber properties for the composites are given in Tables 2 and 3. We choose to use the data for composites made from same type of fibers, AS4. Therefore, the composites have the same value of σ_0. The Weibull distribution parameter, m, is assumed to be 6, which is a typical value used in the literature [*12*].

The parameter σ_0 is determined in the following way. For the AS4/2220-3 system, E_f/G_m = 195.8, which should result in a σ_f/σ_0 equal to 0.558, based on our prediction of strength shown in Fig. 9. Here σ_f represents the stress of the fibers when the composite failed and was published as 3.18 GPa. Therefore, we have $\sigma_0 = \sigma_f/0.559 = 3.18/0.559 = 5.69$. The measured tensile strength values [*14*] for material systems APC2(AS4/PEEK), AS4/8551-7, and AS4/2220-1 are normalized by the σ_0 and are shown in Fig. 9. The predicted strength values match reasonably well with the experimental data.

One objective of this research is to address the question of optimal design for tensile-controlled performance of composites. Equations 8 and 11 indicate that, for given Weibull parameters m and σ_0, the strength value is inversely proportional to λ_i. λ_i, as depicted in

TABLE 2—*Fiber properties.*

Fiber	Manufacturer	Tensile Modulus	Description
AS4	Hercules	235 GPa	Carbon

TABLE 3—*Matrix properties.*

Resin	2220-1	2220-3	8551-7	PEEK
Shear modulus	1.2 GPa	1.1 GPa	1.2 GPa	1.4 GPa

Eq 11, is more sensitive to stress concentration factor, C_i, than to the ineffective length, δ_i, since m is generally larger than 3. Therefore, the stress concentration factor tends to cause more rapid perturbations of tensile strength values, especially for large values of m.

The above fact is very important in discussing the optimal value of tensile strength. Let us look at strength as a function of material properties E_f/G_m shown in Fig. 6. The present analysis considers stress concentration effects and thus should provide more accurate predictions of strength as compared to those calculated from bundle theory concepts. Nonetheless, the shape of the strength curve remains the same for either methodology. This is due to the fact that the stress concentration factor is virtually unchanged for different ratios of E_f/G_m. Therefore, in this case, the strength is highly dependent on the ineffective length.

The existing analyses in the literature (e.g., Hedgepeth and Van Dyke [11], Carman and Reifsnider [15]) suggest that, at least for polymeric composites which have a large ratio of E_f/G_m, the stress concentration is virtually unchanged for different values of E_f/G_m. Therefore, strength variation is a monotonic function of E_f/G_m controlled by the corresponding variation of ineffective length, δ_i. This is true for regular (homogeneous) fiber spacing, since the micromechanical analyses are based on this assumption.

However, in the vast majority of composites, the fibers are irregularly spaced, resulting in local regions of effective fiber volume fraction less than that of the composite global average. Hence, one would expect the strength in these composites to be less than composites containing regular fiber spacing. In Fig. 10, we present strength predictions for composites with irregular fiber spacing. The composites have a global fiber volume fraction of 0.7 and $m = 8$. The effective fiber volume fraction (resulting from irregular fiber spacing in certain local regions) is a function of E_f/G_m, as shown in Fig. 11. In the strength calculations for Fig. 10, all quantities other than the total number of fibers (Eq 38), such as ineffective length δ_i and η, are calculated using the effective fiber volume fraction. It is shown clearly in the present model that, for certain irregular fiber spacing and material properties, there exists a maximum value of strength for polymeric-based composites. We wish to point out that the remarkable resemblance between the two curves, strength and effective volume fraction (resulting from irregular fiber spacing), as a function of material property E_f/G_m, as shown in Figs. 10 and 11. This clearly indicates the great influence of fiber spacing on the strength of a composite.

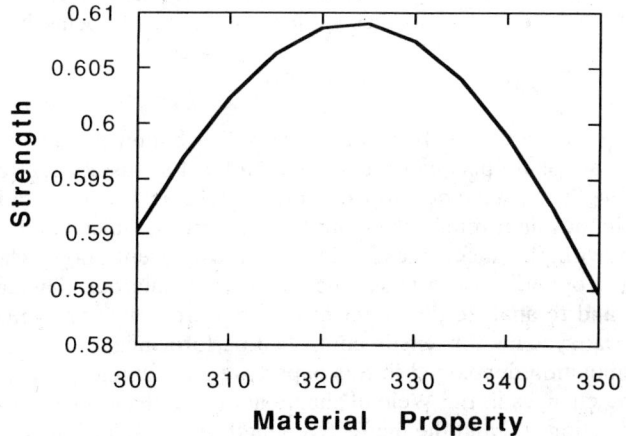

FIG. 10—*Strength prediction for composites with irregular fiber spacing.*

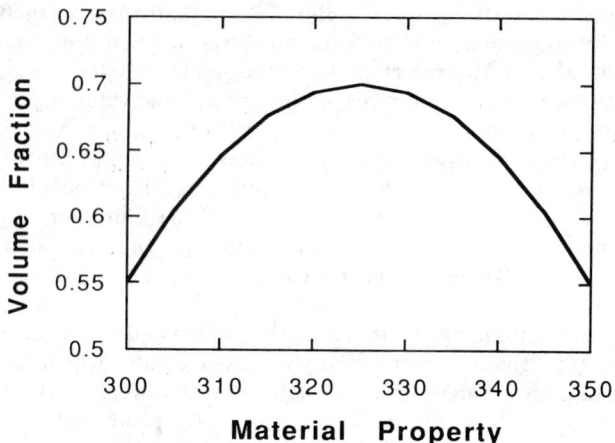

FIG. 11—*Effective volume fraction (resulting from irregular fiber spacing) as a function of material property* E_f/G_m, *used in the calculations of Fig. 10.*

The above example shows how important fiber spacing is to the tensile strength of a composite. The fiber spacing in a composite is not easy to control in the manufacturing process of a composite, nor is it easy to determine for a given composite specimen. Therefore, it is of great importance to control or isolate the effects of fiber spacing when the influence of other micromechanical properties is studied. For instance, to answer the question of how interfacial bonding influences the optimal value of tensile strength, it is critical to exclude the effects of different fiber spacings. As shown in Fig. 10, optimal value can be achieved due to both fiber spacing variations and material property selections if the proper physical parameters are determined and controlled.

The modified local stress expressions developed in this paper involve the stress concentration factors in neighboring unbroken fibers and the ineffective length. This suggests that, compared to bundle theory formulations (Eq 1), our analysis can be applied to a general class of material systems, including those with small E_f/G_m such as ceramic composites. However, it should not be regarded as conclusively determined for small E_f/G_m ratios. Further work is clearly needed to adequately address the issues of the model accuracy for this range of values. These subjects are currently being investigated.

Conclusions

Based on micromechanics analyses of stress redistribution due to multiple fiber fractures, and the relationship between the associated local stress concentration factors and the ineffective length, we have developed a micromechanical representation of the tensile strength of continuous fiber reinforced composite materials, which combines the theory of bundle strength with the mechanics of local stress concentration. The representation provides a unique opportunity to study the influence of micromechanical parameters on tensile strength and to analyze the effect of material properties on the optimal design of composite material systems for tensile-controlled performance.

Using the formulation developed in this paper, we have studied how tensile strength is influenced by the changes in the Weibull shape factor, m, the stress concentration factor, C, the volume fraction, v_f, and the ineffective length, δ, as well as the associated material properties such as E_f and G_m. The introduction of the stress concentration effect into the analysis provides strength values well below those predicted by the bundle theory and in good agreement with the data considered. The magnitude of the stress concentration was found to be most significant for large values of the Weibull shape factor m, i.e., the details of the local stress state are most important when the strength of the fibers has a small statistical variation.

The ineffective length was demonstrated to control the shape of the strength curve as a function of E_f/G_m, at least for polymeric composites. In other words, for such materials, an optimal tensile strength design with regard to material properties is generally achieved by constructing composites with the smallest ineffective length. The present model defines the conditions under which the ineffective length dominates tensile strength (as in typical polymeric composites) or local stress concentrations dominate strength (as in ceramic composite systems). Moreover, the model defines conditions under which both effects contribute significantly to strength and has shown good correlation with experimental data when various material parameters are varied. Hence, the analysis is a strong foundation for the design of continuous fiber reinforced composite systems for optimum tensile strength.

Strength considerations become more complicated with the introduction of irregular fiber spacing into the analysis. This, in conjunction with the fact that virtually no quantitative description is present in the literature, complicates this formidable task. However, we have presented a methodology which addresses the issues of irregular fiber spacing in composite materials and its effects on optimal design for tensile strength. This analysis provides designers with adequate tools to chose not only constitutive materials of the composite, but also processing techniques which yield various fiber spacing and arrangements. Therefore, the fiber spacing of composites, although difficult to control in manufacturing processes, can be considered as an important contributor to tensile strength considerations and to other macromechanical properties of engineering importance.

Although the characteristics of a fiber/matrix interface or interphase were not explicitly involved in the present analysis, the effects are implicitly included in the ineffective length, which is seen to have a large influence on strength. Future work will concentrate on including the interface/interphase region explicitly and on improving our mechanics representation of the local stress state for small E_f/G_m ratios so that this and other related problems can be directly addressed.

Acknowledgments

The authors gratefully acknowledge the support of the National Science Foundation Center for High Performance Polymeric Adhesives and Composites and the Virginia Institute for Material Systems, which provided the opportunity to pursue this research direction. They also gratefully acknowledge significant support from the Air Force Office of Scientific Research under Grant No. AFOSR-89-0216, which provided a strong experimental support for the philosophy which is at the base of the model which we developed.

References

[1] Rosen, B. W., "Tensile Failure of Fibrous Composites," *AIAA Journal,* Vol. 2, No. 11, 1964.
[2] Tsai, S. W. and Hahn, H. T., *Introduction to Composite Materials,* Technomic Publishing Co., Westport, CT, 1980, pp. 407–415.
[3] Batdorf, S. B., "Tensile Strength of Unidirectionally Reinforced Composites—I," *Journal of Reinforced Plastics and Composites,* Vol. 1, 1982.
[4] Weibull, W., "A Statistical Distribution Function of Wide Application," *Journal of Applied Mechanics,* September 1951.
[5] Holister, G. S. and Thomas, C., *Fiber Reinforced Materials,* Elsevier, New York, 1966.

[6] Carrara, A. S. and McGarry, F. J., "Matrix and Interface Stresses in a Discontinuous Fiber Composite Model," *Journal of Composite Materials,* Vol. 2, 1969, p. 222.

[7] Barker, R. M. and MacLaughlin, T. F., "Stress Concentration Near a Discontinuity in Fibrous Composites," *Journal of Composite Materials,* Vol. 5, 1971.

[8] Agarval, B. D. and Bansal, R. K., "Plastic Analysis of Fiber Interactions in Discontinuous Fiber Composites," *Fiber Science and Technology,* Vol. 10, 1977.

[9] Chen, C. H., "Tension of a Composite Bar with Fiber Discontinuities and Soft Inter-fiber Materials," *Fiber Science and Technology,* Vol. 6, 1973.

[10] William, R. S. and Reifsnider, K. L., "Strain Energy Release Rate Method for Predicting Failure Modes in Composite Materials," *Fracture Mechanics (11th Conference), ASTM STP 677,* American Society for Testing and Materials, Philadelphia, 1979.

[11] Hedgepeth, J. M. and Van Dyke, P., "Local Stress Concentration in Imperfect Filamentary Composite Materials," *Journal of Composite Materials,* Vol. 1, 1968, p. 294.

[12] Fukuda, H., "Statistical Strength of Unidirectional Composites with Random Fiber Spacing," *Composite '86: Recent Advances in Japan and the United States,* Proceedings of the Japanese-U.S. CCM III, Tokyo, 1986, pp. 307–314.

[13] Harlow, D. G. and Phoenix, S. L., "Probability Distribution for the Strength of Composite Materials I: Two-Level Bonds," *International Journal of Fracture,* Vol. 17, 1981, pp. 347–372.

[14] Coquill, S. L. and Adams, D. F., "Mechanical Properties of Several Neat Polymer Matrix Materials and Unidirectional Carbon Fiber-Reinforced Composites,"*NASA Contractor Report 181805, 1989.

[15] Carman, G. and Reifsnider, K. L., "Micromechanics of Fiber Fracture with Regards to Tensile Strength," *Composite Materials: Fatigue and Fracture, Fourth Volume, ASTM STP 1156,* American Society for Testing and Materials, Philadelphia, 1993.

Fatigue of Polymer Matrix Composites

Richard L. Wolterman,[1] John M. Kennedy,[1] and Gary L. Farley[2]

Fatigue Damage in Thick, Cross-Ply Laminates with a Center Hole

REFERENCE: Wolterman, R. L., Kennedy, J. M., and Farley, G. L., "**Fatigue Damage in Thick, Cross-Ply Laminates with a Center Hole**," *Composite Materials: Fatigue and Fracture, Fourth Volume, ASTM STP 1156,* W. W. Stinchcomb and N. E. Ashbaugh, Eds., American Society for Testing and Materials, Philadelphia, 1993, pp. 473–490.

ABSTRACT: An experimental investigation was conducted to characterize the response of three composite material systems to long-term cyclic loading. Thick cross-ply laminates of uniwoven AS/4 carbon fabric were produced using a brittle matrix with and without stitching and a tough-matrix material. Quasi-static tension and compression tests were conducted on specimens with and without circular holes to determine strength, modulus, and failing strain. These tests showed that the measured static mechanical properties were insensitive to the type of matrix material because the laminate response was dominated by the 0° fibers. The stitched specimens had significantly lower static compressive strengths.

A series of fatigue tests (tension-tension, compression-compression, and tension-compression) were conducted to study the influence of an open hole on damage and residual strength. These tests showed that the matrix material and stitching influenced the fatigue behavior of the composite. The tough-matrix material developed less damage than the corresponding brittle-matrix specimens and survived one million cycles of fatigue loading at all stress levels. The stitched composite developed less longitudinal damage than the unstitched materials, but it developed more transverse damage. Both the brittle-matrix and the stitched materials failed prior to one million cycles of tension-compression fatigue loading. The residual strength of the materials was affected by the matrix material and the damage state in the specimen.

KEY WORDS: resin matrix composites, stitching, fatigue life, fatigue damage, residual strength, tension, compression

The use of composite materials in aerospace and aeronautical structures has typically resulted in improved structural performance, but at a higher fabrication cost as compared to conventional metallic structures. As experience designing and fabricating composite structures has increased, new failure mechanisms, such as delamination-induced failures [1–4], have been identified. Delamination-induced failures can be produced by foreign object impact, stress concentrations due to holes, deficient manufacturing processes, and ply dropoffs. These failure mechanisms have limited allowable design strain levels of the material, which has resulted in reduced weight savings and increased fabrication cost. Development of materials with higher interlaminar strength to overcome these deficiencies has progressed along two paths, the development of "toughened" matrices and of three-dimensional (3-D) fiber architectures.

The improvements in interlaminar strength through tough matrices have been achieved

[1]Graduate student and associate professor, respectively, Department of Mechanical Engineering, Clemson University, Clemson, SC 29634-0921.

[2]Research engineer, Army Aerostructures Directorate, MS 190, NASA Langley Research Center, Hampton, VA 23665-5225.

at an increase in matrix material cost. The alternate approach to improved interlaminar toughness is through the use of 3-D fiber architecture. Different facets of textile technology, such as 3-D weaving, braiding, stitching, and knitting, can be used to add through-the-thickness fibers to the preform. Although textile composite technology is still in its infancy, there exists a potential to produce damage-tolerant structures at lower cost than conventional metallic structures.

Textile processes have been used to manufacture near-net-shape fiber preforms which, when processed into composite materials, are very damage tolerant [5–7]. However, to confidently design and build composite structures requires an understanding of the relationship between mechanical properties of the constituent materials, the fiber architecture, and the behavior of the composite structure [8].

Often the criteria governing the design of a structure require that material properties, strength or stiffness, remain above a specified critical value throughout its lifetime. When a structure is subjected to fatigue loading, its mechanical response may change with time. If there is degradation of mechanical properties or fatigue-related damage, the safety factor for the structure could decrease below 1.0 and the structure could fail at load levels substantially below the original design conditions. Therefore, static properties cannot be solely used for the design of structures subjected to cyclic loads. Traditionally, advanced composites have been viewed as fatigue resistant, but as stiffer and stronger matrix materials and advanced fiber architectures are developed, the fatigue response may become the limiting factor [9]. Thus, the behavior of the material to cyclic load conditions must be characterized.

The objective of this research program is to investigate the cyclic load behavior of three different cross-ply composite materials. All materials used the same in-plane fiber (AS4). The baseline material used a brittle matrix (3501-6), and the second used a tough matrix (8551-2). The third material used the same brittle matrix as the first except the third material also contained through-the-thickness stitching. Quasi-static loading tests were conducted on all materials to determine baseline tensile and compressive mechanical properties. Fatigue tests were conducted at three stress ratios using specimens with an open hole in the center of the specimen. The mechanisms of damage initiation and the extent of damage growth were recorded. Also, the relationship between the residual strength and stiffness of the material after one million loading cycles to the damage state in the specimen was investigated.

Materials and Equipment

Three composite material systems were investigated. The baseline system was AS4/3501-6.[3] The two damage-tolerant materials were AS4/8551-2[3] and AS4/3501-6 with through-the-thickness stitching. The stitching fiber was Toray 900 carbon stitching fiber. The AS4 fiber has a failing strain of approximately 1.6% and an elastic modulus of 248 GPa. The elastic modulus of the 3501-6 matrix is 4.4 GPa and has a failure strain of 1.1%. The tough 8551-2 matrix has an elastic modulus of approximately 3.4 GPa and a failing strain of approximately 2.0%. The carbon stitching yarn is designed especially for stitching and has a 228-GPa elastic modulus.

The composite laminates were produced from a nine-layer stack of dry uniwoven fabric with fiber orientations of $[0/90/0/90/0/90/0/90/0]_T$. Within a uniwoven layer, all of the carbon fibers were oriented in the warp direction and held together with a fine denier glass fill yarn. The glass yarn comprised less than 1% of the total fiber weight. The uniwoven fabric

[3]Product designation of Hercules, Inc.

consisted of 13 warp yarns per inch with each yarn containing 21 000 filaments. The nine-layer stack of dry uniwoven material was stitched through the thickness using a modified lock stitch having a stitch pitch of 3/cm and a stitch row spacing of 6.4 mm. Stitching was performed in orthogonal directions forming 6.4 by 6.4 mm stitch cells. Resin was infiltrated into the stitched and unstitched preforms using a resin transfer molding (RTM) technique, and the materials were cured in an autoclave according to the curing cycle prescribed by the resin supplier. All composite panels were C-scanned to determine laminate quality. Specimens were cut from regions of the panel that appeared to be of high quality. The C-scans of the stitched panels showed no defects. The specimens without stitching were 7.0 mm thick, whereas the stitched specimens were nominally 7.6 mm thick. The stitched specimens were thicker than the unstitched specimens due to the through-the-thickness fibers and the additional resin required to impregnate those fibers. The fiber volume fraction of all materials was approximately 60% as measured by the ASTM Standard Test Method for Fiber Content of Resin-Matrix Composites by Matrix Digestion (ASTM D 3171–76) [10].

Specimens were cut from the panels such that the 0° plies were in the loading direction of the coupon. Tension specimens were 25 mm wide and 250 mm long. Short-block-compression (SBC) specimens were 38 mm wide and 44 mm long. Static and fatigue notched specimens were 38 mm wide and 200 mm long with a 9.5-mm-diameter hole in the geometric center of the specimen. The specimens were cut using an abrasive cutting wheel. The 9.5-mm-diameter holes were cut using a diamond-coated hole saw.

All tests were conducted in a 250-kN servo-hydraulic closed-loop testing machine under load control. Test machine actuator position and load, far-field longitudinal strain and displacement of a 25-mm gage length spanning the hole were recorded with an automated digital data acquisition system.

The initiation and accumulation of damage were monitored using X-radiography techniques. An X-ray opaque dye made from a solution of zinc oxide was injected into the cracks and delaminations along the free edges of the specimen and inside the hole. The dye was allowed to wet the surface and infiltrate the cracks and delaminations for approximately 10 min. A piece of photographic film was placed behind the specimen and exposed to radiation. The radiographs indicated the size and planar location of damage.

Experimental Program

The experimental test matrix consisted of both quasi-static and fatigue tests, as presented in Table 1. Two static tension and two static compression tests were performed to determine the elastic modulus, ultimate strength, and failing strain for each material. Longitudinal and transverse strain gages were mounted on both sides of the static test specimens. The tension specimens had a 150-mm test section between the test machine grips. The short block compression specimens had opposite and adjacent edges ground parallel and perpendicular, respectively, to avoid eccentrically loading the specimen. The compression tests were run using a short-block-compression test fixture (Fig. 1). Horizontal clamping bolts were incorporated into this test fixture to prevent specimen end brooming.

Static tension and compression tests were run on open-hole specimens to determine strength loss due to the hole. A ring gage and mounting fixture were used on the brittle- and tough-matrix materials, and an extensometer was used on the stitched material to measure the displacement in the loading direction of a 25-mm section spanning the hole (Fig. 2). In Fig. 2, aluminum fixtures with knife edges are attached at ±12.5 mm in the longitudinal direction from the center of the hole. The ring gage measures the displacement

TABLE 1—*Experimental program.*

Test Type	Specimen Type	Loading Rate or Frequency	σ_{max}, MPa	σ_{min}, MPa
Static tension	Unnotched	90 N/s	...	...
Static tension	Open hole	90 N/s	...	...
Static tension	Unnotched	90 N/s	...	...
Static compression	Open hole	90 N/s	...	...
Tension-tension fatigue	Open hole	10 Hz	220	22
Compression-compression fatigue	Open hole	10 Hz	−134	−13
Tension-compression fatigue	Open hole	10 Hz	220	−134
Residual tension (after tension-tension fatigue)	Open hole	90 N/s	...	...
Residual compression (after compression-compression and tension-compression fatigue)	Open hole	90 N/s	...	...

over the 25-mm gage section by measuring the displacement between the two aluminum fixtures. The length of the test section of the open-hole specimens was 100 mm. Load, actuator displacement, and ring gage displacement data were collected throughout the tests. To prevent damage to the fixture, the ring gage was not used for the quasi-static open-hole compression tests; instead, actuator displacement was used to measure hole compliance.

Fatigue tests were conducted on open-hole specimens subjected to sinusoidal loading at a frequency of 10 Hz. The objective was to determine the influence of cyclic load, stress ratio, and load level on the damage. Data were collected periodically to monitor changes in hole elongation which signified changes in the damage state of the material around the hole. Tension-tension fatigue tests were run at a stress ratio, R, of 0.1 with the maximum stress at 44% of the ultimate tensile strength of the statically tested open-hole specimens.

FIG. 1—*Photograph of short block compression fixture with specimen mounted in one end.*

FIG. 2—*Ring gage and mounting fixture for displacement across the hole.*

The compression-compression fatigue tests were run with a maximum compressive stress of 44% of the ultimate compressive strength of the open-hole specimen with $R = 10$. The third set of fatigue tests were run under reversed cyclic loads at $R = -0.64$. This stress ratio corresponded to a tension-compression stress range for which the maximum applied stress equalled that in the tension fatigue tests and the minimum applied stress equalled that in the compression fatigue experiments.

The maximum stress level was chosen such that the tension and compression fatigue specimens could survive one million load cycles, yet also show the initiation and development of damage. The cyclic load tests were halted at one million load cycles, and the residual strength and stiffness were measured by quasi-static loading of the specimen to failure. The initiation and growth of damage were monitored throughout the fatigue tests using X-ray radiography techniques and changes in hole elongation. Radiographs were produced prior to testing, as well as periodically during the tests. The initial radiograph of each specimen gave an indication of the damage due to manufacturing and machining.

Only one test was conducted on each type of open-hole specimen because of the limited quantity of material.

Results and Discussion

Static Tests of Unnotched Specimens

The mechanical properties (shown in Table 2) from quasi-static tension tests of the brittle-matrix (AS4/3501-6) and tough-matrix (AS4/8551-2) composite materials were similar in magnitude. The elastic stiffness and strength of the stitched material (AS4/3501-6 with through-the-thickness stitching) were about 20% lower than the properties of the unstitched materials. The lower tensile stiffness and strength can be partially attributed to the stitching. Because of the high percentage of 0° plies, the stress-strain response of all three materials was nearly linear to failure.

There was little visually detectable damage in the three materials until the applied load was near the ultimate load. The brittle-matrix and stitched materials had similar failure modes, that is extensive longitudinal matrix splitting and fiber breakage throughout the entire test section. The extensive longitudinal matrix splitting is due to the energy released when the specimen failed. The specimen fabricated from the tough-matrix material failed

TABLE 2—*Static properties.*

Property	Brittle Matrix	Tough Matrix	Stitched
Tension			
Modulus, GPa	71.77	75.91	56.03
Strength, MPa	739.0	765.4	639.8
Ultimate strain	0.0103	0.0101	0.0103
Compression			
Modulus, GPa	60.86	59.92	53.48
Strength, MPa	−564.8	−569.2	−361.5
Ultimate strain	−0.0093	−0.0095	−0.0063
Open-hole tension			
Far-field ultimate stress, MPa	492.2	460.2	458.7
Opening displacement at failure, mm	0.780	0.544	0.775
Open-hole compression			
Far-field ultimate stress, MPa	−319.2	−303.2	−242.2
Opening displacement at failure, mm	−0.330	−0.3251[a]	−0.215[b]

[a]Determined from testing machine displacement.
[b]Determined from extensometer.

transverse to the load direction without extensive longitudinal matrix splitting. The tough-matrix material was stronger and had a higher failing strain than the brittle-matrix material. Consequently, more load was transferred by the matrix into fibers adjacent to broken fibers than in the brittle-matrix material in which the matrix failed in shear between fibers or fiber tows.

The stiffness and strength measured from the short-block-compression tests of all materials were similar except the strength of the stitched material was about 30% lower. The stress-strain response was linear to failure. For the unstitched materials, the compressive elastic modulus was as much as 20% lower than the tensile elastic modulus. This difference is probably due to the waviness in the carbon tows produced by the warp yarns in the uniwoven material. The low strength of the stitched material was similar to data reported by Farley et al. [11] and is attributed to the inclusion of the through-the-thickness fibers. The through-the-thickness fibers create in-plane and out-of-plane fiber waviness, create large resin pockets, and significantly reduce the volume of matrix material within individual in-plane tows. Each of these textile process attributes can contribute to lower compression strength [11]. The failure mode in the unstitched specimens was similar. Delaminations developed between plies which precipitated instability-related failures of individual plies or ply groups. The stitching prevented extensive delamination in the stitched specimens. Failure occurred by transverse shearing on a 45° plane normal to the surface of the specimen.

Open-Hole Static Tests

Quasi-static tension and compression tests were conducted on open-hole specimens to determine the reduction in strength caused by a stress concentrator and to establish baseline data for the fatigue tests. The experimental far-field strength and predicted

TABLE 3—*Comparison of open-hole static strength with net-section stress prediction.*

		Strength, MPa		
Test Type	Material	Static	Prediction	Experimental, Predicted
Open-hole tension	Brittle	492.2	554.3	0.888
Open-hole tension	Ductile	460.2	574.1	0.802
Open-hole tension	Stitched	458.7	479.9	0.956
Open-hole compression	Brittle	319.2	423.6	0.754
Open-hole compression	Ductile	303.2	426.9	0.710
Open-hole compression	Stitched	242.2	271.2	0.893

strengths are listed in Table 3. The experimental unnotched strengths of these laminates, given in Table 2, were used to predict the notched strength. The predicted strengths assume the specimen fails due to a uniform stress in the reduced section (laterally adjacent to the hole) equal to the ultimate tensile stress. This prediction implies that there is no stress concentration due to the hole; it is also an upper bound on notched strength. The failure stress of the notched specimens was significantly below the predicted values, which implies that there was a stress concentration near the hole. The stress concentration in the stitched materials was more severe in compression than in tension because the compression failure was initiated by delamination. The difference between predicted and experimental strength of the stitched material was less than 11% in both the tension and compression load cases.

The tensile-stress hole-elongation curves for all three materials were nonlinear, as depicted in Fig. 3. The nonlinear response was due to local damage around the hole and nonlinear matrix material response. However, the data are not sufficient to explicitly quantify nonlinear response mechanisms. The brittle-matrix materials had about 35% higher hole elongation than the tough-matrix material. The brittle-matrix materials failed with extensive longitudinal matrix cracking and fiber breakage throughout the test section. In contrast, the tough-matrix material failed at the hole, primarily through fiber breakage. Stitching had little effect on the hole elongation response of the material in tension.

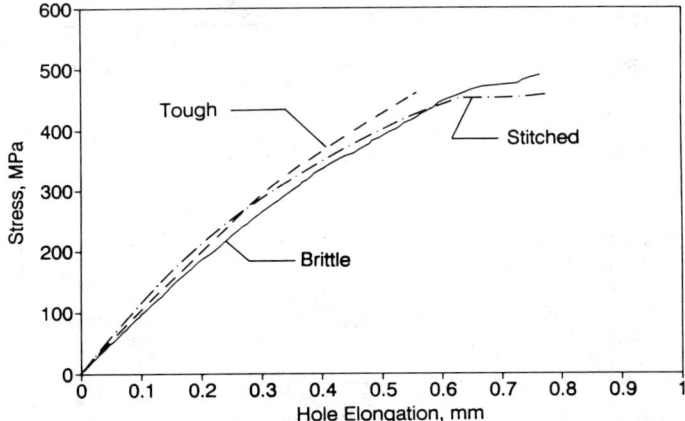

FIG. 3—*Hole elongation for the three materials under static tension loading.*

The tensile results for the unstitched materials demonstrate the effect of matrix failure strain on the suppression of damage around a hole and the effect of damage on ultimate strength. Other researchers [9,12–14] have shown that the tensile strength of composites with notches or holes depends on the noncritical damage state at the ends of the notch. Delamination, longitudinal splitting, and matrix yielding reduce the stress concentration in primary load-carrying (0°) plies near the edges of the hole. The tough-matrix material, which has a high failing strain, can suppress the initiation and growth of damage. Consequently, the resulting failure is caused by the stress concentration due to the hole. The brittle-matrix material develops significantly more damage than the tough-matrix material, making the region more compliant. For uniaxial loading, Goree [14] showed that large shear stresses develop in the matrix material between fibers which are cut by the hole and those that are not. Matrix failures in the form of longitudinal splits develop due to these shear stresses. The longitudinal splits isolate the unbroken fibers from the hole, reducing the stress concentration. Complimentary experimental work [15] confirmed the existence of the large shear stresses using brittle coatings. Additional experimental results using thermography [9] show that longitudinal splitting produces significant stress redistribution in surface plies of carbon/epoxy laminates.

The notched tensile strength of the stitched specimen was within 7% of the notched-tensile strength of the brittle-matrix material. The failure modes were also the same, that is axial splitting in the 0° plies resulted in net-section failure adjacent to the hole. Axial splitting caused a similar stress hole-elongation response as that exhibited by the unstitched brittle-matrix material. However, as discussed, the predominant noncritical damage mode at the edges of the hole is axial splitting, and the stitches are unlikely to suppress this damage mode. The stitches with respect to this damage mode act very much like 0° and 90° plies.

Compressive stress-hole elongation of all three materials was nonlinear (Fig. 4). The total hole elongation of the stitched material was about 35% less than that of the other materials. The response of the unstitched materials was virtually identical. Similar response in this laminate is expected for different matrix materials because the dominant damage mode prior to ultimate failure was longitudinal splitting and not delamination. At failure, extensive delamination developed between plies in the brittle-matrix specimen. The failure mode of the tough-matrix material was localized, with ply groups failing in

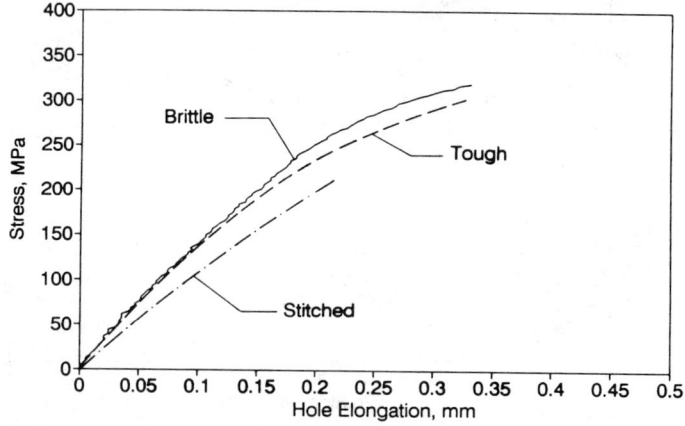

FIG. 4—*Hole elongation for the three materials under static compression loading.*

transverse shear with minimal delamination between these ply groups. In laminates which tend to delaminate in the net section adjacent to the notch, the matrix toughness would have a significant effect on the open-hole compressive strength. The stitched specimen failed in a transverse shear mode with no significant delamination near the failure surface and with no extensive axial splitting at the sides of the hole. The stitching suppressed the damage mode observed in the unstitched materials, but the waviness produced by the stitching caused kinking or bending in the surface plies. The lower slope of the stress hole-elongation curve (Fig. 4) of the stitched specimen as compared to the unstitched materials can be attributed to surface layer kinking, as is the low strength.

Fatigue Test Results

Tension-Tension

The tension-tension fatigue tests were conducted at stress levels that were 44% of the open-hole static-tensile strength with $R = 0.1$. All specimens survived one million cycles of fatigue loading. The most damage developed in the brittle-matrix materials. The radiographs taken throughout the test (Fig. 5) show that the brittle-matrix material developed axial splits, transverse cracking in the 90° plies, and delaminations along the axial splits. The damage in the tough-matrix material consisted of axial splitting and transverse cracking in the 90° plies with little delamination between plies. The stitched material developed delaminations around the hole, and the delamination propagated about one half of a hole diameter in the transverse direction.

After one million cycles of fatigue loading, specimens were statically loaded to failure in tension. A comparison of the open-hole strength with no fatigue loading, the residual strength of the tension-tension fatigue specimens, and the open-hole prediction given in Table 3 is presented in Fig. 6. For all three materials, the residual strength of the fatigued specimens was greater than the strength of the unfatigued open-hole specimen of the same material. The strength increase was due to the reduction in stress concentration around the hole caused by the fatigue-induced damage [14]. The strength increase was small for the brittle-matrix materials. Based on the prediction, there was, however, little potential for the strength to increase because the prediction is an upper bound on the notched strength. The residual strength of the tough-matrix material was 25% greater than the ultimate tensile strength of the unfatigued specimen. The residual strength equalled the predicted strength, which indicates that the fatigue damage rendered the specimen notch insensitive.

Total hole elongation in the brittle-matrix fatigued specimen was unchanged relative to the static test results, as seen in Fig. 7. The lower slope of the curve for the fatigued specimen, as compared to the unfatigued specimen, is an indication of the fatigue damage. The nearly linear response of the fatigued specimen indicates that little additional damage developed in the residual strength test until the specimen was near the failure load. Also, the failure mode in the fatigue specimen was identical to that of the specimen which was not fatigued. These results, along with the fact that the residual strength was only 6% higher than the strength of the unfatigued specimen, indicate that the fatigue damage in the brittle-matrix specimen was not greatly different than the damage which develops during a quasi-static tension test on an unfatigued specimen. Consequently, the damage which develops in the quasi-static tension test and the fatigue test produces about the same redistribution of stress in the primary load-carrying plies.

Hole elongation in the static and residual strength tests of the tough-matrix material are shown in Fig. 8. The total hole elongation was much greater after fatigue loading, indicating that the region around the hole was more compliant due to fatigue damage. After

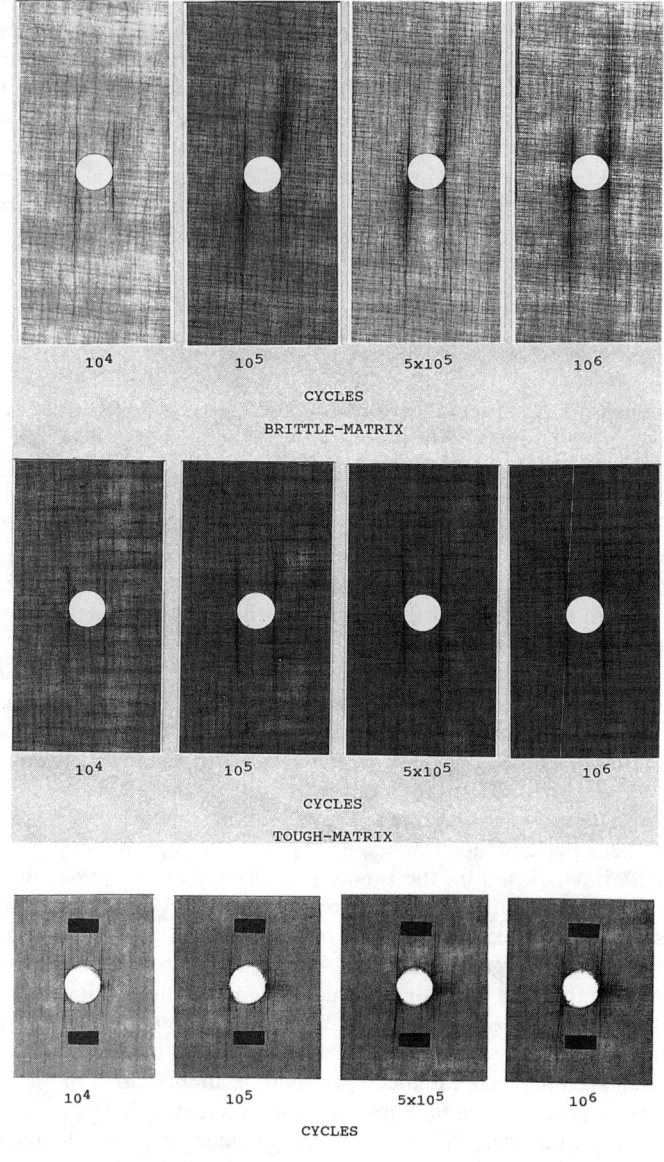

FIG. 5—*Radiographs of all three materials during the tension-tension fatigue tests.*

fatigue loading, the total hole elongation of the tough-matrix material was about the same as that of the brittle-matrix materials. There was a significant change in the tensile failure mode due to tension-tension fatigue loading, as compared to the failure mode of the unfatigued specimen. The failure mode changed from a somewhat self-similar failure initiating at the hole to extensive axial splitting with no defined failure plane. The failure mode after fatigue loading resembled that of the brittle-matrix materials.

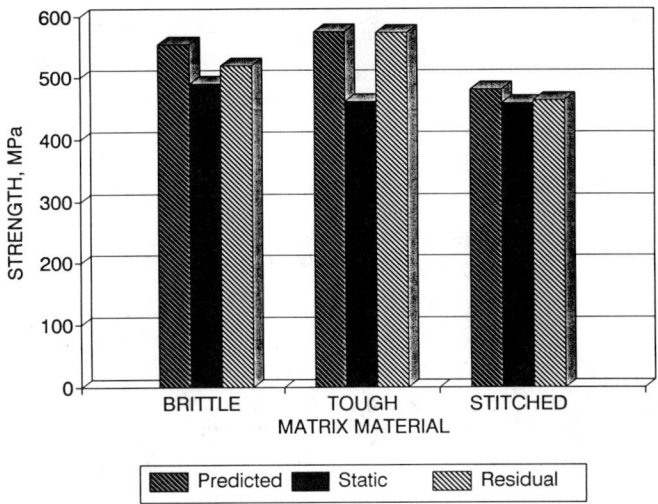

FIG. 6—*Effect of tension-tension fatigue loading on open-hole tensile strength.*

Fatigue loading had little effect on the hole compliance and the total hole elongation of the stitched material. Consequently, the residual strength equalled the static strength and the failure mode was unchanged.

Compression-Compression

The compression-compression fatigue tests were run at a maximum stress of 44% of the static open-hole compressive strength of the materials with $R = 10$. No specimens failed before reaching one million cycles of fatigue loading. The radiographs taken during the fatigue tests (Fig. 9) show that there was much less damage around the hole due to the compression-compression loading than due to the tension-tension loading. The limited damage which accumulated during the compression fatigue tests may be attributed to the

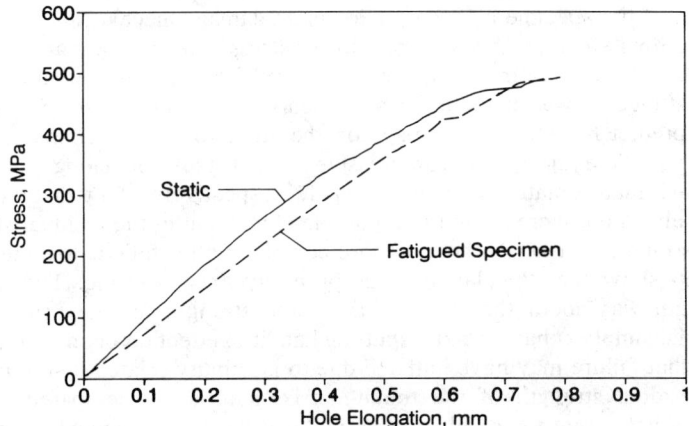

FIG. 7—*Effect of tension-tension fatigue loading on the stress-hole elongation response of the unstitched brittle-matrix material.*

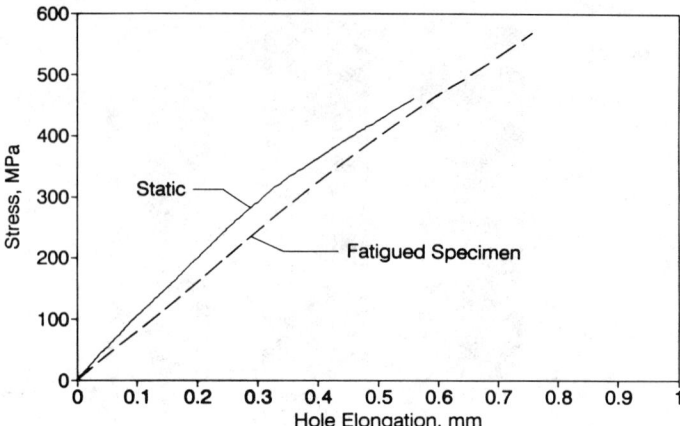

FIG. 8—*Effect of tension-tension fatigue loading on the stress-hole elongation response of the tough-matrix material.*

small stress range at which the tests were conducted. However, this applied stress range corresponds to an approximate far-field strain range of 0.0022, which produces maximum axial strains near the hole equal to 0.0055. The brittle- and tough-matrix materials developed axial splits, which grew in length during the fatigue tests, but there was little or no delamination. Limited axial splitting developed in the stitched material, and some delamination developed around the hole. There were several stitches at the edge of the hole which were partially cut by the hole-drilling process which may have been the initiators of the delamination.

The residual strength, the static strength of open-hole specimens, and the predictions from Table 3 of the compression open-hole fatigue specimens are compared in Fig. 10. The strength of the brittle-matrix and stitched materials was lower after fatigue loading, but the strength reduction was minimal. The residual strength of the tough-matrix material was higher than the static-compressive strength.

In compression, the damage around the hole can increase, decrease, or have no effect on the strength of the specimen. If the predominant damage mechanism is axial splitting, then the strength will increase because of a lower stress concentration at the hole [14]. If delamination develops, creating sublaminates near the hole, then the sublaminates will buckle or fail at much lower loads than a specimen without damage [16]. Consequently, it is difficult to predict the effect of damage on the strength of this laminate, particularly when there is so little damage. The compressive stress versus hole elongation plots of the brittle- and tough-matrix materials (Figs. 11 and 12, respectively) show that fatigue loading resulted in a substantial increase in the total hole elongation in the residual strength test. These data also indicate that the hole is more compliant after fatigue loading. The radiographs (Fig. 9) show that the damage was primarily axial splitting. The fact that the residual strength was about the same as the static strength for the brittle-matrix and stitched materials implies that the axial splitting had little effect on open-hole compressive strength, and that failure may have initiated due to instability-related failure mechanisms. The higher residual strength of the tough-matrix material as compared to the static strength may indicate that the axial splitting reduced the stress concentration in primary load-carrying plies and that the high strength of the matrix suppressed delamination and the instability-related failure. This conclusion is reasonable because axial splitting will

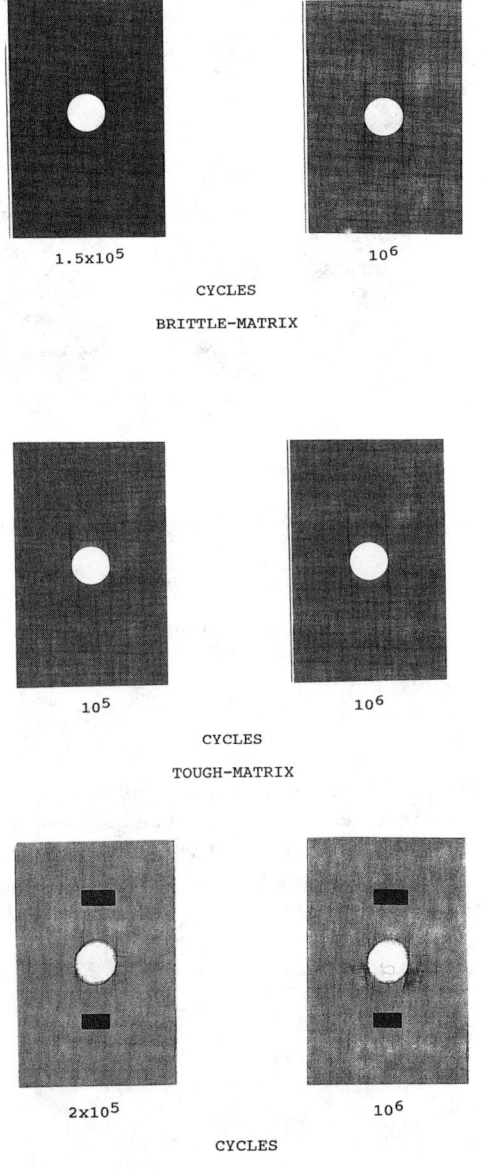

FIG. 9—*Radiographs of all three materials during the compression-compression fatigue tests.*

reduce the stress concentration at the hole, but it will not drastically affect instability failure modes associated with compressive loading.

Even though there was delamination around the hole in the stitched specimen due to fatigue loading, the decrease in strength after fatigue loading was very small. This implies that the stitching suppressed additional delamination growth and controlled sublaminate

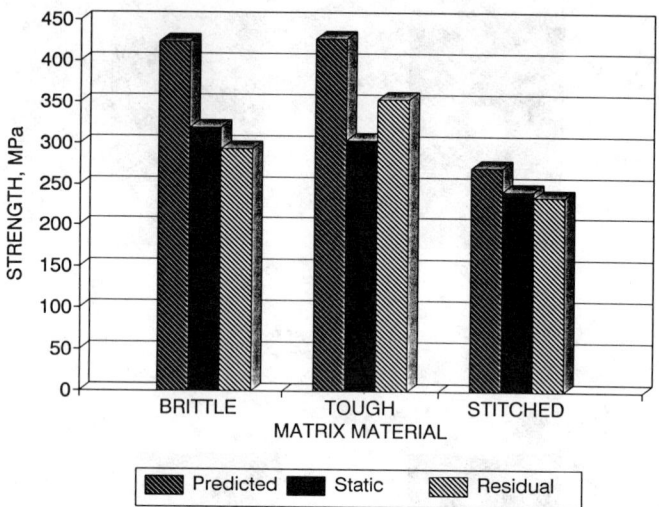

FIG. 10—*Effect of compression-compression fatigue loading on open-hole compressive strength.*

buckling during the residual strength test. Additional experiments and analysis are required to develop a precise understanding of the effect of stitching on notch sensitivity of composites loaded in compression.

Tension-Compression

The objective of the tension-compression fatigue tests was to demonstrate the behavior of the three materials under reversed cyclic loading and to compare it with the material's response at other stress ratios and cyclic load conditions. The specimens were cycled between the maximum loads of the tension and compression fatigue tests. The brittle-matrix materials, both stitched and unstitched, failed during the fatigue tests. The

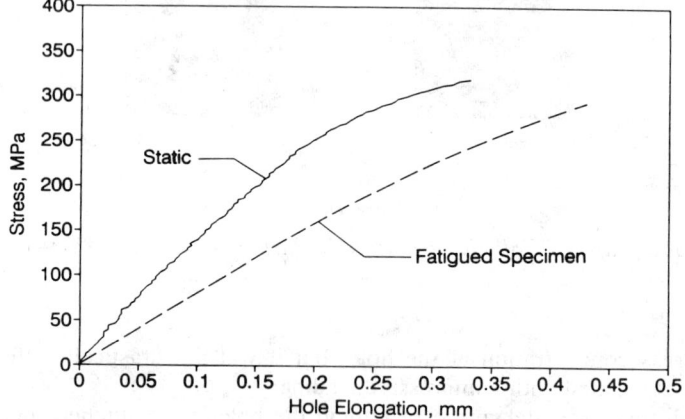

FIG. 11—*Effect of compression-compression fatigue loading on the stress-hole elongation response of the unstitched brittle-matrix material.*

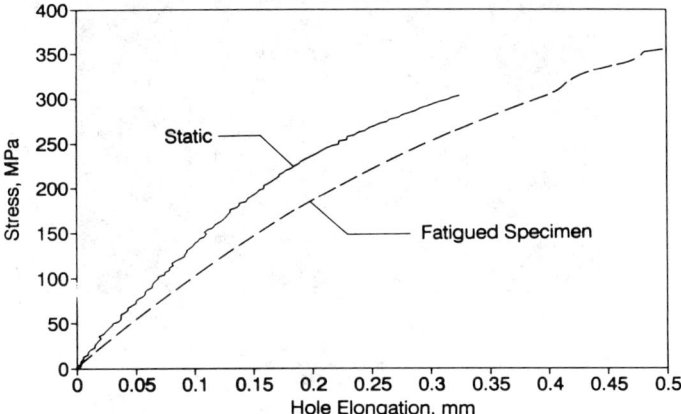

FIG. 12—*Effect of compression-compression fatigue loading on the stress-hole elongation response of the tough-matrix material.*

unstitched specimen failed at 600 000 cycles, and the stitched specimen failed at 7600 cycles. The tough-matrix specimen survived a million cycles of fatigue loading and was then statically loaded to failure in compression.

Figure 13 shows radiographs of the brittle and tough materials at several points in the fatigue tests; the stitched specimen failed before radiographs were made. Much more damage developed in the specimens during the tension-compression fatigue tests than in either the tension or the compression fatigue tests. There was extensive longitudinal damage and transverse cracking in 90° plies in both materials. Delamination around the hole and along the axial splits developed in both materials, but the delamination was much more extensive in the brittle-matrix material. The radiographs of the tough-matrix material showed the development of matrix cracks and only a slight indication of delamination at 600 000 cycles, the number of cycles at which the brittle-matrix specimen failed.

When the brittle-matrix specimen failed in compression at 600 000 cycles, there was extensive delamination over the entire length of the specimen. The failure mode of the tough-matrix specimen during the residual compression test was the same as the failure mode of the open-hole compression test on the unfatigued specimen. The failure was essentially transverse to the hole with some delamination between groups of plies. The stitched specimen failed in tension at 7600 cycles. Different 0° plies failed at locations throughout the entire test section, which was similar to the failure mode of the static open-hole tension stitched specimen.

The residual strength of the tension-compression tough-matrix specimen was greater than the static open-hole compressive strength and the residual strength of the compression fatigue tough-matrix specimens, as shown in Fig. 14. The strength was still below the prediction, which indicates that there is still a stress concentration due to the hole or that delamination is offsetting the positive effect on strength of the axial splitting.

The response of the tension-compression specimens, as compared to that of the tension-tension and compression-compression specimens, indicates the importance of stress range and stress ratio on the fatigue response of composites. The type of fatigue loading did not change the fundamental damage modes, but the tension-compression loading produced much more damage than the tension-tension and compression-compression load cases. The damage in the brittle-matrix and stitched materials was sufficient to cause failure during the tension-compression fatigue tests.

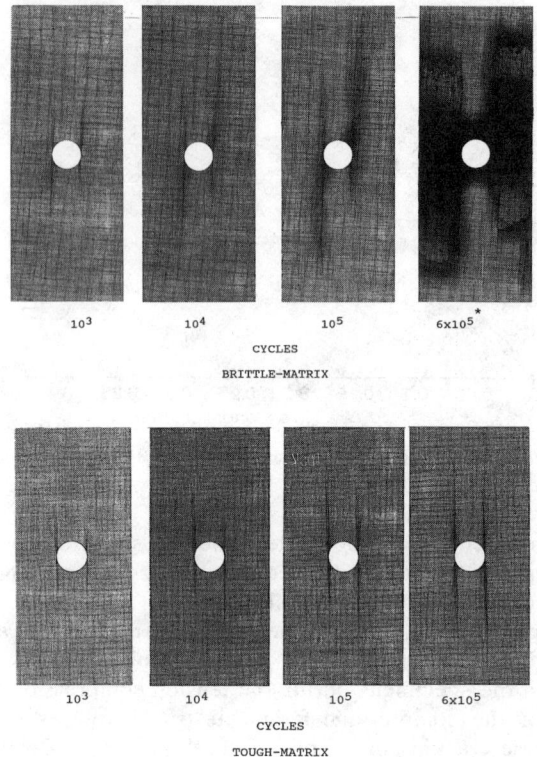

FIG. 13—*Radiographs of the brittle and tough-matrix materials during the tension-compression fatigue tests. Radiograph was made after specimen failed.*

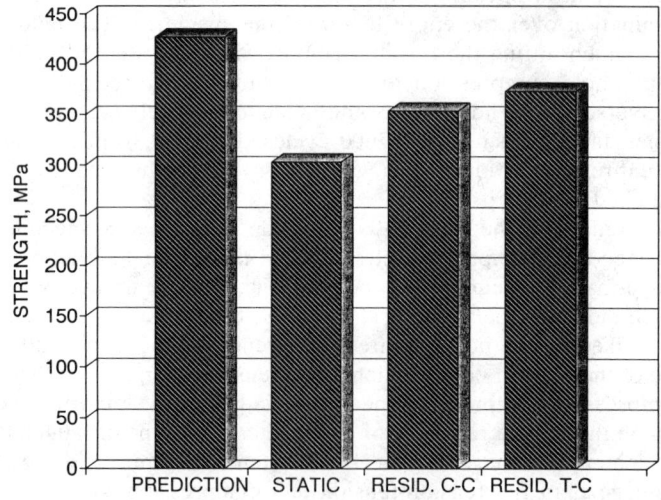

FIG. 14—*Comparison of the residual compressive strength of the tough-matrix material.*

Conclusions

The static and fatigue response in tension and compression of a cross-plied laminate was studied. Tension-tension, compression-compression, and tension-compression fatigue tests were performed on open-hole specimens. The three materials evaluated had a brittle matrix, a tough matrix, and a brittle matrix with through-the-thickness stitching. Based on the results of this study, the following conclusions were developed:

1. The static mechanical properties were insensitive to the matrix material because the high percentage of 0° plies dominated the response. The stitches degraded the tension and compression static properties more than 20%.

2. The fatigue response depended on matrix properties and stitching. The stress ratio and loading range greatly influenced the damage mechanisms, and damage accumulated during one million load cycles. For the test types conducted herein, the tough-matrix material was the most durable; the stitched material was the least durable.

3. All materials demonstrated an increase in residual tensile strength due to tension-tension fatigue loading. The residual compressive strength of the brittle and stitched materials decreased, but the residual strength of the tough-matrix material increased. Radiographs and hole elongation indicated that fatigue damage affected the residual strength.

4. Tension-compression fatigue proved to be the most severe loading.

5. Stitching altered the compressive failure mode and the predominant fatigue damage mechanisms around an open hole.

References

[1] *Composite Materials: Testing and Design, 4th Conference, ASTM STP 617,* American Society for Testing and Materials, Philadelphia, 1977.

[2] *Test Methods and Design Allowables for Fibrous Composites, ASTM STP 734,* C. C. Chamis, Ed., American Society for Testing and Materials, Philadelphia, 1981.

[3] *Composite Materials: Fatigue and Fracture, ASTM STP 907,* H. T. Hahn, Ed., American Society for Testing and Materials, Philadelphia, 1986.

[4] *Composite Materials: Mechanics, Mechanical Properties, and Fabrication,* K. Kawata and T. Akasaka, Eds., Japan Society for Composite Materials, 1982.

[5] Dexter, H. B. and Funk, J. G., "Impact Resistance and Interlaminar Fracture Toughness of Through-The-Thickness Reinforced Graphite/Epoxy," *Proceedings,* AIAA/ASME/ASCE/AHS 27th Structures, Structural Dynamics and Materials Conference, AIAA Paper No. 86-1020-CP, AIAA, Washington, DC, 19–21 May 1986.

[6] Dow, M. B. and Smith, D. L., "Properties of Two Composite Materials Made of Toughened Epoxy Resin and High-Strain Graphite Fiber," NASA TP-2826, National Aeronautics and Space Administration, Washington, DC, July 1988.

[7] Simonds, R. A., Stinchcomb, W., and Jones, R. M., "Mechanical Behavior of Braided Composite Materials," *Composite Materials: Testing and Design, ASTM STP 972,* J. D. Whitcomb, Ed., American Society for Testing and Materials, Philadelphia, 1988, pp. 438–453.

[8] Stinchcomb, W. W. and Reifsnider, K. L., "Damage Accumulation Concepts for Fatigue Loaded Composite Laminates," *Mechanics of Composite Materials—1983,* G. J. Dvorak, Ed., American Society of Mechanical Engineers, New York, pp. 143–148.

[9] Bakis, C. E., Simonds, R. A., Vick, L. W., and Stinchcomb, W. W., "Matrix Toughness, Long-Term Behavior, and Damage Tolerance of Notched Graphite Fiber-Reinforced Composite Materials," *Composite Materials: Testing and Design (Ninth Volume), ASTM STP 1059,* S. P. Garbo, Ed., American Society for Testing and Materials, Philadelphia, 1990, pp. 349–370.

[10] *ASTM Standards and Literature References for Composite Materials,* 1st ed., American Society for Testing and Materials, Philadelphia, 1987, pp. 41–44.

[11] Farley, G. L., "A Mechanism Responsible for Reducing Compression Strength of Through-the-Thickness Reinforced Composite Materials," *Journal of Composite Materials,* Vol. 26, No. 12, 1992, pp. 1784–1795.

[12] Poe, C. C. Jr., "A Unifying Strain Criterion for Fracture of Fibrous Composite Laminates," *Engineering Fracture Mechanics,* Vol. 17, No. 2, pp. 153–171.

[*13*] Kennedy, J. M., "Fracture of Hybrid Composite Laminates," *Proceedings,* 24th AIAA/ASME/ ASCE/AHS Structures, Structural Dynamics and Materials Conference, AIAA Paper No. 83-0804-CP, May 1983, pp. 68–78.

[*14*] Goree, J. G. and Kaw, A. K., "Shear-Lag Analysis of Notched Laminates with Interlaminar Debonding," NASA CR-3798, National Aeronautics and Space Administration, Washington, DC, May 1984.

[*15*] Goree, J. G. and Jones, W. F., "Fracture Behavior of Unidirectional Boron/Aluminum Composite Laminates," NASA CR-3753, National Aeronautics and Space Administration, Washington, DC, December 1983.

[*16*] Whitcomb, J. D., "Mechanics of Instability-Related Delamination Growth," *Composite Materials: Testing and Design (Ninth Volume), ASTM STP 1059,* S. P. Garbo, Ed., American Society for Testing and Materials, Philadelphia, 1990, pp. 215–230.

T. Kevin O'Brien[1] and Steven J. Hooper[2]

Local Delamination in Laminates with Angle Ply Matrix Cracks, Part I: Tension Tests and Stress Analysis

REFERENCE: O'Brien, T. K. and Hooper, S. J., "**Local Delamination in Laminates with Angle Ply Matrix Cracks, Part I: Tension Tests and Stress Analysis,**" *Composite Materials: Fatigue and Fracture, Fourth Volume, ASTM STP 1156,* W. W. Stinchcomb and N. E. Ashbaugh, Eds., American Society for Testing and Materials, Philadelphia, 1993, pp. 491–506.

ABSTRACT: Quasi-static tension tests were conducted on AS4/3501-6 graphite epoxy $(0_2/\theta_2/-\theta_2)_s$ laminates, where θ was 15, 20, 25, or 30°. Dye penetrant-enhanced X-radiography was used to document the onset of matrix cracking in the central $-\theta°$ plies and the onset of local delaminations in the $\theta/-\theta$ interface at the intersection of the matrix cracks and the free edge. Edge micrographs taken after the onset of damage were used to verify the location of the matrix cracks and local delaminations through the laminate thickness.

A quasi-3D finite element analysis was conducted to calculate the stresses responsible for matrix cracking in the off-axis plies. Laminate plate theory indicated that the transverse normal stresses were compressive. However, the finite element analysis yielded tensile transverse normal stresses near the free edge. Matrix cracks formed in the off-axis plies near the free edge where in-plane transverse stresses were tensile and had their greatest magnitude. The influence of the matrix crack on interlaminar stresses is also discussed.

KEY WORDS: composite material, graphite epoxy, delamination, matrix crack

Local delaminations that form at the intersection of matrix cracks in off-axis angle plies and the free edges of a composite laminate have been observed in unnotched, notched, and tapered laminates [1–3]. The accumulation of local delaminations through the laminate thickness redistributes the strain in the 0° plies, resulting in fatigue failures [4]. Previously, local delamination onset in fatigue from 45° matrix ply cracks was predicted using a simple closed-form solution for the strain energy release rate associated with this mechanism [3,4]. However, in order to demonstrate the general applicability of this technique, local delaminations from matrix cracks of arbitrary orientation must be studied. Such solutions could then be applied to arbitrary layups, such as the ones resulting from a rotated straight edge analysis for delamination around an open hole [5], to predict fatigue life.

In this study, tension tests were conducted on $(0_2/\theta_2/-\theta_2)_s$ AS4/3501-6 graphite/epoxy laminates, where θ was 15, 20, 25, and 30°. Dye penetrant-enhanced X-ray radiography was used to document the onset of matrix cracking and the onset of local delaminations at the intersection of the matrix cracks and the free edge. A quasi-3D finite element analysis was conducted to calculate the stresses responsible for matrix cracking in the off-axis

[1]U.S. Army Aerostructures Directorate, NASA Langley Research Center, Hampton, VA 23681-0001.

[2]Wichita State University, Wichita, KS 67208.

plies. These stresses were compared to stresses calculated from laminated plate theory. The influence of the matrix crack on interlaminar stresses is also discussed.

Experiments

Materials

Panels were manufactured in an autoclave at NASA Langley Research Center from Hercules AS4/3501-6 graphite epoxy prepreg using the standard manufacturer's procedure with a maximum temperature of 177°C (350°F) in the cure cycle. One panel was produced for each of the $(0_2/\theta_2/ -\theta_2)_s$ layups, where θ was 15, 20, 25, or 30°. Panels 305 mm (12 in.) square were manufactured and then cut into 20 individual coupons, 127 mm (5 in.) long by 25 mm (1.0 in.) wide. Each specimen had a nominal ply thickness of 0.124 mm (0.0049 in.) and a fiber volume fraction of 66.5%.

Experimental Procedure

Coupons were coated with a thin film of water-based typewriter correction fluid on either edge to act as a brittle coating for detecting the onset of matrix cracking and delamination. Then specimens were placed in the hydraulic grips of an MTS hydraulic load frame such that each specimen had a gage length between the grips of approximately 76 mm (3.0 in.). A 25-mm (1-in.) gage length extensometer was mounted on the specimen midway between the grips to measure nominal strain. Specimens were loaded in tension at a rate of 13 mm/min. (0.5 in./min). During the loading, a plot of load versus nominal strain was generated using an X-Y recorder. Specimens were loaded until the first audible indication of cracking was heard. All tests had linear load versus nominal strain plots up to this point when the loading was stopped. To enhance the ability to detect damage, the load was first reduced by approximately 5 to 10% to prevent the possibility of the formation of further damage, or damage growth due to creep, and this reduced load was maintained on the specimen for 10 to 15 min while the dye penetrant was added and the X-ray radiograph was taken. The edge was examined visually with a magnifying glass while the specimen was under load to determine if a matrix crack or delamination had formed. In addition, a zinc iodide solution (60 g ZnI_2, 8 mL H_2O, 10 mL isopropyl alcohol, and 3 mL Kodak Photo-Flo 200) that was opaque to X-rays was placed on the edge using a hypodermic needle. An X-ray radiograph was taken of the specimen width by exposing the Type-55 Polaroid film, mounted directly behind each 12-ply laminate, for 60 s to an X-ray beam with an intensity of 10 mA and a voltage of 15 kV. After the onset of cracking was detected, specimens were removed from the load frame and the specimen edges were polished. Photographs of the polished edges were taken through an optical microscope to examine the damage on the edge.

Experimental Results

In all cases, the radiograph taken at the first audible indication of the onset of damage showed at least one, and in some cases up to ten, matrix cracks in the central $-\theta°$ plies (Fig. 1). The $\theta = 15$ and 20° laminates (Figs. 1a and 1b) typically had only one or two matrix cracks, whereas the $\theta = 25$ and 30° laminates (Figs. 1c and 1d) typically had two or more matrix cracks. The dark squares in the center of the laminate are thin aluminum blocks that were bonded to the specimen surface to hold the knife edges of the extensometer. The radiographs also showed a triangular-shaped shaded region bounded by the free edge and the matrix crack, indicating that a local delamination was present (Fig. 1).

Figure 2 shows photographs of the edge of laminates that were tested under a monotonically increasing tension load until the onset of damage. As shown in Fig. 2, local delaminations were present in both $\theta/-\theta$ interfaces, extending in the same direction from curved matrix cracks in the central $-\theta°$ plies. For all the specimens tested, the radiographs and edge micrographs indicated that both the $-\theta°$ ply matrix crack and the $\theta/-\theta$ interface delaminations were present. There was no evidence of the presence of matrix cracking without the delamination, nor was there any evidence of the presence of delamination without a matrix crack.

Stress Analysis

Laminated Plate Theory

Laminated plate theory [6] was used to calculate the in-plane stresses in the central $-\theta°$ plies for $(0/\theta/-\theta)_s$ graphite epoxy laminates using lamina properties listed in Table 1. The in-plane stresses were transformed into the lamina coordinate system, and the stresses normal to the fiber direction, σ_{22}, and the shear stresses along the fiber direction, τ_{12}, were determined. These stress components contribute directly to the formation of matrix ply cracks.

Figure 3 shows calculated values of σ_{22} and τ_{12} as a function of θ for $(0/\theta/-\theta)_s$ graphite laminates subjected to a total (mechanical plus thermal) axial strain of 0.01, assuming a ΔT of $-156°C$ ($-280°F$). This ΔT corresponds to the difference in the cure temperature of $177°C$ ($350°F$) and the room temperature of $21°C$ ($70°F$) at which the laminates were tested. These laminated plate theory results indicate that for $(0/\theta/-\theta)_s$ graphite laminates, where θ is between 15 and 30°, relatively high shear stresses are present along the fiber direction but the transverse normal stresses are compressive. For $\theta > 30°$, the shear stresses are also high, but the transverse normal stresses are tensile.

The stresses plotted in Fig. 3, which were calculated assuming a ΔT of $-156°C$ ($-280°F$), correspond to laminates that are completely dry. However, the epoxy matrix will absorb moisture from the air. This absorbed moisture in the matrix creates swelling stresses in the laminate. These swelling stresses tend to relax the residual thermal stresses that result from cooling the laminate after it is cured. If enough moisture is absorbed such that the residual thermal stresses are completely relaxed, the stresses in the laminate would be the same as if there were no thermal contribution ($\Delta T = 0$). Figures 4 and 5 shows how the σ_{22} and τ_{12} stresses vary between a $\Delta T = -156°C$ ($-280°F$) and a $\Delta T = 0$. The $\Delta T = 0$ case also indicates that when θ is between 15 and 30°, relatively high shear stresses are present along the fiber but the transverse normal stresses are compressive.

Finite Element Analysis

A quasi-3D finite element code [7] was used to calculate the in-plane and interlaminar stresses near the free edges of $(0/\theta/-\theta)_s$ laminates subjected to the same loading specified above. The model was constructed assuming a unit ply thickness, h, in a laminate whose total width was $40h$. Figure 6 shows the region of the mesh, near the free edge, used to model a representative cross section of the laminate normal to the direction of the applied tensile load. Because of symmetry, only one quarter of the cross section was modeled. The model consisted of 108 eight-noded quadrilateral isoparametric elements having a total of 367 nodes, each with three degrees of freedom. The smallest element at the free edge

Fig. 1a. RADIOGRAPH OF $(0_2/15_2/-15_2)_s$ LAMINATE AFTER ONSET OF DAMAGE

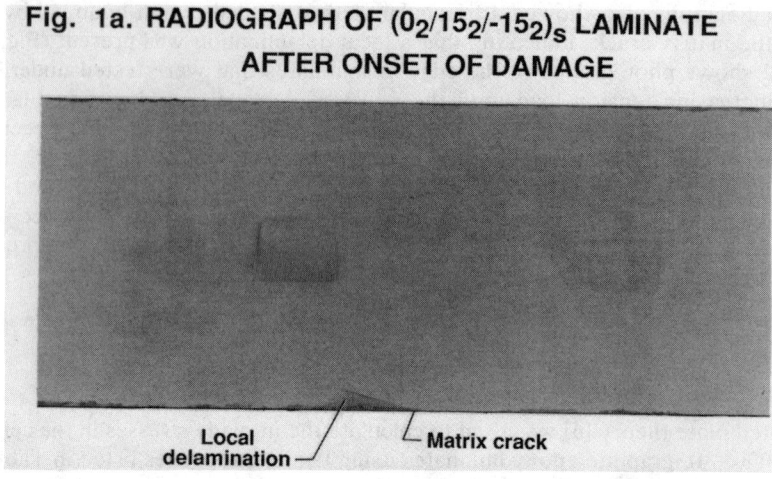

Local delamination —   Matrix crack

Fig. 1b. RADIOGRAPH OF $(0_2/20_2/-20_2)_s$ LAMINATE AFTER ONSET OF DAMAGE

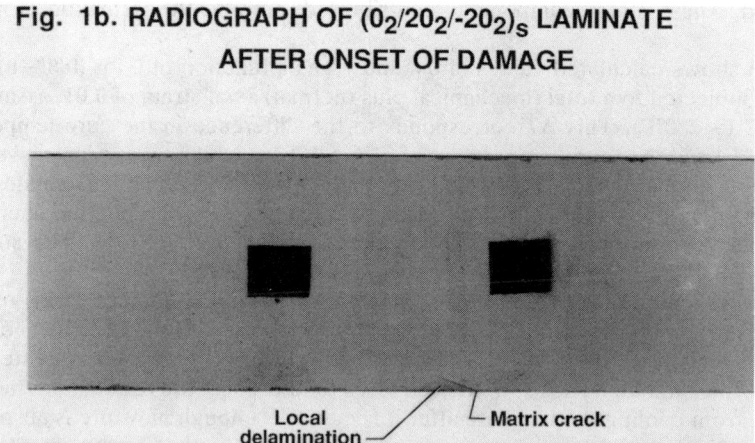

Local delamination —   Matrix crack

FIG. 1—*Dye penetrant-enhanced X-radiographs of $(0_2/\theta_2/-\theta_2)_s$ graphite epoxy laminates taken after the onset of damage in static tests.*

was equal to one eighth of a ply thickness, corresponding to a distance of one sixteenth of a ply thickness between the node on the free edge and the nearest midside node. Because the graphite epoxy material being modeled has approximately 20 fibers through the thickness of a single ply, no attempt was made to refine the mesh further because the size of the element used was on the same order as the fiber diameter. If the mesh were to be further refined, the assumption that the material was a homogeneous anisotropic continuum would no longer be valid.

The in-plane stresses in the central $-\theta°$ plies were calculated at nodes along the midplane of the laminate thickness. These in-plane stresses were transformed into the lamina coordinate system, and the stresses normal to the fiber direction, σ_{22}, and the shear stresses along the fiber direction, τ_{12}, were plotted in the vicinity of the straight free edge.

Fig. 1c. RADIOGRAPH OF $(0_2/25_2/-25_2)_s$ LAMINATE AFTER ONSET OF DAMAGE

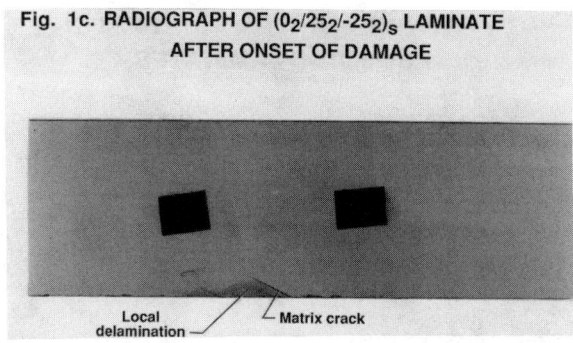

Local delamination — / — Matrix crack

Fig. 1d. RADIOGRAPH OF $(0_2/30_2/-30_2)_s$ LAMINATE AFTER ONSET OF DAMAGE

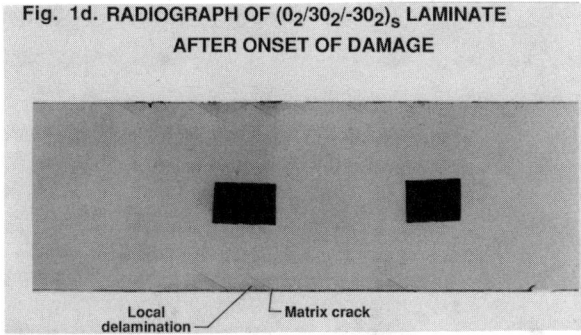

Local delamination — / — Matrix crack

FIG. 1—*Continued.*

These stress distributions were used to obtain a qualitative assessment of the stresses responsible for matrix cracking in the off-axis plies.

In-plane Normal Stresses

Figure 7 shows the distribution across the laminate width of the in-plane normal stress, σ_{22}, for $(0/\theta/-\theta)_s$ graphite laminates subjected to a total (mechanical plus thermal) axial strain of 0.01, assuming a ΔT of $-156°C$ ($-280°F$), where $\theta = 15$, 30, and 45°. The values in the interior, corresponding to $(b - y)/h = 5$, agree with the values calculated from laminated plate theory (Fig. 3), where the normal stresses in the 45° case are in tension and the normal stresses in the 15 and 30° cases are in compression. However, near the free edge, $(b - y)/h = 0$, the magnitude of the tensile stresses increases for the 45° case and the sign of the normal stresses changes from compression to tension for the 15 and 30° cases. The results for the 15 and 30° cases were expected to bound the results for the 20 and 25° cases, and hence, these layups were not analyzed.

The reversal in sign of the transverse normal stresses for the 15 and 30° cases is not limited to the particular applied loading shown in Fig. 7. For example, Fig. 8 shows the distribution across the laminate width of the in-plane normal stress, σ_{22}, for a $(0/15/-15)_s$ laminate subjected to total axial strains of 0.005 and 0.01 and a ΔT of $-156°C$ ($-280°F$). Figure 9 shows the distribution across the laminate width of the in-plane normal stress, σ_{22}, for a $(0/15/-15)_s$ laminate subjected to a total axial strains of 0.01 with a $\Delta T =$

Fig. 2a. EDGE MICROGRAPHS OF $(0_2/\theta_2/-\theta_2)_s$
AS4/3501-6 LAMINATES

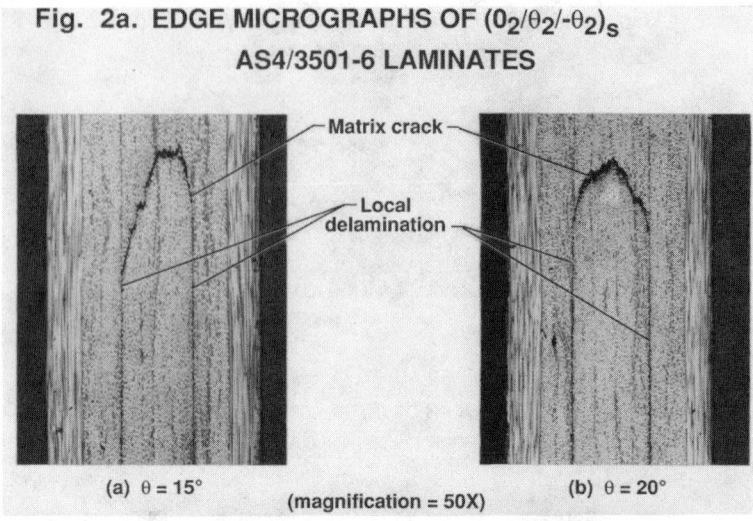

(a) $\theta = 15°$ (magnification = 50X)   (b) $\theta = 20°$

Fig. 2b. EDGE MICROGRAPHS OF $(0_2/\theta_2/-\theta_2)_s$
AS4/3501-6 LAMINATES

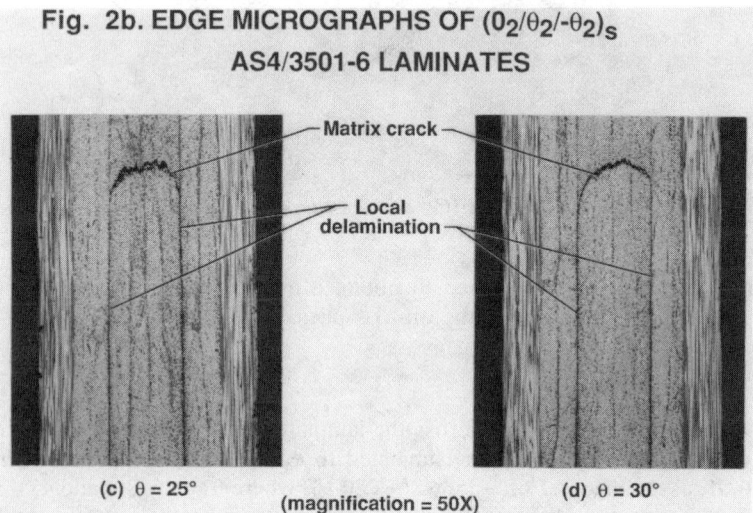

(c) $\theta = 25°$ (magnification = 50X)   (d) $\theta = 30°$

FIG. 2—*Photographs of the edge of $(0_2/\theta_2/-\theta_2)_s$ graphite epoxy laminates following static tests.*

TABLE 1—$(0_2/\theta_2/-\theta_2)_s$ *AS4/3501-6 graphite epoxy lamina properties.*

$(0_2/\theta_2/-\theta_2)_s$ AS4/3501-6 Graphite/Epoxy
E_{11} = 135 GPa (19.5 $\times$ 10^6 psi)
E_{22} = 11 GPa (1.6 $\times$ 10^6 psi)
G_{12} = 5.8 GPa (0.847 $\times$ 10^6 psi)
ν_{12} = 0.301
α_1 = $-0.41 \times 10^6/°C$ $(-0.23 \times 10^{-6}/°F)$
α_2 = 26.8 $\times 10^{-6}/°C$ $(14.9 \times 10^{-6}/°F)$

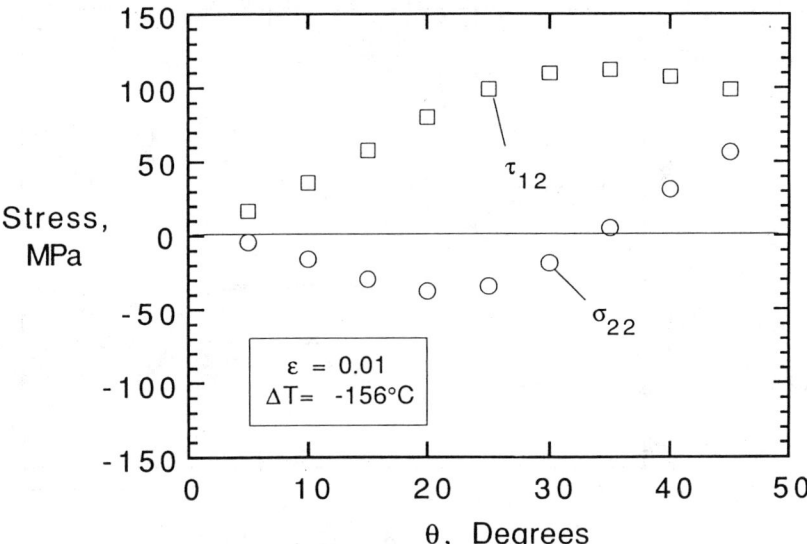

FIG. 3—*In-plane transverse normal stress, σ_{22}, and shear stress, τ_{12}, in $-\theta°$ ply calculated from laminated plate theory as a function of θ in $(0/\theta/-\theta)_s$ graphite epoxy laminates.*

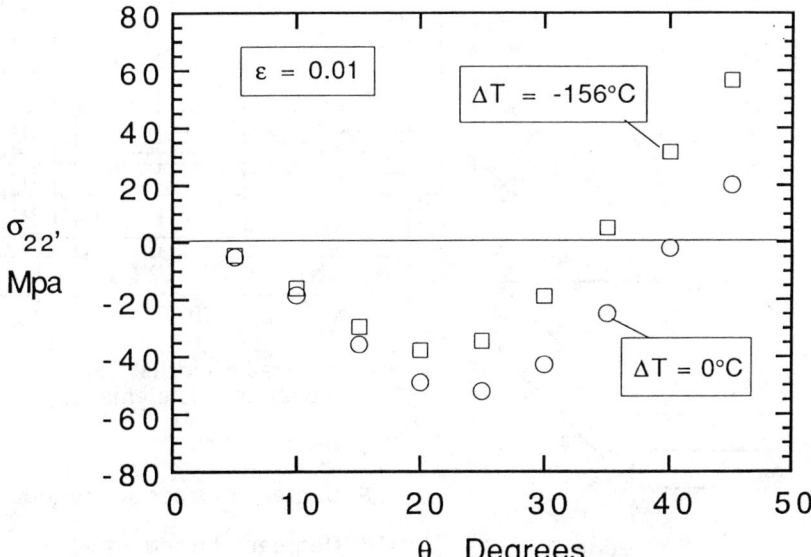

FIG. 4—*In-plane transverse normal stress, σ_{22}, in $-\theta°$ ply calculated from laminated plate theory as a function of θ in $(0/\theta/-\theta)_s$ graphite epoxy laminates.*

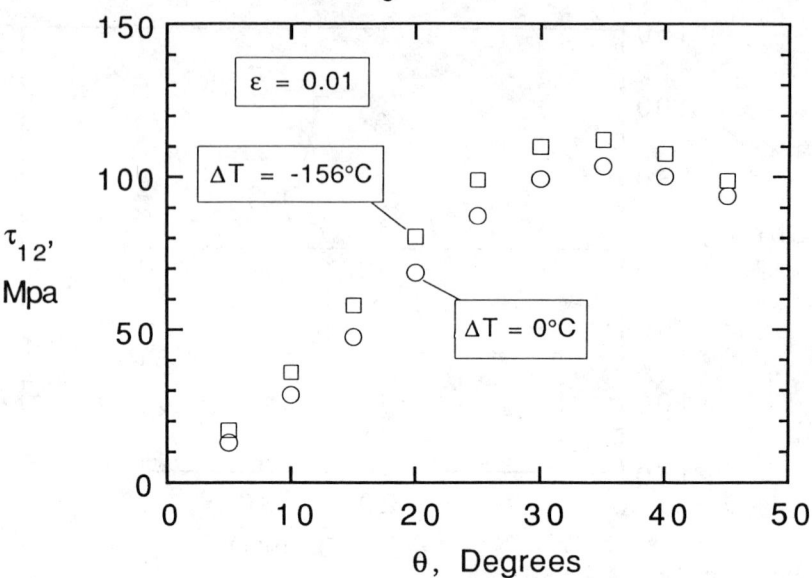

FIG. 5—*In-plane shear stress*, τ_{12}, *in* $-\theta°$ *ply calculated from laminated plate theory as a function of* θ *in* $(0/\theta/-\theta)_s$ *graphite epoxy laminates.*

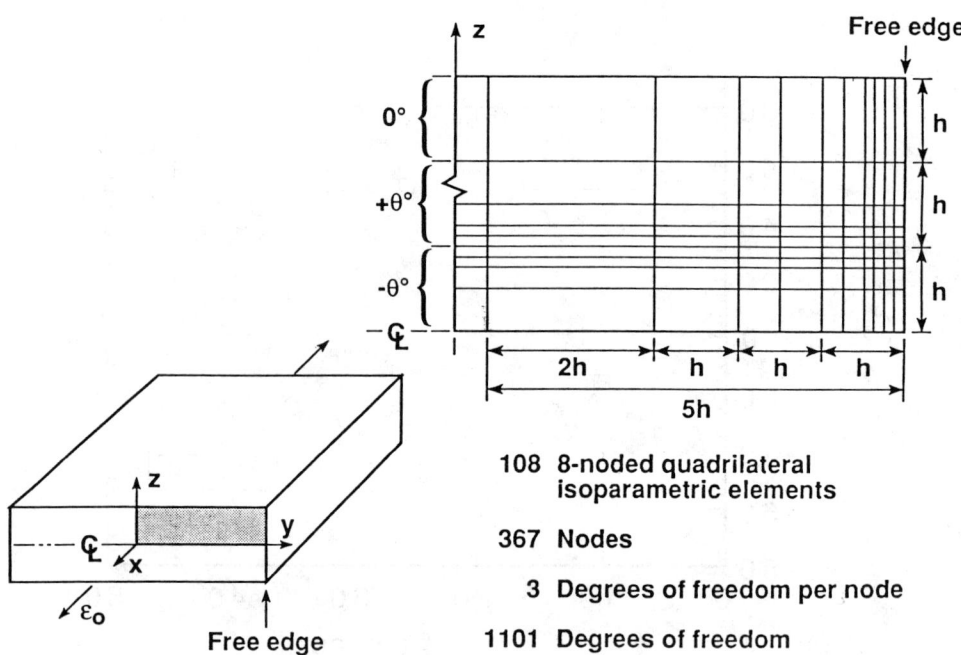

108 8-noded quadrilateral isoparametric elements

367 Nodes

3 Degrees of freedom per node

1101 Degrees of freedom

FIG. 6—*Schematic of quasi-3D finite element mesh near the free edge of a* $(0/\theta/-\theta)_s$ *laminate.*

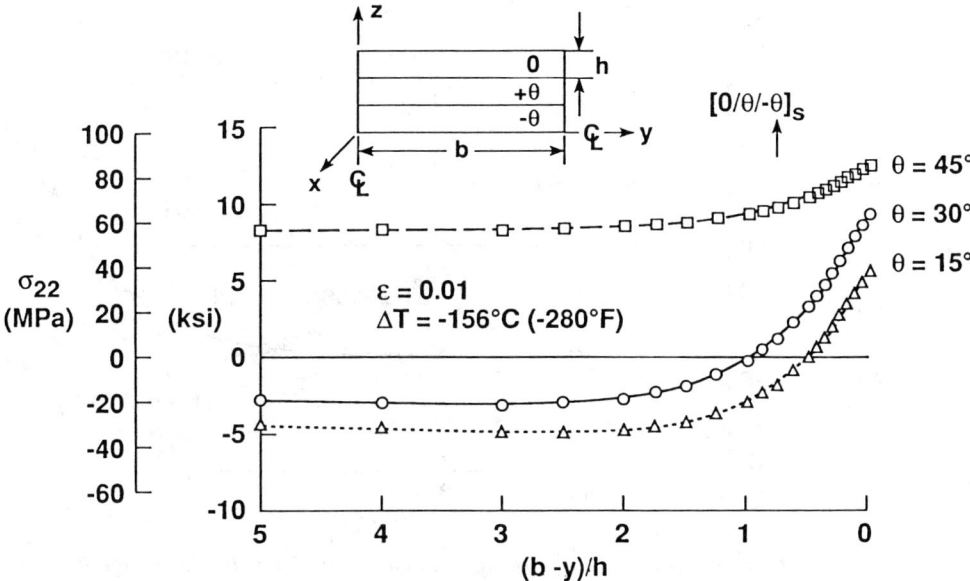

FIG. 7—*In-plane transverse normal stress, σ_{22}, distribution near the free edge in $-\theta°$ ply of $(0/\theta/-\theta)_s$ graphite epoxy laminates.*

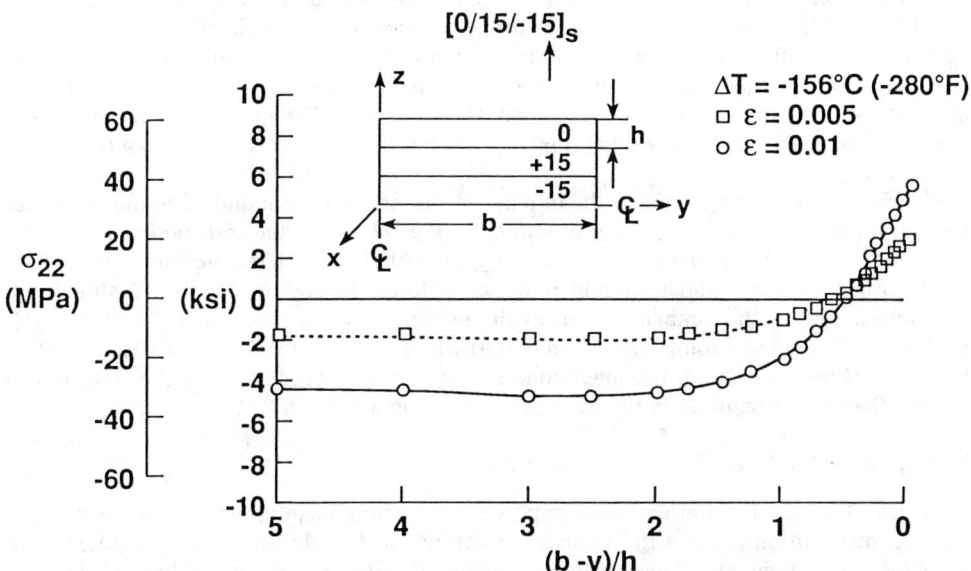

FIG. 8—*In-plane transverse normal stress, σ_{22}, distribution near the free edge in $-15°$ ply of a $(0/15/-15)_s$ graphite epoxy laminate.*

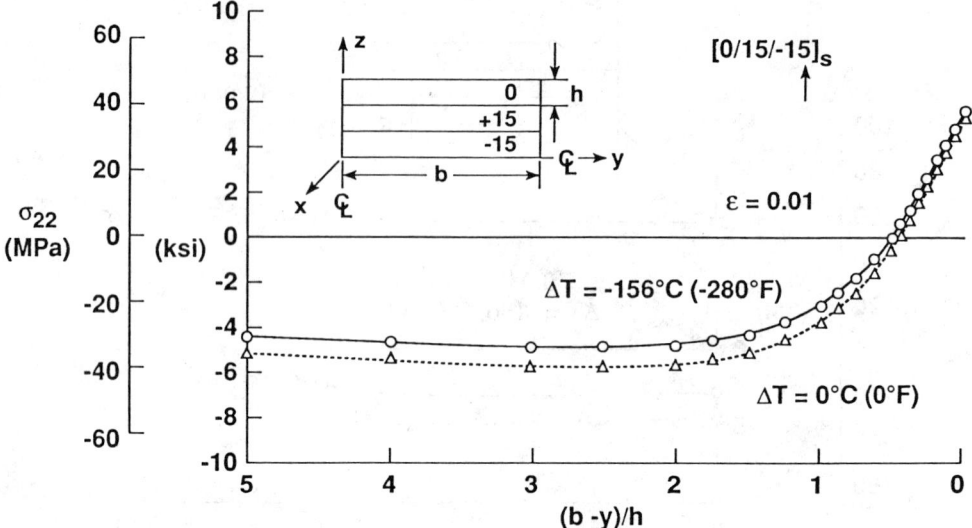

FIG. 9—*In-plane transverse normal stress, σ_{22}, distribution near the free edge in $-15°$ ply of a $(0/15/-15)_s$ graphite epoxy laminate for ambient and dry conditions.*

$-156°C\ (-280°F)$ and a $\Delta T = 0$. In both Figs. 8 and 9, the transverse normal stress is also compressive in the interior of the laminate width and becomes tensile at the free edge.

In-plane Shear Stresses

Figure 10 shows the distribution across the laminate width of the in-plane shear stress, τ_{12}, for $\theta = 15$, 30, and 45°. The values in the interior, corresponding to $(b - y)/h = 5$, agree with the values calculated by laminated plate theory (Fig. 3). In the interior of the laminate width, the 30° case has the largest shear stresses and the 15° case has the lowest shear stresses. However, near the free edge, $(b - y)/h = 0$, the magnitudes of these shear stresses change. Near the free edge, the 15° case has larger shear stresses than the 30 and 45° cases.

The change in the magnitude of the in-plane shear stresses is not limited to the particular applied loading shown in Fig. 10. For example, Fig. 11 shows the distribution across the laminate width of the in-plane shear stress, τ_{12}, for a $(0/15/-15)_s$ laminate subjected to total axial strains of 0.005 and 0.01 and a ΔT of $-156°C\ (-280°F)$. Figure 12 shows the distribution across the laminate width of the in-plane shear stress, τ_{12}, for a $(0/15/-15)_s$ laminate subjected to a total axial strain of 0.01 with a $\Delta T = -156°C\ (-280°F)$ and a $\Delta T = 0$. In both Figs. 11 and 12, the magnitude of the in-plane shear stress at the free edge is greater than the magnitude in the interior of the laminate width.

Interlaminar Stresses

Figure 13 shows the interlaminar normal stress, σ_{zz}, and the interlaminar shear stress, τ_{xz} and τ_{yz}, distributions across the laminate width in the 15/−15 interface of a $(0/15/-15)_s$ graphite epoxy laminate. This analysis indicates that the magnitude of the interlaminar shear stress, τ_{xz}, is very high near the free edge, and that the interlaminar normal stress, σ_{zz}, is compressive. Other authors have obtained similar results and have concluded that

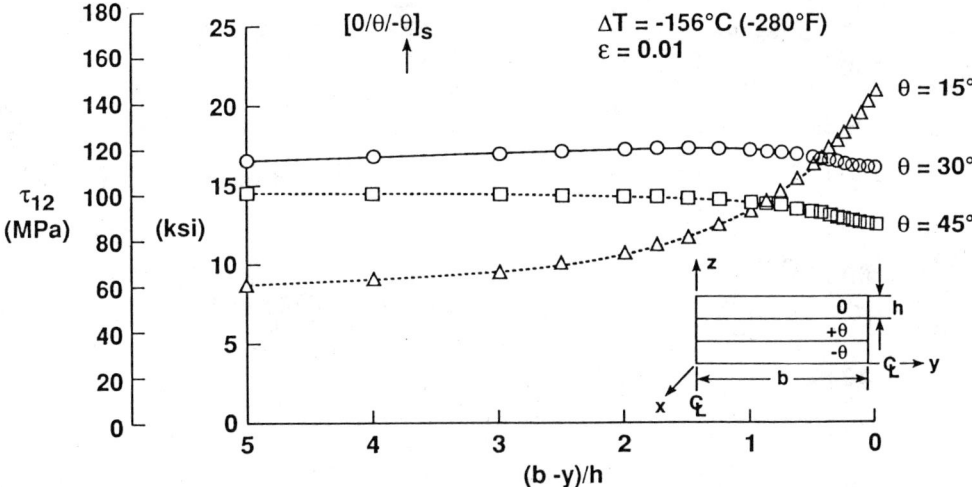

FIG. 10—*In-plane shear stress, τ_{12}, distribution near the free edge in $-\theta°$ ply of $(0/\theta/-\theta)_s$ graphite epoxy laminates.*

edge delaminations will form in the $15/-15$ interfaces of these laminates due to the high interlaminar shear stress, τ_{xz} [8–10]. However, the results in Fig. 13 do not take into account the matrix cracking that may be present in the $-15°$ plies.

The presence of matrix cracks may alter the interlaminar stress state at the free edge, resulting in a change in sign, from compression to tension, of the interlaminar normal stress, σ_{zz} [11,12]. For example, Fig. 14 shows the interlaminar normal stresses, calculated

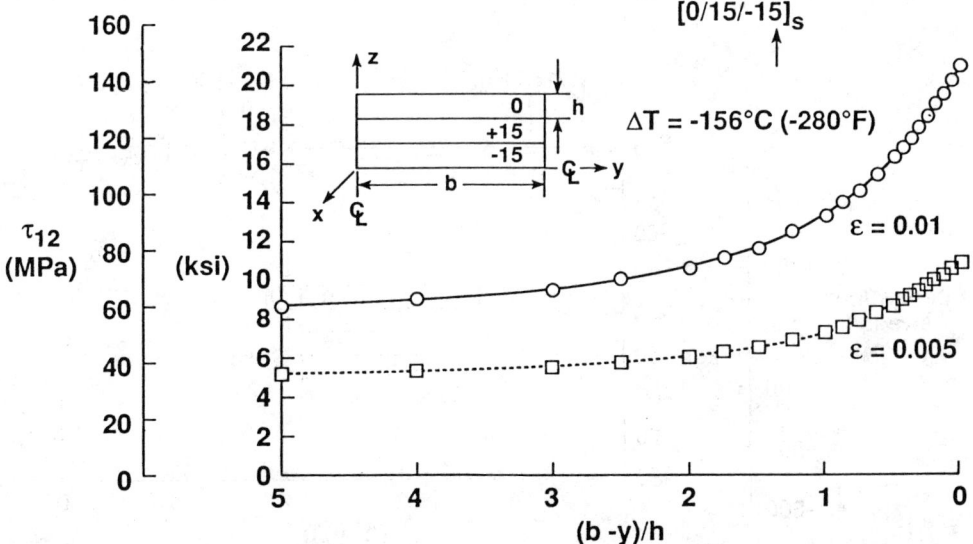

FIG. 11—*In-plane shear stress, τ_{12}, distribution near the free edge in $-15°$ ply of a (0/ 15/-15)_s graphite epoxy laminate.*

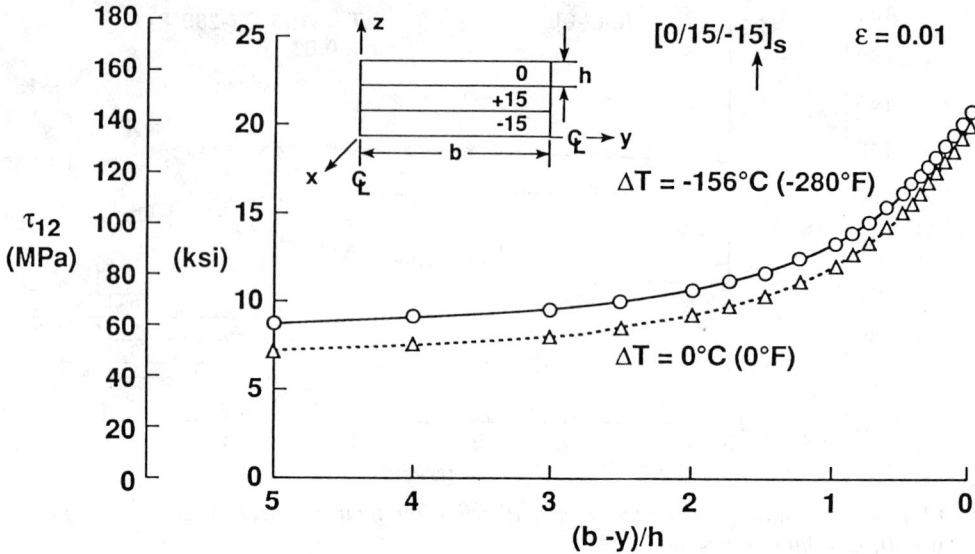

FIG. 12—*In-plane shear stress, τ_{12}, distribution near the free edge in* $-15°$ *ply of a* $(0/15/-15)_s$ *graphite epoxy laminate for ambient and dry conditions.*

from a 3D finite element analysis, in the $15/-15$ interfaces of a $(0/15/-15)_s$ graphite epoxy laminate at the intersection of the straight free edge and a matrix crack in the $-15°$ ply. The σ_{zz} stresses near the free edge in the $15/-15$ interface that were compressive before the introduction of the matrix crack (Fig. 13) become very large tensile stresses once the matrix crack is present. A similar reversal in σ_{zz} stresses was found [12] when $15°$ matrix cracks were modeled in $(15/90/-15)_s$ glass epoxy laminates. The length of the $15°$ matrix

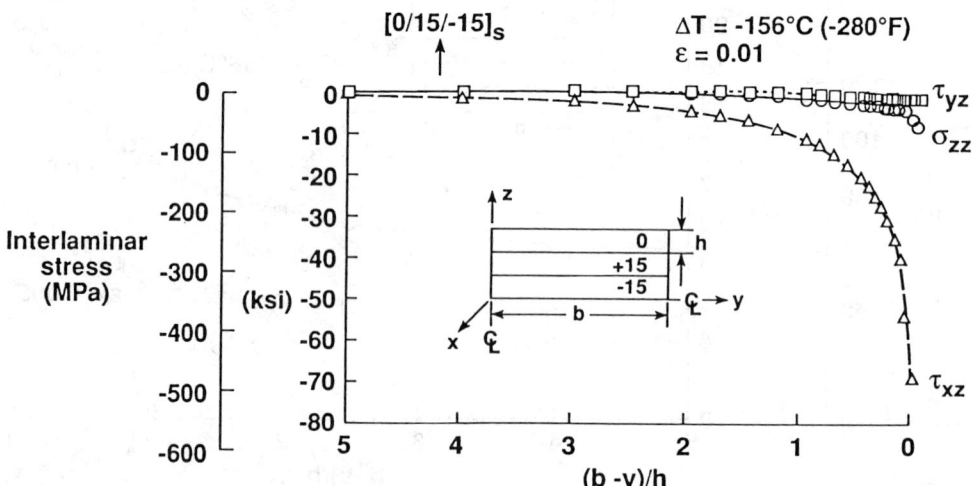

FIG. 13—*Interlaminar stress distributions near the free edge in the* $15/-15$ *interface of a* $(0/15/-15)_s$ *graphite epoxy laminate.*

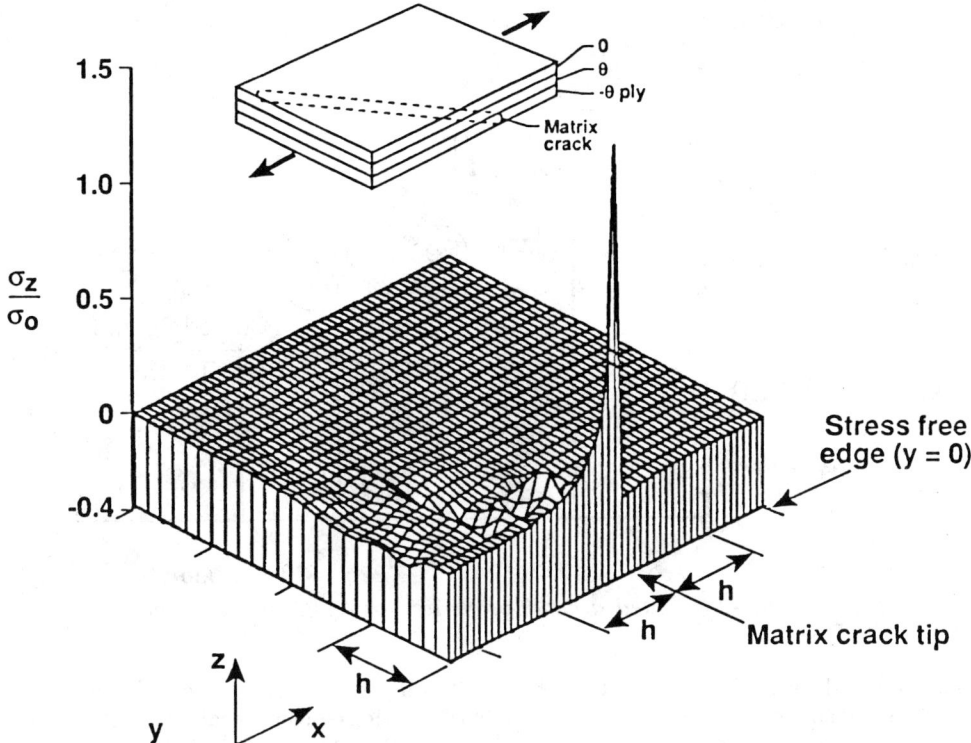

FIG. 14—*Interlaminar normal stress distribution from 3D finite element analysis* [11] *near the free edge in the 15/−15 interface of a (0/15/−15)ₛ graphite epoxy laminate.*

crack influenced the magnitude of the σ_{zz} stresses, but the reversal of sign from compression to tension was always present when a matrix crack of any length was introduced [*12*]. The matrix cracks modeled in Refs *11* and *12* were straight cracks through the ply thickness, normal to the ply interface. However, as shown in Figs. 2*a* through 2*d*, the matrix cracks are typically curved, intersecting the ply interface at an oblique angle. Although the matrix crack curvature through the ply thickness observed in Figs. 2*a* through 2*d* may be predicted [*11*], a 3D analysis simulating these crack configurations is needed to quantify the influence of this curvature on the interlaminar normal stresses.

The interlaminar shear stresses, τ_{xz}, that were present before the introduction of the angle ply matrix crack also increase in magnitude locally near the intersection of the free edge and the matrix crack [*11,12*]. However, as will be discussed later, this increase in magnitude of τ_{xz} may not have as dramatic an effect on the onset of local delamination as the change in the sign of the interlaminar normal stress.

Discussion

The finite element results indicate that matrix cracks may initiate in the central $-\theta$ plies at the edge of the $(0/\theta/-\theta)_s$ laminates where the in-plane transverse tensile stresses exist and have their greatest magnitude. This is especially true for brittle epoxy matrix composites, such as AS4/3501-6, where the transverse tensile strength is typically on the order of

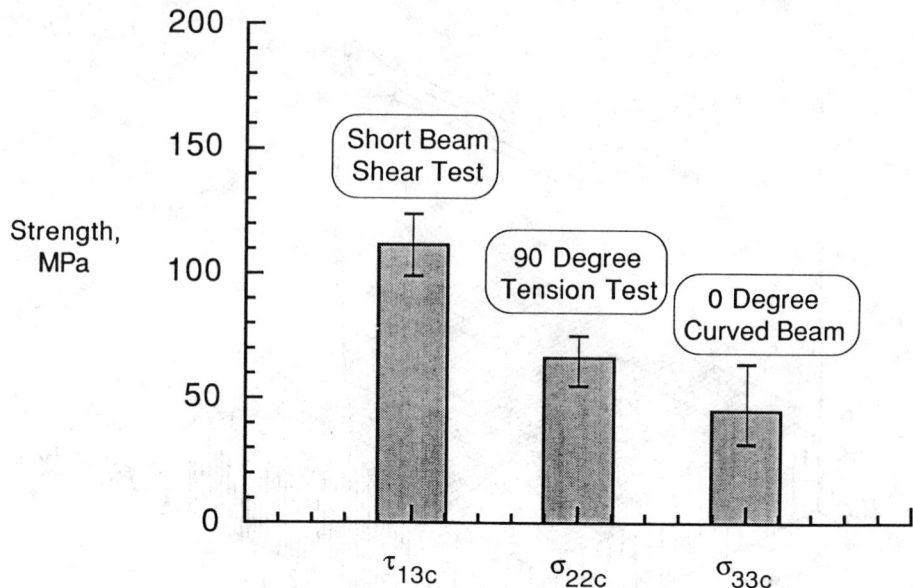

FIG. 15—*Matrix-dominated strength properties for AS4/3501-6 graphite epoxy.*

one half the shear strength. For example, Fig. 15 compares the relative magnitudes of the interlaminar shear strength, τ_{13c}, measured using a short beam shear test [13], to the transverse tensile strength, σ_{22c}, measured using a 90° tension test [14], and the interlaminar tensile strength, σ_{33c}, measured using a 0° curved beam bending test [14]. The matrix-dominated tensile strengths are approximately one half of the interlaminar shear strength for AS4/3501-6.

The analytical results indicate that the use of in-plane stresses calculated from laminated plate theory to predict the onset of matrix cracking, often referred to as "first ply failure" in phenomenological failure criteria, may be unconservative because these laminate theory stresses represent minimum values in the interior of the laminate. The first "failure" of a ply, however, occurs near the free edge (Fig. 1) where in-plane transverse tensile stresses exist and have their greatest magnitude (Fig. 7).

The presence of matrix cracks may alter the interlaminar stress state at the free edge, resulting in a change in sign, from compression to tension, of the interlaminar normal stress, σ_{zz} [11,12]. Furthermore, these laminates do not delaminate uniformly along the length of the laminate near the free edges as typically observed for edge delamination [15–17]. Instead, the delaminations are localized and bounded by angle ply matrix cracks and the free edge, as shown in Figs. 1 and 2, and in Fig. 3 of Ref 9. Hence, the local tensile interlaminar normal stresses near the free edge resulting from the formation of matrix cracks, in addition to the high interlaminar shear stresses that are present near the free edge, will contribute significantly to the formation of local delaminations in these laminates.

Conclusions

Quasi-static tension tests were conducted on AS4/3501-6 graphite epoxy $(0_2/\theta_2/-\theta_2)_s$ laminates, where θ was 15, 20, 25, and 30°. Dye penetrant-enhanced X-ray radiography was used to document the onset of matrix cracking in the central $-\theta°$ plies and local delamination onset in the $\theta/-\theta$ interface at the intersection of the matrix cracks and the free edge.

A quasi-3D finite element analysis was conducted to calculate the stresses responsible for matrix cracking in the off-axis plies. In contrast to the results of laminated plate theory, which indicated that the transverse normal stresses were compressive, the finite element analysis yielded tensile transverse normal stresses near the free edge. Hence, the analytical results indicated that the use of in-plane stresses calculated from laminated plate theory to predict the onset of matrix cracking, often referred to as "first ply failure" in phenomenological failure criteria, may be unconservative because these laminate theory stresses represent minimum values in the interior of the laminate. The first "failure" of a ply, however, occurs near the free edge where in-plane transverse tensile stresses exist and have their greatest magnitude.

References

[1] Reifsnider, K. L. and Talug, A., "Analysis of Fatigue Damage in Composite Laminates," *International Journal of Fatigue*, Vol. 3, No. 1, January 1980, pp. 3–11.

[2] Kress, G. R. and Stinchcomb, W. W., "Fatigue Response of Notched Graphite Epoxy Laminates," *Recent Advances in Composites in the United States and Japan, ASTM STP 864*, American Society for Testing and Materials, Philadelphia, 1985, pp. 173–196.

[3] Murri, G. B., O'Brien, T. K., and Salpekar, S. A., "Tension Fatigue of Glass/Epoxy and Graphite/Epoxy Tapered Laminates," *Proceedings*, 46th AHS Annual Forum, Vol. 1, May 1990, pp. 721–734 (also in NASA TM 102628, April 1990, NASA, Washington, DC).

[4] O'Brien, T. K., Rigamonti, M., and Zanotti, C., "Tension Fatigue Analysis and Life Prediction for Composite Laminates," *International Journal of Fatigue*, Vol. 11, No. 6, November 1989, pp. 379–394.

[5] O'Brien, T. K. and Raju, I. S., "Strain Energy Release Rate Analysis of Delamination Around an Open Hole in a Composite Laminate," *Proceedings*, 25th AIAA Structures, Dynamics, and Materials Conference, Palm Springs, May 1984, AIAA-84-0961, AIAA, New York, pp. 526–536.

[6] Jones, R. M., *Mechanics of Composite Materials*, McGraw-Hill, Washington, DC, 1975.

[7] Raju, I. S., "Q3DG—A Computer Program for Strain Energy Release Rates for Delamination Growth in Composite Laminates," NASA CR 178205, NASA, Washington, DC, November 1986.

[8] Soni, S. R. and Kim, R. Y., "Delamination of Composite Laminates Stimulated by Interlaminar Shear," *Composite Materials: Testing and Design (Seventh Conference), ASTM STP 893*, American Society for Testing and Materials, Philadelphia, 1986, pp. 286–307.

[9] Lagace, P. A. and Brewer, J. C., "Studies of Delamination Growth and Final Failure under Tensile Loading," *Proceedings*, ICCM VI, London, Vol. 5, Elsevier, London and New York, July 1987, pp. 262–273.

[10] Brewer, J. C. and Lagace, P. A., "Quadratic Stress Criterion for Initiation of Delamination," *Journal of Composite Materials*, Vol. 22, No. 4, December 1988, pp. 1141–1155.

[11] Salpekar, S. A. and O'Brien, T. K., "Analysis of Matrix Cracking and Local Delamination in $(0/\theta/-\theta)_s$ Graphite Epoxy Laminates under Tension Load," *Proceedings*, ICCM VIII, Honolulu, July 1991, SAMPE, Covina, CA.

[12] Fish, J. C. and O'Brien, T. K., "Three-Dimensional Finite Element Analysis of Delamination from Matrix Cracks in Glass-Epoxy Laminates," *Composite Materials: Testing and Design (10th Volume), ASTM STP 1120*, American Society for Testing and Materials, Philadelphia, 1992.

[13] Sun, C. T. and Kelly, S. R., "Failure in Composite Angle Structures, Part I: Initial Failure," *Journal of Reinforced Plastics and Composites*, Vol. 7, May 1988, pp. 220–232.

[14] Martin, R. H. and Jackson, W. C., "Damage Prediction in Curved Composite Laminates," *Composite Materials: Fatigue and Fracture, Fourth Volume, ASTM STP 1156*, American Society for Testing and Materials, Philadelphia, 1993.

[15] O'Brien, T. K., Johnston, N. J., Raju, I. S., and Morris, D. H., "Comparisons of Various Configurations of the Edge Delamination Test for Interlaminar Fracture Toughness," *Toughened Composites, ASTM STP 937*, American Society for Testing and Materials, Philadelphia, 1987, pp. 199–221.

[*16*] O'Brien, T. K., "Characterization of Delamination Onset and Growth in a Composite Laminate," *Damage in Composite Materials, ASTM STP 775,* American Society for Testing and Materials, Philadelphia, 1982, pp. 140–167.

[*17*] O'Brien, T. K., "Mixed-mode Strain Energy Release Rate Effects on Edge Delamination of Composites," *Effects of Defects in Composite Materials, ASTM STP 836,* American Society for Testing and Materials, Philadelphia, 1984, pp. 125–142.

T. Kevin O'Brien[1]

Local Delamination in Laminates with Angle Ply Matrix Cracks, Part II: Delamination Fracture Analysis and Fatigue Characterization

REFERENCE: O'Brien, T. K., "**Local Delamination in Laminates with Angle Ply Matrix Cracks, Part II: Delamination Fracture Analysis and Fatigue Characterization,**" *Composite Materials: Fatigue and Fracture, Fourth Volume, ASTM STP 1156,* W. W. Stinchcomb and N. E. Ashbaugh, Eds., American Society for Testing and Materials, Philadelphia, 1993, pp. 507–538.

ABSTRACT: Constant amplitude tension-tension fatigue tests were conducted on AS4/3501-6 graphite epoxy $(0_2/\theta_2/-\theta_2)_s$ laminates, where θ was 15, 20, 25, or 30°. Fatigue tests were conducted at a frequency of 5 Hz and an R-ratio of 0.1. Dye penetrant-enhanced X-radiography was used to document the onset of matrix cracking in the central $-\theta°$ plies and the subsequent onset of local delaminations in the $\theta/-\theta$ interface at the intersection of the matrix cracks and the free edge.

Two strain energy release rate (G) solutions for local delamination from angle ply matrix cracks were derived: one for a uniform delamination front across the laminate width, and one for a triangular-shaped delamination area that extended only partially into the laminate width from the free edge. Plots of maximum cyclic G versus the number of cycles to local delamination onset (G_{max} versus N) were generated to assess the accuracy of these G solutions. The influence of residual thermal and moisture stresses on G were also quantified.

KEY WORDS: composite material, graphite epoxy, delamination, matrix crack, fatigue

In Part I of this study [*1*], the damage observed in quasi-static tension tests on $(0_2/\theta_2/-\theta_2)_s$ AS4/3501-6 graphite/epoxy laminates, where θ was 15, 20, 25, and 30°, was documented. In addition, the stresses responsible for the matrix cracking and associated local delamination observed were calculated using finite element analyses.

In Part II of this study, constant amplitude, tension-tension fatigue tests were conducted at a frequency of 5 Hz and an R-ratio of 0.1 on $(0_2/\theta_2/-\theta_2)_s$ AS4/3501-6 graphite/epoxy laminates, where θ was 15, 20, 25, and 30°. Dye penetrant–enhanced X-ray radiography was used to document the onset of matrix cracking and the subsequent onset of local delaminations at the intersection of the matrix cracks and the free edge as a function of the number of fatigue cycles. Two strain energy release rate (G) solutions were introduced for local delamination from matrix cracks. These G solutions were used to correlate the onset of local delamination in fatigue between layups with different off-axis angle plies.

[1]U.S. Army Aerostructures Directorate, NASA Langley Research Center, Hampton, VA 23681-0001.

Experiments

Materials

As documented in Part I [*1*], panels were manufactured in an autoclave at NASA Langley Research Center from Hercules AS4/3501-6 graphite epoxy prepreg using the standard manufacturer's procedure with a maximum temperature of 177°C (350°F) in the cure cycle. One panel was produced for each of the $(0_2/\theta_2/-\theta_2)_s$ layups, where θ was 15, 20, 25, or 30°. Panels 305 mm (12 in.) square were manufactured and then cut into 20 individual coupons, 127 mm (5.0 in.) long by 25 mm (1.0 in.) wide. Each specimen had a nominal ply thickness of 0.124 mm (0.0049 in.) and a fiber volume fraction of 67%.

Experimental Procedures

All static tests were conducted using the procedure outlined in Part I of this study [*1*]. Fatigue specimens were placed in the hydraulic grips of a hydraulic load frame such that each specimen had a gage length between the grips of approximately 76 mm (3.0 in.). A 25-mm (1-in.) gage length extensometer was mounted on the specimen midway between the grips to measure nominal strain. As specimens were loaded quasi-statically to the mean load, a plot of load versus nominal strain was generated using an X-Y recorder. The extensometer was then removed, and the specimens were loaded at a cyclic frequency of 5 Hz and an *R*-ratio of 0.1. For each layup, ten fatigue tests were run at maximum load levels below the critical load required to initiate damage in the static tests. The cyclic loading was stopped at fixed cyclic intervals, and a zinc iodide solution that was opaque to X-rays was placed on the edge using a hypodermic needle. A radiograph was taken of the specimen width, with the specimen loaded at the mean load, to document the presence of matrix cracking and delamination.

Table 1 lists the maximum cyclic loads and the schedule used to obtain radiographs for each cyclic load level. For example, tests run at maximum cyclic loads of 5000 and 6000 lb were interrupted every 1000 cycles to obtain a radiograph of the specimen. Tests run at a maximum cyclic load of 3500 lb were interrupted every 1000 cycles for the first 10 000 cycles; every 2000 cycles between 10 000 and 20 000 cycles; every 5000 cycles between 20 000 and 50 000 cycles; and every 10 000 cycles thereafter, to obtain a radiograph of the specimen. Loading was continued until a radiograph clearly indicated the presence of a local delamination associated with a matrix crack near the free edge. After the fatigue specimens were removed from the hydraulic grips, the edges were polished and examined

TABLE 1—*X-ray schedule for $(0_2/\theta_2/-\theta_2)_s$ fatigue tests.*

P_{max}, N (lb)	X-ray Taken Every N Cycle	Between N_1 and N_2	
		Cycles	Cycles
26 690 (6000)	1 000	1	5 000
22 240 (5000)	1 000	1	5 000
17 790 (4000)	1 000	1	10 000
	2 000	10 000	20 000
15 570 (3500)	1 000	1	10 000
	2 000	10 000	20 000
	5 000	20 000	50 000
	10 000	50 000	110 000
13 340 (3000)	10 000	1	100 000
	50 000	100 000	1 000 000

in the light microscope. Photographs of the polished edges were taken through an optical microscope to document the damage observed on the straight edge.

Initially, some coupons were coated with a thin film of water-based typewriter correction fluid on either edge. After a few fatigue tests were performed, however, it was apparent that the onset of matrix cracking and local delamination was difficult to detect visually on the edge with this brittle coating. Hence, most of the fatigue coupons were tested without coating the edges. It was also easier to see the formation of the matrix cracks in the radiographs if there was no coating on the edge because the dye penetrant used to reveal the damage was more likely to accumulate near a coated edge.

Experimental Results

Static Tests—The results from the quasi-static tests were documented in Part I of this study [1]. The radiographs showed triangular-shaped shaded regions bounded by free edge and matrix cracks, indicating that a local delamination was present. Edge micrographs showed that local delaminations were present in both $\theta/-\theta$ interfaces, extending in the same direction from curved matrix cracks in the central $-\theta°$ plies.

Fatigue Tests—The fatigue specimens had the same type of damage seen in the radiographs and edge micrographs of specimens taken after the quasi-static tests [1]. However, unlike the quasi-static tests, the sequence of damage was evident in the series of radiographs taken during the cyclic loading. For example, Fig. 1 shows radiographs taken immediately after the onset of matrix cracking (Fig. 1a) and the subsequent onset of local delamination from the matrix crack (Fig. 1b), for a $(0_2/20_2/-20_2)_s$ laminate that was subjected to a maximum cyclic load of 15 570 N (3500 lbs). Similar radiographs were obtained for the other layups tested in fatigue. At the number of cycles corresponding to local delamination onset, all of the specimens had at least one, and in most cases, several matrix cracks in the central $-\theta°$ plies. A few matrix cracks were also observed in the $+\theta°$ plies for some laminates. Table 2 lists the average number of matrix cracks in the gage section measured from the radiographs of laminates taken just after the onset of local delamination. Local delaminations were present in the $\theta/-\theta$ interfaces near some, but not all, of these matrix cracks. By the time the fatigue tests were terminated, matrix cracks had also

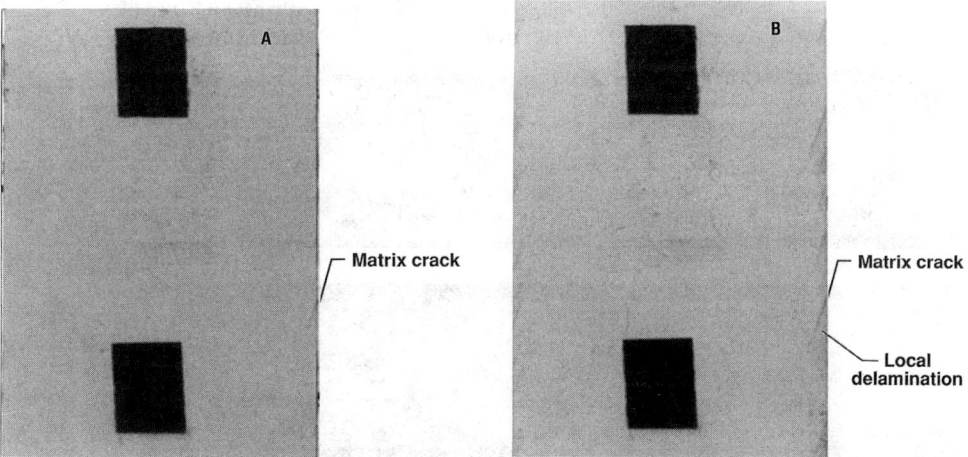

FIG. 1—*Dye penetrant-enhanced X-radiographs of* $(0_2/20_2/-20_2)_s$ *specimens taken after the onset of* (a) *matrix cracking, and* (b) *local delamination in fatigue.*

TABLE 2—*Average number of matrix cracks at local delamination onset in $(0_2/\theta_2/-\theta_2)_s$ AS4/3501-6 graphite epoxy laminates.*

P_{max}, N (lb)	$\theta°$			
	15	20	25	30
Static test	1.35	1.57	2.67	5.25
17 790 (4000)	14.2	9.7	11.0	20.7
15 570 (3500)	12.0	10.3	13.7	23.3
13 340 (3000)	9.0	14.0	15.7	32.0

developed in the $+\theta°$ plies of many laminates. Delaminations in the $\theta/-\theta$ interfaces often terminated at these $+\theta°$ matrix ply cracks (Fig. 2).

Figures 3 to 6 show the cumulative number of cycles to the onset of matrix cracking and subsequent local delamination onset for each layup. The loads at onset of damage under a monotonically increasing load, as observed during the quasi-static tests performed in Part I of this study [1], are shown on the ordinate. Figures 7 through 10 show the cumulative number of cycles to local delamination onset minus the number of cycles to onset of matrix cracking. Hence, Figs. 7 through 10 represent the number of cycles to delamination onset once the matrix crack is present.

Strain Energy Release Rate Analysis

Uniform Local Delamination

In Ref 2, an equation was derived for the strain energy release rate associated with a uniform local delamination growing from a 90° matrix crack (Fig. 11a) in a multidirectional

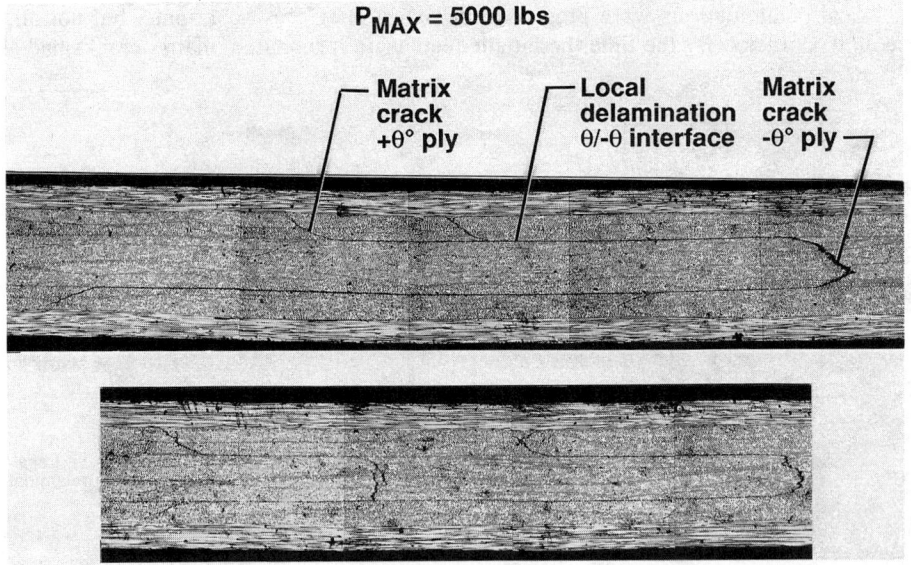

FIG. 2—*Edge micrographs of $(0_2/20_2/-20_2)_s$ laminate taken after the completion of cyclic loading showing local delaminations arresting at $+20°$ matrix cracks.*

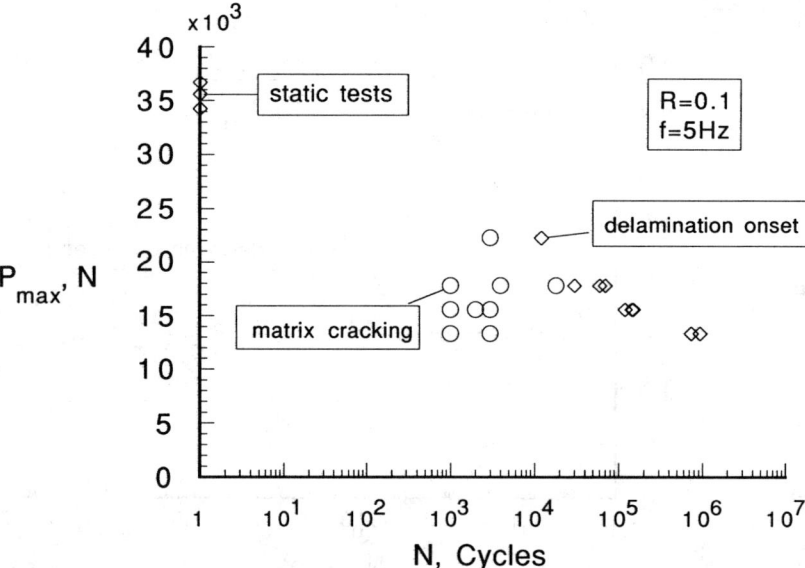

FIG. 3—*Maximum cyclic load as a function of the number of cycles to matrix cracking and subsequent delamination onset in $(0_2/15_2/-15_2)_s$ graphite epoxy laminates.*

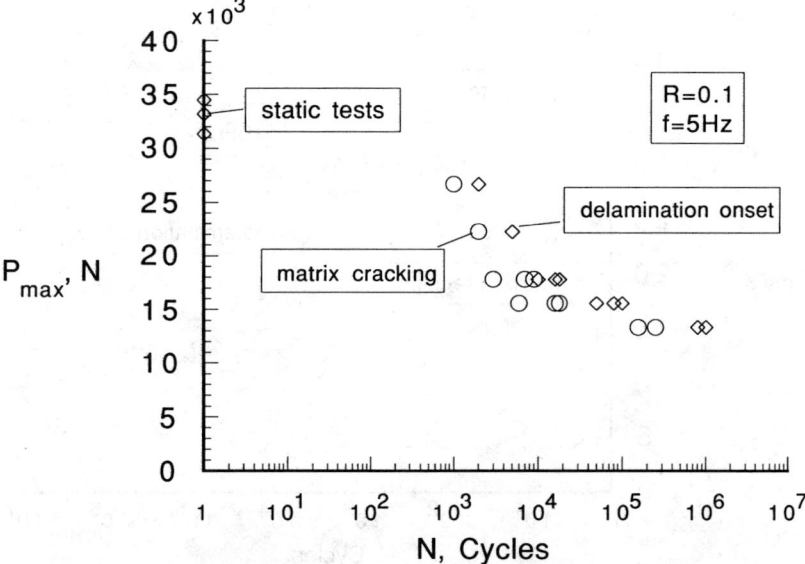

FIG. 4—*Maximum cyclic load as a function of the number of cycles to matrix cracking and subsequent delamination onset in $(0_2/20_2/-20_2)_s$ graphite epoxy laminates.*

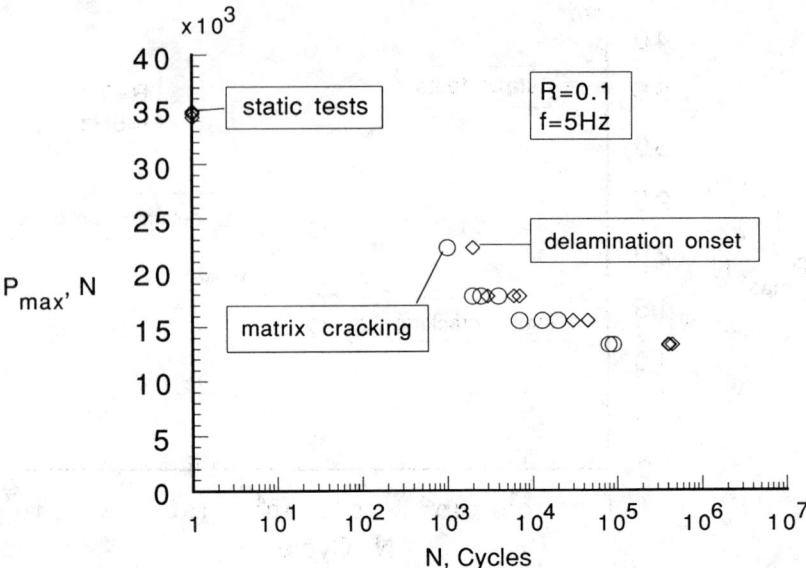

FIG. 5—*Maximum cyclic load as a function of the number of cycles to matrix cracking and subsequent delamination onset in $(0_2/25_2/-25_2)_s$ graphite epoxy laminates.*

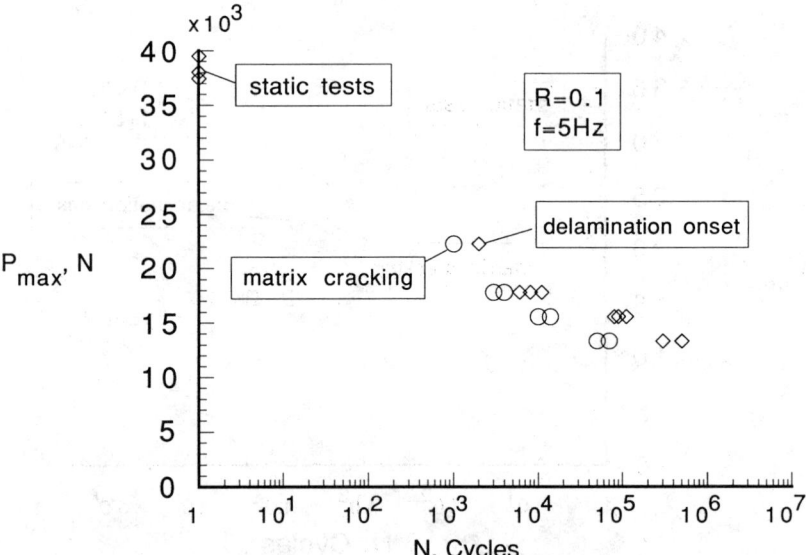

FIG. 6—*Maximum cyclic load as a function of the number of cycles to matrix cracking and subsequent delamination onset in $(0_2/30_2/-30_2)_s$ graphite epoxy laminates.*

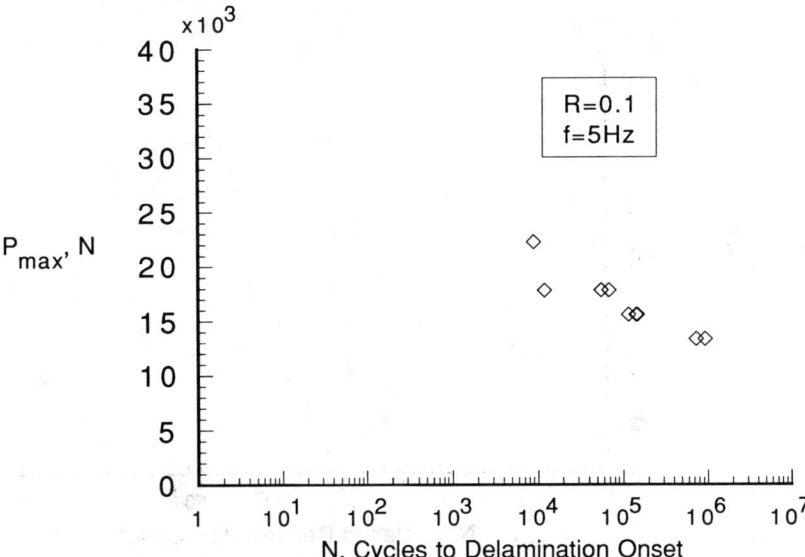

FIG. 7—*Maximum cyclic load as a function of the number of cycles to delamination onset in $(0_2/15_2/-15_2)_s$ graphite epoxy laminates.*

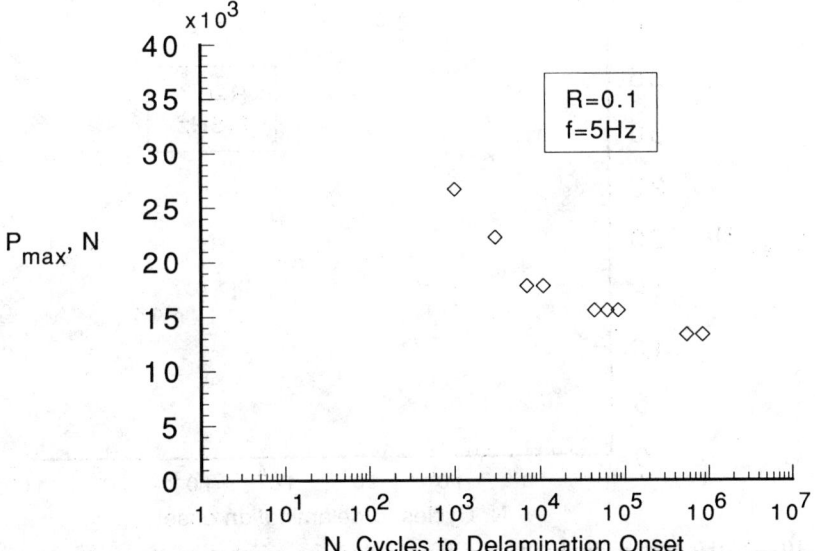

FIG. 8—*Maximum cyclic load as a function of the number of cycles to delamination onset in $(0_2/20_2/-20_2)_s$ graphite epoxy laminates.*

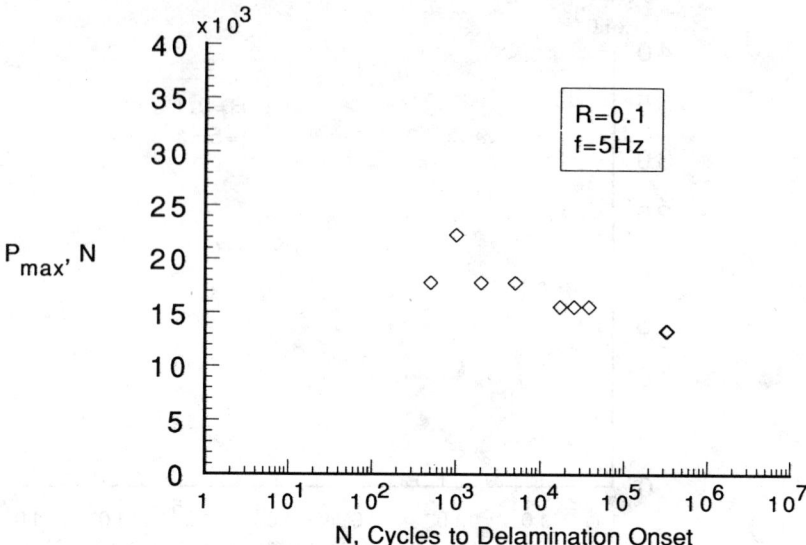

FIG. 9—*Maximum cyclic load as a function of the number of cycles to delamination onset in $(0_2/25_2/-25_2)_s$ graphite epoxy laminates.*

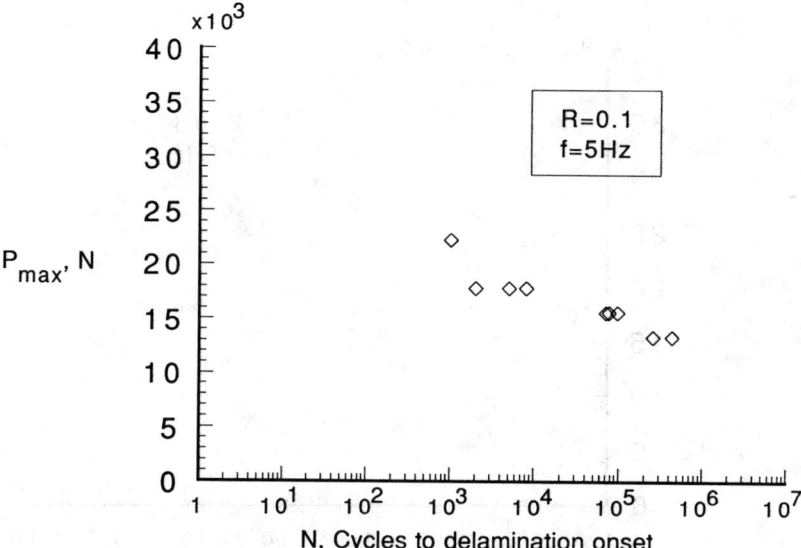

FIG. 10—*Maximum cyclic load as a function of the number of cycles to delamination onset in $(0_2/30_2/-30_2)_s$ graphite epoxy laminates.*

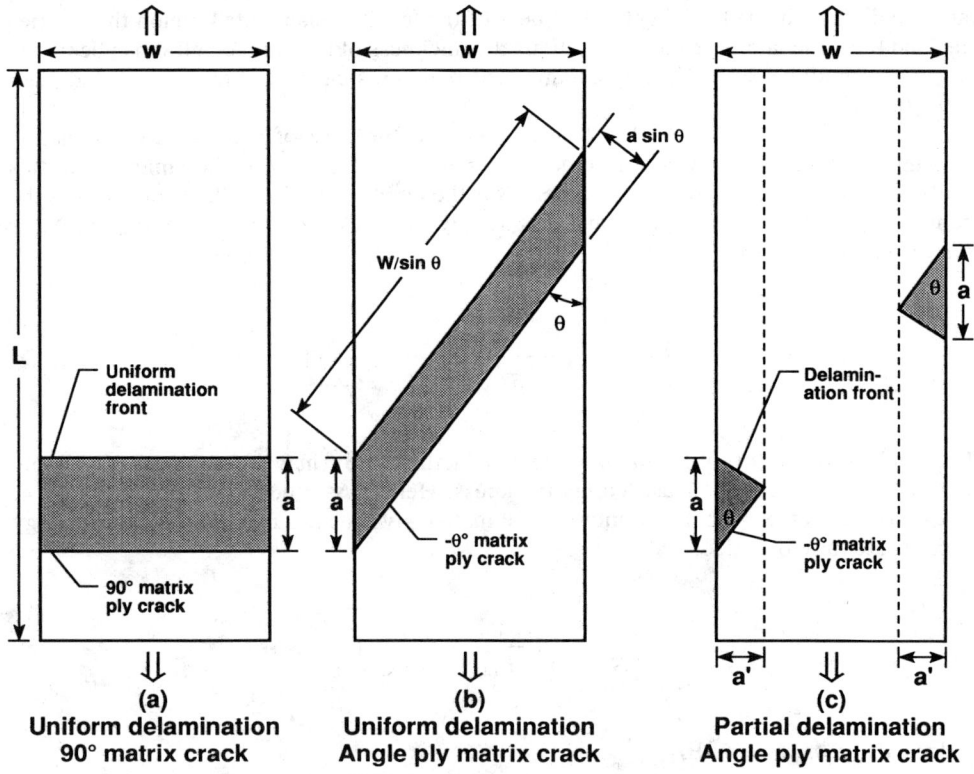

FIG. 11—*Local delamination models.*

laminate using the classical definition for G as

$$G = \frac{P^2}{2} \frac{dC}{dA} \tag{1}$$

where P is the applied mechanical load on the laminate, and dC is the incremental change in laminate compliance associated with an incremental change in surface area, dA, created by the local delamination. This solution assumed that the composite consisted of a laminated region and a locally delaminated region in series, resulting in the following expression

$$G = \frac{P^2}{2mw^2} \left(\frac{1}{t_{LD}E_{LD}} - \frac{1}{tE_{LAM}} \right) \tag{2}$$

where w is the laminate width and m is the number of delaminations growing from the matrix crack. The parameter, m, would have a value of 2 if the cracked off-axis ply is in the interior of the laminate, corresponding to local delamination on either side of the matrix crack. However, m would have a value of 1 if the cracked off-axis ply was a surface ply. In Eq 2, t is the laminate thickness, E_{LAM} is the laminate modulus (as calculated from laminated plate theory), t_{LD} is the thickness of the locally delaminated region that carries the load (i.e., the laminate thickness minus the thickness of the cracked off-axis plies), and E_{LD} is the modulus of the locally delaminated region (as calculated from laminated plate theory).

As shown in Appendix A, Eq 2 also applies to the case of a uniform delamination growing from an angle ply matrix crack as shown in Fig. 11b. For the laminates in this study, $t = Nh$ and $t_{LD} = (N - n)h$, where N is the number of plies in the laminate, n is the number of cracked off-axis plies, and h is the ply thickness. Therefore, noting that $N_x = P/w$, Eq 2 may be rewritten as

$$G = \frac{N_x^2}{2mh}\left(\frac{1}{(N - n)E_{LD}} - \frac{1}{NE_{LAM}}\right) \tag{3}$$

The twelve-ply $(0_2/\theta_2/-\theta_2)_s$ laminates tested in this study may be analyzed as six-ply laminates with twice the measured ply thickness. Hence, $N = 6$, $n = 2$, and $N - n = 4$. Then, noting that $m = 2$ for an interior $-\theta$ matrix ply crack, a normalized strain energy release rate may be calculated as

$$\frac{Gh}{N_x^2} = \frac{1}{4}\left(\frac{1}{4E_{LD}} - \frac{1}{6E_{LAM}}\right) \tag{4}$$

Influence of Coupling on E_{LD}

The value of E_{LD} in Eq 4 depends on whether or not the locally delaminated region consists of symmetric or asymmetric sublaminates. For example, the locally delaminated region may be thought of as a $(0/\theta)_s$ symmetric sublaminate. In this case, the modulus of the locally delaminated region is simply

$$E_{LD} = \frac{1}{(N - n)h(A_{11})^{-1}} \tag{5}$$

where $(A_{11})^{-1}$ is the first term of the inverse extensional stiffness matrix for the $(0/\theta)_s$ sublaminate. However, as noted in Ref 3, if the locally delaminated region consists of two two-ply asymmetric sublaminates, then the presence of bending-extension coupling should be reflected in the modulus calculation. This may be accomplished using the laminated plate theory by prescribing an applied mechanical strain, ϵ_x, setting the applied N_y, N_{xy}, and M_y equal to 0, and setting the κ_x and κ_{xy} equal to 0. The N_x that results from this analysis is derived in Appendix B and may be used to determine E_{LD} as

$$E_{LD} = \frac{2}{(N - n)h}\left[A_{11} + \left(\frac{D_1}{D}\right)A_{12} + \left(\frac{D_2}{D}\right)A_{16} + \left(\frac{D_3}{D}\right)B_{12}\right] \tag{6}$$

where

$$D_1 = \begin{vmatrix} -A_{12} & A_{26} & B_{22} \\ -A_{16} & A_{66} & B_{26} \\ -B_{12} & B_{26} & D_{22} \end{vmatrix}$$

$$D_2 = \begin{vmatrix} A_{22} & -A_{12} & B_{22} \\ A_{26} & -A_{16} & B_{26} \\ B_{22} & -B_{12} & D_{22} \end{vmatrix}$$

$$D_3 = \begin{vmatrix} A_{22} & A_{26} & -A_{12} \\ A_{26} & A_{66} & -A_{16} \\ B_{22} & B_{26} & -B_{12} \end{vmatrix}$$

and

$$D = \begin{vmatrix} A_{22} & A_{26} & B_{22} \\ A_{26} & A_{66} & B_{26} \\ B_{22} & B_{26} & D_{22} \end{vmatrix}$$

Any difference between Eqs 5 and 6 reflects the constraint imposed on the two-ply asymmetric sublaminate by the undamaged laminate and the grips (if the sublaminate extends into the grips). Because of this constraint, the sublaminate is not able to develop a κ_x or κ_{xy} curvature. However, under tension loading, a nonzero κ_y curvature may develop away from the grips in the sublaminate. For the type of sublaminates that form as a result of edge delamination [3–5], this influence of bending-extension coupling may be significant. However, for the $(0/\theta)_T$ sublaminates that constitute the locally delaminated region in the $(0/\theta/-\theta)_s$ laminates with $-\theta$ matrix cracks, the influence of bending-extension coupling is small. This small influence is shown in Fig. 12, where the normalized G of Eq 4 is calculated as a function of θ for $(0/\theta/-\theta)_s$ laminates with $-\theta$ matrix cracks using the material properties in Table 3. The normalized G values determined by calculating E_{LD} from Eq 5, assuming a symmetric $(0/\theta)_s$ sublaminate, are nearly identical to the normalized G values determined by calculating E_{LD} from Eq 6 using the loading prescribed above in the laminated plate theory analysis. However, unlike many of the sublaminates formed by edge delamination, the $(0/\theta)_T$ sublaminates that constitute the locally delaminated region in the $(0/\theta/-\theta)_s$ laminates with $-\theta$ matrix cracks will also exhibit shear-extension coupling. This shear-extension coupling may have more influence on the calculated E_{LD} than the bending extension coupling.

The influence of shear-extension coupling on the calculated E_{LD} for the two-ply asymmetric sublaminate may be determined using laminated plate theory by prescribing the loading outlined above, but setting $\gamma_{xy} = 0$ instead of prescribing $N_{xy} = 0$. The N_x that results from this analysis is derived in Appendix B and may be used to determine E_{LD} as

$$E_{LD} = \frac{2}{(N-n)h}\left[A_{11} - \left(\frac{A_{12}D_{22} - B_{12}B_{22}}{A_{22}D_{22} - B_{22}B_{22}}\right)A_{12} + \left(\frac{A_{12}B_{22} - A_{22}B_{12}}{A_{22}D_{22} - B_{22}B_{22}}\right)B_{12} \right] \quad (7)$$

This E_{LD} value reflects the shear constraint imposed on the locally delaminated region by the undamaged laminated region. The shear constraint results in a greater apparent stiff-

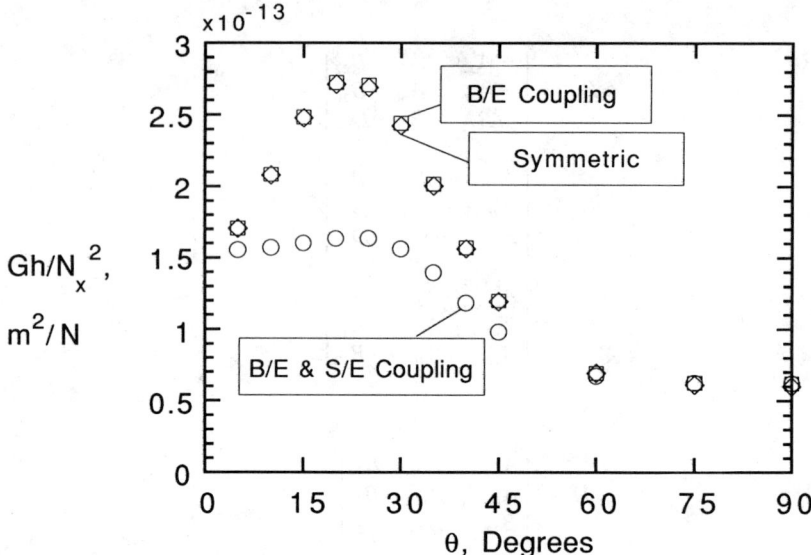

FIG. 12—*Normalized strain energy release rates for* $\theta/-\theta$ *uniform local delaminations in* $(0/\theta/-\theta)_s$ *graphite epoxy laminates as a function of* θ.

ness, E_{LD}, and hence, a correspondingly lower strain energy release rate. Figure 12 shows the normalized G of Eq 4, calculated as a function of θ for $(0/\theta/-\theta)_s$ laminates with $-\theta$ matrix cracks, where the normalized G values were determined by calculating E_{LD} using Eq 7. This loading assumes that the locally delaminated region consists of two 2-ply $(0/\theta)_T$ asymmetric sublaminates with bending-extension and shear-extension coupling. As shown in Fig. 12, the normalized G values determined by calculating E_{LD} using Eq 7 are significantly lower than the normalized G values determined by calculating E_{LD} using Eqs 5 and 6 for θ between 15 and 30°.

Uniform Local Delamination versus Edge Delamination

As noted earlier, the delaminations that form in the $\theta/-\theta$ interfaces of $(0/\theta/-\theta)_s$ laminates are localized and bounded by $-\theta°$ ply matrix cracks. Hence, the physical evidence dictates that these delaminations should be analyzed as local delaminations. However, it

TABLE 3—*AS4/3501-6 Graphite epoxy lamina properties.*

AS4/3501-6 Graphite/Epoxy
E_{11} = 135 GPa (19.5 × 10^6 psi)
E_{22} = 11 GPa (1.6 × 10^6 psi)
G_{12} = 5.8 GPa (0.847 × 10^6 psi)
ν_{12} = 0.301
α_1 = −0.41 × 10^{-6}/°C (−0.23 × 10^{-6}/°F)
α_2 = 26.8 × 10^{-6}/°C (14.9 × 10^{-6}/°F)
β_1 = 0.00/weight %
β_2 = 5.56 × 10^{-3}/weight %

may be enlightening to calculate the strain energy release rate, G, for edge delamination [4,5] in the $\theta/-\theta$ interfaces of the $(0/\theta/-\theta)_s$ laminates and compare that to the corresponding G for local delamination. The strain energy release rate, G, for edge delamination is given by

$$G = \frac{\epsilon^2 t}{2} (E_{LAM} - E^*) \tag{8}$$

where ϵ is the nominal applied strain, t is the laminate thickness, E_{LAM} is the laminate modulus, and E^* is the modulus of the laminate once it is totally delaminated, as determined by the rule of mixtures

$$E^* = \frac{\sum\limits_{i=1}^{M} E_i t_i}{t} \tag{9}$$

where M is the number of sublaminates formed by the edge delamination, and E_i and t_i are the modulus and thickness of the i sublaminates, respectively. For edge delamination in the $\theta/-\theta$ interfaces of the $(0/\theta/-\theta)_s$ laminates, Eq 9 becomes

$$E^* = \frac{2E_{(0/\theta)} + E_{(-\theta)}}{3} \tag{10}$$

The strain energy release rate calculated in Eq 8 will depend on the assumptions used to determine the $(0/\theta)$ and $(-\theta)$ sublaminate moduli in Eq 10, which determines the delaminated modulus, E^*, in Eq 9.

To compare the different possibilities, Eq 8 was rearranged to yield a normalized G as

$$\frac{G}{\epsilon^2 t} = \frac{1}{2} (E_{LAM} - E^*) \tag{11}$$

Figure 13 shows the normalized strain energy release rate, $G/\epsilon^2 t$, for edge delamination in the $\theta/-\theta$ interfaces of the $(0/\theta/-\theta)_s$ laminates as a function of θ where the sublaminate moduli were determined assuming they were either symmetric, asymmetric with bending-extension coupling only (B/E coupling), or asymmetric with shear-extension coupling in addition to bending-extension coupling (B/E and S/E coupling). Unlike the comparison for local delamination, there is a significant difference in the normalized G for edge delamination between the symmetric sublaminate case and the bending-extension coupling case. This difference results from the effect of bending-extension coupling on the modulus of the $-\theta$ sublaminate in Eq 10, which does not contribute to G for local delamination. Furthermore, when the influence of shear-extension coupling is included, the normalized G for edge delamination becomes very small for all values of θ.

In order to compare normalized G values for both local and edge delamination, Hooke's law in the form $N_x = Nh\epsilon E_{LAM}$, where $N = 6$, was substituted into Eq 4 to yield

$$\frac{G}{\epsilon^2 t} = \frac{3E_{LAM}^2}{2} \left(\frac{1}{4E_{LD}} - \frac{1}{6E_{LAM}} \right) \tag{12}$$

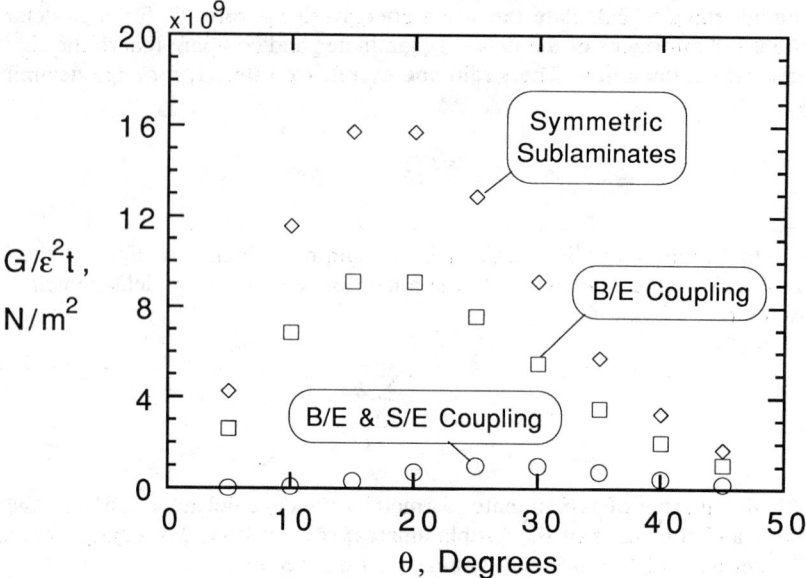

FIG. 13—*Normalized strain energy release rates for* $\theta/-\theta$ *edge delaminations in* $(0/\theta/-\theta)_s$ *graphite epoxy laminates as a function of* θ.

Figure 14 shows G normalized by $\epsilon^2 t$ for edge delamination and uniform local delamination, as calculated by Eqs 11 and 12, respectively, as a function of θ for delamination in the $\theta/-\theta$ interfaces of the $(0/\theta/-\theta)_s$ laminates. The sublaminate moduli were determined assuming they were asymmetric, with shear-extension coupling in addition to bending-extension coupling. Figure 14 indicates that, for the same applied strain, the strain energy release rate associated with a uniform local delamination from a matrix crack exceeds the strain energy release rate associated with edge delamination for all values of θ.

Partial Local Delamination

The local delamination growing from an angle ply matrix crack modeled in Fig. 11*b* has a uniform delamination front across the laminate width. However, the local delaminations observed in the experiments have triangular-shaped delamination areas that extend only partially into the laminate width from the free edge [*1*]. Therefore, an analysis was developed to account for the fact that the local delaminations in the $(0/\theta/-\theta)_s$ laminates did not extend across the entire laminate width, but instead were limited to the extent of the matrix crack in the $-\theta°$ ply. The model was based on the assumption that the regions near the free edges containing the partial local delaminations act as a system of sublaminates in series, just as was assumed for the uniform local delamination, but this series system was assumed to be in parallel with the undamaged laminate in the interior of the laminate width (Fig. 11*c*).

Figure 15 shows three possible ways of visualizing the delamination growing from the matrix crack in the $-\theta°$ ply. Figure 15*a* shows a triangular-shaped delamination that intersects the free edge at 90°. Figure 15*b* shows a triangular-shaped delamination that intersects the matrix crack at 90°. Finally, Fig. 15*c* shows a triangular-shaped delamination

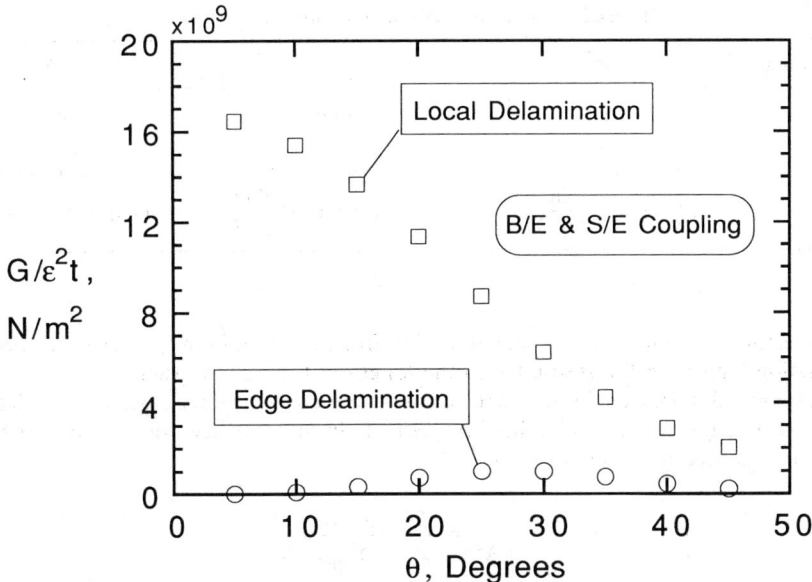

FIG. 14—*Comparison of normalized strain energy release rates for* θ/−θ *edge and uniform local delaminations in* (0/θ/−θ)$_s$ *graphite epoxy laminates as a function of* θ.

that intersects the free edge at an angle of 2θ. Eventually, if the local delamination in the θ/−θ interface reaches a +θ° matrix crack and arrests, the local delamination will appear to be an isosceles triangle with +θ and −θ angles intersecting the straight free edge. The relationship between the matrix crack length, d, and the delamination length along the free edge, a, is different for the three cases shown in Fig. 15. These relationships are given in

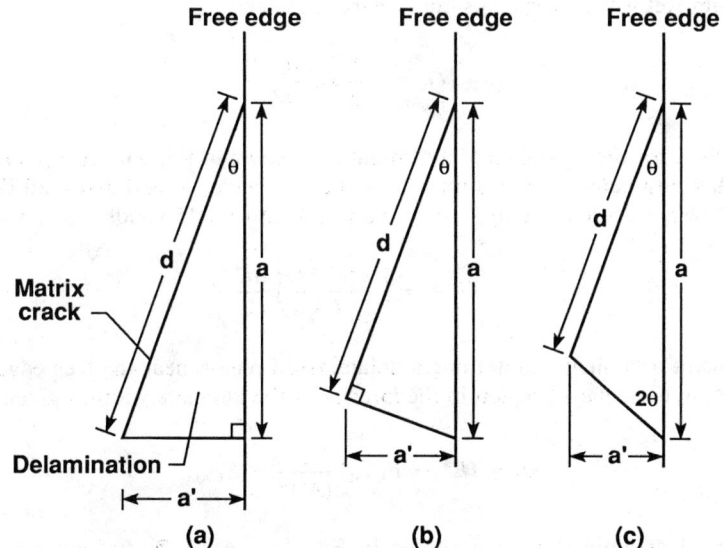

FIG. 15—*Partial local delamination geometries.*

TABLE 4—*Partial local delamination geometries.*

Configuration	d	a'	k
Fig. 15a	$\dfrac{a}{\cos\theta}$	$a\tan\theta$	$\tan\theta$
Fig. 15b	$a\cos\theta$	$a\cos\theta\sin\theta$	$\cos\theta\sin\theta$
Fig. 15c	$a\left(\cos\theta - \dfrac{\sin\theta}{\tan 3\theta}\right)$	$a\sin\theta\left(\cos\theta - \dfrac{\sin\theta}{\tan 3\theta}\right)$	$\sin\theta\left(\cos\theta - \dfrac{\sin\theta}{\tan 3\theta}\right)$

Table 4. Hence, the shape of the delamination front must be known in order to relate the delamination length on the free edge to the length of the matrix crack.

For a local delamination from a matrix crack, the compliance, S, where $S = 1/E$, for a laminate with a given local delamination area, A, in an interface whose total area is A^*, may be represented as (Appendix A)

$$\frac{A}{A^*} = \frac{S - S_{\text{LAM}}}{S^* - S_{\text{LAM}}} \tag{13}$$

where $S^* = 1/E^*_{\text{LD}}$, and $E^*_{\text{LD}} = (t_{\text{LD}}E_{\text{LD}})/t$ as defined in Ref 2. Assuming a partial local delamination is present at a matrix crack on both free edges (Fig. 11c), and noting that $a' = ka$, where k is given in Table 4 for the three geometries shown in Figs. 15a through 15c, then $A = maa' = mka^2$ and $A^* = 2mLka$. Substituting for A and A^* in Eq 13 for any of these three geometries yields

$$S = \frac{1}{2}\left(\frac{1}{t_{\text{LD}}E_{\text{LD}}} - \frac{1}{tE_{\text{LAM}}}\right)\left(\frac{t}{L}\right)a + \frac{1}{E_{\text{LAM}}} \tag{14}$$

recalling from Ref 4 that, for a system in parallel,

$$G = -\frac{1}{2}V\epsilon^2\frac{dE}{dA} \tag{15}$$

and noting that, for the partial local delaminations shown in Fig. 11c, $A = mka^2$ and $dA = 2mkada$. The parameter, k, is a function of the ply angle, θ, and assumed delamination geometry (Table 4). Substituting $V = wLt$ and dA into Eq 15 yields

$$G = -\frac{1}{2}\frac{wt\epsilon^2}{2mk}\left(\frac{L}{a}\right)\frac{dE}{da} \tag{16}$$

The modulus of a laminate containing a delaminated region near the free edges loaded in parallel with an undamaged region in the interior of the laminate width is given by Ref 4 as

$$E = (E^* - E_{\text{LAM}})\frac{A'}{(A')^*} + E_{\text{LAM}} \tag{17}$$

For the partial delamination case shown in Fig. 11c, $A' = 2mLa'$ and $(A')^* = mwL$. Substituting these expressions in Eq 17, and noting that $a' = ka$, where k is given in Table

4 for the three geometries in Figs. 15a to 15c, yields

$$E = (E^* - E_{\mathrm{LAM}}) \frac{2k}{w} a + E_{\mathrm{LAM}} \tag{18}$$

Unlike E^* calculated for edge delamination in Eq 9, E^* in Eq 18 is a function of the local delamination length along the free edge, a. Differentiating Eq 18 with respect to a, and substituting into Eq 16 yields

$$G = \frac{\epsilon^2 t}{2m} \left(\frac{L}{a} \right) \left[E_{\mathrm{LAM}} - E^* - a \frac{dE^*}{da} \right] \tag{19}$$

Substituting $E^* = 1/S$, where S is given by Eq 14, into Eq 19 and differentiating yields

$$G = \frac{\epsilon^2 t}{2m} \left(\frac{L}{a} \right) \left[E_{\mathrm{LAM}} - \frac{1}{(Ba + C)} + \frac{Ba}{(Ba + C)^2} \right] \tag{20}$$

where

$$B = \frac{1}{2} \left(\frac{1}{t_{\mathrm{LD}} E_{\mathrm{LD}}} - \frac{1}{t E_{\mathrm{LAM}}} \right) \left(\frac{t}{L} \right) \tag{21}$$

and

$$C = \frac{1}{E_{\mathrm{LAM}}} \tag{22}$$

Although the relationship between the matrix crack length, d, and the delamination length along the free edge, a, is different for the three cases shown in Fig. 15, Eq 20 is independent of k, and hence, $G(a/L)$ is independent of the local delamination geometry assumed. However, as noted previously for Eq 2, G in Eq 20 will depend on the assumptions made when determining E_{LD} in Eq 21. Therefore, B in Eq 21, and hence G in Eq 20, will depend on whether Eqs 5, 6, or 7 are used to calculate E_{LD}. For the remainder of this paper, it will be assumed that the sublaminates that constitute the locally delaminated region experience both bending-extension and shear-extension coupling. Hence, Eq 7 will be used to calculate E_{LD}.

Equation 20 may be rearranged to yield

$$G = \frac{\epsilon^2 t}{2m} \left(\frac{D}{C \left[D \left(\frac{a}{L} \right) + C \right]} + \frac{D}{\left[D \left(\frac{a}{L} \right) + C \right]^2} \right) \tag{23}$$

where $D = BL$. When $a/L = 0$, then Eq 20 becomes

$$G = \frac{\epsilon^2 t}{2m} \left(\frac{2D}{C^2} \right) \tag{24}$$

Substituting for D and C, normalizing by $\epsilon^2 t$, and noting that $t_{LD} = 4h$ and $t = 6h$, Eq 24 becomes

$$\frac{G}{\epsilon^2 t} = \frac{3E_{LAM}^2}{2} \left(\frac{1}{4E_{LD}} - \frac{1}{6E_{LAM}} \right) \qquad (25)$$

which is identical to Eq 12 for the uniform local delamination.

Figures 16 through 19 compare normalized G values calculated using Eq 23 for partial local delamination, as a function of delamination length along the free edge normalized by the laminate length, a/L, to normalized G values calculated from Eq 12 for uniform local delamination and from Eq 11 for edge delamination modeled in the $\theta/-\theta$ interfaces of $(0/\theta/-\theta)_s$ laminates, where $\theta = 15, 20, 25,$ and $30°$. The uniform local delamination and edge delamination solutions are independent of delamination length, but for the partial local delamination solution, G decreases with increasing delamination length along the free edge. Both local delamination solutions yield greater normalized G values than the edge delamination solution for all delamination lengths.

Figure 20 shows G normalized by $\epsilon^2 t$, as a function of a/L up to $a/L = 0.1$, calculated using Eq 20 for partial load delaminations in the $\theta/-\theta$ interfaces of $(0/\theta/-\theta)_s$ laminates, where $\theta = 15, 20, 25,$ and $30°$. Normalized G values were only plotted for a/L values up to 0.1 because the magnitudes for small matrix crack lengths, and hence small delamination lengths, were of primary interest for comparing to experimental data. In each case, G decreases slightly with delamination length, unlike the solutions for uniform local delamination which were independent of delamination length. If the partial delamination geometry can be anticipated and linked to the matrix crack length as shown in Figs. 15a through 15c, then both the laminate toughness as a function of fatigue cycles and a critical matrix crack length would be needed to predict the onset of a partial local delamination in

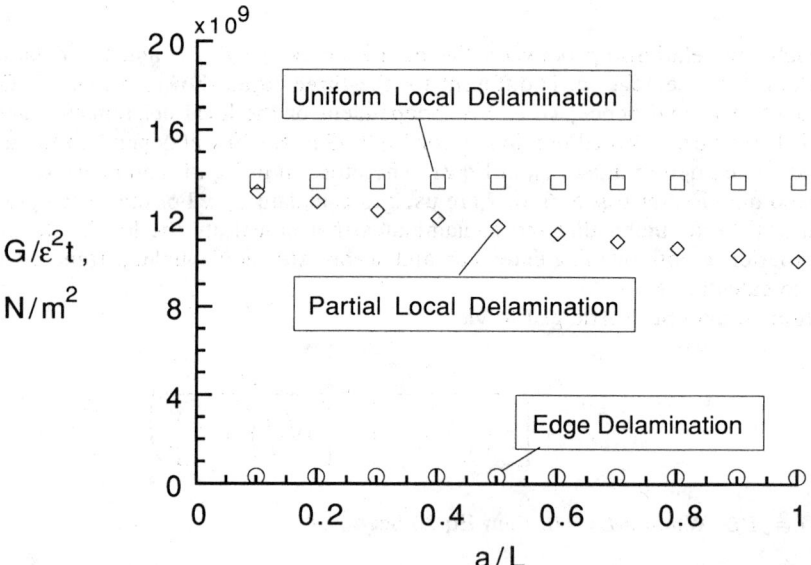

FIG. 16—*Normalized* G *for local delamination and edge delamination as a function of normalized delamination length at the free edge in a (0/15/−15)$_s$ graphite epoxy laminate.*

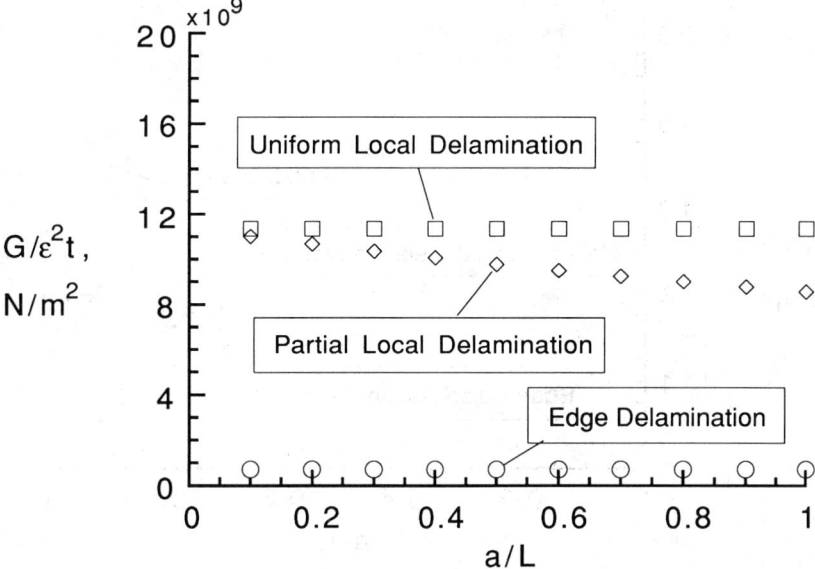

FIG. 17—*Normalized* G *for local delamination and edge delamination as a function of normalized delamination length at the free edge in a (0/20/ − 20)$_s$ graphite epoxy laminate.*

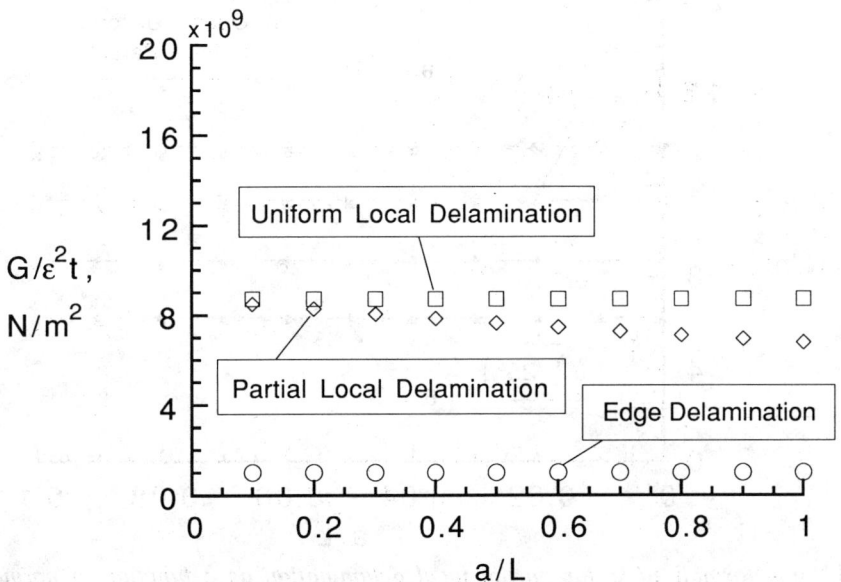

FIG. 18—*Normalized* G *for local delamination and edge delamination as a function of normalized delamination length at the free edge in a (0/25/ − 25)$_s$ graphite epoxy laminate.*

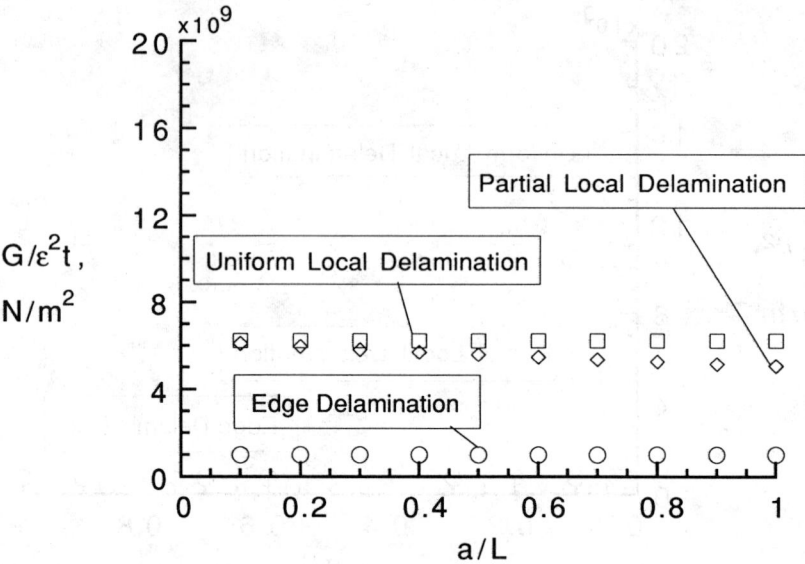

FIG. 19—*Normalized G for local delamination and edge delamination as a function of normalized delamination length at the free edge in a (0/30/−30)ₛ graphite epoxy laminate.*

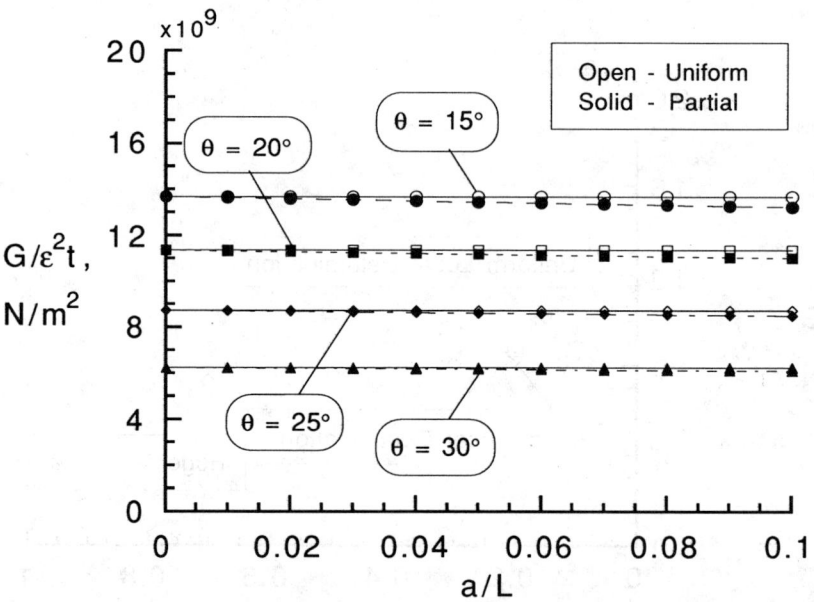

FIG. 20—*Normalized G for partial local delamination as a function of normalized delamination length at the free edge in (0/θ/−θ)ₛ graphite epoxy laminates.*

fatigue. However, as shown in Fig. 20, if the critical matrix crack length is sufficiently small, and hence, the corresponding delamination length along the free edge is small, the difference between the uniform and partial delamination solutions will be insignificant. In this case, G may be calculated from Eq 2 to compare to experimental data, even though the local delamination is not uniform across the laminate width.

Thermal/Moisture Contribution to G

The strain energy release rate for local delamination will be influenced by the presence of residual thermal and moisture stresses in the laminate. A uniform local delamination G solution that includes the effect of the mechanical load, the residual thermal stresses that exist in the laminate after it cools down from the curing temperature, and the hygroscopic stresses that exist in the laminate due to the presence of absorbed moisture, was recently developed [6]. Figures 21 through 24 show the contribution of mechanical load, M, temperature, T, and hygroscopic, H, moisture weight gain to G for a uniform local delamination in the $\theta/-\theta$ interfaces of $(0/\theta/-\theta)_s$ laminates (where $\theta = 15$, 20, 25, and 30°, respectively) subjected to a mechanical load of 3000 lb/in. and a ΔT of $-156°C$ ($-280°F$). As shown in Figs. 21 through 24, when the contribution of the residual thermal stress is included, G^{M+T} will increase above the G^M value obtained from Eq 2 for an applied mechanical load alone. However, when the presence of absorbed moisture is included, as may occur from exposure to the ambient air (Appendix C), the thermal stresses will relax and reduce the strain energy release rate, G^{M+T+H}, with increasing percentage moisture weight gain, ΔH.

Similar behavior was observed for edge delamination [7]. However, the magnitude of the thermal and moisture effects is much more significant for edge delamination. The

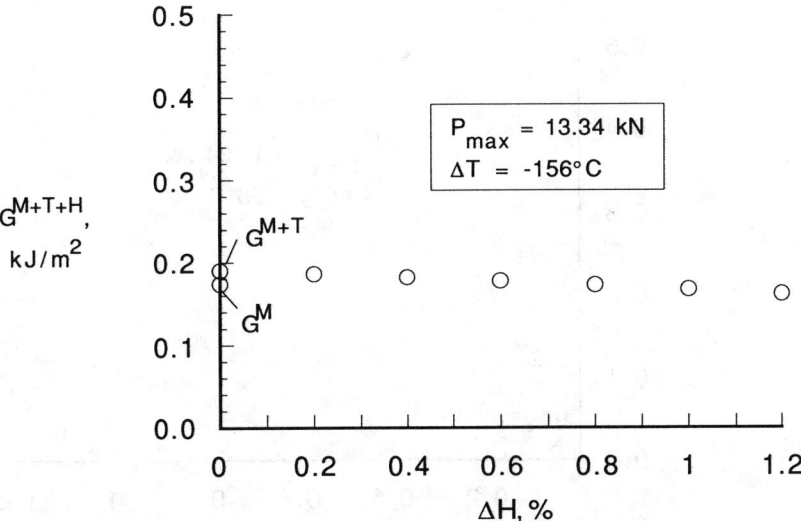

FIG. 21—*Influence of residual thermal and moisture stresses on the strain energy release rate for local delamination from* $-15°$ *ply cracks in* $(0/15/-15)_s$ *graphite epoxy laminates.*

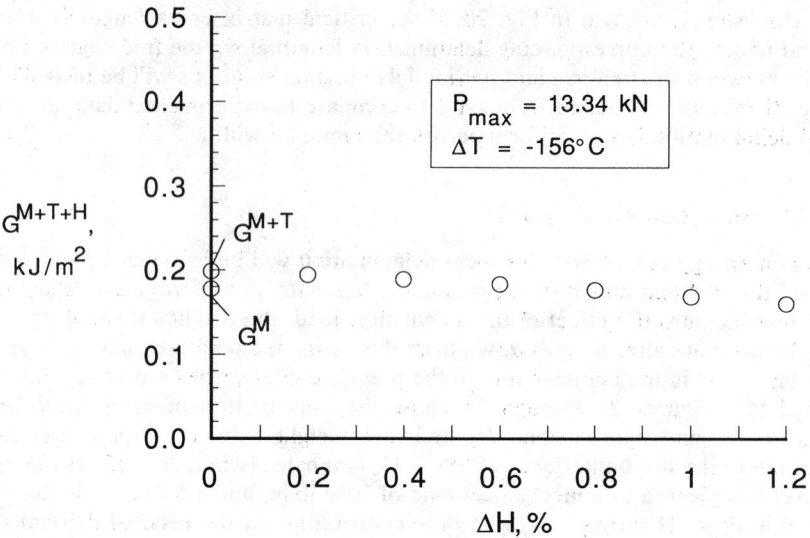

FIG. 22—*Influence of residual thermal and moisture stresses on the strain energy release rate for local delamination from* $-20°$ *ply cracks in* $(0/20/-20)_s$ *graphite epoxy laminates.*

laminates tested in this study had a ΔH between 0.2 and 0.3% (Appendix C). For the $(0/\theta/-\theta)_s$ laminates shown in Figs. 21 through 24, this corresponds to a 5.8 to 7.7% increase in G^{M+T+H} compared to the G^M solution for mechanical loading only. The difference in G^{M+T+H} and G^M is even less at higher mechanical load levels (Table 5).

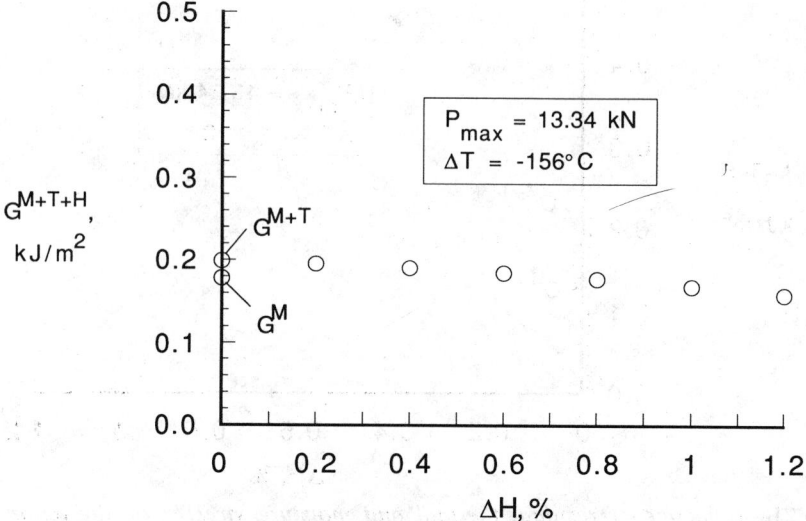

FIG. 23—*Influence of residual thermal and moisture stresses on the strain energy release rate for local delamination from* $-25°$ *ply cracks in* $(0/25/-25)_s$ *graphite epoxy laminates.*

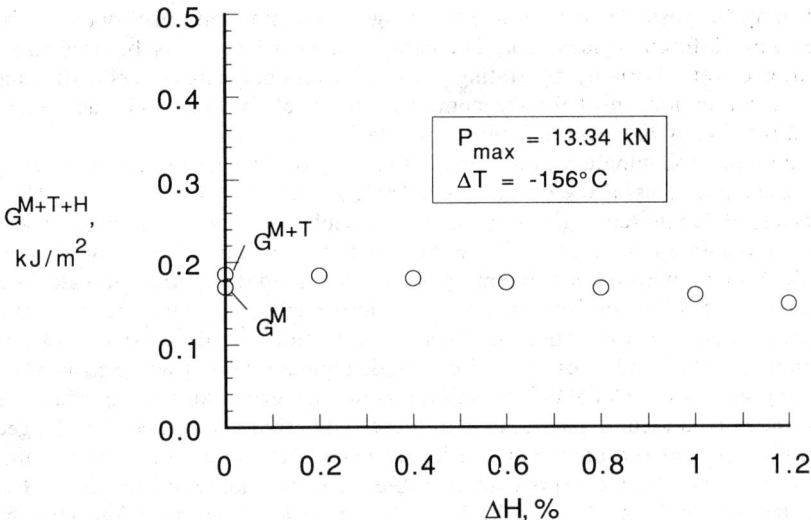

FIG. 24—*Influence of residual thermal and moisture stresses on the strain energy release rate for local delamination from* $-30°$ *ply cracks in* $(0/30/-30)_s$ *graphite epoxy laminates.*

Correlation of Analysis and Experiments

Ideally, the closed form solutions for local delaminations from matrix cracks should be verified and calibrated by comparing them to detailed finite element solutions that sum the G components for the three fracture modes to obtain a total G. Such comparisons have been made for edge delamination [3–5]. However, to date only the case of uniform local delamination from 90° matrix cracks has been verified [8]. Part of the difficulty is that the modeling for local delamination requires a full 3D analysis, with the matrix crack geometry modeled accurately. The solutions for non-90° matrix cracks that have been attempted to date [9,10], have assumed that the angle ply matrix cracks may be represented as straight cracks through the laminate thickness that are normal to the $\theta_2/-\theta_2$ interface. However, as seen in micrographs of the edges for the $(0_2/\theta_2/-\theta_2)_s$ laminates (Fig. 2), these angle ply matrix cracks are curved, intersecting the ply interface at oblique angles. This difference in actual versus assumed matrix crack geometry through the laminate thickness may have a significant effect on the G components calculated from a finite element analysis for total delamination.

Although the fatigue behavior of the $(0_2/\theta_2/-\theta_2)_s$ laminates cannot currently be pre-

TABLE 5—*Percentage difference between* G^M *and* G^{M+T+H} *for* $\Delta H = 0.3\%$.

P_{max}, N (lb)	$\theta°$			
	15, %	20, %	25, %	30, %
13 340 (3000)	5.8	7.5	7.7	7.0
15 570 (3500)	5.0	6.6	6.8	6.1
17 790 (4000)	4.3	5.8	6.2	5.8
22 240 (5000)	3.7	4.9	5.1	4.9

dicted, it may be possible to correlate the fatigue delamination onset behavior between laminates with different values of θ. The data from each layup may be used to generate delamination onset criteria by generating plots of maximum cyclic G versus the number of cycles to delamination onset for the composite material. A similar characterization has been used previously for edge delamination data [5,11].

Laminate and sublaminate moduli for the $(0_2/\theta_2/-\theta_2)_s$ AS4/3501-6 graphite epoxy laminates were calculated using the properties in Table 3 in laminated plate theory. The moduli of the locally delaminated regions, E_{LD}, were calculated using Eq 7 to incorporate the influence of bending-extension and shear-extension coupling. These moduli were then used in Eq 2, along with the maximum cyclic loads in the fatigue tests, to calculate strain energy release rates for uniform local delaminations corresponding to the measured number of cycles to delamination onset for each layup. In addition, the matrix crack length, d, corresponding to the number of cycles for local delamination onset was measured from the X-ray radiographs where local delamination was first observed for each specimen. Examination of the partial local delaminations in these radiographs indicated that the geometry shown in Fig. 15c was reasonable for the layups tested. Therefore, measured matrix crack lengths, d, were used in the expression in Table 4 corresponding to Fig. 15c to determine critical delamination lengths along the free edge, a, at local delamination onset for each layup.

Table 6 shows the range of delamination lengths along the free edge, a, normalized by the gage length, L, measured just after local delamination onset in fatigue. These critical delamination lengths were used, along with the maximum cyclic loads in the fatigue tests, to calculate strain energy release rates from Eq 20 for partial local delaminations corresponding to the measured number of cycles to delamination onset for each layup. Hence, plots of G_{max} versus the number of cycles to delamination onset, N, were generated for each layup assuming either uniform, or partial, local delamination onset from a $-\theta°$ matrix ply crack.

Figure 25 shows the G_{max} versus N plot for both the uniform and partial local delamination models. The maximum cyclic G levels for the uniform and partial local delamination models are very similar because the experimentally determined critical delamination lengths were small (Table 6; Fig. 20). In addition, for any particular maximum cyclic load, the maximum cyclic G levels for all values of θ also agree fairly well, indicating that the assumptions of shear-extension and bending-extension coupling reflected in Eq 7 were accurate (Fig. 12). Finally, the number of cycles to delamination onset at each cyclic G level appear to be greater for the 15 and 20° cases than for the 25 and 30° cases. These differences in the number of cycles to delamination onset may be due to a difference in either the mixed-mode ratio, or the extent of matrix cracking prior to delamination onset, for the individual cases. Neither of these effects are incorporated in the current analyses.

Figure 26 shows the G_{max} versus N plot for the uniform local delamination models with

TABLE 6—*Range of critical normalized delamination lengths at the free edge of $(0_2/\theta_2/-\theta_2)_s$ graphite epoxy laminates measured after local delamination onset in fatigue.*

θ°	a/L
15	0.034–0.106
20	0.036–0.090
25	0.029–0.063
30	0.019–0.062

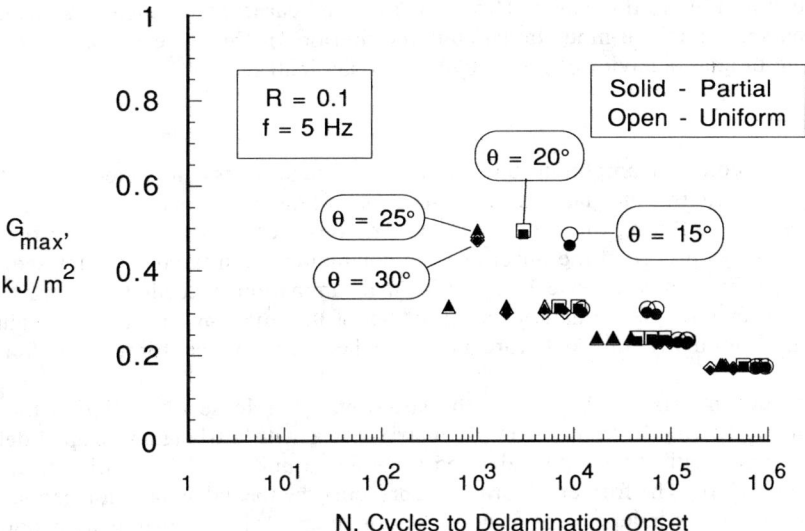

FIG. 25—*Maximum cyclic G as a function of cycles to local delamination onset for* $(0_2/\theta_2/-\theta_2)_s$ *graphite epoxy laminates.*

mechanical loading only, G^M, and with mechanical, thermal, and hygroscopic loading combined, G^{M+T+H}. For any particular maximum cyclic load, the maximum cyclic G^{M+T+H} is only slightly greater than the maximum cyclic G^M because of the relatively small contribution of the residual thermal and moisture stresses to G (Table 5).

The G_{max} versus N plots in Figs. 25 and 26 provide some insight into the accuracy of the

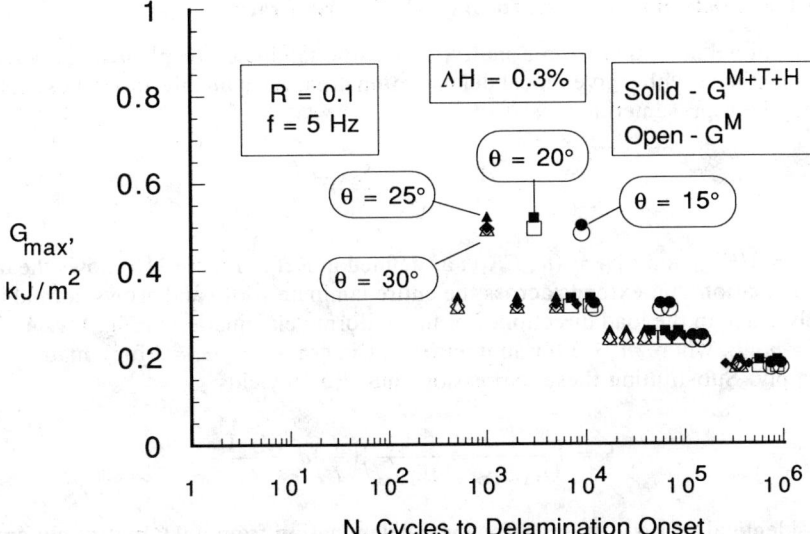

FIG. 26—*Maximum cyclic G^M and G^{M+T+H} as a function of cycles to local delamination onset for* $(0_2/\theta_2/-\theta_2)_s$ *graphite epoxy laminates.*

G solutions derived in this paper. However, as noted earlier, a detailed analysis of the G components and a mixed-mode fatigue failure criterion for this material may be needed to predict the fatigue behavior of these $(0_2/\theta_2/-\theta_2)_s$ laminates.

Conclusions

1. For the constant amplitude tension-tension fatigue tests conducted on AS4/3501-6 graphite epoxy $(0_2/\theta_2/-\theta_2)_s$ laminates the sequence of damage was evident in the series of radiographs taken during the cyclic loading. The onset of matrix cracking in the central $-\theta°$ plies always preceded the onset of local delaminations in the $\theta/-\theta$ interface.

2. The strain energy release rate for local delamination exceeded the strain energy release rate for edge delamination for all values of θ, indicating that local delaminations will initiate from matrix cracks before an edge delamination would form in the $(0_2/\theta_2/-\theta_2)_s$ laminates.

3. For small matrix crack lengths, the strain energy release rate solution for a local delamination growing from an angle ply matrix crack with a triangular-shaped delamination area agrees with the solution derived assuming a uniform delamination front across the laminate width. The former accurately represents the local delamination geometry, but the latter solution is independent of delamination length. The uniform delamination solution may be used to characterize local delamination onset.

4. The influence of residual thermal and moisture stresses on G for local delamination from matrix cracks in the $(0_2/\theta_2/-\theta_2)_s$ AS4/3501-6 laminates is relatively small and may be neglected.

APPENDIX A

Uniform Local Delamination from an Angle Ply Matrix Crack

For a local delamination from a matrix crack, the laminate compliance, S, where $S = 1/E$, for a laminate with a given local delamination area, A, in an interface whose total area is A^*, may be represented as

$$\frac{A}{A^*} = \frac{S - S_{LAM}}{S^* - S_{LAM}} \tag{26}$$

where $S^* = 1/E_{LD}^*$, and $E_{LD}^* = (t_{LD}E_{LD})/t$ as defined in Ref 2. Figure 11a shows the uniform local delamination that extends across the entire laminate width and grows normal to a 90° matrix ply crack in the load direction. For the uniform delamination in Fig. 11a, $A = mwa$, and $A^* = mwL$, where $m = 2$ for an interior matrix crack and $m = 1$ for a matrix crack in a surface ply. Substituting these expressions into Eq 26 yields

$$S = \left(\frac{1}{t_{LD}E_{LD}} - \frac{1}{tE_{LAM}}\right)\left(\frac{t}{L}\right)a + \frac{1}{E_{LAM}} \tag{27}$$

which is identical to Eq 7 in Ref 2 for local delamination from a 90° matrix ply crack.

Figure 11b shows the uniform local delamination that extends across the entire laminate width and grows from a $-\theta°$ matrix ply crack normal to the load direction. For the uniform delamination in Fig. 11b, $A = (mw/\sin \theta) (a \sin \theta) = mwa$, and $A^* = mwL$. Substituting these expressions into Eq 26 also yields Eq 27.

For the 90° matrix crack–induced local delamination in Fig. 11a, $dA = mwda$. Similarly, for the $-\theta°$ matrix crack–induced local delamination in Fig. 11b, $dA = (mw/\sin \theta) d(a \sin \theta) = mwda$. In Ref 2, G is defined as

$$G = \frac{1}{2} V\sigma^2 \frac{dS}{dA} \tag{28}$$

Substituting for $V = wLt$ and $dA = mwda$ and differentiating Eq 27 with respect to a yields

$$G = \frac{P^2}{2mw^2} \left(\frac{1}{t_{LD}E_{LD}} - \frac{1}{t_{LAM}E_{LAM}} \right) \tag{29}$$

for uniform local delaminations from either a 90° or a $-\theta°$ matrix crack.

APPENDIX B

Analysis of Moduli for Asymmetric Sublaminates

In laminated plate theory, the load and moment resultants are related to the midplane strains and curvatures as

$$\left(\frac{N}{M} \right) = \begin{bmatrix} A & B \\ B & D \end{bmatrix} \left(\frac{\epsilon}{\kappa} \right) \tag{30}$$

where A, D, and B represent the extensional, bending, and bending-extension coupling matrices, respectively. By prescribing an applied mechanical strain, $\epsilon_x = \epsilon_0$, setting the applied N_y, N_{xy}, and M_y equal to 0, and setting κ_x and κ_{xy} equal to 0, Eq 30 becomes

$$\begin{Bmatrix} N_x \\ 0 \\ 0 \\ M_x \\ 0 \\ M_{xy} \end{Bmatrix} = \begin{bmatrix} A & B \\ B & D \end{bmatrix} \begin{Bmatrix} \epsilon_0 \\ \epsilon_y \\ \gamma_{xy} \\ 0 \\ \kappa_y \\ 0 \end{Bmatrix} \tag{31}$$

The three equations whose left hand side are zero may be written as

$$-A_{12}\epsilon_0 = A_{22}\epsilon_y + A_{26}\gamma_{xy} + B_{22}\kappa_y$$

$$-A_{16}\epsilon_0 = A_{26}\epsilon_y + A_{66}\gamma_{xy} + B_{26}\kappa_y \tag{32}$$

$$-B_{12}\epsilon_0 = B_{22}\epsilon_y + B_{26}\gamma_{xy} + D_{22}\kappa_y$$

where $A_{ij} = A_{ji}$. This set of three nonhomogeneous equations may be solved for ϵ_y, γ_{xy}, and κ_{xy} using Cramer's rule. Hence,

$$\epsilon_y = \frac{D_1}{D}\epsilon_0, \; \gamma_{xy} = \frac{D_2}{D}\epsilon_0, \; \kappa_y = \frac{D_3}{D}\epsilon_0 \tag{33}$$

where

$$D_1 = \begin{vmatrix} -A_{12} & A_{26} & B_{22} \\ -A_{16} & A_{66} & B_{26} \\ -B_{12} & B_{26} & D_{22} \end{vmatrix}$$

$$D_2 = \begin{vmatrix} A_{22} & -A_{12} & B_{22} \\ A_{26} & -A_{16} & B_{26} \\ B_{22} & -B_{12} & D_{22} \end{vmatrix}$$

$$D_3 = \begin{vmatrix} A_{22} & A_{26} & -A_{12} \\ A_{26} & A_{66} & -A_{16} \\ B_{22} & B_{26} & -B_{12} \end{vmatrix}$$

and

$$D = \begin{vmatrix} A_{22} & A_{26} & B_{22} \\ A_{26} & A_{66} & B_{26} \\ B_{22} & B_{26} & D_{22} \end{vmatrix}$$

These expressions for ϵ_y, γ_{xy}, and κ_{xy} may be substituted into the first equation of Eq 31 to yield

$$N_x = \left[A_{11} + \left(\frac{D_1}{D}\right)A_{12} + \left(\frac{D_2}{D}\right)A_{16} + \left(\frac{D_3}{D}\right)B_{12} \right]\epsilon_0 \tag{34}$$

Hence, N_x may be obtained explicitly as a function of the terms in the A, B, and D matrices and the prescribed axial strain, ϵ_0. Then noting that

$$E_{LD} = \frac{2N_x}{(N - n)h\epsilon_0} \tag{35}$$

yields

$$E_{LD} = \frac{2}{(N - n)h}\left[A_{11} + \left(\frac{D_1}{D}\right)A_{12} + \left(\frac{D_2}{D}\right)A_{16} + \left(\frac{D_3}{D}\right)B_{12} \right] \tag{36}$$

By prescribing the loading outlined above, but setting $\gamma_{xy} = 0$ instead of prescribing that $N_{xy} = 0$, i.e., by prescribing an applied mechanical strain, $\epsilon_x = \epsilon_0$, setting the applied N_y

and M_y equal to 0, and setting γ_{xy}, κ_x, and κ_{xy} equal to 0, Eq 30 becomes

$$\begin{Bmatrix} N_x \\ 0 \\ N_{xy} \\ M_x \\ 0 \\ M_{xy} \end{Bmatrix} = \begin{bmatrix} A & B \\ B & D \end{bmatrix} \begin{Bmatrix} \epsilon_0 \\ \epsilon_y \\ 0 \\ 0 \\ \kappa_y \\ 0 \end{Bmatrix} \tag{37}$$

The two equations whose left hand side are zero are

$$0 = A_{12}\epsilon_0 + A_{22}\epsilon_y + B_{22}\kappa_y \tag{38}$$

$$0 = B_{12}\epsilon_0 + B_{22}\epsilon_y + D_{22}\kappa_y \tag{39}$$

where $A_{ij} = A_{ji}$ and $B_{ij} = B_{ji}$. Multiplying Eq 38 by $-D_{22}$, multiplying Eq 39 by B_{22}, and then adding the two equations yields

$$\epsilon_y = \left(\frac{B_{12}B_{22} - A_{12}D_{22}}{A_{22}D_{22} - B_{22}B_{22}} \right) \epsilon_0 \tag{40}$$

Furthermore, multiplying Eq 38 by $-B_{22}$, multiplying Eq 39 by A_{22}, and adding the two equations yields

$$\kappa_y = \left(\frac{A_{12}B_{22} - A_{22}B_{12}}{A_{22}D_{22} - B_{22}B_{22}} \right) \epsilon_0 \tag{41}$$

Finally, substituting Eqs 40 and 41 into the first equation of Eq 37 yields

$$N_x = \left[A_{11} - \left(\frac{A_{12}D_{22} - B_{12}B_{22}}{A_{22}D_{22} - B_{22}B_{22}} \right) A_{12} + \left(\frac{A_{12}B_{22} - A_{22}B_{12}}{A_{22}D_{22} - B_{22}B_{22}} \right) B_{12} \right] \epsilon_0 \tag{42}$$

Hence, N_x may be obtained explicitly as a function of the terms in the A, B, and D matrices and the prescribed axial strain, ϵ_0. Then recalling Eq 35:

$$E_{LD} = \frac{2N_x}{(N - n)h\epsilon_0}$$

and substituting the expression for N_x from Eq 42 yields

$$E_{LD} = \frac{2}{(N - n)h} \left[A_{11} - \left(\frac{A_{12}D_{22} - B_{12}B_{22}}{A_{22}D_{22} - B_{22}B_{22}} \right) A_{12} + \left(\frac{A_{12}B_{22} - A_{22}B_{12}}{A_{22}D_{22} - B_{22}B_{22}} \right) B_{12} \right] \tag{43}$$

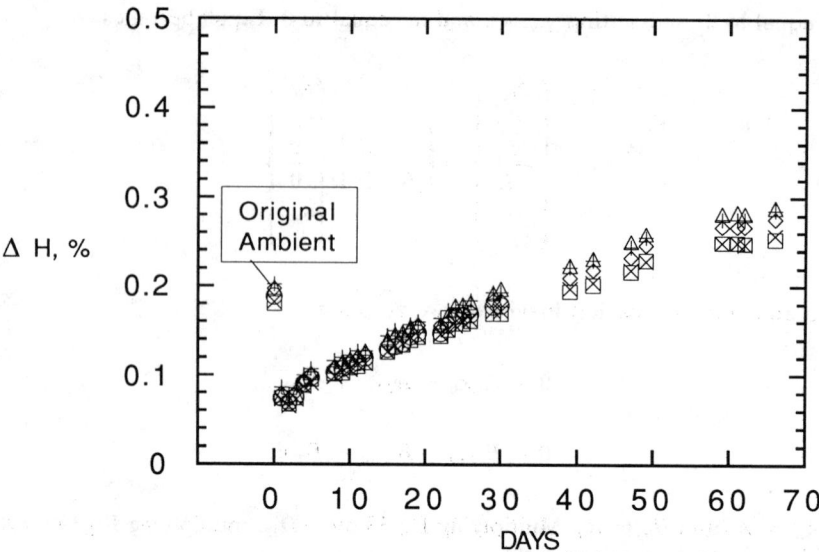

FIG. 27—*Percentage moisture weight gain as a function of the number of days since drying.*

APPENDIX C

Moisture Content Determination

Just prior to testing the $(0_2/\theta_2/-\theta_2)_s$ laminates, two control coupons of each layup were weighed, dried, and then weighed again to determine ΔH for the laminates in the ambient environment. These dried control specimens were then weighed periodically to identify their moisture content by weight as a function of time.

Figure 27 shows the percentage moisture content by weight, ΔH, as a function of the number of days from the time the specimens were dried. There was very little difference in the percentage moisture weight gain for the different layups. The percentage difference in the original weight measured just before drying and the weight measured immediately after the specimens were dried was approximately 0.2% for all the layups. This value is shown in Fig. 27 as the original ambient condition. Because all of the fatigue data were generated within a period of 30 days, the percentage moisture weight gain for all the laminates that were tested in fatigue was assumed to be between 0.2 and 0.3%.

References

[1] O'Brien, T. K. and Hooper, S. J., "Local Delamination in Laminates with Angle Ply Matrix Cracks: Part I, Tension Tests and Stress Analysis," this publication.

[2] O'Brien, T. K., "Analysis of Local Delaminations and Their Influence on Composite Laminate Behavior," *Delamination and Debonding of Materials, ASTM STP 876*, American Society for Testing and Materials, Philadelphia, 1985, pp. 282–297.

[3] O'Brien, T. K., Johnston, N. J., Raju, I. S., and Morris, D. H., "Comparisons of Various Configurations of the Edge Delamination Test for Interlaminar Fracture Toughness," *Toughened*

Composites, ASTM STP 937, American Society for Testing and Materials, Philadelphia, 1987, pp. 199–221.

[4] O'Brien, T. K., "Characterization of Delamination Onset and Growth in a Composite Laminate," *Damage in Composite Materials, ASTM STP 775,* American Society for Testing and Materials, Philadelphia, 1982, pp. 140–167.

[5] O'Brien, T. K., "Mixed-mode Strain Energy Release Rate Effects on Edge Delamination of Composites," *Effects of Defects in Composite Materials, ASTM STP 836,* American Society for Testing and Materials, Philadelphia, 1984, pp. 125–142.

[6] O'Brien, T. K., "Residual Thermal and Moisture Influences on the Strain Energy Release Rate Analysis of Local Delaminations from Matrix Cracks," NASA TM 104077, 1991, Submitted to the ASTM *Journal of Composites Technology and Research,* Vol. 14, No. 2, Summer 1992, pp. 86–94.

[7] O'Brien, T. K., Raju, I. S., and Garber, D. P., "Residual Thermal and Moisture Influences on the Strain Energy Release Rate Analysis of Edge Delamination," *Journal of Composites Technology and Research,* Vol. 8, No. 2, Summer, 1986.

[8] Salpekar, S. A. and O'Brien, T. K., "Combined Effect of Matrix Cracking and Free Edge on Delamination," *Composite Materials: Fatigue and Fracture, 3rd Volume, ASTM STP 1110,* American Society for Testing and Materials, Philadelphia, 1991, pp. 287–311.

[9] Fish, J. C. and O'Brien, T. K., "Three-Dimensional Finite Element Analysis of Delamination from Matrix Cracks in Glass-Epoxy Laminates," *Composite Materials: Testing and Design, 10th Volume, ASTM STP 1120,* American Society for Testing and Materials, Philadelphia, 1992.

[10] Salpekar, S. A. and O'Brien, T. K., "Analysis of Matrix Cracking and Local Delamination in $(0/\theta/-\theta)_s$ Graphite Epoxy Laminates under Tension Load," *Proceedings,* ICCM VIII, Honolulu, July 1991, SAMPE, Covina, CA, p. 28-G.

[11] Adams, D. F., Zimmerman, R. S., and Odom, E. M., "Frequency and Load Ratio Effects on Critical Strain Energy Release Rate G_c Thresholds of Graphite/Epoxy Composites," *Toughened Composites, ASTM STP 937,* American Society for Testing and Materials, Philadelphia, 1987, pp. 242–259.

G. C. Scrivner[1] and W. S. Chan[2]

Effects of Stress Ratio on Edge Delamination Characteristics in Laminated Composites

REFERENCE: Scrivner, G. C. and Chan, W. S., "**Effects of Stress Ratio on Edge Delamination Characteristics in Laminated Composites,**" *Composite Materials: Fatigue and Fracture, Fourth Volume, ASTM STP 1156,* W. W. Stinchcomb and N. E. Ashbaugh, Eds., American Society for Testing and Materials, Philadelphia, 1993, pp. 538–551.

ABSTRACT: The effects of stress ratio on delamination onset behavior under cyclic loading was investigated in AS4/3502 $[+/-25_2/90]_s$ graphite/epoxy laminates. For the purpose of this study, delamination onset was defined as a delamination 5% of the laminate width. The resulting data showed definite changes in delamination behavior between stress ratios (R ratio) of 0.1 and 0.5.

For maximum amplitude stresses below about 80% of the static delamination stress, delamination onset occurred later at $R = 0.5$ than for $R = 0.1$. Also for the corresponding G_{max}, delamination growth rates at delamination onset were slower for $R = 0.5$ than for $R = 0.1$. This was reflected in changes of power law exponent from 10.2 for $R = 0.1$ to 31.6 for $R = 0.5$ and the constant from 1.10×10^{-2} for $R = 0.1$ to 254 for $R = 0.5$.

Fractographic analysis suggested that the failure mode, which appeared to be predominantly Mode I, remained unchanged; however, the density of shallow hackles was greater for $R = 0.1$ than for $R = 0.5$. This presumably was due to a more rapid coalescence of microcracks which was consistent with the larger crack opening displacements for $R = 0.1$.

KEY WORDS: composites, damage tolerance, durability, edge delamination, graphite/epoxy, fatigue, stress ratio

Unlike metals, the fatigue behavior of composites appears to be affected by the stress ratio (R ratio). Studies by Zimmerman, Adams, and Odom [1] showed that a change in stress ratio from 0.1 to 0.5 delayed the onset of delamination. Mandell and Meier [2] showed that the time to failure was influenced by the stress ratio. Also, Chan and Wang [3] showed that a change in stress ratio from 0.1 to 0.5 decreased the delamination growth rate.

The determination of delamination onset under fatigue conditions is difficult. Methods used to detect delamination onset under static conditions such as detection of changes in stiffness or visual observation of edge cracks involve continuous monitoring. However, continuous monitoring of a fatigue test is impractical; therefore, delamination onset criteria that employ sampling are used.

In the work of Zimmerman et al. [1] delamination onset was defined as a 5% decrease in the dynamic laminate stiffness. A minicomputer sampled the laminate stiffness according to a logarithmic schedule, and the test was stopped upon detection of a 5% stiffness

[1]Graduate student, Material Science and Engineering, University of Texas at Arlington, Arlington, TX 76019.

[2]Associate professor, Center for Composite Materials, Department of Mechanical and Aerospace Engineering, University of Texas at Arlington, Arlington, TX 76019.

decrease between the previous and current sample. The delamination width corresponding to a 5% stiffness decrease is large, particularly in a matrix-dominated laminate; therefore, this detection method missed the actual delamination onset.

In this study, the stress ratio effect was measured with delamination onset defined as a delamination width which was 5% of the laminate width. This allows measurements to be made much closer to the actual delamination onset; however, this method is more time-consuming than the method used by Zimmerman et al [1]. Previously, Subramanian and Chan [4] used a 3% delamination width to define delamination onset to characterize fatigue behavior at a stress ratio of 0.1 with good results.

Experimental

A $[+/-25_2,90]_S$ AS4/3502 graphite/epoxy laminate was chosen for study. The material properties of AS4/3502 are given in Table 1. This laminated configuration was advantageous for study because delamination is confined to the $-25/90$ interface. The mismatch of Poisson's ratio is at maximum at the $-25/90$ interface, and the resulting interlaminar stresses were expected to be large. Delamination behavior was characterized under static conditions prior to fatigue testing.

Coupon Preparation

Coupons were prepared from 152 by 330-mm (6 by 13-in.) panels which were cured using a 177°C (350°F) cure cycle. The panel was trimmed 12.7 mm (1/2 in.) on each edge. The resulting 127 by 305-mm (5 by 12-in.) panels were C-scanned to determine the presence of voids. The panels were then cut into five coupons. Each coupon was approximately 25.4 by 305-mm (1 by 12 in.). A Micromatic wafering saw with a liquid-cooled diamond blade was used for all trimming and cutting operations. Initially, the coupons were X-rayed to determine any edge cracking that may have developed during cutting. The first five coupons were consistently free of edge cracks, so afterwards the edges were visually inspected using a ×9 magnifier. Afterwards, 25.4 by 38.1-mm (1 by 1.5-in.) fiberglass end tabs were bonded to the coupon. Figure 1 shows a typical configuration of the test coupons.

Static Tests

Three static tests were made in which two coupons were loaded to failure and the third was loaded to 60% of the ultimate stress. The coupons were loaded in tension at a rate of 1.27 mm/min (0.05 in./min). Mechanical tests (static and fatigue) were made using an MTS servohydraulic test machine with a 100-kip load cell installed. Strain gages were installed on the static test coupons, and the stress-strain response was recorded using a Daytronic data acquisition system. A level was used to align the coupon vertically in the load grips.

TABLE 1—*Material constants for AS4/3502 graphite epoxy lamina.*

AS4/3502 Graphite/Epoxy
$E_{11} = $ 124.8 Gpa
$E_{22} = $ 11.2 Gpa
$G_{12} = $ 6.11 Gpa
$\nu_{12} = $ 0.29
$t_{ply} = 1.27 \times 10^{-4}$ m
Fiber volume = 66.8%

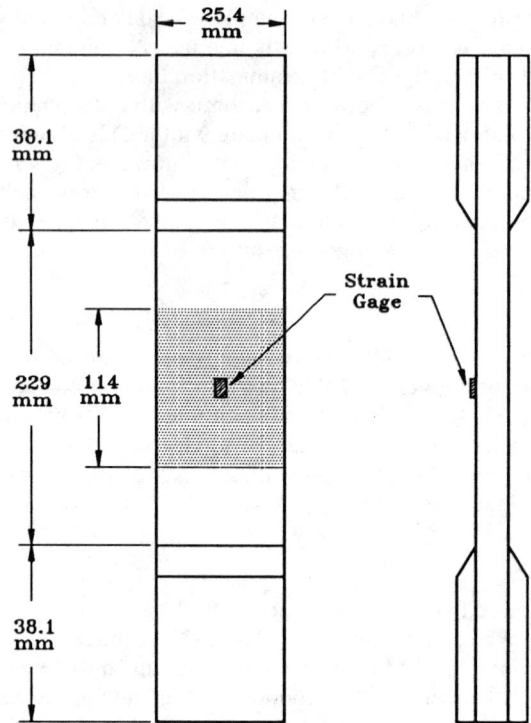

FIG. 1—*Test coupon configuration. Shaded area indicates the place where radiographs were taken.*

Load grip alignment was checked using a plumb bob. During the test, the coupon edge was visually monitored over the entire coupon length with a ×9 magnifier for edge cracks which indicated delamination onset. The load at which edge cracks appeared was the delamination onset load.

TABLE 2—*Test matrix for fatigue tests.*

Stress Ratio	Stress Amplitude[a]	Number of Coupons
0.1	85	1
0.1	80	4
0.1	75	2
0.1	70	3
0.1	65	1
0.5	85	2
0.5	80	3
0.5	75	3
0.5	70	2

[a]Stress amplitude is expressed in percent of the static edge delamination stress.

TABLE 3—*Inspection intervals for delamination onset.*

Applied Cycles	Interval
1–1 000	Continuous
1 001–10 000	1 000
10 001–100 000	10 000
100 001–1 000 000	100 000

Fatigue Tests

The test matrix for fatigue tests is given in Table 2. Fatigue tests were made using load control at a frequency of 5 Hz. The coupon edges were visually monitored over the entire coupon length using a ×9 magnifier for edge cracks according to the schedule in Table 3. After edge cracks were detected, the coupon edges were treated with a zinc iodide solution and X-rayed by using a Hewlett-Packard cabinet X-ray machine. Since the graphite/epoxy material is transparent to X-ray, application of the zinc iodide was allowed to penetrate into the damage area. The test coupon revealed the exact extent of delamination as a dark patch and the matrix cracks as dark transverse lines when exposed to the X-ray. All the test coupons were exposed to X-rays generated at 8 kV for 10 min. Afterwards, radiographs were made at 10 000, 100 000, 500 000 and 1 000 000 cycles. The delamination width was determined by measuring to the nearest millimetre the radiograph image at three locations on each side and then taking an average. The delamination width was then normalized to the laminate half-width.

Test Results

Static Tests

The static test results are given in Table 4. The strain at delamination onset, ϵ_{delam}, as determined from the visually observed delamination onset stress, σ_{delam}, correlated well to the decrease in modulus shown by the stress-strain curve as demonstrated by Fig. 2. The calculated modulus using the values in Table 1 was 73.8 GPa and was in good agreement with the average measured value of 71.0 GPa. The critical strain energy release rate, G_C, was calculated from ϵ_{delam} using [5]

$$G_C = \frac{\epsilon_{delam}^2}{2} t(E_{LAM} - E^*) \qquad (1)$$

TABLE 4—*Static test results.*

Coupon	Delamination Onset			
	Stress, MPa	Strain, μm/m	Modulus, GPa	Ultimate Stress, MPa
1	213	3016	72.0	487
2	201	2875	70.9	487
3	213	3031	70.3	⋯
Average	209	2974	71.0	487

where

t = laminate thickness,

E_{LAM} = modulus in the loading direction for laminate without a delamination, and

E^* = modulus in the loading direction for laminate completely delaminated along an interface.

E^* was calculated by partitioning the laminate along the two 25/90 interfaces into two $[+/-25_2]_S$ laminates and a $[90_2]$ laminate and using rule of the mixture

$$E^* = \frac{(2E' + 8E'')}{10} \qquad (2)$$

where

E' = modulus of $[90_2]$ sublaminate in the loading direction, and

E'' = modulus of $[+/-25_2]_S$ sublaminate in the loading direction.

The calculated value of E^* was 52.1 GPa. Using the average ε_{delam} in Table 4, G_C was calculated as 134 J/m².

The micrographs of a static coupon edge (Fig. 3) show that the delamination was confined to a single $-25/90$ interface although the delamination would cross the 90° plies to the opposite interface. A scanning electron micrograph (SEM) of the fracture surface (Fig. 4) shows signs of resin cleavage indicative of a high degree of Mode I failure [6] in addition to fiber-resin debonding.

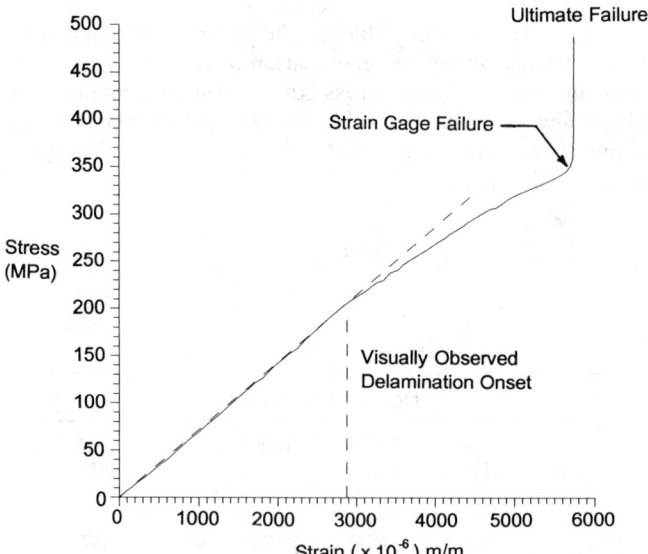

FIG. 2—*Typical load-strain curve showing simultaneous modulus decrease with delamination onset.*

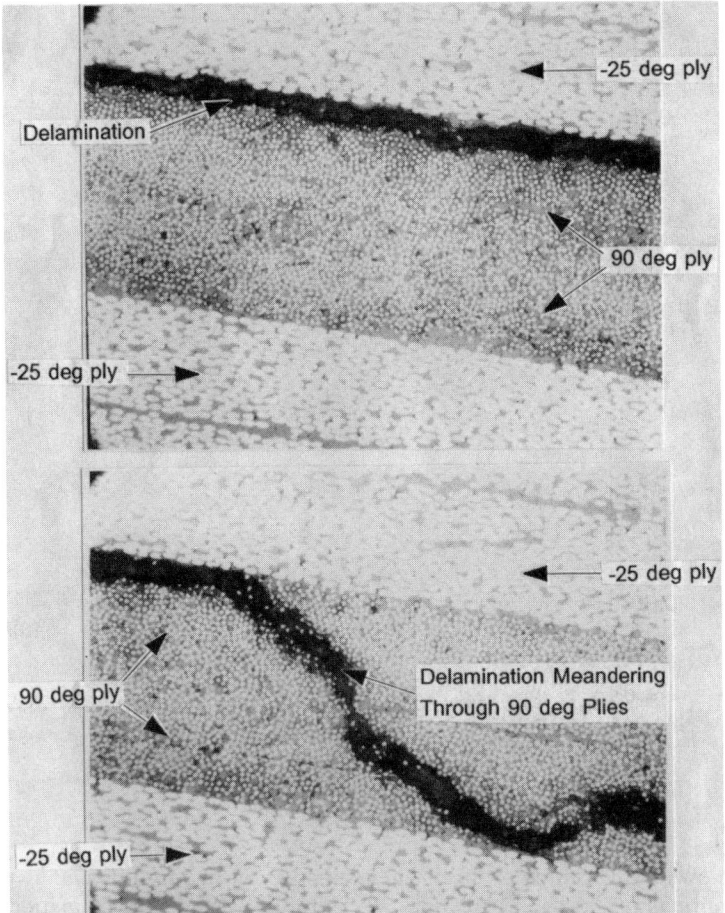

FIG. 3—*Static coupon edge (magnification ×200).*

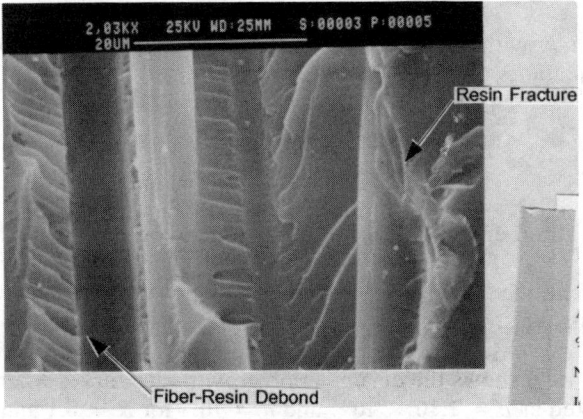

FIG. 4—*Delamination fracture surface under static conditions (magnification ×2030).*

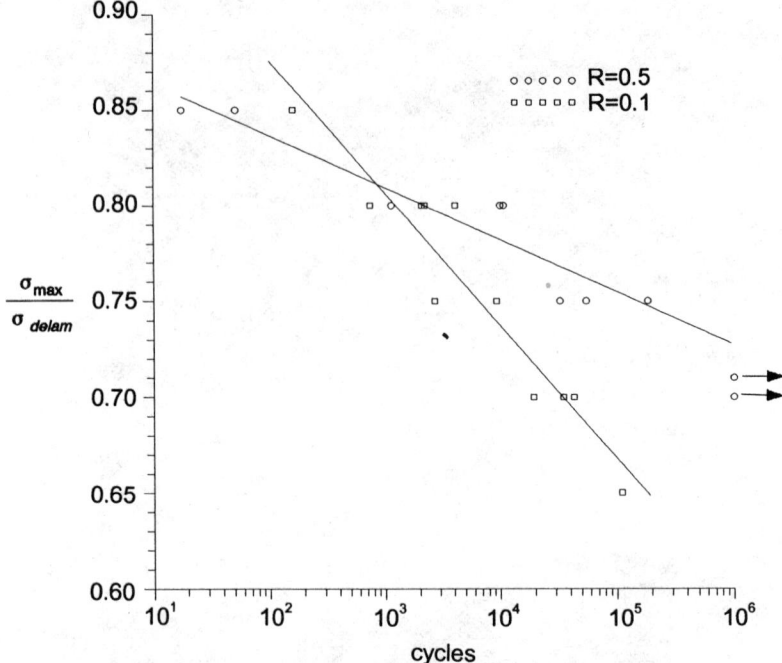

FIG. 5—*Variation of 5% delamination onset with maximum stress.*

Fatigue Tests

Figure 5 shows the variation of the 5% delamination onset with maximum stress amplitude, σ_{max}, normalized to σ_{delam}, for both stress ratios. The 5% delamination onset was calculated from the least-squares fit of the normalized delamination width, $\bar{a}$, versus fatigue cycles, N, data. Below 0.8 σ_{delam}, the 5% delamination onset occurs later at $R = 0.5$ as compared to $R = 0.1$. This behavior is similar to that observed by Zimmerman et al. [1].

The delamination growth rate at 5% delamination onset, $d\bar{a}/dN$, was least-squares fitted to a power law equation

$$\frac{d\bar{a}}{dN} = c \left(\frac{G_{\text{max}}}{G_c} \right)^n \tag{3}$$

where G_{max} is the strain energy release rate corresponding to the maximum stress amplitude.

It should be mentioned that, in the propagation power law, Poursartip [6] used the delamination resistance strain energy release rate, G_R, instead of the critical strain energy release rate, G_C. The reason for using G_C in the power law is that G_C is a constant value and G_R is not. Figure 6 shows the delamination growth rate curves where the constants in Eq 3 were calculated as $c = 1.10 \times 10^{-2}$ and $n = 10.1$ for $R = 0.1$ and $c = 254$ and $n =$

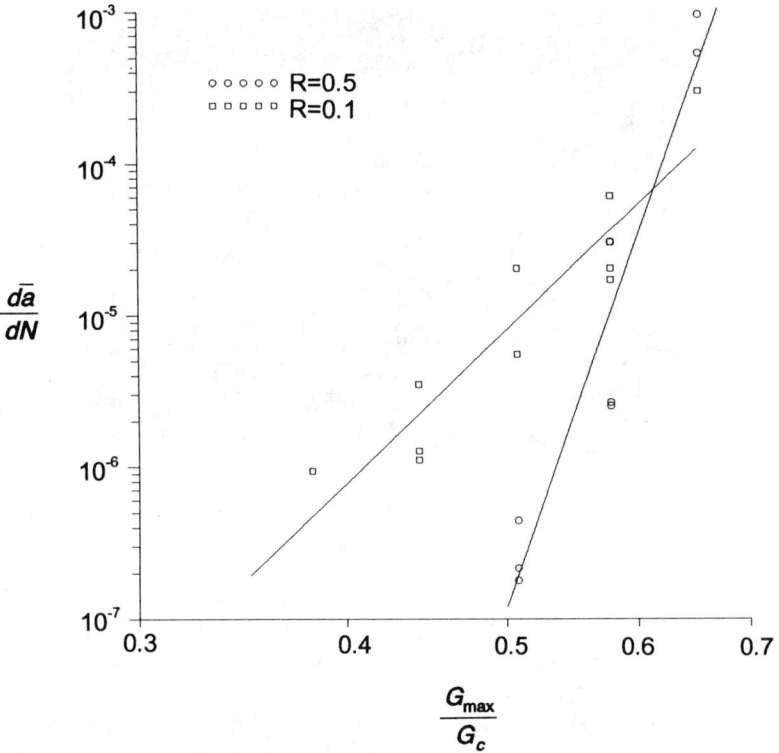

FIG. 6—*Delamination growth rate curves.*

31.0 for $R = 0.5$. The trend is similar to that shown in Fig. 5. The growth rates at high stress levels are approximately equal; however, at lower stress levels, the growth rates are slower for $R = 0.1$.

To determine the failure modes involved, SEMs of the fracture surface of coupons were tested at maximum stress levels of 0.85 σ_{delam} and 0.75 σ_{delam} for both stress ratios. The fracture surface for 0.85 σ_{delam} at $R = 0.1$ is shown in Fig. 7 and at $R = 0.5$ in Fig. 8. Although the left portion of Fig. 7 shows an area of resin cleavage, fiber-resin debonding appeared to be the primary failure mechanism. As with the case of $R = 0.1$, a large degree of debonding is present in Fig. 8; however, shallow hackles can be seen between fibers. Figures 9 and 10 show the fracture surfaces for 0.75 σ_{delam} at stress ratios of 0.1 and 0.5, respectively. In these figures, rows of hackles which are about a fiber diameter wide appear in addition to the fiber-resin debonding. Also, the density of hackles is greater for $R = 0.1$ and presumably is due to a more rapid coalescence of microcracks. This is consistent with the larger crack opening displacements for $R = 0.1$. The failure mode for both stress ratios appeared to be Mode I dominated.

Hackles are usually associated with a significant degree of Mode II failure in brittle resins [7]. However, Arcan et al. [8] observed hackles under Mode I failure of AS4/3501-6 and theorized how hackles formed under Mode I conditions. The region ahead of a fiber-resin debond experiences a shear stress, τ_{yz}, besides a normal stress, σ_{zz}. The interaction of these stresses produces a principle stress off-normal to the fiber. This off-normal stress

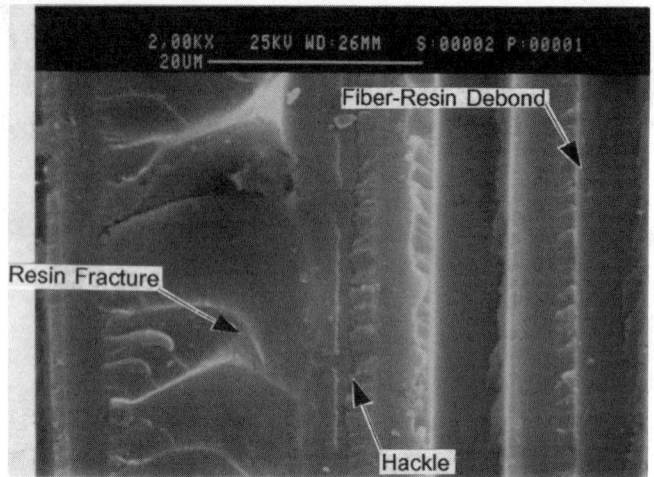

FIG. 7—*Delamination fracture surface under fatigue conditions of 0.85 σ_{delam} and R = 0.1 (magnification $\times 2000$).*

creates a resin fracture inclined to the fiber. In fact, the rows of hackles in Figs. 8 and 9 are probably the resin side of a fiber-resin debond in which this mechanism occurred.

The same pattern of edge cracking which was present in the static case was also observed from coupons tested at both stress ratios. This pattern of edge cracking may be consistent with the presence of a large degree of fiber-resin debonding. Lee [9] observed a similar pattern of edge cracking in coupons with a low fiber-matrix adhesion. The lack of a drop in load, as shown in Fig. 2, was also characteristic of low fiber-matrix adhesion.

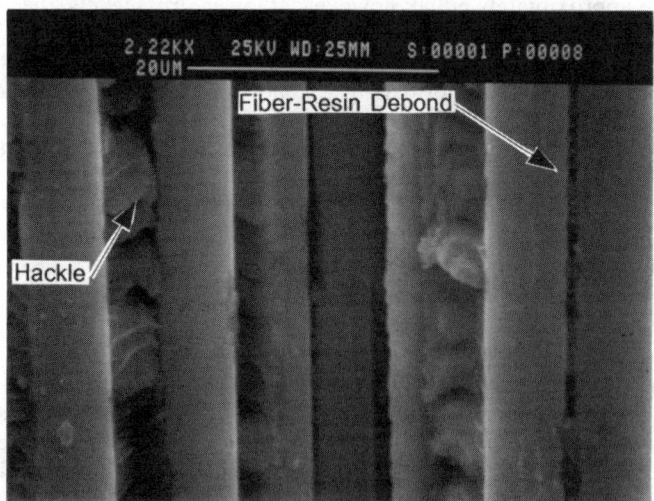

FIG. 8—*Delamination fracture surface under fatigue conditions of 0.85 σ_{delam} and R = 0.5 (magnification $\times 2220$).*

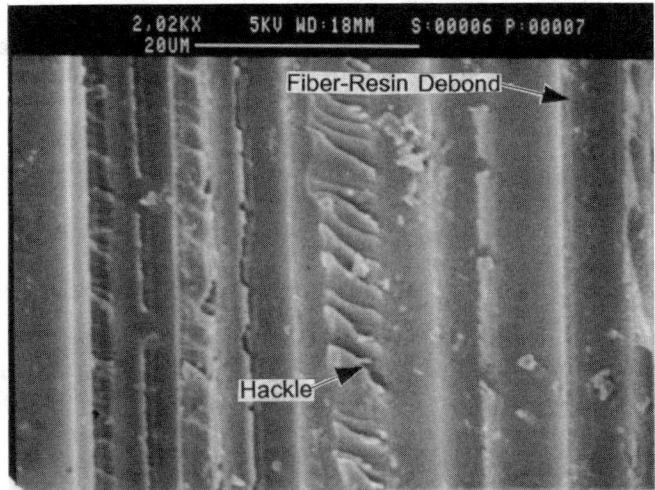

FIG. 9—*Delamination fracture surface under fatigue conditions of 0.75 σ_{delam} and* R = *0.1 (magnification ×2020).*

Four additional coupon tests were conducted at a stress ratio of 0.1. Two coupons were tested at an initial maximum stress level of 0.70 σ_{delam} for half the 5% delamination onset life and then at a maximum stress level of 0.8 σ_{delam} until the delamination size of the test coupon reached 5% of the coupon width. The other two coupons were tested in the same manner, except the maximum stress level was originally at 0.75 σ_{delam} and then at 0.85 σ_{delam}. The effect of changing the stress amplitude was to delay the 5% delamination onset

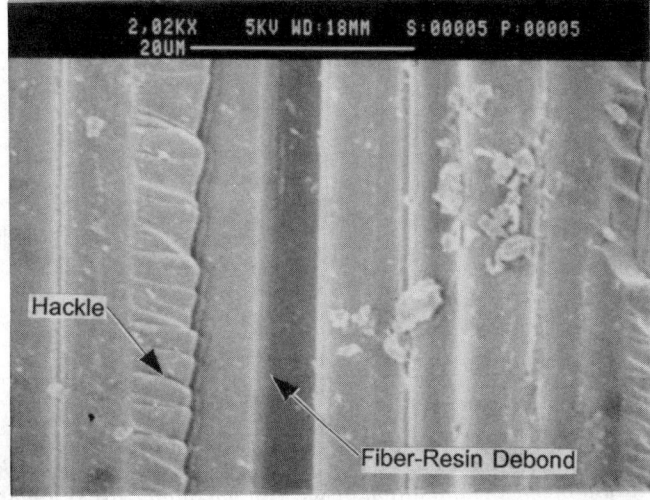

FIG. 10—*Delamination fracture surface under fatigue conditions of 0.75 σ_{delam} and* R = *0.5 (magnification ×2020).*

and to reduce the delamination growth rate. Figure 11 shows the variation of maximum stress amplitude with the 5% delamination onset, and Fig. 12 shows the delamination growth rate curve. Apparently in both situations, the original curve for a stress ratio of 0.1 was shifted. Figure 13 shows an SEM of the fracture surface in which the initial stress was 0.75 σ_{delam}. Upon comparison with Fig. 6, more resin fracture and less fiber-resin debonding occurred.

Discussion

A finite element analysis was also conducted to confirm experimental observation. A quasi-three-dimensional finite element model [10] was used to calculate strain energy release rate for a given delamination. This model has been successfully employed to investigate delamination characteristics in various laminates. Finite element results indicate that the Mode III component of the total strain energy release rate is negligible compared to the components of Modes I and II. Figure 14 shows the results of the normalized total strain energy release rate and its Mode I component. The ratio of G_I/G_T is also shown in the figure. It is seen that the Mode I component contributes 68% of the total G. This high-percentage Mode I component contributes a predominant Mode I failure as observed in fractographic analysis.

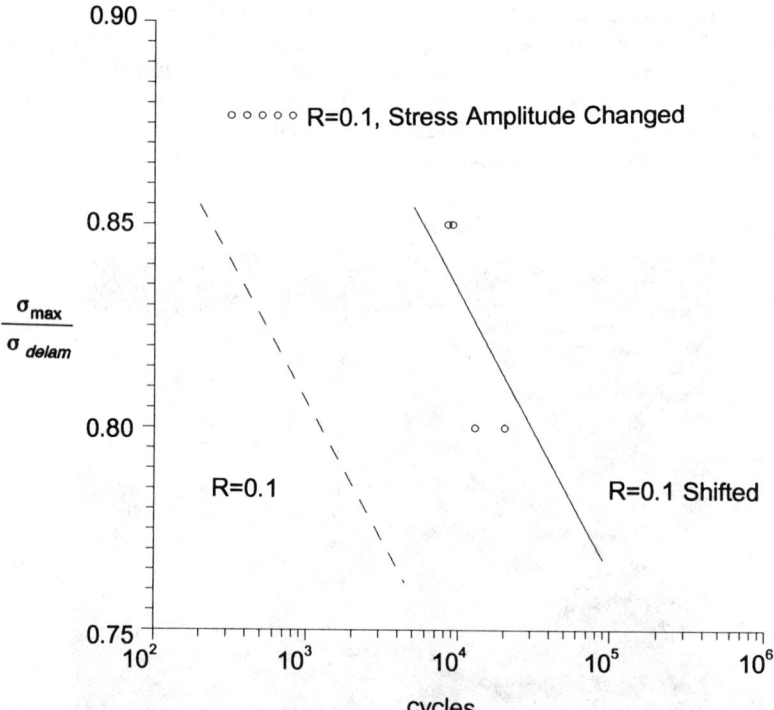

FIG. 11—*Variation of 5% delamination onset with maximum stress under conditions of changing maximum stress amplitude.*

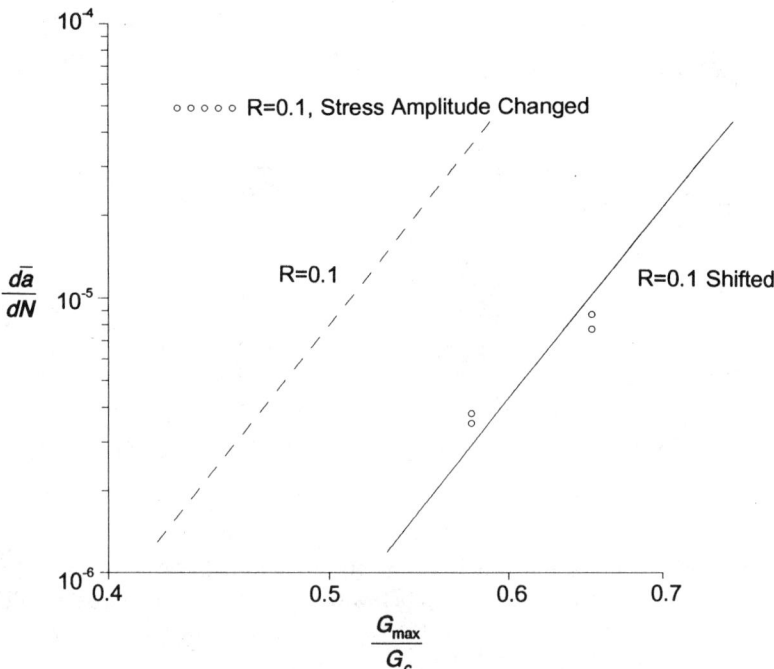

FIG. 12—*Delamination growth rate curves under conditions of changing maximum stress amplitude.*

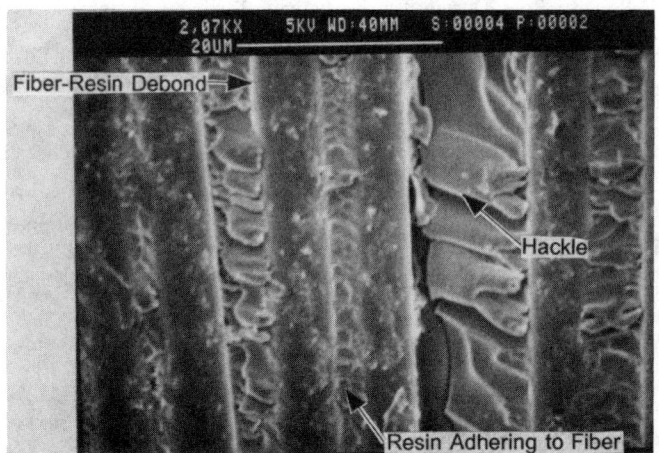

FIG. 13—*Delamination fracture surface under fatigue conditions of 0.85 σ_{delam}, R = 0.1 and changing maximum stress amplitude (magnification ×2070).*

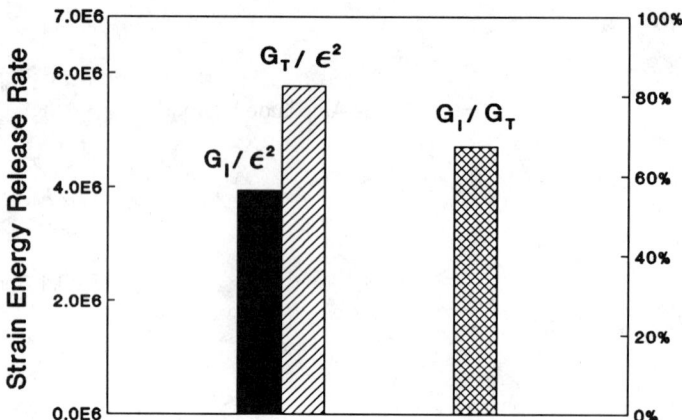

FIG. 14—*Strain energy release rate normalized by square of applied strain.*

Conclusions

Stress ratio had a significant effect on the delamination onset and the delamination growth rate at maximum stress amplitudes below 0.80 σ_{delam}. Apparently, the failure mode was unaffected by stress ratio. However, crack opening displacements, which were a result of stress ratio, did affect the density of hackles observed.

Acknowledgement

The authors would like to thank George Law of the Fort Worth Division of the General Dynamics Corporation for providing the panels used in this study.

References

[1] Zimmerman, R. S., Adams, D. F., and Odom, E. M., "Load Ratio and Frequency Effects on Strain Energy Release Rate During Tensile Fatigue Testing Utilizing the Edge Delamination Test," *Toughened Composites ASTM STP 937*, ASTM, Philadelphia, 1986.

[2] Mandell, J. F. and Meier, U., "Effects of Stress Ratio, Frequency, and Loading Time in the Tensile Fatigue of Glass-Reinforced Epoxy," *Long Term Behavior of Composites, ASTM STP 873*, American Society for Testing and Materials, Philadelphia, 1983, pp. 55–77.

[3] Chan, W. S. and Wang, A. S. D., "Free Edge Delamination Characteristics in S2/CE9000 Glass/Epoxy Laminates Under Static and Fatigue Loads," *Composite Materials: Fatigue and Fracture, Second Volume, ASTM STP 1012*, ASTM, Philadelphia, 1989, pp. 270–295.

[4] Subramanian, S. and Chan, W. S., "Frequency Effect of Delamination Characteristics of Laminates With and Without an Interleaf," presented at the International Conference on Composite Materials, VIII, Honolulu, 15–19 July 1991, pp. 28-O-1 to 28-O-20.

[5] O'Brien, T. K., "Characterization of Delamination Onset and Growth in a Composite Laminate," *Damage in Composite Materials, ASTM STP 775*, K. L. Reifsnider, Ed., American Society for Testing and Materials, Philadelphia, 1982, pp. 140–167.

[6] Poursartip, A. "The Characterization of Edge Delamination Growth in Laminates under Tensile Loading," *Toughened Composites, ASTM STP 937*, ASTM, Philadelphia, 1986.

[7] Hibbs, M. I. and Bradley, W. L., "Correlations Between Micromechanical Failure Processes and the Delamination Toughness of Graphite/Epoxy Systems," *Fractography of Modern Engineering Materials, Composites and Metals, ASTM STP 948*, J. E. Masters and J. J. Au, Eds., American Society for Testing and Materials, Philadelphia, 1988, pp. 68–97.

[8] Arcan, L., Arcan, M., and Daniel, I. M., "SEM Fractography of Pure and Mixed-Mode Interlaminar Fractures in Graphite/Epoxy Composites," *Fractography of Modern Engineering*

Materials: Composites and Metals, ASTM STP 948, J. E. Masters and J. J. Au, Eds., American Society for Testing and Materials, Philadelphia, 1987, pp. 41–67.

[9] Lee, S. M., "Failure Mechanisms of Edge Delamination of Composites," *Journal of Composite Materials*, Vol. 24, November 1990, pp. 1200–1212.

[*10*] Chan, W. S. and Ochoa, O. O., "Delamination Characterization of Laminates Under Tension, Bending and Torsion Load," *Computational Mechanics*, Vol. 6, No. 5/6, 1990, pp. 393–405.

R. E. Swain,[1] *C. E. Bakis,*[2] *and Kenneth L. Reifsnider*[3]

Effect of Interleaves on the Damage Mechanisms and Residual Strength of Notched Composite Laminates Subjected to Axial Fatigue Loading

REFERENCE: Swain, R. E., Bakis, C. E., and Reifsnider, K. L., "**Effect of Interleaves on the Damage Mechanisms and Residual Strength of Notched Composite Laminates Subjected to Axial Fatigue Loading,**" *Composite Materials: Fatigue and Fracture, Fourth Volume, ASTM STP 1156*, W. W. Stinchcomb and N. E. Ashbaugh, Eds., American Society for Testing and Materials, Philadelphia, 1993, pp. 552–574.

ABSTRACT: The effect of interleaving on the fatigue damage mechanisms and residual tensile strength of two graphite/epoxy material systems subjected to cyclic axial loads is investigated. The two materials, AS4/985 and AS4/1808, have stacking sequences of $[0/45/90/-45]_{s2}$ and $[0/-45/90/45]_{s4}$, respectively. All specimens have centrally located through-holes. The 16-ply specimens are subjected to tensile fatigue loads, while the 32-ply specimens are subjected to fully reversed fatigue loads. Interleaving these laminates at every interface with a thin (on the order of 10% of the ply thickness), tough, thermoplastic film results in an expected increase in mass and decrease in laminate stiffness, but also in higher virgin tensile strengths, lower normalized residual tensile strengths, and slightly altered damage patterns and rates of damage growth in both laminate types. The presence of the interleaves alters the baseline virgin tensile load-to-failure and damage patterns to a greater extent in the 1808 material than in the 985 material.

KEY WORDS: composite material, fatigue, notch, damage, residual strength, toughness, interleaf

Engineers are increasingly turning to composite materials in situations where the stiffness, strength, electromagnetic, or thermomechanical properties of a material must be tailored for special applications. By choosing the constituent materials wisely, one can "engineer" the material for optimal characteristics or properties. An example of intelligent engineering of composites that is particularly relevant in this paper is the "toughening" of composite laminates for improved "damage resistance" and "damage tolerance." For our discussion, "damage resistance" is simply defined as the resistance of the material to damage initiation during loading, and "damage tolerance" as the ability of the material to retain strength in the presence of damage. Our discussion will be limited to polymer matrix composites (PMCs) reinforced with high-modulus graphite fibers.

Due to what the authors perceive as a continuing need for an increased understanding of

[1] Graduate research associate, Dept. of Engineering Science and Mechanics, Virginia Polytechnic Institute and State University, Blacksburg, VA 24061. Presently senior research engineer at Smith & Nephew Richards Inc., Memphis, TN 38116.
[2] Assistant professor, Dept. of Engineering Science and Mechanics, Penn State University, University Park, PA 16802.
[3] Alexander F. Giacco professor, Dept. of Engineering Science and Mechanics, Virginia Polytechnic Institute and State University, Blacksburg, VA 24061.

damage mechanisms and their effect on residual strength, this investigation was conceived to test the efficacy of various "toughening" schemes, particularly interleaving with tough thermoplastic films, on damage development and residual strength during the cyclic fatigue of notched laminates. In order to elucidate the various methods and mechanisms of composite laminate toughening, we preface our discussion with a review of the relevant literature.

Quantification of Damage Resistance and Damage Tolerance

Meaningful measures of damage resistance and damage tolerance for composites are practically impossible to define because of the diverse mechanisms of damage that appear in these materials (e.g., matrix cracks, delamination, fiber fracture), the different effects of each damage mechanism on various components of strength (e.g., tensile, compressive), and the complicating effect of stress concentrators (e.g., cut outs, ply drops, manufacturing flaws). Consequently, a given material system can appear to be damage tolerant in certain loading conditions and environments, yet damage intolerant in others. For example, it has been shown that incorporating a resin matrix with a higher elongation to failure (higher ductility) may lessen the extent of impact-induced delamination in and improve the post-impact compressive strength of certain laminates [1–3]. Higher ductility matrix materials also have been reported to improve damage resistance during the cyclic loading of unnotched laminates [4,5], although there does not seem to be a consensus on this matter [6]. In any case, improvements in damage resistance and damage tolerance sometimes evident in toughened unnotched laminates do not seem to translate to notched laminates during reversed cyclic loading [7–9]. That is, while a notched laminate with a ductile matrix may incur a smaller fatigue-damaged area, the type of damage that does occur may be different than that seen in brittle matrix materials, and much more deleterious to the laminate's residual strength. It is widely held that composites with comparatively brittle matrix materials undergo relatively more matrix damage near the notch, thereby relieving the local in-plane stress concentration and increasing the residual tensile strength: the so-called "wear-in," notch "blunting," or material "softening" effect [7–17]. In contrast, composites with comparatively ductile matrix materials experience relatively more fiber fracture due to the continuing high stress concentration associated directly with the initial notch geometry or with previously broken load-bearing fibers near the notch. Kress [12] and Bakis [7,16] have provided some experimental measurements of (surface) stress redistribution in notched laminates during fatigue, but there clearly remains much more to be learned about the three-dimensional and microscopic nature of damage-induced stress redistribution before the engineering community can feel confident about predicting the fatigue response of new, "damage tolerant" material systems.

The composites engineer would be wise to consider the intended use of the material before selecting not only candidate materials for optimal "toughness," but also the test method to screen the materials. For instance, several investigators have noticed that the increased resistance of ductile matrix composites (versus their brittle matrix counterparts) to delamination onset during quasistatic and low-cycle fatigue loadings vanishes when the material is subjected to increasingly longer-term cyclic loading histories [18–20]. If the engineer was faced with a long-term loading situation, the selection of alternative methods of delamination prevention might be prudent.

Realizing that delamination often leads to fatigue failure in cyclically loaded laminates, the authors have chosen to focus on interleaved material systems that—based on short-term experimental screening methods such as interlaminar fracture toughness, damage assessment after impact, and compressive strength after impact—might offer better damage resistance and damage tolerance than noninterleaved material systems. The goal is to see if the promising results of the short-term tests (reported elsewhere) carry over to long-term tests. A brief review of the interleaving concept follows.

Interleaving as a Laminate Toughening Concept

Delamination has been termed "the most prevalent life-limiting failure mode in advanced composite structures" [21]. In direct correlation to the prevalence of delamination in composite structures is the prevalence of delamination studies in the technical literature. Much effort has been expended in the analysis of delamination initiation and propagation, and, in turn, much knowledge has been gained. It is the successful application of this knowledge that has allowed composite scientists to develop schemes to prevent or suppress delamination.

Interleaving—the selective introduction of discrete, compliant, adhesive layers between composite plies—seeks to improve the laminate's resistance to delamination in a number of ways. First, the presence of the low-modulus layer reduces the magnitude of the interlaminar stresses and may, in fact, locally strengthen the ply interface [22]. Then, if delamination does occur, the tough adhesive layer requires much more energy than a conventional laminate to propagate the delamination.

Research on interleaved laminates within the past several years has focused mainly on their ability to resist delamination in loading modes that frequently initiate delamination. Tests on interleaved composite laminates subjected to Mode I [1,21,23–28] and Mode II [1,25–28] fracture toughness tests, as well as static in-plane loading [21,22,26,29–38], multispan shear loading [39], and impact [1,23–26,34,40–43], have shown that the interleaves have performed their task of suppressing delamination damage surprisingly well.

The literature is replete with data espousing the merit of interleaved composite design as a means to significantly increase delamination and impact damage resistance. What has yet to be carefully explored is the effect interleaving has on the composite laminate's in-service mechanical response. Though improvements with regard to the above damage modes have been recorded, the effect that interleaving has on the in-plane response of laminates in instances where delamination is no longer critical is still largely unknown.

The response of interleaved laminates to compressive loading [26,33,34,38], fatigue loading [30,36], and creep, among others, has yet to be substantially investigated, with some noted exceptions. Chan [30] has studied the effect of interleaving on laminates which are prone to delaminate by inserting strips of adhesive layers between critical interfaces. He subjected the laminates with and without interleaves to static tensile and fatigue loading. He concluded that the delay in delamination growth produced in the interleaved specimens resulted in a significant increase in the fatigue life for matrix-dominated laminates. It was also felt that delamination was never really arrested, only that the growth rate was reduced. Subramanian [36] followed up this work by investigating the effect that frequency had on the delamination initiation and propagation of tensile-fatigued specimens. In studies conducted on delamination-prone laminates of IM6/3501-6 with and without interleaves of American Cyanamid FM 300 film adhesive placed at critical interfaces, it was concluded that the delamination growth rate in the interleaved laminates was lower at low load levels when compared to baseline laminates; at higher load levels, however, the growth rates were found to be nearly equal for all frequencies tested.

Experimental Procedure

Cured panels of the various material systems were ultrasonically C-scanned to ensure the lack of manufacturing flaws. Prior to machining, the uncut panels were stored in a desiccator. After machining, the specimens were stored and tested in an air-conditioned laboratory environment. Straight cuts were machined with diamond abrasive wheels, and circular holes were drilled with diamond abrasive core drills. In all machining operations. the tool and specimen were cooled using a commercial synthetic lubricant. During drilling, the back of the specimen was supported with plastic to prevent delaminations near the back surface.

The following notation for ply angles has been adopted: positive angles are measured counterclockwise from the vertical (loading) axis. For ease of interpretation, the conventional numbering scheme for coordinate quadrants has been maintained, i.e., the "first" quadrant defines the upper right quadrant (which encompasses ply angles of 0° to −90°), while the "second" quadrant lies in the upper left quadrant. Stresses in the discussion are computed by dividing the load by the specimen's gross cross-sectional area.

Mechanical testing of the 32-ply specimens was performed on 89 and 222 kN capacity, servohydraulic load frames equipped with hydraulic wedge grips. All of the 16-ply specimens were tested on the smaller of the two load frames. The gripped ends of the specimens were not tabbed, but instead were wrapped with one or two layers of single-sided abrasive paper (abrasive side inward) to prevent grip-induced damage during fatigue tests. All fatigue tests were carried out in load control using a 10-Hz sinusoidal wave form. Stress ratios were $R = 0.1$ for the tensile fatigue tests, and $R = -1$ for the reversed fatigue tests. The test method for reversed cyclic loads detailed in Ref 44 was used in the present investigation. According to this method, no anti-buckling supports are used since they may interfere with the natural development of damage. To preclude gross column buckling, a rather short unsupported length of 63.5 mm (2.5 in.) was chosen for the reversed loading tests in this investigation.

Strains in tension and compression (where applicable) were measured with 25.4-mm (1.0-in.) gage length extensometers centered on the notch as described in Ref 44. Stiffness was calculated by dividing the applied stress by the measured strain. Due to high strain gradients around the hole and the different hole diameters used in different specimen types, actual stiffness values measured with the fixed gage length extensometers reflected material as well as geometric constraints and, therefore, could not be compared from one specimen geometry to the next. Consequently, the reported stiffness data were normalized by the stiffness measured on the first cycle for a particular specimen. Dynamic stiffness was measured continuously with digital peak detectors, although actual data were recorded statically on an as-needed basis in order to construct smooth curves of stiffness versus cycles. These stiffness curves, together with X-ray radiographs, were used to estimate the fraction of fatigue lifetime consumed prior to a residual strength measurement. A zinc iodide liquid penetrant was applied to the damaged specimens prior to radiographic inspection [44]. Two radiographic views of each specimen were taken: one where the X-ray beam passes through the thickness of the specimen (flatwise view) and one where the beam passes through the width of the specimen (edge view).

Residual strength tests were carried out by quasi-statically loading the specimens to failure using a ramp function that lasted between 30 s and 1 min. The loading rate for these tests was appreciably lower than that typified by the sinusoidal cyclic loads during the fatigue tests.

Description of 16-Ply AS4/985 Specimens

Prepreg tape comprised of Hercules' AS4 fiber (possessing a fiber areal weight of 145 g/m²) bound in American Cyanamid's Cycom 985 matrix [24,33] was laid up into several 16-ply, quasi-isotropic panels. The panels were laminated in two configurations. The baseline panel possessed a $[0/45/90/-45]_{s2}$ stacking sequence. The other panel possessed the identical orientation; however, thin adhesive interleaves, measuring 12.7 μm (0.0005 in.) each in thickness, were co-cured in the panel between each interface and on one of the outside surfaces (see Fig. 1). The adhesive material constituting the interleaf is a proprietary thermoplastic film termed "Film C" in Ref 26. Throughout this study, baseline (or

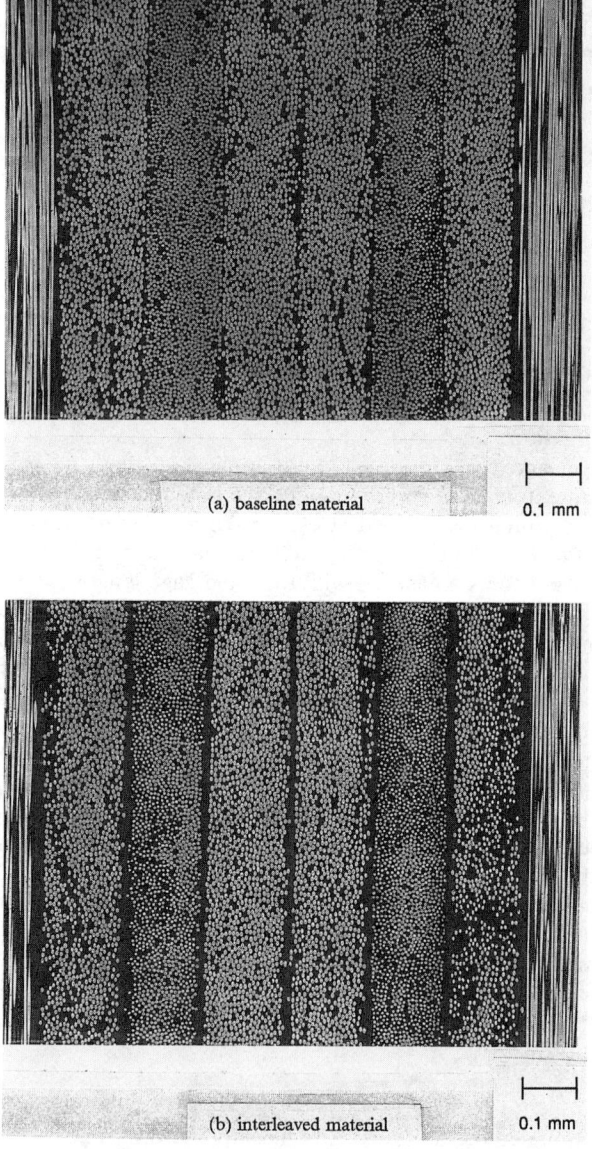

(a) baseline material   0.1 mm

(b) interleaved material   0.1 mm

FIG. 1—*Edge photomicrograph of AS4/985 baseline and interleaved material. View of symmetric [0/45/90/-45] sublaminate: (a) baseline material, (b) interleaved material.*

noninterleaved) laminates will be referred to as "B" laminates while interleaved laminates will be designated with an "I."

Specimens measuring a nominal 139.7 mm (5.5 in.) in length and 25.4 mm (1.0 in.) in width were cut from the B and I panels. On the average, the thickness of the B specimen was 2.306 mm (0.0908 in.) while the addition of the interleaves augmented the thickness of the I specimen to 2.461 mm (0.0969 in.)—an increase of 6.7% over the baseline thickness. A 6.35-mm (0.25-in.)-diameter hole was centered in each specimen. The average mass of the B specimen was found to be 12.32 g, while the average mass of the I specimen was 13.06 g—an increase of 6% over the baseline mass.

Description of 32-ply AS4/1808 Specimens

Panels of Hercules' AS4 fiber (again, possessing a fiber areal weight of 145 g/m^2) in American Cyanamid's Cycom 1808 matrix were laid up in a 32-ply, quasi-isotropic configuration. The Cycom 1808 resin system represents an improvement in toughness over the Cycom 985 system. Good comparative data for the Cycom resins for neat and composite performance have been published [24,33,34]. The baseline 1808 material possessed a [0/−45/90/45]$_{s4}$ layup. Another 1808 panel possessed the identical orientation; however, thin adhesive interleaves, measuring 12.7 μm (0.0005 in.) each in thickness, were co-cured into the panel between each interface and on one of the outside surfaces. The adhesive material constituting the interleaf was the same film (Film C) and thickness (12.7 μm) as used in the AS4/985 panels. Results from fracture toughness tests and post-impact compression strength performed using AS4/1808/Film C laminates are found in Refs 25 and 26. Other AS4/1808 toughness data have been published [1,24–26,33,34,39].

The width of the B and I specimens was a nominal 38.1 mm (1.5 in.). The length of the I specimens was 165.1 mm (6.5 in.), while the B specimens' length was 152.4 mm (6.0 in.). The average thickness of the 1808 B specimens was 4.587 mm (0.1810 in.), while the average thickness of the 1808 I specimen was 5.027 mm (0.1979 in.); this represents an increase of 9.3% over the 1808 baseline thickness. A 9.525-mm (0.375-in.)-diameter hole was centered in each specimen. Unfortunately, the mass of these specimen types was not recorded.

Comparative Response of AS4/985

Static Virgin Strength

In order to ascertain the fatigue stress levels to be employed, static strengths were determined from the virgin, notched material. Three specimens of each laminate type were loaded to failure at a rate of 4448 N/s (1000 lb/s). The individual data and the average values are located in Table 1. Static strength data for virgin, unnotched specimens are located in Table 1 of Ref 35.

Several issues demand amplification. First, specimens B-1 through B-3 revealed an average gross section strength of 349 MPa (50.7 ksi). This value was compared to the average strength of 344 MPa (49.9 ksi) obtained from the I specimens. Since these values seemed reasonably close, a fatigue load level applicable to both specimen types was determined from these values. It was soon discovered that the strengths established by specimens B-1 and B-2 represented the high end of the statistical scatter, since a number of subsequent specimens failed during their initial stiffness measurements. Three such specimens were incorporated into the virgin strength sample size, thus reducing the strength value from 349 to 331 MPa (48.0 ksi). The average gross section strength for the B material, as it turned out, represented 100% of the maximum applied cyclic stress level. If

TABLE 1—*Virgin notched strength of AS4/985 baseline and interleaved specimens, individual and averaged data.*

Specimen ID	Failure Load, kN	Gross Area, cm²	Strength, MPa
B–1	21.5	0.588	366
B–2	21.4	0.575	373
B–3	17.9	0.581	309
		Avg of three =	349
B–6	18.2	0.578	315
B–7	19.0	0.581	326
B–8	17.6	0.595	296
		Avg of six =	331
I–1	20.1	0.626	321
I–2	22.8	0.610	374
I–3	21.1	0.625	338
		Avg of three =	344

B specimens survived early stiffness measurements, they would respond in a reproducible manner to the high applied cyclic loads.

It is interesting to note that the average gross section strength of the three I specimens is 4.0% higher than the average of the six B specimens. This increase becomes more significant if one recalls that the thickness (and, therefore, the gross section area) is greater in the I laminate by 6.7% due only to the presence of the *low-modulus* interleaf—material that is contributing an insignificant amount to the load-carrying capability of the material. It should be noted, however, that if one determines the specific strength (strength/mass) of these materials, the B specimen's average of 26.8 MPa/g is 1.7% higher than the I specimen's average of 26.3 MPa/g.

The initial stiffness of 13 notched I and B specimens was averaged. The average B stiffness is 37.7 GPa (5.47 Msi), while the I stiffness value is 34.9 GPa (5.06 Msi)—a decrease of 7.5%. The average unnotched stiffness of the B specimen (sample size is $n = 3$) is 46.4 GPa (6.73 Msi), while the I specimen value ($n = 3$) is 44.1 GPa (6.39 Msi)—a decrease of 5.1% due to the presence of the interleaves.

Fatigue Results

Cyclic stress levels were originally targeted to be 90% of the average ultimate gross section strength of the two materials. The emergence of significant scatter retroactively adjusted the maximum stress level for the B specimens to be 100%. The maximum applied load level was kept at a constant of 19.3 kN (4330 lb) between the two materials in order to facilitate a one-to-one comparison.

Stiffness Versus Life—The normalized stiffness versus normalized life curves for each material are located in Fig. 2. At least one specimen of each material displayed the familiar "three stiffness fatigue regimes," a sudden decrease in stiffness at early life, a steady, gentle decay of stiffness throughout the middle life, then a cataclysmic degradation at late life [12,45,46]. The repeatability and relatively low spread in the stiffness data enabled the authors to use stiffness degradation as an indication of remaining life.

Residual Strength and its Relationship to Damage Development—Fatigue data detailing the response of the B and I specimens at "early," "middle," and "late" life are presented in Table 2. Figure 3 is a plot of the residual strength data for the I and B materials during their fatigue lifetime, normalized by their respective average virgin strengths. The residual strengths of the two laminate types reveal their greatest disparity at "middle" life—

defined as a stiffness degradation of 13 to 15% as determined from Fig. 2. An explanation for the differences seen in the middle-life residual strengths is sought from the damage states.

A B specimen experiencing a 15% stiffness decay (after 30 000 cycles) is shown in Fig. 4. Delamination is clearly evident in this specimen, emanating from the hole and from four quadrants at the edge of the specimen. The tangential 0° cracks have extended to nearly the length of the gage section. These cracks serve as the major axis for the semi-elliptical shape created by the delaminations. Notice that the delaminations have increased in size about the hole and edge unsymmetrically. The largest delaminations appear in the first and third quadrants—the same quadrants that delamination first appeared in the early-life radiographs. Three 0° longitudinal cracks are located at the delamination front in the edge delamination located in the third quadrant; one such crack lies in the middle of the ligament between the first and fourth quadrants.

The damage state of an I specimen at middle life (15% stiffness degradation at 70 000 cycles) is shown in Fig. 5. The tangential 0° cracks have not grown to the same extent as those shown in Fig. 4, yet they have developed in a skewed position. The tangential crack length and skewed position was nominally the same for each of the three middle-life I specimens. These longitudinal cracks also serve as a major axis for the growth of the delaminations. Yet, an important difference exists in the respective delamination growth patterns. Some delamination in the I material has formed into four "lobes" around the hole. This pattern is reproducible in the I specimens at middle life and differentiates itself from the B specimens in that extensive delamination does not occur in the area of maximum axial stress along each side of the hole. The lobes extend unsymmetrically, with the lobes in the second and fourth quadrants growing larger and generally outside of the longitudinal cracks; the other two lobes tend to grow somewhat symmetrically about the longitudinal cracks (seen also in Fig. 4). The position and extent of edge delamination appears nearly identical to the edge delaminations seen in Fig. 4. Another major difference between the two damage states is the density of off-axis cracking. If these are comparable middle-life damage states, then the I material has suffered a greater amount of 90° cracking, and, as a consequence, −45° and 45° cracking than the B material (unfortunately, this is difficult to detect in the reproductions of the radiographs). This observation is somewhat surprising since, initially, the individual plies of the I laminate are experiencing a slightly lower strain (CLT calculations indicate the difference to be less than 0.5% of the baseline strain) than the plies of the B laminate since each specimen endures the same global load and the interleaves carry a modicum of this load. Perhaps this phenomenon indicates the role that the interleaves play on the constraint of the plies, since the effectively "thicker" 90° ply would be expected to fail at a lower load [47]. Yet this effect is counter-balanced by the lower crack density expected in the I material due to a greater shear-lag effect [31].

The residual *net* section strengths of the B specimens at middle life approach and exceed the unnotched strength of virgin B specimens (found, on average, to be 624 MPa [35]). This result forces one to conclude that the effect of the center notch is no longer felt with respect to residual strength, i.e., the material responds as if symmetric ligaments were being loaded. The residual strengths of the I laminates have increased from their early-life values only to reach the early-life values of the B specimens. A correlation between an increase in the residual notched strength of quasi-isotropic laminates and an increase in length of tangential 0° cracks has been reported [12]. Though the tangential 0° cracks are of differing lengths in the middle-life regime, early-life radiographs reveal no significant difference in crack length (and extent of matrix damage, for that matter), yet the early-life residual strengths of the B specimens are at least 10% greater than the early-life residual strengths of the I specimens.

Any disparity seen in late-life residual strengths between the two laminate types must be

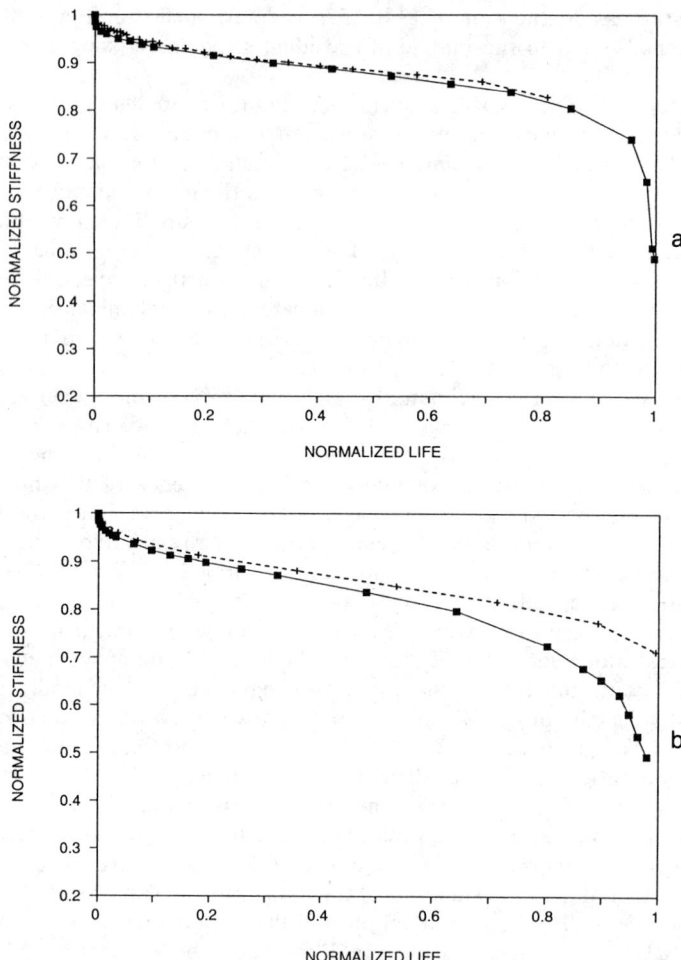

FIG. 2—*Comparison of normalized stiffness versus normalized life for AS4/985 baseline and interleaved material:* (a) *two individual baseline specimens, w = 28.8, d = 20.3;* (b) *two individual interleaved specimens, w = 28.8, d = 20.5;* (c) *one baseline and one interleaved specimen, w = 28.8, d = 20.6.*

interpreted with regard to their proximity to failure. Yet, both laminates reveal an identical sequence towards fracture. In each case, delaminations (occurring along the outer 0/45 interfaces) from the straight edge and the hole edge meet near the center of each ligament. The convergence of these delaminations effectively divides the laminate into sublaminates, reduces the stiffness, unbalances the laminate, and causes the interior stresses to rise dramatically.

Discussion of Results

The temptation exists to compare the behavior of the B and I specimens on the basis of cyclic life. This is bolstered by the fact that no B specimen survived over 90 000 cycles while no I specimen failed prior to 115 000 cycles, keeping in mind the large statistical

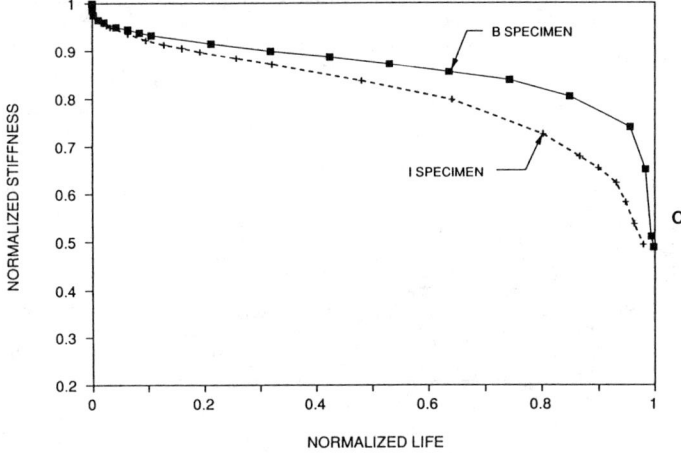

FIG. 2—*Continued.*

spread that attends fatigue testing. Yet, since the materials were compared at a constant maximum load level, one must conclude that the stress in the B laminate was somewhat greater than the stress in the I laminate. Since the applied stresses were extremely high with regards to the materials' initial strength, one may suspect that subtle differences in the applied stress levels as a percentage of ultimate strength could account for the magnitude in the difference between their respective lives. This trend may reverse itself as damage is introduced, however. Figure 2*c* indicates that the I laminate experiences a

TABLE 2—*Residual strength data from fatigue testing of AS4/985 baseline and interleaved specimens. Tested at R = 0.1, maximum applied load of 19.3 kN, cycled at 10 Hz.*

	Specimen ID	Cycles	Estimated % of Life	Fraction of Initial Stiffness	Residual Strength, MPa	Fraction of Initial Strength
B early	9	1 000	<3%	0.96	413	1.25
	17	1 000	<3%	0.97	409	1.24
	18	1 000	<3%	0.97	408	1.23
B middle	13	25 000	50%	0.87	481	1.45
	15	40 000	50%	0.87	489	1.48
	16	30 000	50%	0.85	452	1.37
B late	11	74 700	>80%	0.78	399	1.21
I early	6	500	<1%	···	371	1.08
	7	1 000	1%	0.95	361	1.05
	14	3 000	<3%	0.97	373	1.08
I middle	8	74 700	50%	0.84	407	1.18
	11	60 000	50%	0.85	427	1.24
	12	70 000	50%	0.85	408	1.19
I late	4	155 700	>95%	<0.48	300	0.87
	9	110 000	80%	0.76	380	1.10
	10	116 900	>95%	0.50	345	1.00

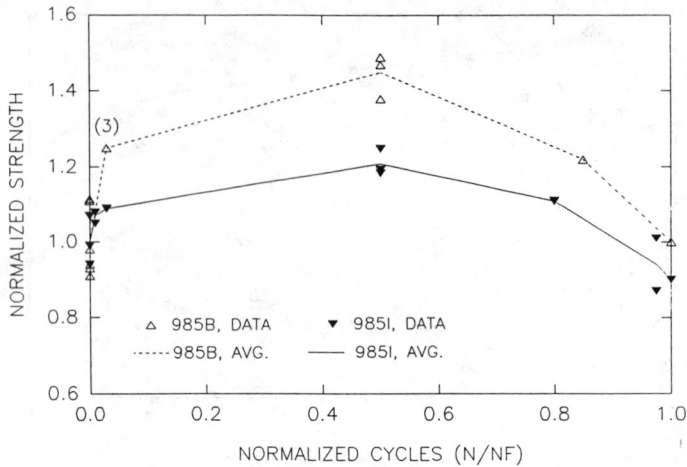

FIG. 3—*Normalized residual strength versus normalized life for AS4/985 baseline and interleaved material, individual and averaged data.*

greater change in stiffness throughout its lifetime than the B laminate. Recall, too, that the early and middle-life residual strength of the baseline laminate was larger than the interleaved laminate. Thus, though the B specimens initially responded to a stress level commensurate with its initial strength, this applied level quickly becomes a lower percentage of the specimens' ultimate strength, lower than the ratio of applied stress to ultimate

FIG. 4—*X-ray radiograph of baseline specimen at 30 000 cycles (15% stiffness decay).*

FIG. 5—*X-ray radiograph of interleaved specimen at 70 000 cycles (15% stiffness decay).*

strength in the I laminate. With these notions in mind, perhaps one may begin to truly compare the respective lives attained by these two materials.

Comparisons between damage states at equal stages of life—stages determined by the quantity of stiffness decline—revealed that the extent of delamination was not significantly different. Perhaps this finding is not very significant if one believes that the extent of delamination largely determines the amount of stiffness degradation. Yet, when compared on a cycle-to-cycle basis, the extent of delamination in the I material was obviously less. It has been observed in this study that while the threshold of delamination initiation was nearly the same for the two materials, the rate of propagation in the I laminate was less than the rate in the B laminate by approximately one half (based on damage states at equal cycles and by respective lives attained). This trend was echoed by Chan [*30*] in fatigue studies of delamination-prone laminates interleaved with adhesive strips. He also concluded that the disparity in the propagation rates between interleaved and noninterleaved laminates increases as the applied stress decreases.

If the applied stress levels in the two materials are considered approximately equal throughout their lives, then the interleaf appears to extend the life of the material at this high applied load level. A surprising observation, however, is the great increase in residual strength occurring in the B laminates when compared to the moderate increase experienced by the I laminates. Perhaps it is the relative position and extent of the delamination with respect to the regions of highest axial stress that dictate to what extent the critical elements (the 0° plies) may endure load prior to failure. Local stress concentrations are transmitted to the 0° plies via the interfaces. If the 0° interfaces delaminate in critical, high-stress regions, then 0° fibers would be free to deform without any proximate constraint. At this applied load level, delamination controls the life of the specimen; inhibiting the growth rate of the delamination extends the life of the specimen. Yet, apparently, it is this same delamination that controls the residual strength of the specimen during the early and middle stages of the lifetime by blunting the notch. Though the extent of the delamination is similar at comparable levels of stiffness loss, it may be the *position* and the extent of delamination with regards to position that determines the residual strength of these two materials at this applied load level. One can conclude that the residual strength of early and middle-life specimens is governed by a blunted notch effect, while the strength of late-life specimens is governed by the debilitating effect of large-scale delaminations.

Comparative Response of AS4/1808

Static Virgin Strength

The individual data and the average values for the static tensile and compressive virgin notched strength of the B and I specimens are found in Table 3. The average tensile load to failure of the I material is 12% higher than the average tensile load to failure of the B material. The observation of an increase in load to failure—in excess of the contribution made by the interleaves themselves—for the I laminates over the B laminates was echoed in the results of the AS4/985 material and in other unnotched material systems [22,29–32]. The difference in the average tensile strength between the two materials is minor, however. This is due, again, to the fact that the interleaf increases the cross-sectional area of the laminate, while contributing little to the load-carrying capacity. Though the specific calculations were not made, one should expect the specific strength of the B specimen to be on the order of 5% higher than the specific strength of the I laminate.

The compressive static properties of the I specimens do not reveal any significant differences over their tensile properties, while it is obvious from this limited sample size that the B material exhibits an unusual increase in compressive performance. One might have expected a decrease in compressive behavior in the I material due to the presence of low modulus support surrounding the 0° plies; this is not evident in the data. No obvious

TABLE 3—*Virgin notched strength of AS4/1808 baseline and interleaved specimens, tensile and compressive response, individual and averaged data.*

	Specimen ID	Failure Load, kN	Gross Area, cm²	Strength, MPa
Tensile	B–1–10	47.6	1.73	275
	B–1–12	47.6	1.73	275
	B–2–14	46.7	1.76	265
		Avg of three =		272
Tensile	I–1	52.0	1.92	272
	I–2	54.3	1.90	285
		Avg of two =		279
Compressive	B–1–9	55.6	1.75	318
	B–1–11	53.4	1.75	304
		Avg of two =		311
Compressive	I–3	51.6	1.93	267
	I–4	55.6	1.92	289
		Avg of two =		278

explanation, however, attends the increase in baseline performance. If, indeed, the AS4/ 1808 notched system possesses superior static compressive performance over tensile performance, then the statement may be made that the interleaf has a slight degrading effect on its compressive performance.

Fatigue Results

The baseline material was tested as part of a large study examining the differences between high load fatigue and low load fatigue in notched composites subjected to fully reversed ($R = -1$) fatigue loading [48]. Due to the limited amount of I material available to the authors, a direct comparison of the B and I laminates was not possible. The notion of testing the I material at a "middle" load level was settled upon. If the interleaf did not affect the response, then one would expect the test results to nestle in between the two extremes of behavior exhibited by the respective high/low response of the B material. If the interleaving did affect the response, one would expect the test results to tend to one of the extremes or appear uncharacteristic of an AS4/1808, quasi-isotropic B laminate under $R = -1$ cyclic fatigue. The high load level (a level which induces failure between 10 000 and 70 000 cycles) in the B material was considered to be 65% (203 MPa) of the ultimate compressive strength (UCS) of the B material. The low load level (the lowest level inducing failure before run-out occurs) in the B material was considered to be 50% (156 MPa) of the UCS of the B material. With these numbers established, the "middle" load level of the I material was chosen to be 60% (168 MPa) of the UCS of the I material.

Stiffness versus Life—A plot of the normalized stiffness versus normalized life for each load level is located in Fig. 6. The high load and low load response of the B material is the upper and lower curve, respectively. The low load response is highlighted by the three large stiffness reductions. These reductions correlate to the growth of large-scale delaminations. The overall stiffness reduction indicates the preponderance of matrix damage in the low load response. The high load curve reveals a steady degradation in stiffness

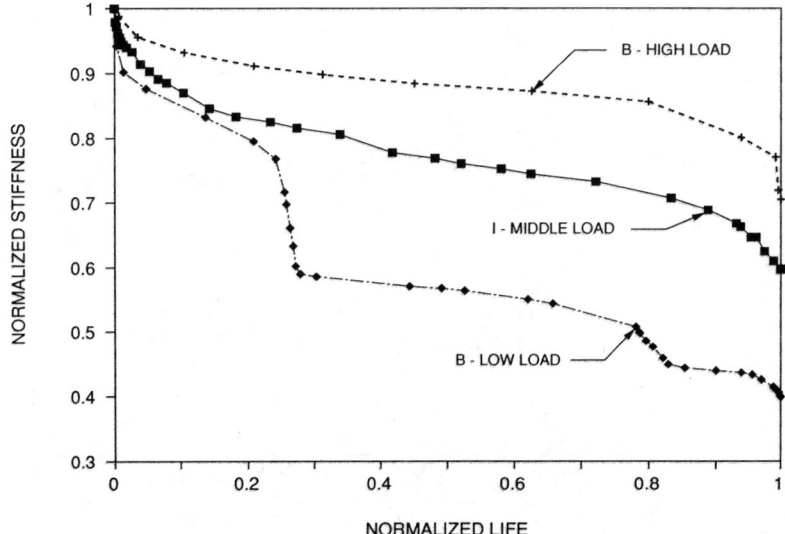

FIG. 6—*Comparison of normalized stiffness versus normalized life for AS4/1808 baseline and interleaved material fatigued at three distinct load levels.*

until near failure, where the stiffness decreases rapidly. This indicates that the matrix damage is not as extensive as the low load case. The curve corresponding to the I response nestles between the two extremes. The slope of the curve is quite similar to the high load response of the B material through 90% of normalized life. The first 5% and final 5% of the normalized life is similar, instead, to the low load response, exhibiting comparable wear-in and wear-out. Therefore, one might be tempted to conclude from the comparative stiffness curves that the response of the I material falls naturally between the two load level extremes of the B material.

Residual Strength and Its Relationship to Damage Development—A plot of the residual *tensile* (unfortunately, compressive residual strengths for the I material were not obtained) strength data for the I and B materials during their fatigue lifetime, normalized by their respective strengths, is located in Fig. 7. Early-life residual strengths increase for each system on the order of 10%. Yet differences in tensile residual strength are clearly visible at middle life. Again, it is necessary to appeal to the respective damage states in order to elucidate these differences.

The distinct damage mechanisms affecting each laminate are clearly illustrated in the radiographs taken at "middle" life (see Figs. 8 through 10). Figure 8 shows the damage state of a high load B specimen. Delaminations are growing both laterally and longitudinally from the center hole. The largest delaminations occur at the outermost 0/−45 interfaces. As this delamination grows, the 0° ply splits. This sequence gradually rids the outer two 0° plies of their compressive load-carrying capability, causing the interior plies to assume the brunt on this half of the load cycle. In the radiograph of the low load B specimen (Fig. 9), the 0/−45 interface has completely delaminated from the right side of the specimen. The density of matrix cracking in this specimen is much higher; the stiffness results gave every indication that this would be the case. Exterior 0° ply fracture is apparent from Fig. 9b. Again, as the exterior plies fail, either by delamination alone or by 0° ply fracture followed by delamination, a larger onus is placed on the interior of the

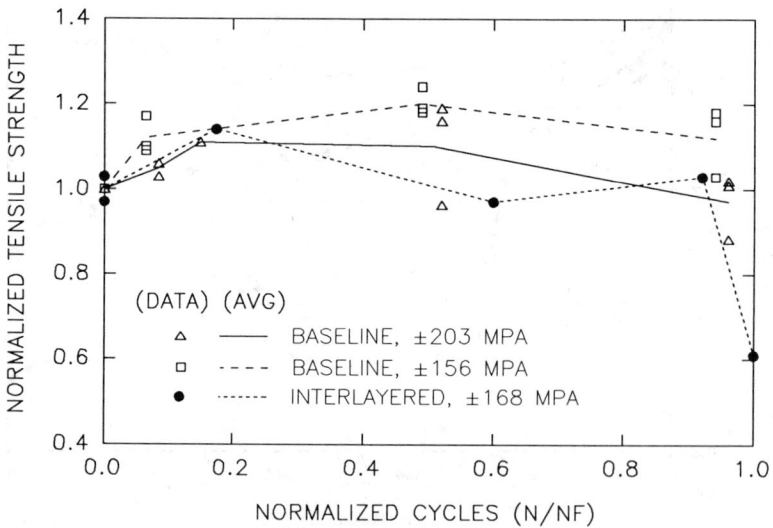

FIG. 7—*Normalized residual strength versus normalized life for AS4/1808 baseline and interleaved material fatigued at three distinct load levels; individual and averaged data.*

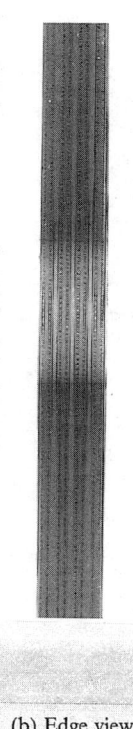

(a) Front view   (b) Edge view

FIG. 8—*Baseline specimen under high load fatigue at "middle" life* (n = *8800 cycles):* (a) *front view,* (b) *edge view.*

laminate. Notice that no fractures are yet apparent in the interior 0° plies. The damage state of the I specimen is documented in Fig. 10. This damage state shares characteristics of both the high and low B response. Delaminations are present, and their presence (at least between the outermost 0/ − 45 interfaces) is accompanied by 0° longitudinal cracking; this is reminiscent of the high load B response. A high density of matrix damage also attends this stage of life, like that found in the low load B response. A *very* intense zone of damage is seen at 90° on each side of the hole. Unseen from both the flatwise and edge radiographs (for *all* of these 1808 specimens) is the level of outer 0° ply fracture that occurs in a step-wise fashion along the path of the major − 45° cracks emanating from the hole. These fractures occur during the compressive excursion of the load cycle. The most dramatic difference between the I damage state and the B damage states at both high and low loads is seen in Fig. 10*b.* Though the radiograph lacks the optimal definition, bands of fracture appear in the interior of the laminate where the 0° plies are located. It is likely that this damage is located at the hole boundary and is compression-induced.

It is this particular difference in the I and B damage states at this stage in life that likely accounts for the variations in residual tensile strength between the two materials (see Fig. 7). The I residual strength has approached its virgin value, no longer revealing a reduction in the notch stress concentration. Both B specimens still reveal an increase in their respective tensile residual strengths, with the low load specimen showing a larger increase of the two. However, one data point representing a middle-life, high load B specimen is significantly lower than other comparative points. Radiographs of this low strength specimen reveal less delamination and more 0° fiber fracture than that seen in similar specimens

(a) Front view   (b) Edge view

FIG. 9—*Baseline specimen under low load fatigue at "middle" life (n = 1 052 000 cycles): (a) front view, (b) edge view.*

at this point in the lifetime, even though the stiffness drop is similar. This information, combined with additional information in Ref *48*, leads to the conclusion that the high load level for the B specimens is at or slightly below a threshold for a transition in the mode of damage formation and propagation. In this case, the transition would be from a damage pattern characterized by both transverse 0° fiber fracture and longitudinal delamination growth to one characterized by relatively less delamination and more fiber fracture, similar to the transition seen by Simonds [8,9] in notched AS4/PEEK specimens with the same stacking sequence. Simonds noted that the residual tensile strength of specimens experiencing the fiber fracture-dominated damage pattern was significantly less than that of delamination-dominated specimens—sufficiently so, in fact, to cause tensile failures during fully reversed fatigue loading rather than the usual compressive failures seen with lower load levels.

The late-life damage states under the three load levels all reveal essentially the same failure characteristics. One face of the laminate is fractured (as in Fig. 9*b*), which causes an imbalance in the applied load. This outer-ply fracture contributes to the final failure differently in the two laminate types, however. The high and low load baseline specimens fail in a compression/instability mode. Obviously, their respective compressive strengths decay at a greater rate than their tensile strengths; this is shown in Ref *48*. The interleaved specimen, on the other hand, tested at this "middle" load level, fails in *tension*.

Discussion of Results—The dearth of interleaved specimens prevented a direct comparison of B and I material performance subjected to R = −1 fatigue. Instead, the I

(a) Front view   (b) Edge view

FIG. 10—*Interleaved specimen under middle load fatigue at "middle" life* (n = *110 000 cycles): (a) front view, (b) edge view.*

material was exposed to a "middle" load level in order to compare its behavior to the baseline laminates subjected to high and low load extremes. The normalized stiffness versus normalized life curves, and, for the most part, the damage states as recorded by X-ray radiography revealed that the I materials' response settled in between the two load level extremes. Differences in the residual tensile strength of the I and B laminates arose during and after "middle" life that could be traced to the development of damage, seen especially in the edge radiographs. Residual compressive strength tests on these 32-ply specimens may shed light on the relative importance of fiber fracture and delamination on the fatigue failure mode of the AS4/1808 material system. Such results, given in Refs *48* and *49* for B specimens, indicate rather similar residual compressive strengths for the two fully reversed cyclic load levels investigated. In both cases the decline was monotonic throughout the lifetime, culminating in a compressive fatigue failure mode. Questions remain whether or not the compressive strength of the I material degrades in the same fashion as the B material, and how the I material's compressive strength depends on load level.

Summary and Conclusions

The most graphic means of visualizing the effect that interleaves have on the residual strength and damage mechanisms of center-notched, cyclically loaded laminates is to study the differences obtained over the baseline performance. In the 16-ply, center-notched AS4/985 subjected to high load, $R = 0.1$ fatigue, the damage sequence was as follows: delamination originated at the center hole and at the two specimen edges and propagated inward. The predominant delaminations occurring at the outer 0/45 interfaces eventually converged in the center of the ligament, created unsymmetric sublaminates, and caused the interior stresses to rise precipitously, all leading to imminent failure. The early presence of delamination at the hole boundary relieved the geometric stress concentration and allowed the net section residual strength of the B laminate to attain its unnotched virgin strength. Even at applied stresses approaching 100% of the virgin notched strength, 0° fiber fracture was not a relevant damage mechanism.

The interleaves were responsible for slightly altering the above course of events in the 16-ply laminates. While it was surprising how relatively ineffective the interleaves were at suppressing the presence of delamination, obviously the rate of delamination propagation was influenced. At this extremely high fatigue load level, though the failure sequence was essentially the same, by slowing the rate of delamination propagation (the life-limiting mechanism in this specific scenario) the fatigue life of the I laminates was extended. This was achieved at the expense of substantial gains in residual strength, however. Though delamination was present early in the fatigue life, the position and extent of the delamination may have precluded optimal stress concentration relief.

The baseline performance of the 32-ply, center-notched AS4/1808 material was studied at two $R = -1$ fatigue load extremes which, in turn, were responsible for creating two distinct responses. Though direct comparisons between interleaved and baseline performance are tenuous at best due to the test matrix adopted, some general statements can be made. Delamination is, again, a life-limiting failure mode, yet its presence is influenced by the presence of 0° fiber fracture. This outer ply fiber fracture combines with the delamination and gradually rids the outer plies of their load-carrying capability, leading to dramatically increased stresses in the interior plies and, in some instances, unsymmetric sublaminates. The creation of unsymmetric sublaminates is particularly detrimental to compressively loaded specimens because of the accelerated growth of delamination on the side of the laminate experiencing relatively greater compression. The two load level extremes in the baseline material differ in their severity of fiber fracture growth versus delamination growth. These differing mechanisms may account for the differences seen in the tensile residual strength throughout the fatigue life.

In the 1808 specimens, the interleaves again alter the damage sequence and material behavior. Though exhibiting both outer ply fiber fracture and delamination, the interior of the laminate responds to these damage mechanisms differently. Interior 0° ply fracture was present early on in the fatigue lifetime in some specimens, and the presence of this fracture seemed to be more deleterious to the residual tensile strength than the residual compressive strength since fatigue failures in the interleaved specimens were tensile.

In general, the 1808 material system responded similarly to the 985 system to interleaving. That is, in addition to the expected increase in mass and decrease in laminate stiffness, interleaving resulted in slightly higher virgin tensile strengths, lower normalized residual tensile strengths during fatigue, and altered damage patterns. Compared to their effect on the 985 material, the interleaves had a somewhat stronger influence on fatigue damage in the 1808 material, but a noticeably smaller influence on residual tensile strength of those same specimens at the early-, middle-, and late-life measurement points. It is difficult to explain this curious result at this time because of the confounding effects of matrix toughness and stress ratio that need to be separated by testing specimens with identical thicknesses.

The results obtained herein suggest several possible schemes to improve laminate static and fatigue performance through interleaving. Interleaving at every interface seems superfluous. The placement of interleaves at critical interfaces should produce similar improvements in performance while markedly reducing the laminate's mass and increasing the laminate's stiffness. Perhaps the selective placement of strip- or annular-shaped interleaves could replace the full-width interleaves used within this study. Interleaves placed a small distance away from the notch could allow the notch-blunting effect to occur unabated, yet still attempt to contain the life-limiting growth of delaminations to a small area near the notch. The employment of these schemes and the study of other pertinent variables remain as a topic for future research.

Acknowledgments

The authors gratefully acknowledge the support of the Air Force Office of Scientific Research through a contract monitored by Lt. Col. George K. Haritos. The authors also wish to acknowledge the American Cyanamid Company and Dr. John E. Masters for supplying the material employed in this study.

References

[1] Masters, J. E., "Characterization of Impact Damage Development in Graphite/Epoxy Laminates," *Fractography of Modern Engineering Materials: Composites and Metals, ASTM STP 948*, J. E. Masters and J. J. Au, Eds., American Society for Testing and Materials, Philadelphia, 1987, pp. 238–258.

[2] Chang, I. Y., "PEKK as a New Thermoplastic Matrix for High Performance Composites," *Proceedings, 33rd International SAMPE Symposium and Exhibition*, G. Carrillo, E. D. Hewell, W. D. Brown, and P. Phelan, Eds., 7-10 March 1988, Anaheim, CA, Society for the Advancement of Material and Process Engineering, Covina, CA, pp. 194–205.

[3] Wedgewood, A. R., Su, K. B., and Narin, J. A., "Toughness Properties and Service Performance of High Temperature Thermoplastics and Their Composites," *Proceedings, 19th International SAMPE Technical Conference*, T. Lynch, J. Persh, T. Wolf, and N. Rupurt, Eds., 13-15 Oct. 1987, Crystal City, VA, Society for the Advancement of Material and Process Engineering, Covina, CA, pp. 454–460.

[4] Baron, C., Schulte, K., and Harig, H., "Influence of Fibre and Matrix Failure Strain on Static and Fatigue Properties of Carbon Fibre-Reinforced Plastics," *Composites Science and Technology*, Vol. 29, 1987, pp. 257–272.

[5] Baron, C. and Schulte, K., "Fatigue Damage Response of CFRP with Toughened Matrices and Improved Fibres," *Proceedings, International Conference on Composite Materials VI/ECCM II*, F. L. Matthews, N. C. R. Buskell, J. M. Hodgkinson, and J. Morton, Eds., Vol. 4, Elsevier Applied Science, New York, 1987, pp. 65–75.

[6] Curtis, P. T., "An Investigation of the Tensile Fatigue Behaviour of Improved Carbon Fibre Composite Materials," *Proceedings, International Conference on Composite Materials VI/ECCM II*, F. L. Matthews, N. C. R. Buskell, J. M. Hodgkinson, and J. Morton, Eds., Vol. 4, Elsevier Applied Science, New York, 1987, pp. 54–64.

[7] Bakis, C. E., Simonds, R. A., Vick, L. W., and Stinchcomb, W. W., "Matrix Toughness, Long-Term Behavior, and Damage Tolerance of Notched Graphite Fiber-Reinforced Composite Materials," *Composite Materials: Testing and Design (Ninth Volume), ASTM STP 1059*, S. P. Garbo, Ed., American Society for Testing and Materials, Philadelphia, 1990, pp. 349–370.

[8] Simonds, R. A., Bakis, C. E., and Stinchcomb, W. W., "Effects of Matrix Toughness on Fatigue Response of Graphite Fiber Composite Laminates," *Composite Materials: Fatigue and Fracture, Second Volume, ASTM STP 1012*, P. A. Lagace, Ed., American Society for Testing and Materials, Philadelphia, 1989, pp. 5–18.

[9] Simonds, R. A. and Stinchcomb, W. W., "Response of Notched AS4/PEEK Laminates to Tension/Compression Loading," *Advances in Thermoplastic Matrix Composite Materials, ASTM STP 1044*, G. M. Newaz, Ed., American Society for Testing and Materials, Philadelphia, 1989, pp. 133–145.

[10] Reifsnider, K. L., Stinchcomb, W. W., and O'Brien, T. K., "Frequency Effects on a Stiffness-Based Fatigue Failure Criterion in Flawed Composite Specimens," *Fatigue of Filamentary Composite Materials, ASTM STP 636*, K. L. Reifsnider and K. N. Lauraitis, Eds., American Society for Testing and Materials, Philadelphia, 1977, pp. 171–184.

[11] Bishop, S. M. and Dorey, G., "The Effect of Damage on the Tensile and Compressive

Performance of Carbon Fibre Laminates," *Characterization, Analysis, and Significance of Defects in Composite Materials,* AGARD-CP-355, 1983, pp. 10.1–10.10.

[12] Kress, G. R. and Stinchcomb, W. W., "Fatigue Response of Notched Graphite/Epoxy Laminates," *Recent Advances in Composites in the United States and Japan, ASTM STP 864,* J. R. Vinson and M. Taya, Eds., American Society for Testing and Materials, Philadelphia, 1985, pp. 173–196.

[13] Bakis, C. E. and Stinchcomb, W. W., "Response of Thick, Notched Laminates Subjected to Tension-Compression Cyclic Loads," *Composite Materials: Fatigue and Fracture, ASTM STP 907,* H. T. Hahn, Ed., American Society for Testing and Materials, Philadelphia, 1986, pp. 314–334.

[14] Morton, J., Kellas, S., and Bishop, S. M., "The Effect of Environment on the Fatigue Damage Development in Notched Carbon Fibre Composites," *Proceedings, International Conference on Composite Materials VI/ECCM II,* F. L. Matthews, N. C. R. Buskell, J. M. Hodgkinson, and J. Morton, Eds., Vol. 4, Elsevier Applied Science, New York, 1987, pp. 139–149.

[15] Kellas, S., Morton, J., and Bishop, S. M., "Fatigue Damage Development in a Notched Carbon Fibre Composite," *Composite Structures,* Vol. 5, 1986, pp. 143–157.

[16] Bakis, C. E., Yih, H. R., Stinchcomb, W. W., and Reifsnider, K. L., "Damage Initiation and Growth in Notched Laminates Under Reversed Cyclic Loading," *Composite Materials: Fatigue and Fracture, Second Volume, ASTM STP 1012,* P. A. Lagace, Ed., American Society for Testing and Materials, Philadelphia, 1989, pp. 66–83.

[17] Maier, G., Ott, H., Protzner, A., and Protz, B., "Notch Sensitivity of Multidirectional Carbon Fibre-Reinforced Polyimides in Fatigue Loading as a Function of Stress Ratio," *Composites,* Vol. 18, No. 5, November 1987, pp. 375–380.

[18] O'Brien, T. K., "Fatigue Delamination Behavior of PEEK Thermoplastic Composite Laminates," *Proceedings, First Technical Conference of the American Society for Composites,* Technomic Publishing Co., Lancaster, PA, 1986, pp. 404–420.

[19] O'Brien, T. K., Murri, G. B., and Salpekar, S. A., "Interlaminar Shear Fracture Toughness and Fatigue Thresholds for Composite Materials," *Composite Materials: Fatigue and Fracture, Second Volume, ASTM STP 1012,* P. A. Lagace, Ed., American Society for Testing and Materials, Philadelphia, 1989, pp. 222–250.

[20] Mall, S., Yun, K. -T., and Kochhar, N. K., "Characterization of Matrix Toughness Effect on Cyclic Delamination Growth in Graphite Fiber Composites," *Composite Materials: Fatigue and Fracture, Second Volume, ASTM STP 1012,* P. A. Lagace, Ed., American Society for Testing and Materials, Philadelphia, 1989, pp. 296–310.

[21] Browning, C. E. and Schwartz, H. S., "Delamination Resistant Composite Concepts," *Composite Materials: Testing and Design (Seventh Conference), ASTM STP 893,* J. M. Whitney, Ed., American Society for Testing and Materials, Philadelphia, 1986, pp. 256–265.

[22] Soni, S. R. and Kim, R. Y., "Analysis of Suppression of Free-Edge Delamination by Introducing Adhesive Layer," *Proceedings, International Conference on Composite Materials VI/ECCM II,* F. L. Matthews, N. C. R. Buskell, J. M. Hodgkinson, and J. Morton, Eds., Vol. 5, Elsevier Applied Science, New York, 1987, pp. 219–230.

[23] Krieger, R. B. Jr., "The Relation Between Graphite Composite Toughness and Matrix Shear Stress-Strain Properties," *Proceedings, Technology Vectors: 29th National SAMPE Symposium,* Society for the Advancement of Material and Process Engineering, Covina, CA, April 1984, pp. 1570–1584.

[24] Hirshbuehler, K. R., "A Comparison of Several Mechanical Tests Used to Evaluate the Toughness of Composites," *Toughened Composites, ASTM STP 937,* N. J. Johnston, Ed., American Society for Testing and Materials, Philadelphia, 1987, pp. 61–73.

[25] Masters, J. E., "Correlation of Impact and Delamination Resistance in Interleafed Laminates," *Proceedings, International Conference on Composite Materials VI/ECCM II,* F. L. Matthews, N. C. R. Buskell, J. M. Hodgkinson, and J. Morton, Eds., Vol. 3, Elsevier Applied Science, New York, 1987, pp. 96–107.

[26] Masters, J. E., "Development of Composites Having Improved Resistance to Delamination and Impact," Interim Technical Report No. 4, Contract No. F33615-84-C-5024, The American Cyanamid Company, Stamford, CT, October 1985.

[27] Ishai, O., Rosenthal, H., Sela, N., and Drukker, E., "Effect of Selective Interleaving on Interlaminar Fracture Toughness of Graphite/Epoxy Composite Laminates," *Composites,* Vol. 19, No. 1, January 1988, pp. 49–54.

[28] Sela, N., Ishai, O., and Banks-Sills, L., "The Effect of Adhesive Thickness on Interlaminar Fracture Toughness of Interleaved CFRP Specimens," *Composites,* Vol. 20, No. 3, May 1989, pp. 257–264.

[29] Chan, W. S., Rogers, C., and Aker, S., "Improvement of Edge Delamination Strength of Composite Laminates Using Adhesive Layers," *Composite Materials: Testing and Design (Seventh Conference), ASTM STP 893,* J. M. Whitney, Ed., American Society for Testing and Materials, Philadelphia, 1986, pp. 266–285.

[30] Chan, W. S., "Delamination Arrester—An Adhesive Inner Layer in Laminated Composites," *Composite Materials: Fatigue and Fracture, ASTM STP 907,* H. T. Hahn, Ed., American Society for Testing and Materials, Philadelphia, 1986, pp. 176–196.

[31] Sun, C. T. and Jen, K. C., "On the Effect of Matrix Cracks on Laminate Strength," *Proceedings, First Technical Conference of the American Society for Composites,* Technomic Publishing Co., Lancaster, PA, 1986, pp. 352–367.

[32] Lagace, P. A., Weems, D. B., and Brewer, J. C., "Suppression of Delamination via an Interply Adhesive Layer," *Composites '86: Recent Advances in Japan and the United States,* S. Umekama and A. Kobayashi, Eds., proceedings of the Japan-U.S. CCM-III, Tokyo, 1986, Japan Society for Composite Materials, pp. 323–330.

[33] Evans, R. E. and Masters, J. E., "A New Generation of Epoxy Composites for Primary Structural Applications: Materials and Mechanics," *Toughened Composites, ASTM STP 937,* N. J. Johnston, Ed., American Society for Testing and Materials, Philadelphia, 1987, pp. 413–436.

[34] Hirshbuehler, K. R., "An Improved 270° F Performance Interleaf System Having Extremely High Impact Resistance," *SAMPE Quarterly,* Vol. 17, No. 1, October 1985, pp. 46–49.

[35] Swain, R. E., Reifsnider, K. L., and Vittoser, J., "Investigation of Damage in Composite Laminates Using the Incremental Strain Test," *Composite Materials: Testing and Design (Ninth Volume), ASTM STP 1059,* S. P. Garbo, Ed., American Society for Testing and Materials, Philadelphia, 1990, pp. 390–403.

[36] Subramanian, S., "Frequency Effect on the Fatigue Development Onset and Growth Characteristics of Laminated Composites," Master's thesis, Dept. of Materials Science, The University of Texas at Arlington, August 1990.

[37] Altus, E. and Ishai, O., "The Effect of Soft Interleaved Layers on the Combined Transverse Cracking/Delamination Mechanisms in Composite Laminates," *Composite Science and Technology,* Vol. 39, January 1990, pp. 13–27.

[38] Lubowinski, S. J., Guynn, E. G., Elber, W., and Whitcomb, J. D., "Loading Rate Sensitivity of Open-Hole Composite Specimens in Compression," *Composite Materials: Testing and Design (Ninth Volume), ASTM STP 1059,* S. P. Garbo, Ed., American Society for Testing and Materials, Philadelphia, 1990, pp. 457–476.

[39] Williams, J. G., "The Multi-Beam Shear Test Method for Studying Composite Transverse Shear Failure Characteristics," *Proceedings,* AIAA/ASME/ASCE/AHS 26th Structures, Structural Dynamics and Materials Conference, Orlando, April 1985, ASCE, New York.

[40] Masters, J. E., Courter, J. L., and Evans, R. E., "Impact Fracture and Failure Suppression Using Interleafed Composites," *International SAMPE Symposium and Exhibition Series,* Vol. 31, Las Vegas, Society for the Advancement of Material and Process Engineering, Covina, CA, April 1986, pp. 844–858.

[41] Sun, C. T., "Suppression of Delamination in Composite Laminates Subjected to Impact Loading," *11th Annual Mechanics of Composite Review,* Air Force Materials Laboratory, Dayton, OH, October 1986, pp. 106–113.

[42] Sun, C. T. and Rechak, S., "Effect of Adhesive Layers on Impact Damage in Composite Laminates," *Composite Materials: Testing and Design (Eighth Conference), ASTM STP 972,* J. D. Whitcomb, Ed., American Society for Testing and Materials, Philadelphia, 1988, pp. 97–123.

[43] Gandhe, G. V., "Impact Response of Interleaved Composite Materials," masters thesis, College of Engineering, Virginia Polytechnic Institute and State University, Blacksburg, VA, December 1988.

[44] Bakis, C. E., Simonds, R. A., and Stinchcomb, W. W., "A Test Method to Measure the Response of Composite Materials Under Reversed Cyclic Loads," *Test Methods for Design Allowables for Fibrous Composites, Second Volume, ASTM STP 1003,* C. C. Chamis, Ed., American Society for Testing and Materials, Philadelphia, 1989, pp. 180–193.

[45] Reifsnider, K. L., Schulte, K., and Duke, J. C., "Long-Term Fatigue Behavior of Composite Materials," *Long-Term Behavior of Composites, ASTM STP 813,* T. K. O'Brien, Ed., American Society for Testing and Materials, Philadelphia, 1983, pp. 136–159.

[46] Reifsnider, K. L. and Stinchcomb, W. W., "Stiffness Change as a Fatigue Damage Parameter for Composite Laminates," *1983 Advances in Aerospace Structures, Materials, and Dynamics— AD-06,* U. Yuceoglu, R. L. Sierakowski, and D. A. Glasgow, Eds., American Society of Mechanical Engineers, New York, 1983, pp. 1–6.

[47] Crossman, F. W. and Wang, A. S. D., "The Dependence of Transverse Cracking and

Delamination on Ply Thickness in Graphite/Epoxy Laminates," *Damage in Composite Materials, ASTM STP 775*, K. L. Reifsnider, Ed., American Society for Testing and Materials, Philadelphia, 1982, pp. 118–139.

[48] Bakis, C. E., "Fatigue Behavior of Notched Carbon Epoxy Laminates During Reversed Cyclic Loads," Ph.D. dissertation, College of Engineering, Virginia Polytechnic Institute and State University, Blacksburg, VA, August 1988.

[49] Reifsnider, K. L., Stinchcomb, W. W., Bakis, C. E., and Yih, R. Y., "The Mechanics of Micro-Damage in Notched Composite Laminates," *Damage Mechanics in Composites*, AD Vol. 12, A. S. D. Wang and G. K. Haritos, Eds., ASME, New York, 1987, pp. 65–72.

Mehran Elahi,[1] *Ahmad Razvan,*[1] *and Kenneth L. Reifsnider*[1]

Characterization of Composite Material's Dynamic Response Using Load/Stroke Frequency Response Measurement

REFERENCE: Elahi, M., Razvan, A., and Reifsnider, K. L., "**Characterization of Composite Material's Dynamic Response Using Load/Stroke Frequency Response Measurement,**" *Composite Materials: Fatigue and Fracture, Fourth Volume, ASTM STP 1156,* W. W. Stinchcomb and N. E. Ashbaugh, Eds., American Society for Testing and Materials, Philadelphia, 1993, pp. 575–588.

ABSTRACT: A new experimental technique was developed to characterize damage development in composite materials. The technique does not require interruption of the test. It utilizes the frequency response measurement of load/stroke signals to characterize fatigue damage in terms of parameters such as phase and gain. Test frame and frequency dependency of the method was investigated. Center-notch quasi-isotropic unidirectional specimens were fatigue cycled at various load levels. From the fatigue test results, it was found that gain is related to the total damage and that phase may be related to the rate of damage in the specimen. Results also indicated that the method was sensitive to the applied load level and the material systems. This technique might be able to overcome problems involved in fatigue damage characterization of unidirectional laminates during high-temperature testing of composites where extensive splitting and elevated temperatures limit the use of extensometers.

KEY WORDS: phase lag, gain, unidirectional laminate, quasi-isotropic, fatigue, stiffness, dynamic response

Failure modes of composite materials are complicated and generally different from those of metals. Unlike most isotropic materials where a single crack is the dominant mode of failure, in composites fatigue damage is extensive and is spread throughout the specimen volume. In composites microcracks may be initiated at an early stage of loading, but the materials can sustain the load until final failure after many additional cycles of loading [1,2].

Fatigue failure can occur if the residual strength of material degrades to the level of applied load [3]. This degradation can be caused by matrix cracking [4], delamination [5], fiber fracture [6], and interfacial debonding [7]. Any combination of these may be responsible for fatigue damage, which may result in reduced fatigue strength and stiffness [1–7]. Material properties, specimen geometry, stacking sequence, waveform type, loading waveform frequency, loading mode, loading rate, time, and temperature are some of the variables critical in any fatigue study or service environment [8–12]. Variation in any of these variables could result in different damage evolution mechanisms and processes.

The state of damage is related to the three most important engineering material characteristics: stiffness, strength, and life. In general, stiffness is related to damage in a deter-

[1]Graduate research assistant, research associate, and Alexander Giacco professor, respectively, Materials Response Group, Engineering Science and Mechanics Dept., Virginia Polytechnic Institute and State University, Blacksburg, VA 24061-0219.

ministic fashion. Therefore, stiffness loss, which is often large and easily measured, can often provide a basis for the characterization of fatigue damage or rate of damage development [13–15]. The most common technique for measuring stiffness change is by means of an interrupted fatigue test. Fatigue-cycled specimens are typically stopped at various stages of life and loaded quasi-statically to measure their stiffness. This procedure is repeated during the life of the specimen. The end result is a relationship between the static stiffness degradation and the stage of life. Due to the static nature of this measurement technique, important information is neglected using this method. Every time the test is stopped, the initial conditions to this forced vibration problem are altered [16]. In order to fully understand the fatigue behavior of materials as a function of stiffness change, it is desirable to monitor the dynamic response of the specimen continuously in real time.

Several investigators have utilized dynamic stress-strain signals in order to characterize fatigue damage development by measuring phase and stiffness where the phase values are obtained either from direct measurement at the zero-crossing point of signals [16] or by constructing a hysteresis loop [17–18]. This procedure requires use of an extensometer or an attached strain gage to measure strain. In high-temperature or high-frequency fatigue tests, there is always the possibility of extensometer slippage [16], and it is difficult to attach any strain measurement device to a specimen inside a furnace. Another shortcoming of an extensometer procedure is in the fatigue testing of unidirectional materials. These fatigue tests may cause a great deal of matrix splitting on the specimen surface which, in turn, limits the use of an extensometer by disrupting the position of the extensometer contact points.

So et al. [19] developed a free-vibration technique for the measurement of material damping under periodic chirp excitation. Using an experimental arrangement similar to that of So [19], in the present case a new experimental technique is developed to resolve the aforementioned problems. Based on this technique, a new approach to the interpretation of fatigue behavior is proposed. This method utilizes the load and stroke signals from a servohydraulic test machine to measure quantities such as phase lag and gain[2] for measuring damping and compliance, respectively. Even though a thorough understanding of these dynamic parameters is not at hand, our preliminary results indicate a good correlation between stiffness degradation, phase, and gain response measured in this fashion. If this is true, this technique could be a valuable tool for the dynamic evaluation of composite parts under conditions in which a change of strength, stiffness, and life must be monitored and interpreted in terms of durability, damage tolerance, etc.

Experimental Procedure

In order to characterize the phase and gain functionals, four different servohydraulic test frame units were used. This was in conjunction with two different material systems (described below). As listed in Table 1, Units 3 and 4 have the same load frames, while having different hydraulic grips. Units 2 and 3 have the same hydraulic grips but different load frames. Unit 1 was unique and did not have any common features with the other units. It is believed that through such a test arrangement the effect of gripping, load frame characteristics, signal conditioning, as well as amplifiers could easily be identified and determined.

Different material systems were fatigue cycled (tension-tension, $R = 0.1$) under load-

---

[2]Gain is defined as the ratio of the magnitude (amplitude) of a steady-state sinusoidal output relative to the causal input; the length of a phasor from the origin to a point of the transfer locus in a complex plane [22].

TABLE 1—*MTS servohydraulic test frames used in the study.*

Load Frame unit	Controller Model Number	Load Frame Model Number	Grip Assembly Model Number	Load Capacity kN/kip
MTS 1	458	309.21	641.36	250/55
MTS 2	413	312.21	647	100/22
MTS 3	448.82	380.10	647	100/22
MTS 4	436	380.10	641.35	100/22

controlled conditions using MTS servohydraulic testing machines at a frequency of 10 Hz under a sinusoidal waveform. Notched specimens made of Hercules AS4 fiber and American Cyanamid's Cycom 985 matrix with a stacking sequence of $[0/45/90/-45]_{s2}$ were used. Thin adhesive interlayers measuring 12.7 μm (0.0005 in.) in thickness were placed between each layer and on one of the outside surfaces. These specimens had an average length of 139.7 mm (5.5 in.), width of 25.4 mm (1.0 in.), thickness of 2.461 mm (0.0969 in.), and a 6.35-mm (0.25-in.)-diameter hole at the center [20].

Also, eight-ply unidirectional coupons made of Hexcel's graphite/epoxy prepreg with a layer of release cloth (non-TFE-fluorocarbon coated) embedded at the middle ply were used in this investigation. Specimens had an average length of 152.4 mm (6 in.), width of 25.4 mm (1 in.), and thickness of 1.27 mm (0.05 in.). Glass epoxy tabs with 30° tapering were used to reduce the stress concentration due to gripping of the specimen. Tabs and the tab-section surfaces of the specimen were lightly sandblasted for cleaning as well as to increase the mechanical friction between the adjoining surfaces. Cyanamid's FM-300K adhesion film was used for binding the tabs to the specimens. The tabs were then cured at 177°C (350°F) for 1 h in a hot press at 75 psi (517 kPa) [21].

Dynamic data acquisition was performed using an HP 9000/PC-315 computer in-line with an HP-3852A/HP-3853A data acquisition control unit as well as an HP 3562-A dynamic signal analyzer (Fig. 1). A software routine was developed for real time dynamic response analysis. Phase lag and gain response of the load/stroke signals from the MTS servohydraulic load frame were plotted in conjunction with load, stroke, and temperature data. To avoid disk storage overflow, data were sampled according to their relative change with respect to the previous events.

Working in the frequency domain, all measurements were made at the system's excitation frequency in a linear resolution mode. First, by using the cross spectrum function, the fundamental frequency was determined. Then the phase lag and gain measurements were made at this frequency using a frequency response measurement function (Figs. 2 and 3).

Results and Discussion

The frequency response measurement, often called the "transfer function," is defined as the ratio of a system output to its input and yields both gain and phase as a function of frequency. In the HP 3562-A, the signal on Channel 1 is assumed to be the system's input (load), and the signal on Channel 2 is assumed to be its output (stroke). The frequency response is calculated as the ratio of the cross spectrum to the Channel 1 (load) signal power spectrum [22]

$$H(f) = \frac{G_{xy}}{G_{xx}}$$

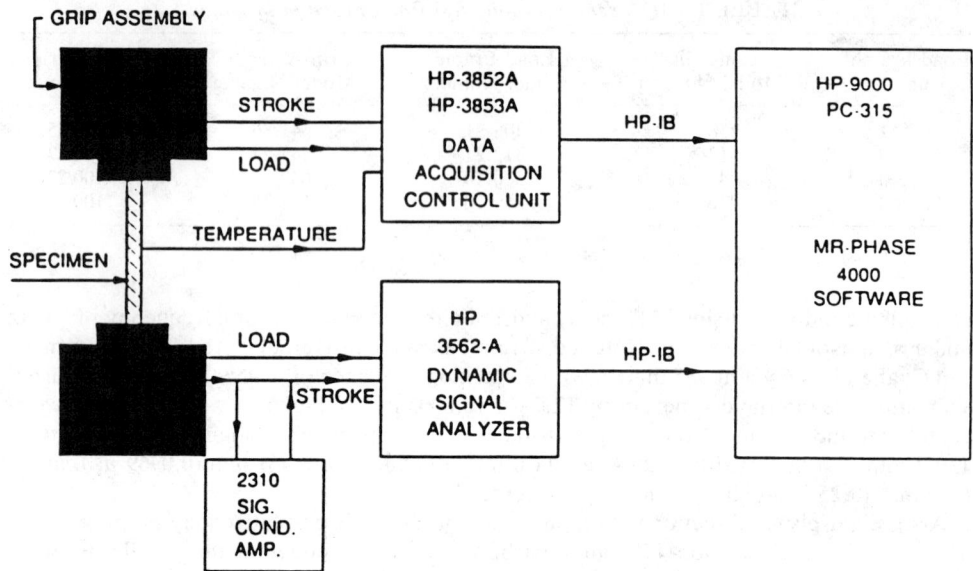

FIG. 1—*Schematic diagram of the experimental setup.*

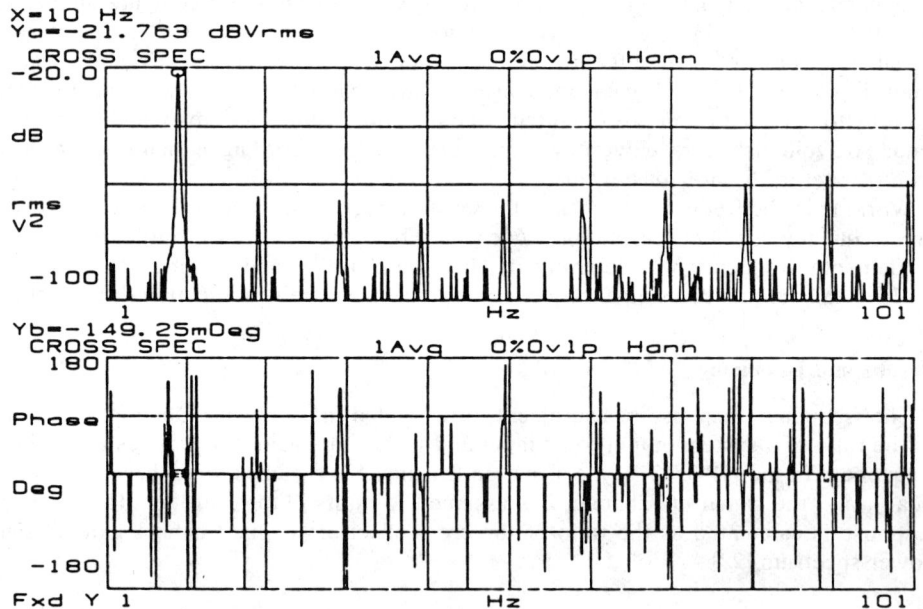

FIG. 2—*Cross spectrum measurement of an aluminum specimen fatigued at a frequency of 10 Hz and 40% σ_{yt}.*

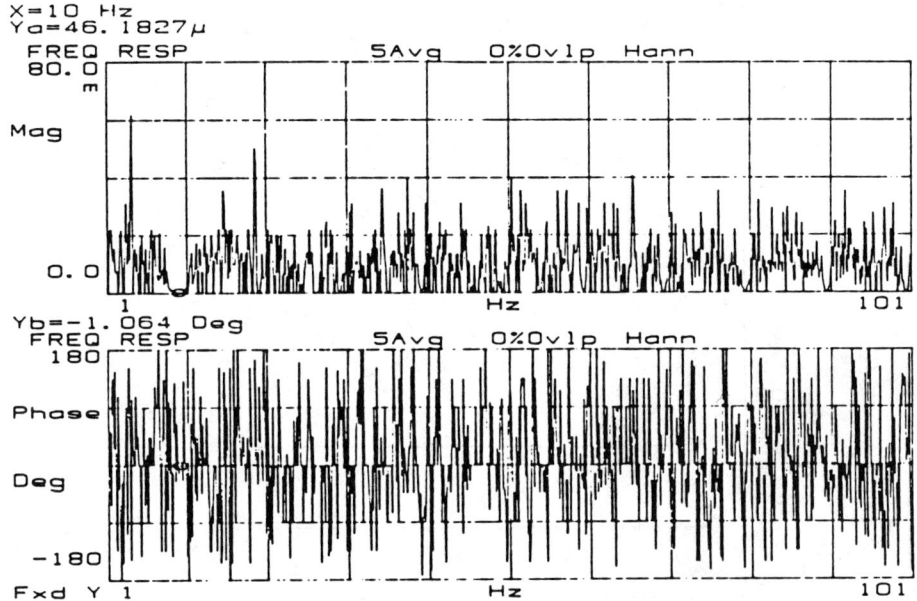

FIG. 3—*Frequency response measurement of an aluminum specimen fatigued at a frequency of 10 Hz and 40% σ_{yt}.*

where

G_{xy} = cross spectrum, and
G_{xx} = Channel 1 power spectrum.

and where

$$G_{xy} = F_x F_y^*$$

where

F_y^* = Channel 2 linear spectrum's complex conjugate, and
F_x = Channel 1 linear spectrum.

and where

$$G_{xx} = F_x F_x^*$$

* Denotes two different F_x.

The dependency of gain and phase on parameters such as hydraulic test frame, excitation frequency, and the applied load level were investigated. To obtain the effect of test frame and frequency on phase and gain, a steel specimen was cycled at 40% of its tensile yield strength with a fatigue ratio of 0.1. It is observed that gain is not influenced by frequency but that its base level is somewhat influenced by the test frame (Fig. 4). As is

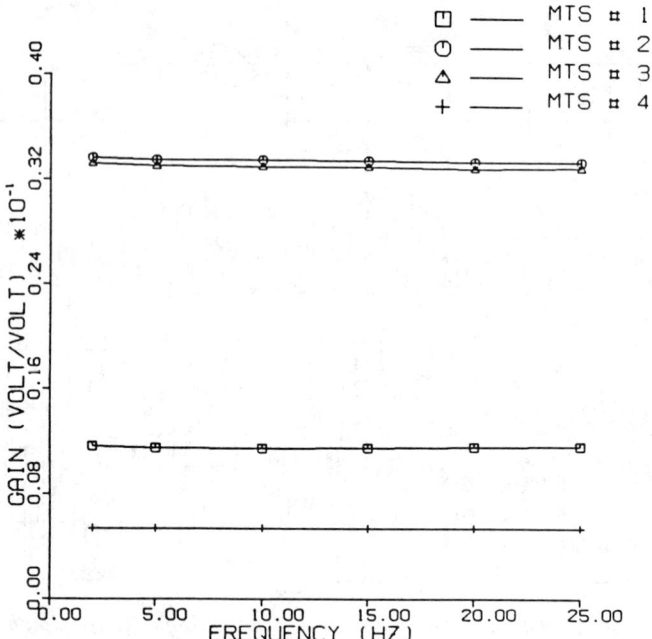

FIG. 4—*Influence of test frame and frequency on gain, based on fatigue test of a steel specimen cycled at 40%* σ_{yt}.

illustrated in Fig. 4, units with identical grip assemblies (Units 2 and 3) have identical gain responses. This is in comparison with Units 1 and 4, which have different grip assemblies.

Figure 5 shows the effect of excitation frequency on the phase lag response as a function of test frame. As can be seen, phase lag is influenced linearly by the test frequency for Test Frame 2. This linear dependency was no longer observed after removal of the load-conditioning filter from the circuitry of the controller unit. The information on the circuitry of the unit is not available to the authors at this moment, but it is believed this problem is due to the inherent nonlinear nature of filters. This filter was removed on all the test frames except Test Frame 1; the filter could not be accessed on that unit. It can be seen that the phase log response for Test Frame 1 stands apart from the others. This could be a filter problem for this unit. The variation between the results obtained for Test Frames 2, 3 and 4 could be related to having different unsupported length in the specimen. It should be noted that the negative value of the phase is due to the fact that in the hydraulic test frame the displacement signal leads the load signal, where (theoretically) the displacement signal is generally assumed to be lagging behind the load signal. The phase values were normalized by a normalization factor of − 1.

Using a 100-KN (22-kip) servohydraulic load frame (MTS 3), an interlayered center-notched specimen was cycled (tension-tension) to failure at 78% of its ultimate tensile strength (UTS). Using the experimental setup in Fig. 1, the gain and phase lag response versus normalized life was plotted (Figs. 6 and 7). Also, using a contact thermocouple, the temperature variation was recorded as a function of cycles (Fig. 8).

Gain and temperature curves (Figs. 6 and 8) show three distinct regions resembling the stiffness degradation in this material system as discussed by others [23–25]. The first region occurs during the early life of the specimen where rapid increases in gain and

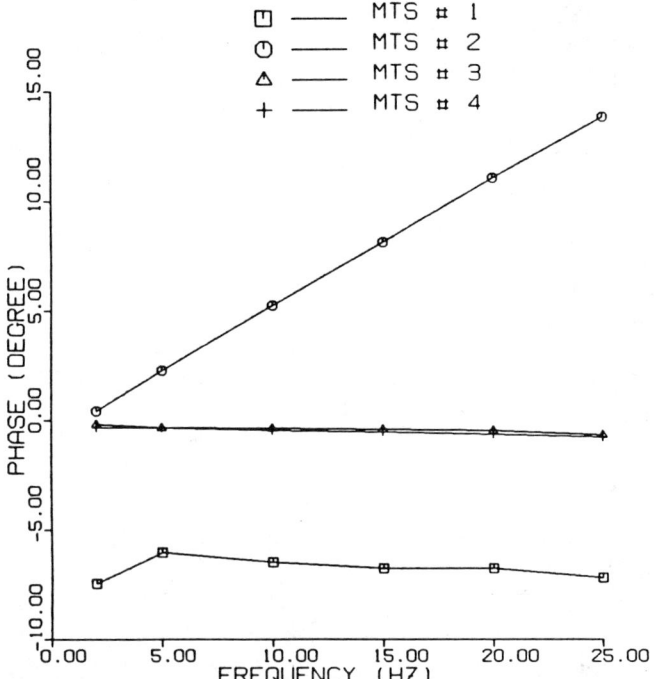

FIG. 5—*Influence of test frame and frequency on phase, based on fatigue test of a steel specimen cycled at 40% σ_{yt}.*

temperature are seen. It is observed that this region is associated with matrix cracking in the 0° direction and with the start of delamination around the notch. This, in turn, created friction surfaces which gave rise to temperature increases. The second region occurs during the middle portion of the life of the specimen. It indicates a slow rate of increase in gain and temperature. This region is observed to be related to greater delamination and 0° matrix cracking around the notch. This stage is also the start of delamination on the specimen edges due to the edge effect. These also create more friction surfaces which result in temperature increase. The third region occurs during the late life of the specimen. It shows a very rapid rate of increase in gain and temperature. It was observed that during this period of life, the delaminations on either side of the notch start to grow until they finally meet. There is also a great amount of fiber fracture during the final periods all the way to failure. The temperature fluctuations in that region may be due to disrupted contact between the specimen and thermocouple tip.

Normalized stiffness and 1/gain are plotted against normalized life in Fig. 9. The similarity of these two curves indicates that gain is related to the total damage. Comparison of gain/cycle and phase/cycle curves (Figs. 6 and 7) suggest that the phase might be related to the rate of change of gain. To test this hypothesis, the gain/cycle curve was differentiated with respect to cycles. Figure 10 shows a plot of the slope of the gain versus cycles. This figure is very similar to the phase/cycle curve in Fig. 7. This observation suggests that phase lag may be related to the damage rate in this case, an important parameter which is otherwise difficult to measure.

Operating at 90% of the UTS, another interlayered specimen was fatigue cycled to obtain phase, gain, and temperature response. To investigate the effect of load level, these

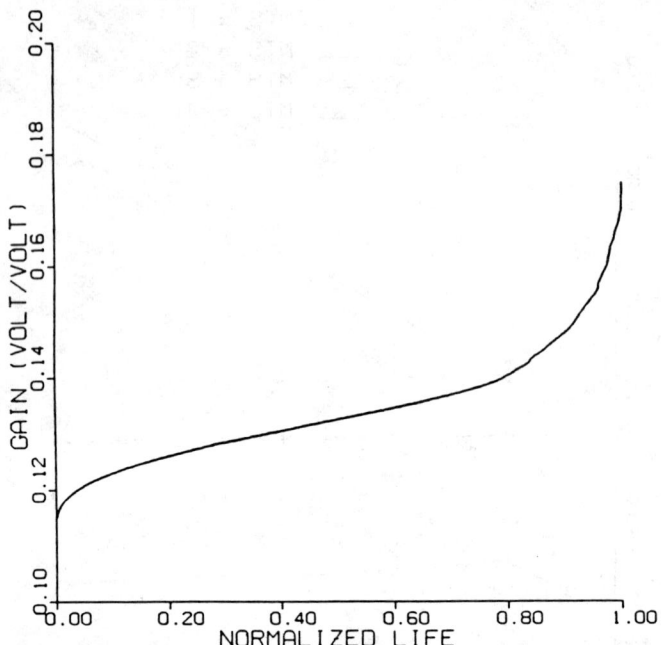

FIG. 6—*Gain response versus normalized life of AS4/985 [0/45/90/ −45]$_{s2}$ interlayered laminate at 78% σ$_u$ and frequency of 10 Hz.*

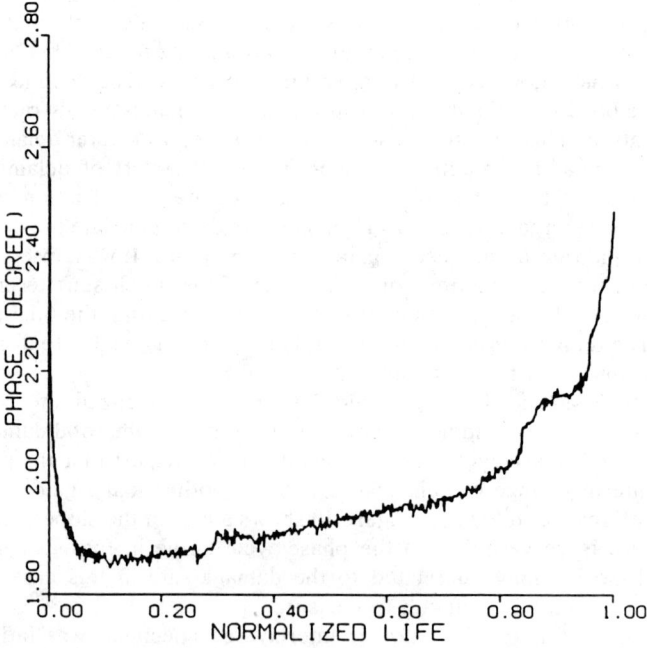

FIG. 7—*Phase response versus normalized life of AS4/985 [0/45/90/ −45]$_{s2}$ interlayered laminate at 78% σ$_u$ and frequency of 10 Hz.*

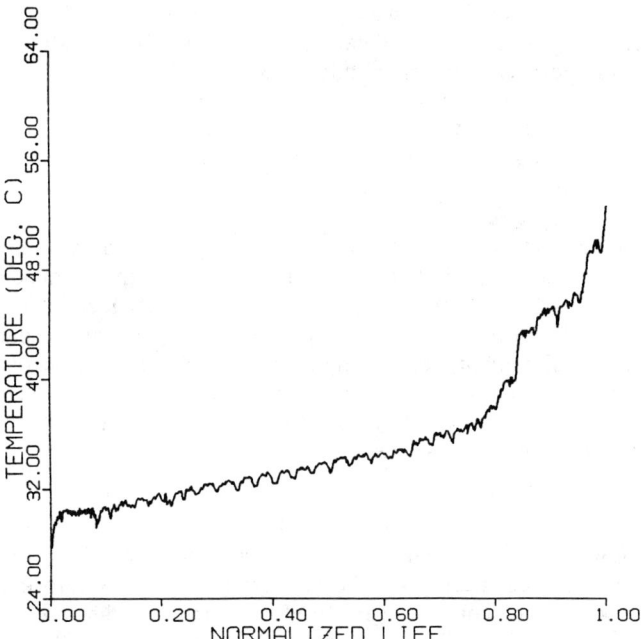

FIG. 8—*Temperature response versus normalized life of AS4/985 [0/45/90/−45]$_{s2}$ interlayered laminate at 78% σ_u and frequency of 10 Hz.*

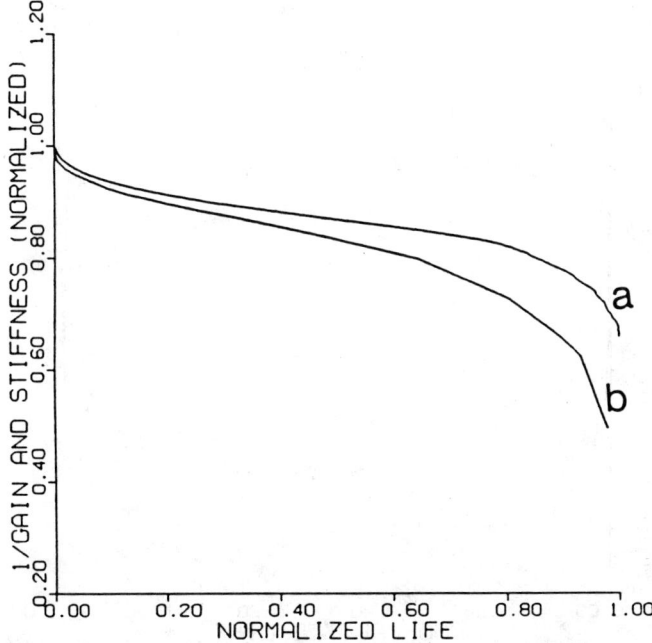

FIG. 9—*AS4/985 [0/45/90/−45]$_{s2}$ laminate at* (a) *normalized 1/gain, and* (b) *normalized stiffness versus normalized life.*

values were plotted versus cycles along with data obtained from the 78% UTS load level test (Figs. 11 and 12). These figures indicate that load level has a distinct effect on these damage parameters such that the magnitude as well as shape of phase and gain versus cycle curves are altered, but the basic features of the curves remain unchanged. Next, the unidirectional laminates were fatigued at 60, 65, and 70% UTS. The gain and phase were plotted versus cycles (Figs. 13 and 14). These plots show the laminate response under various applied load levels. The fluctuations as well as various jumps illustrated in the figures are due to the longitudinal matrix splittings and fiber fracture. This was verified visually where the audible sound of matrix splitting was followed by sudden jumps in the real time plots. The nature of damage in unidirectional laminates makes it difficult to obtain any quantitative measurement of damage parameters such as residual strength and stiffness. It is believed that this problem could be solved using the present technique.

Even though the results might suggest an order in relative magnitude of gain and phase lag response in these material systems, no comments as to the exact nature for this behavior is available at the present time. Further analysis of this technique is currently underway and will be reported at a later date.

Conclusion

It has been shown that the dynamic load and stroke signals from the controllers of standard servohydraulic test frames can be interpreted with a waveform analysis device during the cyclic loading of composite materials in a manner that provides information about dynamic compliance and phase lag that is directly related to the damage development processes in those materials. It is further shown that the measured parameters can provide quantitative information about the level of damage and the rate of damage devel-

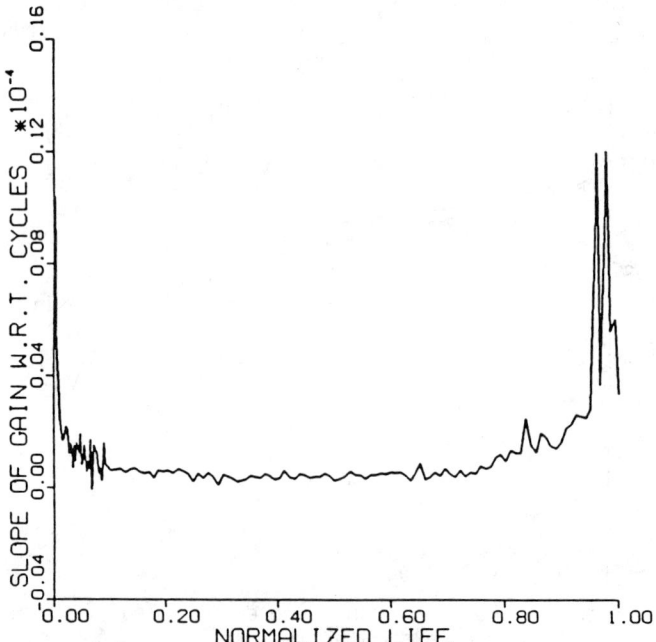

FIG. 10—*Slope of gain w/r to cycles versus normalized life of AS4/985 [0/45/90/−45]$_{s2}$ laminate at 78% σ_u and frequency of 10 Hz.*

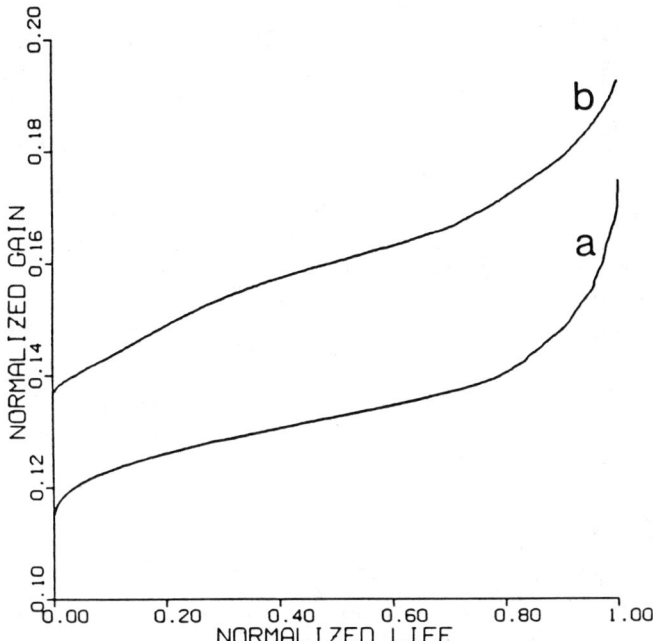

FIG. 11—*Normalized gain versus normalized life of AS4/985 [0/45/90/ − 45]$_{s2}$ laminate at* (a) *78%* σ$_u$ *and* (b) *90%* σ$_u$.

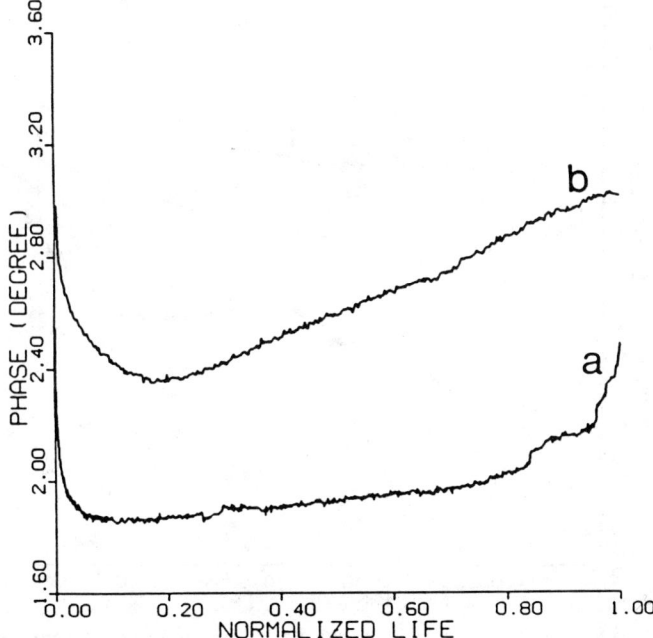

FIG. 12—*Phase versus normalized life of AS4/985 [0/45/90/ − 45]$_{s2}$ laminate at* (a) *78%* σ$_u$ *and* (b) *90%* σ$_u$.

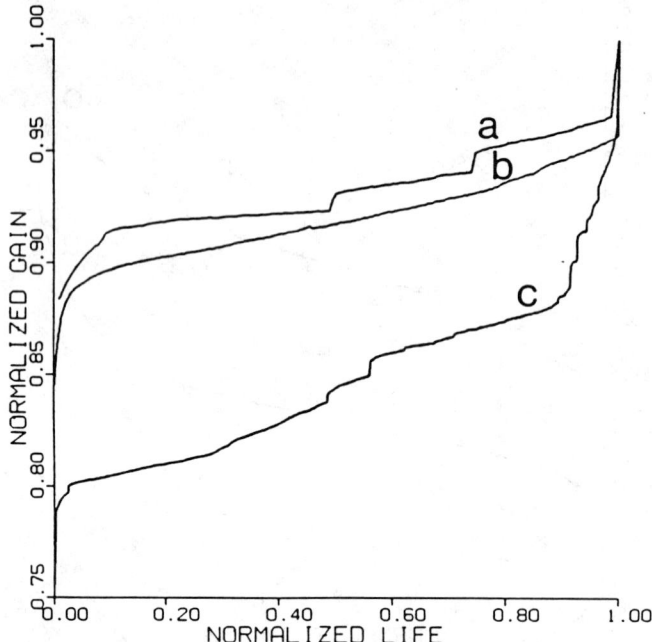

FIG. 13—*Normalized gain versus normalized life of unidirectional laminate at* (a) *60%,* (b) *65%, and* (c) *70%* σ_u.

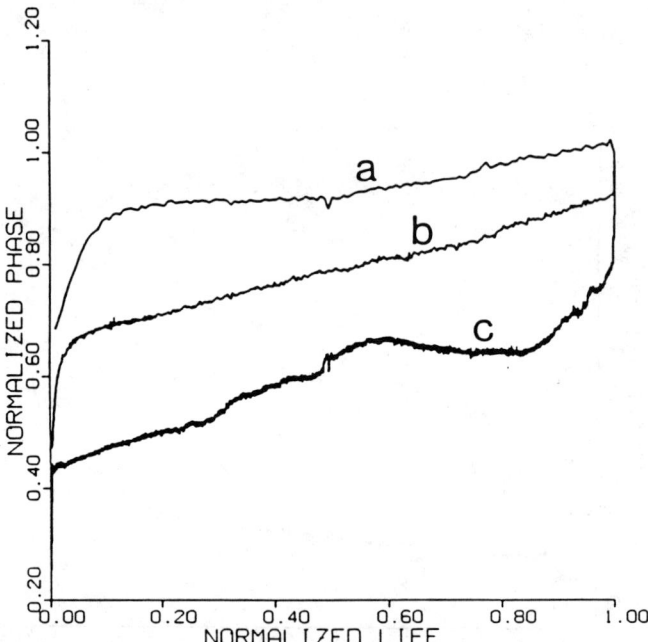

FIG. 14—*Normalized phase versus normalized life of unidirectional laminate at* (a) *60%,* (b) *65%, and* (c) *70%* σ_u.

opment during such tests, a particularly valuable result since the quantitative measurement of damage rate during testing is otherwise very difficult. Major advantages of the method include the use of standard, easily obtained test equipment and test information (from standard servohydraulic test system controller signals), the elimination of need for contact methods for strain measurement (which are difficult to use in high-temperature tests), and the ease of automated data retrieval devices to recover the test information.

Continuing efforts include the study of specific relationships between damage events and measured parameter changes, study of the damage rate measurement capabilities of the method, and use of the device for high-temperature dynamic characterization of composite material systems.

Acknowledgments

The authors gratefully acknowledge the support of the Air Force Office of Scientific Research under Grant Number 85.0087 and the Virginia Institute for Material Systems.

References

[1] Lorenzo, L. and Hahn, H. T., "Fatigue Failure Mechanisms in Unidirectional Composites," *Composite Materials: Fatigue and Fracture, ASTM STP 907,* American Society for Testing and Materials, Philadelphia, 1986, pp. 210–232.

[2] Rotem, A., "Fatigue and Residual Strength of Composite Laminates," *Engineering Fracture Mechanics,* Vól. 25, No. 516, 1986, pp. 819–827.

[3] Tsai, S. W. and Hahn, H. T., *Introduction to Composite Materials,* Technomic Publishing Co., Inc., Lancaster, PA, 1980.

[4] Talreja, R., *Fatigue of Composite Materials,* Chapter 5, Technomic Publishing Co. Inc., Lancaster, PA, 1987.

[5] Herakovich, C. T., "On the Relationship Between Engineering Properties and Delamination of Composite Materials," *Journal of Composite Materials,* Vol. 15, July 1981, pp. 338–348.

[6] Razvan, A. and Reifsnider, K. L., "Fiber Fracture and Strength Relationship in Unidirectional Graphite/Epoxy Composite Materials," *Theoretical and Applied Fracture Mechanics,* Vol. 16, 1991, pp. 81–89.

[7] Piggott, M. R., "The Interface—An Overview," *Proceedings,* 36th International SAMPE Symposium, 15–18 April 1991, Society for the Advancement of Material and Process Engineering, Covina, CA, pp. 1773–1786.

[8] Herakovich, C. T., "Influence of Layer Thickness on the Strength of Angle-Ply Laminates," *Journal of Composite Materials,* Vol. 16, May 1982, pp. 216–227.

[9] Pagano, N. J. and Pipes, R. B., "The Influence of Stacking Sequence on Laminate Strength," *Journal of Composite Materials,* Vol. 5, January 1971, pp. 55–57.

[10] Razvan, A., Bakis, C. E., Wagnecz, L., and Reifsnider, K. L., "Influence of Cyclic Load Amplitude on Damage Accumulation and Fracture of Composite Laminates," *Journal of Composite Technology and Research,* Vol. 10, No. 1, Spring 1988, pp. 3–10.

[11] Dan-Jumbo, E., Zhou, S. G., and Sun, C. T., "Load-Frequency Effect on Fatigue Life of IMP6/APC-2 Thermoplastic Composite Laminates," *Advances in Thermoplastic Matrix Composite Materials, ASTM STP 1044,* G. M. Newaz, Ed., American Society for Testing and Materials, Philadelphia, 1989, pp. 113–132.

[12] Curtis, D. C., Moore, D. R., Slater, B., and Zahlan, N., "Fatigue Testing of Multi-Angle Laminates of CF/Peek," *Composites,* Vol. 19, No. 6, November 1988.

[13] O'Brien, T. K. and Reifsnider, K. L., "Fatigue Damage: Stiffness/Strength Comparisons for Composite Materials," *Journal of Testing and Evaluation,* Vol. 5, No. 5, 1977, pp. 384–393.

[14] Camponeschi, E. T. and Stinchcomb, W. W., "Stiffness Reduction as an Indicator of Damage in Graphite/Epoxy Laminates," *Composite Materials: Testing and Design (Sixth Conference), ASTM STP 787,* I. M. Daniel, Ed., American Society for Testing and Materials, Philadelphia, 1982, pp. 225–246.

[15] Hahn, M. T., "Fatigue Behavior and Life Prediction of Composite Laminates," *Composite Materials: Testing and Design, ASTM STP 674,* S. W. Tsai, Ed., American Society for Testing and Materials, Philadelphia, pp. 383–417.

[16] Lifshitz, J. M., "Deformational Behavior of Unidirectional Graphite/Epoxy Composite Under

Compressive Fatigue," *Journal of Composite Technology and Research,* Vol. 11, No. 3, Fall 1989, pp. 99–105.

[17] Sims, G. D. and Bascombe, D., "Continuous Monitoring of Fatigue Degradation in Composites by Dynamic Mechanical Analysis," *Proceedings,* Sixth International Conference on Composite Materials, Second European Conference on Composite Materials, ICCMVI/ECCMII, F. L. Matthews, N. C. R. Buskell, J. M. Hodgkinson, J. Morton, Eds., Vol. 4, pp. 4.161–4.171.

[18] Renz, R., Altstadt, V., and Ehrenstein, G. W., "Hysteresis Measurement for Characterizing the Dynamic Fatigue of R-SMC," *Journal of Reinforced Plastics and Composites,* Vol. 7, September 1988, pp. 413–433.

[19] So, C. K., Lai, T. C., and Tse, P. C., "The Measurement of Material Damping by Free-Vibration Technique with Periodic Excitation," *Experimental Techniques,* May/June 1990, pp. 41–42.

[20] Swain, R. E., Bakis, C. E., and Reifsnider, K. L., "Effect of Interleaves on the Damage Mechanisms and Residual Strength of Notched Composite Laminates," *Composite Materials: Fatigue and Fracture, Fourth Volume, ASTM STP 1156,* W. W. Stinchcomb and N. E. Ashbaugh, Eds., American Society for Testing and Materials, Philadelphia, 1992.

[21] Razvan, A. and Reifsnider, K. L., "Fiber Fracture and Strength Degradation in Unidirectional Graphite/Epoxy Composite Materials," *Theoretical and Applied Fracture Mechanics,* Vol. 16, No. 1.

[22] "The Fundamentals of Signal Analysis," Application Note 243, Hewlett Packard Co., Palo Alto, CA, 1989.

[23] Bakis, C. E. and Stinchcomb, W. W., "Response of Thick, Notched Laminates Subjected to Tension-Compression Cyclic Loads," *Composites Materials: Fatigue and Fracture, ASTM STP 907,* H. T. Hahn, Ed., American Society for Testing and Materials, Philadelphia, 1986, pp. 314–334.

[24] Reifsnider, K. L. and Stinchcomb, W. W., "A Critical-Element Model of the Residual Strength and Life of Fatigue-Loaded Composite Coupons," *Composite Materials: Fatigue and Fracture, ASTM STP 907,* H. T. Hahn, Ed., American Society for Testing and Materials, Philadelphia, 1982, pp. 50–62.

[25] Bakis, C. E., "A Test Method to Measure the Response of Composite Materials Under Reversed Cyclic Loads," *Test Methods and Design Allowables for Fibrous Composites, Second Volume, ASTM STP 1003,* American Society for Testing and Materials, Philadelphia, 1989.

Fatigue of Ceramic Matrix, Metal Matrix, and Specialty Composites

M. Mirdamadi,[1] W. S. Johnson,[2] Y. A. Bahei-El-Din,[3] and M. G. Castelli[4]

Analysis of Thermomechanical Fatigue of Unidirectional Titanium Metal Matrix Composites

REFERENCE: Mirdamadi, M., Johnson, W. S., Bahei-El-Din, Y. A., and Castelli, M. G., **"Analysis of Thermomechanical Fatigue of Unidirectional Titanium Metal Matrix Composites,"** *Composite Materials: Fatigue and Fracture, Fourth Volume, ASTM STP 1156,* W. W. Stinchcomb and N. E. Ashbaugh, Eds., American Society for Testing and Materials, Philadelphia, 1993, pp. 591–607.

ABSTRACT: Thermomechanical fatigue (TMF) data have been generated for a Ti-15V-3Cr-3Al-3Sn (Ti-15-3) material reinforced with SCS-6 silicon carbide fibers for both in-phase and out-of-phase testing. Significant differences in failure mechanisms and fatigue life are noted for the in-phase and out-of-phase testing. The purpose of the research reported in this paper is to apply a micromechanics model to analysis of the data. The analysis predicts the stresses in the fiber and in the matrix material during the thermal and mechanical cycling by calculating both the thermal and mechanical stresses and their time-dependent behavior. The rate-dependent behavior of the matrix was characterized and was used to calculate the constituent stresses in the composite. The predicted 0° fiber stress range was used to explain the composite failure. It was found that for a given condition, temperature, loading frequency, and time at temperature, the 0° fiber stress range may control the fatigue life of the unidirectional composite.

KEY WORDS: silicon-carbide fibers, isothermal fatigue, stress range, interface, residual stresses, in phase, out of phase, nonisothermal

Nomenclature

A, A^*, n, n^*	Curve-fitting constants
E	Young's modulus
H	Overstress
k, p	Temperature-dependent experimentally determined parameters
R	Stress ratio (S_{min}/S_{max})
S	Composite stress
S_{min}	Minimum stress
S_{max}	Maximum stress
T	Temperature
$\dot{\epsilon}$	Strain rate

[1]National Research Council research associate, NASA Langley Research Center, Mail Stop 188E, Hampton, VA 23665-5225.

[2]Senior scientist, NASA Langley Research Center, Mail Stop 188E, Hampton, VA 23665-5225.

[3]Research associate professor, Department of Civil Engineering, Rensselaer Polytechnic Institute, Troy, NY 12180. On leave from Cairo University, Giza, Egypt.

[4]Research engineer, Sverdrup Technology, Inc., NASA Lewis Research Center, Mail Stop 49-7, Cleveland, OH 44135.

$\dot{\epsilon}^T$ Thermal strain rate
$\dot{\epsilon}^e$ Elastic strain rate
$\dot{\epsilon}^{in}$ Inelastic strain rate
α Coefficient of thermal expansion
$\Delta\sigma_f$ Fiber stress range

Titanium matrix composites, such as Ti-15V-3Cr-3Al-3Sn (Ti-15-3) reinforced with continuous silicon carbide (SCS-6) fibers, are being considered for use on hypersonic aerospace vehicles. The structural materials in such vehicles require high-temperature capability, light weight, and high stiffness. However, due to the significant difference in the coefficient of thermal expansion (CTE) of the fiber and matrix, cyclic stresses and strains are produced in the fiber and matrix whenever the temperature is cycled. Thermal cycling could result in internal damage which may affect the macromechanical behavior of the composite. Furthermore, at temperatures above 480°C, the titanium matrix exhibits thermo-viscoplastic behavior. Since projected operating temperatures are expected to be well into the creep range, the rate-dependent behavior of the matrix material must be considered in the thermomechanical fatigue (TMF) evaluation of the composite behavior.

The objectives of this research are: (1) to use an analytical method that was developed under contract with NASA Langley Research Center to evaluate TMF and nonisothermal fatigue data of unidirectional SCS-6/Ti-15-3 composites under in-phase and out-of-phase loading, and (2) to determine the critical constituent stresses (strains) that control laminate fatigue behavior under TMF and nonisothermal fatigue loading, taking into account the rate-dependent behavior of the matrix.

The constitutive equations of a unidirectional ply are defined using the vanishing fiber diameter (VFD) model [1]; these constitutive equations are then used in a laminated plate analysis to model a laminate [2]. The constitutive behavior of the matrix is described with a thermo-viscoplastic model developed by Bahei-El-Din et al. [3].

Among some of the recent work on the fatigue of SCS-6/Ti-15-3, Johnson et al. [4] and Pollock and Johnson [5] studied the static isothermal tensile behavior and isothermal fatigue behavior of a variety of layups at room temperature and at an elevated temperature of 650°C. These laminates were made with fugitive binders. The composite used in the room temperature research [4] was in the "as-fabricated" condition. The laminates used in the 650°C tests [5] were aged for 1 h at 650°C in air prior to testing. All layups had a fiber volume fraction of 0.325. The isothermal fatigue tests were conducted under load control at 10 Hz with an R ratio (S_{min}/S_{max}) of 0.1.

At room temperature, fatigue failure was observed in the fibers but not in the matrix, except for one long-life test. It was suggested that the 0° plies played a dominant role in controlling fatigue life. In particular, the authors suggested that the stress range in the 0° fibers controlled life. They showed that the room temperature fatigue data of several layups could be collapsed to a narrow band when plotted as a function of the stress range in the 0° fiber, as shown in Fig. 1. At 650°C, Pollock and Johnson [5] observed that a high cyclic strain resulted in a short life due to multiple fiber failure with no matrix fatigue cracking. On the other hand, low cyclic strains resulted in long lives with extensive matrix cracking prior to fiber failure. Thus, the cyclic life of the composite at 650°C was also governed by the cyclic stress range in the 0° fibers; however, more scatter was found in the 0° fiber stress range at 650°C than was found in the room temperature data.

Castelli et al. [6] studied the TMF behavior of a nine-ply unidirectional SCS-6/Ti-15-3 laminate in a temperature range from 93 to 538°C. The TMF tests were conducted under load control with $R = 0.05$ and a cycle time of 180 s. The composites used in their study

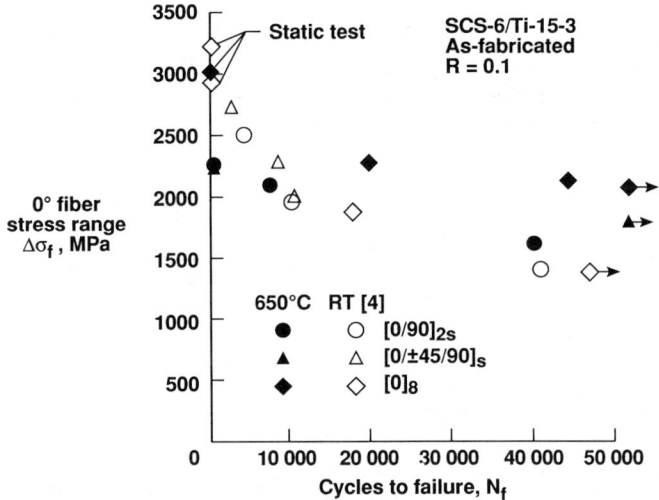

FIG. 1—*The stress range in the 0° fiber versus number of cycles to failure* [5].

were fabricated from matrix foil and plies of unidirectional SiC fibers held in place by fugitive binders and had a fiber volume fraction of 0.34. The composite was aged at 700°C for 24 h in an argon environment prior to testing. Castelli observed that TMF greatly reduced cyclic fatigue lives when compared with isothermal and in-phase nonisothermal data. The failure modes under TMF test conditions were significantly different from those observed under comparable isothermal test conditions. Out-of-phase TMF loadings displayed a matrix-dominated failure, where cracking and crack initiation sites were found entirely at surface and near-surface locations. In-phase TMF loadings resulted in ductile matrix failure and fiber pullout across the entire fracture surface, with extensive fiber breakage in the absence of matrix cracking.

Isothermal and in-phase and out-of-phase load-controlled nonisothermal fatigue experiments were performed by Gabb et al. [7] on a unidirectional SCS-6/Ti-15-3 composite between 300 and 550°C. The composite sheets used in their study were fabricated from matrix foil and plies of unidirectional SiC fibers held in place by Ti-6Al-4V weaving. Unlike the true TMF tests where the load and the temperature are cycled simultaneously, in the nonisothermal tests the load was cycled at a constant temperature followed by a temperature cycle at zero load. The fatigue test cycles used by Gabb are shown in Fig. 2. Under in-phase loading conditions, the load was cycled to a maximum at the maximum temperature of 550°C, followed by a thermal cycle from 550 to 300°C at zero load. Under out-of-phase loading conditions, the load was cycled to a maximum at the minimum temperature of 300°C, followed by a thermal cycle from 300 to 550°C at zero load. Gabb found that, based on the cyclic stress range, both in-phase and out-of-phase nonisothermal fatigue lives were shorter than the isothermal fatigue lives at 300 and 550°C.

Analytical Method—VISCOPLY

The VISCOPLY code developed by Bahei-El-Din is a micromechanics analysis based on constituent properties. The program combines the vanishing fiber diameter (VFD) model [2], a thermo-viscoplastic theory [3], and the laminated plate theory to evaluate the rate-dependent, high-temperature response of symmetric fibrous composite laminates. The

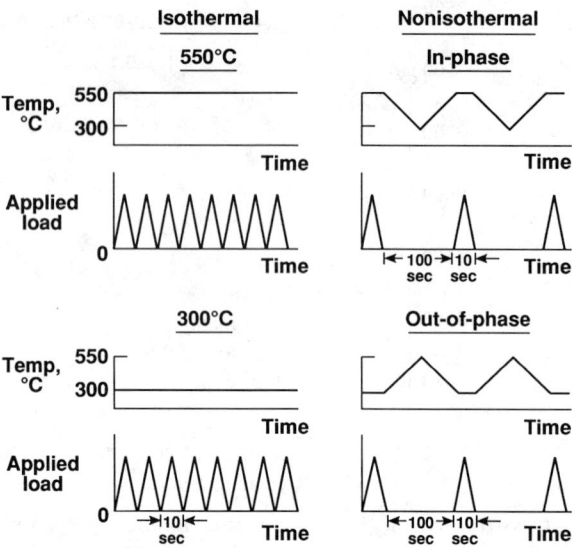

FIG. 2—*The waveform in fatigue test cycles* [7].

viscoplastic theory models isotropic materials where fiber and matrix are characterized individually, and both can be viscoplastic. The VFD model accounts for orthotropic behavior. The VISCOPLY program accepts any combination of thermal and mechanical loads. The mechanical loads consist of in-plane axial stress or strain rates, in-plane transverse stress rate, in-plane shear stress rate, and out-of-plane normal stress rate. The program can run sequential jobs which allows any order and rate of load and temperature. The program computes fiber and matrix stresses and strains and the overall composite response under thermomechanical loading conditions and can simulate fabrication and actual tests performed on composite specimens.

The viscoplastic theory used in the VISCOPLY program was developed by Bahei-El-Din et al. [3] for high-temperature, nonisothermal applications. The theory reduces to the formulation given by Eisenberg and Yen [9] at low temperatures and isothermal conditions. The theory used in VISCOPLY assumes that the elastic response is rate-independent and that inelastic strains develop when the current stress point lies outside an equilibrium yield surface. When the material is subjected to quasi-static loading, the yield surface may harden kinematically and isotropically, causing the stress point to fall on the yield surface. The inelastic strain in this case can be found with a rate-independent plasticity theory. The quasi-static stress-strain curve is referred to as the equilibrium curve and corresponds to the theoretical lower bound of the dynamic response. If the material is loaded at a finite stress or strain rate and the yield stress is exceeded, the stress point may fall outside the initial yield surface. The inelastic strain rate is described using a power law function of the overstress, H, which is defined as the difference between the current stress and the equilibrium stress.

The total strain rate is given by Ref 3 as the sum of the elastic strain rate, $\dot{\epsilon}^e$, the thermal strain rate, $\dot{\epsilon}^T$, and the inelastic strain rate, $\dot{\epsilon}^{in}$. Under uniaxial loading, the strain rate is given by

$$\dot{\epsilon} = \dot{\epsilon}^e + \dot{\epsilon}^T + \dot{\epsilon}^{in} = \dot{\sigma}/E(T) + m(T)\dot{T} + k(T)H^{p(T)} \tag{1}$$

where

$$m(T) = \alpha(T) - (\sigma/E^2(T))dE(T)/dT, \text{ and}$$
$$H = \sigma - \sigma^*.$$

Here E is Young's elastic modulus, T is the temperature, α is the coefficient of thermal expansion, and σ^* is the equilibrium stress defined by the quasi-static stress-strain curve.

The parameters k and p in Eq 1 must be determined experimentally and are temperature dependent. When isotropic hardening and thermal recovery effects are not significant, these parameters can be found from two monotonic uniaxial tests. However, if isotropic hardening and thermal recovery effects are significant, the behavior becomes more complex, and evaluation of the quasi-static curve is more difficult [3]. In the present paper, we determined the material parameters, k and p, assuming that isotropic hardening and thermal recovery effects were negligible for the Ti-15-3 matrix. Determination of the material parameters, k and p, will be presented in the experimental section of this paper. The fiber was assumed to be isotropic and to remain elastic, although the model has the capability to model the fiber as a viscoplastic material with transverse orthotropic properties. Loss of composite stiffness as a result of matrix cracking may be approximated by reducing the modulus of the matrix in the model.

Material and Experimental Procedure

In order to analyze the TMF data using the thermo-viscoplastic model [3], several material parameters were required. Strain-controlled and load-controlled tests at specific temperatures and loading rates were conducted to evaluate the necessary material parameters. Previously generated test data were also used to validate the model.

Material

The Ti-15-3 matrix material is a metastable beta strip alloy. The alloy is weldable, ageable, and relatively insensitive to corrosive environments [8]. Long exposures at elevated temperatures can lead to the precipitation of an alpha phase, which may alter the macroscopic behavior of the material [8]. Therefore, the material in the present study was subjected to a heat treatment prior to testing. The matrix alloy was obtained in a 500 by 500 by 5-mm panel and cut into specimens 5 mm thick, 13 mm wide, and 152 mm long. The specimens could be subjected to desired test loads without buckling.

The specimens were heat treated at 650°C for 1 h followed by an air quench. This heat treatment was based on differential scanning calorimetry (DSC) tests conducted by Pollock and Johnson [5]. The DSC measures the heat flow into a sample as a function of time. Since the heat flow varies with metallurgical instabilities such as phase changes, it was used to determine the time necessary to reach a steady-state heat flow during an isothermal hold.

Test Procedures

The tests necessary to determine the equilibrium curves and the material parameters, k and p, were conducted in a 100-kN servo-hydraulic test frame with water-cooled grips. The specimens were heated using a 5-kW induction generator. The temperature distribution along the gage section of the specimen was measured with six thermocouples (three on each face of the specimen, measured from center, 15 mm apart on one face and 20 mm apart on the other face). The induction coil was adjusted to minimize the temperature gradient within the gage section to a differential of less than 10°C.

In all tests, the temperature of each specimen was monitored and controlled with a K-type thermocouple spot welded to the center of the specimen and a microprocessor-controlled induction generator. Axial strains were measured on the edge of the specimen by a high-temperature, water-cooled extensometer with a 25-mm gage length. The strain-controlled tests used to determine the equilibrium curves were conducted with a strain rate of 1×10^{-4} mm/mm/s. The load-controlled monotonic tests used to determine k and p were conducted at a constant rate of 2.56 MPa/s. These rates were representative of loading rates expected in actual service conditions.

Tests were recorded on a X-Y plotter and a strip chart recorder. Material property data was determined at the following temperatures: room temperature, 316, 482, 566, and 650°C. Strain-controlled tests on Ti-15-3 showed that once the maximum stress was reached, the stress dropped (stress softening) with further straining, approaching a constant stress level. This behavior was observed consistently at strain rates between 1×10^{-4} and 1×10^{-2} mm/mm/s when the temperature was above 316°C. However, under cyclic straining, the cyclic stress-strain response stabilized rapidly after the first cycle. Therefore, in all tests, the matrix material was initially cycled four times using a fully reversed triangular waveform to stabilize the softening behavior. The strain amplitude was chosen to avoid buckling of the specimen.

To define the equilibrium curves, 5-min hold periods were imposed at predetermined strain levels as illustrated in Fig. 3a. The equilibrium curve was then approximated by a power law $[\sigma^* = A^*(\epsilon^{in})^{n^*}]$ fit to the calculated values of stress and inelastic strain, using the method of least squares. The stress-control test was conducted with a constant stress rate. Given the equilibrium curve, the overstress, H, is determined as illustrated in Fig. 3b. The stress and inelastic strain determined from the monotonic stress-controlled tests were also fit to a power law $[\sigma = A(\epsilon^{in})^{n}]$ using the method of the least squares. The resulting equation was differentiated with respect to time to determine the inelastic strain rate. Once ϵ^{in} and H were known, it was a simple matter to find k and p using a log-log plot of inelastic strain rate and corresponding overstress values as illustrated in Fig. 3c.

In the present paper, we determined the material parameters, k and p, as described above for the Ti-15-3 matrix at room temperature, 482, 566, and 650°C.

Results and Discussion

In this section the experimental and analytical results are presented. First, the constituent properties are presented. Second, the matrix and composite behavior are analyzed using VISCOPLY and compared with observed experimental behavior. Last, VISCOPLY is used to analyze the TMF data performed by Castelli et al. [6] and nonisothermal fatigue data of Gabb et al. [7], and results are compared with isothermal fatigue data of Pollock and Johnson [5] and Gabb et al. [7].

Constituent Properties

The fiber properties used in this study are shown in Table 1. As mentioned, the fiber was assumed to remain elastic and to have no temperature dependence. The Ti-15-3 matrix properties were determined from tests conducted at room temperature and the elevated temperatures of 316, 482, 566, and 650°C.

Determination of the constants, k and p, at the elevated temperatures of 566 and 650°C was easily done as described earlier. However, at room temperature and 482°C, determination of the material constants, k and p, required some manipulation in the curve fitting of

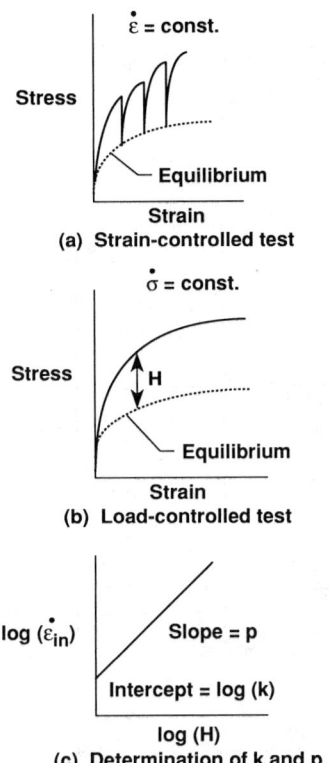

FIG. 3—*Equilibrium stress-strain curve and determination of parameters* k *and* p.

the load-controlled tests. After the initial hardening, the load-controlled ($\dot{\sigma}$ = 2.56 MPa/s) stress-strain curve approached zero slope. Thus, the calculated value of overstress, H, was nearly constant with increasing values of the inelastic strain rate, $\dot{\epsilon}^{in}$, resulting in nonconverging values of the constants, k and p (Fig. 3c). The exponent n used in the power law for the experimental data was adjusted to give reasonable values of k and p and still accurately fit the experimental results. At 316°C, the stress relaxation was minimal (as in room the temperature tests), and no load-controlled test was conducted at this temperature. Therefore, the same values of k and p found at room temperature were used with the elastic modulus, the yield stress, and the equilibrium curve found at 316°C. The validity of this assumption was verified using strain-controlled test results at 316°C. The elastic constants and yield stress as a function of temperature are given in Table 2. The curve-fitting parameters used for the equilibrium and overstress curves and the parameters k and p are also listed in Table 2.

The equilibrium curve and the parameters k and p were used in VISCOPLY to recon-

TABLE 1—*SCS-6 fiber properties* [4].

SCS-6	
Longitudinal modulus	400 GPa
Poisson's ratio	0.25
Coefficient of thermal expansion	4.86×10^{-6}/°C

TABLE 2—*Ti-15-3 matrix properties.*

Temperature, °C	Elastic Modulus, E_m, GPa	Poisson's Ratio,[a] V_m	CTE[b] $\times 10^6$, α_m, m/m/°C	Yield Stress, Y_m, MPa	Equilibrium[c] Curve Constants		Stress-Strain[d] Curve Constants		Inelastic Strain[e] Rate Parameter	
					A^*, MPa	n^*	A, MPa	n	k, MPa^{-p}/s	p
25	91.800	0.36	8.48	772	1180.87	0.05	1371.14	0.05	3.374×10^{-13}	4.37
316	80.440	0.36	9.16	492	769	0.05	...	...	3.374×10^{-13}	4.37
482	72.240	0.36	9.71	142	273	0.08	1089	0.13	4.53×10^{-9}	5.68
566	64.400	0.36	9.98	20.48	116.81	0.18	660	0.16	5.53×10^{-17}	5.35
650	53.000	0.36	10.26	10.20	36.33	0.14	452	0.165	1.17×10^{-14}	4.72

[a]See Ref 4.
[b]See Ref 8.
[c]Equilibrium stress-inelastic strain curve, $\sigma^* = A^*(\epsilon^{in})^n$.
[d]Stress-inelastic strain curve at 2.56 MPa/s, $\sigma = A(\epsilon^{in})^n$.
[e]Constitutive equation, $\dot{\epsilon}_{in} = kH^p$.

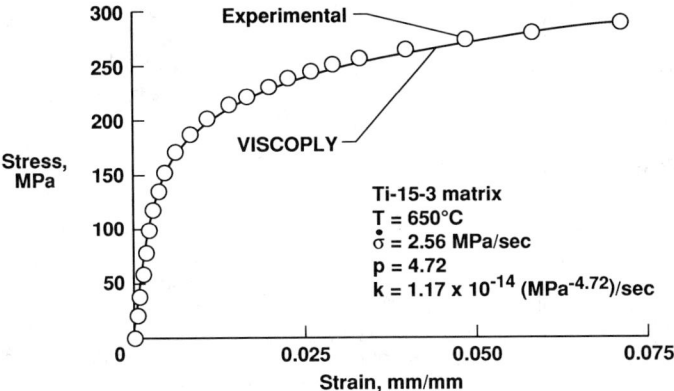

FIG. 4—*Best fit approximation to the experimental stress-strain data for Ti-15-3 at 650°C.*

struct the load-controlled stress-strain curve for the matrix. Figures 4 and 5 show the best-fit approximation of the experimental results. VISCOPLY was used to predict the stress relaxation of the matrix at 482 and 650°C. Figure 6 compares the prediction and experimental results for 650°C. The prediction and the observed experimental behavior agree well at 650°C. Figure 7 compares the prediction and experimental results at 482°C. Poor agreement between the prediction and the experimental data was found at this temperature. The discrepancy between the prediction and the experiment may be attributed to isotropic hardening and thermal recovery effects which are not accounted for in the constitutive model implemented in VISCOPLY.

Laminate Behavior

The constituent properties given in Table 1 and Table 2 were used to predict the laminate behavior for the isothermal tests conducted in Ref 5. Figure 8 compares the first

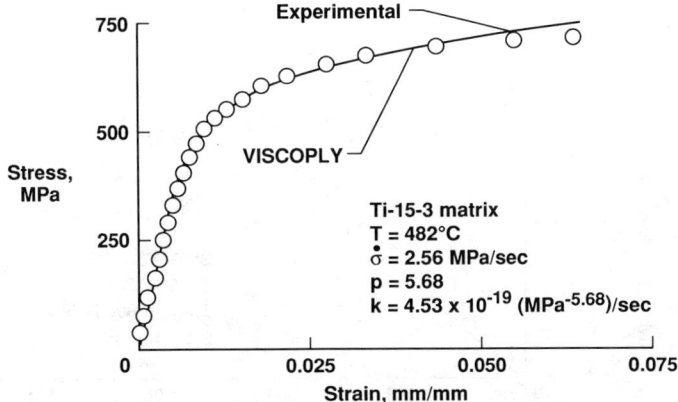

FIG. 5—*Best fit approximation to the experimental stress-strain data for Ti-15-3 at 482°C.*

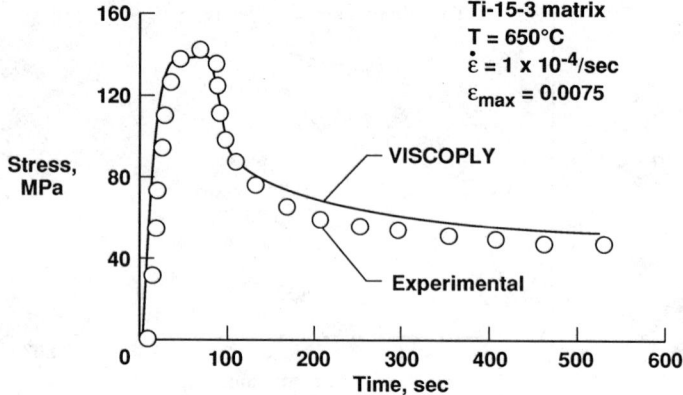

FIG. 6—*Prediction of time-dependent matrix response at 650°C.*

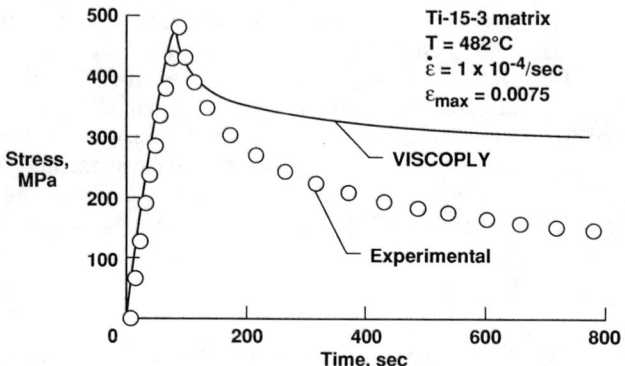

FIG. 7—*Prediction of time-dependent matrix response at 482°C.*

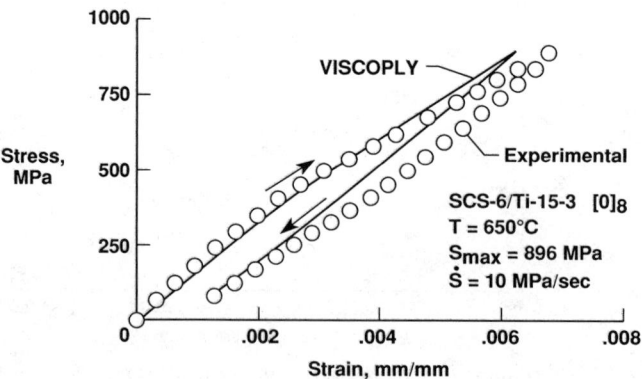

FIG. 8—*Prediction of cyclic stress-strain composite response.*

cycle of an experimental stress-strain curve for a $[0]_8$ laminate at 650°C with the prediction using VISCOPLY. The loading rate was 10.34 MPa/s. As seen in the figure, VISCOPLY captured the fundamental aspects of the loading and unloading response of the composite. Note that the analysis was based on the constituents properties only, and no adjustments (e.g., reducing fiber modulus) were made to improve the predictions.

Thermomechanical Response

In this section, the TMF test conducted by Castelli et al. [6] and nonisothermal fatigue conducted by Gabb et al. [7] were analyzed using VISCOPLY. In the analysis, the TMF and nonisothermal waveforms were cycled until the fiber stress range did not change with further cycling (the fiber stress usually stabilized within five load cycles). The results will also be compared with the isothermal fatigue data of Pollock and Johnson [5] and Gabb et al. [7].

Castelli, Bartolotta, Ellis—In-phase and out-of-phase TMF tests were conducted by Castelli et al. [6] in the temperature range from 93 to 538°C using a sinusoidal waveform for both load and temperature with a cycle time of 180 s. For simplicity of computation, a simplified waveform, shown in Fig. 9, was used in VISCOPLY for the analysis of the data from Ref 6. In-phase TMF corresponds to maximum stress at maximum temperature; out-of-phase TMF corresponds to maximum stress at minimum temperature. The fatigue lives for in-phase and out-of-phase loading, as a function of the applied maximum stress S_{max}, are shown in Fig. 10. The TMF lives under in-phase loading conditions were considerably shorter than those under out-of-phase loading conditions.

The VISCOPLY program was used to analyze the observed difference in TMF lives. Predictions were made using the constituent properties given in Table 1 and Table 2. The fiber properties given in Table 1 were assumed to remain constant throughout the TMF cycle. The matrix properties were assumed to be the same from room temperature to 150°C. It was assumed that any residual stresses that developed during the fabrication of the composites would be relieved due to relaxation at absolute temperatures greater than one half of the melting temperature of the matrix [10]. Therefore, a temperature change of 555°C was used to simulate the cooldown during the fabrication process. The stress-strain responses under in-phase and out-of-phase loading are shown in Figs. 11 and 12, respectively. Also shown in the Figs. 11 and 12 are the VISCOPLY predictions. Under in-phase loading TMF conditions, the VISCOPLY predicted the creep ratcheting behavior of the composite. For the out-of-phase TMF loading, VISCOPLY predicted an almost elastic

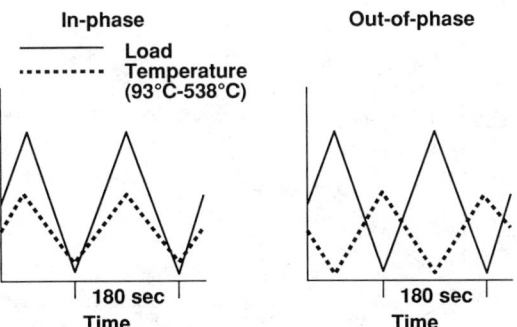

FIG. 9—*Waveform used to analyze the in-phase and out-of-phase TMF data.*

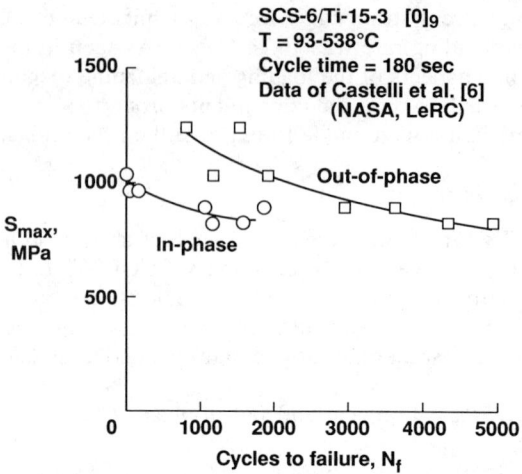

FIG. 10—*The maximum applied stress as a function of cycles to failure under TMF* [6].

response which was not found experimentally; however, the minimum and maximum strains were predicted accurately.

Figure 13 shows the predicted stresses in the fiber and matrix in a TMF cycle under in-phase and out-of-phase loading. As shown in the figure, under in-phase loading the peak stress in the fiber is higher than for the out-of-phase loading. These predictions are consistent with the fact that in the case of in-phase TMF loading, the load-carrying capacity of the matrix is greatly reduced at elevated temperatures; therefore, more load is carried by the fibers than in the case of out-of-phase loading. Figure 13 also shows that the matrix stress is higher during the out-of-phase loading than during in-phase loading. This is consistent with the observations made by Castelli et al. [6] that specimens tested in out-of-phase loading developed extensive matrix cracking. Conversely, the in-phase loading resulted in no significant matrix cracking prior to failure.

The predicted fiber stress ratio ($\sigma_{min}/\sigma_{max}$) under both in-phase and out-of-phase TMF loading conditions was 0.07. Figure 14 shows the resulting predicted fiber stress range versus observed fatigue lives. As shown in the figure, the in-phase and out-of-phase TMF

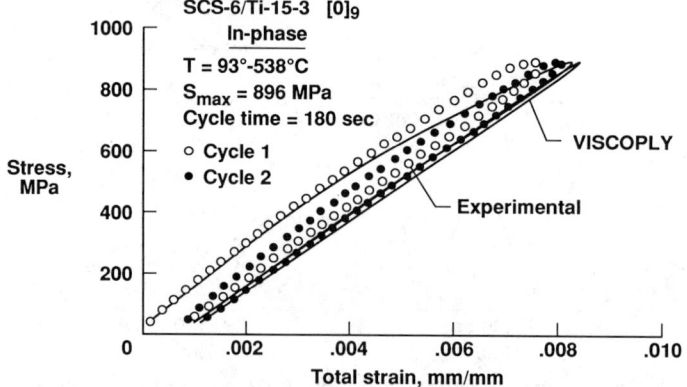

FIG. 11—*Prediction of composite response under in-phase TMF.*

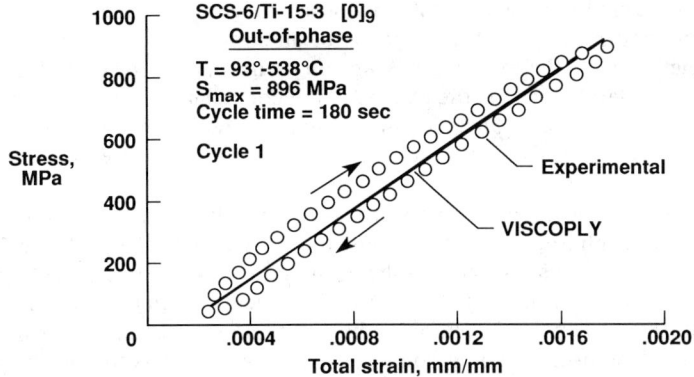

FIG. 12—*Prediction of composite response under out-of-phase TMF.*

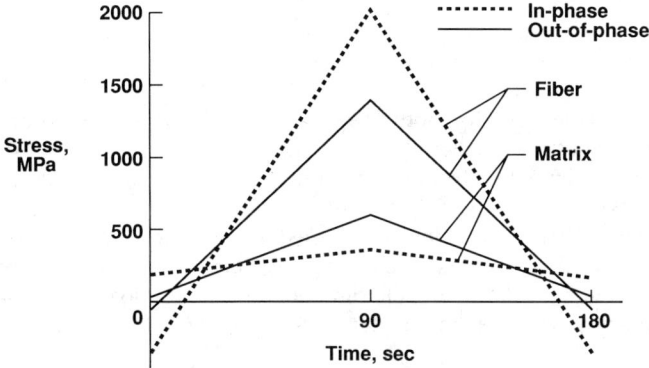

FIG. 13—*VISCOPLY analysis of in-phase and out-of-phase TMF. T = 93°-538°C; S_{max} = 896 MPa.*

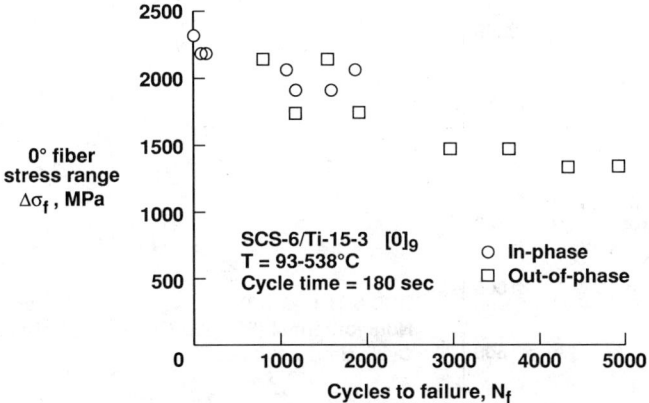

FIG. 14—*The stress range in the 0° fiber as a function of cycles to failure.*

data collapsed when plotted against the fiber stress range, indicating that the 0° fiber stress range apparently controls the TMF response of unidirectional composites.

As mentioned earlier, the matrix cracking can be accounted for by reducing the matrix modulus in the model. However, the change in the composite stiffness as a result of matrix cracking under out-of-phase TMF loading conditions was not thoroughly investigated, and, therefore, the analysis was based on the initial measured elastic modulus of the matrix.

Gabb, Gayda, and MacKay—As described previously, isothermal and nonisothermal in-phase and out-of-phase tests were conducted by Gabb et al. [7] using the waveform shown in Fig. 2. The nonisothermal in-phase and out-of-phase experimental data of Gabb et al. [7] was analyzed using VISCOPLY. The calculated fiber stress ratio, $\sigma_{min}/\sigma_{max}$, varied from 0.01 to 0.1 under in-phase loading conditions and from -0.1 to -0.05 under out-of-phase nonisothermal loading conditions. Figure 15 shows the observed fatigue life as a function of the predicted 0° fiber stress range. As seen from the figure, the fatigue data collapsed to a narrow band, indicating that fatigue failure of the unidirectional composite is apparently governed by the 0° fiber stress range.

Comparisons

The observed fatigue lives for both the Castelli et al. [6] and Gabb et al. [7] data are plotted as a function of the predicted 0° fiber stress range in Fig. 16. As shown, the TMF and nonisothermal fatigue data from the two references separate into two bands due to the different testing conditions. The data from Ref 6 consistently had shorter lives than data from Ref 7. The TMF tests conducted by Castelli et al. [6] had a cycle time of 180 s, whereas nonisothermal fatigue tests conducted by Gabb et al. [7] had a mechanical cycle time of 10 s followed by a thermal cycle time of 100 s at zero load. This may indicate that the slower frequency or more time at temperature lowers the fatigue strength of the 0° fibers.

The 0° fiber stress range was also used to compare the fatigue lives from the isothermal tests [5,7] with the TMF [6] and nonisothermal fatigue tests [7], as shown in Fig. 17. The isothermal fatigue tests of Pollock and Johnson [5] were conducted at 650°C with fre-

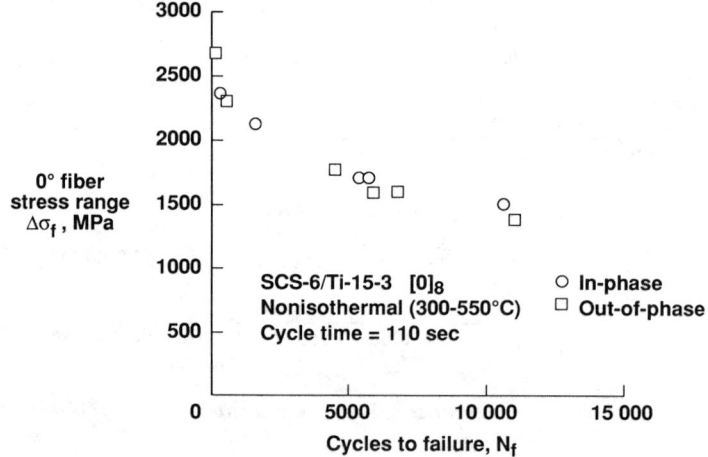

FIG. 15—*The stress range in the 0° fiber as a function of cycles to failure.*

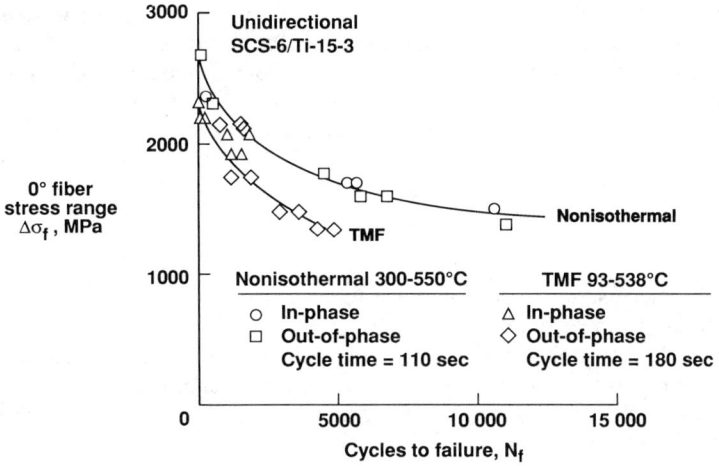

FIG. 16—*The stress range in the 0° fiber as a function of cycles to failure.*

quency of 10 Hz (cycle time of 0.1 s), and the isothermal fatigue tests of Gabb et al. [7] were conducted at 300 and 550°C with a frequency of 0.1 (cycle time of 10 s). For these isothermal tests, the 0° fiber stress range was calculated from the composite strain range by multiplying the strain range by the modulus of the fiber. The 0° fiber stress range for the TMF, isothermal fatigue, and nonisothermal fatigue conditions are shown in Fig. 17. As seen in the figure, the isothermal fatigue lives were distinctly greater than and separate from the TMF and nonisothermal fatigue lives. The isothermal fatigue lives at 650°C with a frequency of 10 Hz were greater than the fatigue lives at either 300 or 550°C with a frequency of 0.1 Hz. However, the isothermal fatigue lives at 300°C were higher than those at 550°C. Thus, for the isothermal fatigue tests at a given frequency, the higher the temperature the shorter the fatigue life. The TMF and nonisothermal tests were conducted at much lower frequencies than the isothermal tests; thus, the TMF and nonisothermal

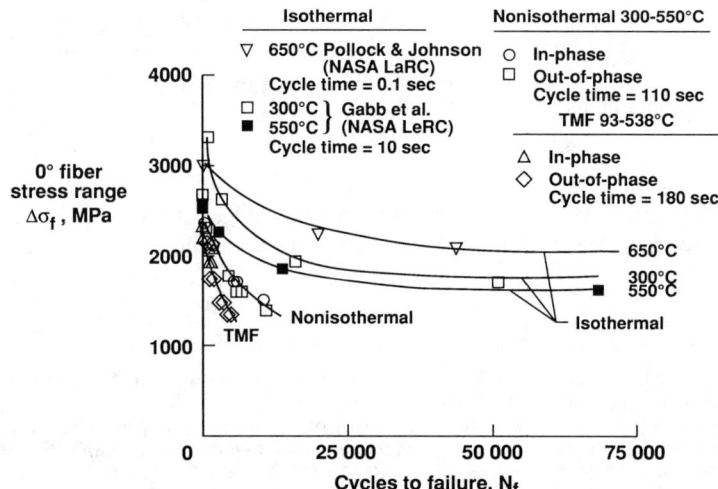

FIG. 17—*The stress range in the 0° fiber as a function of cycles to failure.*

fatigue specimens were exposed to high temperatures for longer periods of time. The longer exposure time at a higher temperature lowered the fatigue strength of the 0° fiber. This increased exposure at elevated temperature could produce more reaction between the fiber and matrix, lowering the static strength as reported by Jeng et al. [11] and lowering fatigue resistance as reported by Naik, Johnson, and Pollock [12]. Therefore this difference in fatigue life may be more affected by length of the exposure to elevated temperature than the type of loading.

Conclusions

VISCOPLY was used to analyze the rate-dependent behavior of the Ti-15-3 matrix alloy. The stress-strain response of the SCS-6 fiber reinforced Ti-15-3 unidirectional composite under TMF and nonisothermal fatigue loading conditions was also predicted. The predicted 0° fiber stress range was used to explain the observed fatigue behavior of unidirectional composites under isothermal, nonisothermal, and TMF loading conditions. The following conclusions were made:

1. The VISCOPLY program reasonably predicted the stress relaxation of the matrix at 650°C; however, the prediction at 482°C was poor. This may be due to hardening and thermal recovery which was not accounted for in the VISCOPLY code.
2. The VISCOPLY program predicted the stress-strain response of the unidirectional composite under in-phase TMF loading reasonably well. In the case of the out-of-phase loading, the VISCOPLY predicted a nearly elastic stress-strain response which was not seen in the experimental data; however, the minimum and maximum strains were accurately predicted. The VISCOPLY program has proven to be a useful tool for analyzing TMF of metal matrix composites.
3. Isothermal and nonisothermal fatigue data for the SCS-6/Ti-15-3 composite were analyzed using 0° fiber stress range. The results indicated that the fatigue strength of the 0° is controlled by a combination of temperature, loading frequency, and time at temperature. Furthermore, within a given set of parameters (i.e. temperature, loading frequency, and time at temperature) the stress range in the 0° fiber controls fatigue life.
4. Predictions were made using constituent properties only and no adjustments were made to the constituent properties to fit any specific set of experimental data. However, it may be appropriate to adjust constituent properties to account for observed damage mechanisms in the composite.

Acknowledgment

The first author, M. Mirdamadi, gratefully acknowledges the support extended by the National Research Council (NRC), Washington, DC, through their Associateship program.

References

[1] Dvorak, G. J. and Bahei-El-Din, Y. A., "Plasticity Analysis of Fibrous Composites," *Journal of Applied Mechanics,* Vol. 49, 1982, pp. 327–335.
[2] Bahei-El-Din, Y. A., "Plasticity Analysis of Fibrous Composite Laminates Under Thermomechanical Loads," *Thermal and Mechanical Behavior of Ceramic and Metal Matrix Composites, ASTM STP 1080,* J. M. Kennedy, H. H. Moeller, and W. S. Johnson, Eds., American Society for Testing and Materials, Philadelphia, 1990, pp. 20–39.
[3] Bahei-El-Din, Y. A., Shah, R. S., and Dvorak, G. J., "Numerical Analysis of the Rate-

Dependent Behavior of High Temperature Fibrous Composites," *AMD,* Vol. 118: *Mechanics of Composites at Elevated and Cryogenic Temperatures,* S. N. Singhal, W. F. Jones, C. T. Herakovich, and T. Cruse, Eds., 1991, pp. 67–78.

[4] Johnson, W. S., Lubowinski, S. L., and Highsmith, A. L., "Mechanical Characterization of Unnotched SCS_6/Ti-15-3 Metal Matrix Composites at Room Temperature," *Thermal and Mechanical Behavior of Metal Matrix and Ceramic Matrix Composites, ASTM STP 1080,* J. M. Kennedy, H. H. Moeller, and W. S. Johnson, Eds., American Society for Testing and Materials, Philadelphia, 1990, pp. 193–218.

[5] Jeng, S. M., Yang, C. J., Alassoeur, P., and Yang, J. M., "Deformation and Fracture Mechanisms of Fiber-Reinforced Titanium Alloy Matrix Composites," *Proceedings,* Eighth International Conference on Composite Materials (ICCM/8), S. W. Tsai and G. S. Springer, Eds., Society for the Advancement of Material and Process Engineering (SAMPE), Covina, CA, 1991, pp. 25-C-1, 25-C-12.

[6] Castelli, M. G., Bartolotta, P. A., and Ellis, J. R., "Thermomechanical Fatigue Testing of High Temperature Composites: Thermomechanical Fatigue Behavior of SiC(SCS-6)/Ti-15-3," *Composite Materials: Testing and Design (Tenth Volume), ASTM STP 1120,* G. C. Grimes, Ed., American Society for Testing and Materials, Philadelphia, 1991.

[7] Gabb, T. P., Gayda, J., and MacKay, R. A., "Isothermal and Nonisothermal Fatigue Behavior of a Metal Matrix Composite," *Journal of Composite Materials,* Vol. 24, June 1990.

[8] Rosenberg, H. W., "Ti-15-3: A New Cold-Formable Sheet Titanium Alloy," *Journal of Metals,* Vol. 35, No. 11, 1986, pp. 30–34.

[9] Eisenberg, M. A. and Yen, C. F., "A Theory of Multiaxial Anisotropic Viscoplasticity," *ASME Journal of Applied Mechanics,* Vol. 48, June 1981, pp. 276–284.

[10] Dieter, E. D., *Mechanical Metallurgy,* 2nd ed., McGraw-Hill, New York, 1976, pp. 452–453.

[11] Jeng, S. M., Yang, C. J., Alassoeur, P., and Yang, J. M., "Deformation and Fracture Mechanisms of Fiber-Reinforced Titanium Alloy Matrix Composite," to be presented at ICCM-VIII, Honolulu, HI, 1991.

[12] Naik, R. A., Johnson, W. S., and Pollock, W. D., "Effect of High Temperature Cycle on the Mechanical Properties of Silicon Carbide/Titanium Metal Matrix Composites," *Proceedings of the ASC on High Temperature Composites,* Technomics, Lancaster, PA, 1989, pp. 94–103.

Michael J. Verrilli[1] *and Timothy P. Gabb*[1]

High-Temperature Tension-Compression Fatigue Behavior of a Unidirectional Tungsten Copper Composite

REFERENCE: Verrilli, M. J. and Gabb, T. P., "**High-Temperature Tension-Compression Fatigue Behavior of a Unidirectional Tungsten Copper Composite,**" *Composite Materials: Fatigue and Fracture, Fourth Volume, ASTM STP 1156,* W. W. Stinchcomb and N. E. Ashbaugh, Eds., American Society for Testing and Materials, Philadelphia, 1993, pp. 608–619.

ABSTRACT: The high-temperature fatigue behavior of a $[0]_{12}$ tungsten fiber-reinforced copper matrix composite was investigated. Specimens having fiber volume percentages of 10 and 36 were fatigued under fully reversed, strain-controlled conditions at both 260 and 560°C. The fatigue life was found to be independent of fiber volume fraction because fatigue damage preferentially occurred in the matrix. Also, the composite fatigue lives were shorter at 560°C as compared to 260°C, due to changes in mode of matrix failure. On a total strain basis, the fatigue life of the composite at 560°C was the same as the life of unreinforced copper, indicating that the presence of the fibers did not degrade the fatigue resistance of the copper matrix in this composite system. Comparison of strain-controlled fatigue data to previously generated load-controlled data revealed that the strain-controlled fatigue lives were longer because of mean strain and mean stress effects.

KEY WORDS: metal matrix composites, fatigue, high temperature, tungsten fibers, copper matrix, strain control

The projected use of metal matrix composites at elevated temperatures will subject these composites to thermal and thermomechanical cycles, thus requiring an understanding of composite behavior under these conditions. These types of loadings are expected to be severe because of the inherent difference of the coefficient of thermal expansion between the matrix and fiber. As part of a program to study the fatigue of metal matrix composites, the fatigue behavior of tungsten fiber-reinforced copper has been investigated.

Previous work on this composite system [1,2] focused on load-controlled fatigue under high-temperature isothermal and thermomechanical fatigue (TMF) conditions. During the tension-tension load-controlled tests performed, the cyclic mean strains of the $[0]_4$, 0.1 fiber volume fraction composite increased significantly during the fatigue cycling, resulting in hysteresis loops which never stabilized. To gain a better understanding of the fatigue failure processes, the current effort concentrated on the fully reversed strain-controlled fatigue behavior of this same composite system. Isothermal fatigue tests were performed as a first step in the characterization of overall fatigue behavior. This paper reports the results of these isothermal tests.

The specific objectives of this work were to compare the strain-controlled low-cycle fatigue behavior of two fiber volume fractions of a tungsten-copper composite at two

[1]Research engineer and research metallurgist, respectively, NASA Lewis Research Center, 21000 Brookpark Rd., Mail Stop 49-7, Cleveland, OH 44135.

temperatures and also to compare this data with previously generated load-controlled fatigue data.

Material and Specimens

The material studied, copper reinforced with tungsten fibers, is a candidate material for rocket nozzle liner applications. This composite was manufactured in the form of 12-ply panels using an arc spray technique [3]. The matrix was oxygen-free high-conductivity (OFHC) copper, and the fibers were G.E. 218 CS continuous tungsten wire of 200-μm diameter. The composite plates contained unidirectional tungsten fibers arranged in a square array, with the center of the fiber axes located at the corners of the square.

Specimens were machined from unidirectional composite plates having fiber volume percents (v/o) of 10 and 36. All specimens had the fibers oriented parallel to the loading axis. Figure 1 shows the two specimen geometries employed. The specimen shown in Fig. 1a was used for all experiments conducted at 260°C and for the 10 v/o composite tested at 560°C. The specimen geometry of Fig. 1b, which has a shorter (17.78 mm) straight section and a bigger transition radius between the grip end and straight section, was used to test the 36 v/o composite at 560°C. As the fiber spacing was smaller in the 36 v/o composite than in the 10 v/o composite, the 36 v/o plate was thinner (3.91 mm) than the 10 v/o plate (6.99 mm). Thus, the second specimen geometry was necessary to minimize the possibility of buckling of the thinner 36 v/o composite at the higher temperature.

Test Procedures

Fatigue tests were conducted using a 90-kN servohydraulic test system fitted with an environmental chamber. Strain was utilized as the control variable in fatigue experiments using an R-ratio (R = minimum strain/maximum strain) of −1. A triangular strain wave-

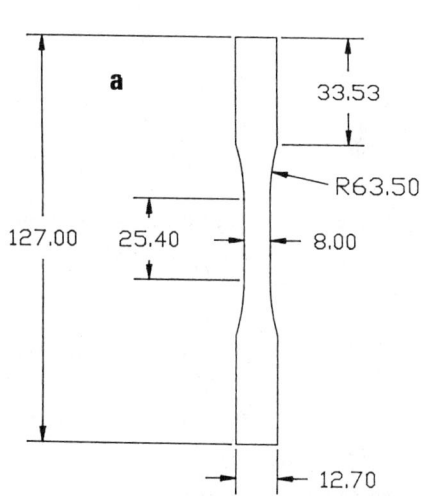

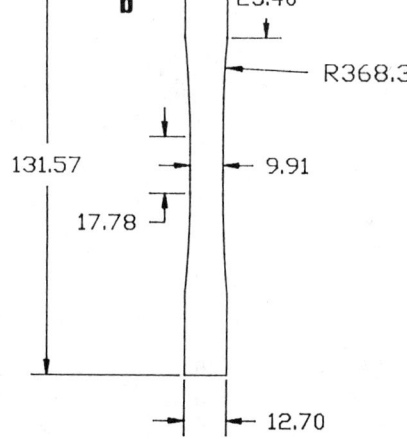

FIG. 1—*Composite specimens: (a) geometry used for all 10 v/o material and 36 v/o material tested at 260°C; (b) geometry used for 36 v/o material tested at 560°C. The 10 v/o specimens were 6.99 mm thick and the 36 v/o specimens were 3.91 mm thick.*

form was employed for test control, using a strain rate of 0.002 mm/mm/s. Strain was measured and controlled using a 12.7-mm gage length high-temperature extensometer. Strain control was employed to enable life comparisons to be made on a constant strain basis and to eliminate the overall extension caused by strain ratchetting in the load-controlled experiments [1,2]. The failure criteria was a 25% decrease of the tensile stress from the maximum value attained during the test. It was observed that the thinner 36 v/o specimens would buckle at 560°C when a large fatigue crack was present. Defining failure in the chosen manner generally ended the tests before specimen buckling.

The fatigue tests were conducted at temperatures of 260 and 560°C. Specimens were heated by induction heating. The temperature variation along the specimen straight section was less than 5°C. Temperature was measured and controlled using a Type K thermocouple wrapped around and spring-loaded against the specimen gage section. A vacuum of less than 6×10^{-6} torr was used as the test environment to minimize specimen oxidation.

In order to determine the mechanisms of specimen failure, fracture surfaces and polished sections of failed specimens were examined with optical and scanning electron microscopy.

Results

The results of fatigue tests on the 10 and 36 v/o composite specimens at both 260 and 560°C are summarized in Table 1. The stresses shown are the stabilized values as measured at half life.

Cyclic Stress-Strain Behavior

Typical high-temperature stress-strain behavior of 10 and 36 v/o specimens are shown in Fig. 2. Shown are the first fatigue cycle and a cycle near failure. As indicated by the open width of the hysteresis loops, the composite experienced inelastic strains. The change of

TABLE 1—*Fatigue test results of the W/Cu composite.*

Specimen Number	Fiber Volume Fraction	Test Temperature, °C	Strain Range, mm/mm	Cycles to Failure	Stress Range, MPa
10-04	0.1	260	0.0123	625	599
10-03	0.1	260	0.0102	1710	559
10-05	0.1	260	0.0093	3317	555
10-06	0.1	260	0.0085	4726	522
10-07	0.1	260	0.0085	4309	523
10-13	0.1	560	0.0099	334	347
10-08	0.1	560	0.0085	733	330
10-09	0.1	560	0.0077	1370	313
10-10	0.1	560	0.0067	3770	283
10-14	0.1	560	0.0055	8050	247
40-08	0.36	260	0.0142	327	1087
40-07	0.36	260	0.0112	1200	1032
40-05	0.36	260	0.0093	1680	941
40-13	0.36	260	0.0085	6800	846
40-14	0.36	260	0.0077	7820	864
40-18	0.36	560	0.0077	1470	739
40-15	0.36	560	0.0057	8900	632

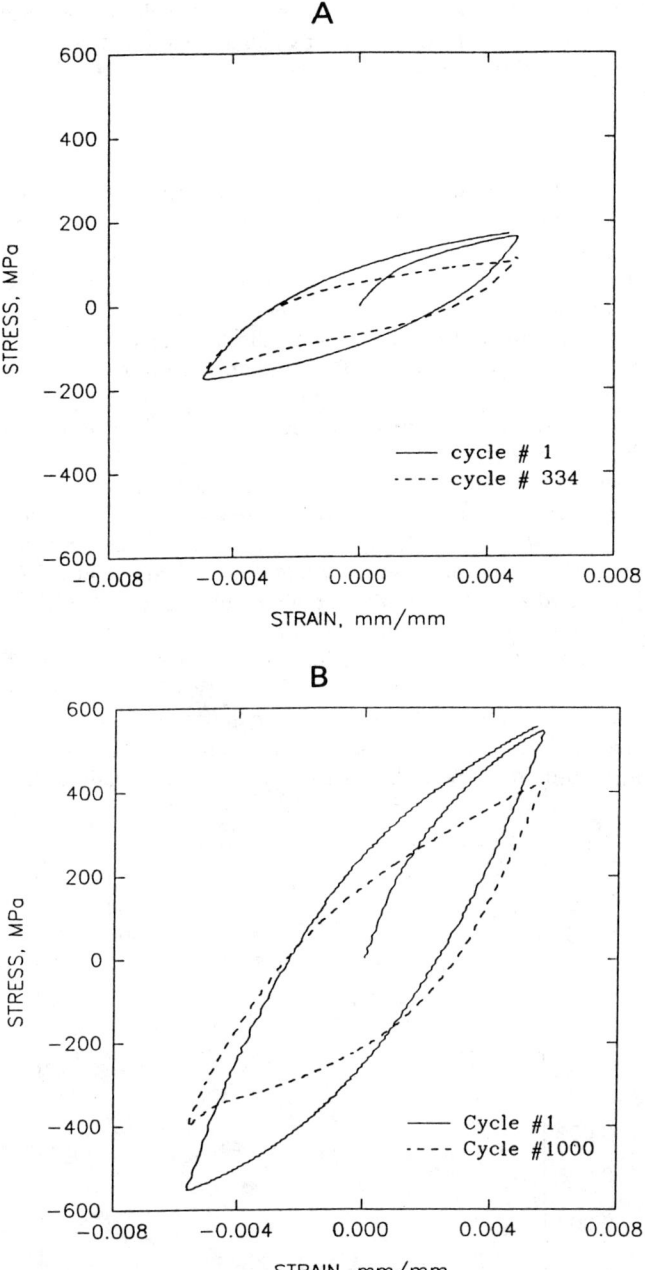

FIG. 2—*Stress-strain behavior of W-Cu, showing the first hysteresis loop and one near failure:* (a) *10 v/o W-Cu tested at 560°C;* (b) *36 v/o W-Cu tested at 260°C.*

the shape of the loops as the test progressed is due to cyclic deformation and damage accumulation. Near failure, the hysteresis loops had a cusp in the compressive portion of the stress-strain curve, which is believed caused by matrix cracking in the specimen.

Fatigue Life

A comparison of the fatigue life as a function of fiber volume fraction and test temperature is given in Fig. 3. The lives are compared on a total strain basis. At a given temperature, fatigue life does not vary between the test tested fiber volume fractions. However, fatigue life is inversely temperature dependent. The fatigue lives at 260°C on a total strain basis are about five times longer than those at 560°C for the strain regime tested.

The 560°C composite lives are compared with the lives of OFHC copper at 538°C, as reported by Conway et al. [4] in Fig. 4. The copper tests were conducted under strain control at the same strain rate as the composite tests. Different test techniques were employed in the two studies, as will be discussed later. The data shown are for both fully annealed and fully hardened copper because one would expect the degree of work hardening of the as-received composite matrix to fall between these two bounds. Even though the copper data were generated at a slightly lower temperature, their fatigue lives are about the same as the 560°C composite data. In contrast, a strain-based comparison of matrix and composite fatigue lives of a brittle/ductile metal matrix composite, SiC/Ti, at elevated temperatures [5] revealed that the life of the composite was less than that of the matrix.

The copper and composite life data are compared on a stress range basis in Fig. 5. The advantage of the composite in fatigue load-carrying capability is shown. For a given fatigue life, the stress-carrying capability of the 10 v/o W-Cu is about four times greater and the 36 v/o composite is about ten times greater than that of the copper.

The 10 v/o composite lives are compared with fatigue lives of the same material tested in a tension-tension, load-control mode [2] in Fig. 6. Under the load-controlled test condi-

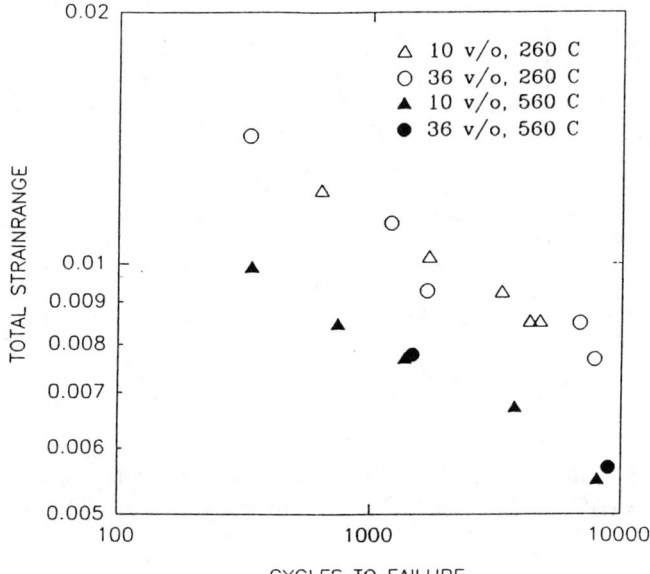

FIG. 3—*Total strain range as a function of fatigue life for W-Cu at 260 and 560°C.*

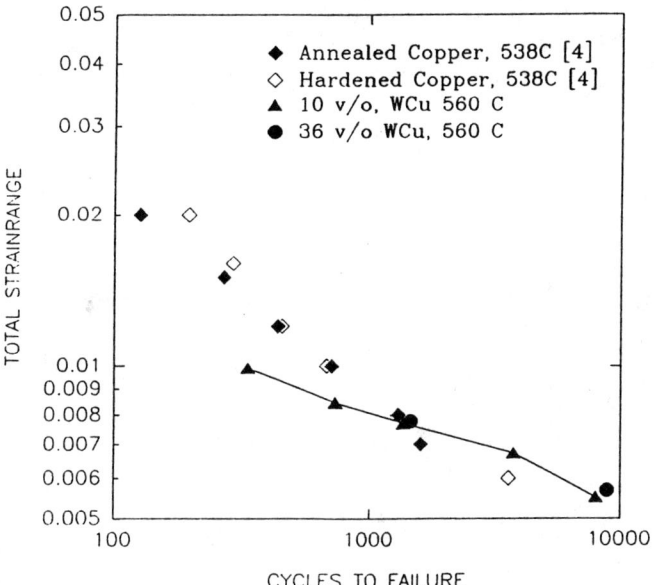

FIG. 4—*Total strain range as a function of fatigue life for W-Cu tested at 560°C and OFHC copper tested at 538°C [4].*

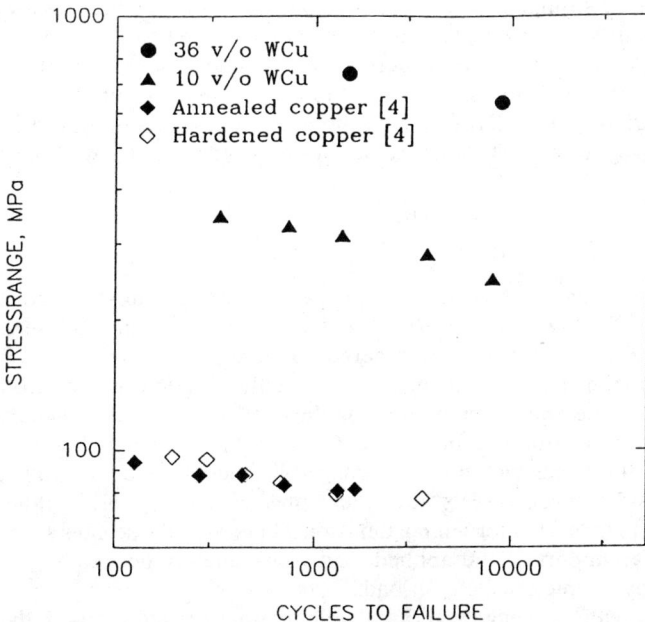

FIG. 5—*Strain range as a function of fatigue life for W-Cu tested at 560°C and OFHC copper tested at 538°C [4].*

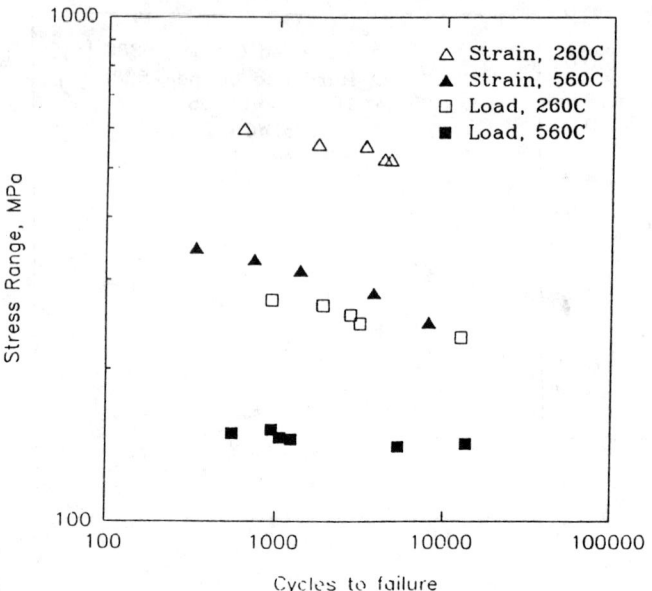

FIG. 6—*Stress range as a function of fatigue life for 10 v/o W-Cu fatigue tested under load and strain control.*

tions, the composites continuously ratchetted, resulting in fatigue strains and hysteresis loops which never stabilized. Therefore, the life comparison of the two test control modes was made on a stress range basis. The stress ranges shown for the strain-controlled tests were those measured at one half fatigue life. For each temperature, lives of the strain-controlled tests are at least two orders of magnitude longer than those of the load-controlled tests. Although not shown here, the lives were also compared on a stress range versus time-to-failure basis. Even though the cycle frequency employed during the load-controlled test program was slower by a factor of ten, the data showed the same trends as seeen in Fig. 6.

Fractography and Metallography

Fatigue-cracked regions of 10 and 36 v/o specimens tested in strain control at 260 and 560°C are shown in Figs. 7a and 7b. The fatigue cracking and failure modes did not significantly vary with volume fraction here and were similar to previous load-controlled results [1,2]. Fatigue damage was again not usually initiated at the strong fiber-matrix interfaces. The fatigue cracks initiated along the specimen surface in general, particularly at corners. The cracks initiated in the matrix along the specimen sides and corners and also at fibers on the edges which were damaged during specimen machining.

In previous load-controlled tests, the cracks propagated only through the matrix, growing around and then past intervening tungsten fibers [1]. These fibers, left bridging the crack and locally supporting the applied load, subsequently necked to 70% reduction in area and failed by simple tensile overload.

Fractographic evidence suggested that the fibers fracture differently in the crack growth process in strain-controlled tests. Metallographic examinations of secondary cracks away from the fracture surfaces indicate cracking still preferentially occurred in the matrix as in

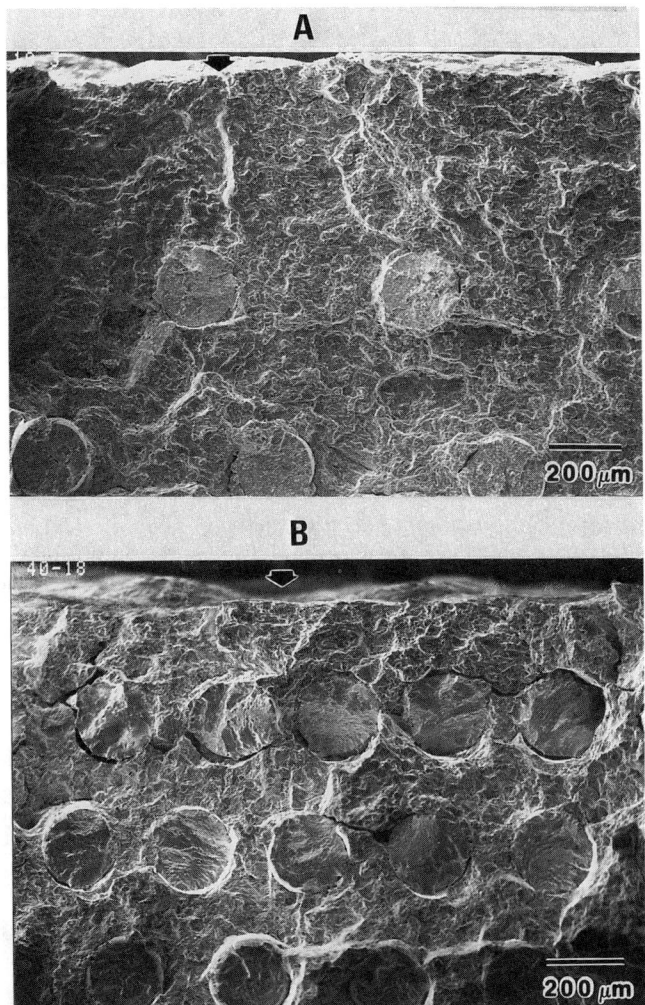

FIG. 7—*Fatigue cracked regions:* (a) *10 v/o specimen tested at 260°C;* (b) *36 v/o specimen tested at 560°C. Cracks initiated at specimen sides at indicated locations.*

load-controlled tests. These cracks still grew around the impeding fibers. The remaining crack-bridging fibers subsequently necked and fractured behind the crack tip as shown in Fig. 8. Unlike the case for the load-controlled specimens, the fibers fractured at about the same elevation as the matrix, but the fiber and matrix fracture surface topologies were not continuous. The fibers were necked only 1 to 14% in fatigue-cracked regions of the strain-controlled specimens. The lower reduction in area is in some part due to the constant maximum strain amplitude imposed in these tests. However, the strain-controlled fiber failure mode differed substantially from the earlier tensile overload failure of the load-controlled tests, as evidenced by the lower reduction in area.

The matrix failed in fatigue principally by formation of cavities at grain boundaries, as shown in Fig. 9, as in previous load-controlled tests. In tests at 260°C, the cavitation was mixed with minor secondary transgranular microvoid coalescence. The intergranular cavi-

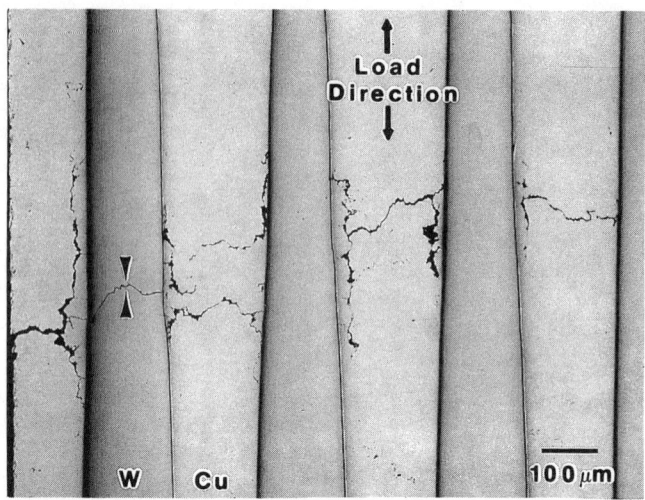

FIG. 8—*Metallographic longitudinal section showing a copper (Cu) matrix crack initiated at the specimen corner which grew around the tungsten (W) fiber. The outermost tungsten fiber subsequently necked and fractured at the indicated location. 36 v/o specimen tested at 260°C.*

tation was more severe at 560°C and clearly predominated. The fatigue-cracked regions sometimes had significant compressive damage as evidenced by surface flattening, especially near crack initiation points.

In summary, fractographic analyses indicated surface crack initiation and preferential

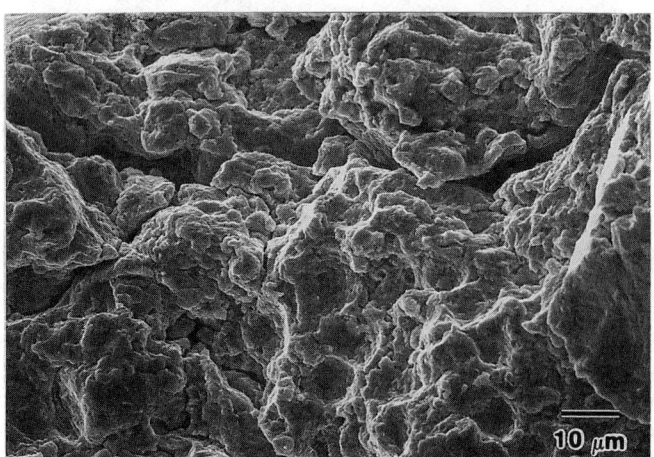

FIG. 9—*Predominant intergranular cavitation fatigue failure in the Cu matrix. 36 v/o specimen tested at 560°C.*

matrix cracking occurred for specimens of each volume fraction tested at each temperature. The matrix failed in fatigue predominantly by cavitation at grain boundaries.

Discussion

The composite and copper fatigue lives are the same at the elevated temperature. This may be related to the preferential fatigue cracking in the copper matrix of the composite. The matrix failed principally by formation of grain boundary cavities, the same fatigue damage mechanism observed in stand-alone copper [6]. The data suggest that fatigue cracking of the matrix played the major role in controlling the failure of the composite and that cracking of the tungsten fibers was less important. The coincident fatigue lives may indicate that an axial-strain driven failure process is operative here. For isostrain conditions employed in this study, the axial strain in the matrix would be equivalent for both composite volume fractions as well as unreinforced copper.

The composite lives may actually be better than the copper lives under equivalent testing conditions. Different test techniques were employed in the testing of the copper [4] and the composite. The copper specimens had a cylindrical cross section, not a square cross section as the composite specimens did. Cracks in the composite initiated at corners and fibers damaged during specimen machining, probably yielding lower lives than one would obtain testing a cylindrical specimen with no corners or machining-damaged fibers. Also, the failure criteria used for the composite tests was not separation into two pieces as used in the copper tests but a drop in tensile load. As most composite specimens did not break before the load drop occurred, the composites would be expected to have even longer lives if allowed to fatigue to separation. As an example, one 10 v/o composite specimen was tested using a strain range of 0.927% at 260°C. The life, as defined by the 25% drop in tensile load failure criteria, was 1680 cycles; however, the specimen was allowed to accumulate 63 389 cycles before the test was stopped with the specimen still in one piece. Hence, the composite data given in Fig. 4 probably represent a lower bound, and the composite may have better fatigue resistance than the stand-alone copper when compared on a strain basis.

Although no copper data were available for comparison to the composite data at the lower temperature, the fatigue life of copper is expected to be similar to that of the composite, as the failure of the composite at 260°C was also controlled by matrix cracking.

Matrix cracking appears to dominate composite fatigue life. Because of the matrix cracking in the composite, fatigue lives were independent of fiber volume fraction at each temperature. The microscopic examination suggested that the failure mechanism of the fibers was independent of temperature. Therefore, the shorter composite fatigue lives at 560°C as compared to 260°C can be largely attributed to changes in matrix failure between the temperatures. The observed temperature dependency on the degree of matrix damage indicates, as shown by others [7], that the cavitation rates at 560°C were higher than at 260°C.

Load-controlled fatigue lives were shorter than the strain-controlled lives, probably due to the accumulation of much larger mean strains and the higher mean stress. The composite tested under the load-controlled conditions experienced large mean strains, up to 14% at 260°C [2]. The load-controlled tests were conducted using a load R-ratio of 0.05, whereas the strain-controlled tests were conducted under fully reversed ($R_\varepsilon = -1$) conditions. This produced tensile mean stresses in the load-controlled tests, which can degrade fatigue life [8], in the range of 110 to 137 MPa at 260°C and around 70 to 76 MPa at 560°C. In contrast, little or no mean stress was generated during the strain-controlled tests.

Although this discussion is based on 10 v/o composite data, the same trends are expected for the 36 v/o composite as well, since the life differences are due to tensile mean stresses and mean strains imposed by the test method.

Fatigue tests of metal matrix composites are typically conducted using a zero-tension, load-controlled waveform. The data shown in this report imply that one should take great care in using this data for design purposes. For example, if the composite experiences compression in service, then the use of tension-tension, load-controlled data for design analysis may be overly conservative.

The tungsten fiber/copper matrix composite system presents unique opportunities to support the development of fatigue damage accumulation and life prediction models. Both the fibers and the matrix have good ductility and fatigue resistance over the temperature range from 260 to 560°C. The fiber-matrix interface bond is very strong and does not fail prematurely in isothermal fatigue. In isothermal fatigue over the temperature range from 260 to 560°C, this composite failed by a single general fatigue failure mechanism. This mechanism involved matrix cracking through the formation of cavities at grain boundaries, a process which is well understood and documented in monolithic materials including copper [7]. A model based on this matrix failure mechanism was very successful in predicting fatigue life of 10 v/o load-controlled tests at 560°C in previous work [9]. The present results suggest such an approach would also be useful for isothermal strain-controlled fatigue of this composite in these test conditions.

Conclusions

1. For this composite tested at these temperatures, fatigue life is independent of fiber volume fraction because fatigue damage preferentially occurs in the matrix.
2. When compared on a total strain basis, fatigue lives of unreinforced copper and the composite were the same at 560°C. Thus, the presence of fibers does not degrade the fatigue resistance of the copper matrix in this composite system.
3. Strain-controlled fatigue lives are longer than those generated using a load-controlled test method, presumably due to mean strain and mean stress effects.

References

[1] Kim, Y. S., Verrilli, M. J., and Gabb, T. P., "Characterization of Failure Processes in Tungsten Copper Composites Under Fatigue Loading Conditions," *Proceedings, ISTFA 1989—International Symposium for Testing and Failure Analysis: Advanced Materials*, Los Angeles, CA, 6–10 Nov. 1989, ASM International, Metals Park, OH.
[2] Verrilli, M. J., Kim, Y. S., and Gabb, T. P., "High Temperature Fatigue Behavior of Tungsten Copper Composites," *Fundamental Relationships Between Microstructures and Mechanical Properties of Metal Matrix Composites*, P. K. Liaw and M. N. Gungor, Eds., TMS, Warrendale, PA, 1990, pp. 479–495.
[3] Ammon, R. L. and Buckman, R. W., "Fabrication of High Strength, High Conductivity Tungsten Fiber Reinforced Copper Composites for Application in Space Power Systems," *Proceedings, Sixth Symposium on Space Nuclear Power Systems*, Albuquerque, NM, Orbit Book Co., Melbourne, FL, January 1989.
[4] Conway, J. B., Stentz, R. H., and Berling, J. T., "High Temperature, Low-Cycle Fatigue of Copper-Base Alloys in Argon: Part I—Preliminary Results for 12 Alloys at 1000°F (538°C)," NASA CR-121259, January 1973.
[5] Gayda, J., Gabb, T. P., and Freed, A. D., "The Isothermal Fatigue Behavior of a Unidirectional SiC/Ti Composite and the Ti Alloy Matrix," *Fundamental Relationships Between Microstructures and Mechanical Properties of Metal Matrix Composites*, P. K. Liaw and M. N. Gungor, Eds., TMS, Warrendale, PA, 1990, pp. 497–514.
[6] Page, R. and Weertman, J. R., "Investigation of Fatigue-Induced Grain Boundary Cavitation by Small Angle Neutron Scattering," *Scripta Metallurgica*, Vol. 14, 1980, pp. 773–777.

[7] Weertman, J. R., "Fatigue-Induced Cavitation in a Single-Phase Material," *Canadian Metallurgy Quarterly,* Vol. 18, 1979, pp. 73–81.
[8] Smith, K. N., Watson, P., and Topper, T. H., "A Stress-Strain Function for the Fatigue of Metals," *Journal of Materials,* Vol. 5, No. 2, 1970, pp. 767–778.
[9] Kim, Y. S., Verrilli, M. J., and Gabb, T. P., "A Model for Predicting High Temperature Fatigue Failure of a W/Cu Composite by Creep Cavity Growth," *HITEMP Review 1990,* NASA CP-10051, October 1990.

Kin Liao,[1] Thomas J. Dunyak,[2] Wayne W. Stinchcomb,[3] and Kenneth L. Reifsnider[4]

Monitoring Fatigue Damage Development in Ceramic Matrix Composite Tubular Specimens by a Thermoelastic Technique

REFERENCE: Liao, K., Dunyak, T. J., Stinchcomb, W. W., and Reifsnider, K. L., "**Monitoring Fatigue Damage Development in Ceramic Matrix Composite Tubular Specimens by a Thermoelastic Technique**," *Composite Materials: Fatigue and Fracture, Fourth Volume, ASTM STP 1156,* W. W. Stinchcomb and N. E. Ashbaugh, Eds., American Society for Testing and Materials, Philadelphia, 1993, pp. 620–636.

ABSTRACT: Tubular specimens made of borosilicate glass reinforced by chopped carbon fibers for high-temperature applications were designed and fabricated by an injection molding process. Specimens were cyclically loaded in tension-tension at maximum stresses of 60, 70, and 85% ($R = 0.1$) of the ultimate tensile strength at room temperature to one million cycles and then loaded to failure in quasistatic tension. A thermoelastic technique known as SPATE (stress pattern analysis by thermal emission) was used to characterize fatigue damage development in these tubular specimens. Qualitative SPATE results were shown to be related to surface crack initiation and growth in the specimens, circumferentially and radially. Damage initiation sites and subsequent growth of cracks as well as residual strengths were found to be primarily influenced by the local manufacturing related microstructure of individual specimens and secondarily influenced by fatigue damage.

KEY WORDS: ceramic matrix composite material, fatigue damage, thermoelastic technique

Advanced ceramic matrix composite (CMC) materials have an increasing number of potential engineering applications because of their desirable properties at elevated temperatures. For instance, they are candidate materials in structures such as heat exchangers, regenerators, turbines, and heat engines, to name only a few applications. Concurrent with needs for new and improved CMC material systems are needs for reliable mechanical test methods, nondestructive evaluation (NDE) techniques, and mechanistic, performance prediction methodologies. Destructive and nondestructive test data are required to develop relationships between manufacturing processes, properties of materials, and performance of CMC components. These relationships are an essential part of an overall capability to describe and predict long-term performance.

In a previous study [1], a number of NDE techniques were examined for their capability to detect defects and damage development in CMC tubular specimens during quasistatic

[1]Graduate research assistant, Materials Response Group, Engineering Science and Mechanics Dept., Virginia Polytechnic Institute and State University, Blacksburg, VA 24061-0219.

[2]Formerly graduate project assistant, Materials Response Group, Engineering Science and Mechanics Dept., Virginia Tech; currently, engineer, Engine Materials Testing Laboratory, General Electric Aircraft Engines, Cincinnati, OH 45215-6301.

[3]Professor, Materials Response Group, Engineering Science and Mechanics Dept., Virginia Polytechnic Institute and State University, Blacksburg, VA 24061-0219.

[4]Alexander Giacco Professor, Materials Response Group, Engineering Science and Mechanics Dept., Virginia Polytechnic Institute and State University, Blacksburg, VA 24061-0219.

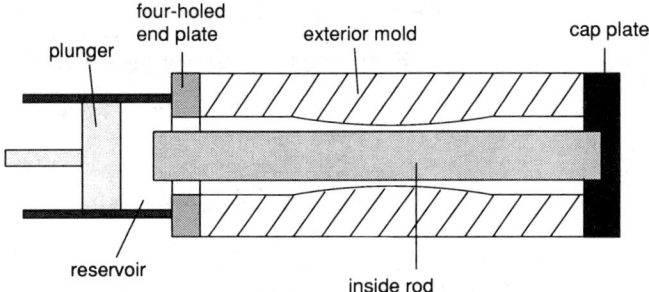

FIG. 1—*Injection molding assembly.*

tension tests. To characterize surface-related fatigue damage initiation and development in the tubular specimens in situ, a study was carried out using a nondestructive, thermoelastic technique known as SPATE (stress pattern analysis by thermal emission). The results of the study are reported herein.

Specimen Fabrication

The fabrication process has a direct influence on the properties and performance of the specimens. An understanding of that process and the microstructure of the specimens produced in such a process is necessary for the interpretation of test results and subsequent analytical modeling. The specimens were fabricated by the United Technologies Research Center (UTRC) with an injection molding process and consisted of a borosilicate glass matrix reinforced with chopped graphite fibers. The details of the fabrication process are considered proprietary by UTRC and cannot be described in detail in this paper. Only a brief summary of the process can be provided.

A schematic diagram of the injection molding assembly is shown in Fig. 1. Glass beads and chopped fibers are thoroughly mixed and added to the reservoir. The mixture is heated to produce a viscous molding compound and then pushed from the reservoir through four symmetrically positioned holes in the end plate and into the mold. The assembly is then cooled prior to removing the interior rod and exterior mold. The exterior mold is a continuous graphite structure and is machined in half in order to be removed. The surface finish on the outside of the tube was very smooth, and no additional machining was

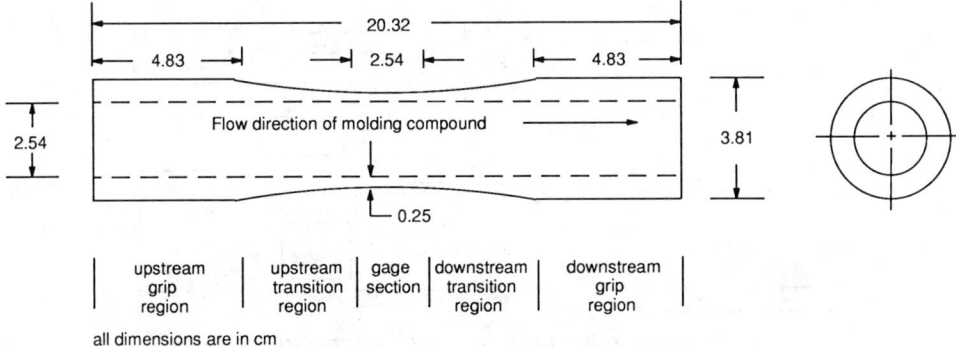

FIG. 2—*Specimen configuration.*

TABLE 1—*Physical properties of tubes.*

Specimen	Glass Type	Fiber Type	Fiber Length, cm	Fiber volume, %	Bulk Density, g/cc	Porosity, %
258-89	CGW 7070	HMU-PVA-3k	0.94	27	1.98	1.97
259-89	CGW 7070	HMU-PVA-3k	0.94	27	1.98	2.48
260-89	CGW 7070	HMU-PVA-3k	0.94	27	1.98	4.65

required except in the grip regions. The inside rod was also made of graphite and removed by machining. The outside of the grip regions was then machined in order to obtain proper gripping tolerances. A schematic diagram of an as-fabricated tube is shown in Fig. 2. Before gripping, two 5.1-cm-long cylindrical end plugs made of graphite with diameters slightly less than 2.54 cm were inserted in the two grip regions of the specimen to provide support for those regions and prevent grip-related damage. Some of the physical properties of the specimens are listed in Table 1.

As a result of the injection molding process, the fiber distributions and directions varied throughout the finished specimen. Detailed identification of the fiber orientations was difficult, but important to an understanding and interpretation of mechanical performance. In order to accurately reference the various parts of the tube, the tube is divided into five regions (Fig. 2). The terms "upstream" and "downstream" are based on the flow direction of the molding compound and are required since the fiber distribution in the upstream transition region is significantly different than the distribution in the downstream transition region. These differences have a major effect on the damage initiation, development, and failure location in the specimens.

Three tubular CMC specimens were selected for the fatigue damage development study in order to simulate some of the performance-limiting events associated with complex engineering components. X-ray radiography was used to examine the state of the as-fabricated specimens. A detailed description of the method used can be found elsewhere [1]. The fiber distributions and orientations are determined from the radiographs. A schematic diagram of the microstructure determined by X-ray radiography is shown in Fig. 3.

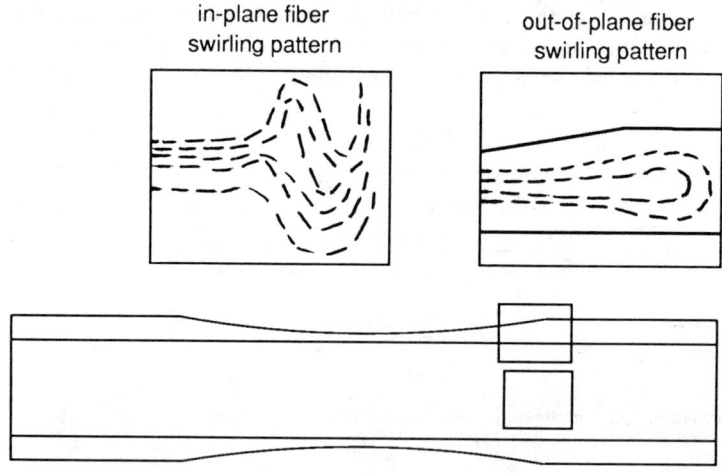

FIG. 3—*Microstructure in the downstream transition region.*

For all the three specimens, the flow pattern in the upper grip region due to the four injection points in the end plate are very apparent. The effect of these injection points extends all the way through the gage section and even into the downstream transition region. The fibers in the upstream grip region, the upstream transition region, and especially the gage section tend to be nearly aligned along the axis of the tube. However, the fibers appear to be swirled throughout the downstream transition and grip regions. This swirling is the result of the flow pattern as the viscous material exits the gage section in the molding process and the back pressure when the molding compound reaches the end of the molding cavity. Two different kinds of fiber swirling were identified, as shown in Fig. 3. The in-plane swirling was identified by X-ray radiography, while the out-of-plane swirling can only be shown after longitudinal sectioning of a specimen. These fiber swirling patterns in the specimens are extremely critical to the strength of the specimens.

Test Facility

A servohydraulic test frame was developed under a corporative effort between the Materials Response Group and the Instron Corporation. The test frame is a biaxial, two-post system rated for 4.448E5 N (100 kip) axial and 5.65E3 N/m (50 kip-in.) torsional load. A high-stiffness frame was required to reduce the amount of twist in the crosshead under rated loads. The stiffness of the frame is rated at 8.76E9 N/m (5.0E6 lb/in.) axial and 4.29E5 N/m/degree (3.8E6 in.-lb/degree) torsional at 101.6 cm separation between the crosshead and base. This high stiffness is a basic requirement for the testing of CMC materials. Twist tends to introduce a bending moment into the specimen under torsional loading, and even a slight amount of bending can be very detrimental to the performance of a ceramic composite component. The load cell and actuators are rated at 2.224E5 N (50 kip) axial and 2.824E3 N/m (25 kip-in.) torsional. A pair of hydraulic grips was specially designed and developed for testing the CMC tubular specimens.

The SPATE Technique

It has long been known that if a material is stressed under adiabatic conditions within its elastic range, it will experience a small, reversible temperature change resulting from the dilational deformation of the material. Thompson [2] has shown that for a homogeneous, isotropic, linear elastic material, such a thermoelastic effect can be described by the following relationship

$$\Delta T = -\frac{\alpha}{C_\sigma \rho} T_0 (\Delta \sigma_1 + \Delta \sigma_2 + \Delta \sigma_3) \qquad (1)$$

where

$$\Delta T = \text{change in temperature,}$$
$$\alpha = \text{coefficient of thermal expansion,}$$
$$C_\sigma = \text{specific heat at constant stress,}$$
$$\rho = \text{density,}$$
$$T_0 = \text{initial temperature, and}$$
$$\Delta \sigma_i (i = 1,2,3) = \text{change in principal stresses.}$$

It follows from Eq 1 that for materials with a positive thermoelastic constant, $\alpha/C_\sigma \rho$, a positive change in the sum of the principal stresses (i.e., a change from a lower stress level to a higher stress level) will result in a decrease in the temperature of the material and vice versa. For anisotropic materials, a similar but more complicated expression was given by Biot [3].

If cyclic load is applied to the material at a rate such that no heat transfer occurs between the material and its surroundings, i.e., an adiabatic condition is maintained, the temperature of the material will vary with the applied load at the loading frequency. Based on this thermoelastic principle, a SPATE system was developed and marketed by Ometron Inc. The system (SPATE 9000) consists of an infrared photon detector unit, a correlator, and two computers. It is capable of measuring the full field, minute temperature changes on a specimen surface at the loading frequency of the specimen with a temperature resolution of 0.001 K by detecting the change in infrared radiation photon emittance of the specimen surface in a pointwise, raster-like manner. The functions of the computer and correlator are to control the scan activities of the system and condition the measured photon emittance such that only the sinusoidal temperature variation occurring at the same frequency as the applied load is recorded. Any other temperature change that is not occurring at the loading frequency is rejected. The peak-to-peak change in photon emittance, $\Delta\Phi$, is related to changes in temperature, ΔT, through the relationship [4–6]

$$\Delta\Phi = 3eBT_0^2\Delta T \tag{2}$$

where

e = emissivity of the prepared test surface, and
B = Stefan-Boltzmann constant.

Equation 2 is obtained by differentiating the Stefan-Boltzmann law of photon emission. Substituting Eq 1 into Eq 2 results in

$$\Delta\Phi = -\frac{3eB\alpha}{C_\sigma\rho} T_0^3(\Delta\sigma_1 + \Delta\sigma_2 + \Delta\sigma_3) \tag{3}$$

The detected change in infrared photon emittance, $\Delta\Phi$, can be converted to a change in sum of principal stresses, as can be seen from Eq 3.

Experimental Procedures

Three tubular specimens were loaded to failure in monotonic tension at room temperature. The loading rate was 4448 N/min (1000 lbs/min). Nominal monotonic ultimate tensile strengths of the tubes were 82.8, 83.8, and 87.6 MPa. The mean ultimate tensile strength was 84.7 MPa, with a scatter of +2.9 MPa and −1.9 MPa. Each of the specimens failed at the downstream transition regions where fiber swirling was found.

Three specimens were chosen for examination by the SPATE technique during cyclic loading. Specimens 260-89, 259-89, and 258-89 were cyclically loaded ($R = 0.1$) at room temperature at 60, 70, and 85%, respectively, of the mean ultimate tensile strength (UTS), as determined from the monotonic tests. The loading frequency was 5 Hz.

To prepare for a SPATE scan, a thin layer of flat black spray paint was evenly applied to the specimen surfaces to obtain uniform surface emissivity in the infrared spectrum and to reduce the possibility of reflected heat sources being modulated at the test frequency [4]. A black backdrop was mounted behind the specimen to eliminate thermal noise coming from the vicinity of the specimen.

Since all the failures in the monotonic tests occurred within the downstream transition region of the specimens, it was anticipated that fatigue damage is more likely to be initiated in this region than in other regions of the specimen. Therefore, a small area on the downstream transition region of each specimen was randomly identified to be scanned using SPATE. The distance between the specimen and the SPATE scan unit was fixed at about 38 cm. At this distance, the diameter of each measurement spot was 0.064 cm. The time that the scan unit spent on each measurement spot, the sample time, was set at 0.4 s. Since only the occurrence and development of fatigue damage was of interest (but not the actual stress values), a calibration factor of unity was used for all the scans.

A widely accepted practice in using SPATE to detect fatigue damage is to scan the specimen surface at a lower load level than the actual cyclic load so that no further damage occurs during the scan period. However, for the current CMC specimens, damage development is not detectable at low load levels. It is believed that crack openings at lower load levels were not wide enough for SPATE resolution. A few scans for Specimens 260-89 (cyclic stress = 60% UTS) were made at cyclic scan stress of up to 50% UTS, with no cracks being detected. When the scan stress was increased to 60% UTS, abnormal SPATE signals were readily observed, which later were shown to be from the cracked regions. In order to enhance detectability of cracks and compare the effect of the cyclic scan load on the SPATE signal, the specimens were scanned periodically at their cyclic load; i.e., the scan stress for Specimens 258-89, 259-89, and 260-89 were 85, 70, and 60% UTS, respectively.

Two different kinds of SPATE scans were made for each specimen, a "normal scan," and a "detailed scan." For a normal scan, the distance between the center of two measurement spots was 0.04 cm, while for a detailed scan that distance was reduced to 0.02 cm. Usually, a detailed scan was made between several normal scans. The scan times for each normal scan and detailed scan were about 25 and 90 min, respectively. After each specimen was cyclically loaded to one million cycles, it was unloaded and was examined by a dye penetrant method to detect surface cracks. The specimens were loaded to failure afterward in monotonic tension to measure their residual strengths.

Results and Discussion

Four selected SPATE scan images for each of the three specimens are shown in Figs. 4 through 6. The bar at the right of each of the images represents *qualitative* SPATE signal strength, since a calibration factor of unity was used. The gray/black colored areas are cracked regions. The result of a dye penetrant test for Specimen 260-89, performed after one million cycles, is shown in Fig. 7. The cracked regions shown by the dye penetrant test approximately matched those shown in the SPATE images in Fig. 4d. This strongly confirms the cracked regions recorded by the SPATE images.

In all the SPATE images shown, the stress patterns in regions without cracks were "truncated" in order to enhance visibility of the cracked regions. This was done by defining a qualitative maximum *display* signal which is lower than the actual recorded maximum value. Also, the color associated with the display maximum was adjusted to white so that regions with signal strength above the defined display maximum was displayed in that color. The thermal emission pattern of the regions without cracks was found to be highly nonuniform. Bakis and Reifsnider [6] have shown that, for a two-phase composite laminate, the thermoelastic emission pattern strongly depends on the orientation of the surface ply. Also, the nonuniform thermoelastic response of the surface ply is a combined effect of the fiber and matrix phases. For short fiber reinforced CMC material,

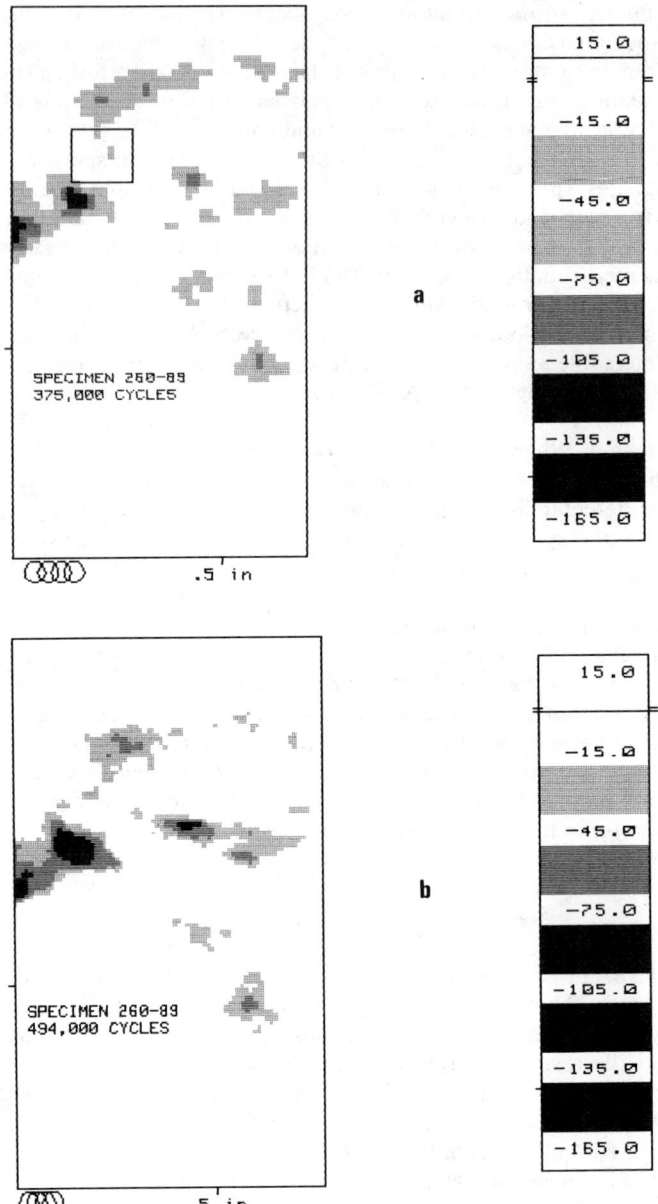

FIG. 4a–d—*SPATE image of Specimen 260-89 at the end of* (a) *375 000* (b) *494 000* (c) *744 000* (d) *976 800 cycles. Longitudinal crack linkage appeared in the boxed region.*

the highly nonuniform thermal emission pattern is also believed to be caused by the nonuniform mixing of the ceramic matrix and the reinforcing chopped fibers.

Surface crack initiation and development in the specimens are readily observed in the SPATE images. Surface cracks initiated in the downstream transition region in all the

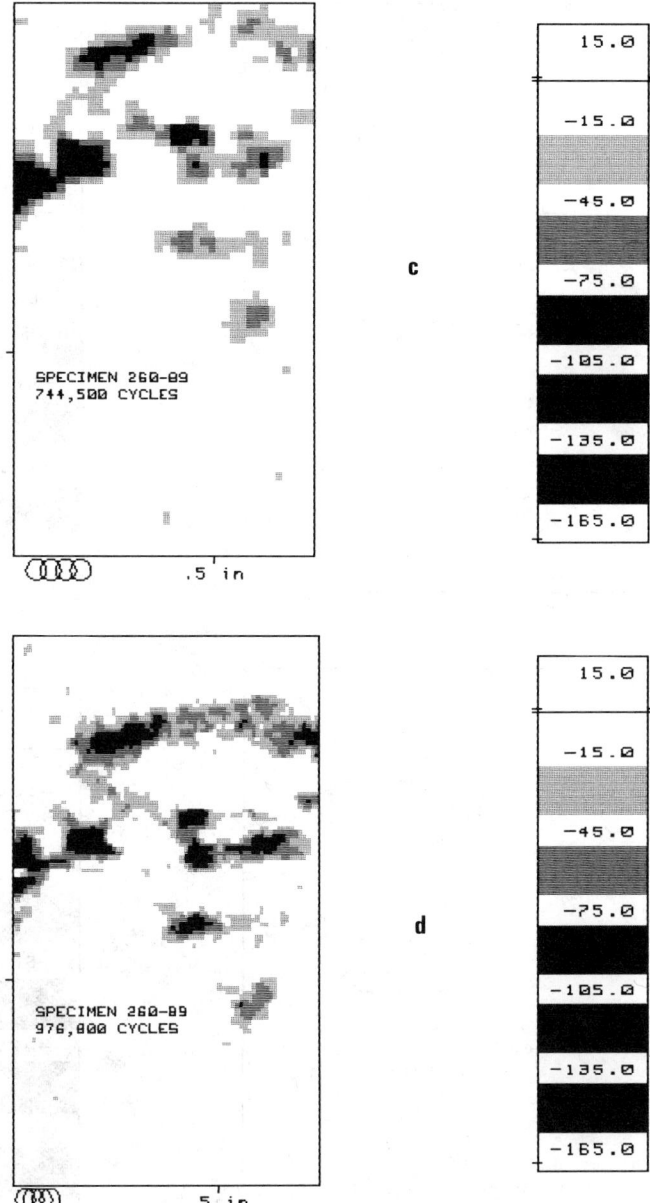

FIG. 4a–d—Continued.

three specimens and extended around the circumference in a direction perpendicular to the longitudinal axis of the specimens. This is easily noticed in Figs. 5a through 5d, where a small crack about 0.13 cm long (shown in the boxed region in Fig. 5a) first appears near the center of the image at 286 000 cycles. It developed around the circumference of the specimen and extended to a 1.0-cm-long crack at one million cycles. Small cracks also

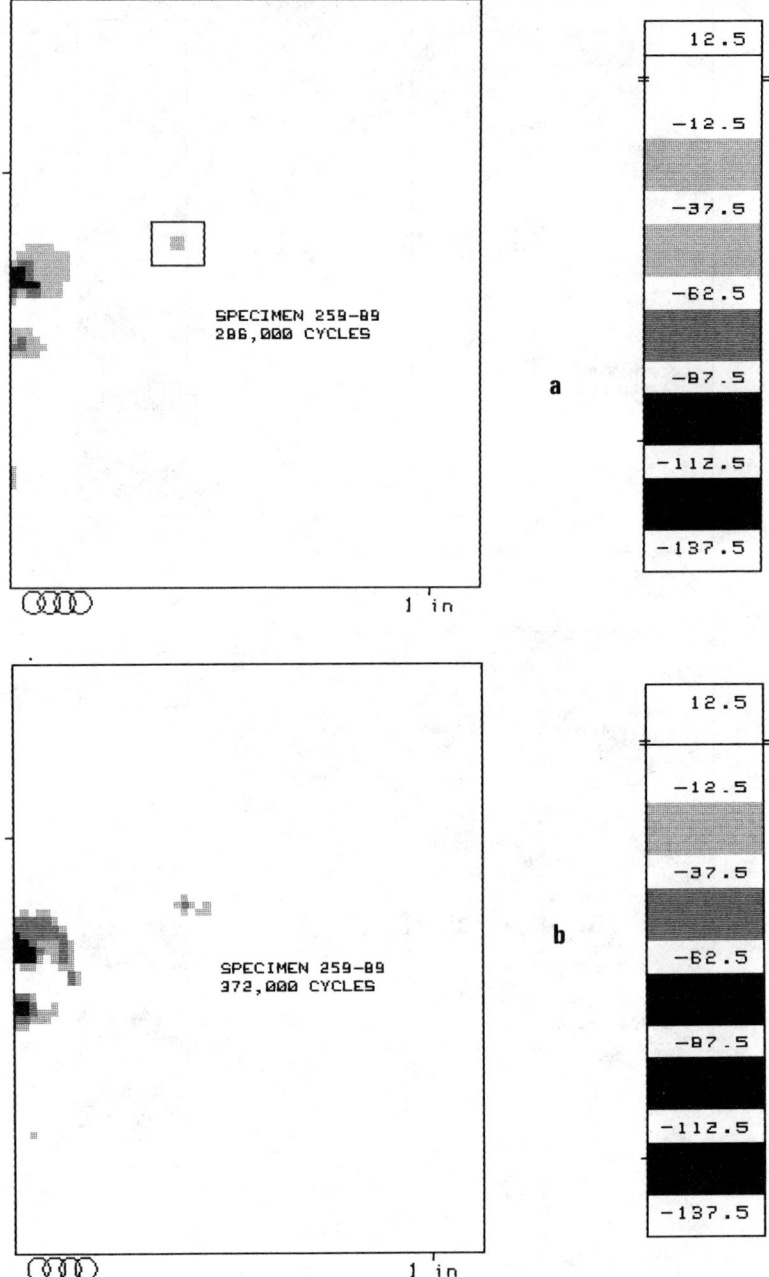

FIG. 5a–d—SPATE image of Specimen 259-89 at the end of (a) 286 000 (b) 372 000 (c) 811 000 (d) 990 000 cycles. Circumferential crack extension appeared in the boxed region.

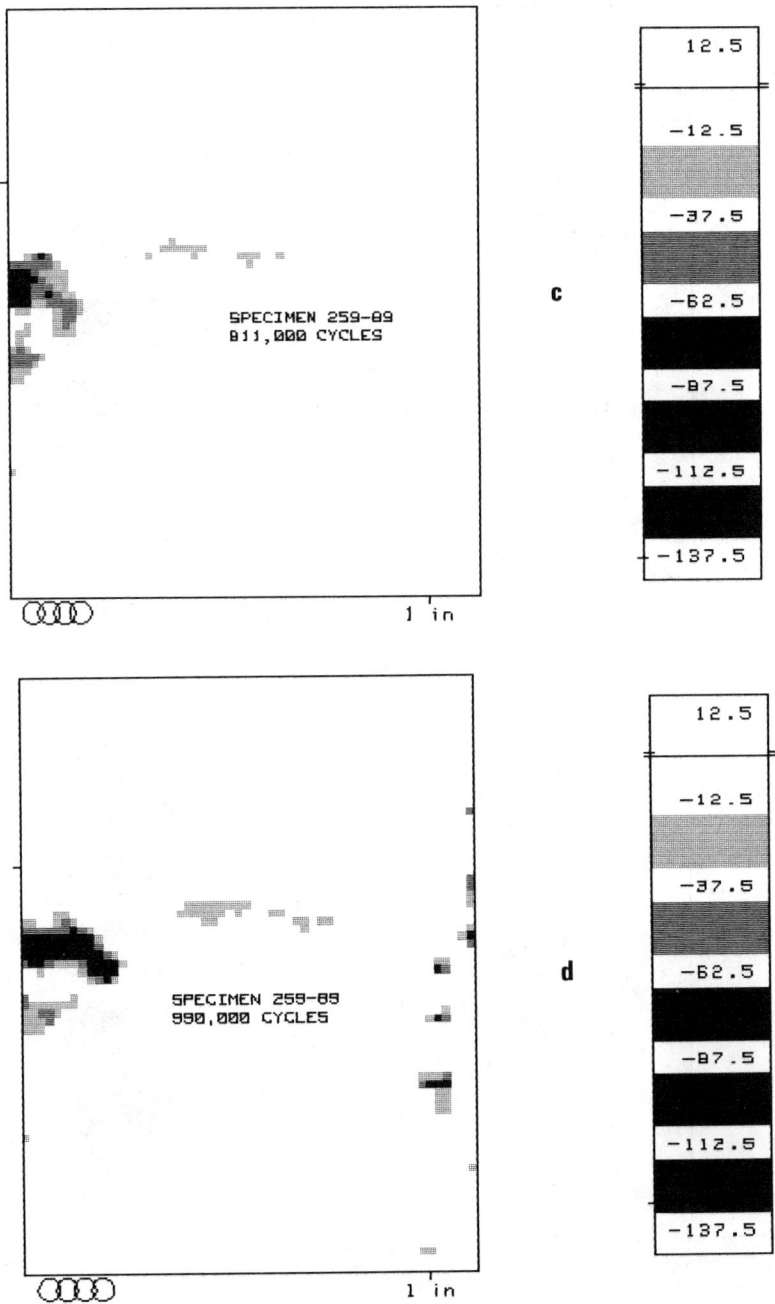

FIG. 5a–d—Continued.

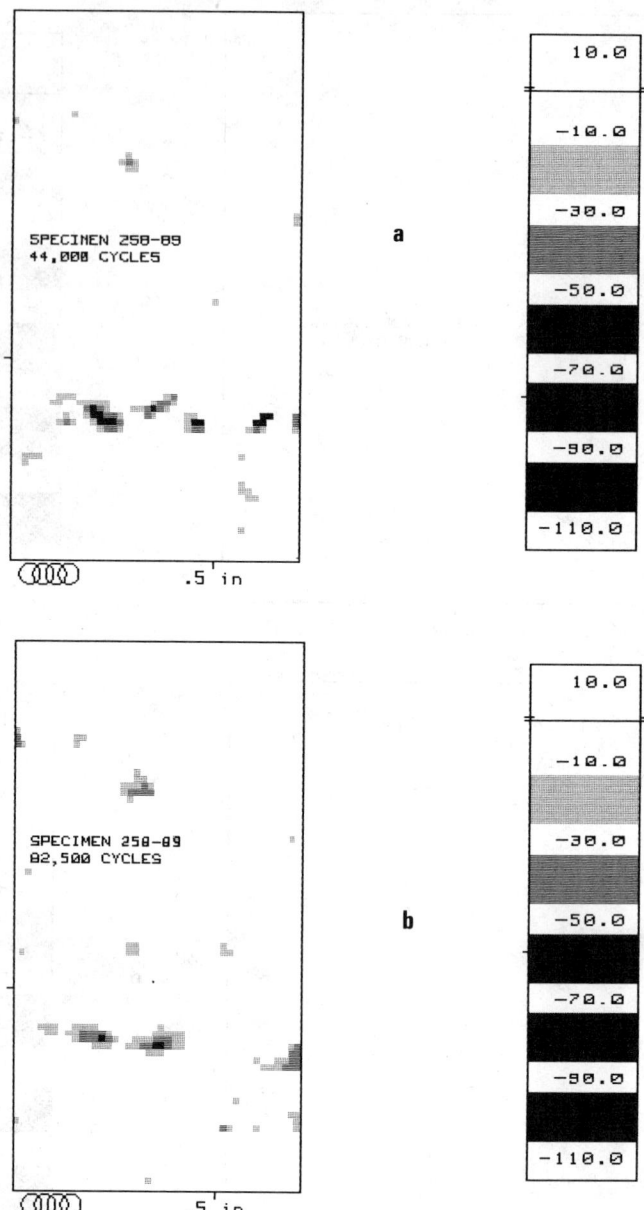

FIG. 6a–d—SPATE image of Specimen 258-89 at the end of (a) 44 000 (b) 82 500 (c) 334 300 (d) 556 000 cycles.

tend to link together around the circumference, and they may also do so longitudinally (shown in the boxed area in Fig. 4a), as seen in Figs. 4a through 4d.

In regions without cracks, the qualitative SPATE signals are positive, while in cracked regions, they are negative. The coefficient of thermal expansion of the specimen surface,

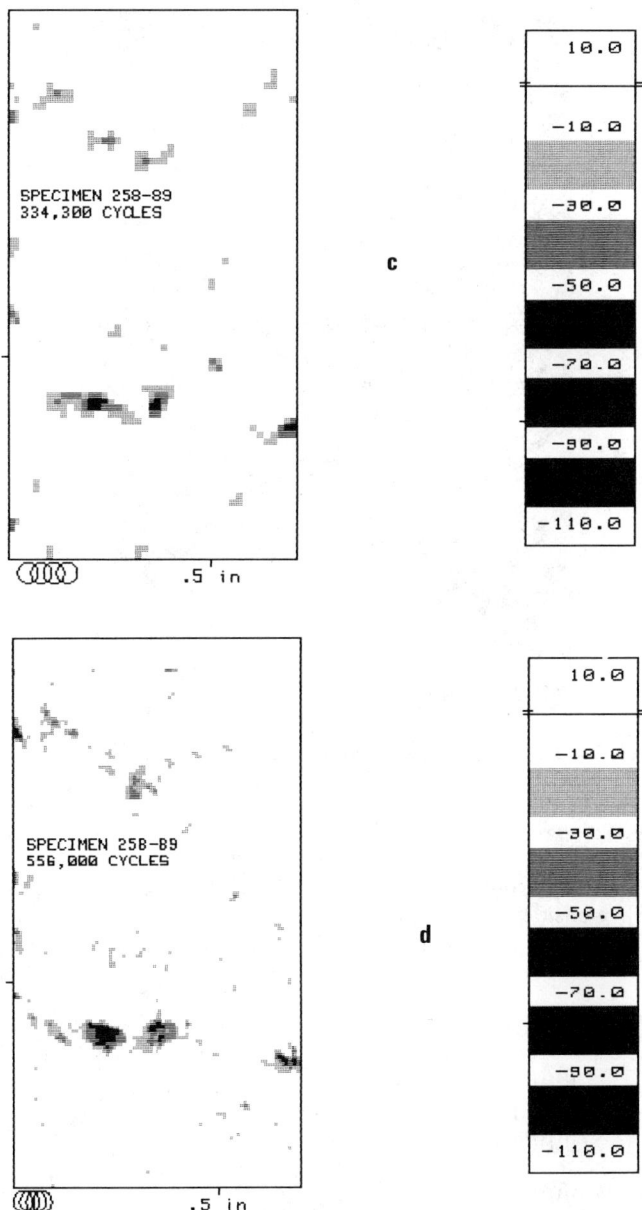

FIG. 6a–d—*Continued*.

which is predominantly glass matrix material, is positive. For materials with positive coefficients of thermal expansion, a positive change in thermal emission will occur from the high peak to the low peak of a load cycle when the materials are cyclically loaded in tension-tension, as predicted by the thermoelastic theory (Eq 1). This is observed in regions without cracks.

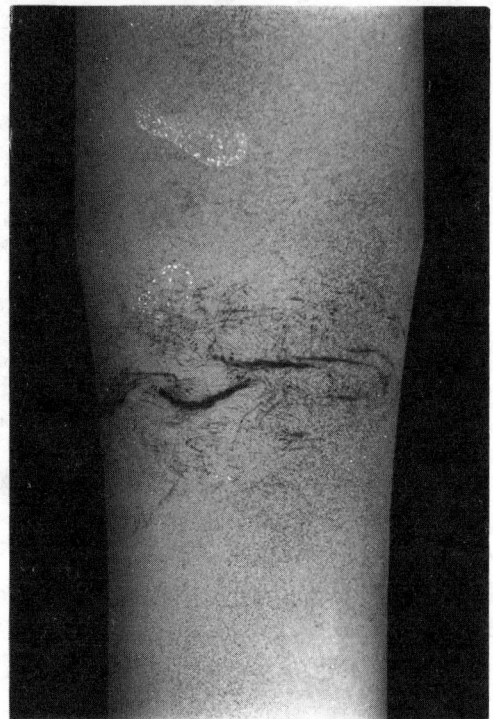

FIG. 7—*Result of a dye penetrant test for Specimen 260-89.*

In the cracked regions, the observed negative SPATE signal may be explained by the concept of cavity emission. It is well known that the thermal radiation from a cavity of any shape is higher than that from a flat surface made of the same material and at the same temperature [7,8]. Consider the behavior of a crack during a tension-tension load cycle (Fig. 8), assuming for the moment that there is no crack extension. At the high peak of a

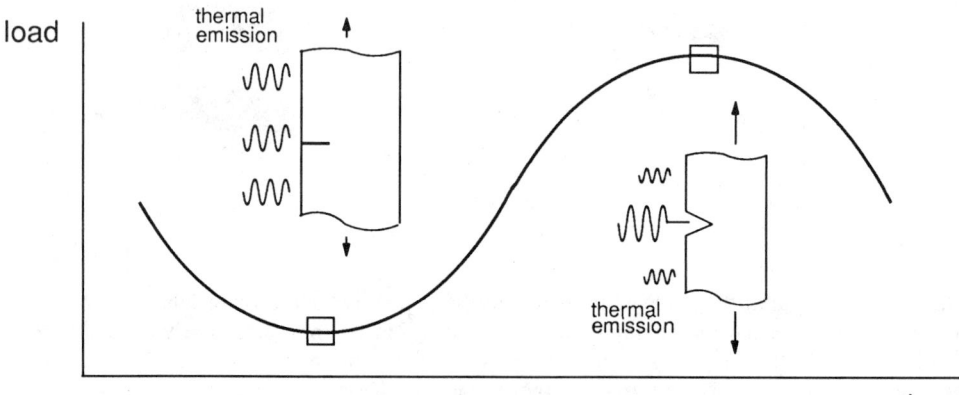

FIG. 8—*Thermal emission of a cracked region under cyclic load.*

load cycle, a crack opens up. The free surfaces of the opened crack act like a cavity, emitting higher thermal radiation than those emitted from the rest of the specimen surface. When the specimen is at the low peak of the load cycle, the crack is closed due to the elastic response of the material containing residual stresses, so that the thermal emission from the crack region is essentially the same as from the rest of the specimen surface. The net change in thermal emission in the cracked region, due to a change of load from high peak to low peak, is therefore negative, which coincides with the observed results.

This phenomenon can be explained further, qualitatively, using a fracture mechanics approach. It has been shown that the effective emissivity of a cavity is related to the ratio of the dimension of the opening of the cavity to its surface area such that the effective emissivity decreases with an increase in that ratio for all values of initial open wall emissivity [7,8]. For simplicity, assuming that the material is isotropic and a surface crack in the tubular specimen is semi-elliptically shaped, the maximum crack opening displacement, COD_{max}, is [9]

$$COD_{max} = \frac{4\sigma a}{E} \tag{4}$$

where

σ = remote stress,
a = depth of the crack, and
E = elastic modulus.

The total surface area of the crack, A_c, is

$$A_c = \pi ab \tag{5}$$

where b is one half the crack length. We now define $r_{d/a}$ to be the ratio of COD_{max} to A_c, such that

$$r_{d/a} = \frac{COD_{max}}{A_c} = \left(\frac{4}{\pi E}\right)\left(\frac{\sigma}{b}\right) \tag{6}$$

Equation 6 states that the ratio of maximum crack opening displacement to the total surface area of the crack, $r_{d/a}$, will decrease with increase of crack length, b, but will increase with the applied stress, σ. This implies that, at a constant stress amplitude, σ, and an increase in crack length with fatigue cycles, the effective emissivity of the cracked region will increase due to a decrease of the ratio $r_{d/a}$. Equation 6 is used here to illustrate the basic relationship between the emissivity of a crack (represented by $r_{d/a}$), crack geometry, and remote stress. In the cracked region, as will be discussed later, fibers are essentially parallel to the circumference of the tube due to swirling. Qualitatively, Eq 6 can be used in this case without the consideration of fiber reinforcement. If the effect of fiber reinforcement is to be considered, as shown by Marshall et al. [10,11], the basic relationship between $r_{d/a}$, σ, and b still have the same fundamental form as in Eq 6.

For each of the scan images of Specimen 258-89, a point in a cracked region was chosen and the SPATE signals were plotted against log of cycles (Fig. 9). It is shown in Fig. 9 that the SPATE signal from the same point in a cracked region increases (in its negative value) with an increase in fatigue cycles. Note that each data point shown in the figure is an

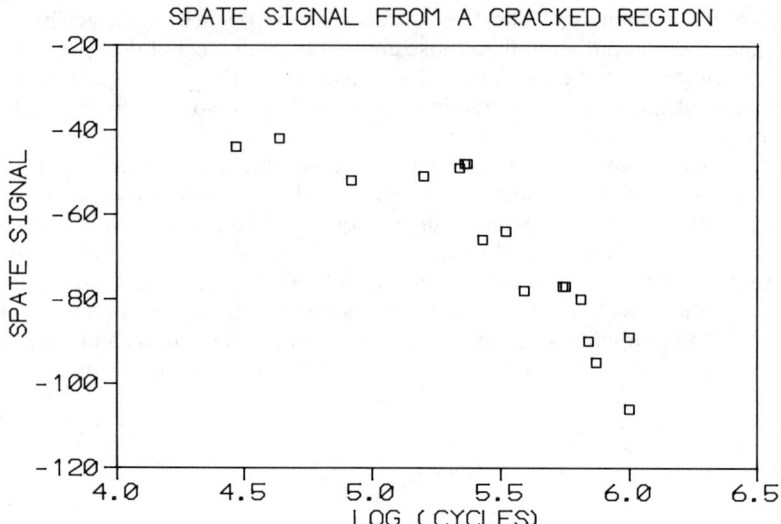

FIG. 9—*SPATE signal from a point in a cracked region.*

averaged value. This is obtained by using a smoothing mechanism of the SPATE system over the entire image such that each data point is averaged with nine of its neighboring data points. If we apply the concept of cavity emission, as the fatigue crack extends into the specimen, the effect of cavity emission increases; thus, the net change in thermal emission also increases. Qualitatively, the data shown in Fig. 9 coincide with the implication of Eq 6.

In general, information obtained from a SPATE scan can be related to surface crack growth in the circumferential longitudinal, and radial directions. The rate of circumferential crack growth can simply be measured from the SPATE image, as shown by Stanley and Leaity [12,13]. However, for the CMC specimens considered, such information cannot be directly related to the residual strength of the specimen as in the case of isotropic materials.

All three specimens failed at the downstream transition region in monotonic tension tests conducted after the completion of one million cycles. The residual strengths of Specimens 258-89, 259-89, and 260-89 after one million cycles were 101.4, 85.5, and 75.2 MPa (14.7, 12.4, and 10.9 ksi), respectively, based on the gage section area. The results are very interesting when these numbers are compared to the mean quasistatic strength of 84.7 MPa (12.3 ksi). Apparently, the residual strength of the specimens loaded under high cyclic load is higher than those loaded under low cyclic loads. Examination of the fracture surfaces of the specimens suggested that the residual strengths were strongly influenced by local microstructure in the failure region.

An examination of local microstructure in the failure region provides some explanation for the strength data. The orientation of fiber bundles on the fractured surfaces for the specimens were complicated. A scanning electron micrograph (SEM) of the fractured surface of Specimen 258-89 is shown in Fig. 10. In general, fiber bundles in the downstream transition region are oriented perpendicularly, obliquely, or aligned with the longitudinal axis of the tube due to swirling in that region. More perpendicularly oriented fiber bundles (and hence less aligned fibers) were found on the fractured surface of Specimen 260-89. This specimen, with the lowest residual strength, fractured along a major swirled

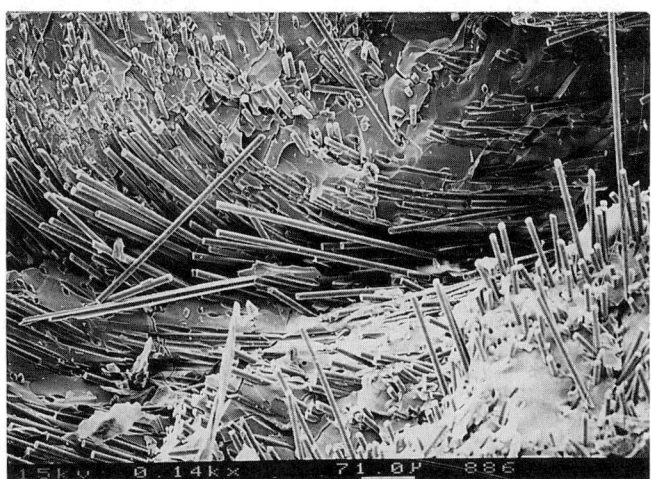

FIG. 10—*Scanning electron micrograph of fractured surface of Specimen 258-89.*

fiber bundle oriented perpendicularly to the longitudinal axis. The fracture surface of Specimen 258-89, with the highest residual strength and under the highest cyclic load, contained more fractured fiber bundles which were aligned with the longitudinal axis of the tube. The fracture surface and residual strength of Specimen 259-89 were found to be intermediate cases between the high and low cyclic load specimens.

Summary

The SPATE technique is an NDE method capable of monitoring surface-related damage initiation and development in CMC materials in situ during cyclic loading. It is shown that SPATE signals can be related to crack extension, circumferentially, longitudinally, and radially. The critical damage region in the CMC tubes produced for this study was the downstream transition region where cracks initiate and develop following fiber swirling directions. The residual strengths of the specimens were primarily controlled by the local manufacturing related microstructure of the specimens.

Acknowledgments

The authors gratefully acknowledge the support of this work by Martin Marietta Energy Systems, Inc. for the U.S. Department of Energy under Subcontract No. 19X-SA946C, WBS Element VPI-1. The cooperative participation of the United Technology Research Center and the Instron Corporation are very much appreciated. The able assistance of Tom Baxter is also acknowledged.

References

[1] Dunyak, T. J., Reifsnider, K. L., and Stinchcomb, W. W., "An Examination of Selected NDE Methods for Ceramic Composite Tubes," CCMS-89-18, Virginia Tech Center for Composite Materials and Structures, Blacksburg, VA, 1989.
[2] Thompson, W., "On the Dynamic Theory of Heat," *Transactions of the Royal Society,* Vol. 20, 1853, pp. 261–283.
[3] Biot, M. A., "General Theory of 3-D Consolidation," *Journal of Applied Physics,* Vol. 12, 1941, pp. 155–164.

[4] Bakis, C. E. and Reifsnider, K. L., "Adiabatic Thermoelastic Measurements," *Manual on Experimental Methods for Mechanical Testing of Composites,* Society for Experimental Mechanics, Bethel, CT, 1990.

[5] Oliver, D. E., "Stress Pattern Analysis by Thermal Emission," Chapter 14, *Handbook of Experimental Mechanics,* A. S. Kobayashi, Ed., Prentice-Hall, Englewood Cliffs, NJ, 1986.

[6] Bakis, C. E. and Reifsnider, K. L., "The Adiabatic Thermoelastic Effect in Laminated Fiber Composites," *Journal of Composite Materials,* Vol. 25, July 1991, pp. 809–830.

[7] Simon, I., *Infrared Radiation,* D. Van Nostrand Company, Inc., New York, 1966.

[8] Bramson, M. A., *Infrared Radiation: A Handbook for Applications,* Plenum Press, New York, 1968.

[9] Broek, D., *Elementary Engineering Fracture Mechanics,* 3rd ed., Martinus Nijhoff Publishers, The Hague, Netherlands, 1983.

[10] Marshall, D. B., Cox, B. N., and Evans, A. G., "The Mechanics of Matrix Cracking in Brittle-Matrix Fiber Composites," *Acta Metallurgica,* Vol. 33, No. 11, 1985, pp. 2013–2021.

[11] Marshall, D. B. and Evans, A. G., "The Tensile Strength of Uniaxially Reinforced Ceramic Fiber Composites," *Fracture Mechanics of Ceramics,* R. C. Bradt, et al. Eds., Vol. 7, Plenum Publishing, New York, 1986.

[12] Stanley, P. and Chan, W. K., "The Determination of Stress Intensity Factors and Crack-tip Velocities from Thermoelastic Infrared Emissions," Paper C262/86, *Proceedings,* Vol. 1, *International Conference on Fatigue of Engineering Materials and Structures,* Sheffield, September 1986, Mechanical Engineering Publications for the Institution of Mechanical Engineers, London, pp. 105–114.

[13] Leaity, G. P. and Smith, R. A., "The Use of SPATE to Measure Residual Stresses and Fatigue Crack Growth," *Fatigue and Fracture of Engineering Materials and Structures,* Vol. 12, No. 4, 1989, pp. 271–282.

Stefanie E. Stanzl-Tschegg,[1] Maria Papakyriacou,[2] Herwig R. Mayer,[3] Jaap Schijve,[4] and Elmar K. Tschegg[5]

High-Cycle Fatigue Crack Growth Properties of Aramid-Reinforced Aluminum Laminates

REFERENCE: Stanzl-Tschegg, S. E., Papakyriacou, M., Mayer, H. R., Schijve, J., and Tschegg, E. K., "**High-Cycle Fatigue Crack Growth Properties of Aramid-Reinforced Aluminum Laminates,**" *Composite Materials: Fatigue and Fracture, Fourth Volume, ASTM STP 1156,* W. W. Stinchcomb and N. E. Ashbaugh, Eds., American Society for Testing and Materials, Philadelphia, 1993, pp. 637–652.

ABSTRACT: The fatigue crack growth properties of the laminated fiber-reinforced composite material Arall[6] (Arall® 1 and Arall® 2) have been studied at low cyclic stress amplitudes and high numbers of cycles. For the experiments, a high-frequency (21 kHz) ultrasound resonance fatigue machine has been used ($R = -1$).

In order to determine the threshold stress intensity (for Arall 1), tests were performed by increasing the stress amplitude stepwise. The resulting threshold stress intensity increases with increasing crack length. In addition, tests with constant cyclic stress amplitude were carried out (for Arall 1 and Arall 2). For these, the fatigue crack propagation rates decrease with increasing crack length and become zero at a defined crack length if the applied constant stress amplitude is lower than a defined "critical" stress amplitude. Fiber bridging, causing extensive crack closure, is the reason for this effect. At cyclic stresses higher than this "critical" stress, cracks continue to grow, though with reduced speed.

Damaging effects like delamination owing to failure of the resin in addition influence the fatigue crack growth behavior of Arall at high cyclic stress amplitudes.

KEY WORDS: fiber-reinforced composite material, laminated material, fatigue thresholds, threshold stress intensity, crack growth, crack arrest, crack closure, bridging, delamination, fiber fracture

Nomenclature

ϵ — Strain
ϵ_{max} — Maximum strain
a — Total crack length including notch depth, mm
c — Crack length emanating from the notch, mm
c_L — "Limiting" crack length emanating from the notch, mm
E — Modulus of elasticity, GPa
R — $\sigma_{min}/\sigma_{max}$

[1]Professor, University of Agriculture of Vienna, Institute for Meteorology and Physics, Türkenschanzstrasse 18, A-1180 Wien, Austria.
[2]Research assistant, University of Agriculture of Vienna, Institute for Meteorology and Physics, Türkenschanzstrasse 18, A-1180 Wien, Austria.
[3]Assistant professor, University of Vienna, Institute for Solid State Physics, Boltzmanngasse 5, A-1090 Wien, Austria.
[4]Professor, Delft University of Technology, Faculty of Engineering, 2600 GB Delft, Netherlands.
[5]Professor, Technical University, Institute for Applied and Technical Physics, Karlsplatz 13, A-1040 Vienna, Austria.
[6]Registered trademark, Aluminum Company of America, Alcoa Center, PA 15069.

n	Paris exponent for crack growth
$\Delta a/\Delta N$	Crack growth rate, m/cycle
σ_{max}	Maximum stress amplitude, MPa
σ_{min}	Minimum stress amplitude, MPa
σ_m	Mean stress amplitude, MPa
$\Delta\sigma$	$\sigma_{max} - \sigma_{min}$
$\sigma_{max,th}$	Maximum threshold stress amplitude, MPa
$\sigma_{max,eff,th}$	Maximum effective threshold stress amplitude, MPa
$\sigma_{max,br}$	Maximum bridging stress amplitude, MPa
K_{max}	Maximum stress intensity, MPa$\sqrt{m}$
K_{min}	Minimum stress intensity, MPa$\sqrt{m}$
ΔK	$K_{max} - K_{min}$
$K_{max,th}$	Maximum threshold stress intensity, MPa$\sqrt{m}$
$K_{max,eff}$	Maximum effective stress intensity, MPa$\sqrt{m}$
$K_{max,nom}$	Maximum nominal stress intensity, MPa$\sqrt{m}$
$K_{max,br}$	Maximum bridging stress intensity, MPa$\sqrt{m}$

An attempt is made nowadays to replace the materials of several fatigue critical aircraft parts, for example, lower wing skin, fuselage, and tail skin, by new materials. The new material should give not only good fatigue properties, but also a low density in order to reduce the component weight. In addition, this material is expected not to be susceptible to notches and cracks. Such a material is the laminated fiber-reinforced hybrid composite material, Arall® [1] (name is an acronym for Aramid fiber Reinforced ALuminum Laminates). It can be used for a lower wing skin, for a fuselage skin, and as crack stopper bands in a pressurized fuselage.

Static tests, fatigue tests, and tests on damping, impact, and acoustic fatigue and studies of environmental influences have been performed, especially in Delft [2,3] and by Alcoa [1,3–6]. Usually the loading direction was parallel to the fiber direction, and a notch was introduced perpendicular to the fibers in most studies of the fatigue crack growth behavior, the damage, and crack closure mechanisms. Tests with an Arall® 2 panel fixed to the lower wing skin of a Fokker 27 revealed a 33% weight savings relative to the aluminum design, slow crack growth rates, and no through cracks [1,7]. Similar results were presented at the 32nd International SAMPE Symposium of Alcoa [4]. The lifetime of unstretched Arall material was ten times and that of post-stretched Arall more than 1000 times longer than for 7075-T6 alloy. The fatigue crack growth rate decreased continuously with increasing stress intensity factor range ΔK in post-stretched Arall. It should be explained here that Arall, after the bonding cycle in the autoclave, carried an unfavorable residual stress system with tensile stress in the aluminum-alloy layers and compression stress in the fiber layers. However, if the as-cured Arall is post-stretched, the residual stress system is reversed into compression in the metal sheets and tension in the fibers. This favorable residual stress system is present in the commercial Arall® 1 grade. Additional experiments with constant-amplitude loading [2,8] and flight-simulation loading [2] have amply confirmed the superiority of the laminated material.

S-N curves obtained on riveted lap joints of Arall, 2024-T3, and 7075-T6 showed lifetimes three to twenty times longer than the Arall joints [2]. Center-cracked tensile panels, reinforced with Arall strips, showed a longer lifetime than panels with 2024-T3 or 7075-T6 strips, but more important, the Arall strips did not fail, contrary to the 2024-T3 and 7075-T6 strips [9]. Marissen [8] studied the dependence of fatigue crack growth and found that the crack growth rate decreased with increasing crack length until crack arrest occurred at low enough amplitudes.

The classical mechanism to explain crack closure is based on plastic deformation left in the wake of the crack. Another mechanism is based on the surface roughness-induced crack closure. The predominant mechanism to explain slow crack growth in Arall is crack bridging by unbroken fibers in the wake of the crack [10]. The Aramid fibers with their high stiffness "bridge" the crack in the metal layers. This implies a restraint on crack opening. The resulting fiber forces are transmitted by shear deformation of the resin from the fiber layer to the aluminum layers. Thus the effective stress intensity at the crack tip of the metallic part of the laminate is considerably reduced [4,5,11]. The better fatigue and crack growth properties of Arall in comparison to the properties of aluminum alloys should be attributed mainly to this effect. It is of interest to correlate the reduction of cyclic stress intensity with the length of the bridging zone behind the crack tip, where the fibers are still unbroken [10].

The main damaging mechanism, besides fatigue cracking of the metallic and resin layers are delamination [2,8] and fiber fracture [2,10]. Delamination occurs in areas of high shear stress. Crack-bridging forces of the fibers cause high shear forces in the resin rich layer between the fibers and the aluminum alloy layer. It leads to shear deformation and damage of the resin. As a result, delamination will occur. According to Marissen [8], delamination increases exponentially with the resin shear stress and the energy release rate associated with delamination.

Fiber fracture can occur in Aramid fibers, especially in as-cured Arall if σ_{min} is low or compressive. If cracks occur in the Al-alloy layers, some cracking also occurs in the resin. If the crack is opened under a positive stress, some "fiber pullout" will occur. After reversing the loading, the fibers are loaded in compression. The locally unsupported fibers will buckle, and cyclic buckling leads to fiber damage and ultimately to fiber failure. It can be prevented by avoiding compression in the fibers, i.e., by a sufficiently positive σ_{min} or by post-stretching [12]. In our tests with $\sigma_m = 0$ ($R = -1$), fiber failure should be expected in the as-cured Arall and probably not in the post-stretched Arall.

Crack propagation in Arall has been measured down to rates of 1×10^{-11} m/cycle [8,10] in the literature until now. However, it is of great interest to know the material crack growth resistance at still lower cyclic stress intensity values. For measurements of such low crack growth rates, high numbers of cycles, i.e., more than 10^7 cycles, are necessary. Measurements with conventional testing machines need extremely long testing times and thus are very expensive. With an ultrasound fatigue testing machine working at a frequency of 21 kHz, the experiments need approximately one hundred times shorter testing times [13].

The influence of the testing frequency has been studied on Arall only for the range of 0.02 to 10 Hz until now [2,12]. In these tests, which were performed at $R = 0$, little or no frequency influence on prestrained Arall laminate specimens could be detected. However, three to four times higher crack growth rates were found in the "as cured" Arall® specimens at the testing frequencies of 0.02 and 0.2 Hz instead of 10 Hz. Ultrasonic frequency results have only been compared with conventional frequency tests for metallic materials so far. No real frequency influence on the fatigue crack growth behavior could be detected if temperature and environmental influences were excluded [14].

In this work, fatigue crack growth measurements in the threshold regime and fatigue experiments at constant stress values have been performed on Arall 1 and Arall 2 fiber-reinforced composite material with an ultrasound device in order to characterize the fatigue crack growth properties and the relevant parameters at low crack growth rates. The damaging mechanisms of crack propagation, delamination, and fiber fracturing have been studied in addition by means of light microscopy and with a scanning electron microscope (SEM).

Material and Experimental Procedure

Material and Experimental Details

Arall is a laminated, partly metallic and partly synthetic composite material. The metal part consists of aluminum alloy sheets and the synthetic part (prepreg) of an epoxy-phenol resin matrix which contains Aramid fibers. Metal and prepreg layers are alternately glued together. Arall may be produced with various numbers of layers, different layer thickness, fiber orientation, different aluminum alloys and pretreatments.

In this work, Arall 1 and Arall 2 [1,3–5] have been tested. For comparison purposes, the aluminum alloy 2024-T3 has been used. The Arall 1 used contains three sheets of the aluminum alloy 7075-T6 and two layers of prepreg. Its total (nominal) thickness is 1.3 mm, where the aluminum alloy laminates are 0.3 mm thick, and the prepreg layers are 0.2 mm thick. The fiber orientation is parallel to the rolling direction of the aluminum alloy. The weight fraction of the fibers is 50% of the prepreg. The distribution of the fibers in the resin is not homogeneous; there is a fiber-poor part in the vicinity of the Al-sheets and a fiber-rich part in the middle [2]. Arall 1 is post-stretched (0.4%) after the bonding cycle. That means that it carries a residual compressive stress of 69 MPa in the metal layers [3,6] and a residual tensile stress in the fibers.

The metallic base material of Arall 2 is the aluminum alloy 2024-T3. Arall 2 is not post-stretched, and as a result a residual tensile stress of 34.5 MPa occurs in the metal layers [3,6] and a residual compressive stress in the fibers. Mechanical properties and additional data of Arall 1 and Arall 2 with the 3/2 layer configuration and a nominal thickness of 1.3 mm, are listed in Table 1 [4,6]. The properties of the alloys 7075-T6 and 2024-T3 with a thickness of 1.6 mm are listed for comparison.

Fatigue loading was performed with an ultrasound resonance device working at a frequency of 21 kHz. The specimens must have specified dimensions in order to vibrate in resonance. They are loaded longitudinally, with a mean load equal to zero ($R = -1$). Vibration amplitude and frequency are controlled with the aid of a magnet-coil assembly

TABLE 1—*Mechanical properties and additional data of Arall 1, Arall 2, 7075-T6 and 2024-T3.*

Property	Arall 1[a]	Arall 2[a]	7075-T6[b]	2024-T3[b]
Tensile ultimate strength, MPa	800	717	572	455
0.2% offset tensile yield strength, MPa	641	359	510	359
Tensile total strain to failure, %	1.9	2.5	12	19
Tensile elastic modulus, GPa	68	64	71	72
0.2% offset compressive yield strength, MPa	372	262	503	304
Compressive elastic modulus, GPa	70	67	72	74
Shear modulus, GPa	17	17	27	28
Density, g/cm³	2.29	···	2.79	···

[a]3/2 laminate, 1.3 mm thick.
[b]Aluminum alloy sheet, 1.6 mm thick.

and a PID controller during the test. From the displacement measurements, the strains, ϵ, in the center of the specimens are calculated; they may be obtained also directly by measurements with micro-strain gages. The stresses, σ_{max}, are calculated from the strains according to Hooke's law

$$\sigma_{max} = E \times \epsilon_{max} \tag{1}$$

where E = modulus of elasticity (for values of E, see Table 1).

In order to keep the specimen temperature constant, the damping heat is removed by forced air cooling and by introducing periodic pauses in between the load pulses. One pulse consisted of 500 cycles in the tests of this work. All testing parameters, like vibration amplitude, number of cycles during one pulse, and length of pauses are controlled by a computer; the resulting data, like number of cycles, frequency, pregiven and actually performed amplitudes are recorded and stored by the computer.

For the tests of this work, flat specimens with an edge notch, as shown in Fig. 1, have been used. The notch was introduced perpendicular to the fiber and load direction. The specimens were polished according to ASTM's recommendations in order to make easier the observation of cracking and delamination. Crack lengths, a, were measured with the aid of a video camera and a macro-optic revealing a magnification of 130 on the TV screen. With this, a minimum crack increment of 10 μm could be detected.

Measuring Procedure

All experiments were performed at low stress amplitudes ranging from 30 to 90 MPa. In the first series of experiments, threshold stress amplitudes for crack propagation were determined. The following measuring procedure was chosen. The specimen is stressed at a defined low constant amplitude first (30 MPa in the tests of this work) and crack propagation is observed. The crack growth rate becomes gradually smaller with increasing crack length until the crack stops. Loading is continued for additional 2.5×10^7 cycles at this crack length in order to make sure that a threshold has been really obtained. This crack length is called "limiting crack length," c_{L1} by the authors. Then the stress is increased by 5 to 7% until the crack begins to grow again. The stress amplitude at which the crack grows again is called by the authors "threshold stress amplitude," $\sigma_{max,th1}$ (c_{L1}). It is plotted versus the previously defined "limiting crack length" in Fig. 2. At a certain length the crack stops to grow (for constant stress loading at $\sigma = \sigma_{max,th1}$). This length is a new "limiting crack length," c_{L2}. The same procedure is repeated until a total crack length a, equal to 4 mm, is obtained.

In a second series of experiments the stress amplitudes were kept constant during the whole test, and the fatigue crack propagation rates were measured at these values. To obtain $\Delta a/\Delta N$ versus c curves, crack increments of ~80 to 150 μm (10 to 20 mm on the TV screen) were evaluated (Figs. 4 and 5). The stress amplitudes were 50, 60, 70, 80, and 90 MPa for Arall 1 and 30, 40, 50, and 60 MPa for Arall 2. For comparison purposes, the crack propagation curve of the aluminum alloy 2024-T3 was determined at 60 MPa (Fig. 3).

A third test series served to characterize the nature of material damage. For this, constant stress amplitude tests at two stress levels were performed. Crack path, fiber fracture, and delamination in the different layers of these specimens were microscopically studied.

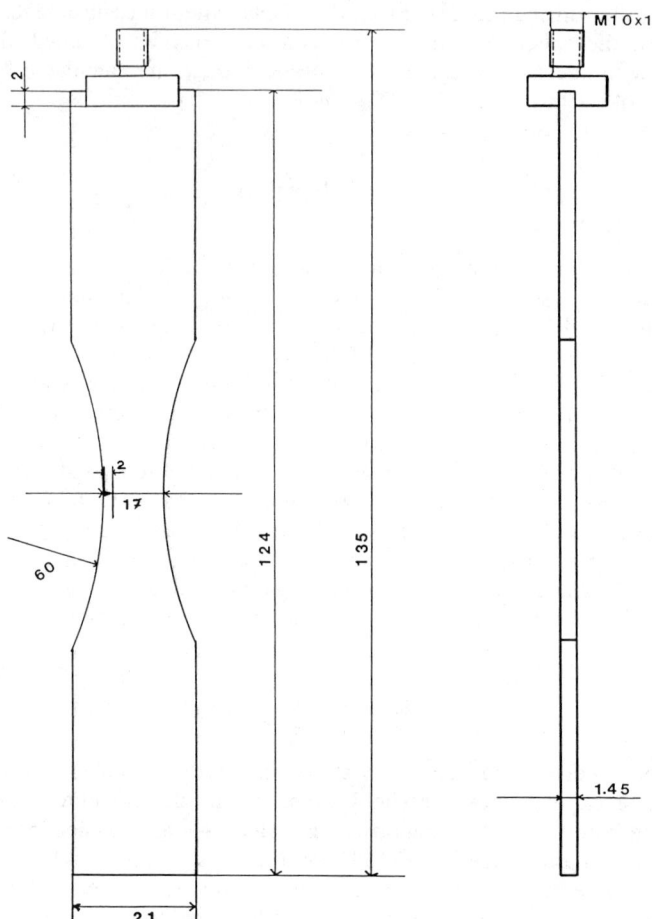

FIG. 1—*Specimen shape for fatigue crack growth and fatigue delamination tests in Arall 1, Arall 2, and 2024-T3 alloy.*

Experimental Results

Influence of Crack Length on Fatigue Crack Propagation

Figure 2 shows the relationship between limiting crack length, c_L, and threshold stress amplitude, $\sigma_{max,th}$ (c_L). "Limiting crack length" means that the crack stops growing at this length at a certain stress amplitude. "Threshold stress amplitude," $\sigma_{max,th}$ (c_L) of a defined crack length (the "limiting crack length") is the stress amplitude at which the crack starts to grow again. Only such data were introduced into Fig. 2, which resulted from a single crack originating from the notch root. To obtain a single crack, it was necessary to prepare a sharp notch with a radius of less than approximately 100 μm. Crack branching occurred if the radius was larger than approximately 100 μm. Figure 2 demonstrates that the stress amplitude has to be increased with increasing crack length in order to keep the crack growing, which appears to be in contrast to linear elastic fracture mechanics (LEFM's)

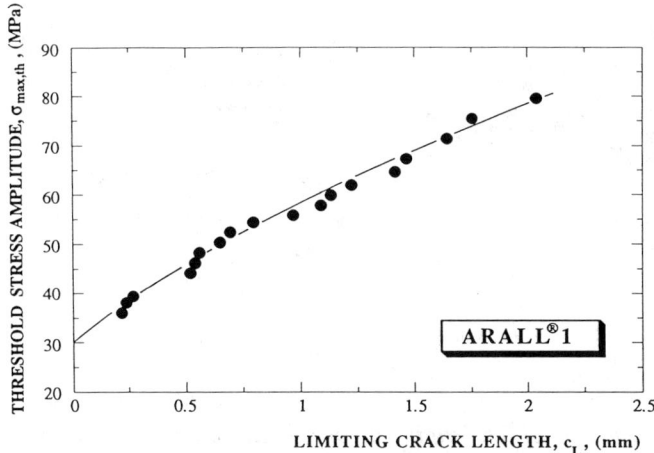

FIG. 2—*Influence of crack length on the threshold stress amplitude, $\sigma_{max,th}$, in Arall 1; the "cyclic threshold stress amplitude," $\sigma_{max,th}$ is the minimum stress necessary for a crack of length, c, to propagate.*

principles. In general the applied stress value has to be lowered for a growing crack in single-edge notched specimens in order to obtain constant stress intensity values.

In Figure 3, the fatigue crack growth rates in Arall 1, Arall 2, and aluminum alloy 2024-T3 are plotted versus the crack length c (c is the length of a crack emanating from the notch) for a constant stress amplitude, $\sigma_{max} = 60$ MPa. According to LEFM's principles, the crack growth rates, $\Delta a/\Delta N$, should increase exponentially with the square root of the crack length for "long" cracks and for constant stress amplitudes. LEFM's corresponding behavior is observed for the Alloy 2024-T3 and for Arall 2 for crack lengths longer than approximately 1.0 and 1.5 mm respectively, but not for Arall 1.

To study the crack length dependence of fatigue crack growth in Arall, additional experiments have been performed at different constant stress values. The results for Arall

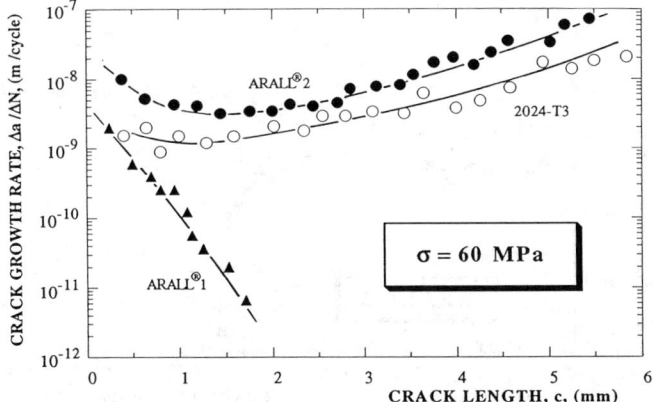

FIG. 3—*Influence of crack length, c, on fatigue crack growth rates on Arall 1, Arall 2, and 2024-T3 at a cyclic stress amplitude, σ_{max}, of 60 MPa.*

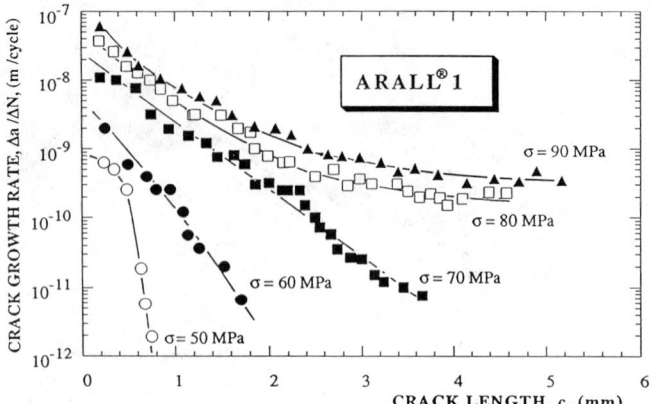

FIG. 4—*Influence of crack length, c, on fatigue crack growth rates of Arall 1 at constant stress amplitudes* σ_{max} = 50, 60, 70, 80, and 90 MPa.

1 are shown in Fig. 4. These results demonstrate that crack propagation stops at a specified length in specimens of 17 mm width up to stress values of 70 MPa. For σ_{max} = 50 MPa, the crack stops at a length of approximately 0.75 mm, for σ_{max} = 60 MPa at 1.8 mm, for σ_{max} = 70 MPa at 3.7 mm, which shows that the critical crack length for crack arrest increases with increasing stress amplitude. In addition, it may be seen from Fig. 4 that the crack growth rate decreases more the lower the stress amplitude is. For stress amplitudes σ_{max} = 80 and 90 MPa, crack arrest is not observed any longer, but still decreasing crack propagation rates with increasing crack lengths are seen. After the crack has grown to a length of approximately 4 mm, it still continues to grow, with a constant crack growth rate of approximately 2×10^{-10} m/cycle for σ_{max} = 80 MPa and 4×10^{-10} m/cycle for σ_{max} = 90 MPa.

The results of similar experiments with constant stress amplitudes for Arall 2 are shown in Fig. 5. Crack arrest is observed within 1.5×10^7 cycles at a stress amplitude of 30 MPa

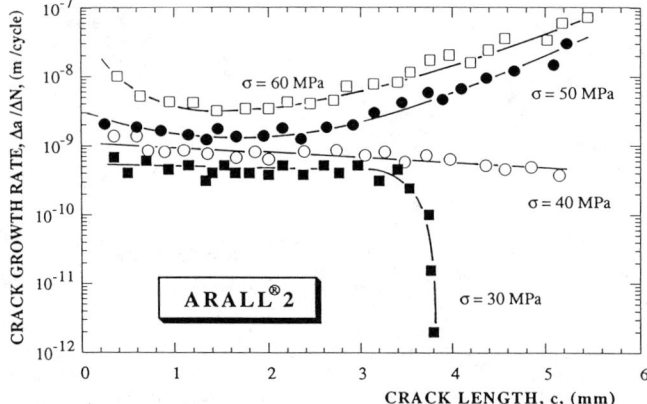

FIG. 5—*Influence of crack length, c, on fatigue crack growth rates of Arall 2 at constant stress amplitudes* σ_{max} = 30, 40, 50, and 60 MPa.

at a crack length of approximately 3.8 mm. A slowly decreasing crack growth rate with increasing crack length is found for $\sigma_{max} = 40$ MPa, whereas increasing crack growth rates result for higher stresses. For these stresses a behavior, as expected from LEFM's principles, is observed.

Comparing Arall 1 and Arall 2 yields the following typical differences:

Above a defined "critical" stress (of 40 MPa), the fatigue crack growth rate of Arall 2 increases with increasing crack length, whereas no such "critical" stress exists up to 90 MPa for Arall 1 for the tested crack lengths and specimen dimensions. The difference of Arall 1 and Arall 2 may be further seen clearly at a stress amplitude of 60 MPa (Fig. 3). While the fatigue crack growth rates of Arall 2 depend on the crack length in a similar manner as that of the metallic part 2024-T3 (increasing crack growth rate with growing crack length), the crack growth rate of Arall 1 strongly decreases with increasing crack length and becomes zero at a crack length of only 1.8 mm.

Characterization of Fatigue Damage in Arall

Crack Front within the Layers and Delamination—Arall 2 specimens which were fatigued with a constant amplitude of 40 MPa were analyzed microscopically afterwards. The crack path in the different layers, delamination, and fiber fracturing have been studied.

The crack length, c, within the different layers through the specimen thickness of Arall 2 are shown in Fig. 6. This specimen was loaded at a stress amplitude of 40 MPa. The photographs of the crack corresponding to the indicated layers are added. The following results are found to be characteristic. The fatigue crack length is highest in the outer aluminum alloy layers (7.1 mm); it is shorter in the fiber-poor part of the epoxy layer (6.1 mm), shorter in the fiber-rich part (5.5 mm), and shortest in the fibers (4.9 mm). It is longer, however, in the center aluminum alloy sheet and almost as long as in the fiber-poor part (5.6 mm). This means that the "bridging zone" is 2.2 mm for the outer aluminum sheet and 0.7 mm for the center aluminum sheet. (The bridging zone is that part of the crack where the fibers are still unbroken, thus bridging the crack in the metallic layer.) The crack is 21% shorter in the center-aluminum alloy layer than in the outer sheet.

A SEM picture of the crack tip within the resin and the fibers is shown in Fig. 7. For this, the outer metallic layer has been removed completely and the resin has been solved away partly, as may be seen in the picture. The resin in this picture is completely broken. The fibers are broken on the right side of this photo and not broken on the left side (bridging zone).

Around the crack tip, delamination takes place. The shape of the delaminated area (brighter color) is shown in Fig. 8, which comes from a test on a notched Arall 1 specimen, which was fatigue loaded with a low cyclic amplitude.

In another test, an unnotched Arall 1 specimen was stressed at a high amplitude of 136 MPa. Fracture took place after 7.94×10^7 cycles. Both outer aluminum alloy layers failed, the upper layer at three places, so that a piece of several millimetres completely broke away from this layer (Fig. 9). In the center-aluminum alloy sheet, three cracks are visible, which, however, do not go through the whole layer. The outer aluminum alloy layers are delaminated in the vicinity of the crack. At the inner side of the delaminated outer aluminum alloy layers, some adhering fibers are visible. It is assumed that delamination owing to crack propagation takes place in the transition area between the fiber-rich and fiber-poor part of the resin [15].

The smaller crack length in the center aluminum alloy layer may be explained by bridging of the unbroken fibers. For the center aluminum alloy layer, fiber bridging is

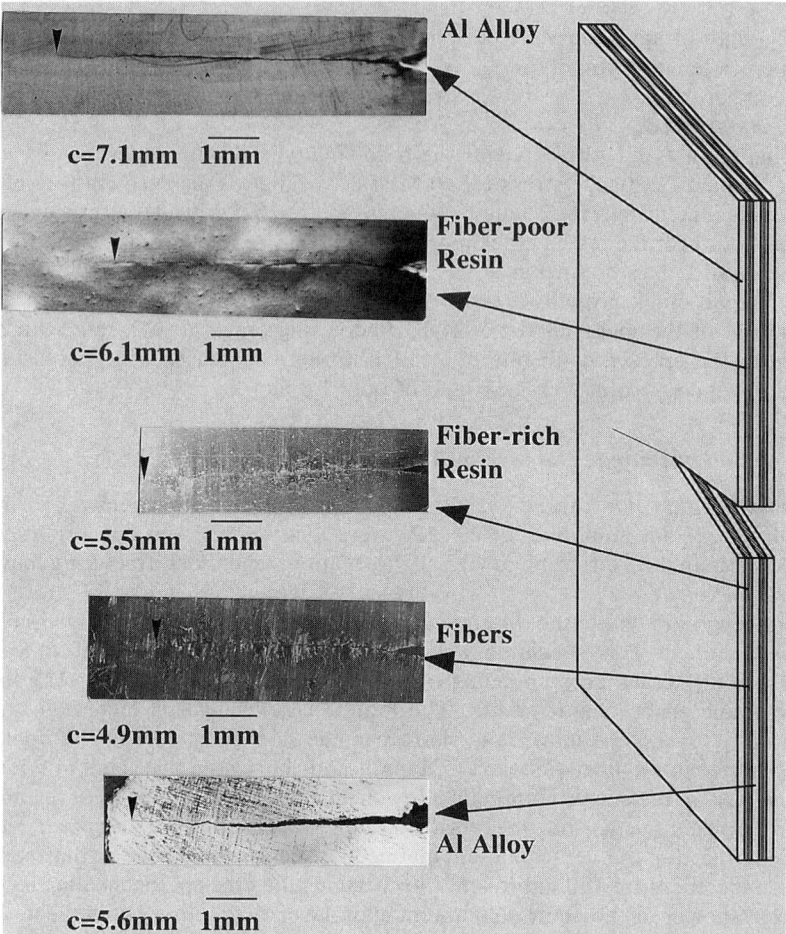

FIG. 6—*Crack front of Arall 2 at constant stress amplitudes; photos of crack length within the different layers of a specimen loaded with a constant stress amplitude of 40 MPa.*

effective at both sides, whereas fibers are acting at one side of the outer aluminum alloy sheets only.

Discussion

Materials which obey LEFM's principles are characterized by a cyclic threshold stress intensity, which is a material constant independent of the applied load or stress or crack length. This implies that the stress amplitude, which is necessary for crack growth, has to be lowered with $1/\sqrt{a}$ (with a = crack length) if the stress intensity value is kept constant; the threshold stress intensity, ΔK_{th}, does not depend on a.

Contrary to this, the stress has to be increased with increasing crack length in order to keep the crack growing in Arall 1 (Fig. 2), i.e., the post-stretched Arall with the favorable residual stress system. This "threshold stress" increases somewhat less than linearly with the crack length, for specimen and crack dimensions as tested in this work.

FIG. 7—*Crack tip in the fibers of the same Arall 2 specimen as shown in Fig. 6 ($\sigma_{max} = 40$ MPa); unbroken fibers are at the left side and broken fibers are at the right side.*

This result may be explained by the influence of unbroken fibers in the wake of the crack tip. These fibers reduce crack opening during the tension part of cycling, thus reducing the cyclic stress intensity at the crack tip to a lower effective value. The number of fibers becoming effective in this way increases linearly with increasing crack length, so that a linear dependence of the bridging stress on the crack length should be expected. Delamination which favors fiber fracturing, however, probably reduces this positive fiber bridging effect with increasing crack length and stress amplitude. As a consequence a lower-than-linear increase of the bridging stress with crack length may be expected. On the other hand, the maximum effective threshold stress amplitude $\sigma_{max,eff,th}$ ($\sigma_{max,eff,th}$ is the threshold stress amplitude for crack growth in the Al alloy) is proportional to $1/\sqrt{a}$. The slope of the curve in Fig. 2 may be explained by adding up $\sigma_{max,br}$ and $\sigma_{max,eff,th}$. According to the results of Fig. 2 it cannot be decided clearly whether the dependence of the threshold stress on c is linear or less than linear. The threshold stress intensities for different "limiting" crack lengths are derived from Fig. 2, as shown in Fig. 10. Extrapolation of the resulting curve to the crack length, c, equal to zero ($a = 2$ mm = notch depth) reveals a

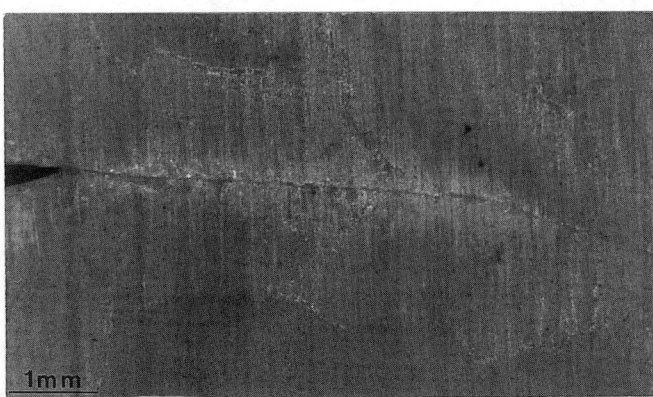

FIG. 8—*Shape of delamination zone around the crack within the prepreg layer of an Arall 1 specimen.*

FIG. 9—*Delamination in an Arall 1 specimen at a stress amplitude of 136 MPa.*

threshold stress intensity value of the compound without influence of the fiber bridging. This value is 2.71 MPa$\sqrt{m}$ if a less-than-linear dependence of $\sigma_{max,th}$ on c is assumed from Fig. 2 (according to Fig. 2, $\sigma_{max,th}$ is 30.5 MPa for $c = 0$), and it is 2.95 MPa$\sqrt{m}$ if $\sigma_{max,th}$ increases linearly with c ($\sigma_{max,th}$ is 33.2 MPa for $c = 0$). These values are similar to the threshold stress intensity K_{max} of 2.44 MPa$\sqrt{m}$ for $R = 0.1$ for the alloy 7075-T6, as reported by Bucci et al. [4]. Differences may be explained by load history influences on the bridging capability.

Extrapolation of the $K_{max,th}$ versus c curve to a crack length, c equal to zero makes it possible to determine stress intensity values which characterize the bridging capability of the material for each crack length approximately. These values are called $K_{max,br}$ in the

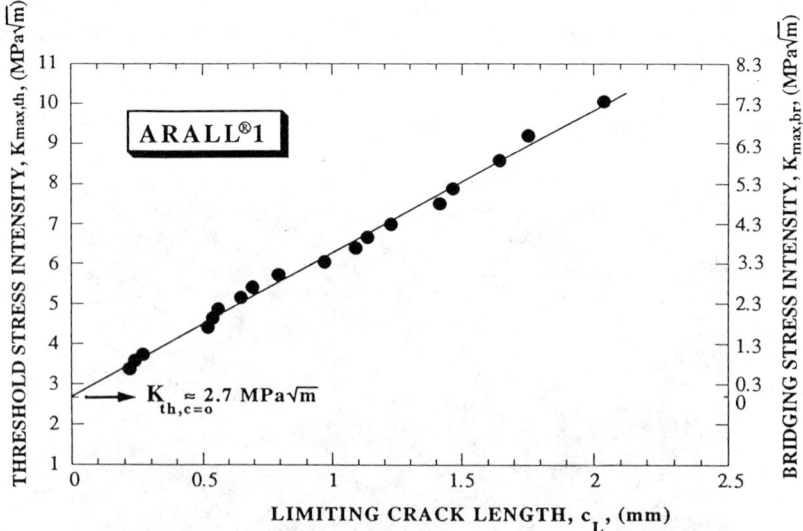

FIG. 10—*Influence of fatigue crack length, c, on the threshold stress intensity, $K_{max,th}$, (left y-axis) in Arall 1. Right y-axis: bridging stress intensity, $K_{max,br}$.*

following and are determined from

$$K_{\max,br}(c) \approx K_{\max,th}(c) - K_{\max,th}(c = 0) \tag{2}$$

The resulting values are indicated on the right y-axis of Fig. 10.

The results of Figs. 4 and 5 show the influence of crack lengths on the fatigue crack growth rates of Arall 1 and Arall 2 at different constant stress amplitudes and may be characterized and explained in the following way. For Arall 1 (Fig. 4) essentially two types of curves are obvious. Up to stresses of approximately 70 MPa, crack growth retardation with increasing crack length and finally crack arrest at a specified crack length takes place. At stresses equal to 80 and 90 MPa, retardation is observed up to a crack length, c, of approximately 4 mm; for further increasing crack lengths the crack growth rate remains constant. From these results it is concluded that the fibers in the wake of the crack tip ("bridging zone") remain unbroken up to a crack length, c, of 4 mm. Beyond this length the fibers break more or less continously with growing crack length such that a wake length with unbroken fibers of approximately 4 mm behind the crack remains constant. The difference of nominal and effective cyclic stress intensity at the crack tip characterizes the bridging cyclic stress intensity. Values up to $\sim$7.3 MPa$\sqrt{m}$ (at a cyclic stress value of 80 MPa) (Fig. 10) have been found in the stress raising test of Figs. 2 and 10.

For higher stress amplitudes like $\sigma_{\max}$ = 80 and 90 MPa, the nominal cyclic stress intensity, $K_{\max,nom}$ obviously becomes high enough at a certain crack length so that the crack growth impeding (bridging) cyclic stress intensity is balanced according to

$$K_{\max,eff} \approx K_{\max,nom} - K_{\max,br} \tag{3}$$

and crack growth is possible. This behavior is observed at a crack length, c, of approximately 4 mm for $\sigma_{\max}$ = 80 and 90 MPa (Fig. 4). Assuming that the maximum bridging stress intensity value, $K_{\max,br}$, is a constant implies that the crack growth rate should increase for further increasing crack lengths. The width of the specimens used in this work, however, is too small to verify this prediction. Further crack growth has to be expected at higher stresses for specimens or components with larger dimensions than those tested in this work [8].

In order to quantify influences of delamination processes on the fatigue crack growth behavior, additional work is performed and will be published later.

Arall 2 with the unfavorable residual stress system shows (Fig. 5) a third typical course of the $\Delta a/\Delta N$ versus c curve (crack growth rates versus crack length at constant stress values). At stresses above 40 MPa, the crack growth rates increase with increasing crack length. The increase is proportional to $(\sqrt{a})^n$ at crack lengths, c larger than 1.5 mm according to LEFM's principles. The $(\Delta a/\Delta N)$ versus (c) curve of 2024-T3 is similar to that of Arall 2 for 60 MPa (Fig. 3), with a similar decrease of the crack growth rate up to a defined crack length, which characterizes the "short crack" length. The results of Fig. 5 imply that the bridging stress intensity is small enough to be overcome by stresses above 40 MPa in Arall 2.

2024-T3 has superior fatigue crack growth properties than 7075-T6. Since the opposite behavior has been observed for the composite materials in this work, the main reason for the better fatigue crack growth properties of Arall 1 must originate from its higher fiber-bridging capability. This is obviously achieved by the pretreatment of Arall 1, which leads to the above-mentioned favorable residual stress system in the fibers as well as in the Al-alloy sheets. The residual compression load in the aluminum alloy sheet is equal to 69 MPa after 0.4% post-stretching and thus reduces the effective load and the crack opening capability during the tension cycle. This reduces the crack opening displacement and the crack growth rates in the Al-alloy on one side and, more importantly, reduces pullout of the fibers on the other side. The tension preload of the fibers leads to reduced effective compressive stresses during the compressive portion of cycling loading, so that buckling of pulled out fibers does not occur. Therefore fiber failure is avoided or reduced, thus resulting in a high bridging capability and therefore better fatigue crack growth properties.

When comparing Arall 2 with 2024-T3, consider that the stress is calculated on the nominal thickness and for Arall 2 only part of that thickness (0.9/1.3) is Alloy 2024-T3. That part carries an unfavorable residual load (tension), which amounts 34.5 MPa. This residual tensile stress adds during cyclic loading with a stress amplitude of 60 MPa in the metal sheet, and a residual compressive stress in the fibers reduces their bridging capability. All these facts contribute to the superior crack growth properties of 2024-T3 in comparison with Arall 2 at an applied stress amplitude of 60 MPa. Comparison of these materials for stresses below approximately 40 MPa, however, shows clearly the superiority of the fiber-reinforced laminate material Arall 2.

Comparison of the high-frequency measurements as performed in this work with results obtained at conventional frequencies, however, is difficult, as R value, notch shape and length, specimen dimension, size and kind of loading, Arall configuration and treatment etc. [2,4,8,10,12,15], have been different in most cases. The R values were positive (or zero) in all tests performed at low frequencies in contrast to the R value ($R = -1$) of this work, and the crack lengths, which strongly influence the crack growth rates, usually are not reported in the literature. In addition, the near-threshold crack growth behavior has not been tested at conventional frequencies in detail as done in this work. Nevertheless, one result may be mentioned.

The crack growth rate in Arall 2 at 10 Hz, $R = 0$, σ_{max} ($= \Delta\sigma$) $= 120$ MPa, crack length of 3.5 mm and a notch depth of 1.5 mm, is reported to be approximately 5×10^{-8} m/cycle [12]. For a testing frequency of 21 kHz (this work, Fig. 5) the crack growth rate for Arall 2 was found to be 1×10^{-8} m/cycle at $R = -1$, $\sigma_{max} = 60$ MPa ($\Delta\sigma = 120$ MPa) and a crack length of 3.5 mm.

This result indicates that no pronounced frequency effect exists for at least 10 Hz to 21 kHz. Therefore it seems to be useful to apply the high frequency fatigue method for studies used at least in the threshold regime in order to save testing times.

Summary and Conclusions

The fatigue crack growth properties of Arall 1 (fibers under residual tension stress and metallic sheets under residual compression stress) and Arall 2 (fibers under residual compression and sheets under residual tension) have been studied in the threshold regime. It has revealed the following features:

1. Threshold stress and stress intensity of Arall 1 increase with increasing crack length. The threshold stress intensity increases approximately linearly in the crack length regime of 0.2 to 2.1 mm (specimen width is 17 mm, notch depth is 2 mm).
2. Fatigue crack propagation tests on Arall 1, Arall 2, and Alloy 2024-T3 with constant stress amplitudes lead to the following results: The crack growth rates decrease with increasing crack length in Arall 2 at stress amplitudes below 40 MPa. At higher stresses, however, an increase of the fatigue crack growth rates with increasing crack lengths is observed, which is similar to that in the metallic material (2024-T3 alloy). In Arall 1, on the other hand, the crack growth rates decrease with increasing crack

length at *all* constant stress amplitudes (up to 90 MPa, as used in this work). The positive influence of the fibers is attributed to their crack bridging capability. Tension prestressing of the fibers (plus compressive prestressing of the metallic layer) intensifies this effect. The post-stretching treatment, which introduces a favorable residual stress system, obviously is the reason for the superiority of Arall 1 over Arall 2: fiber failure is avoided and thus crack bridging is effective at much higher stresses and for longer cracks.

3. The results of this work show that fiber-induced crack bridging increases with increasing crack length (number of fibers increasing with increasing crack length) up to a maximum value. If this crack closure stress is overcome by the applied load, further crack propagation is possible. Based on the results of Fig. 5 it is assumed that the bridging closure stress obtains a maximum constant value at some specified crack length. The effective driving force for crack propagation is given then by the applied stress minus this bridging stress.

Acknowledgments

The authors thank Dr. Vasudevan and ALCOA for having supplied the testing material for this work, Arall 1 and Arall 2, for research purposes. Financial support by the Bundesministerium für Wissenschaft and Forschung, Wien is also gratefully acknowledged.

References

[1] *Arall Letter,* Vol. 1, No. 1, Aluminum Company of America, Alcoa Center, PA, September 1987.
[2] Vogelesang, L. B., Gunnink, J. W., and Schijve, J., "Arall Laminate Research Project, Review 1978–1987 by Posters," Delft University of Technology, Faculty of Aerospace Engineering, Delft, Netherlands.
[3] Bucci, R. J., Mueller, L. N., Vogelesang, L. B., and Gunnink, J. W., "Arall Laminates, Properties and Design Update," *Proceedings,* 33rd International SAMPE Symposium, Society for the Advancement of Materials and Process Engineering, Anaheim, CA, 7–10 March 1988.
[4] Bucci, R. J., Mueller, L. N., Schulz, R. W., and Prohaska, J. L., "Arall Laminates—Results from a Cooperative Test Program," *Proceedings,* 32nd International SAMPE Symposium, Society for the Advancement of Materials and Process Engineering, Anaheim, CA, 6–9 April 1987.
[5] Alcoa Arall Laminate Sheet, Alcoa Aerospace Technical Fact Sheet, Aluminum Company of America, Sheet and Plate Division, Bettendorf, IA.
[6] Bucci, R. J., Mueller, L. N., Gregory, M. A., and Bentley, R. M., "Arall Laminates Scale-up from R&D to Flying Articles," *Proceedings,* 1988 USAF Conference, San Antonio, TX, 29 Nov.–1 Dec. 1988, Alcoa Laboratories, Alcoa Center, PA.
[7] Van Veggel, L. H., Jongebreur, A. A., and Gunnink, J. W., "Damage Tolerance Aspects of an Experimental ARALL F-27 Lower Wing Skin Panel," *Proceedings,* 14th Symposium of the International Committee on Aerospace Fatigue, Ottawa, 8–12 June 1987, Engineering Materials Advisory Services, Cradley Heath, Warley, West Midland, United Kingdom, 1987, pp. 465–502.
[8] Marissen, R., "Fatigue Crack Growth in Arall—A Hybrid Aluminum-Aramid Composite Material (Crack Growth Mechanisms and Quantitative Predictions of the Crack Growth Rates)," Delft University of Technology, Faculty of Aerospace Engineering, Delft, Netherlands, June 1988.
[9] Schijve, J., "Crack Stoppers and ARALL Laminates," *Engineering Fracture Mechanics,* Vol. 37, 1990, pp. 405–421.
[10] Ritchie, R. O., Weikang, Y., and Bucci, R. J., "Fatigue Crack Properties in Arall Laminates: Measurement of the Effect of Crack-Tip Shielding from Crack Bridging," *Engineering Fracture Mechanics,* Vol. 32, No. 3, 1989, pp. 361–377.
[11] Marissen, R., "Fatigue Mechanisms in Arall, a Fatigue Resistant Hybrid Aluminum-Aramid

Composite Material," *Proceedings,* 3rd International Conference on Fatigue and Fatigue Thresholds, University of Virginia, Charlottesville, VA, 28 June–3 July 1987.

[*12*] Roebroeks, G. H. J. J., "Towards GLARE, the Development of a Fatigue Intensitive and Damage Tolerant Aircraft Material," Delft University of Technology, Delft, Netherlands, 1991.

[*13*] Stanzl, S. E., Czegley, M., Mayer, H. R., and Tschegg, E. K., "Fatigue Crack Growth Under Combined Mode I and Mode II Loading," *Fracture Mechanics: Perspectives and Directions, ASTM STP 1020,* R. P. Wei and R. P. Gangloff, Eds., American Society for Testing and Materials, Philadelphia, 1989, pp. 479–496.

[*14*] Stanzl, S. E., "Fatigue Testing at Ultrasonic Frequencies," *Journal of the Society of Environmental Engineers,* Vol. 25, No. 1, 1986, pp. 11–16.

[*15*] Marissen, R., "Fatigue Crack Growth in Aramid Reinforced Aluminum Laminates (ARALL), Mechanisms and Predictions," Report DFVLR-FB 84-37, DFVLR, Institut für Werkstofforschung, Köln, Germany, 1984.

Author Index

Subject Index